# Marine Chemistry in the Coastal Environment

**Thomas M. Church,** EDITOR
*University of Delaware*

A special symposium
sponsored by the
Middle Atlantic Region
at the 169th Meeting
of the American Chemical
Society, Philadelphia,
Penn., April 8-10, 1975

ACS SYMPOSIUM SERIES **18**

AMERICAN CHEMICAL SOCIETY
WASHINGTON, D. C. 1975

Library of Congress CIP Data

Marine chemistry in the coastal environment.
(ACS symposium series; 18. ISSN 0097-6156)

Includes bibliographical references and index.

1. Chemical oceanography—Congresses. 2. Coasts—Congresses. 3. Estuarine oceanography—Congresses. 4. Waste disposal in the ocean—Congresses.

I. Church, Thomas M., 1942- II. American Chemical Society. Middle Atlantic Region. III. Series: American Chemical Society. ACS symposium series; 18.

GC111.2.M37 551.4'601 75-28151
ISBN 0-8412-0300-8 ACSMC8 18 1-710



PRINTED IN THE UNITED STATES OF AMERICA

ACS Symposium Series

**Robert F. Gould,** *Series Editor*

# FOREWORD

The ACS SYMPOSIUM SERIES was founded in 1974 to provide a medium for publishing symposia quickly in book form. The format of the SERIES parallels that of the continuing ADVANCES IN CHEMISTRY SERIES except that in order to save time the papers are not typeset but are reproduced as they are submitted by the authors in camera-ready form. As a further means of saving time, the papers are not edited or reviewed except by the symposium chairman, who becomes editor of the book. Papers published in the ACS SYMPOSIUM SERIES are original contributions not published elsewhere in whole or major part and include reports of research as well as reviews since symposia may embrace both types of presentation.

# CONTENTS

APPLICATIONS AND RESOURCES IN MARINE CHEMISTRY



ORGANIC AND BIOLOGICAL MARINE CHEMISTRY

# PREFACE

Traditionally, marine chemists have followed the descriptive leads of oceanographers in using basic chemical processes to unravel the more simple chemical processes that drive the complex electrolyte called the sea. In the open ocean this approach has been successful because the system is essentially one dimensional, the compositions are temporally invarient, and chemical processes are in rather steady state. In the coastal environment, however, the marine chemist loses most of those advantages since he is dealing with a multi-dimensional system that includes time and where tidal action drives a system whose dynamic processes are rarely in equilibrium. Compounding these problems are the interactions with fresh water supplies, geological solids, immense biological productivity, and not least, the catabolism of human technocracy.

Since the pioneering work of Ketchum and co-workers some 20 years ago, coastal oceanography has suffered an embarrassing gap of knowledge because attentions have been turned largely to the apparently simpler chemical processes of the open and deep sea. The price of this lapse has now closed in on man who is being forced to rely ever more heavily on the resources of the coastal marine environment. Separate cities located traditionally at the fall line of estuarine river systems have turned into a contiguous sprawl of industrialized metropolises. Estuaries, such as in the Middle Atlantic Region, have become open conduits of raw material import and waste export. As the volumes needed to fuel this technocracy escalated and man became less willing to tolerate spoiling of the land, the coastal environment bore the brunt that today threatens its viable existence. As a result, in this decade, the marine chemist has been called to help devise means to prudently manage and revitalize the coastal marine resource.

This symposium is the first attempt to collate some of the most recent advances in the field of marine chemistry in the coastal environment. The sessions follow a theme which looks beyond simple descriptions to those processes that integrate basic chemistry, tracer techniques, geochemistry, pollutant fates, waste disposal, resource applications, and biological dynamics in the coastal ocean. The goal has been to show that instead of merely describing problems, there is greater potential in applying the tools of modern oceanography and chemistry to progressive concepts that give sophisticated insight to help solve these problems.

The papers touch upon several disciplines which a purer chemist might even construe as irrelevant. The intended purpose was to demonstrate that sound chemical knowledge is not enough; realistic solutions must consider the coastal regions as geological, biological, and anthropological entities as well. Hopefully this symposium also reflects the growing recognition of the chemist for the validity of research in this area. However, continued federal support must order national priorities to guarantee prudent use and protection of the coastal environment.

I would like to gratefully acknowledge the session officers who composed the symposium committee: R. H. Wood, O. P. Bricker, G. R. Helz, M. G. Gross, R. J. Huggett, and J. H. Sharp. Their energies were largely responsible for successfully collating a distinguished program, and their continuing enthusiasm helped me immeasurably in organizing the symposium and completing this volume. The College of Marine Studies at the University of Delaware donated much of its resources, including the secretarial assistance of Sandra Northrup. Lastly, we are all sincerely indebted to the Middle Atlantic Region of the American Chemical Society for sponsoring this timely forum on marine chemistry in the coastal environment.

THOMAS M. CHURCH

University of Delaware
Newark, Del.
July 21, 1975

1

# Chemical Equilibrium in Seawater

R. M. PYTKOWICZ, E. ATLAS, and C. H. CULBERSON

School of Oceanography, Oregon State University, Corvallis, Ore. 97331

The concept of chemical equilibrium can be applied to the oceans in three ways; it enters the study of the geochemical control of the oceanic composition, for fast reactions equilibrium constants are used to calculate the concentrations of species present in seawater, and for slow reactions departures from equilibrium are useful for kinetic studies.

Seawater differs from the solutions that are usually examined by chemists because of the large number of solutes and, at times, of suspended particles that are present in the oceans. Also, one must consider the concurrent effects of purely chemical, hydrographic, biological, and geological processes upon the composition of seawater. One must consider as well the gravitational field and pressure, temperature, and compositional gradients that result from the extent and depth of the oceans.

The study of the equilibrium chemistry of the oceans is facilitated because the major ions are present in almost constant proportions. This makes seawater an ionic medium which can be characterized by one compositional parameter, the chlorinity or the salinity. The constancy of the relative composition is not always present in estuaries because river water, which mixes with seawater, has its major ions in proportions which differ from those in seawater. Still, it will be shown that equilibrium data obtained for seawater can often be applied to estuarine waters.

In this work we will first examine briefly the composition of seawater. Then, we will outline some major aspects of equilibria as applied to the oceans. Next, we will consider to what extent these equilibrium considerations are relevant to estuaries and finally, we will examine some topics on the control of the oceanic composition. We will not attempt a comprehensive coverage of the subject but will emphasize basic concepts and methods.

## The Composition of Seawater

All the naturally occurring elements and many of their compounds find their way into the oceans through rivers, ground waters, aerial transport, and submarine volcanism. A few major constituents, shown in Table I, account for over 90% by weight of the solutes present in seawater (1).

Table I. Major Constituents of Seawater of 34.3‰ Salinity (19‰ Chlorinity)

| Constituent | ppm | Constituent | ppm | Constituent | ppm |
|---|---|---|---|---|---|
| $Cl^-$ | 18,971 | $Ca^{2+}$ | 403.9 | $B(OH)_3$ | 25.6 |
| $Na^+$ | 10,555 | $K^+$ | 391 | $Sr^{2+}$ | 7.7 |
| $SO_4^{2-}$ | 2,657 | $HCO_3^-$ | 142 | $F^-$ | 1.3 |
| $Mg^{2+}$ | 1,268 | $Br^-$ | 65.9 | | |

These major constituents are present in almost constant proportions in the oceans (1, 2) indicating that the mixing time of the oceans, which is of the order of 1,000 years, is fast relative to the input rates and to the reactivity of the constituents (3-5). The constant relative composition also led to the definitions of chlorinity and salinity which are presented in standard oceanographic texts, e.g., Riley and Chester (6). It is important to realize that the chlorinity can be used to represent the extent of mixing of river water with seawater in an estuary. The defined salinity loses its meaning, however, when the oceanic proportions of the major ions are altered in estuarine waters or in the pore waters of submarine sediments.

The minor elements vary greatly in time and space (1) and are in general quite important because of their participation in chemical, biological, and geological processes. These elements are examined elsewhere in this volume.

The general geochemical control of the chemical composition of the oceans in terms of equilibria, steady states, and fluxes between natural reservoirs is a vital topic. Its study yields insights into the chemical history of seawater and helps us understand the potential impact of man-made perturbations upon the environment (5,7-11). This subject will be examined briefly at the end of this work, after a critical review of the main reactions which influence the local distributions of chemical species in seawater. The first topic related to local equilibria will be the dissociation of weak acids.

## Dissociation of Weak Acids

The dissociation constants of weak acids and bases can be used for the calculation of the concentrations of molecules and ions of oceanographic interest because the dissociation reactions are fast relative to competing biological and geological processes and reach equilibrium. Carbonic acid is an especially important weak electrolyte because of the roles it plays in life and in the formation of limestones and dolomites. Also, it serves as a model system for the study of other electrolytes which are present in the oceans.

The use of thermodynamic dissociation constants is not recommended for careful work in seawater because it requires the estimate of the total activity coefficients of single ions (12). These coefficients are conventional quantities and, therefore, their accuracy is unknown. Still, thermodynamic constants may be and are often used for fast low accuracy estimates by marine geochemists when constants measured in actual seawater are not available.

Buch et al. (13) introduced to oceanography the so-called apparent dissociation constants measured directly in seawater. In their present form, for a generic acid HA, these constants are defined by (14)

$$K' = \frac{ka_H(A^-)_T}{(HA)} \tag{1}$$

Parentheses represent concentrations and T refers to total (free plus ion-paired) quantities. $a_H$ is the conventional hydrogen ion activity defined, for example, on the NBS scale. k is a constant, within the reproducibility of pH data, which is cancelled out between the determination and the application of $K'$ (14). Thus, k does not affect the accuracy with which $(A^-)_T$ and HA can be determined. Apparent constants have been shown to be invariant for many processes of oceanographic interest (14).

Another definition of dissociation constants, which differs from $K'$ only in that $(H^+)_T$ is used instead of $ka_H$, was proposed by Hansson (15). $(H^+)_T$ can be determined either with tris buffers prepared in seawater (16) or with conventional NBS buffers (17). We will return to alternate definitions of the pH later.

The apparent dissociation constants and the constants defined by Hansson for carbonic acid and for boric acid were determined in seawater at atmospheric pressure by several workers (13,15,18-20). Disteche and Disteche (21), Culberson et al. (22), and Culberson and Pytkowicz (23) extended the measurements to the high pressures which are encountered in the deep oceans. Culberson and Pytkowicz (23) produced correction tables for the increase in pH that occurs when deep seawater samples are brought to shipboard.

The carbon dioxide system in seawater can be described by four equations in six unknowns once the apparent dissociation constants are known and the borate contribution is subtracted from the alkalinity (24). One can, therefore, completely specify the system by means of any two relevant measurements. This has led to a proliferation of methods, each with its own advantages and drawbacks. One can measure the pH and the titration alkalinity which can be obtained accurately by a single acid addition (17). Alternatively one can determine the titration alkalinity and the total $CO_2$ by a Gran titration (25), the $pCO_2$ and the total $CO_2$ gasometrically, and so on. The choice of a method depends upon the quantity of primary interest.

In situ pH probes, such as those developed by Manheim (26) and by Grasshoff (27), are of special interest for time series measurements in estuaries. They are adaptable for probing pore waters of sediments. This eliminates possible shifts in mineral-seawater equilibria that may occur when sediments are brought to different temperatures and pressures in the laboratory.

The apparent dissociation constants of phosphoric acid in seawater, needed for the study of the formation of apatites and phosphorites, were determined by Kester and Pytkowicz (28). $K'_1$ for hydrogen sulfide was measured by Goldhaber (29). Culberson et al. (17) determined the dissociation constants of hydrofluoric acid and of bisulfate ions. Culberson and Pytkowicz (30) measured the ionization of water in seawater.

It should be emphasized that total activity coefficients and, therefore, apparent constants, which are related to thermodynamic ones by these coefficients, depend upon the major ion composition of the medium. Thus, one should use these quantities with care in estuaries and in the pore waters of sediments because the major ion composition of these media may differ from that of oceanic waters.

## pH of Seawater

It was mentioned earlier that the NBS buffer scale, when applied to seawater, yields a quantity $ka_H$ and that k is cancelled out in practice, within the reproducibility of pH data. Thus, it is the reproducibility rather than the accuracy of pH data which is relevant to oceanographic measurements. We found in our laboratory that careful pH measurements in seawater are reproducible to within ± 0.01 pH units between different electrodes and that the reproducibility appears to be limited by the liquid junction potentials.

Hansson (16) proposed a $pmH_T$ scale based upon buffers prepared in seawater. Bates and Macaskill (31) also used such buffers but determined $pmH_F$ instead of $pmH_T$. $(H^+)_T$ and $(H^+)_F$, where the subscripts F and T refer to the free and to the total concentrations, are related by (17)

$$(H^+)_T = (H^+)_F + (HSO_4^-) + (HF) \quad (2)$$

The NBS, the $pmH_T$, and the $pmH_F$ scales can be related by means of the data presented by Culberson et al. (17).

It is possible that the use of buffers prepared in seawater, may reduce the liquid junction potential and, therefore, increase the reproducibility of pH data. On the other hand, the preparation of seawater buffers and of reference electrodes by individual investigators may introduce systematic errors.

With regard to the general control of the oceanic pH, Sillen (7) suggested that clay-seawater interactions exert a primary pH-statting role. Pytkowicz (5) concluded, however, that the carbon dioxide system is the major pH buffering agent in the present oceans.

## Solubilities of Minerals

The most intensely studied salt in seawater has been calcium carbonate because of its biological and geological importance. In addition, work on carbonates has yielded concepts and techniques which are applicable to the interactions of other salts and of solids in general with seawater.

Wattenberg (32) first determined the solubility of calcite in seawater at atmospheric pressure. He used the stoichiometric solubility product

$$K'_{SP} = (Ca^{2+})_{T,S}(CO_3^{\ 2-})_{T,S} \tag{3}$$

where the subscript S refers to the concentration at saturation. The use of stoichiometric solubility products obviates the need to estimate the activity coefficients of single ions. These products remain essentially constant at a given salinity for processes, such as the dissolution and the precipitation of carbonates in the oceans, which have only a slight effect upon the major ion composition of seawater. There have been many measurements of the solubility of carbonates in seawater at atmospheric pressure since Wattenberg (32), the most recent ones being those of McIntire (33) and of Ingle et al. (34).

Pytkowicz and co-workers (35-39) extended the results on carbonates to high pressures by means of potentiometric measurements while Berner (40), Millero and Berner (41), and Duedall (42) used the partial molal volume approach.

Several interesting features emerged from the study of the solubility of calcium carbonate in seawater, which may be relevant to other solids. Some of these features are; the metastable supersaturation of near-surface waters, the hysteresis and lack of reproducibility of solubility data for calcite, the factors that control the crystal form that precipitates from seawater and the diagenetic alteration of sediments, the kinetic behavior of carbonate materials at depth, the cycling of carbonates in nature, the pH buffering of seawater, and the distribution and fluxes of the carbon dioxide system in nature.

The metastability of the calcium carbonate supersaturation in near-surface waters results from the inhibition of nucleation and growth by magnesium ions, organic matter, and phosphate ions (43-46). The results of Pytkowicz (43, 45) showed that the inorganic precipitation of carbonates in the oceans can only occur in a few special environments and that the removal of carbonates from seawater is primarily biogenic.

Weyl (47) observed that the solubility of calcite in seawater undergoes hysteresis-type effects which were not observed to any significant extent for aragonite. He attributed his results to the adsorption of magnesium in surface coatings. Hysteresis due to surface coatings may be the cause of the poor reproducibility observed in calcite solubility data. The presence of hydrated phases, which may affect the solubility behavior of calcite, has recently been proposed (48).

The organic and inorganic factors that control the crystal form of calcium carbonate that is precipitated inorganically from seawater were studied by Kitano (49) and Kitano et al. (50). Although most of the carbonate sedimentation in seawater is biogenic (43,45), these researches may lead to insights into processes within the body fluids of organisms. The form of calcium carbonate that precipitates from normal seawater, to which a solulbe carbonate salt has been added to supersaturate it, is aragonite Calcareous organisms produce primarily a series of magnesian calcites although some aragonite is also formed. The mechanisms which control the crystal forms of calcium carbonate in the shells of calcareous organisms have not yet been elucidated.

Aragonite in sediments is converted gradually to the more stable calcite (51) and magnesian calcites tend to be diagenetically altered to purer calcites and to dolomite (52,53). Chave et al. (54) found that the stability of carbonates in seawater increases in the order; high magnesian calcite, aragonite, low magnesian calcite, pure calcite, dolomite, but Berner (46) recently concluded that calcites with 2-7% mole-fraction of $MgCO_3$ are thermodynamically stable in seawater.

Intermediate oceanic waters are undersaturated in the North Pacific Ocean. This is the result of the high concentration of carbon dioxide present there because the waters are old and extensive oxidation of organic matter has taken place in them. All deep oceanic waters are undersaturated as the result of the effects of high pressures and low temperatures upon the solubility of calcium carbonate (36-39, 55).

Degree of saturation data indicate that carbonate sediments can persist until burial while exposed to undersaturated waters (55,56). This conclusion is confirmed by the fact that the lysocline, the depth at which calcareous tests first show signs of dissolution, is well above the carbonate compensation depth, which marks a sudden decrease in the carbonate content of sediments (57,58). Morse and Berner (59) concluded from their kinetic results that the large increase in dissolution rate at the compensation depth corresponds to a change in the mechanism of solution.

Dissolution of carbonates at depth led to studies of the carbon dioxide-carbonate cycles within and through the oceans as well as to studies of the factors which control the carbon dioxide components and the pH (5,8,9,60).

Other minerals with solubilities that have been determined in seawater are calcium phosphates (61), silica (62,63), the least soluble compounds of a series of trace metals (64), and clays (65).

In solubility work as well as in the case of dissociation constants it is important, before applying equilibrium data obtained in seawater to estuaries and to pore waters to ascertain that there have been no large changes in the major ion proportions.

## Ion Association

This is an important topic because the formation of ion-pairs, by affecting the distributions of solute sizes and charges, modifies most physico-chemical properties of seawater, including solubility equilibria.

Garrels and Thompson (66) pioneered work on the formation of ion-pairs in seawater. They were forced to make a large number of assumptions because needed data was not available to them. Even so, some features of their results were confirmed by subsequent investigations. Kester and Pytkowicz (67) and Pytkowicz and Hawley (68) used potentiometric methods to determine the concentrations of free ions and of ion-pairs in seawater and were able to avoid most of the assumptions made by Garrels and Thompson (66).

There still are some contradictions in the results of different investigators regarding the interactions among the major ions of seawater. In addition, further work is needed to characterize possible triple ions and the effects of temperature, salinity, and pressure upon ion-pairing (68).

## Activity Coefficients

We will dwell at some length upon this topic because much of the material presented here has not been published before. Activity coefficients are valuable because they provide insights into solvent-ion and ion-ion interactions. Also, they are useful for stoichiometric computations made when apparent dissociation constants and stoichiometric solubility products are not available.

One may use free or total mean activity coefficients and free or total single ion activity coefficients, depending upon the problem under consideration. Free coefficients are those determined in solutions which there are no specific ionic interactions. In seawater they are constructs which correspond to the free ions if the ion-pairing model is used (68) or to long-range ionic interactions, of the Debye-Hückel type (69), if the specific interaction model of Brønsted (70) and Guggenheim (71) is employed (72). Total activity coefficients are those obtained in solutions in which ion-pairs or short-range specific interactions occur and are related to the free ones by (12)

$$a = f_F(F) = f_T(T) \quad (4)$$

a is the activity, f the activity coefficient, while (F) and (T) represent the concentration of free ions and the stoichiometric (total) concentration.

The mean free activity coefficient can be obtained from data in chloride solutions if such solutions are indeed unassociated (67). This must be done at the same effective ionic strength as that of the seawater of interest. The effective ionic strength includes the effects of ion-pairing and yields an extended ionic strength principle for non-specific interactions in moderately concentrated multi-electrolyte solutions, as was demonstrated by Pytkowicz and Kester (73). Others prefer to obtain mean free coefficients by assuming that the Debye-Hückel equation is valid at the ionic strength of seawater, with a hydration correction (74) or without it (72).

Total mean activity coefficients can be measured directly in the associated solutions of interest or can be calculated from free coefficients coupled to ion-pair or specific interaction terms.

Corresponding free and total activity coefficients of single ions are conventional quantities which depend upon non-thermodynamic assumptions and are, therefore, of unknown accuracy. Still, they are useful for many of the oceanographic and geochemical computations which are based upon thermodynamic equilibrium constants.

Next, we will calculate activity coefficients by the mean salt method coupled to ion-pair models and will compare them to those obtained by other investigators.

The free activity coefficients of single ions, shown in Table II, were obtained by us in two steps. First, an interpolation equation was used to obtain values of the mean activity coefficients intermediate to those compiled by Harned and Owen (75) and by Robinson and Stokes (76). Then, the activity coefficients of the single ions were obtained by the mean-salt method (66). This method depends upon the validity of the MacInnes (77) convention, $(f_K)_F = f_{Cl}$, which is based upon transference numbers. This convention cannot be verified unambiguously because the activities of single ions cannot be measured. The mean salt

Table II. Free Activity Coefficients of Potassium, Sodium, Calcium, Magnesium and Chloride Versus the Ionic Strength at 25 °C, Obtained by the Mean-Salt Method

| Ionic Strength (molal) | $K^+$ | $Na^+$ | $Ca^{2+}$ | $Mg^{2+}$ | $Cl^-$ |
|---|---|---|---|---|---|
| 0.00 | 1.000 | 1.000 | 1.000 | 1.000 | 1.000 |
| 0.05 | 0.816 | 0.826 | 0.473 | 0.486 | 0.816 |
| 0.10 | 0.769 | 0.787 | 0.392 | 0.408 | 0.769 |
| 0.15 | 0.739 | 0.765 | 0.351 | 0.368 | 0.739 |
| 0.20 | 0.718 | 0.750 | 0.325 | 0.343 | 0.718 |
| 0.25 | 0.701 | 0.739 | 0.307 | 0.326 | 0.701 |
| 0.30 | 0.687 | 0.732 | 0.294 | 0.313 | 0.687 |
| 0.35 | 0.676 | 0.726 | 0.284 | 0.305 | 0.676 |
| 0.40 | 0.666 | 0.721 | 0.277 | 0.298 | 0.666 |
| 0.45 | 0.657 | 0.718 | 0.271 | 0.293 | 0.657 |
| 0.50 | 0.650 | 0.716 | 0.266 | 0.290 | 0.650 |
| 0.55 | 0.643 | 0.714 | 0.263 | 0.288 | 0.643 |
| 0.60 | 0.637 | 0.712 | 0.260 | 0.286 | 0.637 |
| 0.65 | 0.631 | 0.712 | 0.258 | 0.285 | 0.631 |
| 0.70 | 0.626 | 0.711 | 0.256 | 0.285 | 0.626 |
| 0.75 | 0.622 | 0.711 | 0.255 | 0.286 | 0.622 |
| 0.80 | 0.618 | 0.711 | 0.254 | 0.286 | 0.618 |
| 0.85 | 0.614 | 0.711 | 0.253 | 0.288 | 0.614 |
| 0.90 | 0.610 | 0.712 | 0.253 | 0.289 | 0.610 |
| 0.95 | 0.607 | 0.712 | 0.253 | 0.291 | 0.607 |
| 1.00 | 0.603 | 0.713 | 0.253 | 0.293 | 0.603 |

method should not be used for anions unless they do not associate with potassium ions.

Measured mean activity coefficients reflect two types of hydration effects; the direct effect of ion-solvent interactions upon chemical potentials and the changes in concentration and ionic strength due to the removal of water of hydration from the bulk solution. The first effect is taken into account automatically in the mean-salt method as the method is based upon measured mean coefficients. We made a correction for the second effect by means of the equation

$$I_A = I(1 + 0.018h\, I_A/n) \tag{5}$$

$I_A$ is the corrected ionic strength, h is the hydration number, and n is 1 for 1-1 salts and 3 for 1-2 salts. The effect of the change in the ionic strength upon activity coefficients was found to be negligible.

The data in Table II can be used for estuarine and for pore waters as free activity coefficients at a given ionic strength are insensitive to the composition of the solution (67, 68, 73, 78). Furthermore, we calculated from the Harned rule coefficients of Robinson and Bower (79) that the mean activity coefficients of NaCl and $CaCl_2$, in going from a pure NaCl to a pure $CaCl_2$ solution at 0.75 ionic strength, only change by 1.8 and by 0.1% respectively.

Bates et al. (74) used a hydration equation to estimate the free activity coefficients of single ions. They assumed that the Debye-Hückel equation can be used to describe the effects of non-specific electrostatic interactions for ionic strengths up to I = 6 and that chloride ions are not hydrated. Our results, shown in Table II, and those of Bates et al. (74) differ by about 3% for potassium and sodium and by up to 16% for calcium and magnesium at I = 1.0. It is not possible to decide which set of results is more accurate because, as was mentioned earlier, the activity coefficients of single ions are conventional.

Marine chemists often need activity coefficients expressed as a function of the salinity. The coefficients in Table II were expressed in terms of the salinity, as is shown in Table III, by means of the ionic strength-salinity relationship of Lyman and Fleming (80). We found that the use of the effective ionic strength (73) instead of the conventional one has a negligible effect upon the activity coefficients.

The total activity coefficients of the single ions obtained in this work for $K^+$, $Na^+$, $Ca^{2+}$ and $Mg^{2+}$, and shown in Table IV, were calculated from the free ones, Equation 4 in the form $f_T = f_F(F)/(T)$, and the speciation model of Pytkowicz and Hawley (68). The coefficient of sulfate was obtained by the method of Kester and Pytkowicz (67) while those of bicarbonate and carbonate ions were calculated from the ratios of the apparent dissociation constants of carbonic acid to the thermodynamic ones, the activity coefficient of carbonic acid, and the activity of water in seawater. The total mean activity coefficients were then calculated from those for the single ions and are shown in Table V. The results of Berner (81) and of van Breemen (82) were estimated by procedures akin to those of Garrels and Thompson (66).

Whitfield (72) assumed that non-specific ionic interactions

Table III. Free Activity Coefficients of Single Ions at 25°C Versus the Salinity

| Salinity (‰) | $K^+$ | $Na^+$ | $Ca^{2+}$ | $Mg^{2+}$ | $Cl^-$ |
|---|---|---|---|---|---|
| 25 | 0.648 | 0.715 | 0.265 | 0.289 | 0.648 |
| 26 | 0.646 | 0.714 | 0.264 | 0.288 | 0.646 |
| 27 | 0.643 | 0.714 | 0.263 | 0.288 | 0.643 |
| 28 | 0.641 | 0.713 | 0.261 | 0.287 | 0.641 |
| 29 | 0.637 | 0.712 | 0.260 | 0.286 | 0.637 |
| 30 | 0.635 | 0.712 | 0.259 | 0.286 | 0.635 |
| 31 | 0.633 | 0.712 | 0.258 | 0.285 | 0.633 |
| 32 | 0.630 | 0.711 | 0.257 | 0.285 | 0.630 |
| 33 | 0.628 | 0.711 | 0.257 | 0.285 | 0.628 |
| 34 | 0.626 | 0.711 | 0.256 | 0.285 | 0.626 |
| 35 | 0.625 | 0.711 | 0.255 | 0.285 | 0.625 |
| 36 | 0.623 | 0.711 | 0.255 | 0.285 | 0.623 |
| 37 | 0.620 | 0.711 | 0.254 | 0.286 | 0.620 |
| 38 | 0.618 | 0.711 | 0.254 | 0.286 | 0.618 |
| 39 | 0.617 | 0.711 | 0.254 | 0.287 | 0.617 |
| 40 | 0.615 | 0.711 | 0.254 | 0.287 | 0.615 |

Table IV. Total Activity Coefficients of Single Ions at about 35‰ Salinity and 25°C

| Reference | $K^+$ | $Na^+$ | $Ca^{2+}$ | $Mg^{2+}$ | $Cl^-$ | $SO_4^{2-}$ |
|---|---|---|---|---|---|---|
| This work | 0.618 | 0.695 | 0.225 | 0.254 | 0.625 | 0.084 |
| (81) | 0.624 | 0.703 | 0.237 | 0.252 | 0.630 | 0.068 |
| (82) | 0.620 | 0.695 | 0.228 | 0.254 | 0.630 | 0.090 |
| (72) | 0.617 | 0.650 | 0.203 | 0.217 | 0.686 | 0.122 |
| (83) | 0.630 | 0.680 | 0.214 | 0.234 | 0.658 | 0.108 |

| | $HCO_3^-$ | $CO_3^{2-}$ |
|---|---|---|
| This work | 0.501 | 0.030 |

can be described by the modified Debye-Hückel equation with Bå = 1. He did not introduce a hydration correction even though both hydration effects mentioned earlier should be taken into consideration. The specific short range interactions were accounted for by additive interaction terms between cations and anions. The good agreement between his calculated mean coef-

Table V. Total Mean Activity Coefficients at about 35‰ Salinity and 25°C

| Reference / Salt | This work | (81) | (82) | (72) | (83) | (84) | Experimental | |
|---|---|---|---|---|---|---|---|---|
| KCl | 0.621 | 0.627 | 0.625 | 0.650 | 0.644 | 0.639 | | |
| NaCl | 0.659 | 0.666 | 0.662 | 0.668 | 0.666 | 0.669 | 0.672 | (86) |
| $CaCl_2$ | 0.445 | 0.455 | 0.449 | 0.457 | 0.453 | | | |
| $MgCl_2$ | 0.463 | 0.464 | 0.465 | 0.467 | 0.466 | 0.467 | | |
| $K_2SO_4$ | 0.318 | 0.299 | 0.326 | 0.360 | 0.355 | 0.345 | | |
| $Na_2SO_4$ | 0.343 | 0.323 | 0.352 | 0.372 | 0.368 | 0.366 | 0.378 | (87) |
| $CaSO_4$ | 0.137 | 0.127 | 0.143 | 0.157 | | | | |
| $MgSO_4$ | 0.146 | 0.131 | 0.151 | 0.163 | 0.151 | 0.158 | | |
| $KHCO_3$ | 0.56 | | | | | | | |
| $NaHCO_3$ | 0.59 | | | | | | | |
| $Ca(HCO_3)_2$ | 0.39 | | | | | | | |
| $Mg(HCO_3)_2$ | 0.40 | | | | | | | |
| $K_2CO_3$ | 0.23 | | | | | | | |
| $Na_2CO_3$ | 0.24 | | | | | | | |
| $CaCO_3$ | 0.083 | | | | | | 0.090 | (*) |
| $MgCO_3$ | 0.087 | | | | | | | |

(*) Obtained from the ratio of the stoichiometric to the thermodynamic solubility product of calcium carbonate.

ficients and experimental data indicates that hydration enters inadvertently into the interaction terms. Leyendekkers (83) included interactions between ions of the same charge types.

The total activity coefficients calculated by the various methods are in rough agreement. A comparison of Tables IV and V shows that the low value of $(f_{\pm Na_2SO_4})_T$ obtained in this work relative to that of Whitfield (72) is primarily due to the contribution of sulfate ions. Our value of $(f_{SO_4})_T$ was obtained without resort to the mean-salt method and is essentially based upon potentiometric measurements. This indicates an incompatibility between the experimental result of Kester and Pytkowicz (67) obtained with glass and ion-exchange electrodes and that of Platford (87) which was measured with amalgam electrodes. Robinson and Wood (84) calculated the total mean activity coefficients by an extension of the method of Friedman (85).

## Other Equilibria

The equilibria involving the complexation of trace metals have recently been reviewed by Stumm (88) and are further examined elsewhere in this volume. Kester (89) presented a review of gas-seawater interactions. Many other aspects of marine chemistry and geochemistry have been described in Horne (90), Riley and Chester (6), Berner (81), Garrels and Mackenzie (91), Broecker (92), Goldberg (93), and Pytkowicz (5). We will, therefore, limit this work to the topics already described and to a look at the equilibrium chemistry of estuarine waters followed by a brief examination of the control of the oceanic composition.

## Equilibrium in Estuaries

Little appears to have been done concerning equilibria in estuaries. We will examine, therefore, how one may ascertain whether equilibrium constants measured in seawater can be applied to estuaries. This will be illustrated by considering the dissociation reactions of carbonic acid.

The composition of rivers depends upon the geological nature of their drainage basins (91). For illustrative purposes we will, therefore, discuss the effects of mixing seawater with a hypothetical river of a world-average composition which is shown in Table VI. The relative composition of river water differs from that of seawater and their mixing yields in principle estuarine waters in which the proportions of the major ions

Table VI. Concentration Parameters for Mixtures of River Water and Seawater. The Concentrations are Expressed in Molality x $10^3$. $I_T$ is the Total Ionic Strength.

| Cl‰ / SW% | 19.0 / 100 | $15.20_1$ / 80 | $11.40_2$ / 60 | $7.60_4$ / 40 | $3.80_6$ / 20 | .0078 / 0 |
|---|---|---|---|---|---|---|
| $Na^+$ | 476.0 | 378.1 | 281.7 | 186.6 | 92.78 | .274 |
| $K^+$ | 10.37 | 8.247 | 6.157 | 4.096 | 2.063 | .0588 |
| $Mg^{+2}$ | 54.06 | 42.98 | 32.05 | 21.27 | 10.65 | .1687 |
| $Ca^{+2}$ | 10.44 | 8.372 | 6.330 | 4.318 | 2.332 | .374 |
| $Cl^-$ | 544.4 | 440.4 | 328.1 | 217.3 | 108.0 | .2200 |
| $SO_4^{-2}$ | 28.68 | 22.80 | 17.01 | 11.30 | 5.671 | .1166 |
| $HCO_3^-$ | 2.122 | 1.883 | 1.646 | 1.414 | 1.184 | .9575 |
| $I_T$ | .7078 | .5626 | .4196 | .2785 | .1393 | .002074 |

differ from those in the oceans. Thus, estuarine waters cease to be ionic media of constant composition, the simplifying feature which renders apparent constants useful. Fortunately, however, river water is so dilute that estuarine waters tend to retain the major ion proportions of seawater down to fairly low chlorinities.

The extent of mixing in an estuary depends upon a large number of factors such as the seasonally variable river flow, the tidal cycle, the wind state, the topography, and the relative temperatures of the river water and the seawater. One must also consider in actual estuaries the inflow of ground and melt waters. The conventional salinity depends upon the relative composition of seawater and, therefore, the chlorinity and the total salt content are better compositional variables than the salinity to ascertain the extent of mixing at any given location.

In Table VII we present the ratios of the concentrations of the major ions in seawater and in estuarine water at the same chlorinities. It can be seen that the concentrations remain similar in the two media down to low values of the chlorinity.

Next, let us consider by how much equilibrium constants determined in seawater may differ from those in estuarine waters at the same chlorinities. The example will be hypothetical because we are considering a world-average river and

Table VII. Ratios of Concentrations and of the Ionic Strength in Seawater to Those in the Estuarine Water at the Same Chlorinity

| % SW | 100 | 80 | 60 | 40 | 20 |
|---|---|---|---|---|---|
| $Na^+$ | 1.00 | 1.0000 | .9999 | .9997 | .9993 |
| $K^+$ | 1.00 | .9981 | .9964 | .9919 | .9786 |
| $Mg^{+2}$ | 1.00 | .9993 | .9781 | .9958 | .9889 |
| $Ca^{+2}$ | 1.00 | .9909 | .9742 | .9478 | .8721 |
| $Cl^-$ | 1.00 | ----- | ----- | ----- | ----- |
| $SO_4^{-2}$ | 1.00 | .9990 | .9975 | .9943 | .9852 |
| $I_T$ | 1.00 | .999 | .9982 | .9961 | .9897 |

because apparent dissociation constants have not yet been determined at the low chlorinities which may be found near the heads of estuaries.

Apparent dissociation constants are related to the corresponding thermodynamic constants by total activity coefficients. These coefficients in turn are related to the free ones by $f_T = f_F(F)/(T)$ where $f_F$ reflects the ionic strength effect and $(F)/(T)$ represents the ion-pair effect.

The ionic strength effect can be calculated from the data in Table II and from the free activity coefficients of bicarbonate and carbonate ions (94). The effect of ion-pairing can be calculated from the equations of Pytkowicz and Hawley (68) which relate apparent dissociation constants to stoichiometric association constants and free cation concentrations. For $K'_1$, for example, the equations yield

$$\frac{(K'_1)_{EW}}{(K'_1)_{SW}} = \frac{1 + \Sigma K^*_{MHCO_3}(M^{c+})_{F(EW)}}{1 + \Sigma K^*_{MHCO_3}(M^{c+})_{F(SW)}} \qquad (6)$$

where $K^*$ is the stoichiometric association constant, $M^{c+}$ represents cations, EW refers to estuarine waters and SW refers to seawater. The results of the combined effects are shown in Table VIII.

The association constants and the free ion concentrations

Table VIII. Ratios of $K'_1$ and $K'_2$, the Apparent Dissociation Constants of Carbonic Acid, in Seawater to Those in Estuarine Waters at the Same Chlorinities

| % Seawater | $(K'_1)_{SW}/(K'_1)_{EW}$ | $(K'_2)_{SW}/(K'_2)_{EW}$ |
|---|---|---|
| 100 | 1.000 | 1.000 |
| 80 | 0.9994 | 0.9981 |
| 60 | 0.9986 | 0.9943 |
| 40 | 0.9977 | 0.9877 |
| 20 | 0.9953 | 0.9670 |
| 10 | 0.9914 | 0.9290 |

are well known only at 19‰ chlorinity. Still, one may estimate the maximum and the minimum effects of compositional differences upon the apparent dissociation constants by using values of $K^*$ in Equation 6 at 19‰ chlorinity and at infinite dilution respectively. The association constants at infinite dilution are considerably larger than those at 19‰ and enhance the effect of compositional variations. Only the maximum effects are shown in Table VIII.

It can be seen from the results in Table VIII that $K_1$ and $K_2$ determined for oceanic waters can be used in estuaries supplied by a hypothetical world-average river down to estuarine waters which contain only 20% seawater. This can be done with an error of less than 4%, which is within the limits of uncertainty of measured apparent constants (20).

## Control of the Oceanic Composition

An understanding of the control mechanisms for the chemical composition of the oceans is important because it helps us understand the geochemical history of the oceans and the impact of perturbations upon the chemical nature of seawater. This subject was examined critically by Pytkowicz (5) and will only be mentioned briefly here.

The control of the composition of seawater can be approached from two points of view; that of variations in time and space within the oceans and the broad geochemical view in which the oceans are but one of several linked reservoirs. The first approach consists basically in determining the parameters for and solving the equation

$$\frac{\partial N}{\partial t} = \sum_i \frac{\partial}{\partial i} \left( \frac{A_i}{\rho} \frac{\partial N}{\partial i} - V_i N \right) + R \qquad (7)$$

N is the concentration of a non-conservative constituent, $\frac{A_i}{\rho}$ is the eddy diffusion coefficient, and $V_i$ is the advective velocity. R represents the effects of sources and sinks such as uptake by the biota and the sediments and gas exchange (95). Considerable work has been done with various forms of Equation 7 in the open ocean and in estuaries (5).

The geochemical description of the oceans is not as straightforward because of the complexity of reactions during weathering and sedimentation, and has been the subject of considerable discussion (5). Some of the main sinks for naturally occurring and man-produced chemicals brought into the oceans by rivers, winds, and submarine volcanism are adsorption and exchange reactions on the surface of settling detrital particles, authigenic precipitation of oxides and sulfides, diffusion into the sediments, biogenic settling, and the diagenetic alteration of sediments.

One aspect of removal mechanisms, namely, seawater-clay interactions, has received considerable attention because Sillen (7) proposed that the equilibria for such interactions may control the concentrations of the major ions and the pH of seawater. Clays probably do exert a measure of control by adding and removing some seawater ions. It is doubtful, however, that equilibria are achieved in the open oceans and it is likely that the major ion composition of seawater, which may have remained somewhat constant for a long time, has done so as the result of steady states rather than of equilibria (5,8-10). Also, it has been shown that the primary controls of the pH and of the buffering capacity of seawater in the recent oceans result from the action of the carbon dioxide system. Still, aluminium silicates do play a role because their weathering contributes about 15% of the primary buffering agent, bicarbonate ions, to the oceans (5,8,9,60).

There is some possibility that clay-seawater equilibria may be reached within the pore waters of sediments (11). This should lead, if first order models are used as guidelines for thought (5), to solutions of the type

$$B = (k_{AB}/k_e)A + B_e \qquad (8)$$

A and B are the contents of a given constituent in the weathering and in the oceanic reservoirs, $B_e$ is the value of B at equilibrium, and $k_{AB}$ and $k_e$ represent the rate constants for the river and the exchange fluxes. It can be seen from Equation 8 that equilibrium in pore waters would lead to an eventual steady state B in the oceans different from the equilibrium value $B_e$ unless the river input was much smaller than the flux into the sediments. It can also be seen that the term $B_e$ would be absent if the pore waters were not at equilibrium with the sediments. Thus equations governing the removal of pollutants could be quite different in the presence or in the absence of equilibria.

It is important for the modelling of fluxes in the presence of man-made perturbations to determine and use actual rate laws for the transfer of chemicals between natural reservoirs. As an example, Donaghay, Small and Pytkowicz (unpublished results) made preliminary laboratory studies of the transfer of increasing amounts of carbon dioxide in seawater to marine phytoplankton. This is a vital topic because the oceanic biota can take up a considerable fraction of the carbon dioxide released to the atmosphere and the oceans by the combustion of fossil fuels and, thus, attenuate the greenhouse effect.

The experiments were performed in chemostats, with an Isochrysis galbana culture, while bubbling air containing various amounts of carbon dioxide through the system. The biomass went from 7 mg/l for a seawater with a normal carbon dioxide content to 16 mg/l at five times normal to 13 mg/l at ten times normal carbon dioxide. Thus, a tenfold increase in carbon dioxide can double the biomass. The decrease from 16 to 13 mg/l may perhaps reflect the effect of the decreasing pH upon biological activity, a hypothesis that is under study. By generalizing the equations and the extrapolations of Pytkowicz (96) to include the marine phytoplankton, one finds that the atmospheric carbon dioxide may increase in a century by a factor of ten instead of fourteen, the value obtained in the absence of the marine biota. Of course, this is only a preliminary result since a single phytoplankton species was used. Also nutrient limitation, the land biota, and biotic adaptation were not considered, and the prediction of the future consumption of fossil fuels (96) was pessimistic. It illustrates, however, the importance of determining rate laws for the chemical fluxes in nature.

## Acknowledgement

This work was supported by the Oceanography Section of the

National Science Foundation through Grant DES T2-01631 and by the Office of Naval Research Contract N00014-67-A-0369-0007 under NR083-102.

Abstract

Concurrent chemical, biological and geological processes, in addition to the presence of the gravitational potential and of temperature and pressure gradients must be taken into consideration in the study of chemical equilibria in the oceans. A simplifying feature occurs because the approximate constant ratios of the concentrations of the major ions mean that seawater is an ionic medium of constant relative composition. Estuarine waters differ from normal seawater because of time and spacial variations in composition due to the input of river water. The equilibrium chemistry of estuarine waters has not been studied to any significant extent but it will be shown that equilibrium data for open ocean waters are often applicable to estuaries.

Literature Cited

1. Pytkowicz, R. M., Kester, D. R., "Oceanogr. Mar. Biol. Ann. Rev.," H. Barnes, ed., Vol. 9, pp. 11-60, Allen and Unwin, London, 1971.
2. Culkin, F., "Chemical Oceanography," J. P. Riley and G. Skirrow, eds., Vol. 1, pp. 121-161, Academic, New York, 1965.
3. Barth, T. F. W., "Theoretical Petrology," Wiley, New York, 1952.
4. Goldberg, E. D., "Chemical Oceanography," J. P. Riley and G. Skirrow, eds., Vol. 1, pp. 163-196, Academic, New York, 1965.
5. Pytkowicz, R. M., Earth Sci. Rev., Vol. 11, p. 1, 1975.
6. Riley, J. P., Chester, R., "Introduction to Marine Chemistry," Academic, New York, 1971.
7. Sillen, L. G., "Oceanography," M. Sears, ed., Am. Assoc. Adv. Sci., Washington, D. C., 1961.
8. Pytkowicz, R. M., "The Changing Chemistry of the Oceans," D. Dyrssen and D. Jagner, eds., pp. 147-152, Almqvist and Wiksell, Stockholm, 1972.
9. Pytkowicz, R. M., Schweizer. Zeitsch. Hydrol. (1973) 35, 8.
10. Broecker, W. S., Quatern. Res. (1971) 1, 188.
11. Siever, R., Earth Planet. Sci. Letters (1968) 5, 106.

12. Pytkowicz, R. M., Duedall, I. W., Connors, D. N., Science (1966) 152, 640.
13. Buch, K., Harvey, H. W., Wattenberg, H., Gripenberg, S., Rapp. Cons. Int. Explor. Mer (1932) 79, 1.
14. Pytkowicz, R. M., Ingle, S. E., Mehrbach, C., Limnol. Oceanogr. (1974) 19, 665.
15. Hansson, I., Deep-Sea Res. (1973) 20, 46.
16. Hansson, I., Deep-Sea Res. (1973) 20, 479.
17. Culberson, C., Pytkowicz, R. M., Hawley, J. E., J. Mar. Res. (1970) 28, 15.
18. Buch, K., Acta Acad. Aeboensis Mat. Phys. (1938) 11, 1.
19. Lyman, J., Ph. D. Dissertation, University of California, Los Angeles, 1956.
20. Mehrbach, C., Culberson, C. H., Hawley, J. E., Pytkowicz, R. M., Limnol. (1973) 18, 897.
21. Disteche, A., Disteche, S., J. Electrochem. Soc. (1967) 114, 330.
22. Culberson, C., Kester, D. R., Pytkowicz, R. M., Science (1967) 157, 59.
23. Culberson, C., Pytkowicz, R. M., Limnol. Oceanogr. (1968) 13, 403.
24. Park, P. K., Limnol. Oceanogr. (1969) 14, 179.
25. Dyrssen, D., Sillen, L. G., Tellus (1967) 19, 113.
26. Manheim, F., Stockholm Contrib. Geol. (1961) 8, 27.
27. Grasshoff, K., Bull. Cons. Int. Explor. Mer (1964) 123, 1.
28. Kester, D. R., Pytkowicz, R. M., Limnol. Oceanogr. (1967) 12, 246.
29. Goldhaber, M. B., Ph. D. Dissertation, University of California, Los Angeles, 1974.
30. Culberson, C. H., Pytkowicz, R. M., Mar. Chem. (1973) 1, 309.
31. Bates, R. G., Macaskill, J. B., "Analytical Methods of Oceanography," R. F. Gould, ed., American Chemical Society, Washington, D. C., in press.
32. Wattenberg, H., Wiss. Ergebn. Dt. Atlant. Exped. 'Meteor' (1933) 8, 1.
33. McIntire, W. G., Bull. Res. Bd. Canada No. 200, Ottawa, 1965.
34. Ingle, S. E., Culberson, C. H., Hawley, J. E., Pytkowicz, R. M., Mar. Chem. (1973) 1, 295.
35. Pytkowicz, R. M., Connors, D. N., Science (1964) 144, 840.
36. Pytkowicz, R. M., Limnol. Oceanogr. (1965) 10, 220.
37. Pytkowicz, R. M., Disteche, A., Disteche, S., Earth Planet. Sci. (1967) 2, 430.

38. Pytkowicz, R. M., Fowler, G. A., Geochem. J. (1967) 1, 169.
39. Hawley, J. E., Pytkowicz, R. M., Geochim. Cosmochim. Acta (1969) 33, 1557.
40. Berner, R. A., Geochim. Cosmochim. Acta (1965) 29, 947.
41. Millero, F. J., Berner, R. A., Geochim. Cosmochim. Acta (1972) 36, 92.
42. Duedall, I. W., Geochim. Cosmochim. Acta (1972) 36, 729.
43. Pytkowicz, R. M., J. Geol. (1965) 73, 196.
44. Chave, K. E., Suess, E., Limnol. Oceanogr. (1970) 15, 633.
45. Pytkowicz, R. M., Am. J. Sci. (1973) 273, 515.
46. Berner, R. A., Geochim. Cosmochim. Acta, in press.
47. Weyl, P. K., Stud. Trop. Oceanogr. Miami (1967) 5, 178.
48. North, N. A., Geochim. Cosmochim. Acta (1974) 38, 1075.
49. Kitano, Y., Bull. Chem. Soc. Japan (1962) 35, 1973.
50. Kitano, Y., Kanamori, N., Tokuyama, A., Am. Zool. (1969) 9, 681.
51. Bischoff, J. L., Fyfe, W. S., Am. J. Sci. (1968) 266, 65.
52. Peterson, M. N. A., Von Der Borch, C. C., Bien, G. S., Am. J. Sci. (1966) 264, 257.
53. Clayton, R. M., Jones, B. F., Berner, R. A., Geochim. Cosmochim. Acta (1968) 32, 415.
54. Chave, K. E., Deffeyes, K. S., Weyl, P. K., Garrels, R. M., Thompson, M. E., Science (1962) 137, 33.
55. Edmond, J. M., Deep-Sea Res. (1974) 21, 455.
56. Pytkowicz, R. M., Geochim. Cosmochim. Acta (1970) 34, 836.
57. Berger, W. H., Mar. Geol. (1970) 8, 111.
58. Heath, G. R., Culberson, C. H., Geol. Soc. Amer. Bull. (1970) 81, 3157.
59. Morse, J. W., Berner, R. A., Am. J. Sci. (1972) 272, 840.
60. Pytkowicz, R. M., Geochim. Cosmochim. Acta (1967) 31, 63.
61. Roberson, C. E., M. S. Thesis, University of California, San Diego, 1965.
62. Krauskopf, K. B., Geochim. Cosmochim. Acta (1956) 10, 1.
63. Jones, M. M., Pytkowicz, R. M., Bull. Soc. Royale Sci. Liege (1973) 42, 125.
64. Krauskopf, K. B., Geochim. Cosmochim. Acta (1956) 9, 1.
65. Siever, R., Woodford, N., Geochim. Cosmochim. Acta (1973) 37, 1851.

66. Garrels, R. M., Thompson, M. E., Am. J. Sci. (1962) 260, 57.
67. Kester, D. R., Pytkowicz, R. M., Limnol. Oceanogr. (1969) 14, 686.
68. Pytkowicz, R. M., Hawley, J. E., Limnol. Oceanogr. (1974) 19, 223.
69. Debye, P., Hückel, E., Physik. Z. (1923) 24, 185.
70. Brønsted, J. N., J. Am. Chem. Soc. (1922) 44, 877.
71. Guggenheim, E. A., Philos. Mag. (1935) 19, 588.
72. Whitfield, M., Mar. Chem. (1973) 1, 251.
73. Pytkowicz, R. M., Kester, D. R., Am. J. Sci. (1969) 267, 217.
74. Bates, R., Staples, B. R., Robinson, R. A., Analyt. Chem. (1970) 42, 867.
75. Harned, H. S., Owen, B. B., "The Physical Chemistry of Electrolytic Solutions," Am. Chem. Soc. Monogr. 137, Reinhold, New York, 1958.
76. Robinson, R. A., Stokes, R. H., "Electrolyte Solutions," Butterworths, 2nd ed., London, 1965.
77. MacInnes, D. A., J. Am. Chem. Soc. (1919) 41, 1086.
78. Gieskes, J. M., Z. Physik. Chem. (Frankfurt) (1966) 50, 78.
79. Robinson, R. A., Bower, V. E., J. Res. NBS (1966) 70A(4), 313.
80. Lyman, J., Fleming, R. H., J. Mar. Res. (1940) 3, 135.
81. Berner, R. A., "Principles of Chemical Sedimentology," McGraw-Hill, New York, 1971.
82. van Breemen, N., Geochim. Cosmochim. Acta (1973) 37, 101.
83. Leyendekkers, J. V., Mar. Chem. (1973) 1, 75.
84. Robinson, R. A., Wood, R. H., J. Solut. Chem. (1972) 1, 481.
85. Friedman, H. L., "Ionic Solution Theory," Interscience, New York, 1962.
86. Platford, R. F., J. Mar. Res. (1965) 23, 55.
87. Platford, R. F., Dafoe, T., J. Mar. Res. (1965) 23, 63.
88. Stumm, W., "Chemical Oceanography," J. P. Riley and G. Skirrow, eds., 2nd ed., Vol. pp. , Academic New York, .
89. Kester, D. R., "Chemical Oceanography," J. P. Riley and G. Skirrow, eds., 2nd ed., Vol. , pp. , Academic, New York, .

90. Horne, R. A., "Marine Chemistry: The Structure of Water and the Chemistry of the Hydrosphere," Interscience, New York, 1969.
91. Garrels, R. M., Mackenzie, F. T., "Evolution of Sedimentary Rocks," Norton, New York, 1971.
92. Broecker, W. S., "Chemical Oceanography," Harcourt Brace Jovanovich, New York, 1974.
93. Goldberg, E. D., ed., "The Sea: Ideas and Observations on Progress in the Study of the Sea," Vol. 5, Interscience, New York, 1974.
94. Walker, A. C., Bray, V. B., Johnston, J., J. Am. Chem. Soc. (1927) 49, 1235.
95. Redfield, A. C., Ketchum, B. H., Richards, F. A., "The Sea: Ideas and Observations on Progress in the Study of the Sea," M. N. Hill, ed., Vol. 2, pp. 26-77, Interscience, New York, 1963.
96. Pytkowicz, R. M., Comments on Earth Sci.: Geophys. (1972) 3, 15.

2

# The Physical Chemistry of Estuaries

FRANK J. MILLERO

Rosenstiel School of Marine and Atmospheric Science, University of Miami, Miami, Fla. 33149

In recent years, there have been a number of major advances made on the interpretation of ionic interactions in multicomponent electrolyte solutions (1-8). A number of chemical models have been developed from these recent studies. The models have been applied to seawater solutions in an attempt to understand how the physical chemical properties of seawater are related to the ion-water and ion-ion interactions of the major constituents of seawater (8-21) and how the medium of seawater affects the state of metal ions in seawater (8,9,22-39). In the present paper these models and methods will be briefly reviewed and applied to river water and estuarine water (mixtures of river and seawater). In the next two sections the application of these methods to seawater will be reviewed.

## Physical Chemical Properties of Seawater

One of the most useful generalizations to be developed for multicomponent electrolyte solutions is Young's rule (40,41) and its modifications by Wood and coworkers (42-44) and Millero (8, 9-19). Young's rule for a multicomponent electrolyte solution is given by

$$\Phi = \Sigma E_i \phi_i \qquad (1)$$

where $\Phi$ is any apparent equivalent property (such as volume, expansibility, compressibility, enthalpy and heat capacity), $E_i$ is the equivalent ionic fraction of species *i* ($E_i = e_i/e_T$, where $e_i$ is the equivalent molality of species *i* and $e_T = \Sigma e_i$) and $\phi_i$ is the apparent equivalent property of species *i* at the ionic strength of the solution. The apparent equivalent property of the solution is related to the measured property (P) of seawater by

$$\Phi = (P - P^o)/e_T \qquad (2)$$

where $P^o$ is the physical property of pure water and $e_T$ is the total equivalents of sea salt. (Equation 1) essentially states that as a first approximation the excess properties of mixing sea salts at a constant ionic strength can be neglected. This relationship is a very useful first approximation and has been shown to be applicable to many multicomponent electrolyte solutions. For a solution that contains a number of ionic components it is possible to use Young's rule, in terms of the constituent salts, a number of ways. For example, for the major sea salt ions ($Mg^{2+}$, $Na^+$, Cl and $SO_4^{2-}$), we have

$$\Phi = E_{NaCl}\ \phi_{NaCl} + E_{MgSO_4}\ \phi_{MgSO_4} \tag{3}$$

$$\Phi = E_{Na_2SO_4}\ \phi_{Na_2SO_4} + E_{MgCl_2}\ \phi_{MgCl_2} \tag{4}$$

$$\Phi = E_{NaCl}\ \phi_{NaCl} + E_{Na_2SO_4}\ \phi_{Na_2SO_4} + E_{MgCl_2}\ \phi_{MgCl_2} + E_{MgSO_4}\ \phi_{MgSO_4} \tag{5}$$

Wood and Anderson (42) have suggested that (equation 5) is the better of the forms to use since it includes all the possible plus-minus interaction pairs which make up the major ionic interactions. More recent work by Reilly and Wood (44) has indicated a more general form for Young's rule

$$\Phi = \sum_{MX} E_M\ E_X\ \phi_{MX} \tag{6}$$

where $E_M$ is the equivalent fraction of cation M and $E_X$ is the equivalent fraction of anion X. This equation was arrived at by first mixing all the salts of one cation at constant ionic strength in such a manner as to yield the anion composition of the final mixture. This step is repeated for each cation and then all of the resulting cation solutions were mixed to give the final mixture. Millero (8,9) has elaborated on the application of this procedure to seawater. By making the summation for each cation over all its possible salts, one obtains the equivalent weighted cation contribution

$$\Phi(M\Sigma X_i) = E_M E_{Cl}\ \phi_{MCl} + E_M E_{SO_4}\ \phi_{MSO_4} + E_M E_{HCO_3}\ \phi_{MHCO_3} + E_M E_{Br}\ \phi_{MBr} + E_M E_{CO_3}\ \phi_{MCO_3} + E_M E_{B(OH)_4}\ \phi_{MB(OH)_4} + E_M E_F\ \phi_{MF} \tag{7}$$

The total ionic component for seawater is given by

$$\sum_{MX} E_M\ E_X\ \phi_{MX} = \phi(Na\Sigma X_i) + \phi(Mg\Sigma X_i) + \phi(Ca\Sigma X_i) + \phi(K\Sigma X_i) + \phi(Sr\Sigma X_i) \tag{8}$$

By determining the ionic term with this method one essentially eliminates excess cation-anion interactions (possibly due to ion pair formation). Since seawater contains the nonelectrolyte boric acid as one of its major components (i.e., a dissolved solute greater than one part per million; by weight), Millero (8,9) has used a revised form for Young's rule

$$\Phi = \sum_{MX} E_M E_X \phi_{MX} + E_B \phi_B \qquad (9)$$

where $\phi_B$ is the apparent molal property for boric acid and $E_B$ is the equivalent fraction of boric acid in seawater. This extension of Young's rule to electrolyte-nonelectrolyte solutions has been shown to be reliable for the volume of boric acid - NaCl solutions over a wide concentration range (45). Because salts such as $CaSO_4$ are not soluble at the high ionic strength of seawater, one must estimate its apparent properties by using an additivity method

$$\phi(CaSO_4) = \phi(CaCl_2) + \phi(MgSO_4) - \phi(MgCl_2) \qquad (10)$$

and the apparent properties of soluble salts.

By dividing the apparent property into an infinite dilution term ($\Phi^o$) and one or more concentration terms (b is an empirical constant),

$$\Phi = \Phi^o + S\, I_V^{1/2} + b\, I_V + \ldots \qquad (11)$$

it is possible to simplify the use of Young's rule at various concentrations ($I_V$ is the molar ionic strength). Since $\Phi^o = \Sigma E_M E_X \phi_{MX}$, $S = \Sigma E_M E_X S_{MX}$ are independent of specific ion-ion interactions in seawater, Young's rule need only be applied to b (and higher order terms if needed).

$$b = \sum_{MX} E_M E_X b(MX) + E_V b(B) \qquad (12)$$

Since the apparent equivalent property of seawater is related to the physical properties by (equation 2), it is possible to determine the physical properties of seawater from the estimated $\Phi$

$$P = P^o + \Phi\, e_T \qquad (13)$$

By combining this equation with (equation 11) and noting that $e_T = k\, Cl_V$ and $I_V = k'\, Cl_V$ (where the volume chlorinity $Cl_V = Cl(^o/oo) \times d$, the density, we have

$$P = P^o + A\, Cl_V + B\, Cl_V^{3/2} + C\, Cl_V^{2} \qquad (14)$$

where $A = k\Phi^o$, $B = k(k')^{1/2}S$, $C = k\,k'\,b$. This equation predicts the concentration dependence of seawater solutions diluted with pure water. Since $\Phi^o$ is related to ion-water interactions, S and b are related to ion-ion interactions, any physical property of seawater can be visualized as being equal to

$$P = P^o + \Sigma \text{ ion-water interactions} + \Sigma \text{ ion-ion interactions} \qquad (15)$$

The ion-ion interaction term can be split up into a theoretical Debye-Hückel limiting law term and a term due to deviations from the limiting law

$$\Sigma \text{ ion-ion interactions} = \text{Debye-Hückel term} + \Sigma \text{ deviations from Debye-Hückel} \qquad (16)$$

(equations 11 to 14) have been shown to be valid for predicting and representing the concentration dependence of the thermodynamic (8-19) and transport (8) properties of seawater. Recently we have applied these methods to lakes (46), rivers (47) and estuaries (47).

The application of these methods to the physical chemical properties of seawater can be demonstrated by examining the density. The apparent equivalent volume of seawater is related to the density (d) by

$$\Phi_V = 1000\ (d^o - d)\ /\ e_T d^o + M/d^o \qquad (17)$$

where $d^o$ is the density of water, $e_T$ is the normality (equivalents/liter) and M is the mean equivalent weight of sea salt. By combining this equation with the concentration dependence of $\Phi_V$ (equation 11), we obtain (noting that $e_T = 0.0312803\ Cl_V$ and $I_V = 0.0360145\ Cl_V$)

$$d = d^o + A_V\ Cl_V + B_V\ Cl_V^{3/2} + C_V\ Cl_V^{2} \qquad (18)$$

where $A_V = (M - d^o\ \Phi_V^o) \times 3.12803 \times 10^{-5}$, $B_V = S_V d^o \times 5.93621 \times 10^{-6}$ and $C_V = -\ b_V d^o \times 4.78277 \times 10^{-7}$. The calculations of the $\Phi_V^o$ and $b_V$ values for sea salt using the molal volume data given elsewhere (8,48,50) is demonstrated in Table I.

TABLE I

CALCULATION OF $\Phi_V^o$ AND $b_V$ FOR SEA SALT AT 25°C

| Solute | $E_i$ | $\phi_V^o(i)$ | $E_i\phi_V^o(i)$ | $b_V$ | $E_i b_V(i)$ |
|---|---|---|---|---|---|
| $Na^+$ | 0.77268 | - 1.21 | - 0.935 | 1.078 | 0.833 |
| $Mg^{2+}$ | 0.17573 | -10.59 | - 1.861 | -0.197 | -0.035 |
| $Ca^{2+}$ | 0.03390 | - 8.93 | - 0.303 | 0.242 | 0.008 |

TABLE I continued

| Solute | $E_i$ | $\phi_V^o(i)$ | $E_i\phi_V^o(i)$ | $b_V$ | $E_ib_V(i)$ |
|---|---|---|---|---|---|
| $K^+$ | 0.01684 | 9.03 | 0.152 | 1.129 | 0.019 |
| $Sr^{2+}$ | 0.00030 | - 9.08 | - 0.003 | 0.569 | $0.000_2$ |
| $Cl^-$ | 0.90078 | 17.83 | 16.061 | -1.030 | -0.928 |
| $SO_4^{2-}$ | 0.09318 | 6.99 | 0.651 | 0.134 | 0.013 |
| $HCO_3^-$ | 0.00318 | 24.28 | 0.077 | 0.302 | -0.010 |
| $Br^-$ | 0.00139 | 24.71 | 0.034 | -1.107 | -0.001 |
| $B(OH)_3$ | 0.00054 | 39.22 | 0.021 | 1.300 | 0.001 |
| $CO_3^{2-}$ | 0.00067 | - 1.89 | - 0.001 | -0.780 | -0.001 |
| $B(OH)_4^-$ | 0.00014 | 21.84 | 0.003 | 3.630 | 0.001 |
| $F^-$ | 0.00011 | - 1.16 | - $0.000_1$ | -0.538 | $-0.000_1$ |
| | | | 13.896 | | -0.101 |

Combining these values of $\Phi_V^o$ and $b_V$ with $S_V$ = 2.150 and M = 58.034 (51), we obtain the equation

$$d = 0.997075 + 1.38192 \times 10^{-3}\, Cl_V - 1.27255 \times 10^{-5}\, Cl_V^{3/2} + 4.82143 \times 10^{-3}\, Cl_V^2 \qquad (19)$$

The densities calculated from (equation 19) from 5 to 35$^o$/oo salinity are compared to the measured values (52) in Table II.

TABLE II
COMPARISON OF THE CALCULATED AND MEASURED DENSITIES OF SEAWATER SOLUTIONS AT 25°C

| Salinity | $Cl_V$ | $(d-d^o)10^3$ Meas | $(d-d^o)10^3$ Calc | Δ, ppm |
|---|---|---|---|---|
| 5$^o$/oo | 2.770 | 3.765 | 3.769 | - 4 |
| 10 | 5.561 | 7.511 | 7.518 | - 7 |
| 15 | 8.372 | 11.254 | 11.261 | - 7 |
| 20 | 11.205 | 15.002 | 15.007 | - 5 |
| 25 | 14.058 | 18.757 | 18.756 | 1 |
| 30 | 16.932 | 22.523 | 22.512 | 11 |
| 35 | 19.827 | 26.300 | 26.276 | 24 |
| | | | | mean ± 8 |

The agreement is very good when one considers that the single salt density data at high concentrations is only good to ± 10 x $10^{-6}$g cm$^{-3}$. At present it is not possible to state with certainty that the deviations at high concentrations are due to unreliable single salt data or a failure of Young's rule. Further studies are needed to examine the excess volumes of mixing the major sea salts,

$$\Delta V_m = [\Phi_V(\text{meas}) - \Phi_V(\text{calc})]/e_T \qquad (20)$$

In the low salinity range the estimated densities are in excellent agreement with the measured values since the higher order terms ($B_V\ Cl_V^{3/2}$ and $C_V\ Cl_V^{2}$) are not of great importance.

Properties of Ionic Solutes in Seawater

To obtain an understanding of how the medium of seawater or any multicomponent electrolyte affects an ionic chemical reaction

$$A + B \rightarrow C + D \qquad (21)$$

it is necessary to examine the non-ideal behavior of the reactant species. This non-ideal behavior can be studied by examining the activity coefficient ($\gamma_T$) of the reactant species as well as its pressure $\bar{V} - \bar{V}^o = RT(\partial \ln \gamma_T/\partial P)$ and temperature $\bar{H} - \bar{H}^o = RT^2(\partial \ln \gamma_T/\partial T)$ dependence. The thermodynamic activity of the solute i is related to the concentration and activity coefficient by

$$a_i = [i]_T\ \gamma_T(i) \qquad (22)$$

where the subscript T is used to denote total concentrations. There are three types of models that have been used to examine the non ideal behavior of an electrolyte (1) The specific interaction model (2) The ion pairing model (3) The cluster model. A thorough discussion of these and other models is given elsewhere (4,8,21,29,37-39). The starting point for all of the discussions of ionic interactions is the Debye-Hückel theory. The theory predicts that the mean activity coefficients of an electrolyte is given by (55-57)

$$-\ln\gamma_{\pm}(MX) = A\ Z_M\ Z_X\ I^{1/2}/(1 + B\mathring{a}\ I^{1/2}) \qquad (23)$$

where A and B are constants related to the dielectric constant of the pure water and the temperature (A = 0.509, B = 0.329 at 25°C), $Z_M$ and $Z_X$ are the electrostatic change on cation M and anion X, and $\mathring{a}$ is the ion size parameter. The classical methods (8) that have been used to examine the non-ideal behavior of an electrolyte is to examine the deviations from the Debye-Hückel limiting law (in various extended forms involving one or more arbitrary constants). The difference between this form and the experimental data is attributed to non-coulombic effects.

The Specific Interaction Model. In using the specific interaction model developed by Guggenheim (53) the following equation is used

$$-\log \gamma_{\pm}(MX) = \log \gamma\ (\text{elect}) + \nu\beta_{MX}m \qquad (24)$$

where log $\gamma$ (elect) is given by equation (23) with $B\text{å} = 1.0$, $\nu = (2\nu_M/\nu_X)/(\nu_M + \nu_X)$, ($\nu_i$ is the number of species $i$), $\beta_{MX}$ is a constant and $m$ is the molality. To use the specific interaction model to estimate total activity coefficients, the following equations are used (39)

$$\log \gamma_T(MX) = \log \gamma\ (\text{elect}) + (\nu_M/\nu) \sum_X \beta_{MX}[X]_T + (\nu_X/\nu) \sum_M \beta_{MX}[M]_T \qquad (25)$$

$$\log \gamma_T(M) = (Z_M/Z_X) \log \gamma\ (\text{elect}) + \sum_M \beta_{MX}[X]_T \qquad (26)$$

$$\log \gamma_T(X) = (Z_X/Z_M) \log \gamma\ (\text{elect}) + \sum_M \beta_{MX}[M]_T \qquad (27)$$

The total activity coefficients calculated by using the specific interaction model for the major sea salts and ions of seawater are given in Tables III and IV (37). The estimated values for $\gamma_T$ for the ions given in Table IV (37) are in excellent agreement with the measured values. Pitzer (58) has recently

TABLE III
COMPARISON OF THE CALCULATED STOICHIOMETRIC ACTIVITY COEFFICIENT FOR THE MAJOR SEA SALTS AT 25°C AND 0.7 IONIC STRENGTH

| Salt | Ionic Strength | Ion Pairing | Specific Interaction | Cluster |
|---|---|---|---|---|
| HCl | 0.774 | 0.683 | 0.683 | 0.696 |
| $H_2SO_4$ | 0.201 | 0.379 | --- | 0.386 |
| NaCl | 0.666 | 0.664 | 0.668 | 0.668 |
| $Na_2SO_4$ | 0.349 | 0.366 | 0.370 | 0.374 |
| $MgCl_2$ | 0.480 | 0.463 | 0.466 | 0.463 |
| $MgSO_4$ | 0.121 | 0.158 | 0.157 | 0.160 |
| KCl | 0.624 | 0.624 | 0.647 | 0.641 |
| $K_2SO_4$ | 0.336 | 0.337 | 0.350 | 0.345 |
| $CaCl_2$ | 0.460 | 0.444 | 0.455 | 0.448 |
| $CaSO_4$ | --- | 0.148 | 0.157 | 0.159 |
| $NaHCO_3$ | --- | 0.549 | 0.612 | --- |
| $CaCO_3$ | --- | 0.066 | 0.069 | --- |

TABLE IV
COMPARISON OF THE CALCULATED AND MEASURED TOTAL ACTIVITY COEFFICIENT ($\gamma_T$) FOR THE MAJOR IONS OF SEAWATER AT 25°C AND I = 0.7

| Ion | Measured | Ionic Strength | Ion Pairing Model | Specific Interaction Model |
|---|---|---|---|---|
| $H^+$ | 0.74 | 0.85 | 0.74 | 0.74 |
| $Na^+$ | 0.68 | 0.71 | 0.70 | 0.68 |
| $Mg^{2+}$ | 0.23 | 0.29 | 0.25 | 0.23 |
| $Ca^{2+}$ | 0.21 | 0.26 | 0.22 | 0.21 |
| $K^+$ | 0.64 | 0.63 | 0.62 | 0.63 |
| $Sr^{2+}$ | -- | 0.25 | 0.22 | -- |
| $Cl^-$ | 0.68 | 0.63 | 0.63 | 0.66 |
| $SO_4^{2-}$ | 0.11 | 0.22 | 0.10 | 0.11 |
| $HCO_3^-$ | 0.55 | 0.68 | 0.43 | 0.59 |
| $CO_3^{2-}$ | 0.02 | 0.21 | 0.02 | 0.03 |
| $F^-$ | -- | 0.68 | 0.31 | -- |
| $OH^-$ | -- | 0.65 | 0.11 | 0.56 |
| $B(OH)_4^-$ | 0.26 | 0.68 | 0.38 | -- |

developed a set of equations that extend the specific interaction model to higher concentrations and also account for plus-plus and minus-minus ionic interactions. These equations, however, do not improve the estimates made with the more simpler form developed by Guggenheim (53).

Robinson and Stokes (57) have developed a hydration model to examine the differences between log $\gamma_{\pm}$(meas) and log $\gamma$ (elect)

$$\log \gamma_{\pm}(MX) = \log \gamma \text{ (elect)} - (h/\nu) \ln a_W + \ln \{[1 - (h - \nu)m]/55.51\} \quad (28)$$

where $a_W$ is the activity of water ($a_W = p/p_o$ where p and $p_o$ are the vapor pressure of solution and water). Recently Elgquist and Wedborg (60) have used this method to estimate activity coefficient of $SO_4^{2-}$ at high ionic strengths.

Ion Pairing Model. The most popular method of treating the deviations from the Debye-Hückel theory in concentrated solutions is the ion pairing method. This method assumes that short range electrostatic interactions can be represented by the formation of ion pairs

$$M^+ + X^- = MX^o \quad (29)$$

which has a characteristic association constant

$$K_A = a_{MX}/a_M a_X = \{[MX^o]/[M^+]\ [X^-]\}\ (\gamma_{MX^o}/\gamma_M \gamma_X) \qquad (30)$$

In its original formulation by Bjerrum (61) the ion pairing model was applied by treating the solvent as a continuum. In recent years four classes of structural ion pairs have been distinguished: 1. complexes - when the ions are held in contact by covalent bonds, 2. contact ion pairs - when the ions are in contact and linked electrostatically (with no covalent bonding), 3. solvent-shared ion pairs - when the ions are linked electrostatically and separated by a single water molecule and 4. solvent-separated ion pairs - when the ions are linked electrostatically are separated by more than one water molecule. In applying the ion pairing model one makes the assumption that

$$a_M = [M^+]_F\ \gamma_F(M^+) \qquad (31)$$

where $[M^+]_F$ is the free metal concentration and $\gamma_F(M^+)$ is the activity coefficient of the free ion. Combining this relationship with (equation 22), we have upon rearranging

$$\gamma_T(M^+) = ([M^+]_F/[M^+]_T)\ \gamma_F(M^+) \qquad (32)$$

$$\gamma_T(X^-) = ([X^-]_F/[X^-]_T)\ \gamma_F(X^-) \qquad (33)$$

The fraction of free ions, $[M^+]_F/[M^+]_T$ and $[X^-]/[X^-]_T$, are determined from

$$[M^+]_F/[M^+]_T = 1/(1 + \Sigma\ K_A{*}(i)\ [X^-_i]_F) \qquad (34)$$

$$[X^-]_F/[X^-]_T = 1/(1 + \Sigma\ K_A{*}(i)\ [M^+_i]_F) \qquad (35)$$

where $K_A{*}$ is the stoichiometric association constant

$$K_A{*} = K_A\ (\gamma_M \gamma_X/\gamma_{MX^o}) = [MX^o]/[M^+]_F\ [X^-]_F \qquad (36)$$

The $K_A{*}$ can be determined in an ionic medium or estimated by using infinite dilution $K_A$'s - both methods should agree providing reasonable values can be determined for $\gamma_F$ and the ionic medium is non-reacting. The values of $\gamma_T$ determined by using the ion pairing model for the major sea salts and ions are given in Table III and IV. The ion pairing method predicts values for $\gamma_T$ that are in good agreement with the specific interaction model as well as the experimental results. When applying the ion pairing model it is assumed that various types of ion pairs exist in solution. This so-called speciation of various cation and anion forms is determined by using the equations

$$[MX_i]/[M^+]_T = K_A{*}(i)\ [X_i^-]_F/(1+ \Sigma\ K_A{*}(i)\ [X_i^-]_F) \qquad (37)$$

$$[MX_i]/[X^-]_T = K_A^*(i)\ [M_i^+]_F/(1 + \Sigma\ K_A^*(i)\ [M_i^+]_F) \qquad (38)$$

For the major ions of seawater the forms determined by various workers from these equations are in reasonable agreement (9). The speciation of the major cations in seawater are given in Figure 1. Most of the cations are predominately in the free ion form and the major ion pairs are formed with sulfate. The speciation of the major anions in seawater are given in Figure 2. About 50% of $SO_4^{2-}$, $HCO_3^-$ and $B(OH)_4^-$ are complexed in seawater mainly with $Na^+$ and $Mg^{2+}$ ions; for $CO_3^{2-}$ about 90% is complexed mostly with $Ca^{2+}$ and $Mg^{2+}$; for $F^-$ and $OH^-$ the major ion pairs are formed with $Mg^{2+}$ ions. It should be pointed out that although the examination of the speciation of ions in seawater is a convenient way of examining interaction parameters between cations and anions, one cannot prove or disprove their existence by using thermodynamic data. Other types of physical chemical measurements (for example, Raman Spectra - and ultrasonic absorption) are needed to confirm the existence of ion pairs.

Recently, there have been two studies (31,32) made on the speciation of heavy metals in seawater (Cu, Zn, Cd and Pb). Although the speciation of the heavy metals are quite different (37,61), the fraction of free metals are in reasonable agreement (see Table V). Since it is this fraction that is necessary to calculate $\gamma_T$ for the metal ions, the speciation calculations yield similar (low values) for the total activity coefficient (see Table V).

TABLE V.
CALCULATED TOTAL ACTIVITY COEFFICIENTS ($\gamma_T$) FOR SOME MINOR IONS IN SEAWATER AT 25°C AND I = 0.7

| | (°/oo) Free | | | $\gamma_T$(M) | |
|---|---|---|---|---|---|
| Ion | a | b | $\gamma_F$(M) | a | b |
| Cu | 1.0 | 0.7 | 0.253 | $0.002_5$ | $0.001_8$ |
| Zn | 17.0 | 16.1 | 0.268 | $0.046^5$ | $0.043^8$ |
| Cd | 2.5 | 1.7 | 0.012 | $0.000_3$ | $0.000_2$ |
| Pb | 2.1 | 4.5 | 0.328 | $0.007^3$ | $0.014^2$ |

a) Zirino and Yamamoto (31)
b) Dyrssen and Wedborg (32)

The Cluster Expansion Model. The more recent cluster theory of Friedman (1) makes no attempt to separate the electrical and non-electrical interactions. It also considers the importance of all possible interactions (plus-plus, plus-minus and minus-minus). For example, for the major sea salts ($NaCl + MgSO_4$)

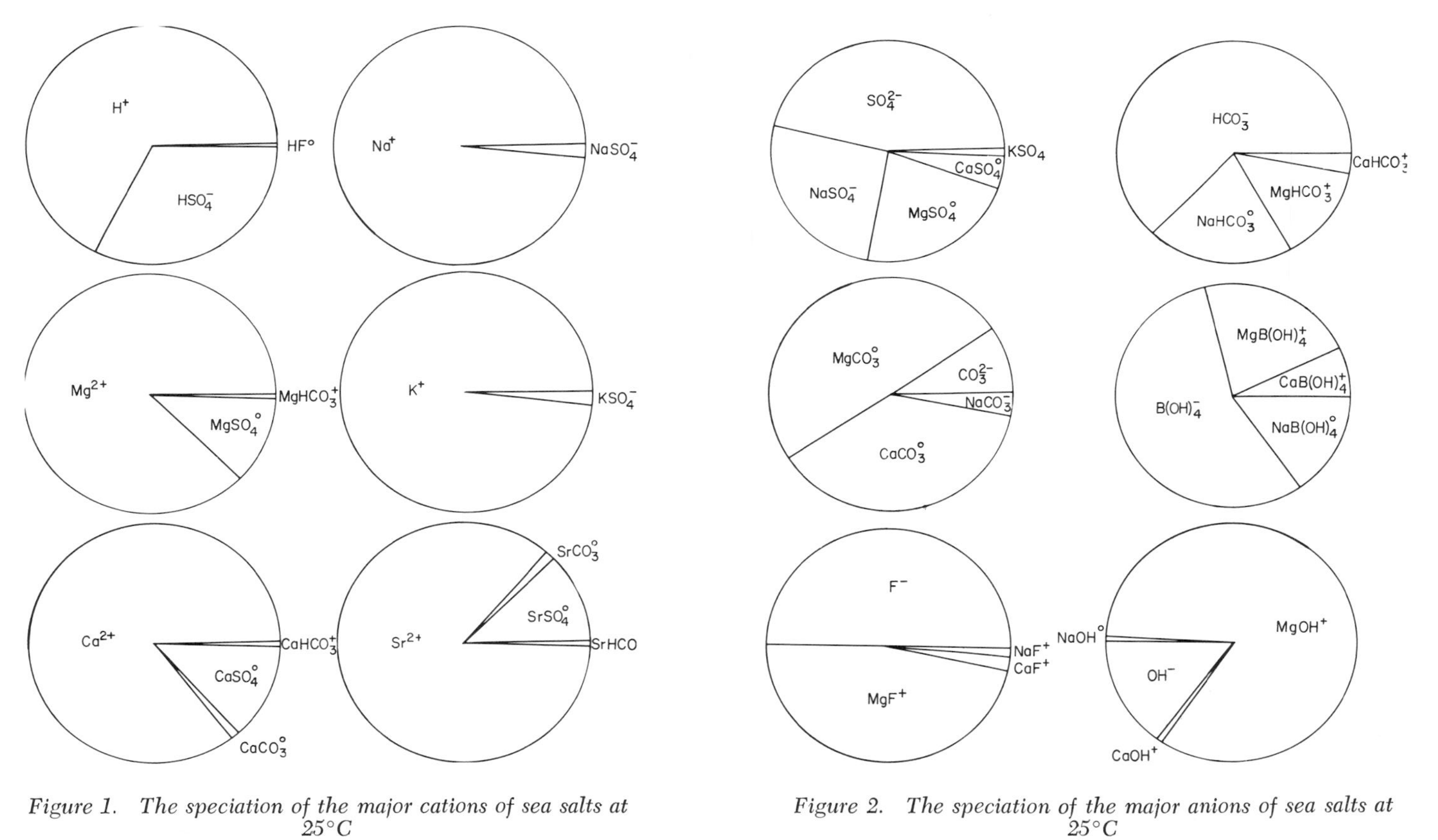


*Figure 1. The speciation of the major cations of sea salts at 25°C*

*Figure 2. The speciation of the major anions of sea salts at 25°C*

there are a number of possible interactions to consider

| Interactions | | Possible Types |
|---|---|---|
| ⊕ | ⊕ | Na-Na, Mg-Mg, Na-Mg |
| ⊖ | ⊖ | Cl-Cl, $SO_4$-$SO_4$, Cl-$SO_4$ |
| ⊕ | ⊖ | Na-Cl, Mg-$SO_4$, Mg-Cl, Na-$SO_4$ |

These interactions can be represented by the cross square diagram

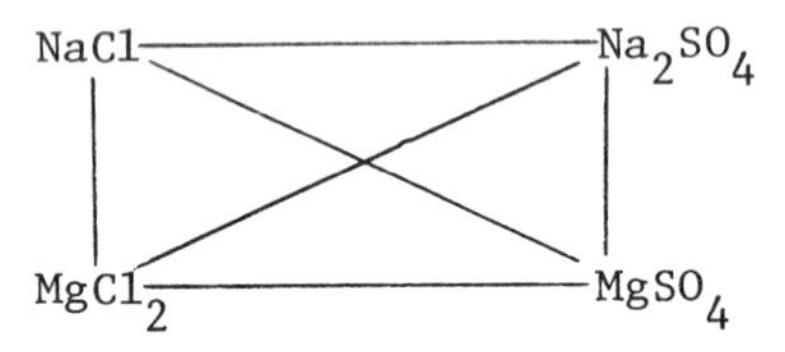


By studying the mixtures along the side of this diagram, one can obtain some information about the ⊕ - ⊕ and ⊖ - ⊖ interactions, by studying the individual salts and the sum around the sides of the diagram ($MgSO_4$ = $MgCl_2$ + $Na_2SO_4$ - 2NaCl) one can study ⊕ - ⊕ interactions. The cross mixtures represent the mixture of simple sea salts. The total activity coefficient of an electrolyte is given by

$$\log \gamma_T(MX) = \log \gamma^o(MX) + \Sigma \oplus - \oplus \text{ terms} + \Sigma \ominus - \ominus \text{ terms} \quad (39)$$

where $\log \gamma^o(MX)$ is the value for MX in itself at the ionic strength of the mixture and the other terms are related to the excess mixing terms. For example, for NaCl in seawater

$$\Sigma \oplus - \oplus = (Na - Mg) + (Na - K) + (Na - Ca) + \ldots \quad (40)$$

$$\Sigma \ominus - \ominus = (Cl - SO_4) + (Cl - HCO_3) + (Cl - Br) + \ldots \quad (41)$$

where the terms in parenthesis are weighted according to the composition of the mixture. Wood and coworkers have applied the cluster theory to the major constituents (20) and some minor constituents of seawater (62). The results for the major constituents shown in Table III are in excellent agreement with the values obtained using the specific interaction model and the ion pairing model. This is not surprising since the ⊕ - ⊕ and ⊖ - ⊖ interaction terms are small for free energies of mixing.

In conclusion, all of the methods yield comparable results for the $\gamma_T$ of the major sea salts. It should be pointed out that due to the lack of thermodynamic data, the specific interaction

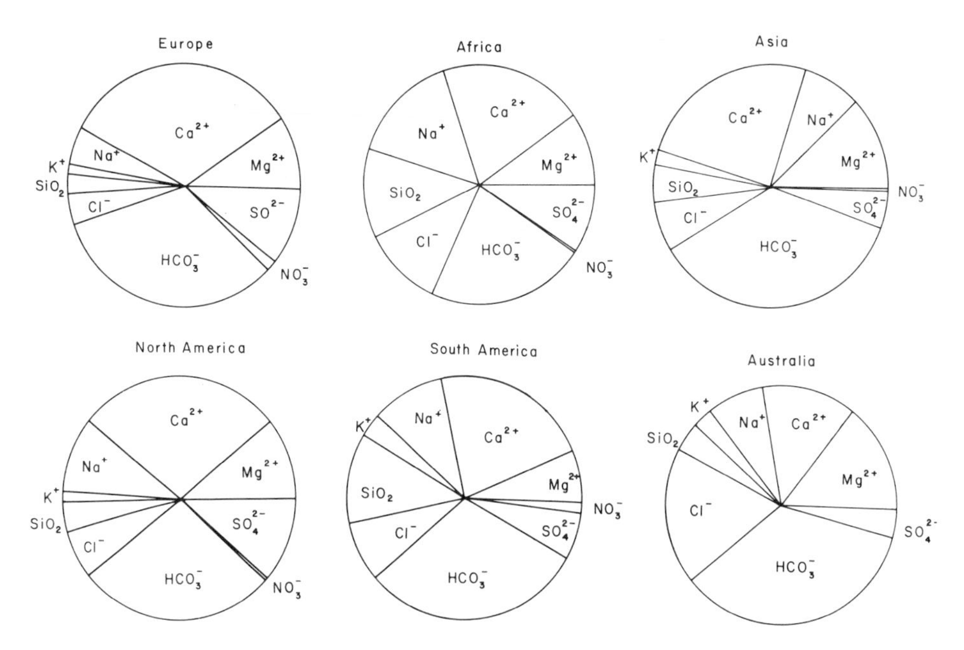


*Figure 3. The major constituents of various rivers*

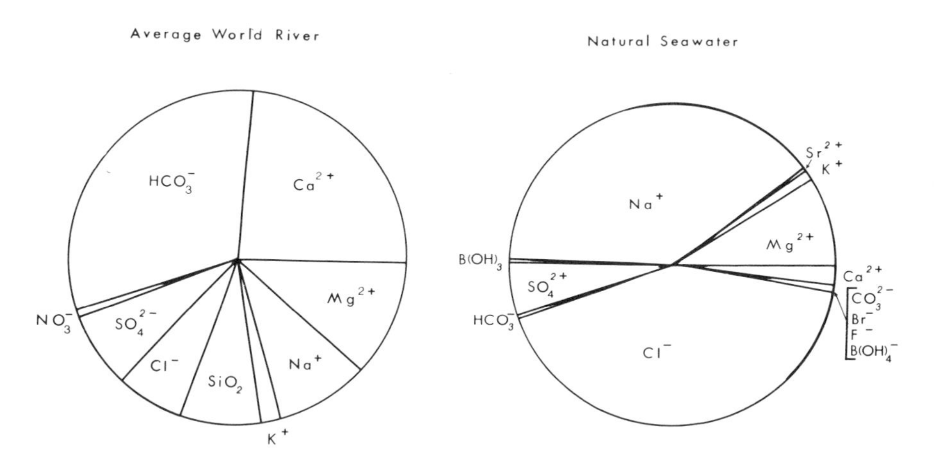


*Figure 4. The equivalent fraction of the cations and anions for world river and sea salts*

model and the cluster model cannot be used to estimate the $\gamma_T$ for trace constituents in seawater. Even though reliable data is available for the chloride and sulfates of trace metals, the interaction parameters for hydroxide and carbonate salts are nonexistent due to solubility limitations at high ionic strengths. Although it is difficult to determine $\gamma_T$ for trace constituents at high ionic strengths, the effect of pressure $(\bar{V} - \bar{V}^o)$ and temperature $(\bar{H} - \bar{H}^o)$ can be determined in various ionic media or estimated with reasonable accuracy (9, 34-36).

## The Composition of Estuarine Waters

An estuary is defined by Pritchard (63) as "a semi-enclosed coastal body of water which has a free connection with the open sea and within which seawater is measurably diluted with fresh water derived from land drainage". Since the composition of the river draining into an estuary can be different, we must first examine the composition of various rivers. Livingstone (64) has tabulated the composition of the major constituents and a comparison of the equivalent fraction of these major constituents are shown in Figure 3. As is quite apparent from Figure 3, although the total solid input of various rivers vary from 70 to 200 parts per million, the equivalent fraction of most rivers are similar. The major cations are $Ca^{2+}$, $Mg^{2+}$ and $Na^+$ and the major anions are $HCO_3^-$, $SO_4^{2-}$ and $Cl^-$. Most of the $Na^+$ and $Cl^-$ is the result of recycled NaCl from seawater aerosols. The $SiO_2$ is predominately in the unionized form $Si(OH)_4$ at the pH of most rivers (7.3 to 8.0). A comparison of the equivalent fraction of the major constituents in "world" river water (a weighted input from all the continents(64)) and seawater are shown in Figure 4. The equivalent fraction of the major cations and anions for "world" river water (64) are compared to seawater in Figure 5. This comparison demonstrates that the major constituents of most rivers ($Ca^{2+}$ and $HCO_3^-$) are quite different from the major constituents of seawater ($Na^+$ and $Cl^-$). One might expect from this comparison that the physical properties of rivers and seawaters (diluted with pure water) and an estuary formed by mixing river water with seawater to be quite different. The composition for one liter of world or average river water is given in Table VI.

TABLE VI
THE COMPOSITION OF ONE LITER OF WORLD OR AVERAGE RIVER WATER

| Species | $g_i$ $10^6$ [a] | $n_i$ $10^3$ [b] | $e_i$ $10^3$ | $I_i$ $10^3$ |
|---|---|---|---|---|
| $Na^+$ | 6.5 | 0.283 | 0.283 | 0.283 |
| $Mg^{2+}$ | 4.1 | 0.169 | 0.337 | 0.674 |
| $Ca^{2+}$ | 15.0 | 0.374 | 0.749 | 1.496 |
| $K^+$ | 2.3 | 0.059 | 0.059 | 0.059 |

TABLE VI continued

| Species | $g_i\ 10^6$ a | $n_i\ 10^3$ b | $e_i\ 10^3$ | $I_i\ 10^3$ |
|---|---|---|---|---|
| $Cl^-$ | 7.8 | 0.220 | 0.220 | 0.220 |
| $SO_4^{2-}$ | 11.2 | 0.117 | 0.233 | 0.466 |
| $HCO_3^-$ | 58.4 | 0.950 | 0.950 | 0.950 |
| $CO_3^{2-}$ | --- | 0.002 | 0.004 | 0.008 |
| $NO_3^-$ | 1.0 | 0.016 | 0.016 | 0.016 |
| $Si(OH)_3O^-$ | --- | 0.005 | 0.005 | 0.005 |
| | | $1/2\Sigma$ = 1.086 | $1/2\Sigma$ = 1.428 | $1/2\Sigma$ = 2.089 |
| $Si(OH)_4$ | 20.5 | 0.213 | 0.213 | --- |
| | $g_T$ = 126.8 | $n_T$ = 1.299 | $e_T$ = 1.641 | $I_T$ = 2.089 |

a) Taken from Livingstone (64). The value for $Na^+$ has been adjusted slightly (6.3 to 6.5) by making the sum of $\Sigma$ equivalents of the cations = the sum of $\Sigma$ equivalents of the anions.
b) The value for $[CO_3^{2-}]$ was determined from $K_2 = 4.69 \times 10^{-11}$ (65) using a pH of 7.7 - $[HCO_3^-]/[CO_3^{2-}] = 425.43$. The value for $[Si(OH)_3O^-]$ was determined from $K = 5 \times 10^{-10}$ (66) using a pH of 7.7 - $[Si(OH)_3O^-]/[Si(OH)_4] = 0.025$.

$g_T = 126.8 \times 10^{-6} g\ l^{-1}$ or $127 \times 10^{-6} g\ kg^{-1}$ (since the density of the river is approximately equal to the density of water, $d^o$ = 0.997075 g $ml^{-1}$ at 25°C (67)). The molarity of the river equals $1.299 \times 10^{-6} mol\ l^{-1}$ and the molar ionic strength equals $2.089 \times 10^{-6} mol\ l^{-1}$. Since the grams of water are equal to $\sim$ 1000.00, the molality and molal ionic strength are equal to the molar values.

In the next section we will examine the use of the methods described earlier to examine the densities of river waters and an estuary formed by mixing river water with seawater (47).

## The Density of River and Estuarine Waters

As was discussed earlier, the physical chemical properties of multicomponent electrolyte solutions, like seawater, can be estimated from binary electrolyte data. Recently, we have used these methods to estimate the physical chemical properties of rivers (47), estuaries (47) and lakes (46). In this section we shall examine the density of rivers and estuarine waters determined by using Young's rule. To determine the density for rivers, one must determine the $\Phi_V{}^o$, $S_V$ and $b_V$ as well as $M_T$, the mean equivalent weight for river salts. A sample calculation of $\Phi_V{}^o$, $b_V$ and $M_T$ for "world" river water is shown in Table VII.

TABLE VII
CALCULATION OF $\Phi_V^o$, $M_T$ AND $b_V$ FOR RIVER SALTS AT 25°C

| Solute | $\Phi_V^o$ | $E_i\phi_V^o$ | $M_i$ | $E_iM_i$ | $b_V(i)$ | $E_ib_V(i)$ |
|---|---|---|---|---|---|---|
| $Ca^{2+}$ | - 8.93 | -4.077 | 20.0400 | 9.150 | 0.242 | 0.110 |
| $Mg^{2+}$ | -10.59 | -2.175 | 12.1525 | 2.496 | -0.197 | -0.040 |
| $Na^+$ | - 1.21 | -0.202 | 22.9898 | 3.839 | 1.078 | 0.180 |
| $K^+$ | 9.03 | 0.325 | 39.1020 | 1.408 | 1.129 | 0.041 |
| $HCO_3^-$ | 24.29 | 14.171 | 61.0172 | 35.597 | 2.122 | 1.238 |
| $SO_4^{2-}$ | 6.99 | 0.993 | 48.0288 | 6.820 | 0.134 | 0.019 |
| $Cl^-$ | 17.83 | 2.391 | 35.4530 | 4.754 | -1.030 | -0.138 |
| $NO_3^-$ | 26.20 | 0.257 | 62.0049 | 0.608 | -1.000 | -0.010 |
| $Si(OH)_4$ | 60.0 [a] | 7.974 | 96.1156 | 12.774 | --- | --- |
| | | $\Phi_V^o$ = 19.657 | | M = 77.446 | | $b_V$ = 1.400 |

a) Taken from reference (68).

A summary of the values of $\Phi_V^o$, $S_V$, $b_V$ and $M_T$ for various rivers as well as seawater are given in Table VIII.

TABLE VIII
THE $\Phi_V^o$, $S_V$, $b_V$, $M_T$ FOR RIVER AND SEA SALTS AT 25°C

| Source | $\Phi_V^o$ | $S_V$ | $b_V$ | $M_T$ |
|---|---|---|---|---|
| North America | 14.934 | 2.612 | 1.417 | 74.672 |
| South America | 25.168 | 2.055 | 1.339 | 80.734 |
| Europe | 14.512 | 2.739 | 1.523 | 76.407 |
| Asia | 18.723 | 2.432 | 1.572 | 77.151 |
| Africa | 22.226 | 2.104 | 1.006 | 75.856 |
| Australia | 23.008 | 2.276 | 1.273 | 78.540 |
| World | 19.657 | 2.371 | 1.400 | 77.446 |
| Seawater | 13.896 | 2.150 | -0.101 | 58.034 |

The infinite dilution $\Phi_V^o$ and the mean equivalent weight for rivers are all larger than the sea salt values. By combining (equation 17) and

$$\Phi_V = \Phi_V^o + S_V I_V^{1/2} + b_V I \qquad (42)$$

we have an equation that can be used to estimate the density of the rivers

$$d = d^o + 10^{-3}(M_T - d^o \Phi_V^o) e_T - 10^{-3} S_V d^o e_T I_V^{1/2} - 10^{-3} b_V d^o e_T I_V \qquad (43)$$

Since $e_T$ and $I_V$ are very small, the last two terms are not needed and the densities can be estimated from infinite dilution apparent molal volumes, $\Phi_V{}^o = \Sigma E_i \phi_V{}^o$. The relative densities of various rivers determined from (equation 43) are given in Table IX.

TABLE IX

THE RELATIVE DENSITIES OF VARIOUS RIVERS AT 25°C

| Source | $g_T$, ppm | $(d - d^o)10^6$ River | $(d - d^o)10^6$ Seawater | Δ, ppm |
|---|---|---|---|---|
| North America | 147.8 | 118.3 | 111.8 | 6.5 |
| South America | 75.1 | 51.7 | 57.0 | -5.3 |
| Europe | 185.4 | 150.3 | 149.6 | 0.7 |
| Asia | 148.8 | 112.8 | 112.5 | -0.3 |
| Africa | 133.8 | 94.7 | 101.2 | -6.5 |
| Australia | 61.4 | 43.5 | 46.8 | -3.3 |
| World | 126.8 | 94.9 | 96.0 | -1.1 |
| | | | | mean ± 3.4 |

Also given in Table IX are the densities of seawater solutions diluted with pure water calculated at the same total solid concentration, related to the salinity by (8,9,51)

$$g_T(SW) = 1.004847\ S(^o/oo) \qquad (44)$$

The calculated densities of the various rivers are in good agreement with those determined from seawater at the same total solid concentration (on the average to $\pm$ 3.4 x $10^{-6}$g cm$^{-3}$). Thus, the densities (as well as other physical chemical properties) of most rivers are equal to seawater at the same total dissolved solids.

The density of mixtures of river salts and sea salts can be estimated from the $\Phi_V$ for the mixture by using Young's rule

$$\Phi_V(\text{Estuary}) = E_R\ \phi_V(R) + E_S\ \phi_V(S) \qquad (45)$$

where $E_R$ and $E_S$ are the equivalent fractions of river and sea salts, $\phi_V(R)$ and $\phi_V(S)$ are the apparent equivalent volume at the ionic strength of the mixtures ($I_T = E_R\ I_R + E_S\ I_S$). The values of $\Phi_V$(Est) for the estuary determined from (equation 45) at various weight fractions of seawater are given in Table X.

TABLE X
THE APPARENT EQUIVALENT VOLUME, MEAN MOLECULAR WEIGHT AND TOTAL NORMALITY OF ESTUARY SALTS AT VARIOUS WEIGHT FRACTIONS OF SEA SALT

| $X_{SW}$ | $E_S$ | $\Phi_V$(Est) | $M_T$ | $N_T$ |
|---|---|---|---|---|
| 0 | 0 | 19.768 | 77.446 | 0.00164 |
| 0.02 | 0.8827 | 14.866 | 63.310 | 0.01370 |
| 0.04 | 0.9392 | 14.644 | 59.215 | 0.02576 |
| 0.06 | 0.9594 | 14.608 | 58.822 | 0.03785 |
| 0.08 | 0.9698 | 14.617 | 58.620 | 0.04995 |
| 0.10 | 0.9762 | 14.641 | 58.496 | 0.06207 |
| 0.20 | 0.9893 | 14.806 | 58.242 | 0.12280 |
| 0.30 | 0.9937 | 14.962 | 58.156 | 0.18388 |
| 0.40 | 0.9960 | 15.098 | 58.112 | 0.24525 |
| 0.50 | 0.9973 | 15.217 | 58.086 | 0.30694 |
| 0.60 | 0.9982 | 15.323 | 58.069 | 0.36895 |
| 0.70 | 0.9988 | 15.418 | 58.057 | 0.43127 |
| 0.80 | 0.9993 | 15.508 | 58.047 | 0.49393 |
| 0.90 | 0.9997 | 15.590 | 58.040 | 0.55689 |
| 1.00 | 1.0000 | 15.671 | 58.034 | 0.62019 |

Also given in this Table are the mean equivalent weight of the estuary salt

$$M_T = E_R M_R + E_S M_S \qquad (46)$$

and total normality

$$N_T = E_R N_R + E_S N_S \qquad (47)$$

The relative densities of the estuary have been calculated from the $\Phi_V$, $M_T$ and N given in Table X, using the equation

$$d - d^o = (M_T - d^o \Phi_V) N_T \qquad (48)$$

and are given in Table XI. Also given in this Table are the densities of seawater diluted with pure water (52) calculated at the same total solid concentration. Over the entire weight fraction of sea salt the densities calculated from the seawater

TABLE XI
THE DENSITIES OF ESTUARINE WATERS AT VARIOUS FRACTIONS OF SEA SALT

| $X_S$ | $g_T$ | $(d - d^o)10^3$ Estuary | $(d - d^o)10^3$ Seawater | Δ, ppm |
|---|---|---|---|---|
| 0 | 0.127 | 0.095 | 0.096 | - 1 |
| 0.02 | 0.827 | 0.622 | 0.622 | 0 |
| 0.04 | 1.529 | 1.149 | 1.148 | 1 |

TABLE XI continued

| $X_S$ | $g_T$ | Estuary | Seawater | Δ, ppm |
|---|---|---|---|---|
| 0.06 | 2.229 | 1.675 | 1.672 | 3 |
| 0.08 | 2.931 | 2.200 | 2.197 | 3 |
| 0.10 | 3.631 | 2.725 | 2.720 | 5 |
| 0.20 | 7.136 | 5.339 | 5.335 | 4 |
| 0.30 | 10.640 | 7.951 | 7.947 | 4 |
| 0.40 | 14.144 | 10.560 | 10.558 | 2 |
| 0.50 | 17.649 | 13.172 | 13.172 | 0 |
| 0.60 | 21.153 | 15.788 | 15.789 | - 1 |
| 0.70 | 24.657 | 18.408 | 18.410 | - 2 |
| 0.80 | 28.161 | 21.034 | 21.035 | - 1 |
| 0.90 | 31.666 | 23.665 | 23.666 | - 1 |
| 1.00 | 35.170 | 26.301 | 26.301 | 0 |
| | | | mean | ± 1.9 |

equation of state are in excellent agreement with the values estimated from (equation 48). Recently, we have made direct measurements on mixtures of artificial river water and standard seawater (47) and found that the measured values agreed with those calculated from the seawater equation at the same total solid concentration to $\pm$ 3 x $10^{-6}$g $cm^{-3}$ from 0 to 35°/oo salinity. From these comparisons, it is clear that the densities (as well as other physical chemical properties) of seawater diluted with pure water can be used in rivers and estuaries as well as lakes and seas (46).

By knowing the river water input of total dissolved solids, the solids of the estuarine salt can be determined from

$$g_T(\text{Estuary}) = g_T(\text{River}) + b\ S(^o/oo) \qquad (49)$$

where S(°/oo) is the conductivity salinity and the constant b is given by

$$b = [35.170 - g_T(\text{River})]\ /\ 35.000 \qquad (50)$$

Recently, we have used these equations to examine the densities of Baltic Sea waters (52). We have found that (see Table XII)

TABLE XII
COMPARISONS OF THE MEASURED DENSITIES OF THE BALTIC WITH THOSE OF SEAWATER DILUTED WITH PURE WATER

| Ours-Cox, et.al., 0°C | | | Ours-Kremling, 0.36°C | | |
|---|---|---|---|---|---|
| S(°/oo) | Uncorr | Corr | S(°/oo) | Uncorr | Corr |
| 9.579 | -53 | 17 | 15.247 | -57 | - 2 |
| 15.541 | -45 | 9 | 20.083 | -44 | - 3 |
| 20.130 | -51 | -10 | 25.013 | -38 | -10 |

TABLE XII continued

Ours-Cox, et.al., 0°C

| $S(^o/oo)$ | Uncorr | Corr |
|---|---|---|
| 20.154 | -32 | 9 |
| 25.439 | -29 | - 2 |
| 29.698 | -25 | -10 |
| 35.004 | - 4 | - 4 |
| 39.232 | 8 | - 3 |
| 40.288 | 9 | - 5 |
| | | ± 7.7 |

Ours-Kremling, 0.36°C

| $S(^o/oo)$ | Uncorr | Corr |
|---|---|---|
| 29.839 | -13 | 1 |
| 31.191 | - 9 | 2 |
| 31.920 | -11 | - 3 |
| 32.319 | - 4 | 4 |
| 35.495 | 3 | 2 |
| 39.232 | - 3 | -14 |
| | | ± 4.5 |

the density measurements of Cox, et.al., (69) and Kremling (70) are in good agreement (± 7 ppm) with our measurements on seawater diluted with pure water if the $g_T$(River) = 0.120 g $kg^{-1}$. This value for the river input is in excellent agreement with the value determined from the composition data of Kremling (71). It is interesting to note that the earlier measurements of Knudsen (72) give a value of $g_T$(River) = 0.073 g $kg^{-1}$ which is in agreement with earlier composition data of Lyman and Fleming (73). It, thus, appears that the river input or the evaporation to precipitation has increased in the Baltic Sea in the last 60 years.

The densities of the Mediterranean and Red Seas were also found (52) to be in agreement with evaporated seawater at the same $g_T$ as determined from (equation 49) with $g_T$(River) = 0.120 g $kg^{-1}$. Thus, these seas behave as an evaporated estuary, not evaporated seawater.

The fact that the densities of rivers, estuaries, lakes, seas and seawater are in good agreement at the same total dissolved solid concentration, brings up a problem concerning the new definition of salinity (74)

$$S(^o/oo) = 1.80655\ Cl(^o/oo) \qquad (51)$$

This new definition causes a difference between the physical chemical properties of seawater diluted with pure water and direct measurements made on natural waters (due to the lack of a river water input term). By using (equation 49) (or its equivalent expressed in chlorinity using (equation 51), or the earlier definition of salinity by Knudsen, these problems could have been avoided. Since the conductivity of seawater diluted with pure water are nearly equivalent to the conductivity of natural water at the same $g_T$ (52), the redefinition of the conductivity ratio ($R_{15}$) in terms of weight diluted or evaporated seawater will yield weight salinities $S(^o/oo)_T$ valid for all natural waters (52).

$$S(^o/oo)_T = 27.25860610\ R_{15} + 19.0618570\ R_{15}^2 - 27.23834557\ R_{15}^3 + 27.09960846\ R_{15}^4 - 14.19791134\ R_{15}^5 + 3.01618532\ R_{15}^6 \quad (52)$$

Due to the sensitivity of conductivity measurements, the conductivities of estuaries at low salinities will deviate considerably from seawater diluted with pure water. The equivalent conductivity of solutions unlike the thermodynamic properties approach values at infinite dilution that are dependent upon the composition. For example, if we estimate the infinite dilution equivalent conductance ($\Lambda = 10^3\ L_{sp}/e_T$, where $L_{sp}$ is the specific conductance) of river salts and sea salts from

$$\Lambda^o = \Sigma\ E_i\ \Lambda^o_i \quad (53)$$

we obtain (see Table XIII) $\Lambda^o = 97.34$ and $127.85\ cm^2\ \Omega^{-1} eq^{-1}$

TABLE XIII
CALCULATION OF THE INFINITE DILUTION EQUIVALENT CONDUCTANCE OF RIVER AND SEA SALTS AT 25°C

| Ion | $\Lambda^o_i$ [a] | River Salt $E_i\ \Lambda^o_i$ | Sea Salt $E_i\ \Lambda^o_i$ |
|---|---|---|---|
| $Na^+$ | 50.10 | 8.367 | 38.711 |
| $Mg^{2+}$ | 53.05 | 10.896 | 9.322 |
| $Ca^{2+}$ | 59.50 | 27.168 | 2.017 |
| $K^+$ | 73.50 | 2.646 | 1.238 |
| $Sr^{2+}$ | 59.47 | --- | 0.018 |
| $Cl^-$ | 76.35 | 10.239 | 68.775 |
| $SO_4^{2-}$ | 80.02 | 11.363 | 7.457 |
| $HCO_3^-$ | 44.50 | 25.961 | 0.142 |
| $Br^-$ | 78.14 | --- | 0.109 |
| $CO_3^{2-}$ | 69.3 | --- | 0.047 |
| $B(OH)_4^-$ | 65.0 | --- | 0.009 |
| $F^-$ | 55.4 | --- | 0.006 |
| $NO_3^-$ | 71.46 | 0.700 | --- |
| | | 97.340 | 127.850 |

a) From Robinson and Stokes (57). The value for $B(OH)_4^-$ has been estimated.

respectively, for river and sea salts (using values of $\Lambda^o$ given in reference (57)). If it is assumed that (equation 53) is valid over a wide concentration range, it is possible to estimate the $\Lambda$ of an estuary formed by mixing river and sea salts. These values of $\Lambda$(Estuary) are shown in Figure 6 (75). At concentrations of $e_T$ greater than 0.122 or $g_T = 7.13\ g\ kg^{-1}$ the

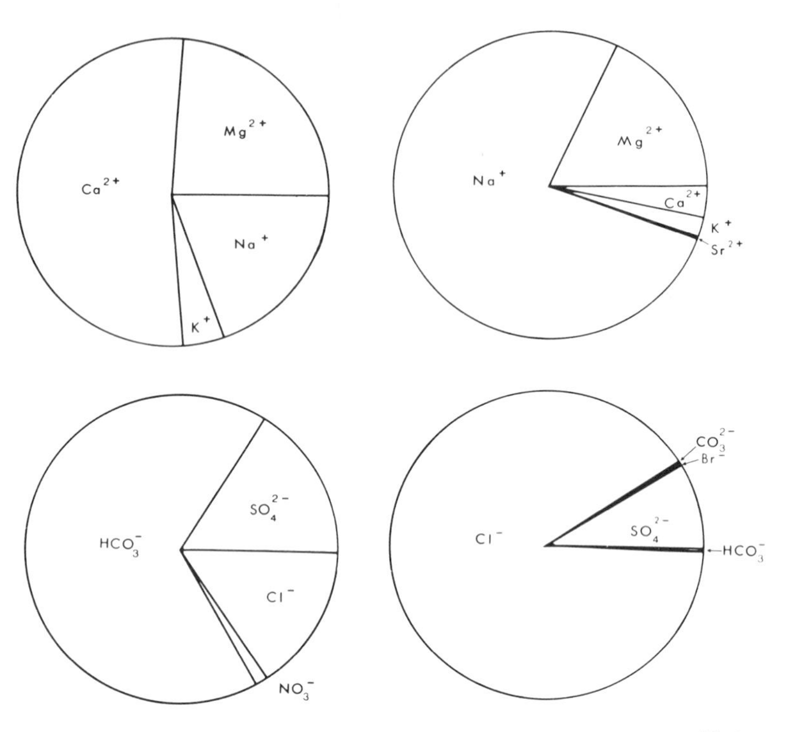


*Figure 5. The equivalent fraction of the major constituents for world river and sea salts*

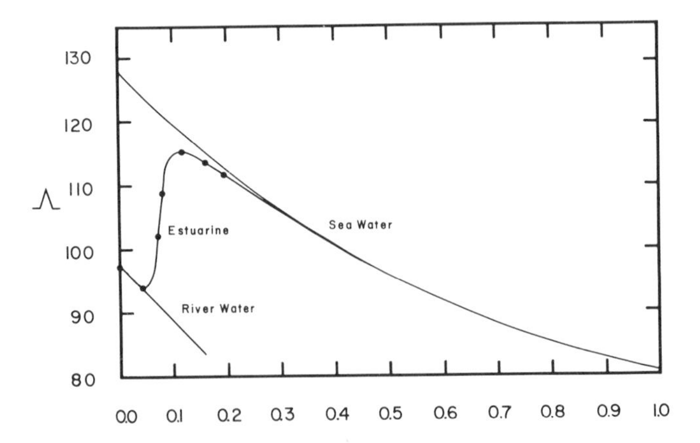


*Figure 6. The equivalent conductance of river water, seawater, and estuarine waters as a function of the square root of normality at 25°C*

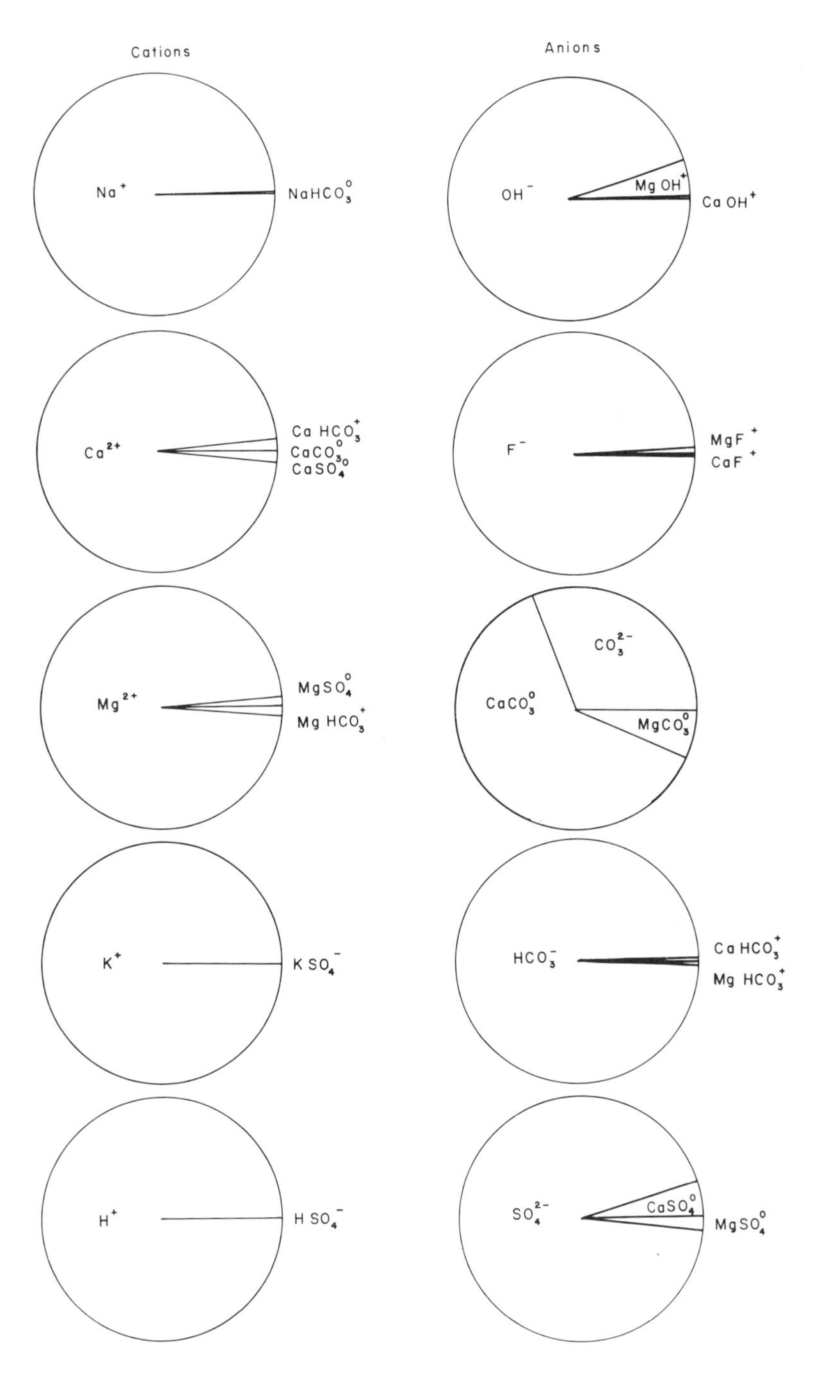


*Figure 7. The speciation of the major cations and anions of river salt at 25°C*

$\Lambda$(Estuary) is approximately equal to the $\Lambda$ of seawater diluted with pure water. Further experimental conductance studies are needed to see if these calculations are correct.

## The Speciation of Major and Minor Ions in River Waters

Since the composition of river waters and seawater are quite different, one might expect chemical reactions in estuaries to be affected by the mixing of the two waters. One of the easiest ways of examining how the composition of rivers affects the properties of ionic solutes is to compare the speciation of the ionic solutes in river waters and seawaters. Since the ionic strength of river water is very low, it is possible to estimate the activity coefficient of free ions and charged ion pairs by using the equation (55-57)

$$\log \gamma_F = -(0.509)\, Z^2\, [I^{1/2}/(1 + I^{1/2}) - 0.3I] \qquad (54)$$

This equation yields values of $\gamma_F$ = 0.951 for monovalent ions or ion pairs and 0.817 for divalent ions or ion pairs at I = 2.095 x $10^{-3}$ mol $kg^{-1}$. The stoichiometric association constants determined from thermodynamic association constants (9) for the major river salts are given in Table XIV.

TABLE XIV
THE STOICHIOMETRIC ASSOCIATION CONSTANTS FOR THE MAJOR RIVER SALTS AT 25°C

| | $K_A^*$ | | | | | |
|---|---|---|---|---|---|---|
| | $SO_4^{2-}$ | $HCO_3^-$ | $CO_3^{2-}$ | $NO_3^-$ | $F^-$ | $OH^-$ |
| $Na^+$ | 4.3 | 1.3 | 2.9 | $0.2_3$ | 0.5 | $0.2_4$ |
| $Mg^{2+}$ | 108 | 13.2 | 1272 | --- | 54 | 310 |
| $Ca^{2+}$ | 137 | 14.5 | 5552 | $1.5_5$ | 9.0 | 15.2 |
| $K^+$ | 7.4 | --- | --- | $0.5_7$ | --- | --- |
| $H^+$ | 78 | --- | --- | --- | --- | --- |

The speciation of the major ionic components of river salt are given in Table XV, and shown in Figure 7. As in seawater, the

TABLE XV
THE SPECIATION OF THE MAJOR CATIONS AND ANIONS IN RIVER WATER AT 25°C

| Cation | % M | % M $SO_4$ | % M $HCO_3$ | % M $CO_3$ | % M OH |
|---|---|---|---|---|---|
| $H^+$ | 99.98 | 0.02 | --- | --- | --- |
| $Na^{2+}$ | 99.83 | 0.05 | 0.12 | --- | --- |
| $Mg^{2+}$ | 97.54 | 1.15 | 1.21 | 0.08 | 0.01 |

TABLE XV continued

| Cation | % M | % M $SO_4$ | % M $HCO_3$ | % M $CO_3$ | % M OH |
|---|---|---|---|---|---|
| $Ca^{2+}$ | 96.89 | 1.45 | 1.32 | 0.33 | 0.01 |
| $K^+$ | 99.92 | 0.08 | --- | --- | --- |

| Anion | % X | % Na X | % Mg X | % Ca X | % K X |
|---|---|---|---|---|---|
| $Cl^-$ | 100.00 | --- | --- | --- | --- |
| $SO_4^{2-}$ | 93.55 | 0.11 | 1.66 | 4.64 | 0.04 |
| $HCO_3^-$ | 99.23 | 0.04 | 0.21 | 0.52 | --- |
| $CO_3^{2-}$ | 31.03 | 0.03 | 6.50 | 62.44 | --- |
| $NO_3^-$ | 99.93 | 0.01 | --- | 0.06 | --- |
| $OH^-$ | 94.64 | 0.01 | 4.83 | 0.52 | --- |
| $F^-$ | 98.79 | 0.01 | 0.88 | 0.32 | --- |

major river cations are predominately in the free form. The anions with the exception of $CO_3^{2-}$ are also predominately in the free ion form. The total activity coefficients for river ions determined from (equations 32 and 33) are given in Table XVI.

TABLE XVI
THE TOTAL ACTIVITY COEFFICIENTS OF THE MAJOR RIVER IONS

| Ion | % Free | $\gamma_F$ | $\gamma_T$ |
|---|---|---|---|
| $H^+$ | 99.98 | 0.951 | 0.951 |
| $Na^+$ | 99.83 | 0.951 | 0.949 |
| $Mg^{2+}$ | 97.54 | 0.817 | 0.797 |
| $Ca^{2+}$ | 96.89 | 0.817 | 0.792 |
| $K^+$ | 99.92 | 0.951 | 0.950 |
| $Cl^-$ | 100.00 | 0.951 | 0.951 |
| $SO_4^{2-}$ | 93.55 | 0.817 | 0.764 |
| $HCO_3^-$ | 99.23 | 0.951 | 0.944 |
| $CO_3^{2-}$ | 31.03 | 0.817 | 0.254 |
| $NO_3^-$ | 99.93 | 0.951 | 0.950 |
| $OH^-$ | 94.64 | 0.951 | 0.900 |
| $F^-$ | 98.79 | 0.951 | 0.939 |

It should be pointed out that values of $\gamma_T$ for river ions determined from the specific interaction model are in good agreement with the values given in Table XVI. The difference between using values of $\gamma_F$ and $\gamma_T$ for ionic equilibria in river waters can be demonstrated by examining the value $\gamma_\pm^2(CaCO_3) = \gamma(Ca) \times \gamma(CO_3)$; which can be used to examine the solubility of $CaCO_3$. The value of $\gamma_\pm^2(CaCO_3)$ determined from free ion activity coefficients (0.667) is 3.3 times larger than the value determined from the total activity coefficients (0.201).

The speciation of the heavy metals Cu, Zn, Cd and Pb in

river waters can also be made by using free ion activity coefficients determined from (equation 54) and the thermodynamic constants tabulated by Zirino and Yamamoto (32). The speciation of these metals in river and seawater are shown in Figure 8. The predominate form of all the heavy metals in river waters is the carbonate complex, while in seawater hydroxide, chloride and carbonate complexes are important. It is interesting to note that a greater percentage of free heavy metal ions exist in seawater than in river waters. The total activity coefficients of the heavy metals in river water are given in Table XVII.

TABLE XVII
THE TOTAL ACTIVITY COEFFICIENTS OF HEAVY METALS IN RIVER WATERS

| Metal | % Free | $\gamma_F$ | $\gamma_T$ |
|---|---|---|---|
| $Cu^{2+}$ | 0.0197 | 0.817 | 0.0002 |
| $Zn^{2+}$ | 1.1691 | 0.817 | 0.0096 |
| $Cd^{2+}$ | 0.9367 | 0.817 | 0.0077 |
| $Pb^{2+}$ | 0.0077 | 0.817 | 0.0001 |

The importance of examining all of the interaction parameters for heavy metal ions is clearly demonstrated by these calculations. The ratio of the free ion activity coefficients for Cu, Zn and Pb in river and seawaters is 2.5 to 3.2, while the value for the total activity coefficients is 0.01 to 0.2. The cadium calculations do not show these effects since the $\gamma_F$ in seawater determined by using the mean salt method from $\gamma_{\pm}(CdCl_2)$ is quite low (0.012). This is due to the strong interactions between $Cd^{2+}$ and $Cl^-$ ions. If the $\gamma_F$ for Cd in seawater had been estimated from (equation 54) ($\gamma_F = 0.34$ (32)), the results would have been similar [$\gamma_F$(River)/ $\gamma_F$(SW) = 2.4 and $\gamma_T$(River)/ $\gamma_T$(SW) = 0.9] to the ratios for Cu, Zn and Pb. These results point out the importance of using the same methods to determine $\gamma_F$ at both low and high concentrations,if one is going to compare values of $\gamma_T$ or ionic equilibria in fresh and saline waters. Although the mean salt method yields more reliable values for $\gamma_F$ at high concentrations, it cannot easily be used at low concentrations (below 0.1 m) since tabulations of experimental data are not available. These differences occur, however, only for the few metals that form complexes with $Cl^-$ ions.

Since organic ligands of unknown concentrations present in river waters could also complex these heavy metals, these calculations must be considered as a first approximation. The calculations do show, however, that large changes in the state of metal ions can occur at the interface between rivers and seawaters. Further work is necessary to examine how the state of ionic solutes influences chemical processes (such as solubility and ion-exchange) that occur in estuaries.

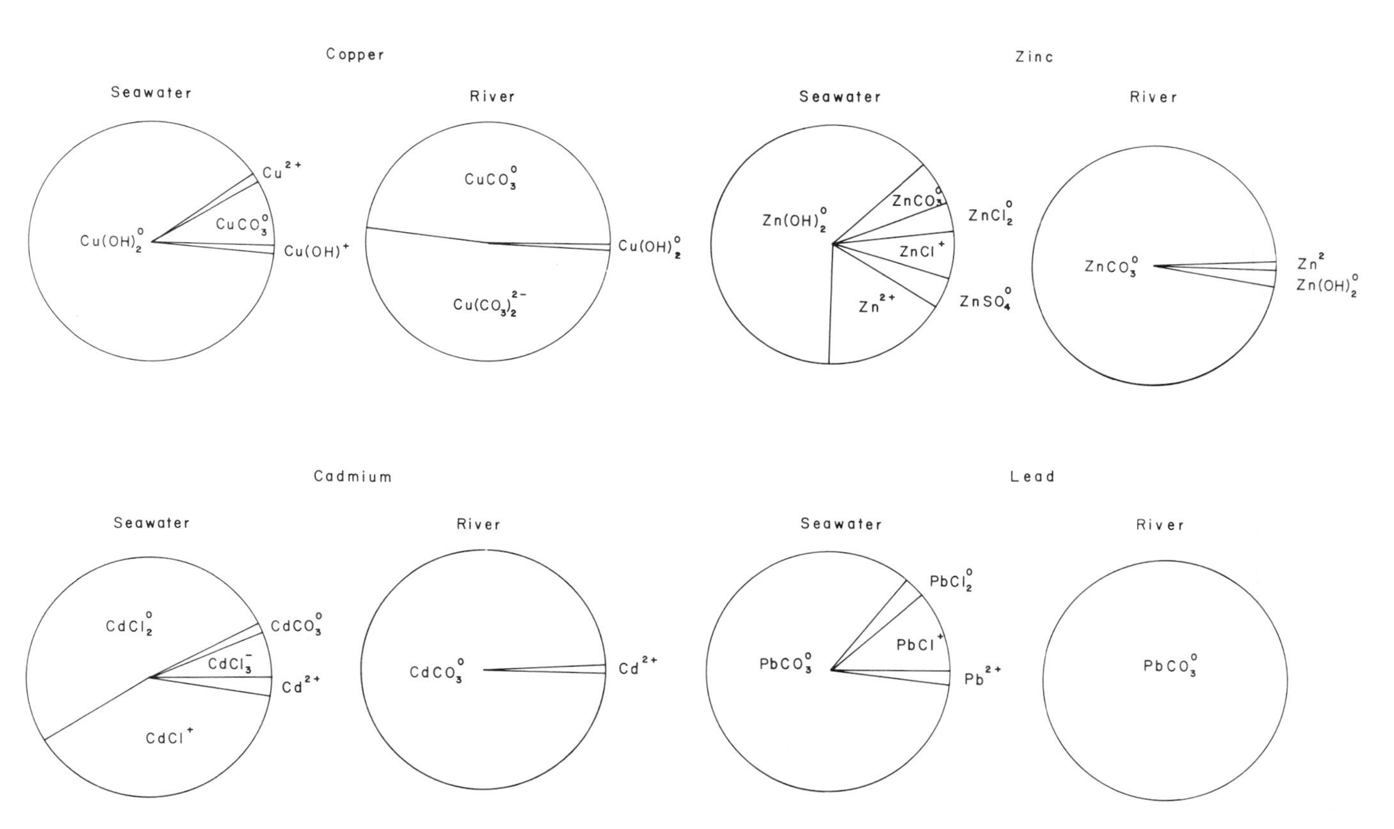


*Figure 8. Comparison of speciation of copper, zinc, cadmium, and lead in river water and seawater at 25°C*

Acknowledgement

The author wishes to acknowledge the support of the Office of Naval Research (Contract ONR - N00014-75-C-0173) and the oceanographic branch of the National Science Foundation (GA-40532) for this study. The author also wishes to thank Mr. Peter Chetirkin for preparing the Figures.

Abstract

In recent years a great deal of progress has been made in developing models to examine the physical chemistry of multicomponent electrolyte solutions like seawater. These models have been used to examine the physical chemical properties of seawater and the state of ionic solutes in seawater. In the present paper these models will be briefly reviewed and applied to estuaries. The physical chemical properties of estuaries have been estimated by mixing river water with seawater and assuming that the excess properties of the resulting mixture are equal to zero. These calculations demonstrate that at the same total solid concentration the physical chemical properties of estuaries are equal to those of seawater diluted with pure water. The state of metal ions in estuaries have also been examined by using the ion pairing model for ionic interactions.

Literature Cited

(1) Friedman, H. L., J. Chem. Phys., (1960), 32, 1134.
(2) Harned, H. S., Robinson, R. S., "Multicomponent Electrolyte Solutions", Pergamon, Oxford (1968).
(3) Scatchard, G., J. Amer. Chem. Soc., (1968), 90, 3124.
(4) Wood, R. H., Reilly, P. J., Ann. Rev. Phys. Chem., (1970), 21, 287.
(5) Millero, F. J., in "Biophysical Properties of Skin", H. R. Elden, ed., Wiley, New York (1971), p. 329.
(6) Scatchard, G., J. Amer. Chem. Soc., (1969), 91, 2410.
(7) Anderson, H. L., Wood, R. H., in "Water", F. Franks, ed., Plenum, New York (1974), p. 119.
(8) Millero, F. J., in "The Sea" Vol. 5, E. D. Goldberg, ed., Wiley, New York (1974), p. 3.
(9) Millero, F. J., Ann. Rev. Earth Planet. Sci., (1974), 2, 101.
(10) Lepple, F. K., Millero, F. J., Deep Sea Res., (1971), 18, 1233.
(11) Millero, F. J., Deep Sea Res., (1973), 20, 101.
(12) Millero, F. J., J. Soln. Chem., (1973), 2, 1.
(13) Millero, F. J., Lepple, F. K., Mar. Chem., (1973), 1, 89.
(14) Millero, F. J., Perron, G., Desnoyers, J. E., J. Geophys. Res., (1973), 78, 4499.

(15) Millero, F. J., in "Structure of Water and Aqueous Solutions", W. A. P. Luck, ed., Verlag Chemie, Germany (1973), p. 513.
(16) Emmet, R. T., Millero, F. J., J. Geophys. Res., (1974), 79, 3463.
(17) Millero, F. J., Naval Rev., (1974), 27, 40.
(18) Leung, W. H., Millero, F. J., J. Chem. Thermodyn., (1975), in press.
(19) Millero, F. J., Duer, W., Oglesby, B., Leung, W. H., J. Soln. Chem., (1975), submitted.
(20) Robinson, R. A., Wood, R. H., J. Soln. Chem., (1972), 1, 481.
(21) Whitfield, M., Deep Sea Res., (1974), 21, 57.
(22) Sillen, L. G., in "Oceanography", M. Sears, ed., A.A.A.S., Washington, D. C., (1961), p. 549.
(23) Garrels, R. M., Thompson, M. E., Amer. J. Sci., (1962), 260, 57.
(24) Kester, D. R., Pytkowicz, R. M., Limnol. Oceanogr., (1969), 14, 686.
(25) Pytkowicz, R. M., Kester, D. R., Amer. J. Sci., (1969), 267, 217.
(26) Lafon, G. M., Ph.D. Thesis, Northwestern University, Evanston, Illinois (1969).
(27) Kester, D. R., Pytkowicz, R. M., Geochim. Cosmochim. Acta, (1970), 34, 1039.
(28) Kester, D. R., Byrne, R. H., Jr., "Ferromanganese Deposits on the Ocean Floor", Lamont, New York (1972), p. 107.
(29) Atkinson, G., Dayhoff, M. O., Ebdon, D. W., in "Marine Electrochemistry", J. B. Berkowitz, et.al., ed., The Electrochem. Soc. Inc., Princeton, (1973), p. 124.
(30) Morel, F., Morgan, J., Environ. Sci. Tech., (1972), 6, 58.
(31) Dyrssen, D., Wedborg, M., in "The Sea" Vol. 5, E. D. Goldberg, ed., Wiley, New York (1974), p. 181.
(32) Zirino, A., Yamamoto, S., Limnol. Oceanogr., (1972), 17, 661.
(33) Stumm, W., Brauner, P. A., in "Chemical Oceanography" Vol. 1, (2nd Ed.), J. P. Riley and G. Skirrow, eds., Academic Press, New York (1975), in press.
(34) Millero, F. J., Limnol. Oceanogr., (1969), 14, 376.
(35) Millero, F. J., Geochim. Cosmochim. Acta, (1971), 35, 1089.
(36) Millero, F. J., Berner, R. A., Geochim. Cosmochim. Acta, (1972), 36, 92.
(37) Millero, F. J., in "The Sea" Vol. 6, E. D. Goldberg, ed., Wiley, New York (1975).
(38) Leyendekkers, J. V., Mar. Chem., (1972), 1, 75.
(39) Whitfield, M., Mar. Chem., (1973), 1, 251.
(40) Young, T. F., Rec. Chem. Progr., (1951), 12, 81.
(41) Young, T. F., Smith, M. B., J. Phys. Chem., (1954), 58, 716.
(42) Wood, R. H., Anderson, H. L., J. Phys. Chem., (1966), 70, 1877.

(43) Reilly, P. J., Wood, R. H., J. Phys. Chem., (1969), 73, 4292.
(44) Reilly, P. J., Wood, R. H., Robinson, R. A., J. Phys. Chem. (1971), 75, 1305.
(45) Ward, G. K., Millero, F. J., J. Soln. Chem., (1974), 3, 431.
(46) Millero, F. J., Earth Planet. Sci. Letters, (1975), in press.
(47) Millero, F. J., Lawson, D. R., J. Geophys. Res., (1975), submitted.
(48) Millero, F. J., Chem. Rev., (1971), 71, 147.
(49) Millero, F. J., in "Water and Aqueous Solutions", R. H. Horne, ed., Wiley, New York (1972), p. 519.
(50) Millero, F. J., Gonzalez, A., J. Chem. Eng. Data, (1975), to be submitted.
(51) Millero, F. J., in "The Oceans Handbook", R. H. Horne, ed., Wiley, New York (1975), in press.
(52) Millero, F. J., Gonzalez, A., Ward, G. K., J. Mar. Res., (1975), submitted.
(53) Guggenheim, E. A., Phil. Mag., (1935), 19, 588.
(54) Scatchard, G., Rush, R. M., Johnson, J. S., J. Phys. Chem., (1970), 74, 3786.
(55) Lewis, G. N., Randall, M., "Thermodynamics", (2nd Ed.), revised by K. S. Pitzer and F. Brewer, McGraw-Hill, New York (1961).
(56) Harned, H. S., Owen, B. B., "The Physical Chemistry of Electrolyte Solutions", ACS Mono. Ser. No. 137, Reinhold, New York (1958).
(57) Robinson, R. A., Stokes, R. H., "Electrolyte Solutions", Butterworths, London (1959).
(58) Pitzer, K. S., J. Phys. Chem., (1973), 77, 268.
(59) Whitfield, M., personal communication, (1975).
(60) Elgquist, B., Wedborg, M., Mar. Chem., (1974), 2, 1.
(61) Millero, F. J., Thallasia Yugoslavia, (1975), in press.
(62) Watson, M. W., Wood, R. H., Millero, F. J., in this book.
(63) Pritchard, D. W., "Estuaries", A.A.A.S. Pub. 83, Washington D. C., (1967), p. 3.
(64) Livingstone, D. A., in "Data of Geochemistry", M. Fleisher, ed., Geol. Sur. Prof. Paper 440-G, p. 641.
(65) Harned, H. S., Scholes, S. R., Jr., J. Amer. Chem. Soc., (1941), 63, 1706.
(66) Chatterjee, B., J. Indian Chem. Soc., (1939), 14, 589.
(67) Kell, G. S., J. Chem. Eng. Data, (1967), 12, 66.
(68) Brewer, P., Bradshaw, A., J. Mar. Res., (1975), in press.
(69) Cox, R. A., McCartney, M. J., Culkin, R., Deep Sea Res., (1970), 17, 679.
(70) Kremling, K., Deep Sea Res., (1972), 19, 377.
(71) Kremling, K., Kieler Meeresf., (1972), 28, 99.
(72) Knudsen, M., "Hydrographische Tabellen", G.E.C., Grod., Copenhagen (1901).

(73) Lyman, J., Fleming, R. H., J. Mar. Res., (1940), 3, 134.
(74) UNESCO, Second Report on Joint Panel on Oceanography Tables and Standards UNESCO Tech. Rept., Mar. Sci. No. 4, (1966).
(75) Thomas, B. D., Thompson, T. G., Utterback, C. L., J. Cons. Int. Explor. Mer., (1934), 9, 28.

# 3

# Redox Reactions and Solution Complexes of Iron in Marine Systems

DANA R. KESTER, ROBERT H. BYRNE, JR., and YU-JEAN LIANG

Graduate School of Oceanography, University of Rhode Island, Kingston, R.I. 02881

The chemistry of iron in natural waters has considerable effect on a variety of environmental processes. Iron plays a unique role in many biological systems due to its ability to form porphyrin molecules which participate in biochemical oxidation-reduction reactions. Changes in the oxidation state of iron in response to environmental changes in redox potential is a significant factor for geochemical processes such as the formation of pyrite ores and ferro-manganese nodules. The chemical behavior of iron in natural waters is also important because of its prevalent use in structural materials and its subsequent deterioration through corrosion reactions. The tendency of iron to form colloidal and particulate phases provides a mechanism for the removal of dissolved trace elements from natural waters by adsorption and coprecipitation. Due to this diverse range of chemical processes, it is important to establish a reliable understanding of the redox reactions of iron in marine waters.

Analytical measurements of the concentration of iron in natural waters provides a basis for evaluating the significance of this element in various chemical systems. The distinction between dissolved and particulate iron has generally been made on the basis of filtration using a 0.5 μm pore size filter. This procedure leads to an operationally defined separation of iron into soluble and solid phases which can be difficult to relate to chemically defined forms such as aqueous ions, solution complexes, and colloidal particles. It is important for this reason to qualify analytical observations according to the separation techniques employed. Table I provides a brief example of the concentrations of iron found in some natural waters. These results do not show the variability which occurs in the amount of iron in natural waters, but they represent the magnitudes which are encountered and they suggest the significance of a removal mechanism for iron as waters pass from fresh to estuarine to oceanic environments.

A second step in defining the chemical processes involving iron is to distinguish between Fe(II) and Fe(III) and to deter-

Table I. Concentrations of iron in some natural waters.

| Environment | Iron (mol/kg) | Form | Reference |
|---|---|---|---|
| River | $1.2 \times 10^{-5}$ | Dissolved and particulate not specified | (1) |
| Estuarine | $5 \times 10^{-6}$ | Particulate | (2) |
| Estuarine (1°/₀₀ Salinity) | $3 \times 10^{-6}$ | Dissolved | (3) |
| Estuarine (20°/₀₀ Salinity) | $6 \times 10^{-7}$ | Dissolved | (3) |
| Oceanic | $2 \times 10^{-8}$ | Dissolved | (4) |
| Oceanic | $3 \times 10^{-9}$ | Particulate | (5) |

mine the tendency of these two oxidation states to form solution complexes and solid phases. At the present time techniques do not exist which can reveal the specific chemical forms in aqueous samples containing less than $10^{-6}$ mol/kg iron. It is therefore necessary to pursue a less direct approach in which a chemical model for iron in natural waters is developed from thermodynamic and kinetic information. Some of the general characteristics of chemical models and their relationship to natural systems were considered by Morgan (6). Previous considerations of iron chemistry in marine systems have examined the importance of hydroxide, phosphate, and organic ligands (7-13). In this paper we will incorporate data recently obtained in our laboratory into a consideration of iron in marine waters giving particular attention to the effects of salinity and pH. It is possible to identify from this model the most significant forms and reactions of iron, and techniques may then be devised to characterize these chemical systems under natural conditions. A chemical model is limited by its inability to account for the effects of biochemical processes on the chemistry of a constituent such as iron. Nevertheless, deviations between the predictions from chemical models and the environmental observations can indicate the existence of additional factors which require consideration.

This paper presents a chemical model for iron in marine systems based on thermodynamic and kinetic information. Emphasis will be placed on the redox reactions of iron because of their unique importance in this metal's environmental chemistry. An Eh-pH diagram (Figure 1) provides a useful orientation for the material to be presented. The two broken lines delineate the stability range for water molecules relative to oxygen and hydrogen gases; all stable aqueous systems fall within these bounds. The redox potential (Eh) of marine systems appears to be controlled by either oxygen or sulfide. Thermodynamically, the oxygen-water reaction should control the Eh of aerated systems (line ABC in Figure 1), but due to kinetic factors it is possible that the oxygen-hydrogen peroxide reaction regulates the Eh in these systems (14, 15). For marine environments devoid of oxygen the sulfate-sulfide reaction (line DEF in Figure 1) controls the Eh. A chemical model for ferrous iron will be developed which applies along line DEF. A similar model for ferric iron applicable to line ABC will be presented next. It will then be possible to examine the redox equilibrium between these two oxidation states, and finally the kinetics of ferrous oxygenation in natural waters will be considered.

## Solution Equilibria of Ferrous Ions

The first step in formulating a chemical model in natural waters is to consider the composition of the medium and the tendency for these components to form metal complexes. Table II lists the molalities of the major constituents in river water (1), ocean water (16, 17) and mixtures of these two waters. This information may be combined with data on the formation of solution complexes (18, 19) to obtain a chemical model for a constituent. The lack of knowledge about the organic constituents in marine waters and of their ability to form iron-organic complexes is a limiting factor in developing a quantitative model. One approach to this problem is to consider the inorganic speciation of iron and then assess the possible significance of organic ligands relative to the inorganic ones. There are two areas in which the present knowledge of inorganic iron complexes is inadequate for marine systems: the possible formation of bicarbonate, carbonate, and sulfide complexes can not be incorporated into the model with existing data; and the possible significance of mixed ligand complexes has not been established. The role of mixed ligand complexes such as chloro-hydroxy species is a chemical problem of special importance to multicomponent natural waters which chemists studying single salt solutions have avoided. Some considerations have indicated that mixed ligand complexes may be important for copper, zinc, and mercury in seawater (20). A chemical model for iron can be developed within these limitations.

Stability constants for hydroxide, chloride, and sulfate complexes of ferrous iron were selected in the following manner. These equilibrium constants have been measured in a variety of ionic media such as sodium perchlorate solutions with ionic

Table II. Molality of constituents and ionic strength (I) parameters in mixtures of surface oceanic seawater and world average river water (RW).

| | Salinity | | | | |
|---|---|---|---|---|---|
| Constituent | RW | 4°/oo | 10°/oo | 20°/oo | 35°/oo |
| $Na^{+}$ | 0.00027 | 0.05426 | 0.13731 | 0.27574 | 0.48614 |
| $Mg^{2+}$ | 0.00017 | 0.00623 | 0.01556 | 0.03110 | 0.05473 |
| $Ca^{2+}$ | 0.00037 | 0.00151 | 0.00327 | 0.00620 | 0.01065 |
| $K^{+}$ | 0.00006 | 0.00123 | 0.00303 | 0.00627 | 0.0106 |
| $Cl^{-}$ | 0.00022 | 0.06306 | 0.15973 | 0.32086 | 0.56577 |
| $SO_4^{2-}$ | 0.00011 | 0.00335 | 0.00833 | 0.01664 | 0.02926 |
| $HCO_3^{-}$ | 0.00096 | 0.00113 | 0.00139 | 0.00182 | 0.00247 |
| $F^{-}$ | 0.00001 | 0.00002 | 0.00003 | 0.00004 | 0.00007 |
| I | 0.0021 | 0.0820 | 0.2051 | 0.4102 | 0.7218 |
| $\sqrt{I}/(1 + \sqrt{I})$ | 0.0434 | 0.2226 | 0.3117 | 0.3904 | 0.4593 |

strengths (I) between 0.1 and 3.0 molal (18, 19). In some cases these equilibrium constants were evaluated at infinite dilution or zero ionic strength. A plot of the logarithm of the stability constant for each complex versus $\sqrt{I}/(1 + \sqrt{I})$ was constructed in order to examine the consistency of the various values and to obtain their ionic strength dependence. The functional form $\sqrt{I}/(1 + \sqrt{I})$ provides a convenient means of interpolating between different ionic strengths and of extrapolating data to infinite dilution (21, 22).

The resulting stability constants are summarized in Table III. The values which have been reported for log $*\beta_2(Fe(OH)_2)$ range from -17.6 to -5.85 for $0 \leq I \leq 1$. Such wide discrepancies may occur between different studies when in one case the formation of related complexes such as $FeOH^{+}$ is taken into account while in another case these complexes are not considered and their effects thus become associated with the $Fe(OH)_2°$. Alternatively, measurements of this constant may be particularly difficult due to the tendency of Fe(II) to oxidize to Fe(III) which undergoes much more extensive hydrolysis. The systematic selection of free energy data by Langmuir (23) provides a consistent set of values for hydroxy species at infinite dilution. Other data for $*\beta_2(Fe(OH)_2°)$ would require that this constant increase by six to twelve orders of magnitude between I = 0 and

Table III. Stability constants selected for the calculation of ferrous complexes in natural waters.

| Constant | Value | Reference |
| --- | --- | --- |
| $^*\beta_1(FeOH^+) = \frac{[FeOH^+][H^+]}{[Fe^{2+}]}$ | $\log {}^*\beta_1 = -8.30 + 3.58\ \sqrt{I}/(1 + \sqrt{I})$ | (23, 24) |
| $^*\beta_2(Fe(OH)_2^\circ) = \frac{[Fe(OH)_2^\circ][H^+]^2}{[Fe^{2+}]}$ | $\log {}^*\beta_2 = -17.6$ | (23) |
| $\beta_1(FeCl^+) = \frac{[FeCl^+]}{[Fe^{2+}][Cl^-]}$ | $\log \beta_1 = 0.79 - 0.73\ \sqrt{I}/(1 + \sqrt{I})$ | (25) |
| $\beta_2(FeCl_2^\circ) = \frac{[FeCl_2^\circ]}{[Fe^{2+}][Cl^-]^2}$ | $\log \beta_2 = 0.21 - 0.29\ \sqrt{I}/(1 + \sqrt{I})$ | (25) |
| $\beta_1(FeSO_4^\circ) = \frac{[FeSO_4^\circ]}{[Fe^{2+}][SO_4^{2-}]}$ | $\log \beta_1 = 2.12 - 2.47\ \sqrt{I}/(1 + \sqrt{I})$ | (26) |
| $\beta_1(HSO_4^-) = \frac{[HSO_4^-]}{[H^+][SO_4^{2-}]}$ | $\log \beta_1 = 1.99 - 1.06\ \sqrt{I}/(1 + \sqrt{I})$ | (27-30) |

1, which is totally contrary to the ionic strength dependence which can be associated with the various species. For example:

$$*\beta_2(Fe(OH)_2°, I = 0.7) = *\beta_2(Fe(OH)_2°, I = 0) \frac{g_{Fe^{2+}}}{g_{Fe(OH)_2°}\, g^2_{H^+}} \quad (1)$$

where the g factors represent the free activity coefficients of each species at I = 0.7. Even though accurate values can not be assigned to individual ionic activity coefficients, limits can be placed on their variation with I such that the quotient of g-values in equation (1) can not be expected to vary by six orders of magnitude between I = 0 and 1. In fact "rule of thumb" values for these activity coefficients (31) suggest that $*\beta_2(Fe(OH)_2°)$ should be relatively insensitive to variations in I; this rationale yielded the value for this constant selected in Table III. The equilibrium constants for ferrous chloride and sulfate complexes were not available over a range of ionic strengths, so their ionic strength dependences were estimated also from a relationship analogous to equation (1) and from information on the typical behavior of activity coefficient variations between I = 0 and 1 (31).

The calculation of ferrous iron speciation is based on the expression for total Fe(II):

$$T(Fe(II)) = [Fe^{2+}] + [FeCl^+] + [FeCl_2°] + [FeSO_4°] + [FeOH^+] + [Fe(OH)_2°] \quad (2)$$

Due to the low concentration of iron in natural waters it is not necessary to consider polynuclear complexes of iron. The fraction of Fe(II) which is uncomplexed may be derived from equation (2) with the definition of stability constants contained in Table III:

$$[Fe^{2+}]/T(Fe(II)) = 1/(1 + \beta_1(FeCl^+)[Cl^-] + \beta_2(FeCl_2°)[Cl^-]^2 + \beta_1(FeSO_4°)[SO_4{}^{2-}] + *\beta_1(FeOH^+)/[H^+] + *\beta_2(Fe(OH)_2°)/[H^+]^2) \quad (3)$$

and similarly the fraction of Fe(II) which occurs as the generic species $FeL_n$ is given by:

$$[FeL_n]/T(Fe(II)) = \beta_n(FeL_n)[L]^n/(1 + \beta_1(FeCl^+)[Cl^-] + \beta_2(FeCl_2°)[Cl^-]^2 + \beta_1(FeSO_4°)[SO_4{}^{2-}] + *\beta_1(FeOH^+)/[H^+] + *\beta_2(Fe(OH)_2°)/[H^+]^2) \quad (4)$$

The square brackets designate the molality of the enclosed

species (e.g., unassociated or free ligands) and the species in parentheses following each beta identifies the particular stability constant as defined in Table III.

The molality of free ligands was based on an ion-pairing model for seawater in which the free chloride was assumed to be equal to the total chloride (32). The free sulfate was 39% of the total sulfate at 35°/oo salinity (32) and this percentage increased to 100% in river water in proportion to the decrease in salinity. The decrease in $[SO_4^{2-}]$ at low pH due to the formation of $HSO_4^-$ was accounted for in the calculations. For the purpose of this chemical model of iron a pH scale was used such that pH = $-\log [H^+]$ so that the pH yields the free $H^+$ directly. This pH scale has a number of advantages for treating seawater equilibria, and it departs from the conventional National Bureau of Standards pH scale by 0.01 pH units (33, 34).

The results of these calculations for ferrous iron over the pH range of 3.0 to 10.0 for river water, 4°/oo salinity, and 35°/oo salinity seawater are shown in Figure 2. In river water below pH 7 the Fe(II) exists primarily as $Fe^{2+}$ whereas between pH 7 and 10 the hydroxide complexes of Fe(II) become significant. At 35°/oo salinity chloro-complexes are a major form of Fe(II) up to a pH of 7, beyond which the $FeOH^+$ species is the predominant form. The effect of salinity on the Fe(II) species distribution at pH 8 is illustrated in Figure 3. There are two factors involved in the effect of salinity on Fe(II): one is the change in concentration of the chloride and sulfate complexing ligands; the second is the change in stability constants with ionic strength. Table IV provides a more convenient quantitative representation of the net effect of these two factors at pH 7 and pH 8. These results provide an inorganic chemical model for iron which is applicable to line DEF in Figure 1 at various salinities.

It is possible to evaluate the potential significance of organic ligands in complexing ferrous iron. Marine waters typically contain a maximum of 1-5 mg of dissolved organic carbon (DOC) per liter (1, 35, 36). Assuming approximately one functional group capable of complexing $Fe^{2+}$ per six carbon atoms and an average DOC concentration of 2 mg/l yields an effective concentration of $2.8 \times 10^{-5}$ molal for the organic ligands. In order for these ligands to complex 20% of the Fe(II) in 35°/oo salinity seawater at pH 8.0 they must have a stability constant on the order of $2.2 \times 10^5$ whereas in river water at pH 7 an Fe(II)-organic stability constant of $9.6 \times 10^3$ would be sufficient to complex 20% of the Fe(II). Stability constants for the formation of $Fe^{2+}$ complexes with the functional groups of amino acids, acetate, citrate, and quinoline carboxylate are generally in the range $1\text{-}6 \times 10^3$ (18, 19) which indicates that these types of ligands could not compete with the inorganic ones for Fe(II) in seawater, but they could be significant in river water.

A second factor to consider in evaluating the effectiveness

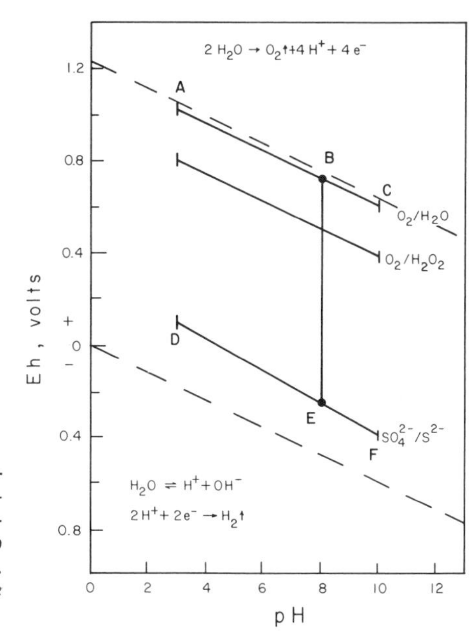


*Figure 1. Eh–pH diagram for natural waters. Line ABC shows the relationship between Eh and pH when the $O_2/H_2O$ controls the Eh. Line DEF is the relationship for the $SO_4^{2-}/S^{2-}$ reaction in marine systems. The intermediate line shows control by the $O_2/H_2O_2$ reaction.*

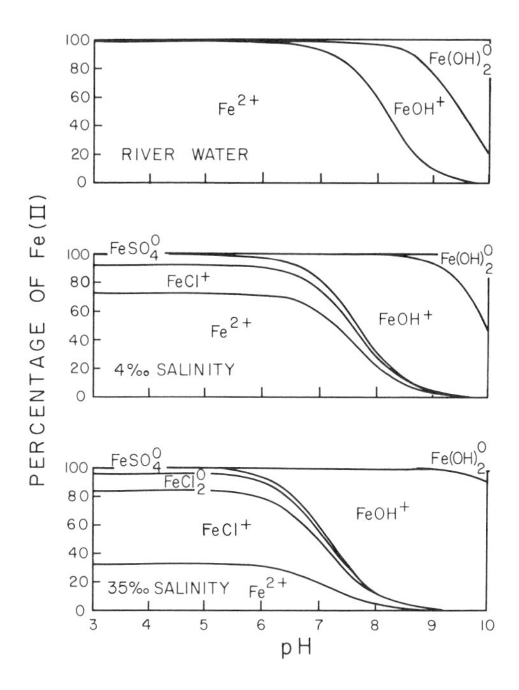


*Figure 2. Distribution of Fe(II) species as a function of pH at three salinities. The vertical distance between curves represents the percentage of the species between the curves.*

Table IV. Percentage distribution of Fe(II) species over the estuarine range of salinity at constant pH.

pH = 7

| Salinity | RW | 4°/oo | 10°/oo | 20°/oo | 35°/oo |
|---|---|---|---|---|---|
| $Fe^{2+}$ | 92.2 | 59.1 | 41.7 | 28.3 | 18.8 |
| $FeCl^{+}$ | 0.1 | 15.8 | 24.3 | 29.0 | 30.3 |
| $FeCl_2^{\circ}$ | -- | 0.3 | 1.4 | 3.6 | 7.2 |
| $FeSO_4^{\circ}$ | 1.0 | 6.2 | 5.4 | 3.6 | 2.1 |
| $FeOH^{+}$ | 6.6 | 18.5 | 27.3 | 35.4 | 41.6 |
| $Fe(OH)_2^{\circ}$ | -- | 0.01 | 0.01 | -- | -- |

pH = 8.0

| Salinity | RW | 4°/oo | 10°/oo | 20°/oo | 35°/oo |
|---|---|---|---|---|---|
| $Fe^{2+}$ | 57.0 | 22.0 | 12.0 | 6.7 | 4.0 |
| $FeCl^{+}$ | 0.07 | 5.9 | 7.0 | 6.9 | 6.4 |
| $FeCl_2^{\circ}$ | -- | 0.1 | 0.4 | 0.9 | 1.5 |
| $FeSO_4^{\circ}$ | 0.6 | 2.3 | 1.6 | 0.8 | 0.4 |
| $FeOH^{+}$ | 40.9 | 69.1 | 78.7 | 84.4 | 87.6 |
| $Fe(OH)_2^{\circ}$ | 1.4 | 0.6 | 0.3 | 0.2 | 0.1 |

of organic ligands in complexing Fe(II) is the competition between $Fe^{2+}$ and $Mg^{2+}$ or $Ca^{2+}$ for the organic functional groups (37, 38). The availability of functional groups for these divalent ions is likely to depend on the hydrogen ion dissociation equilibrium for acid groups. The estimate of organic metal complexing will be maximized by considering that all the functional groups are available for the metal ions at pH 7.0 and 8.0. Table V illustrates that most of the organic ligands capable of complexing $Fe^{2+}$ will be complexed by the much more abundant $Mg^{2+}$ and $Ca^{2+}$ ions in seawater. This effect further decreases the likelihood of significant Fe(II)-organic complexes at pH 8 in seawater. In river water at pH 7 the $Fe^{2+}$ can compete effectively with the $Mg^{2+}$ and $Ca^{2+}$ and about 10% of the Fe(II) could be complexed by organic ligands with a $6 \times 10^3$ stability constant. A definitive evaluation of the importance of Fe(II)-organic complexes will require measurements of β(Fe-org) using naturally derived organic matter from marine systems. Nevertheless the present analysis shows that unless the natural organic matter has a much greater selectivity for $Fe^{2+}$ relative to $Mg^{2+}$ it will not be significant for Fe(II) in seawater, but it could be important for Fe(II) in river water and in coastal waters of low salinity.

Table V. Evaluation of the competition between ferrous, magnesium, and calcium ions for organic ligands in 35‰ salinity seawater and in river water.

$$T(org) = [Fe\text{-}org] + [Mg\text{-}org] + [Ca\text{-}org] + [org]$$

$$\frac{[Fe\text{-}org]}{T(org)} = \frac{\beta(Fe\text{-}org)[Fe^{2+}]}{1 + \beta(Mg\text{-}org)[Mg^{2+}] + \beta(Ca\text{-}org)[Ca^{2+}] + \beta(Fe\text{-}org)[Fe^{2+}]}$$

| | | Seawater, pH = 8.0 | | River Water, pH = 7.0 | |
|---|---|---|---|---|---|
| Metal | (M-org) | [M] | % T(org) | [M] | % T(org) |
| $Fe^{2+}$ | $6 \times 10^3$ | $4 \times 10^{-8}$ | $4 \times 10^{-3}$ | $1 \times 10^{-5}$ | 5.5 |
| $Mg^{2+}$ | $1 \times 10^2$ | $5 \times 10^{-2}$ | 81 | $2 \times 10^{-4}$ | 1.8 |
| $Ca^{2+}$ | 20 | $1 \times 10^{-2}$ | 3 | $4 \times 10^{-4}$ | 0.7 |

## Solution Equilibria of Ferric Ions

An analysis for ferric iron in natural waters may be made similar to the preceding one for Fe(II). Preliminary calculations of this type considering a wide range of ligands ($Cl^-$,

Table VI. Stability constants selected for the calculation of ferric species in natural waters.

| Constant | Value | Reference |
|---|---|---|
| $^{*}\beta_1(FeOH^{2+}) = [FeOH^{2+}][H^+]/[Fe^{3+}]$ | $\log {}^{*}\beta_1 = -2.17 - 1.19\sqrt{I}/(1+\sqrt{I})$ | (23, 40, 41) |
| $^{*}\beta_2(Fe(OH)_2^{+}) = [Fe(OH)_2^{+}][H^+]^2/[Fe^{3+}]$ | $\log {}^{*}\beta_2 = -7.17 + 0.22\sqrt{I}/(1+\sqrt{I})$ | (23, 40, 41) |
| $^{*}\beta_3(Fe(OH)_3^{\circ}) = [Fe(OH)_3^{\circ}][H^+]^3/[Fe^{3+}]$ | $\log {}^{*}\beta_3 = -13.6$ | (23, 40, 42) |
| $\beta_1(FeCl^{2+}) = [FeCl^{2+}]/[Fe^{3+}][Cl^-]$ | $\log \beta_1 = 1.48 - 2.30\sqrt{I}/(1+\sqrt{I})$ | (40, 41, 44, 45) |
| $\beta_2(FeCl_2^{+}) = [FeCl_2^{+}]/[Fe^{3+}][Cl^-]^2$ | $\log \beta_2 = 0.65 - 2.21\sqrt{I}/(1+\sqrt{I})$ | (40, 41, 45) |
| $\beta_2(FeSO_4^{+}) = [FeSO_4^{+}]/[Fe^{3+}][SO_4^{2-}]$ | $\log \beta_1 = 4.04 - 4.02\sqrt{I}/(1+\sqrt{I})$ | (46, 47) |
| $\beta_1(FeF^{2+}) = [FeF^{2+}]/[Fe^{3+}][F^-]$ | $\log \beta_1 = 6.0 - 2.0\sqrt{I}/(1+\sqrt{I})$ | (48, 49) |
| $\beta_1(HF) = [HF]/[H^+][F^-]$ | $\log \beta_1 = 3.18 - 0.65\sqrt{I}/(1+\sqrt{I})$ | (50-52) |

$SO_4^{2-}$, $Br^-$, $F^-$, $B(OH)_4^-$, $HPO_4^{2-}$, $H_2PO_4^-$, and $OH^-$) showed the predominant influence of hydroxide species at pH values greater than 5 (39). During the past several years a program has been carried out in our laboratory to measure the stability constants for ferric hydroxide species which are appropriate to equilibrium calculations in a seawater medium (40-43). Stability constants for the principal inorganic ferric species were obtained as a function of ionic strength in the same manner as for Fe(II). The results are summarized in Table VI. The relative amounts of ferric species were calculated using expressions analogous to equations (3) and (4). The free fluoride ligand was taken to be 50% of the total fluoride in 35°/₀₀ salinity seawater (53) in order to account for fluoride ion-pairing by $Mg^{2+}$, $Ca^{2+}$, and $Na^+$. This percentage increased to 100% in river water in proportion to the decrease in salinity. At low pH values the protonation of $F^-$ was accounted for by use of the $\beta_1(HF)$ listed in Table VI.

The results of Fe(III) species distributions are shown in Figure 4. One of the major differences between the Fe(III) and Fe(II) speciation models is that Fe(III) is not substantially affected by salinity changes for pH values greater than 4 due to the predominance of hydroxide over the other inorganic ligands. Table VII illustrates the competition which exists between hydroxide and the other anions at pH 4 as salinity increases from river water to oceanic values.

Table VII. Percentage distribution of Fe(III) species over the estuarine range of salinity at pH = 4.0.

| Salinity | RW | 4°/₀₀ | 10°/₀₀ | 20°/₀₀ | 35°/₀₀ |
|---|---|---|---|---|---|
| $Fe^{3+}$ | 1.3 | 1.8 | 2.1 | 2.5 | 2.7 |
| $FeCl^{2+}$ | -- | 1.0 | 2.0 | 3.0 | 4.1 |
| $FeCl_2^+$ | -- | -- | -- | 0.2 | 0.4 |
| $FeF^{2+}$ | 9.4 | 10.5 | 10.9 | 10.0 | 11.2 |
| $FeSO_4^+$ | 1.0 | 7.1 | 7.4 | 6.4 | 4.9 |
| $FeOH^{2+}$ | 79.0 | 65.9 | 60.8 | 57.6 | 53.0 |
| $Fe(OH)_2^+$ | 9.1 | 13.6 | 16.7 | 20.4 | 23.6 |
| $Fe(OH)_3^\circ$ | 0.03 | 0.05 | 0.05 | 0.06 | 0.07 |

The requirements for significant organic complexes of Fe(III) relative to the hydroxide species at pH 8 can be formulated in a manner similar to that for Fe(II). This consideration will be made for 35°/₀₀ salinity, because it is essentially independent of the ionic strength and the anion concentrations

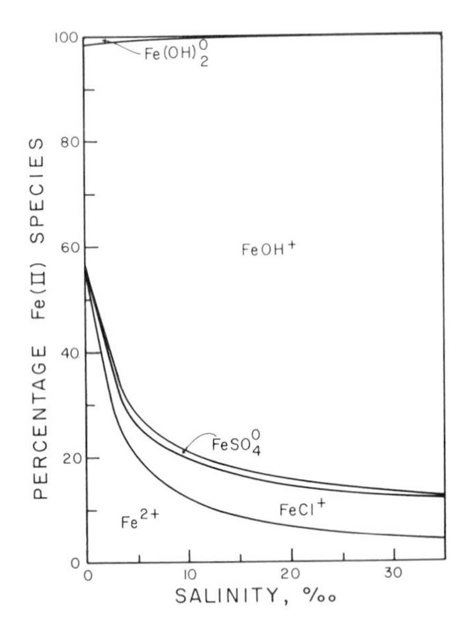


*Figure 3. Effect of salinity on the distribution of Fe(II) species at pH = 8.0*

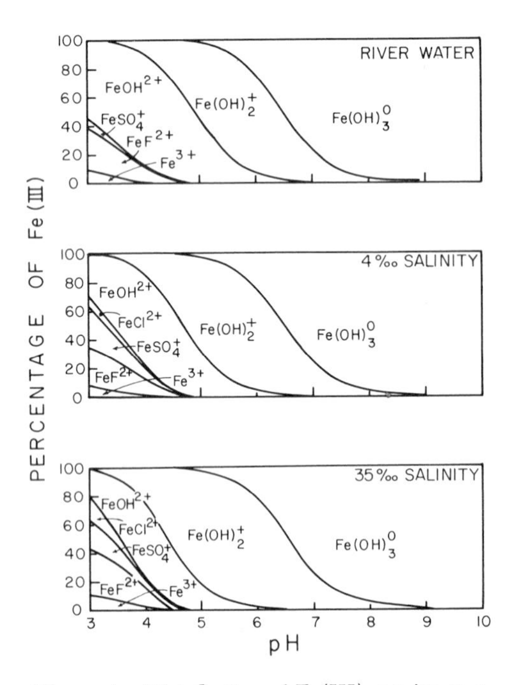


*Figure 4. Distribution of Fe(III) species as a function of pH for three salinities*

other than $OH^-$. Organic ligands such as amino acids (glutamic, aspartic, and alanine), hydroxy acids (citric), and carboxylic acids (propionic) have stability constants with $Fe^{3+}$ of less than $1 \times 10^{12}$. Due to the extensive hydroxide complexing of $Fe^{3+}$ at pH 8 the iron can not compete with the magnesium and calcium ions in seawater for the organic ligands (Table VIII).

Table VIII. Estimation of the ability of ferric ions to compete with the magnesium and calcium ions in seawater at pH = 8.0 for typical organic ligands. The $[Fe^{3+}]$ is based on a $T(Fe(III)) = 1 \times 10^{-7}$.

| Metal | (M-org) | [M] | % T(org) |
|---|---|---|---|
| $Fe^{3+}$ | $1 \times 10^{12}$ | $3.9 \times 10^{-18}$ | $6.3 \times 10^{-5}$ |
| $Mg^{2+}$ | $1 \times 10^{2}$ | $5 \times 10^{-2}$ | 81 |
| $Ca^{2+}$ | 20 | $1 \times 10^{-2}$ | 3 |

These estimates indicate that $3.5 \times 10^{-8}\%$ of the T(Fe(III)) could be complexed by organic matter. In order for Fe(III)-organic complexes to be significant in seawater it would require that the natural organic matter in seawater has a stability constant with iron which is $10^{18}$ times greater than its stability constant with magnesium.

The calculation of ferric iron complexes in seawater and in river water provides a description of the solution equilibria of iron along line ABC in Figure 1. The resulting conclusions are that ferric hydroxide complexes predominate for pH values greater than 4 and that these species are insensitive to changes in salinity. Other ferric complexes are important only under extreme conditions such as acidic mine waters or the microenvironments near surfaces where the pH may decrease below 4. The uncharged $Fe(OH)_3^\circ$ species has been treated as a solution complex in this analysis. The experiments used to determine its stability constant (40, 42) provided an operational characterization of this species such as the fact that it passed through a 0.45 µm pore size filter; it was retained on a 0.05 µm filter (possibly by adsorption); it passed through a dialysis membrane; and it was much more rapidly converted to a filterable form than the bulk portion of precipitated hydrous ferric oxide when the pH was reduced from about 8 to 4. These properties are consistent with what would be expected for an uncharged complex, but they may also represent features of very small colloidal particles as indicated in ultracentrifugation experiments (54). Even though the molecular nature of this $Fe(OH)_3^\circ$ species has not been established, it is possible to differentiate it chemically and physically from other solution complexes and from bulk particulate phases (particle size greater than 0.1-0.5 μm).

Redox Equilibria of Fe(II) and Fe(III)

The preceding considerations of Fe(II) and Fe(III) speciation provide a basis for examining the redox equilibrium between these two oxidation states. These reactions frequently are expressed in terms of standard electrode potentials (relative to the conventional hydrogen scale) in which case the parameter Eh and the following expressions are useful:

$$Fe^{3+} + e^{-} \rightleftharpoons Fe^{2+} \qquad E^{\circ} = +0.771 \text{ volt} \tag{5}$$

$$E^{\circ} = \frac{RT}{F} \ln \frac{a(Fe^{2+})}{a(Fe^{3+})\ a(e)} \qquad RT/F = 0.02569 \text{ volts at } 25^{\circ}C \tag{6}$$

$$\ln \frac{a(Fe^{2+})}{a(Fe^{3+})} = \frac{E^{\circ} - Eh}{RT/F} \quad \text{where } Eh = -\frac{RT}{F} \ln a(e) \tag{7}$$

The a(i) represents the activity of the i constituent.

Equation (7) provides the basis for calculating the relative thermodynamic stability of Fe(II) and Fe(III) as a function of Eh. In order to relate such a calculation to analytical observations it is more useful to consider T(Fe(II)) and T(Fe(III)) than $a(Fe^{2+})$ and $a(Fe^{3+})$. These quantities are related to the activity coefficients of free metal ions, g, and to the speciation of the metal by the following:

$$a(Fe^{2+}) = g(Fe^{2+})\ [Fe^{2+}] \text{ and } a(Fe^{3+}) = g(Fe^{3+})\ [Fe^{3+}] \tag{8}$$

$$T(Fe(II)) = \frac{a(Fe^{2+})/g(Fe^{2+})}{[Fe^{2+}]/T(Fe(II))} \text{ and } T(Fe(III)) = \frac{a(Fe^{3+})/g(Fe^{3+})}{[Fe^{3+}]/T(Fe(III))} \tag{9}$$

The ratios [Fe]/T(Fe) for each oxidation state are obtained from the two preceding sections of this paper, and for the purpose of these calculations it is sufficient to assign "rule of thumb" values to $g(Fe^{2+})$ and $g(Fe^{3+})$ on the basis of solution ionic strength and ionic charge (31). For 35°/₀₀ salinity seawater $g(Fe^{2+}) = 0.25$, $g(Fe^{3+}) = 0.044$ and for river water $g(Fe^{2+}) = 0.915$, $g(Fe^{3+}) = 0.819$.

The resulting calculation of the relative magnitudes of T(Fe(III)) and T(Fe(II)) based on equations (7) and (9) is shown by the solid lines in Figure 5. The redox equilibrium between Fe(II) and Fe(III) does not change much with salinity, but a pH change from 8 to 6 alters the ratio of Fe(III) to Fe(II) by five orders of magnitude.

It is necessary to establish a value for the Eh of natural

waters in order to relate Figure 5 to marine systems. For anoxic environments containing hydrogen sulfide there is general agreement between thermodynamic calculations and Eh measurements using a platinum electrode. The sulfate/sulfide reaction regulates Eh under these conditions and its effect on the Fe(III) and Fe(II) equilibrium is shown by the horizontal diamonds in Figure 5.

For waters containing dissolved oxygen, thermodynamic considerations predict that Eh should be controlled by the $O_2/H_2O$ reaction, but measurements with a platinum electrode yield values which are not well-defined but which are several tenths of a volt lower than predicted. One possible explanation for this observation is that the $O_2/H_2O$ reaction does not occur reversibly on a platinum surface. Another possible factor is that the breaking of the O=O bond in the $O_2/H_2O$ reaction kinetically prevents the attainment of equilibrium and an intermediate redox couple, $O_2/H_2O_2$, may set the Eh of the system (14, 15). It has not been clear to what extent these factors of Eh control at electrode surfaces by dissolved oxygen reflect the mechanism of Eh control for the homogeneous solution reaction between ferrous ions and oxygen. Regardless of which oxygen reaction controls the Fe(II)/Fe(III) couple, these calculations indicate thermodynamically that ferrous iron should yield ferric iron in oxygenated natural waters having pH values greater than 6 and salinities ranging from river to ocean waters.

## Kinetics of Ferrous Oxygenation in Natural Waters

The kinetics of the reaction between Fe(II) and oxygen will determine the extent to which the preceding equilibrium considerations reflect the forms of iron in oxygenated waters. In $NaHCO_3$ solutions with pH = 6.0 to 7.5 Stumm and Lee (55) found that the rate of ferrous oxygenation varied according to:

$$\frac{d\ T(Fe(II))}{d\ t} = k\ T(Fe(II))\ P_{O_2}[OH^-]^2 \qquad (10)$$

The first order dependence of the rate on the partial pressure of oxygen, P, and on T(Fe(II)) appears to be a characteristic feature of the reaction except in highly acidic solutions ($pH < 2$) where the rate is second order with respect to $[Fe^{2+}]$ (56). The second order dependence on $[OH^-]$ persists to a pH = 5, but this pH effect diminishes below this value so that for $pH < 3$ there is very little change in the rate with a further decrease in pH (57). The pH dependence of this reaction is highly significant for natural waters. At pH = 6.0 about a week is required to oxidize 95% of an initial quantity of Fe(II), whereas at pH = 8.0 the reaction is 95% complete in one minute (31).

Evidence has been acquired which shows that the rate of ferrous oxidation differs for various solution complexes of $Fe^{2+}$

(55). The rate increases in the order of the following anions: $ClO_4^-$, $SO_4^{2-}$, $Cl^-$, $PO_4^{3-}$, and $OH^-$. There are several possible mechanisms for the ferrous oxygenation reaction. The Haber-Weiss mechanism has received the most attention (15) and it involves the following rate determining step:

$$FeOH^+ + O_2 + H^+ \rightarrow FeOH^{2+} + HO_2 \quad (11)$$

This reaction would account for the first order dependence on Fe(II) and $O_2$, but it predicts a zero order pH dependence. The $HO_2$ species formed by reaction (11) would react rapidly to form $H_2O_2$ and possibly $H_2O$ with the subsequent electron transfer reactions involving either additional Fe(II) or another reduced substance in the system. Goto *et al.* (58) proposed the following rate determining step to account for the observed dependence on Fe(II), $O_2$, and $OH^-$:

$$FeOH^+ + O_2OH^- \rightarrow Fe(OH)_2^+ + O_2^- \quad (12)$$

The $O_2^-$ species would undergo rapid protonation to $HO_2$ with subsequent reactions as suggested in the Haber-Weiss mechanism. Reaction (12) postulates the existence of an $O_2OH^-$ species which is a hydrated oxygen molecule that has undergone proton dissociation similar to the hydrolysis reactions of transition metal ions. The $O_2OH^-$ species must be quantitatively insignificant compared to ordinary dissolved oxygen even in 2 N NaOH solutions, because the solubility of oxygen in aqueous solution is not pH dependent.

We can suggest two possible rate determining steps which conform to the rate law of equation (10) without invoking a special form of dissolved oxygen. The first one is:

$$Fe(OH)_2^\circ + O_2 \rightarrow Fe(OH)_2^+ + O_2^- \quad (13)$$

The calculations of Fe(II) species indicate that the $Fe(OH)_2^\circ$ complex is less than 0.02% of T(Fe(II)) at pH = 7.0, so that reaction (13) would require the rate constant for $Fe(OH)_2^\circ$ oxygenation to be several orders of magnitude greater than one for an $FeOH^+ + O_2$ reaction. The second possibility is:

$$FeOHH_2O^+ + O_2 \rightarrow Fe(OH)_2^+ + HO_2 \quad (14)$$

in which the rate determining step involves the deprotonation of a hydrated water molecule along with the electron transfer between Fe and $O_2$. Both the deprotonation and the concentration of $FeOH^+$ would be favored by an increase in pH thereby leading to the second order dependence of the rate on $OH^-$. Reaction (14) is more consistent than (13) with the principal Fe(II) species in the neutral pH range.

The role of dissolved organic matter in natural waters is another factor which is significant in the kinetics of the Fe(II)-

Fe(III) reaction. Experiments have shown that organic substances may have several effects on iron reactions (59, 60). Tannic acid, gallic acid, and pyrogallol prevented the oxygenation of Fe(II). Other substances such as glutamic acid, tartaric acid, and glutamine produced a slower rate of oxygenation than was observed in $NaHCO_3$ solutions at the same pH = 6.3. Citric acid produced an initial rate which was greater than the one in $NaHCO_3$. Thus different organic materials can inhibit, retard, or accelerate the rate of ferrous oxygenation. Theis and Singer (60) also found that $10^{-4}$ molar concentrations of numerous organic compounds could reduce $10^{-6}$ to $10^{-5}$ molar concentrations of Fe(III) to Fe(II) in the presence of dissolved oxygen. The humic acid portion of organic material extracted from fresh waters exhibits some of the same characteristics with Fe(II) and Fe(III) as tannic acid. These results have significant implications which require investigation of the kinetics of iron redox reactions in marine systems.

Some initial studies of the rate of ferrous oxygenation in natural waters have been carried out in our laboratory. Two experimental approaches were used. A direct monitoring of the Fe(III) product as a function of time was performed by recording the absorbance at 295 nm of the reaction solution contained in a 10 cm cell. Approximately $5 \times 10^{-6}$ moles of $FeCl_2$ was added to the reaction medium which had a pH between 7.9 and 8.3. The increase in absorbance at 295 nm with time is due to the formation of ferric hydroxy complexes and the scattering of light by colloidal and particulate hydrous ferric oxide. Because the attenuation of light intensity due to absorption and scattering may not have been directly proportional to the Fe(III) concentration a second technique was used in which scattering was not a factor. Samples were allowed to react for various periods of time, and the reaction was quenched by the addition of 1.1 N $HClO_4$ to reduce the pH to 2.5. The UV spectrum of the quenched sample was characteristic of Fe(III) complexes over the wavelength range 240-400 nm. The absorbance at 303 nm of the quenched samples varied linearly with the ferric concentration over the range of 1-8 μM.

An example of the results obtained by the direct monitoring of absorbance at 295 nm is shown in Figure 6. For the reaction:

$$Fe(II) + O_2 \rightarrow Fe(III)$$

the pseudo first order rate law may be written

$$\frac{d\,[Fe(II)]}{d\,t} = k' \{[Fe(II)]_o - [Fe(III)]\} \qquad (15)$$

where [Fe(II)] and [Fe(III)] are the concentrations of reduced and oxidized iron at time t and $[Fe(II)]_o$ is the initial concentration of ferrous iron. The rate parameter k' will be a constant

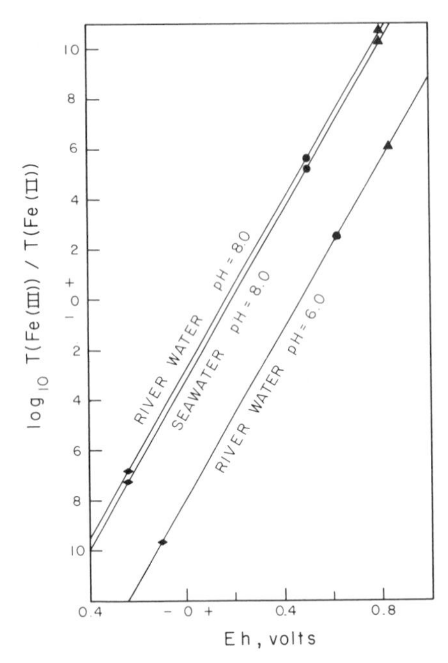


*Figure 5. Variation in the ratio of total Fe(III) to total Fe(II) molality as a function of Eh for three natural water systems.* ▲, *Eh control by the $O_2/H_2O$ reaction;* ●, *Eh control by the $O_2/H_2O_2$ reaction;* ♦, *the control of Eh by the $SO_4^{2-}/S^{2-}$ reaction.*

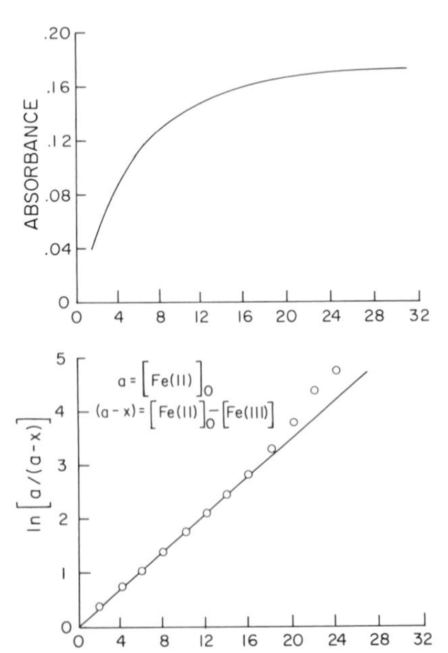


*Figure 6. Rate of Fe(II) oxygenation in Narragansett Bay seawater at pH = 8.08. The upper curve shows the continuously recorded absorbance at 295 nm. The lower plot shows the logarithmic fit which is characteristic of a pseudo first order rate, the slope of which yields k′.*

provided that pH and P are constant (compare equation (15) with (10)). The plot of ln $[Fe(II)]_o/([Fe(II)]_o - [Fe(III)])$ versus time yields a straight line during the initial portion of the reaction (e.g., the first 16 min in Figure 6). Deviations from this line are observed over longer periods of time which may be due to departures in a direct proportionality between [Fe(III)] and absorbance due to coagulation and scattering effects of the hydrous ferric oxide. The value of k' was obtained from the initial linear portion of the data. This rate parameter was indistinguishable from the value obtained by the quenching technique in which scattering of the light was not a factor.

The rate constant k in equation (10) was calculated from k', pH, and P. The results of our measurements in natural seawater from Narragansett Bay and surface Sargasso Seawater are shown in Table IX along with the values obtained by Singer and Stumm (57) in sodium bicarbonate solutions. The $t_{\frac{1}{2}}$ is the time required for the concentration of Fe(II) to be diminished by a factor of two. These preliminary results show that the rate of this reaction is much slower in natural marine waters than would be expected on the basis of measurements in sodium bicarbonate solutions. This observation could reflect the importance of organic matter as pointed out by Theis and Singer (60) or of other constituents in marine waters on the rate of ferrous oxygenation.

Table IX. The rate of ferrous oxygenation in marine waters and in sodium bicarbonate solutions.

| Medium | $k^a$ | $t_{\frac{1}{2}}$ at $P_{O_2} = 0.2094$ | | Source |
|---|---|---|---|---|
| | | pH = 7 (min) | pH = 8 (min) | |
| $NaHCO_3$, pH = 2-6 | $8.0 \times 10^{13}$ | 4 | 0.04 | (57) |
| Narragansett Bay S°/oo = 31.2°/oo, pH = 7.9 - 8.3 | $6 \times 10^{11}$ | 550 | 5.5 | This work |
| Sargasso Seawater S°/oo = 36.0°/oo, pH = 8.2 | $10 \times 10^{11}$ | 330 | 3.3 | This work |

a: The units for k are liter$^2$ mol$^{-2}$ atm$^{-1}$ min$^{-1}$

## Summary of Iron Reactions in Marine Waters

Four aspects of the chemistry of iron have been considered

in this paper. The formation of ferrous solution complexes varies significantly with salinity and pH over the range of conditions encountered in coastal regions. Complexes of ferric iron are primarily hydroxide species which depend on pH and which appear not to vary greatly with salinity. The redox equilibrium between Fe(II) and Fe(III) favors the predominance of Fe(III) if the redox potential is set by either the $O_2/H_2O$ or $O_2/H_2O_2$ reaction, but Fe(II) is the stable oxidation state when the redox potential is set by the $SO_4^{2-}/S^{2-}$ reaction. The rate of oxygenation of Fe(II) in natural waters depends greatly on pH, and preliminary data indicate that this reaction is slower in marine waters than in sodium bicarbonate solutions. Knowledge of these types of chemical processes under environmental conditions provides a basis for predicting the behavior of dissolved iron in marine systems.

This analysis of iron chemistry points to several areas where additional information is needed. Data are lacking for evaluating the significance of bicarbonate and carbonate species of iron and for possible sulfide species of Fe(II). Equilibrium considerations indicate that iron organic complexes are not likely to be stable in marine systems, but this conclusion should be tested by experimental methods and it should be recognized that biochemically synthesized iron organic species may be kinetically, if not thermodynamically, stable. Additional work is required to determine the kinetics of ferrous oxygenation in marine systems. The role of dissolved organic matter appears to be of considerable importance. Observations by Lewin and Chen (61) indicate the presence of Fe(II) in coastal waters immediately after sampling and the quantity of Fe(II) decreases over a period of 30-40 hours upon storage. There may be a steady state concentration of Fe(II) in marine waters due to a balance between its rate of production by biological systems and its rate of oxygenation to Fe(III). The development of analytical techniques to determine $10^{-8}$ molar concentrations of organic and inorganic forms of Fe(II) and Fe(III) in marine waters would be a great asset to further work on the marine chemistry of iron.

Acknowledgements

We appreciate the assistance of Marilyn Maley in the preparation of the manuscript. This work was supported by the Office of Naval Research Contract N00014-68A-0215-0003.

Literature Cited

1. Livingstone, D.A., "Data of Geochemistry, 6th edition, Chapter G," U.S. Geological Survey, Professional Paper 440-G, page G1-G64. U.S. Government Printing Office, Washington, D.C., 1963.
2. Burton, J.D. and Head, P.C., Limnology and Oceanography (1970)

15 (1): 164-167.
3. Boyle, E., Collier, R., Dengler, A.T., Edmond, J.M., Ng, A.C. and Stollard, R.F., Geochimica et Cosmochimica Acta (1974) 38 (11): 1719-1728.
4. Chester, R. and Stoner, J.H., Marine Chemistry (1974) 2 (1): 17-32.
5. Betzer, P.R. and Pilson, M.E.Q., Journal of Marine Research (1970) 28 (2): 251-267.
6. Morgan, J.J., "Equilibrium Concepts in Natural Water Systems" R.F. Gould, editor, pages 1-29, American Chemical Society, Washington, D.C., 1967.
7. Cooper, L.H.N., Journal of the Marine Biological Association, United Kingdom (1948) 27 (2): 314-321.
8. Schindler, P.W., "Equilibrium Concepts in Natural Water Systems," R.F. Gould, editor, pages 196-221, American Chemical Society, Washington, D.C., 1967.
9. Duursma, E.K. and Sevenhuysen, W., Netherlands Journal of Sea Research (1966) 3 (1): 95-106.
10. Hem, J.D. and Cropper W.H., "A Survey of Ferrous-Ferric Chemical Equilibria and Redox Potentials," U.S. Geological Survey, Water-Supply Paper No. 1459-A, pages 1-31, U.S. Government Printing Office, Washington, D.C., 1959.
11. Hem, J.D., Bulletin of the Geological Society of America (1972) 83 (2): 443-450.
12. Morel, F., McDuff, R.E., and Morgan, J.J., "Trace Metals and Metal-Organic Interactions in Natural Waters," P.C. Singer, editor, pages 157-200, Ann Arbor Science Publishers, Ann Arbor, Michigan, 1973.
13. Lerman, A. and Childs, C.W., "Trace Metals and Metal-Organic Interactions in Natural Waters," P.C. Singer, editor, pages 201-235, Ann Arbor Science Publishers, Ann Arbor, Michigan, 1973.
14. Breck, W.G., "The Sea: Ideas and Observations on Progress in the Study of the Sea," Vol. 5, E.D. Goldberg, editor, pages 153-179, Wiley-Interscience, New York, 1974.
15. Parsons, R., "Dahlem Workshop on the Nature of Seawater," E.D. Goldberg, editor, Dahlem Konferenzen, Berlin, 1975, In Press.
16. Pytkowicz, R.M. and Kester, D.R., "Oceanography and Marine Biology: Annual Review," Vol. 9, H. Barnes, editor, pages 11-68, George Allen and Unwin, London, 1971.
17. Carpenter, J.H. and Manella, M.E., Journal of Geophysical Research (1973) 78 (18): 3621-3626.
18. Sillen, L.G. and Martell, A.E., "Stability Constants," Special Publication No. 17, 754 pages, The Chemical Society, London, 1964.
19. Sillen, L.G. and Martell, A.E., "Stability Constants, Supplement No. 1," Special Publication 25, 865 pages, The Chemical Society, London, 1971.
20. Dyrssen, D. and Wedborg, M., "The Sea: Ideas and Observations

on Progress in the Study of the Sea," Vol. 5, E.D. Goldberg, editor, pages 181-195, Wiley-Interscience, New York, 1974.
21. Marshall, W.L., Journal of Physical Chemistry (1967) 73 (11): 3584-3588.
22. Kester, D.R., "Ion Association of Sodium, Magnesium, and Calcium with Sulfate in Aqueous Solution," Ph.D. Thesis, 116 pages, Oregon State University, Corvallis, 1969.
23. Langmuir, D., "The Gibbs Free Energies of Substances in the System $Fe-O_2-H_2O-CO_2$ at 25°C," U.S. Geological Survey, Professional Paper 650-B, pages B180-B184, U.S. Government Printing Office, Washington, D.C., 1969.
24. Komar, N.P., Uchenye Zapiski Khar'kov Univ. (1963) 133 (Trudy Khim. Fak. KhGU., 19), 189 (Cited by reference 18).
25. Olerup, H., Dissertation (1944) University of Lund (Cited by reference 18)
26. Wells, C.F. and Salam, M.A., Journal of the Chemical Society (1968) No. 2, pages 308-315.
27. Dunsmore, H.S. and Nancollas, G.H., Journal of Physical Chemistry (1964) 68 (6): 1579-1581.
28. Marshall, W.L. and Jones, E.V., Journal of Physical Chemistry (1966) 70 (12): 4028-4040.
29. Helgeson, H.C., Journal of Physical Chemistry (1967) 71 (10): 3121-3136.
30. Daly, F.P., Brown, C.W., and Kester, D.R., Journal of Physical Chemistry (1972) 76 (24): 3664-3668.
31. Kester, D.R., "The Oceans Handbook," R.A. Horne, editor, Marcel Dekker, New York, 1975, In Press.
32. Kester, D.R. and Pytkowicz, R.M., Limnology and Oceanography (1969) 14 (5): 686-692.
33. Bates, R.G., "Dahlem Workshop on the Nature of Seawater," E.D. Goldberg, editor, Dahlem Konferenzen, Berlin, 1975, In Press.
34. Hansson, I., Deep-Sea Research (1973) 20 (5): 479-491.
35. Duursma, E., "Chemical Oceanography," Vol. 1, J.P. Riley and G. Skirrow, editors, pages 433-475, Academic Press, New York, 1965.
36. Wagner, F.S., Contributions in Marine Biology, University of Texas at Port Aransas (1969) 14: 115-153.
37. Stumm, W. and Morgan, J.J., "Aquatic Chemistry," 583 pages, Wiley-Interscience, New York, 1970.
38. Stumm, W. and Brauner, P.A., "Chemical Oceanography," Vol. 1, 2nd edition, J.P. Riley and G. Skirrow, editors, Academic Press, London, 1975, In Press.
39. Kester, D.R. and Byrne, R.H., Jr., "Ferromanganese Deposits on the Ocean Floor," D.R. Horn, editor, pages 107-116, Columbia University, Palisades, New York, 1972.
40. Byrne, R.H., Jr., "Iron Speciation and Solubility in

Seawater," Ph.D. Thesis, 205 pages, University of Rhode Island, Kingston, 1974.
41. Byrne, R.H., Jr. and Kester, D.R., "Potentiometric Determination of Ferric Hydroxide Complexes in Aqueous Solutions at 0.7 Ionic Strength," Manuscript in Preparation, 1975.
42. Byrne, R.H., Jr. and Kester, D.R., "Determination of the Solubility of Hydrous Ferric Oxide in Seawater," Manuscript in Preparation, 1975.
43. Kester, D.R., O'Connor, T.P., and Byrne, R.H., Jr., Thallasia Jugoslavica (1975), In Press.
44. Bent, H.E. and French, C.L., Journal of the American Chemical Society (1941) 63:568-572.
45. Rabinowitch, E. and Stockmayer, W.H., Journal of the American Chemical Society (1942) 64: 335-347.
46. Mattoo, B.N., Zeitschrift für Physikalische Chemie (Frankfurt) (1959) 19: 156-167.
47. Willix, R.L.S., Transactions of the Faraday Society (1963) 59 (6): 1315-1324.
48. Paul, A.D., Thesis (1955) University of California at Berkeley, UCRL-2926 (Cited by reference 18).
49. Connick, R.E., Hepler, L.G., Hugus, Z.Z., Jr., Kury, J.W., Latimer, W.M., and Tsao, M., Journal of the American Chemical Society (1956) 78 (9): 1827-1829.
50. Vanderborgh, N.E., Talanta (1963) 15 (10): 1009-1013.
51. Ellis, A.J., Journal of the Chemical Society (1963) No. 9, 4300-4304.
52. Hudis, J. and Wahl, A.C., Journal of the American Chemical Society (1953) 75: 4153-4158.
53. Miller, G.R. and Kester, D.R., Marine Chemistry (1975) In Press.
54. Lengweiller, H., Buser, W., and Feitknecht, W., Helvita Chimica Acta (1961) 44: 796-805.
55. Stumm, W. and Lee, G.F., Industrial and Engineering Chemistry (1961) 53 (2): 143-146.
56. George, P., Journal of the Chemical Society (1954): 4349-4359.
57. Singer, P.C. and Stumm, W., Science (1970) 167 (3921): 1121-1123.
58. Goto, K., Tamura, H., and Nagayama, M., Inorganic Chemistry (1970) 9 (4): 963-964.
59. Theis, T.L. and Singer, P.C., "Trace Metals and Metal-Organic Interactions in Natural Waters," P.C. Singer, editor, pages 303-320, Ann Arbor Science Publishers, Ann Arbor, Michigan, 1973.
60. Theis, T.L. and Singer, P.C., Environmental Science and Technology (1974) 8 (6): 569-573.
61. Lewin, J. and Chen, C-H., Limnology and Oceanography (1973) 18 (4): 590-596.

# 4

# Minor Element Models in Coastal Waters

PETER G. BREWER and DEREK W. SPENCER

Woods Hole Oceanographic Institution, Woods Hole, Mass. 02543

It is now almost twenty years since Richards (1) discussed the state of our knowledge of trace elements in the ocean. In a brief, and somewhat pungent paper he pointed out that by 1956 there had been no valid analyses of deep ocean water for Sb, Ba, Cd, Cs, Ce, Cr, Co, Ga, Ge, La, Pb, Hg, Mo, Ni, Sc, Se, Ag, Tl, Th, W, Sn, V, Y and Zr; that the assumption of a single oceanic concentration level for these elements was doubtful, and that computed residence times for these elements were of limited use in assessing their oceanic chemistry. Since then a great deal of work has been carried out. Recently, one of us was obliged to review (2) what is now known of minor element concentrations in sea water. The task set was simply to review the observational data base; discussions of analytical methods, theoretical models, or calculations of chemical speciation were to be kept to a minimum. The result was disturbing, for it was soon apparent that we are still far from having an adequate set of analytical data on which to base extensive theoretical models of minor element chemical processes in the ocean (3, 4, 5). In this paper, we briefly examine what is known of some minor element concentrations in sea water. We ask what chemical processes may be controlling their distribution, and inquire as to under what circumstances these processes may be revealed through direct observation.

## Analytical Data

It is now true that, of the list of elements given above, some reasonable estimate of the deep water concentrations may be given. A listing of these data, and the calculated residence times are given in Table I. How accurate are these data, and what may be inferred from the residence time calculation? An inspection of the recent literature (2) shows that, of the elements listed by Richards (1), we may now be sure of the deep water abundance of Ba, Cs, Mo, Sc (6, 7, 8, 9). Of these elements, Mo and Cs exhibit essentially conservative behavior, though the concentration of Mo may be perturbed in anoxic systems; Sc shows a small, but statistically significant, increase with depth from $1.42 \times 10^{-11}$ M/l above 2000m. to $2.04 \times 10^{-11}$ M/l below 2000m; Ba exhibits a well documented range of values from ca. $3 \times 10^{-8}$ M/l to ca. $22 \times 10^{-8}$ M/l. The covariance of Ba with Si is strong and a good deal may be said of its oceanic distribution and chemistry. For the rest, data on Sb, Cd, Ce, Cr, Co, Ga, Ge, La, Pb, Hg, Ni, Se, Ag, Tl, Th, W, Sn, V, Y and Zr are possibly correct to an order of magnitude. Sampling problems are severe for Sb, Cd, Cr, Co, Pb, Hg, Ni, Ag, Sn, V, and sample contamination is probably a limiting factor at the present time. Problems of detection limit, and particularly the precision of the determination, are troublesome for Ce, Ga, Ge, La, Se, Tl, Th, W, Y and Zr. The goal of attaining at least one, accurate, deep water value is extraordinarily limited. A more productive effort would be the determination of a detailed vertical profile in a major ocean basin. From this, the correlation of the element with well known oceanic variables (T, S‰, $O_2$, $NO_3$, $PO_4$, $Si(OH)_4$, $\Sigma CO_2$, alkalinity) can be examined. This leads to estimates of biological cycling, and important information on sources, sinks and transport mechanisms.

There is a general feeling that trace metal data are "getting lower all the time" as sampling and analytical methods improve. Has this in fact occurred? Figure 1 shows data taken from references 2 and 10, for several trace elements, plotted against year of publication for the last 40 or so years. It is by no means clear that a definite trend towards lower concentration is observed. Rather, the data show erratic variations, often with a fairly wide (i.e. order of magnitude) range of values for any one author. This may be taken to mean: (i) that the ocean contains inherent trace metal variability over short scales of distance or time, (ii) that analytical problems exist such that the data presented are essentially noise, or (iii) that sampling problems exist to the extent that even the most careful analysis will see only a signal dominated by the large variations of the metal ions coming from, or to, the individual container surfaces.

Hypotheses (ii) and (iii) must contribute greatly to the scatter apparent in the data. Not all samples have been stored in the acidified condition to minimize adsorbtion (11) and sample contamination has been shown to exist in a number of cases (12, 13). Finally, recent data on carefully taken oceanic samples (14, 15) do show a markedly lower abundance for Cu ($1\text{-}3 \times 10^{-9}$ M/l) and for Mn ($5 \times 10^{-10}$ M/kg) than the mean concentration reported over the last 40 years. These data, made possible by the remarkable sensitivity and specificity of graphite furnace atomic absorption spectroscopy, provide the most convincing evidence so far of a history of analytical or sampling problems over the last few decades.

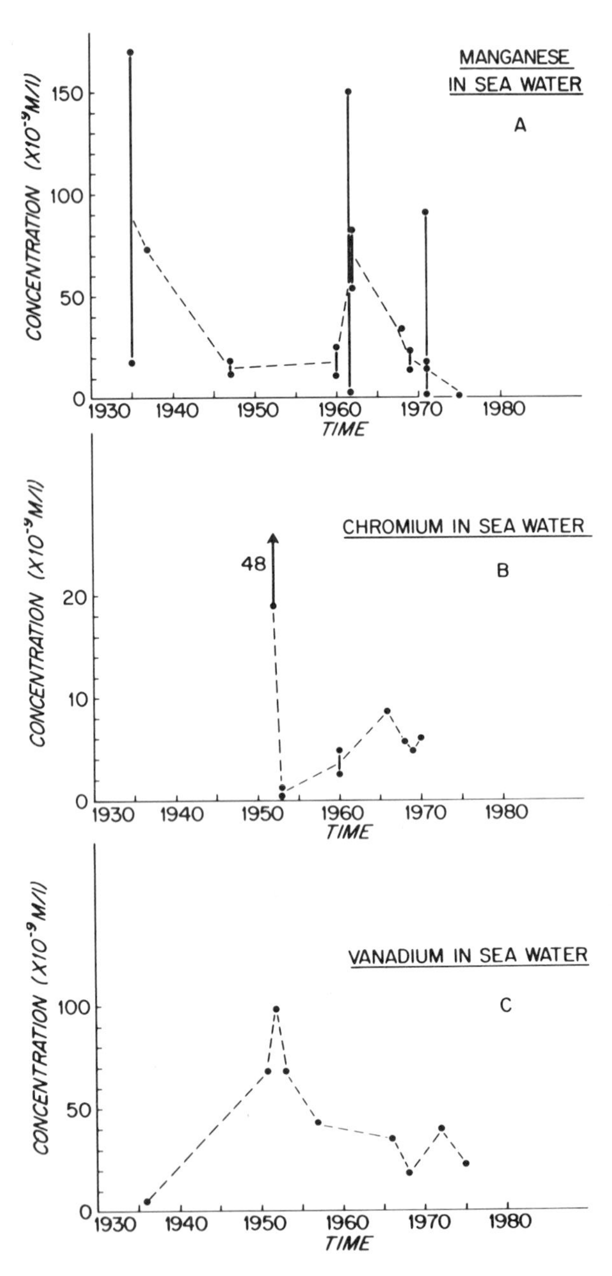


*Figure 1. Estimates of the concentration of three trace elements in seawater with time. Data taken principally from references 2 and 10. No clear trend is apparent.*

In spite of these caveats a substantial body of evidence may be advanced in favor of hypothesis (i). If this is correct then the concept of a residence time, as given in Table I, is not particularly helpful. A residence time of 10,000 years corresponds to approximatley 10 global oceanic stirs, sufficient to smear out all observable concentration gradients. An element with a half-life of this time scale would be quasi-uniformly distributed. However, small scale regional variations have been shown to exist (11) and to be correlated with hydrographic features, or local biological perturbation (16) for Cu, Zn, Fe and Mn. This demonstrates that chemical or biological controls of some kind may be operating on a time scale two or three orders of magnitude shorter than the conventional residence time would suggest.

## MODEL CALCULATIONS

How might we, even crudely, assess this? In the case of biological effects, we can obtain average values for primary productivity, and estimates of the elemental composition of marine phytoplankton are available (17). Assuming that primary productivity is $\sim$ 100 g $C/m^2/yr.$, the depth of the euphotic zone is 100 m, then the growth rate is approximately $2.5 \times 10^{-3}$ g dry weight of phytoplankton/l/yr. In Table II,we show estimates of removal times of some trace elements in surface waters with respect to removal by phytoplankton, assuming either complete uptake and removal by sinking, or 90% recycling of the element in a manner analogous to phosphate. The true result probably lies somewhere between. The values range from $< 1$ to $\sim 10^5$ years. Some of the data are misleading; for instance, the low values for Al and Ti are not due to pronounced metabolic activity. They probably reflect the incorporation of clay particles into the phytoplankton sample.

Let us consider the simplest possible case. It is implicit in the traditional residence time concept that the continents represent a constant source, and the oceans a uniform sink. We incorporate a coastal zone interface connecting the two via horizontal diffusion. Given what we know of concentration gradients and uptake rates (Table II),would non-conservative effects be distinguishable from purely conservative transport by diffusion?

If we consider a one dimensional model of the surface ocean with distance from the coast, a simple case could be given by

$$\frac{dc}{dt} = K \frac{d^2c}{dx^2} - \lambda c \qquad (1)$$

where c is the concentration of a constituent, x is distance from the coast, t is time, $\lambda$ is a first order removal rate constant, and K is the horizontal eddy diffusivity.

TABLE I

The Concentration and Residence Times of Some Minor Elements in Sea Water

| Element | Concentration (M/l) | *Residence Time (Years) |
|---|---|---|
| Be | $6 \times 10^{-11}$ | ? Short |
| Al | $7 \times 10^{-8}$ | $1 \times 10^{2}$ |
| Sc | $2 \times 10^{-11}$ | $4 \times 10^{4}$ |
| Ti | $2 \times 10^{-8}$ | $1 \times 10^{4}$ |
| V | $5 \times 10^{-8}$ | $8 \times 10^{4}$ |
| Cr | $6 \times 10^{-9}$ | $6 \times 10^{3}$ |
| Mn | $5 \times 10^{-10}$ | $2 \times 10^{3}$ |
| Fe | $3 \times 10^{-8}$ | $2 \times 10^{2}$ |
| Co | $8 \times 10^{-10}$ | $3 \times 10^{4}$ |
| Ni | $3 \times 10^{-8}$ | $5 \times 10^{4}$ |
| Cu | $8 \times 10^{-9}$ | $2 \times 10^{4}$ |
| Zn | $8 \times 10^{-8}$ | $2 \times 10^{4}$ |
| Ga | $4 \times 10^{-10}$ | $1 \times 10^{4}$ |
| Ge | $7 \times 10^{-10}$ | ? |
| As | $5 \times 10^{-8}$ | $5 \times 10^{4}$ |
| Se | $3 \times 10^{-9}$ | $2 \times 10^{4}$ |
| Rb | $1.4 \times 10^{-6}$ | $4 \times 10^{6}$ |
| Y | $1 \times 10^{-10}$ | ? |
| Zr | $3 \times 10^{-10}$ | ? |
| Nb | $1 \times 10^{-10}$ | ? |
| Mo | $1 \times 10^{-7}$ | $2 \times 10^{5}$ |
| Ag | $4 \times 10^{-10}$ | $4 \times 10^{4}$ |
| Cd | $1 \times 10^{-9}$ | ? |
| Sn | $8 \times 10^{-11}$ | ? |
| Sb | $2 \times 10^{-9}$ | $7 \times 10^{3}$ |
| I | $5 \times 10^{-7}$ | $4 \times 10^{5}$ |

TABLE I (Continued)

| Element | Concentration (M/1) | *Residence Time (Years) |
|---|---|---|
| Cs | $3 \times 10^{-9}$ | $6 \times 10^{5}$ |
| Ba | $1\text{-}3 \times 10^{-7}$ | $4 \times 10^{4}$ |
| La | $2 \times 10^{-11}$ | $6 \times 10^{2}$ |
| Ce | $1 \times 10^{-10}$ | ? |
| Pr | $4 \times 10^{-10}$ | ? |
| ND | $1.5 \times 10^{-11}$ | ? |
| Sm | $3 \times 10^{-12}$ | ? |
| Eu | $6 \times 10^{-13}$ | ? |
| Gd | $4 \times 10^{-12}$ | ? |
| Tb | $9 \times 10^{-13}$ | ? |
| Dy | $6 \times 10^{-13}$ | ? |
| Ho | $1 \times 10^{-12}$ | ? |
| Er | $4 \times 10^{-12}$ | ? |
| Tm | $8 \times 10^{-13}$ | ? |
| Yb | $3 \times 10^{-12}$ | ? |
| Lu | ? | ? |
| Hf | $4 \times 10^{-11}$ | ? |
| Ta | $1 \times 10^{-11}$ | |
| W | $5 \times 10^{-10}$ | $1 \times 10^{5}$ |
| Re | $4 \times 10^{-11}$ | ? |
| Au | $2 \times 10^{-10}$ | $2 \times 10^{5}$ |
| Hg | $1.5 \times 10^{-10}$ | $8 \times 10^{4}$ |
| Tl | $5 \times 10^{-11}$ | ? |
| Pb | $1 \times 10^{-10}$ | $4 \times 10^{2}$ |
| Bi | $1 \times 10^{-10}$ | ? |
| Ra | $3 \times 10^{-16}$ | ? |
| Th | $4 \times 10^{-11}$ | $2 \times 10^{2}$ |
| U | $1 \times 10^{-8}$ | $3 \times 10^{6}$ |

* Residence Time ($\tau$) is defined as

$$\tau = \frac{A}{dA/dt}$$

where A is the total amount of the element dissolved in the ocean, and dA/dt is the amount introduced to, or removed from, the ocean each year. A steady state is assumed.

TABLE II

Concentration of Some Minor Elements in Phytoplankton and Estimated Residence Times in Surface Waters With Respect to Biological Removal

| Element | *Concentration in Phytoplankton (μg/g dry weight) | †Residence Time (Years) | |
|---|---|---|---|
| | | A (complete removal) | B (90% recycling) |
| Sr | 100 - 700 | $4 \times 10^3$ - $3 \times 10^4$ | $4 \times 10^4$ - $3 \times 10^5$ |
| Ba | 20 - 300 | 13 - 200 | 130 - 2000 |
| Al | 10 - 400 | 1 - 40 | 10 - 400 |
| Fe | 200 - 1500 | 0.8 - 6 | 8 - 60 |
| Mn | 5 - 15 | 26 - 80 | 260 - 800 |
| Ti | 0 - 30 | ⩾ 13 | ⩾ 130 |
| Cr | ∿ 5 | ∿ 40 | ∿ 400 |
| Cu | 1 - 15 | 27 - 800 | 270 - 8000 |
| Ni | 1 - 10 | 40 - 400 | 400 - 4000 |
| Zn | 10 - 100 | 40 - 400 | 400 - 4000 |
| Ag | 0.1 - 1 | 4 - 40 | 40 - 400 |
| Cd | 1 - 5 | 15 - 80 | 150 - 800 |
| Pb | 0 - 10 | ⩾ 40 | ⩾ 400 |
| Hg | 0 - 0.2 | ⩾ 40 | ⩾ 40 |

*Data taken from references 2, 18

†Primary productivity = 100 g $c/m^2/yr$.
Depth of euphotic zone = 100 m
Growth rate = $1 \times 10^{-3}$ g c/l/yr   $2.5 \times 10^{-3}$ g dry weight/l/yr.

Assuming a steady state condition is achieved then $\frac{dc}{dt} = 0$ and a particular solution of the equation (1) is given by

$$C = Co\, e^{-Mx} + \frac{e^{MX}-e^{-MX}}{e^{MXm}-e^{-MXm}} (Cm - Co\, e^{-MXm})$$

where $M = \sqrt{\frac{\lambda}{K}}$, Co is a fixed concentration at X = 0 and Cm is a fixed concentration at X = Xm. Figure 2 shows several solutions of this model where we assume the concentration at the coast to be fixed at 10 units by river input and the concentration in the open ocean, at 200 km from the shore, fixed at a lower level of 0.1 units maintained by a diffusive flux from below. The figure shows solutions for eddy diffusivities (K) of $10^6$ and $10^7$ $cm^2$ $sec^{-1}$ and removal rate constants giving residence times of 0.5, 1 and 5 years. With the precision that is normally reported for trace element determinations, it is apparent that at eddy diffusivities of $10^7$ $cm^2$ $sec^{-1}$ a residence time of even 0.5 year could hardly be distinguished from the straight diffusion model. With a eddy diffusivity of $10^6$ $cm^2$ $sec^{-1}$ a residence time of one year should be detectable but it is doubtful that a five year residence time could be seen.

A somewhat more sophisticated model could consider the case of a coastal zone in which, in addition to a fixed boundary concentration maintained by river input, there occurs an additional input either from an atmospheric flux or flux from the bottom sediments. Such a case could be modelled by

$$\frac{dc}{dt} = K \frac{d^2c}{dx^2} - \lambda C + \alpha = 0 \qquad (2)$$

where $\lambda$ is a first order removal rate constant and $\alpha$ is an input function which decreases with distance from the coast according to

$$\alpha = \alpha_o\, e^{-\mu X}$$

so that the maximum input is in the near coastal zone.
A solution to equation (2) is given by

$$C = C_o\, e^{-\sqrt{\frac{\lambda}{K}} X} + \frac{\alpha_o}{\lambda - \mu^2 K} \left[ e^{-\mu X} - e^{-\sqrt{\frac{\lambda}{K}} X} \right]$$

where, at infinite X, C approaches zero.
Such a model may be a simple approximation for Pb, and other heavy metals that have a significant atmospheric flux, however, we know of no stable trace element data suitable to test this model.

A possible application is to the distribution of $Ra^{228}$. Moore and Sackett (18) and Moore (19) have suggested that the

$Ra^{228}$ distribution in the surface ocean is strongly influenced by input from the continental shelves. This conclusion has been adopted by Kaufman et al. (20) who give substantial amounts of surface data for $Ra^{228}$, including a horizontal traverse from the near shore of Long Island into the Sargasso Sea. These authors modelled the data of this traverse using the simple first order decay - diffusive model given by equation (1) to suggest an eddy diffusion coefficient of about $2 \times 10^6$ $cm^2$ $sec^{-1}$. However, of the 380 km represented by the traverse, about 200 km is on the continental shelf and one would expect that equation (2) would be a more appropriate model.

Figure 3 shows the $Ra^{228}$ data from Kaufman et al. (20) together with three solutions of equation (2). The first solution allows no input from the continental shelf (i.e. sets $\alpha_0 = 0$). This would model the case of a boundary condition maintained by river input and a reasonable fit to the observed data can be obtained with this boundary concentration set at 25 dpm/100 kg. The second solution arbitrarily sets $\alpha_0 = 100$ dpm/100 kg/yr. and $\mu = 0.05$. The boundary concentration, Co, necessary for a reasonable fit to the observed data is about 18 dpm/100 kg. The third solution sets the boundary concentration, Co, to zero and, in this case, a reasonable fit to the data could be obtained with $\alpha = 1300\ e^{-0.1x}$ dpm/100 kg/yr.

If Moore (21) is correct in his statement that the river concentration of $Ra^{228}$ is extremely low ($\sim$ 2 dpm/100 kg) then the third solution is the most appropriate. However, all three models give an agreeable fit to the data and, if the last is most appropriate, then the flux of $Ra^{228}$ from the bottom cannot be maintained over the whole shelf but must be restricted to the immediate 20-30 mile near shore area.

In the shelf region off Long Island the existence of a stable "winter water" cold layer in the outer shelf region at about 50 meters depth may establish a sufficiently stable water column that vertical diffusion of the $Ra^{228}$ input from the sediments may be essentially restricted to the well mixed near shore zone.

It should be pointed out that the fit to the data given by the three models described above was obtained through the use of a radioactive decay constant appropriate to a half-life of 5.75 years. This is the value given in recent compilations of isotopic data (e.g. reference (22)), but it differs by about 20% from the older value (23) of 6.7 years quoted in many papers on the use of $Ra^{228}$ as a geochemical tracer (18, 19, 20, 21). It is interesting to see how stable the models are to perturbations of this kind, since it is certain that trace metal removal rates, as for instance those calculated in Table II, are subject to large errors. A fit of the third model to the data, using a decay constant of 0.103 ($t_{1/2}$ = 6.7 years) gave $\alpha = 850\ e^{-0.1x}$ dpm/100 kg/yr., with $K = 1 \times 10^6$ $cm^2$/sec. The result is numerically different, requiring a diffusion coefficient of about half the previous value, but the conclusions are substan-

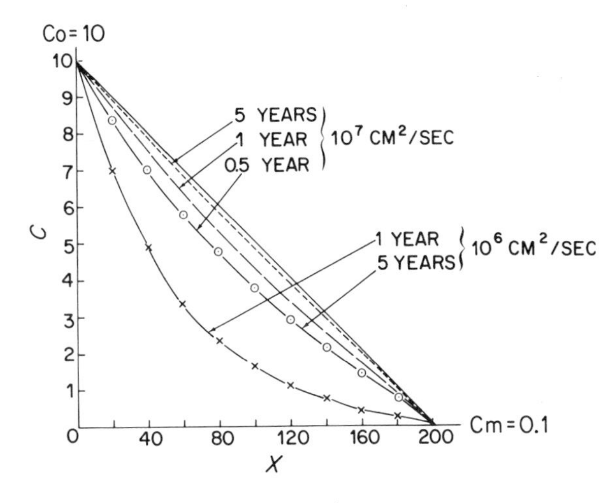


*Figure 2. Simple model of diffusive transport of a non-conservative tracer from a coastal source to an oceanic sink.* $\times$ *is distance from coast. The removal rate is first order, and the concentration gradient Co/Cm = 100. At a diffusion coefficient of* $10^7$ *cm²/sec, a 1 year half-life would produce significant perturbation; at* $10^6$ *cm²/sec* $>$ *5 years half-life could be distinguished.*

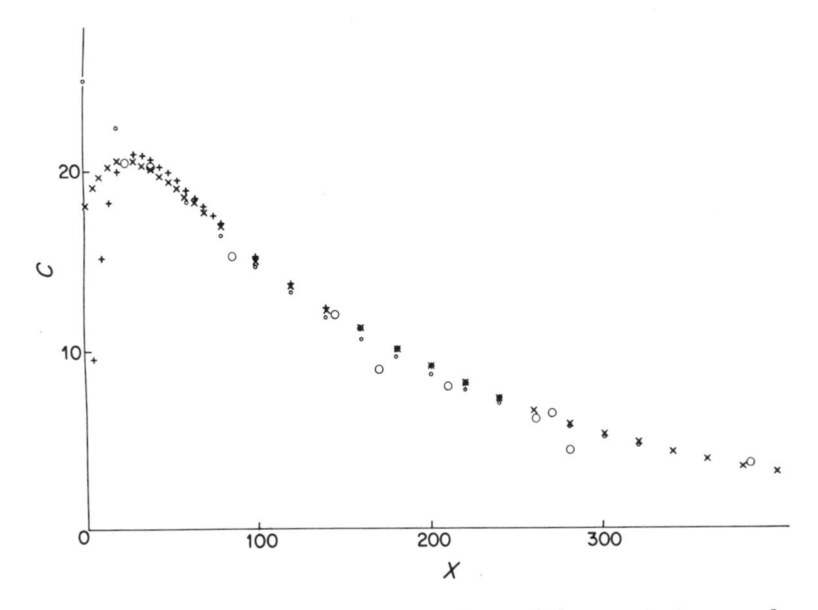


*Figure 3. Application of Equation 2 to the* $Ra^{228}$ *data of Kaufman* et al. *(20).* $\times$ *is distance from coast. Three solutions fit to the data:* ○ = $Ra^{228}$ *data in dpm/100 kg; o* $\alpha_0 = 0$*, Co = 25 dpm/100 kg;* $\times$ $\alpha_0 = 100$ *dpm/100 kg/yr.,* $\mu = 0.05$*, Co = 18; +* $\alpha = 1300\ e^{-0.1x}$ *dpm/100 kg/yr., Co = 0.*

tially the same. If river input is set to zero, then the flux from the sediments is significant only in the 20-30 mile near shore area.

Removal Processes of Shorter Time Scale

Superimposed upon processes such as active biological uptake, and processes of a time scale of months to years, are many short term effects. Precipitation and adsorbtion fall into this category. It is commonly assumed that these processes must take place in the ocean and probably operate on a time scale of minutes to days. Under what circumstances could these processes by investigated by direct observation? We must again consider the rate of the process relative to the rate of diffusion. Equations (1) and (2) dealt with a simple case in which the diffusion coefficient was held to be constant. This is valid over distances of the order of several tens of kilometers. However, horizontal diffusion in the ocean is non-Fickian (24, 25) and over shorter distances this must be taken into account. For Fickian diffusion the variance ($\sigma^2$) of the concentration around a point source is proportional to time (t), permitting the definition of a constant diffusion coefficient. In the ocean the variance is observed to increase faster than t', the physical reason being essentially that large eddies contain proportionately more energy than small ones. Thus, the rate of dispersion increases with time. Okubo (24) has prepared oceanic diffusion diagrams, reducing the results of many experiments, and one of these is reproduced in Figure 4. The slope of all the points is given by

$$\sigma^2 = 0.018\ t^{2.34}$$

Theory would predict a dependence on $t^3$. The relationship between the apparent diffusivity (Ka) and scale length (l), defined as $3\alpha$, is

$$Ka = 0.0103\ l^{1.15}$$

Again, theory would predict a dependence on $l^{4/3}$. Okubo has shown that the 4/3 power law may hold locally.

These data have been obtained from dye experiments, in which a quantity of Rhodamine B dye is released, so as to resemble as closely as possible an instantaneous point source. The initial variance, $\alpha_0{}^2$, may be represented as

$$\alpha_0{}^2 = 6\ P^2 t_0{}^2$$

where P is a diffusion velocity ($\sim$ 1 cm sec$^{-1}$) and $t_0$ is a characteristic growth time from a point source to size $\alpha_0$.

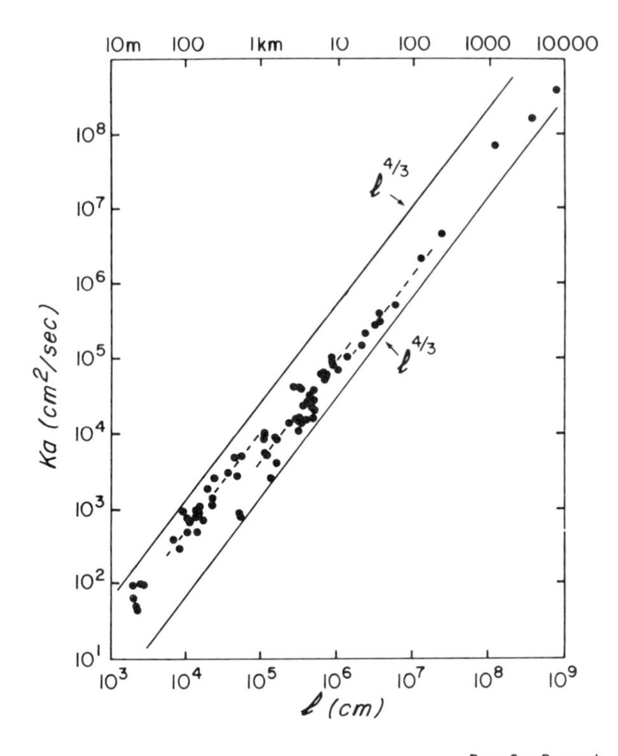


Deep Sea Research

*Figure 4. Oceanic diffusion diagram showing the relationship between apparent diffusivity (Ka) and scale length (l), with the fit of the 4/3 power law locally*

Reliable data on turbulent diffusion may only be obtained at some critical time > $t_0$. A time of 10 $t_0$ is often chosen.

In describing dye diffusion experiments of this kind it is frequently stated that they are of practical importance in connection with the dispersal of pollutants or chemical tracers. If the rate of removal of a tracer is first order (radioactive decay, adsorbtion, precipitation) then a solution to the diffusion equation can be found and, in principle, knowing the diffusion coefficient and measuring the dispension, a removal rate may be calculated from direct observation. So far as we know, this has not been attempted. We can enquire whether it would be possible to construct an experiment in which the controlled release of a trace metal, or other tracer substitute for a dye, in the ocean would provide information on the local removal processes of short time scale. This would be analogous to the problems posed by the dispersal of metals from waste water discharges, or a catastrophic pollution event.

The problem does not appear to be simple. Firstly, the inherent scatter in the oceanic diffusion diagrams is such that it would be impossible to predict the local diffusion coefficient to much better than an order of magnitude. This probably means that measurements of some conservative tracer, released simultaneously with the element of interest, would have to be made in order to provide estimates of local turbulent diffusion. (In fact, the measured variance of a dye patch is independent of any first order decay, the measure of removal being given by the change in peak concentration and mass balance. However, for a tracer with a rapid removal rate, the dispersion would provide an inferior estimate of turbulent diffusion.) An obvious candidate for such a conservative tracer would be Rhodamine B dye. It has been widely used, is harmless, and may be measured with great sensitivity. Unfortunately, it is not clear whether its behavior is truly conservative. Mass balances of 50-70% are often reported (24), the loss being attributed to photochemical decomposition or adsorbtion on to particles. If the latter is a significant factor then the dye would compete for adsorbtion sites with the element of interest.

Secondly, for most minor elements in seawater, there are problems of detection limit and severe problems in generating a significant concentration gradient. The detection limit for Rhodamine B dye is approximately $10^{-8}$ g/kg. For convenience it is customary to use a visual limit of about $10^{-5}$ g/kg for observing the spread of the dye patch by eye and planning sampling strategy. For diffusion experiments of the order of one week in time scale perhaps 50 kg of dye would be used. In this manner concentration gradients of 10-100 from the center of the dye patch to the edge can be maintained. For many chemical elements this would be impossible. It is true that the concentration of minor elements in seawater is not maintained by equilibrium with their least soluble phase, but the natural levels may often not be exceeded grossly before supersaturation does occur. In Table III, we show a comparison of the natural level of some trace metals in seawater with their limiting concentration with respect to the formation of a solid phase (26). The calculated values for Co and Ni would suggest that local concentrations could be raised drastically, but that Cu, Zn and Ba occur naturally within a factor of 10 of their maximum permissible concentration. Kinetic factors may, of course, inhibit precipitation, as in the case of $Mn^{2+}$ where oxidation and precipitation as $MnO_2$ is slow below pH ca. 8.5

Were precipitation to occur then our diffusion experiment would simply reveal the rate of dispersion of the particles. For many problems of practical importance this would be most valuable, and studies of solid phase formation and transport during settling are much needed. Reference to the oceanic diffusion diagram (Figure 3) shows that if precipitation are settling and complete within 24 hours, then the variance of the tracer distribution in the sediments around the source would be of the order of one kilometer. Local advection would elongate the distribution pattern. The actual rate of formation of particles must take place on a much shorter time scale, and may be impossible to observe adequately *in situ*.

If the rate of removal by adsorbtion is of interest then we are limited by the quantity of the solid phase naturally present, and our ignorance of its surface chemical characteristics. The quantity of particulate matter in coastal waters is highly variable ; it can range from > 1 mg/kg to ca. 100 μg/kg. The effective surface area of this material is unknown; a value of 100 $m^2$/gram may be an approximate value for a clay mineral such as illite (26). Let us assume a value of 50 $m^2$/gram for marine suspended matter. At a concentration of 1 mg/kg we thus have 0.05 $m^2$/kg available for adsorbtion. The adsorbtion density of various ions in seawater on to marine particulate matter is unknown. Taking Co as an example of an elemental tracer, then at a concentration of $10^{-7}$ M/l in seawater the adsorbtion density on illite would be $10^{-10}$ moles/$m^2$, and on $\delta$-$MnO_2$ 5 x $10^{-7}$ moles/$m^2$ (26). Using a hypothetical value of $10^{-8}$ moles/$m^2$ for the adsorbtion density on marine particulate matter then we have 5 x $10^{-10}$ moles Co adsorbed per kg of seawater. If we were to add 1 kg (17 moles) of Co to seawater to initiate a tracer experiment, then under the conditions described above, a patch 1 m deep and approximately 10 km radius would form before complete removal by adsorbtion could occur. The limiting factor being the presence of adsorbing sites. A patch of this size would form over the time scale of about one week and would be represented by an effective diffusion coefficient of $\sim 10^5$ $cm^2$ $sec^{-1}$. Reference to a diagram such as Figure 1, with appropriate scale factors, reveals that with an analytical

TABLE III

Natural and Limiting Minor Element Concentration in Seawater with Respect to the Formation of a Solid Phase (26).

| Element | Natural Level (M/l) | Limiting Concentration (M/l) | Solid Phase |
|---|---|---|---|
| Co | $8 \times 10^{-10}$ | $8.5 \times 10^{-6}$ | $CaCO_3$ |
| Ni | $3 \times 10^{-8}$ | $1.1 \times 10^{-2}$ | $Ni(OH)_2$ |
| Cu | $8 \times 10^{-9}$ | $3.4 \times 10^{-7}$ | CuO |
| Zn | $8 \times 10^{-8}$ | $4.4 \times 10^{-7}$ | $ZnCO_3$ |
| Ba | $3 \times 10^{-7}$ | $3.5 \times 10^{-7}$ | $BaSO_4$ |

precision of 1% the removal rate would be scarcely distinguishable from straight diffusive transport of a conservative tracer.

## Conclusions

In writing a simple review paper of this kind, we are forcibly reminded of the acute lack of accurate observational data on the ocean, and in particular of the coastal zone. The admittedly crude estimates of uptake rates given here would tend to suggest that non-conservative effects in trace element chemistry may well be hard to distinguish through direct observation. This should in no way deter good experimental work. It is clear that further advances will not come from casual observation, but from carefully planned expeditions in a defined oceanographic setting with full realization of what the analyst may have to reveal through his data.

Some of the problems discussed here, the coastal-to-oceanic elemental gradients and the adsorbtive properties of marine particulate matter, appear to be of basic importance and should yield to experimental work. If this paper can stimulate some work in these fields, we will count ourselves fortunate.

## Acknowledgements

This work was supported by the U. S. Atomic Energy Commission under Contract No. AT (11-1)-3566, Contribution Number 3539 from the Woods Hole Oceanographic Institution.

## Abstract

The concept of a trace element residence time in the ocean leads to estimates of ca. $10^4$ years. Were this the only factor then trace element gradients would be effectively eliminated. However, some data, and simple uptake calculations, show that gradients in surface waters do occur. Attempts to evaluate this by a simple diffusive model of transport of a non-conservative tracer from a continental source to an oceanic sink through a coastal zone, as in

$$\frac{\partial c}{\partial t} = K \frac{\partial^2 c}{\partial x^2} - \alpha c$$

show that a "half-life" of >> 5 years can probably not be distinguished from truly conservative behavior. The model may be modified to include an exponentially varying input function, as from inshore sediments. This model is applied to published data on $Ra^{228}$ ($t_{1/2}$ = 5.75 years) as an example. Estimates of removal processes of shorter time scale are given. We conclude that basic data on coastal to oceanic elemental gradients, and the adsorptive properties of marine particulate matter, are sadly lacking.

## Literature Cited


(1) Richards, F. A. Geochim. Cosmochim. Acta (1956) 10, 241.
(2) Brewer, P. G. In "Chemical Oceanography" 2nd ed., Academic Press, London, 1975.
(3) Morel, F., Morgan, J. J. Environ. Sci. Technol. (1972) 6, 58.
(4) Craig, H. Earth Planet. Sci. Lett. (1974) 23, 149.
(5) Millero, F. J. In "The Sea", Vol. 6, Interscience. In press.
(6) Bacon, M. P., Edmond, J. M. Earth Planet. Sci. Lett. (1972) 16, 66.
(7) Spencer, D. W., Robertson, D. E., Turekian, K. K. and Folsom, T. R. J. Geophys. Res. (1970) 75, 7688.
(8) Morris, A. W. Deep Sea Res. (1975) 22, 49.
(9) Brewer, P. G., Spencer, D. W., Robertson, D. E. Earth Planet. Sci. Lett. (1972) 16, 111.
(10) Høgdahl, O. T. "The Trace Elements in the Ocean. A Bibliographic Compilation", Central Institute for Industrial Research, Norway, 1963.
(11) Spencer, D. W., Brewer, P. G. Geochim. Cosmochim. Acta (1969) 33, 325.
(12) Schutz, D. F., Turekian, K. K. Geochim. Cosmochim. Acta (1965) 29, 259.
(13) Robertson, D. E. Geochim. Cosmochim. Acta (1970) 34, 553.
(14) Boyle, E., Edmond, J. M. Nature (1975) 253, 107.
(15) Bender, M. Unpublished work.
(16) Morris, A. W. Nature (1971) 233, 427.
(17) Martin, J. H., Knauer, G. A. Geochim. Cosmochim. Acta (1973) 37, 1639.
(18) Moore, W. S., Sackett, W. M. J. Geophys. Res. (1964) 69, 5401.
(19) Moore, W. S. Earth Planet. Sci. Lett. (1969) 6, 437.
(20) Kaufman, A., Trier, R. M., Broecker, W. S., Feely, H. W. J. Geophys. Res. (1973) 78, 8827.
(21) Moore, W. S. J. Geophys. Res. (1969) 74, 694.
(22) Heath, R. L. In "Handbook of Chemistry and Physics", 53rd ed., The Chemical Rubber Co., Ohio, 1972-1973.
(23) Curie, M., Debierne, A., Eve, A. S., Geiger, H., Hahn, O., Lind, S. C., Meyer, St., Rutherford, E., Schweidler, E. Revs. Mod. Phys. (1931) 3, 427.
(24) Okubo, A. Deep Sea Res. (1971) 18, 789.
(25) Csanady, G. T. Turbulent diffusion in the environment, Reidel, 1973.
(26) Murray, J. W., Brewer, P. G. In "Marine Manganese Deposits", Elsevier, in press.

# 5

# The Calculation of Chemical Potentials in Natural Waters. Application to Mixed Chloride–Sulfate Solutions

G. MICHEL LAFON

Department of Earth and Planetary Sciences, The Johns Hopkins University, Baltimore, Md. 21218

A fundamental problem in the study of natural waters is ascertaining the chemical potentials of the dissolved components. The chemical potentials are of particular interest because they provide the most natural expression of equilibrium between the solution and other phases, their gradients are the driving forces of molecular diffusion, and they also characterize the free energy available for non-equilibrium processes such as the precipitation and dissolution of minerals. Because many of the solid phases in contact with natural waters are reactive over relatively short periods of time, departures from equilibrium are often small. For example, the calcite saturation factor in the deep ocean is about 0.6 to 0.7 and decreases below 0.5 only very rarely (1) . The barium concentrations of sediment pore waters from the eastern Pacific Ocean appear to be controlled by equilibrium with barite (2). To study equilibrium and non-equilibrium processes in natural waters, we need to determine the chemical potentials of dissolved components with an accuracy sufficient to detect these small departures from equilibrium (say, 10 per cent or better).

The chemical potentials of dissolved salts can be expressed either in terms of the activities of constituent ions, or more directly in terms of the activities of neutral components.These two approaches have recently been discussed by Leyendekkers (3, 4) and Whitfield (5) , and Millero (6) has reviewed their application to sea water. Results for metal ions, chloride and bicarbonate are in good agreement, but there are serious discrepancies for sulfate and carbonate. Ionic activities cannot be measured directly and must instead be estimated using some arbitrary non-thermodynamic assumption.The models used to estimate ionic activities in mixed-electrolyte solutions (e.g. ion-pairing or specific interaction models) fail in some cases where there are strong interactions between ions (7). Thus, the choice of neutral components which can be investigated experimentally appears preferable.

Many empirical models of the chemical potentials of neutral components in mixed-electrolyte solutions have been proposed, and they are discussed in some detail by Harned and Owen (8), Robinson and Stokes (9) and Harned and Robinson (10). More recently, Friedman (11) has suggested a generalization of Young's (12) Rule, proposing that the excess free energy of a mixed-electrolyte solution be expressed as the sum of the contributions of single-salt solutions at the same ionic strength and of an expansion in the products of the ionic strength fractions of each electrolyte. Using Friedman's approach, Reilly and Wood (13) and Scatchard and colleagues (14,15) have obtained equations for the free energy of arbitrary mixed-electrolyte solutions which appear to be superior to previous treatments.

The purpose of this paper is to show that the equations of Reilly, Wood and Robinson (16) accurately predict the values of chemical potentials in mixed-electrolyte solutions where there are strong ion-ion interactions. The additivity of the interaction terms in the general equations of (16) is demonstrated for systems of the type $NaCl$-$MSO_4$-$H_2O$. Thus, it is likely that these equations predict chemical potentials in natural waters more accurately than has been possible previously.

## Chemical Potentials and Interaction Parameters in mixed-electrolyte solutions.

Using Reilly and Wood's (13) expression for the free energy of an arbitrary mixed-electrolyte solution, Reilly, Wood and Robinson (16) have derived expressions for the osmotic coefficient and for the activity coefficients of components in terms of the experimental properties of single-salt solutions and of interaction parameters. These parameters, denoted here by $g(I,J,K)$ can be obtained from measurements in electrolyte mixtures with a common ion, IJ-IK. Although the general equations of Reilly,Wood and Robinson (RWR) are cumbersome, their mathematical treatment is simple and easily programmed on a computer. Because relatively few single-salt data and interaction parameters are needed for the computation of all the chemical potentials in any mixed-electrolyte solution, the approach of Reilly, Wood and Robinson is particularly well suited to the study of natural waters, especially sea water (17).

Scatchard and colleagues (14,15) have presented a very similar treatment, but have chosen to represent the properties of single-salt solutions and the interaction parameters by power series in the ionic strength. The RWR equations have the advantage that they use experimental quantities directly and that they do not presuppose the functional form of the interaction parameters. Friedman (11) has shown that, for asymmetrically charged mixtures, the interaction parameter approaches ln $I$ in the limit of infinite dilution, which makes a power function representation highly doubtful at low ionic strength.

The RWR expressions for activity coefficients are functions of the interaction parameters and, independently, of their partial derivatives with respect to the ionic strength. The expression for the osmotic coefficient depends solely on the quantity $\partial(Ig)/\partial I$ which can be conveniently denoted by the symbol $w$. To achieve a consistent thermodynamic description of mixtures with a common ion, the values of $g$ obtained from activity coefficients must be related to those of $w$ obtained from the osmotic coefficient. The relation between $g$ and $w$ is discussed in detail elsewhere (18). We simply recall here that, if $w$ can be represented empirically by some function $f$ over the ionic strength interval: $x_1 < I < x_2$, then $g$ must be given by:

$$g = [\int_{x_1}^{I} f(x)\,dx + x_1 f(x_1)]/I \quad . \qquad (1)$$

In particular, if $f$ is the power function: $a_0 + a_1x + a_2x^2 + a_3x^3 + \ldots.$, $g$ is given by:

$$g = C/I + a_0 + (a_1/2)I + (a_2/3)I^2 + (a_3/4)I^3 + \ldots \qquad (2)$$

where C is a constant to be determined from activity coefficient data. The leading inverse term arises from our ignorance of the behavior of $w$ at ionic strengths below $x_1$ and vanishes only when $f$ is also a good representation of $w$ over the interval $(0,x_1)$, which is not generally the case when $f$ is a power function.

It has been shown (16) that the RWR equations accurately predict osmotic and activity coefficients in mixed chloride solutions and that they provide a good first approximation of the trace activity coefficient of $N(ClO_4)_2$ in HCl solutions when the interaction terms are neglected. Before we can apply the general equations to the study of natural waters, it is important to establish that the full equations are valid in the systems of the type MX-NY where marked ion-ion interactions are known to take place. In particular, we wish to test the additive character of the RWR equations which makes it possible to express the properties of complicated mixed-electrolyte solutions in terms of those of single-salt solutions and mixtures with a common ion. A convenient way to test these characteristics is to examine the trace activity coefficients of metal sulfates in sodium chloride solutions.

## The trace activity coefficient of $MSO_4$ in NaCl solutions.

The investigation of the solubility of salts in electrolyte solutions is an important and powerful method for the study of the thermodynamic properties of mixed-electrolytes (8,9). In particular, the solubilities of sparingly soluble sulfate salts in sodium chloride solutions provide an exacting test of the RWR equations. The equilibrium condition for a solid phase of composition $MSO_4.nH_2O$ is:

$$\log K = 2 \log m_{MSO_4} + 2 \log \gamma_{MSO_4} + n \log a_{H_2O}$$

and the activity coefficient of $MSO_4$ is easily obtained from solubility data if we know the activity of $H_2O$ and the equilibrium constant. Because we consider sparingly soluble salts only, the fraction of total ionic strength contributed by the sulfate is very small. Thus, the activity of $H_2O$ does not differ appreciably from its value in solutions of pure NaCl and the activity coefficient determined from equilibrium solubilities is a good estimate of the trace activity coefficient of $MSO_4$ in NaCl solutions.

Using the RWR equations, expressions for the trace activity coefficient of $MSO_4$ in sodium chloride solutions, and activity and osmotic coefficients in $NaCl\text{-}MCl_2\text{-}H_2O$ and $NaCl\text{-}Na_2SO_4\text{-}H_2O$ have been gathered in Table I(equations 3 to 9). The trace activity coefficient of $MSO_4$ in NaCl solutions depends on the properties of pure NaCl, $Na_2SO_4$ and $MCl_2$ solutions and on the sum of the interaction parameters $g(Na,Cl,SO_4)$ and $g(Na,M,Cl)$. Thus, calculations based on solubilities yield an estimate of $[g(Na,Cl,SO_4) + g(Na,M,Cl)]$. These interaction parameters can also be determined quite independently from osmotic and activity coefficients in mixed $NaCl\text{-}Na_2SO_4$ and $NaCl\text{-}MCl_2$ solutions. The RWR equations require that the values of $[g(Na,Cl,SO_4) + g(Na,M,Cl)]$ obtained directly from trace activity coefficients be identical to the sum of the corresponding values obtained from the properties of common-ion mixtures. The comparison of both types of results is therefore a severe test of the general applicability of the RWR equations to multi-electrolyte solutions. Good agreement would strongly imply that chemical potentials in natural waters can be estimated accurately knowing only the properties of single-salt solutions and the interaction parameters characteristic of mixtures with a common ion.

We now examine the solubilities of barite and gypsum in sodium chloride solutions at 25 C and the corresponding trace activity coefficients of $BaSO_4$ and $CaSO_4$. We wish to compare the values of interaction parameters derived from these trace activity coefficients with independent determinations of $g(Na,Cl,SO_4)$, $g(Na,Ba,Cl)$ and $g(Na,Ca,Cl)$.

The thermodynamic properties of $NaCl\text{-}Na_2SO_2\text{-}H_2O$ at 25 C are discussed in detail elsewhere (18). It has been shown that this system is described accurately by equations 4,5 and 6 and that the interaction parameter can be represented by the empirical expression:

$$g(Na,Cl,SO_4) = -0.10037/I - 0.05613 \qquad (10)$$

for ionic strengths between 0.7 and 6.0 $m$.

Trace activity coefficient of $BaSO_4$ in NaCl solutions. Templeton (19) has measured the solubility of barite in sodium chloride solutions at 25 C, and Rosseinsky (20) has made an accurate determination of the activity of $BaSO_4$ at equilibrium with barite. With this equilibrium constant, we have:

Table I. Equations for activity and osmotic coefficients in mixed Na-Cl-M-$SO_4$ solutions at the ionic strength $I$.

1) NaCl - $MSO_4$(trace) - $H_2O$

$$\ln \gamma^{trace}_{MSO_4} = 1.5(1 - \phi^*_{MCl_2} + \ln \gamma^*_{MCl_2}) - 2(1 - \phi^*_{NaCl} + \ln \gamma^*_{NaCl}) + 1.5(1 - \phi^*_{Na_2SO_4} + \ln \gamma^*_{Na_2SO_4}) - 4(1 - \phi^*_{NaCl}) + 1.5\, I[g(Na,M,Cl) + g(Na,Cl,SO_4)] \quad (3)$$

2) NaCl - $Na_2SO_4$ - $H_2O$

$$(2-y)(1-\phi) = y(1 - \phi^*_{Na_2SO_4}) + 2(1-y)(1 - \phi^*_{NaCl}) - y(1-y)I\, w(Na,Cl,SO_4) \quad (4)$$

$$\ln \gamma_{NaCl} = \ln \gamma^*_{NaCl} + y[1 - \phi^*_{NaCl} - 0.5(1 - \phi^*_{Na_2SO_4})] + 0.5\, yI\, [g(Na,Cl,SO_4) + I(1-y)\partial g(Na,Cl,SO_4)/\partial I] \quad (5)$$

$$\ln \gamma_{Na_2SO_4} = \ln \gamma^*_{Na_2SO_4} + (1-y)[1 - \phi^*_{Na_2SO_4} - 2(1 - \phi^*_{NaCl})] + (1-y)I\, [g(Na,Cl,SO_4) + yI\, \partial g(Na,Cl,SO_4)/\partial I] \quad (6)$$

3) NaCl - $MCl_2$ - $H_2O$

$$(2-y)(1-\phi) = y(1 - \phi^*_{MCl_2}) + 2(1-y)(1 - \phi^*_{NaCl}) - y(1-y)\, I\, w(Na,M,Cl) \quad (7)$$

$$\ln \gamma_{NaCl} = \ln \gamma^*_{NaCl} + y[1 - \phi^*_{NaCl} - 0.5(1 - \phi^*_{MCl_2})] + 0.5\, yI[g(Na,M,Cl) + I(1-y)\, \partial g(Na,M,Cl)/\partial I] \quad (8)$$

$$\ln \gamma_{MCl_2} = \ln \gamma^*_{MCl_2} + (1-y)[1 - \phi^*_{MCl_2} - 2(1 - \phi^*_{NaCl})] + (1-y)I\, [g(Na,M,Cl) + yI\, \partial g(Na,M,Cl)/\partial I] \quad (9)$$

Notation: $y$ designates the fraction of the ionic strength $I$ contributed by salts other than NaCl; the starred quantities denote the properties of single-salt solutions at the ionic strength $I$; the parameters $g$ and $w$ are related by: $w = \partial(Ig)/\partial I$.

$$2(\log m_{BaSO_4} + \log \gamma^{trace}_{BaSO_4}) = -9.994 \ . \qquad (11)$$

The values of the trace activity coefficient of $BaSO_4$ in sodium chloride solutions were calculated from equation 11 using the smoothed solubilities tabulated in (19). Activity and osmotic coefficients for pure NaCl and $Na_2SO_4$ solutions were taken from (9), and those for $BaCl_2$ solutions from Robinson and Bower (21). The values of $[g(\text{Na,Cl,SO}_4) + g(\text{Na,Ba,Cl})]$ were then calculated from the trace activity coefficients using equation 3. The results of these calculations are presented in Table III.

The next step is to obtain independent values of $g$(Na,Ba,Cl) from measurements in mixed NaCl-$BaCl_2$ solutions. Robinson and Bower (21) have studied this system by the isopiestic method and they have given information that permits calculation of the activity coefficients of NaCl and $BaCl_2$ in the mixtures by the McKay-Perring method. Reilly, Wood and Robinson (16) have calculated from the isopiestic data the values of the interaction parameter $w$(Na,Ba,Cl). These values are plotted here in Figure 1a and they can be represented empirically by the polynomial:

$$w(\text{Na,Ba,Cl}) = 0.062165 - 0.059424I + 0.017517I^2 - 0.001875I^3 \qquad (12)$$

for ionic strengths between 0.6 and 5.0 $m$. It follows from equation 2 that the interaction parameter $g$(Na,Ba,Cl) is given by

$$g(\text{Na,Ba,Cl}) = A/I + 0.062165 - 0.029712\ I + 0.0058389\ I^2 - 0.0004688\ I^3 \qquad (13)$$

over the same range of ionic strength, where A is an as yet undetermined constant. We obtain the value of A from consideration of the activity coefficient data, by determining directly the best fitting values of $g$(Na,Ba,Cl) in equations 8 and 9 at several ionic strengths. Lanier (22) has measured the activity coefficient of NaCl in mixed NaCl-$BaCl_2$ solutions at 1.0, 3.0 and 5.0 $m$ with a sodium-ion glass electrode - Ag/AgCl electrode couple. Christenson (23) has obtained similar data at 1.0 $m$. The results of the e.m.f. measurements are in good agreement with the isopiestic data except at $I$ = 1.0 $m$ where there is a marked discrepancy between the trace activity coefficients of NaCl.

Values of the activity coefficient of NaCl in the mixed solutions were calculated from the data of (21) at ionic strengths of 2.0, 2.5, 3.0 and 3.5 $m$, and in addition were taken from (22) at 3.0 and 5.0 $m$. Figure 2 illustrates the excellent agreement between both data sets at 3.0 $m$, although it is apparent that the e.m.f. results are systematically slightly higher than the isopiestic ones. Then, the estimates of $g$(Na,Ba,Cl) and its derivative that best fit (in the least squares sense) equation 8 were obtained. No attempt was made at fitting activity coefficients at 1.0 $m$ because of the conflict between e.m.f. and

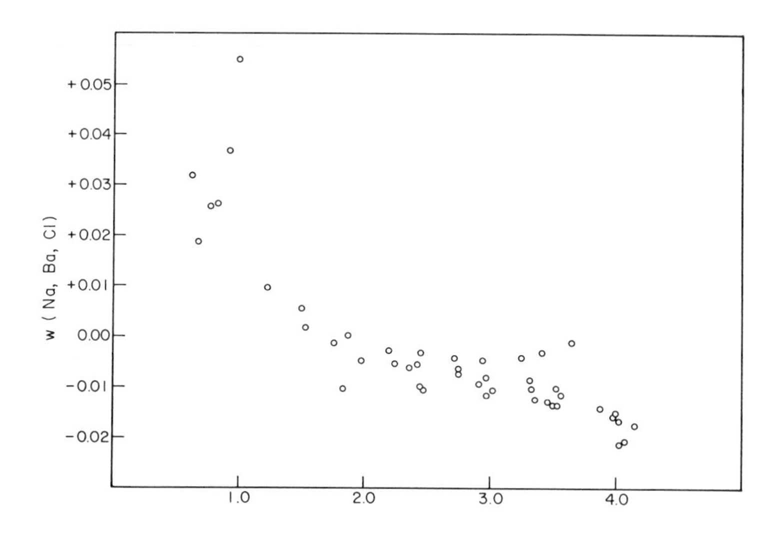


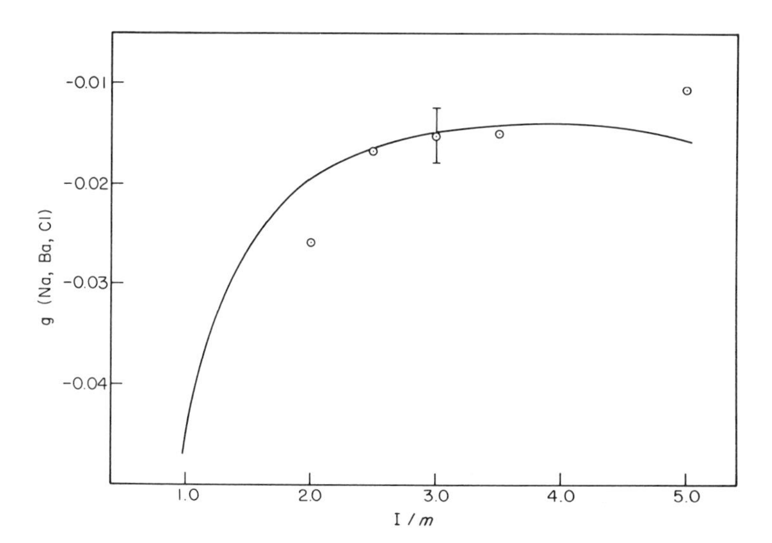


*Figure 1. Interaction parameters in $NaCl–BaCl_2–H_2O$ at 25°C. Top: values of w(Na,Ba,Cl) listed in (21). Bottom: values of g(Na,Ba,Cl) calculated from the activity coefficient of NaCl in mixed $NaCl–BaCl_2$ solutions. The solid curve is computed from Equation 14.*

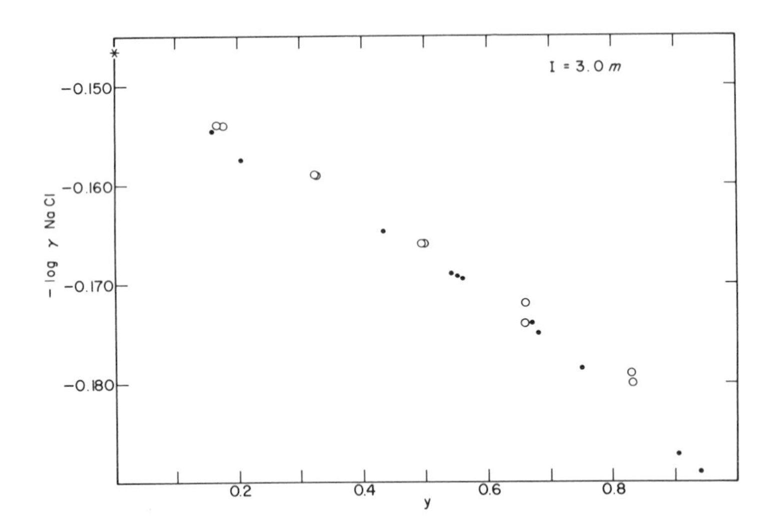


*Figure 2. Activity coefficient of NaCl in mixed NaCl–$BaCl_2$ solutions at 3.0* m *ionic strength.* ○ *from* (22), ● *calculated from* (21).

isopiestic data. The values of the interaction parameter are presented in Table II. Confidence intervals at the 95 per cent level are given for $g$(Na,Ba,Cl) and its derivative at $I$ = 3.0 $m$ only because this is the only ionic strength at which more than one reliable source of data is available. Uncertainties at other ionic strengths are likely to be greater.

Similar calculations were also carried out using the activity coefficient of $BaCl_2$ calculated from the isopiestic data. Results were essentially the same as above and are not presented here to avoid duplication. This is not surprising as the activity coefficient of $BaCl_2$ is derived from the same set of measurements as that of NaCl. This further demonstrates that the properties of common-ion mixtures are well described by the RWR equations.

The data of Table II were used in conjunction with equation 13 to determine the value of the constant A. The value of $g$(Na,Ba, Cl) at 3.0 $m$ , which was obtained from two independent investigations, is likely to be the most accurate and was arbitrarily given a weight of 3 in calculating the average. The values at other ionic strengths were given a weight of unity. The final equation for the interaction parameter is:

$$g(\mathrm{Na,Ba,Cl}) = -0.083615/I + 0.062165 - 0.029712\ I + 0.0058389\ I^2 - 0.0004688\ I^3. \qquad (14)$$

Table II. Values of the interaction parameter $g$(Na,Ba,Cl) calculated from the activity coefficient of NaCl in NaCl-$BaCl_2$-$H_2O$.

| $I$ | $g$(Na,Ba,Cl) | $\partial g$(Na,Ba,Cl)/$\partial I$ |
|---|---|---|
| 2.0 | -0.02585[1] | 0.01571[1] |
| 2.5 | -0.01671[1] | 0.00417[1] |
| 3.0 | -0.01530 ± 0.00277[1,2,3] | 0.00306 ± 0.00274[1,2,3] |
| 3.5 | -0.01503[1] | 0.00135[1] |
| 5.0 | -0.01061[2] | -0.00440[2] |

[1] Derived from activity coefficients calculated from data in (21)
[2] Derived from activity coefficients reported in (22).
[3] The uncertainty quoted is at the 95 per cent confidence level.

A graph of equation 14 is shown in Figure 1b, together with the best fitting estimates of $g$(Na,Ba,Cl). The calculated curve is in excellent agreement with the data points at 2.5, 3.0 and 3.5 $m$. The slight discrepancy observed at 2.0 $m$ is attributed to the small number of observations there, and that at 5.0 $m$ to the systematic difference between e.m.f. and isopiestic measurements already documented at 3.0 $m$. In summary, the osmotic and the two mean activity coefficients in NaCl-$BaCl_2$-$H_2O$ can be represented accurately by equations 7, 8 and 9 where the interaction parameters are given by equations 12 and 14.

Next, we compare the values of [$g$(Na,Cl,$SO_4$) + $g$(Na,Ba,Cl)] calculated from the trace activity coefficient of $BaSO_4$ in sodium chloride solutions with the corresponding independent estimates obtained from equations 10 and 14. Both sets of values are listed in Table III. Agreement between them is excellent up to $I$ = 3.0 $m$ and remains good at higher ionic strengths. Inspection of the original solubility data [Figure 2 of reference (19)] shows that a slightly different smoothing procedure would bring the values at high ionic strength in agreement with the data from common-ion mixtures while remaining in accord with the experimental solubilities. We conclude therefore that the RWR equations predict the solubility of barite in sodium chloride solutions with an accuracy comparable to that of experimental measurements.

Trace activity coefficient of $CaSO_4$ in NaCl solutions. The solubility of gypsum ($CaSO_4 . 2H_2O$) in sodium chloride solutions at 25 C has recently been redetermined by Marshall and Slusher (24). Combining their results with those of previous work, they have obtained a semi-empirical equation that accurately represents experimental solubilities from 0.06 to 6.3 $m$ ionic strength. As the scope of the present paper is limited to the examination of trace activity coefficients, we are restricted to ionic strengths of 3.0 and above in order to keep the contribution of CaSO suitably small. The equilibrium condition is:

Table III. Trace activity coefficient of $BaSO_4$ in NaCl solutions and corresponding interaction parameters.

| $I$ | $\ln \gamma^{trace}_{BaSO_4}$ [1] | $g(Na,Cl,SO_4) + g(Na,Ba,Cl)$ [1] | $g(Na,Cl,SO_4) + g(Na,Ba,Cl)$ [2] |
|---|---|---|---|
| 0.05 | -1.0984 | | |
| 0.1 | -1.3591 | | |
| 0.2 | -1.6417 | | |
| 0.4 | -1.8952 | -0.5703 | |
| 0.6 | -2.0472 | -0.3489 | |
| 0.8 | -2.1678 | -0.2664 | |
| 1.0 | -2.2548 | -0.2175 | -0.2023 |
| 1.5 | -2.4090 | -0.1487 | -0.1496 |
| 2.0 | -2.5188 | -0.1280 | -0.1258 |
| 2.5 | -2.5957 | -0.1133 | -0.1127 |
| 3.0 | -2.6603 | -0.1096 | -0.1045 |
| 3.5 | -2.7209 | -0.1102 | -0.0991 |
| 4.0 | -2.7719 | -0.1116 | -0.0954 |
| 4.5 | -2.8204 | -0.1140 | -0.0930 |
| 5.0 | -2.8666 | -0.1180 | -0.0920 |

[1]Calculated from the solubility of barite in NaCl solutions.
[2]Calculated from activity and osmotic coefficients in the systems $NaCl-BaCl_2-H_2O$ and $NaCl-Na_2SO_4-H_2O$.

$$2(\log \gamma^{trace}_{CaSO_4} + \log m_{CaSO_4} + \log a_{H_2O}) = -4.594 \quad (15)$$

The equilibrium constant at 25 C was calculated from the solubility of gypsum in water and estimates of the activity and osmotic coefficients of $CaSO_4$ in dilute solutions. Although no data for $CaSO_4$ are available, Harned and Owen (8) have pointed out that the activity and osmotic coefficients of the divalent metal sulfates depend very little on the nature of the metal ion at low ionic strength. Therefore, the properties of $ZnSO_4$ solutions were used as good approximations of those of $CaSO_4$ solutions. Note that the equilibrium constant in equation 15 differs markedly from the limiting solubility extrapolated to zero ionic strength by Marshall and Slusher (24). Their extrapolation requires that the activity coefficient of $CaSO_4$ lie above the Debye-Hückel limiting slope at low ionic strength, while we know that the activity coefficient of $ZnSO_4$ approaches this slope from below. Moreover, conductance curves for 2:2 electrolytes are also known to approach the limiting law from below (8), a fact consistent with the widespread assumption of ion-pairing in these solutions. Thus, the limiting solubilities extrapolated by Marshall and Slusher (24) are probably not good estimates of the equilibrium constant for gypsum dissolution.

The results derived from the solubility of gypsum are quite similar to those obtained above from that of barite. Therefore, only sample calculations at 3.0 and 6.0 $m$ are reported here. At these ionic strengths, the logarithms of the molal solubilities are -1.240 and -1.305 respectively [equation 6 of reference (24)]. The corresponding values of the trace activity coefficient of $CaSO_4$ are 0.0982 and 0.1341, leading to values of $[g(Na,Cl,SO_4) + g(Na,Ca,Cl)]$ of -0.0692 at 3.0 $m$ and -0.0495 at 6.0 $m$.

Next, we obtain $g(Na,Ca,Cl)$ from the thermodynamic properties of mixed $NaCl$-$CaCl_2$ solutions in a manner similar to that used for $g(Na,Ba,Cl)$. Robinson and Bower (25) have studied the system $NaCl$-$CaCl_2$-$H_2O$ by the isopiestic method. The osmotic coefficient was calculated from their data at ionic strengths up to 6.0 $m$, using the properties of single-salt solutions listed in (9). The results of these calculations are given in Table IV together with the corresponding values of the interaction parameter $\omega(Na,Ca,Cl)$ which are plotted in Figure 3. They cannot be fitted easily to a power function because of the sharp peak observed near $I = 1.0$ $m$. Recalling that both the $g$ and the $\omega$ parameters for asymmetrical mixtures should behave as $\ln I$ in the limit of infinite dilution (11), we express the $\omega$ interaction parameter as:

$$\omega(Na,Ca,Cl) = 0.07371 \ln I + 0.10814 - 0.07449\, I + 0.007888\, I^2 - 0.0002647\, I^3 \quad (16)$$

for ionic strengths between 0.6 and 6.0 $m$.

We then make direct estimates of the values of $g(Na,Ca,Cl)$ at 3.0 and 6.0 $m$ from the activity coefficient of NaCl in the

Table IV. Osmotic coefficient in $NaCl-CaCl_2-H_2O$ and corresponding values of the interaction parameter $\omega(Na,Ca,Cl)$.

| $m_{NaCl}$[1] | $m_{BaCl_2}$[1] | $I$ | $\phi^2$ | $\omega(Na,Ba,Cl)$[3] |
|---|---|---|---|---|
| 0.3735 | 0.1110 | 0.7065 | 0.9113 | 0.03226 |
| 0.2495 | 0.1968 | 0.8399 | 0.9035 | 0.03371 |
| 0.1268 | 0.2816 | 0.9716 | 0.8960 | 0.03740 |
| 0.4190 | 0.0842 | 0.6716 | 0.9141 | 0.03116 |
| 0.3146 | 0.1561 | 0.7829 | 0.9083 | 0.03675 |
| 0.1832 | 0.2470 | 0.9242 | 0.9002 | 0.03594 |
| 0.6832 | 0.0769 | 0.9139 | 0.9291 | 0.04742 |
| 0.4544 | 0.2297 | 1.1435 | 0.9286 | 0.04785 |
| 0.2241 | 0.3842 | 1.3767 | 0.9269 | 0.04790 |
| 0.5729 | 0.1519 | 1.0286 | 0.9284 | 0.04905 |
| 0.3405 | 0.3072 | 1.2621 | 0.9279 | 0.04829 |
| 0.1271 | 0.4497 | 1.4762 | 0.9275 | 0.09337[4] |
| 0.6887 | 0.2294 | 1.3769 | 0.9455 | 0.04108 |
| 0.4375 | 0.3929 | 1.6162 | 0.9511 | 0.03917 |
| 0.2166 | 0.5367 | 1.8267 | 0.9558 | 0.04315 |
| 0.7891 | 0.1706 | 1.3009 | 0.9438 | 0.04149 |
| 0.5393 | 0.3333 | 1.5392 | 0.9491 | 0.03629 |
| 0.3250 | 0.4723 | 1.7419 | 0.9543 | 0.08215[4] |
| 0.9755 | 0.3348 | 1.9799 | 0.9848 | 0.03662 |
| 0.5078 | 0.6256 | 2.3846 | 1.0063 | 0.03931 |
| 0.2665 | 0.7736 | 2.5873 | 1.0199 | 0.05449 |
| 1.2831 | 0.1656 | 1.7799 | 0.9713 | 0.02436 |
| 0.5750 | 0.6029 | 2.3837 | 1.0056 | 0.03300 |
| 0.3506 | 0.7412 | 2.5742 | 1.0172 | 0.03595 |
| 1.2368 | 0.3258 | 2.2142 | 1.0013 | 0.03580 |
| 0.7203 | 0.6377 | 2.6334 | 1.0293 | 0.03474 |
| 0.2557 | 0.9204 | 3.0169 | 1.0559 | 0.03966 |
| 1.5227 | 0.1600 | 2.0027 | 0.9865 | 0.03395 |
| 1.0256 | 0.4603 | 2.4065 | 1.0133 | 0.03302 |
| 0.5579 | 0.7427 | 2.7860 | 1.0271 | 0.03479 |
| 1.9144 | 0.4751 | 3.3397 | 1.0816 | 0.01749 |
| 1.2728 | 0.8302 | 3.7634 | 1.1271 | 0.03281 |
| 0.6209 | 1.1974 | 4.2131 | 1.1743 | 0.02710 |
| 1.6266 | 0.6524 | 3.5838 | 1.1045 | 0.01907 |
| 0.9691 | 1.0206 | 4.0309 | 1.1510 | 0.01954 |
| 0.3633 | 1.3619 | 4.4490 | 1.1959 | 0.01702 |
| 2.2476 | 0.6360 | 4.1556 | 1.1522 | 0.01995 |
| 1.2284 | 1.1877 | 4.7915 | 1.2254 | 0.01948 |
| 0.3794 | 1.6464 | 5.3186 | 1.2947 | 0.02929 |
| 2.8877 | 0.3111 | 3.8210 | 1.1109 | 0.01070 |
| 1.7012 | 0.9474 | 4.5434 | 1.1933 | 0.01714 |
| 0.8440 | 1.4101 | 5.0743 | 1.2589 | 0.01865 |
| 3.4474 | 0.5772 | 5.1790 | 1.2333 | 0.01858 |

| | | | | |
|---|---|---|---|---|
| 2.1539 | 1.2356 | 5.8607 | 1.3274 | 0.02191 |
| 3.7520 | 0.4410 | 5.0750 | 1.2151 | 0.01261 |
| 2.9372 | 0.8548 | 5.5016 | 1.2710 | 0.01576 |

[1] Taken from (25)
[2] Calculated from data given in (25)
[3] Calculated from equation 7 and the data above.
[4] These points, considered unreliable, were not used in deriving equation 16

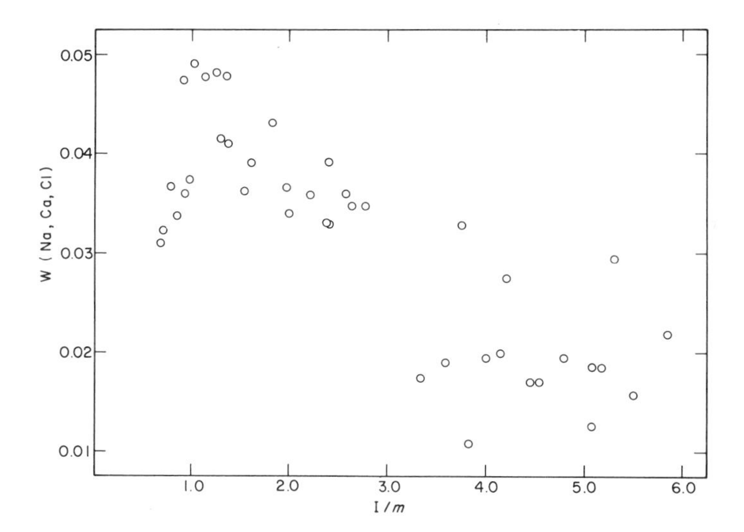


*Figure 3. w(Na,Ca,Cl) at 25°C. These values, calculated from data in (25), are listed in Table IV.*

mixed solutions. The values of the activity coefficient of NaCl were taken from Lanier (22) and also calculated from the data of Robinson and Bower (25). Good agreement was observed between the results of the e.m.f. measurements and those of the isopiestic data. The best fitting estimates of $g$(Na,Ca,Cl) are 0.02559 ± 0.00237 at 3.0 $m$ and 0.02780 ± 0.00256 at 6.0 $m$ (uncertainties quoted are at the 95 per cent confidence level). Using equations 1 and 16, the interaction parameter must be of the form:

$$(\text{Na,Ca,Cl}) = B/I + 0.07371 \ln I + 0.03443 - 0.037246\,I + 0.002629\,I^2 - 0.0000662\,I^3 \qquad (17)$$

where B is a constant. Using the values of $g$ determined above at 3.0 and 6.0 $m$, we find that B is not significantly different from zero, which indicates that equation 16 is also a good approximation of $w$(Na,Ca,Cl) between 0 and 0.6 $m$ ionic strength.

The values of [$g$(Na,Cl,$SO_4$) + $g$(Na,Ca,Cl)] calculated from the data for common-ion mixtures are -0.0640 at 3.0 $m$ and -0.0495 at 6.0 $m$.The corresponding values obtained from the solubility of gypsum are -0.0692 and -0.0570 respectively. Here, again, the agreement between the two methods is excellent, especially when we consider the experimental uncertainties involved at each step of the calculations. We conclude that the solubility of gypsum can be predicted accurately by equation 3 and the interaction parameters given by equations 10 and 17.

## Conclusions

We have shown that the osmotic and activity coefficients in $NaCl$-$BaCl_2$-$H_2O$ and $NaCl$-$CaCl_2$-$H_2O$ can be described accurately by the equations of Reilly, Wood and Robinson (16) where the interaction parameters are given by equations 12, 14, 16 and 17. We have further shown that these interaction parameters can be added to the parameter that characterizes $NaCl$-$Na_2SO_4$-$H_2O$ to provide accurate estimates of the trace activity coefficients of $BaSO_4$and $CaSO_4$ in sodium chloride solutions. Thus, it appears that the additive character of the interaction terms in the general equations of Reilly, Wood and Robinson is valid even for systems where marked specific interactions between ions are known to exist (in this case, between metal ions and sulfate). This strongly suggests that the general RWR equations can be used to estimate chemical potentials in natural waters more accurately than has been possible to date. More specifically, these equations together with appropriate expressions for the interaction parameters are likely to yield significantly improved estimates of the activity coefficients of sulfate and carbonate components in sea water and in continental brines. Finally, the solubilities of sparingly soluble sulfate salts in sodium chloride solutions can be used in conjunction with values of $g$(Na,Cl,$SO_4$) to estimate the interaction parameter $g$(Na,M,Cl) in systems for which no direct data are available.

## Literature Cited


(1) Takahashi T., in "Dissolution of Deep-Sea Carbonates",(1975) Cushman Foundation for Foraminiferal Research, Spec. Publ. 13, p. 11, Washington.
(2) Church T. M. and Wohlgemuth K., Earth Planet. Sci. Letters (1972) 15, 35.
(3) Leyendekkers J. V., Anal. Chem. (1971) 43, 1835.
(4) ——————, Marine Chemistry (1973) 1, 75.
(5) Whitfield M. , Marine Chemistry (1973) 1, 251.
(6) Millero F. J., in "Annual Review of Earth and Planetary Sciences", vol. 2 (1974), p. 101.
(7) Lafon G. M. and Truesdell A. H., ms in preparation.
(8) Harned H. S. and Owen B. B., "The Physical Chemistry of Electrolytic Solutions", 3rd Edition, xxxiii + 803 p., Reinhold Book Corp., New York (1958).
(9) Robinson R. A. and Stokes R. H.,"Electrolyte Solutions", 2nd Edition(revised), xv + 571 p., Butterworths, London,(1970).
(10) Harned H. S. and Robinson R. A., "Multicomponent Electrolyte Solutions", xiii + 110 p., Pergamon Press, Oxford, (1968).
(11) Friedman H. L., J. Chem. Phys. (1960) 32, 1351.
(12) Young T. F. and Smith M. B., J. Phys. Chem. (1954) 58, 716.
(13) Reilly P. J. and Wood R.H.,J. Phys. Chem. (1969) 73, 4292.
(14) Scatchard G., J. Am. Chem. Soc. (1961) 83, 2636.
(15) Scatchard G., Rush R. M. and Johnson J. S., J. Phys. Chem. (1970) 74, 3786.
(16) Reilly P. J., Wood R. H. and Robinson R. A., J. Phys. Chem. (1971) 75, 1305.
(17) Robinson R. A. and Wood R. H., J. Solution Chem. (1972) 1,481
(18) Lafon G. M., ms in preparation.
(19) Templeton C. C., J. Chem. Eng. Data (1960) 5, 514.
(20) Rosseinsky D. R., Trans. Faraday Soc. (1958) 54, 116.
(21) Robinson R. A. and Bower V. E., J. Res. Nat. Bur. Standards (1965) 69A, 19.
(22) Lanier R. D., J. Phys. Chem. (1965) 69, 3992.
(23) Christenson P. G., J. Chem. Eng. Data (1973) 18, 286.
(24) Marshall W. L. and Slusher R., J. Phys. Chem. (1966) 70, 4015
(25) Robinson R. A. and Bower V. E., J. Res. Nat. Bur. Standards (1966) 70A, 313.

# 6

# The Activity of Trace Metals in Artificial Seawater at 25°C

M. W. WATSON and R. H. WOOD
Department of Chemistry and College of Marine Studies, University of Delaware, Newark, Del. 19711
FRANK J. MILLERO
Rosenstiel School of Marine and Atmospheric Science, University of Miami, Miami, Fla. 33149

Considerable interest has been generated recently in the effects of trace metals in the environment. Both the biological and chemical activity of trace metals is determined by the chemical form in which they occur. (1-6) In particular, trace organic chelating agents in seawater play a very important part in changing the effect of trace metals on the environment. (2-4) These chelating agents may solubilize certain trace metals and thus facilitate their transport. If the chelating action is too strong then necessary amounts of trace metals may not be available for use by marine organisms. Conversely, the effect of toxic metals may be reduced by the chelating action. Because of the very small concentrations of both trace metals and chelating agents it is very difficult to learn about the form in which metal ions occur in seawater. For this reason most of the studies have been limited to determining the total concentration of metals in the solution phase. (6)

One way of attacking this problem is to make theoretical calculations of the speciation of the various components in seawater. (7-12) However, the calculation of the interaction of a trace metal ion with a complexing agent requires a knowledge of the activity of the free metal ion in order to perform the calculation. (13) Recently Robinson and Wood (14) have shown that it is possible to use the equations of Reilly, Wood and Robinson (15) to calculate the activity coefficients for the major components of artificial seawater. Similar calculations have been made using both the equations of Scatchard, (16) and the Brønstead-Guggenheim equations either alone (9,10) or with the addition of like-charged interactions (8). All of the results are in substantial agreement and agree with the available experimental results. The purpose of this paper is to extend these calculations to the activity coefficients of the chlorides and sulfates of a variety of trace metals in artificial seawater. Thus these results give

Taken in part from the M.S. Thesis of Mark W. Watson, University of Delaware June 1975.

the activity coefficient of that part of the metal that is uncomplexed by trace inorganic and organic complexing agents. Specifically, the artificial seawater contains sodium, magnesium, potassium, chloride and sulfate ions so interactions of the trace metals with these ions are the only interactions taken into account in the calculations.

## Calculation of Activity Coefficients

The calculations were done using equation A-9 (15) of Reilly, Wood and Robinson which requires knowing the activity and osmotic coefficients of each salt in its own solution at the ionic strength of the mixture and the interaction parameters for each two salt mixture with one ion in common. The activity and osmotic coefficients for all the salts (except $MgSO_4$) were taken from Robinson and Stokes (17). Gardner and Glueckauf (18) and Pitzer (19) have re-evaluated the activity and osmotic coefficients of $MgSO_4$ with similar results. The values of Pitzer were used. The new $\gamma^{\circ}_{\pm}(MgSO_4)$ data resulted in a 2% decrease in the activity of $MgSO_4$ in seawater and much lower changes in the activities of the other salts. The interaction parameters used in the calculations are given in Table 1. All of the interaction parameters involving transition metals and a few of the interactions between major components are not available.

The artificial seawater used for the calculations consisted of a solution of four salts in the ratio 0.4852 m NaCl, 0.0106 m KCl, 0.0367 m $MgCl_2$ and 0.0293 m $MgSO_4$. The resulting solution has a salinity of 35‰, chlorinity of 19.375‰, and a molal ionic strength of 0.7231. The results of the calculation are given in Table II for the trace salts and the major components. The absence of activity and osmotic coefficients for $CaSO_4$ at high concentrations preclude including $Ca^{++}$ in the calculation.

An examination of the basis of the Reilly, Wood and Robinson equation shows that good accuracy is expected for the calculation of "normal" strong electrolytes. On the other hand, it is equally clear that for strongly ion-paired species; for instance, acetic acid ($K_a$ (association) $\simeq 10^5$), the equation will give large errors and an ion-pairing approach using an association constant is needed. As association constants become larger than those for "normal" strong electrolytes, it is not yet clear at what point it is best to go over to an ion-pairing approach. For magnesium sulfate with a $K_a$ of about 200, the equations of Reilly, Wood, and Robinson did not work nearly as accurately as they do for "normal" strong electrolytes. On the other hand, ion-pairing calculations are difficult because of the concentration dependence of the stoichiometric association constant (11). For these reasons, we expect that the activity coefficients for the sulfates of divalent cations calculated in Table II will be somewhat less accurate than the other results. In addition, the very low activity

Table I

Interaction Parameters

| Ion Interactions | $b^{a}_{0,1}$ | $b_{0,2}$ | $b_{0,3}$ | Reference |
|---|---|---|---|---|
| $Na^+$, $Mg^{++}$, $Cl^-$ | 0.0654 | -0.0176 | 0.00191 | b |
| $Na^+$, $K^+$, $Cl^-$ | -0.0253 | -0.00299 | | c |
| $Na^+$, $H^+$, $Cl^-$ | 0.0557 | | | d |
| $Mg^{++}$, $K^+$, $Cl^-$ | 0.0 | | | e |
| $Na^+$, $Mg^{++}$, $SO_4^=$ | 0.03178 | -0.003055 | | b |
| $Na^+$, $K^+$, $SO_4^=$ | -0.0359 | 0.0076 | | f |
| $Mg^{++}$, $K^+$, $SO_4^=$ | 0.0 | | | e |
| $Cl^-$, $SO_4^=$, $Na^+$ | -0.05821 | 0.0 | 0.000439 | b |
| $Cl^-$, $SO_4^=$, $Mg^{++}$ | -0.0368 | | | b |
| $Cl^-$, $SO_4^=$, $K^+$ | -0.0542 | 0.0105 | | f |

a) The interaction parameter g in equation (A-6)[16] is given by $g = b_{0,1} + b_{0,2}\ I/2 + b_{0,3}\ I^2/3$.

b) Wu, Y. C., Rush, R. M., and Scatchard, G., J. Phys. Chem. (1968) 72, 4048; (1969) 73, 2047.

c) Rush, R. M. and Robinson, R. A., J. Tenn. Acad. Sci. (1968) 43, 22.

d) Harned, H. S., J. Phys. Chem. (1959) 63, 1299.

e) Zero used since no data is available.

f) Robinson, R. A., Platford, R. F. and Childs, C. W., J. Soln. Chem. (1972) 1, 167.

Table II

Log $\gamma\pm$ of electrolytes in Artificial Seawater at 35‰ Salinity and 25°C

| Salt | log $\gamma\pm$ | $\gamma\pm$ |
|---|---|---|
| NaCl | -0.176 | 0.667 |
| $Na_2SO_4$ | -0.435 | 0.367 |
| $MgCl_2$ | -0.330 | 0.468 |
| $MgSO_4$ | -0.796 | 0.160 |
| KCl | -0.196 | 0.636 |
| $K_2SO_4$ | -0.463 | 0.345 |
| *$CdCl_2$ | -0.629 | 0.235 |
| *$CdSO_4$ | -1.245 | 0.057 |
| *$CuCl_2$ | -0.339 | 0.458 |
| *$CuSO_4$ | -0.810 | 0.155 |
| *$MnCl_2$ | -0.332 | 0.466 |
| *$MnSO_4$ | -0.799 | 0.159 |
| *$NiCl_2$ | -0.330 | 0.468 |
| *$NiSO_4$ | -0.796 | 0.160 |
| *$ZnCl_2$ | -0.331 | 0.467 |
| *$ZnSO_4$ | -0.797 | 0.160 |

*trace component

coefficient for pure $CdCl_2$, which reflect the strong complex formed between $Cd^{+2}$ and chloride ions, indicates that the activity coefficients calculated for this salt will not be accurate. For the other salts, good accuracy is expected; in fact, Robinson and Wood have shown that the calculated activities of both sodium chloride and sodium sulfate are within experimental error of the measurements that have been made (14).

Unlike the calculations for the major components, the calculated activity coefficients for the transition metal ions in artificial seawater are not expected to be good representations of the activity coefficients in actual seawater. This is because the transition metal ions have very strong interactions with some of the ions excluded from the artificial seawater ($CO_3^=$, $HCO_3^-$, $OH^-$, and trace organics). Indeed, recent calculations using the ion-pairing model show that the expected activity coefficients for the transition metal chlorides and sulfates listed in Table II are lower by a factor of 2 to 10 than the values for artificial seawater listed in Table II. The major reason for these differences are the interactions with ions left out of the artificial seawater. The best estimates on the activities will probably result from using the equations of Reilly, Wood and Robinson to calculate the interactions between the strong electrolytes present in the solution and combining this with calculations of the association between the ions having large association constants (Ka greater than about 200). In this way each method is used where it is the most accurate.

The present results should be useful in the calibration of specific ion electrodes in seawater composed of only the major cations and anions. By adding $OH^-$ and $HCO_3^-$ (as well as other complexing agents) to these mixtures it is possible to study the interactions of trace metals with trace ligands in seawater.

## Calculation of the Freezing Point

Robinson and Wood (14) predicted osmotic coefficient (and thus water activity) for seawater at 25°C using the equations of Reilly, Wood and Robinson (15). The prediction was within 0.1% of the experimental results (20, 21) even for solutions five times more concentrated than seawater. There are two recent measurements of the freezing point of seawater (22,23). The present calculation was made to see if the predictions and experiments at both temperatures were consistent with each other.

The freezing point of the artificial seawater described above was calculated from the predicted osmotic coefficient at 25°C (24) and the heat of dilution of seawater measured by Millero, Hansen and Hoff (25). First the osmotic coefficient at the estimated freezing point is calculated from the equation (17):

$$\int_{1/298}^{1/T_f} 55.5\ \bar{L}_1 d(1/T) = R\ (\Sigma m_i)\int_{1/298}^{1/T_f} d\phi = R(\Sigma m_i)(\phi_{T_f} - \phi_{298})$$

where $\Sigma m_i$ is the total molality of solute ions, $m_i$ is the molality of ion i and the summation is over all the ions, and R is the gas constant, 1.987 cal $K^{-1}$ $mol^{-1}$. Values for $\bar{L}_1$ at 35‰ salinity were plotted vs 1/T and graphically integrated to obtain the area under the curve. This is a small correction since the change in osmotic coefficient is small, $(\phi_{T_f} - \phi_{298}) = -.0122$.

The activity of the water at $T_f$ is given by the following equation (17):

$$\ln a_{H_2O} = \frac{\sum_i m_i W}{10^3} \phi_{T_f}$$

where W is the molecular weight of water. The freezing point depression, Θ, of the solution is then calculated from the equation (17):

$$-\log a_{H_2O} = 2.1 \times 10^{-6}\, \Theta^2 + 4.207 \times 10^{-3}\, \Theta$$

If necessary the calculation is repeated with this new estimate of the freezing point.

Doherty and Kester (22) report for seawater with 35‰ salinity a freezing point of -1.922°C ± .002 while Fujino, Lewis, and Perkin (23) report -1.921 ± .003°C. Our predicted value of -1.927°C agrees to within 0.26% of their result. This agreement is satisfactory since the 25°C data on which the prediction is based (17) has an estimated accuracy of 0.1 to 0.2%.

The agreement between experimental and prediction at 35‰ salinity means freezing points for more dilute natural waters can be predicted with confidence. Similarly previous results indicate that calculation of the activities of the other major components ($Na^{+1}$, $K^{+1}$, $Mg^{+2}$, $Cl^-$, $SO_4^{=}$) should be very accurate. For accurate calculations involving trace transition metals the strong complexes with trace ligands ($OH^-$, $HCO_3^-$, etc.) will have to be included in the calculation.

Acknowledgement. The support of the National Science Foundation (GA-40532 and MPS74-11930), and the office of Naval Research (contract ONR-N00014-75-C-0173) is gratefully acknowledged.

## Summary

The equations of Reilly, Wood and Robinson were used to calculate the trace activity coefficients of the chlorides and sulfates of cadmium, copper, manganese, nickel and zinc in artificial seawater at 25°C. The activity coefficients of the major components of seawater were recalculated using new values for the activity of magnesium sulfate. The freezing point of seawater was predicted from the calculated osmotic coefficient at 25°C and the heat of dilution of seawater.

Literature Cited

1 Barber, R. T., Trace Metals and Metal-Organic Interactions in Natural Waters, Ch. 11. Phillip Singer, ed. Ann Arbor Science Pub., Inc. Ann Arbor, Mich., 1973.
2 Davey, E. W., Gentils, J. H., Erickson, S. J. and Betzor, R., Limnol. Oceanogr. (1970) 15, 486.
3 Steeman-Nielson, E. and Wium-Anderson, S., Mar. Biol. (1970) 6, 93.
4 Davey, E. W., Morgan, M. J., Erickson, S. J., Limnol. Oceanogr. (1973) 18, 993.
5 Martin, J. H., Limnol. Oceanogr. (1970) 15, 756.
6 Millero, F. J., The Sea, Vol. 6 (E. D. Goldberg, ed.), Wiley New York (1975) in press.
7 Garrels, R. M. and Thompson, M. E., Amer. J. Science (1962) 260, 57.
8 Leyendekkers, J. V., Marine Chem. (1972) 1, 75.
9 Whitfield, M., Marine Chem. (1973) 1, 251.
10 Pytkowicz, R. M. and Kester, D. R., Amer. J. Science (1969) 267, 217.
11 Kester, D. R. and Pytkowicz, R. M. Limnol. Oceanogr. (1968) 13, 670; (1969) 14, 686.
12 Whitfield, M., Deep Sea Research (1973) 20, 1.
13 Zirino, A. and Yamanoto, S., Limnol. Oceanogr. (1972) 17, 661.
14 Robinson, R. A. and Wood, R. H., J. Soln. Chem. (1972) 1, 481
15 Reilly, P. J., Wood, R. H. and Robinson, R. A. J. Phys. Chem. (1971) 75, 1305.
16 Scatchard, G., Rush, R. M. and Johnson, J. S., J. Phys. Chem. (1970) 74, 786.
17 Robinson, R. A. and Stokes, R. H., (1970) Electrolyte Solutions, Butterworths, London.
18 Gardner, A. W. and Glueckauf, E., Proc. Roy. Soc. Lond. (1969) A 313, 131.
19 Pitzer, K. S., J. Chem. Soc. Faraday Trans. II (1972) 68, 101.
20 Robinson, R. A. J. Marine Biol. Assoc. U. K. (1954) 33, 449.
21 Rush, R. M. and Johnson, J. S., J. Chem. Eng. Data (1966) 11, 590.
22 Doherty, B. T. and Kester, D. R., J. Marine Res. (1974) 32, 285.
23 Fujino, K., Lewis, E. L. and Perkin, R. G., J. Geophys. Res., (1974) 79, 1792.
24 The change in data for $MgSO_4$ used in the calculation gave a new osmotic coefficient, $\phi = 0.9061$, which is only slightly different from the previous calculation, $\phi = 0.9068$.
25 Millero, F. J., Hansen, L. D. and Hoff, E. V., J. Marine Res. (1973) 31, 21.

# 7

# Methane and Radon-222 as Tracers for Mechanisms of Exchange Across the Sediment–Water Interface in the Hudson River Estuary

D. E. HAMMOND, H. J. SIMPSON, and G. MATHIEU

Lamont–Doherty Geological Observatory of Columbia University, Palisades, N.Y. 10964

One problem in formulating material balances for aquatic systems is predicting the rates of mass transport across the sediment-water and water-atmosphere interfaces. An empirical relationship based on wind speed (1) has been developed to predict the rate of exchange across the upper interface, but exchange across the lower interface is usually calculated by use of reaction-molecular diffusion models. Almost no information exists regarding the accuracy of these models in calculating fluxes. This paper seeks to examine such processes in the Hudson Estuary by examining the distribution of $CH_4$ and $Rn^{222}$, two naturally-occurring tracers which are generated primarily in sediments and migrate into the overlying water column. Due to limitations of space, this paper is a brief preliminary summary of research reported by Hammond (2). Future papers will present a more detailed discussion of the results summarized here.

## Description of the Hudson Estuary

The Hudson Estuary is shown in Figure 1 with several regions of interest labelled. The standard mile point (mp) reference system used in the text is indicated. This index represents statute miles north of the southern tip of Manhattan along the river axis. A distinct difference in bedrock geology has resulted in the development of a broad, shallow channel profile through the quaternary deposits of the Tappan Zee region (mean depth = 5.3m) and a narrow, deep profile through the crystalline rocks of the Hudson Highlands (mean depth 12.8m). A summary of the regional geology can be found in Sanders (3) and references he cites. This distinct morphology change is important to the discussion of methane distribution.

The Hudson has been classified as a partially-mixed estuary (4). Bottom salinities are typically 20-40% greater than surface salinities. The estuary is tidal from the Narrows (mp -8) to the Green Island Dam (mp +154), with an amplitude of about one meter throughout. The 0.1°/₀₀ isohaline penetrates to about mp 25

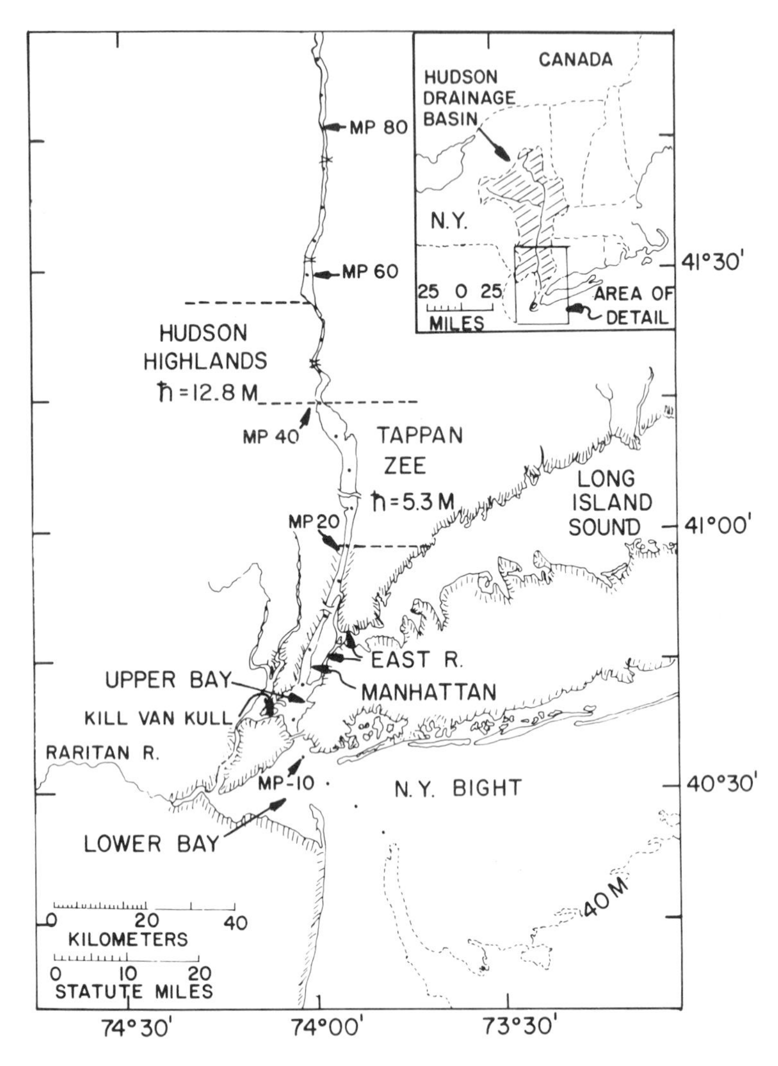


*Figure 1. Map of lower Hudson Estuary. The mile point reference index is abbreviated as MP. The Tappan Zee region has a broad, shallow channel, and the Hudson Highlands has a narrow, deep channel. Large amounts of sewage from New York and New Jersey are discharged to the Upper Bay.*

during periods of high fresh water discharge (spring) and to about mp 60 during periods of low discharge (late summer). Circulation within the estuary has been discussed in terms of a one-layer advection-diffusion model by a number of authors (see 5 and references cited there) and a two-layer advective model (6). On the basis of salinity distribution, the horizontal eddy diffusivity calculated from the first type of model is 3-6 x $10^6$ $cm^2/sec$ (25-50 $km^2/day$).

An estimation of the rate of exchange of dissolved gases across the atmosphere-water interface can be made on the basis of the Lewis and Whitman stagnant film model (7) recently reviewed by Broecker and Peng (8). This model envisions gas flux to be limited by molecular diffusion through a stagnant film of water at the air-water interface. The surface of this film is in equilibrium with the atmosphere ($C_A$) and the base of the film has the same composition as the bulk solution ($C_B$). Film thickness (Z) is a function of wind speed, and the evasive flux of gas C per unit area is then:

$$F_C = \frac{D}{Z}(C_B - C_A) \qquad [1]$$

where D = molecular diffusivity of gas C. Emerson (1) has proposed an empirical calibration curve relating Z to wind speed, and for the mean wind speed of 0.2-0.4 M/sec (5-10 mph) in the New York region, a film thickness Z = 180 ± 60μ is predicted. The mean life of a non-reactive dissolved gas in the water before exchange with the atmosphere will be $\tau = \frac{\bar{h}z}{D}$, and for a mean depth ($\bar{h}$) of 8m, and diffusivity of 2 x $10^{-5}$ $cm^2/sec$, $\tau \sim 8$ days. In this time, with an eddy diffusivity of 30 $km^2/day$, a gas molecule would move approximately $(30 \times 8)^{\frac{1}{2}} \sim 16$ km horizontally from its source before escaping to the atmosphere. Thus, regions showing small horizontal concentration gradients in dissolved gases over this distance can be treated by use of a box model in which exchange with the atmosphere dominates advective transport.

McCrone (9) has found the sediments north of mp 25 to be fairly homogenous in grain size, organic carbon, and cation exchange capacity. The dominant sediment type is a clayey silt with an organic carbon content of 3-10% (dry weight loss on ignition). Near Upper New York Bay (below mp 15), the sand fraction increases, and as a consequence of the large sewage input focused on the Upper Bay, the fraction of organic carbon in these sediments also rises.

South of mp 60, the ratio of drainage basin area to estuary surface area is small. As a consequence, few tributary streams enter this region. Those that do would replace water only once per year in the Hudson Highlands and only once every 4.5 years in the Tappan Zee region. Thus, addition of dissolved gases by tributary streams will be of minor importance. The bedrock characteristics and small drainage areas also indicate that groundwater flow should be quite small, especially in the Tappan Zee region.

Thus, the Hudson appears to be an ideal location for studying processes effecting transport across the sediment-water interface. The sediments (north of the sewage input region) are rather uniform,. the waters are sufficiently well-mixed by tidal movements to largely homogenize dissolved gas distributions, and complications introduced by stream and groundwater inflow should be minimal.

## Methane Distribution

A schematic diagram is shown in Figure 2 to illustrate the expected behavior of methane in the estuarine environment. A sequence of zones may be defined in sediments on the basis of micro-biology and redox potential (10). An oxidized zone overlies one in which sulfate reduction occurs. Methane production occurs below this in anoxic sediments low in sulfate. When *in situ* pressure is less than a few tens of atmospheres, methane concentrations equal to *in situ* saturation may be generated, resulting in formation of a gaseous phase in the sediments. This phase has been identified in Chesapeake Bay sediments as a source of acoustical turbidity (11) and in Hudson Estuary sediments as a cause of low seismic velocities (12). Sediments may flatulate this gas phase (13, 14) into overlying water, where some fraction of the bubbles may dissolve (2, 15) as they rise through the water column.

A second possible mode of transport across the interface is upward diffusion of dissolved methane from the production zone to the overlying water. Some evidence exists that methane may be oxidized by sulfate-reducing bacteria (16) but this process is apparently very slow (14). The oxygenated sediment zone, however, is a more effective barrier and apparently only a small fraction of methane diffusing upward passes through this layer (2).

The time constant for methane oxidation in oxygenated waters of lakes may be only a few hours (17, 18), but in sea water it is greater than 30 days (19). Storage tests on Hudson Estuary waters indicated a time constant greater than two weeks. Since this is greater than the time required for gas exchange, the dominant loss of methane from the water column must be by evasion to the atmosphere.

The distribution of methane in Hudson waters during a warm period (August 1973) and a cold period (March 1974) is shown in Figure 3. Surface and deep samples are averaged. Two trends are clear north of the region of sewage inputs. Concentrations are greater at low temperatures than at high temperatures and greater in deep water columns than in shallow water columns. This distribution is consistent with the conclusion that the most important input of methane to the water column is from bubbles escaping from sediments and partially dissolving as they rise. A model employing a uniform production of bubbles spatially and seasonally has been developed to treat this input quantitatively (2). The temperature dependence of the water column

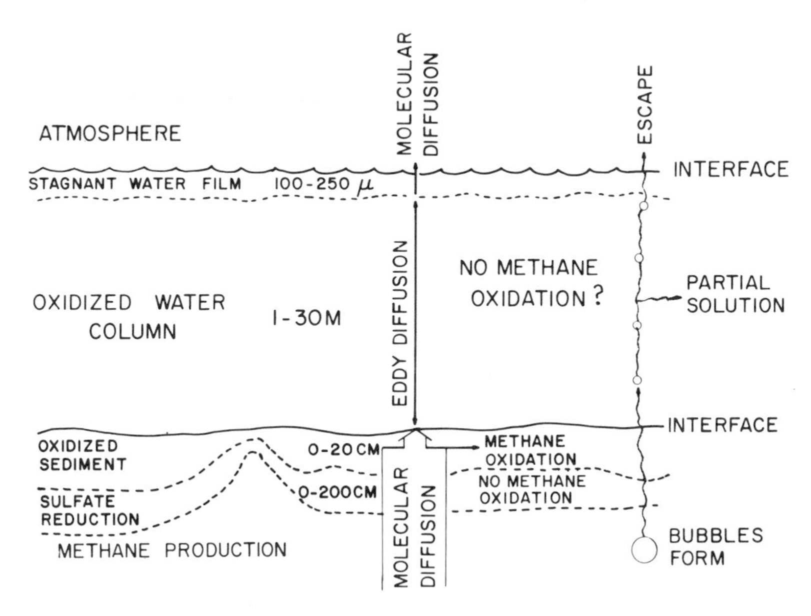


*Figure 2. Biogeochemistry of methane in estuaries*

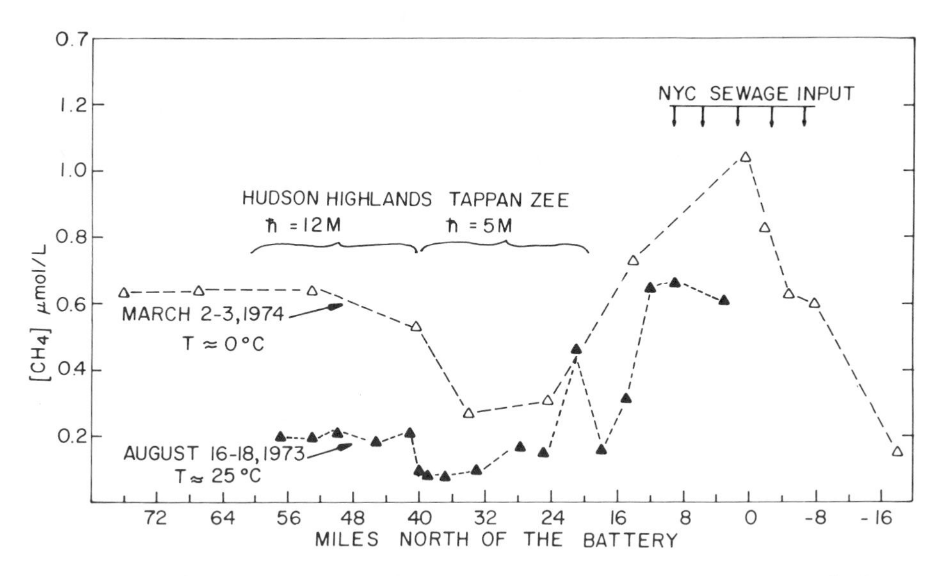


*Figure 3. Methane vs. distance in the Hudson Estuary. Surface and deep samples have been averaged. Note the effect of temperature and mean depth on concentration.*

concentrations is a consequence of increasing methane solubility with decreasing temperature. The mean depth dependence is a consequence of an increase in methane partial pressure within a bubble as hydrostatic pressure increases. A model employing a balance between uniform diffusion and gas exchange would predict no variation with temperature and mean depth.

An alternative model could be developed employing a combination of diffusion, reduced methane oxidation in surface sediments during cooler periods, and increased methane production in less saline regions, but the bubble model requires fewer assumptions. The sharpness of the transition of methane concentrations at the Hudson Highlands-Tappan Zee boundary supports the bubble model. Methane-saturated sediments have been collected and bubbles have been observed breaking at the water surface in the Hudson Estuary (2).

In the region of sewage inputs, advection of methane-rich waters to Lower New York Bay becomes an important loss, particularly in the spring when fresh water discharge is maximal. If advective and evasive losses are summed, the rate of input per unit area from Upper Bay sediments must be ∿ 20 times that upstream of sewage inputs. Inputs have similar seasonal characteristics to those mentioned above (2).

Sandy sediments in the Lower Bay probably supply very little methane. Although the complicated mixing and geometry in this region preclude very reliable modeling, when the assumption is made that flow through the Narrows is the only source of methane, the rate of gas exchange calculated between stations in the Narrows and at the mouth of the Lower Bay during winter months is consistent with the earlier estimate of film thickness (2).

Radon

$Radon^{222}$ is a noble gas with a four day half-life. Its parent, $Radium^{226}$ (1600 year half-life), may be dissolved in the water column, but is primarily bound to or within solid phases. Thus, high concentrations of radon are found in sediment pore waters or in ground waters, and much lower concentrations are found in the water column. In the Hudson, pore water has ∿ 600 dpm/ℓ and overlying water has ∿ 1.4 dpm/ℓ. If Hudson waters were in equilibrium with the atmosphere, the concentration would be ∿ 0.03 dpm/ℓ at 25°C.

Since radon was originally proposed as a tracer for processes in aquatic systems by Broecker (20), it has been used to study mixing rates and gas exchange rates in lakes and the ocean. Figure 4 is an illustration of an ideal radon profile in the open ocean. Throughout most of the water column, radon is in secular equilibrium with its parent. In the surface ocean, a deficiency of radon exists due to a net evasion to the atmosphere. In the deep ocean, an excess is found due to input from the sediments. The shapes of these defficiencies and excesses depends on the characteristics of the mixing processes in the ocean.

In the Hudson Estuary, the surface and bottom zones overlap. Figure 5 is a radon profile collected September 27, 1971 at mp 25. Radon activity is clearly in excess of $Ra^{226}$ activity, indicating that the rate of input from sediments is greater than the rate of evasion to the atmosphere. In this profile, radon increases with depth, but maxima may often be found at mid-depth or at the surface. Vertical variation in radon is typically less than 0.25 dpm/ℓ. The median concentration in 57 surface samples is 1.35 dpm/ℓ and the median concentration in 49 sub-surface samples is 1.45 dpm/ℓ. Analysis of 106 samples has indicated no consistent horizontal gradients between mp 18 and mp 60 (Table I).

Table I

Median Radon Activity (summer)

| Range (mp) | # of Samples | Activity (dpm/ℓ) | σ (dpm/ℓ) |
|---|---|---|---|
| 16-20 | 28 | 1.45 | ± 0.31 |
| 20-40 | 35 | 1.43 | ± 0.26 |
| 40-57 | 33 | 1.40 | ± 0.27 |
| > 57 | 10 | 1.50 | ± 0.35 |
| All Samples | 106 | 1.43 | ± 0.28 |

In the absence of consistent horizontal or vertical gradients, all samples can be grouped, and assuming that dissolved plus suspended $Ra^{226}$ is 0.10 dpm/ℓ (on the basis of 35 samples), the average excess radon (June-September) is $1.33 \pm 0.28$ dpm/ℓ. It is possible that some fraction of this activity in the Hudson Highlands might be supported by groundwater flow, but in the Tappan Zee region, an area of low relief encompassing a small drainage area, groundwater flow could contribute only minor amounts of radon (2). A budget (Table II) can be constructed for the Tappan Zee region (mp 20 - mp 40). Stream inputs are minor (2). Input carried by sediment flatulence of methane bubbles is negligible due to the low $Rn^{222}/CH_4$ ratio in methane-saturated pore water. When these terms are subtracted from the radon losses due to decay in the water column and evasion to the atmosphere, the remaining excess must be supplied from the sediments.

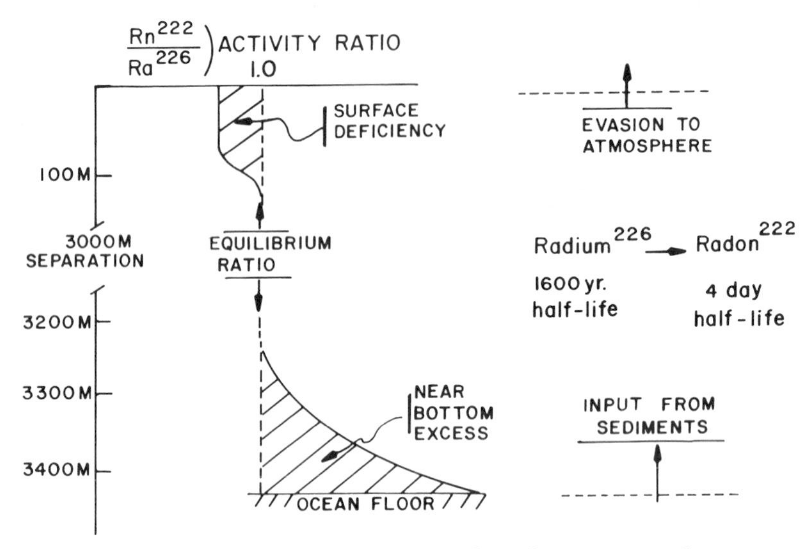


*Figure 4. Deep ocean ideal radon profile. Throughout most of the water column, radon-222 is in equilibrium with its parent. Near the surface, a deficiency exists (shaded area) due to evasion to the atmosphere, and near the bottom an excess exists (shaded area) due to input from sediments. The functionality of these deficiencies and excesses depends on the mixing characteristics in the water column.*

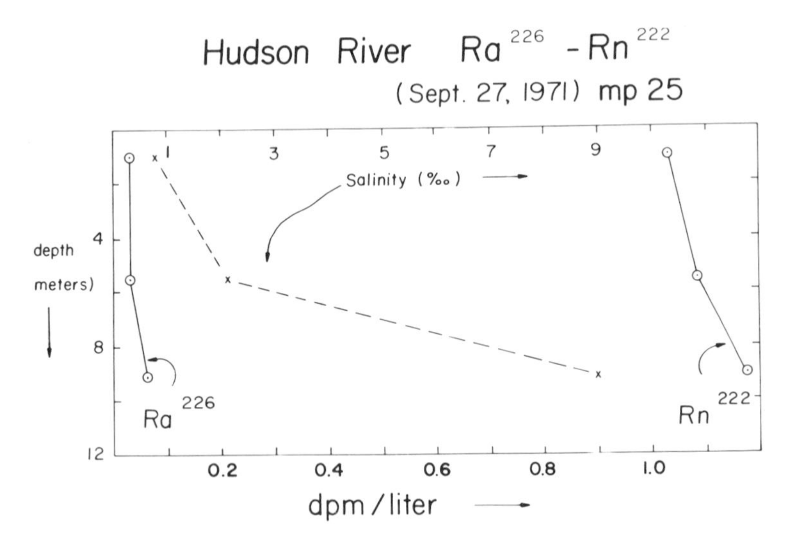


*Figure 5. Hudson River ideal radon profile. Radon-222 activity always exceeds that of its parent. This excess is mixed rather well throughout the water column. The salinity gradient was unusually large when these samples were collected.*

Table II
Summertime Radon Budget (fluxes in atoms/$m^2$-sec) for the Tappan Zee Region (mp 20 -- mp 40)

| | | |
|---|---|---|
| Mean Depth (m) | 5.3 | |
| Median Excess (dpm/ℓ) | 1.33 ± 0.28 | |
| Decay Loss | | -117 ± 25 |
| Evasion to Atmosphere | | -85 ± 30 |
| Stream Input | | 2 ± 5 |
| Flatulent Input | | 0 ± 1 |
| Remainder (Input from Sediment) | | 200 ± 39 |

## Discussion

Broecker (20) has derived an expression for the flux of radon from sediment in terms of the activity supported by its parent, $C_s^{eq}$ (per unit volume of wet sediment). Following this approach, for a uniform diffusivity $D_s$ within the sediment, a balance can be written for production, decay, and diffusion. At steady state,

$$0 = \frac{d}{dx}\left(D_s \frac{dC_s}{dx}\right) + \lambda\,(C_s^{eq} - C_s) \qquad [2]$$

where $\lambda$ is the decay constant of radon.

The solution to [2] is:

$$(C_s^{eq} - C_s) = Me^{-r_1 x} + Ne^{r_1 x} \qquad [3]$$

where $r_1 = \sqrt{\lambda/D_s}$. Applying the boundary conditions

$C_s^{eq} - C_s = 0$ at $x = \infty$

$C_s = C_w$ at $x = 0$
where $C_w$ = overlying water concentration

and differentiating, the flux of radon to the overlying water is:

$$J_{sed} = \sqrt{\lambda D_s}\,(C_s^{eq} - C_w)$$

Measurements of $C_s^{eq}$ were made by making a slurry from a known volume of wet sediment and distilled water. This slurry was sealed in a glass kettle, purged of radon, stored, and the re-growth of radon was measured. For Tappan Zee sediments $C_s^{eq} \approx 0.33 \pm 0.10$ dpm/$cm^3$. For a flux of 200 ± 39 atoms/$m^2$sec, with $\lambda = 2.1 \times 10^{-6}$ $sec^{-1}$ $D_s$ equals $2.7 \pm 1.0 \times 10^{-5}$ $cm^2$/sec. The computed profile is shown in Figure 6 (left side).

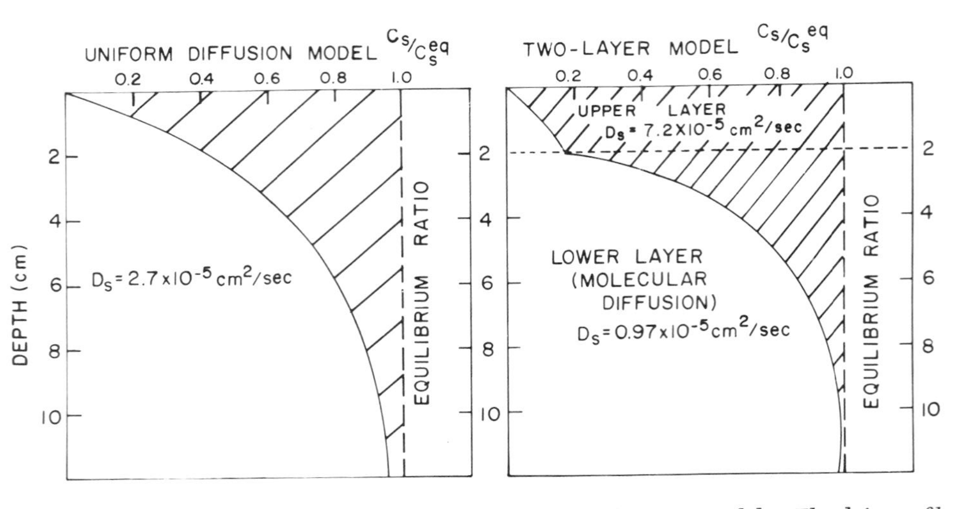


*Figure 6. Radon profiles calculated in sediments for two different models. The left profile illustrates a model in which sediments are stirred by a uniform eddy diffusivity. The right profile is calculated assuming uniform stirring to 2 cm and no stirring below this horizon.*

This value can be compared to the molecular diffusivity in water ($D_m$) measured by Rona (21) at 18°C with a temperature dependence calculated by Peng et al (22), of $D_m = 1.37 \pm 0.14 \times 10^{-5}$ cm$^2$/sec at 25°C. In sediments, this value must be reduced for the effect of tortuosity. In a Pacific red clay, Li and Gregory (23) found $D_s = \frac{D_m}{\theta^2}$ with $\theta = 1.37$ when porosity ($\phi$) was 50%. Since $\phi \approx 0.60$ in Hudson sediment and grain size is slightly larger, $\theta = 1.20 \pm 0.12$ should be a good estimate. Thus, under molecular diffusivity conditions, $D_s = 0.97 \pm 0.30 \times 10^{-5}$ cm$^2$/sec. This value is about 35% of that required to produce the observed flux. Clearly, some process in addition to molecular diffusion is at work.

Sediments in the estuary are characterized by a surface layer of about 2 cm thickness which is lighter in color and soupier than the material which underlies it. The coloration may be due to oxidation and the consistency is probably due to a slightly greater porosity. As an alternative to the model based on uniform diffusivity, a two-layer model may be more appropriate. This model incorporates an upper layer (thickness d) stirred by currents and organisms and a layer below in which molecular diffusion prevails. The equations for such a model in the surface ocean have been developed by Peng et al. (22) and can be applied to sediments as follows. Defining x to represent depth in the upper layer, equation [3] is the solution in the upper layer and:

$$C_s^{eq} - C_s = Pe^{-r_2 y} + Qe^{r_2 y} \qquad [5]$$

applies below, where $r_2 = \sqrt{\lambda\theta^2/D_m}$. Applying boundary conditions:

$$C_s = C_w \qquad \text{at } x = 0$$

$$Me^{-r_1 x} + Ne^{r_1 x} = Pe^{-r_2 y} + Qe^{r_2 y} \qquad \text{at } x = d \quad y = 0$$

$$r_1(Ne^{r_1 x} - Me^{-r_1 x}) = r_2(Qe^{r_2 y} - Pe^{-r_2 y}) \qquad \text{at } x = d \quad y = 0$$

$$C_s^{eq} - C_s = 0 \qquad \text{at } y = \infty$$

the model can be solved (Figure 6 right side). The flux from sediments to the overlying water (x = 0) is

$$J_s = \sqrt{D_s\lambda}\ (M - N) = 200 \text{ atoms/M}^2\text{sec}$$

The model calculated diffusivity will depend on the value of d chosen for the thickness of the mixed zone, but d = 2 cm is reasonable on the basis of the soupiness criteria and indicates $D_s = 7.2 \times 10^{-5}$ cm$^2$/sec. This is about seven times the molecular diffusivity in the layer below and indicates that substances dissolved in interstitial waters may migrate through the upper few centimeters of sediment at rates above their molecular diffusivities. Selection of d = 5 cm would indicate $D_s$ = 3.0 x $10^{-5}$ $cm^2$/sec.

The source of turbulence for this stirring could be small polycheate worms which populate these sediments or it could be stirring by tidal currents. Data collected in winter months, when bioturbation is minimal, may provide an answer to this question. It should be pointed out that alternative models could explain the "enhanced" flux of radon. One possibility is a sediment-reworking model in which localized erosion to several centimeters occurs periodically at random locations. The distribution of shortlived radionuclides in the sediment may be useful in distinguishing between among possible alternatives.

## Conclusions

1. Bubbles of methane which escape from sediments and partially dissolve as they rise through the overlying water column appear to dominate the transport of methane across the sediment-water interface in the Hudson Estuary. Thus, methane is apparently not useful for tracing the process of diffusion through surface sediments in estuaries.

2. A radon excess of 1.33 $\pm$ 0.28 dpm/ℓ exists during summer months in the waters of the Hudson Estuary between mp 16 and mp 60. Most of this excess is supported by input from the sediments at a rate which exceeds evasion to the atmosphere. Radioactive decay of radon in the water column accounts for over half the loss of this sediment input.

3. Summertime radon flux from the sediments is about twice as great as a molecular diffusion model would predict. The mechanisms controlling the flux cannot be uniquely constrained. A model which can account for the flux envisions the sediments to be composed of two layers. In the upper layer, uniform stirring by currents and organisms creates an eddy diffusivity. In the lower layer, molecular diffusion controls transport. Choosing the thickness of the upper layer as 2 cm indicates an eddy diffusivity during summer months approximately seven times greater than molecular diffusivity.

## Acknowledgements

The authors thank J. Goddard, J. Rouen and T. Torgersen for assistance in sample collection and analysis, P. Biscaye, J. Sarmiento, T. Torgersen and S. Williams for reviewing the manuscript, K. Antlitz for typing it, and D. Warner for drafting the figures. The research reported here was supported in part by the Environmental Protection Agency, contract No. R803113-01.

## Abstract

Methane concentrations in Hudson Estuary waters range from ∿ 0.2 μmol/ℓ to ∿ 1 μmol/ℓ. North of mp 20, concentrations are dominated by the balance between input from sediments and evasion to the atmosphere. Seasonal and regional distribution characteristics indicate that transport across the sediment-water interface is dominated by bubbles of methane which are produced in sediment and partially dissolve as they escape and rise through the water column.

Radon-222 is a noble gas with a four day half-life. It is produced primarily in sediments and may migrate into overlying waters where it decays or escapes to the atmosphere. Estimates of the rate of the latter process, combined with measurement of the former, indicates that in the Hudson Estuary migration across the sediment-water interface occurs at twice the rate predicted by a molecular diffusion model developed by Broecker (20). To account for the observed radon flux, a two-layer model developed by Peng et al. (22) can be applied to sediments. The model employs an upper layer in which uniform stirring by currents and organisms creates an eddy diffusivity about seven times the rate of molecular diffusivity in the layer below.



## Literature Cited

1. Emerson, S.R., The Gas Exchange Rate in Small Canadian Shield Lakes, Limnology and Oceanography, (in press), 1975.
2. Hammond, D.E., Dissolved Gases and Kinetic Processes in the Hudson River Estuary, Ph.D. Thesis, Columbia University, 161 pp., 1975.
3. Sanders, J.E., Geomorphology of the Hudson Estuary, in Roels, O.A., ed., Hudson River Colloquium, Annals of the New York Academy of Sciences, vol. 250, pp. 5-38, 1974.
4. Pritchard, D.W., Estuarine Circulation Patterns, Proc. Am. Soc. Civil Engrs., 81, pp. 717/1-717/11, 1955.
5. Simpson, H.J., R. Bopp and D. Thurber, Salt Movement Patterns in the Lower Hudson, Third Symposium on Hudson River Ecology, 1973, Hudson River Environmental Society, New York, 1974.
6. Abood, K.A., Circulation in the Hudson Estuary, in Roels, O.A., ed., Hudson River Colloquium, Annals of the New York Academy of Sciences, vol. 250, pp. 39-111, 1974.
7. Lewis, W.K. and W.G. Whitman, Principles of Gas Absorption, Industrial and Engineering Chemistry, 16, pp. 1215-1220, 1924.
8. Broecker, W.S. and T.H. Peng, Gas Exchange Rates Between Air and Sea, Tellus, 26, pp. 21-35, 1974.
9. McCrone, A.W., The Hudson River Estuary: Sedimentary and Geochemical Properties Between Kingston and Haverstraw, New York, J. Sed. Petrol., 37, pp. 475-486, 1967.

10. Claypool, G.E. and I.R. Kaplan, The Origin and Distribution of Methane in Marine Sediments, in Kaplan, I.R., ed., Natural Gases in Marine Sediments, Plenum Press, New York, pp. 99-139, 1974.
11. Schubel, J.R., Gas Bubbles and the Acoustically Impenetrable, or Turbid, Character of Some Estuarine Sediments, in, Kaplan, I.R., ed., Natural Gases in Marine Sediments, Plenum Press, New York, pp. 275-298, 1974.
12. Worzel, J.L. and C.L. Drake, Structure Section Across the Hudson River at Nyack, New York, from Seismic Observations, Annals of the New York Academy of Sciences, 80, pp. 1092-1105, 1959.
13. Reeburgh, W.S., Observations of Gases in Chesapeake Bay Sediments, Limnology and Oceanography, 14, pp. 368-375, 1969.
14. Martens, C.S. and R.A. Berner, Methane Production in the Interstitial Waters of Sulfate-depleted Marine Sediments, Science, 185, pp. 1167-1169, 1974.
15. Sackett, W.M. and J.M. Brooks, Origin and Distribution of Low-molecular Weight Hydrocarbons in Gulf of Mexico Coastal Waters, (this volume).
16. Nissenbaum, A., B.J. Presley and I.R. Kaplan, Early diagenesis in a Reducing Fjiord, Saanich Inlet, British Columbia. I. Chemical and Isotopic Changes in Major Components of Interstitial Water, Geochimica and Cosmochimica Acta, 36, pp. 1007-1027, 1972.
17. Cappenberg, T.E., Ecological Observations on Heterotrophic, Methane Oxidizing and Sulfate Reducing Bacteria in a Pond, Hydrobiologia, 40, pp. 471-485, 1972.
18. Rudd, J.W.M., R.D. Hamilton and N.E.R. Campbell, Measurement of Microbial Oxidation of Methane in Lake Water, Limnology and Oceanography, 19, pp. 519-524, 1974.
19. Swinnerton, J.W. and V.J. Linnenbom, Gaseous Hydrocarbons in Sea Water: Determination, Science, 156, pp. 1119-1120, 1967.
20. Broecker, W.S., The Application of Natural Radon to Problems in Ocean Circulation, in Ichiye, T., ed., Symposium on Diffusion in Oceans and Fresh Waters, Lamont-Doherty Geological Observatory, Palisades, N.Y., pp. 116-145, 1965.
21. Rona, E., Diffusionsgrösse und Atomdurch-messer der Radiumemanation, Z. Physik Chemie, 92, pp. 213-218, 1917.
22. Peng, T.H., T. Takahashi and W.S. Broecker, Surface radon measurements in the North Pacific Station Papa, Journal of Geophysical Research, 79, pp. 1772-1780, 1974.
23. Li, Y.H. and S. Gregory, Diffusion of ions in sea water and deep-sea sediments, Geochimica and Cosmochimica Acta, 38, pp. 703-714, 1974.

# 8

# Processes Affecting the Transport of Materials from Continents to Oceans

B. N. TROUP

Department of Earth Sciences, Case Western Reserve University, Cleveland, Ohio 44106

O. P. BRICKER

Department of Earth and Planetary Sciences, The Johns Hopkins University, Baltimore, Md. 21218

The major portion of the terrigenous material that reaches the estuarine and shallow marine environments is derived from the continents through weathering processes and transported via river systems. The products of continental weathering and erosion are transported in the form of suspended particulate matter, colloidal material, and dissolved species. The impact that these terrigenous materials have on the shallow water marine environment varies according to the nature and amount of the material and the site of discharge. Some materials introduced by rivers via an estuary may undergo major changes during their residence time in that environment before finally reaching the ocean. Others may be relatively unaffected during transit. The behavior of each material is dictated by its chemical composition and physical properties, and by the chemical, physical, and biological constraints imposed upon it by the estuarine environment. Understanding the chemistry of the estuarine and near-shore marine environment thus entails a knowledge of the chemical inputs into the environment and the nature of the interactions that take place among these materials within the environment.

In river waters and in sea water the species $Na^{+}$, $Mg^{++}$, $Ca^{++}$, $K^{+}$, $Cl^{-}$, $SO_4^{=}$, $HCO_3^{-}$ and $H_4SiO_4$ account for more than 98% of the total dissolved solids. These major elements determine the general chemical character of the waters. The other elements, the trace elements, are present in much smaller quantities, but may nevertheless play important roles in determining the chemical quality of the environment.

The sediments carried by rivers to estuaries and the near-shore marine environment are constituted primarily of the elements O, Si, Al, Ca, Mg, K, Na, and Fe with other elements present in trace amounts.

The continuum of reactions that occur between the waters and solids during the weathering process, transport through river systems and after sedimentation in the estuarine or coastal marine environment, are **major influences** on the composition of the aqueous phase and lead ultimately to the mineral assemblages that are preserved in the sedimentary record.

It may be instructive at this point to examine some of the types of reactions that occur during the sedimentary cycle. The materials of the earth's crust can be separated into two groups on the basis of chemical composition, the silicates and the carbonates. A third group, the evaporite-deposits, is quantitatively unimportant in terms of volume, but can play an important chemical role on a local scale. Chemical weathering of the earth's crust produces two types of products: a) dissolved species, and b) residual solids that are more stable under earth's surface conditions than the original minerals in the rocks being weathered. These residual solids are usually alumino-silicate minerals of the Phyllosilicate group (clays), or hydrous oxides of iron and manganese.

Present information suggests that, once formed in the weathering environment, little happens to the clay minerals other than shifts in the exchangeable ion population until they are deposited in the marine environment. Similarly, the alkali and alkaline earth cations freed in weathering reactions undergo few reactions other than minor involvement in ion exchange reactions with clay minerals and, perhaps with particulate organic matter, during their transit to the sea. In interstitial waters of estuarine and marine sediments, however, reactions occur that involve these elements. For instance, Drever (1) has reported exchange of Mg for Fe in clays in the sediments of the Rio Ameca basin. The clays take up Mg in exchange for iron, and the released iron reacts with sulfide ion in the anoxic interstitial waters to form iron sulfide. Depletion of Mg in the interstitial waters of deep sea sediments has also been reported (2, 3). The interstitial water environment in the sediments appears to be the locale in which significant chemical changes begin to occur.

The behavior of trace elements in estuaries is in marked contrast to the nearly conservative behavior of the major ions. Trace elements participate in a variety of biogeochemical reactions, and an understanding of these reactions is crucial to the understanding of the cycling of trace elements between continents and oceans.

## River Transport

River input is the major source of trace elements to the estuarine and near-shore marine environments. The chemical form of an element in river water to a great extent determines the type of reaction it participates in during its passage through an estuary. Trace elements can be transported by rivers in several different fractions:

A. Dissolved

1. Inorganic - includes the "free" hydrated ion and inorganic ion pairs (i.e., $FeOH^{2+}$, $FeCl_2^+$)

2. Organic - composed of trace elements complexed with dissolved organic matter.

Trace elements in the dissolved fraction are highly reactive and readily participate in all types of estuarine processes.

B. Particulate

1. Adsorbed on the surface of particles - most suspended particles in natural waters are negatively charged and cations are attracted to their surfaces. Additionally, trace elements are adsorbed on ion exchange positions of clay minerals and on hydrous oxides of iron and manganese. Both kinds of adsorbed ions are easily desorbed from particles when the chlorinity of surrounding solution changes.

2. Coprecipitated with iron and manganese in hydrous oxide coatings - during weathering, iron is released to solution and immediately precipitates (at pH's greater than 4-4.5) as a hydrated oxide. This material readily scavenges trace metals either by coprecipitation or sorption. Manganese in solution commonly coprecipitates with iron. These fine-grained, high surface area hydrous oxides coat detrital particles and are transported with them by rivers (4). The coatings are stable in highly oxic surface waters, but are unstable under anoxic conditions and quickly dissolve releasing the contained trace metals.

3. Precipitated on the surface of detrital particles - some trace metals can precipitate in the form of pure phases on surfaces of clay minerals or other detrital particles at low temperatures and pressures (5). These precipitates can dissolve under varying conditions of pH and pE.

4. Bound within the crystalline lattice of sediment particles - these trace metals can only be released to solution under harsh chemical conditions rarely encountered in natural waters.

5. Bound within organic particulates - stable in oxic environments and unstable in anoxic waters and sediments. Rate of decomposition and subsequent release of trace metals depends on the composition of the organic matter and the intensity of bacterial activity.

The chemical reactivity of trace elements is different in each of these fractions; and consequently, the behavior of an element passing through an estuary is strongly influenced by its distribution among these fractions. Gibbs (6) investigated the partitioning of trace metals among some of these fractions in the Amazon River System. His results illustrate the substantial variation in the distribution of four elements among these fractions during their transport in the river (Figure 1). In the Amazon, most of the Fe and Mn are transported as oxide coatings, while most of the Cu and Co reside in the crystalline lattice. In addition, large portions of the total Cu and Mn are in the dissolved state and significant amounts of Fe and Co are bound with organic particulates. Gibbs also reported that the order of dominance of the fractions (i.e., Fe: coating > organic > solution) varied little between rivers but that the absolute percentage of an element in each fraction did change. Thus, since the chemical reactivity of elements varies between the fractions, the amount of each trace metal available for different biological and geochemical reactions in estuaries varies from river to river and must be determined for each element under study.

Several workers have found that trace element concentrations vary greatly between different rivers (7, 8, 9, 10). This observation is not unexpected since the concentration in a river depends on the relative concentration of trace elements being weathered in the drainage basin and on the rate of weathering of the different minerals containing trace elements. Recent studies have also shown that trace metal concentrations in rivers vary greatly as a function of discharge rate, suspended load and time of year.

Carpenter *et. al.* (11) completed an extensive study of the temporal variability of trace metals transported by the Susquehanna River to the Chesapeake Bay. They sampled the river on a weekly basis for a period of approximately 1½ years at a station one mile below the Conowingo Dam. Each sample was separated into three fractions and these fractions were analyzed for trace metals according to the following procedures (11):

1. Settled solids (SS) - 100 liters of river water were allowed to settle for 10-14 days. The supernatant water was removed and the remaining slurry was digested with 1 M HCl and 1.5 M HAc for 48 hours at 60°.

2. Filtered solids (FS) - 50 liters of the supernatant water was filtered through acid-washed 0.2 μm cellulose acetate membrane filters. The filtered solids were then treated identically to the settled solids. The remaining 50 liters of supernatant water was centrifuged to determine the weight concentration of solids.

3. "Soluble" (SOL) - 50 liters of the filtrate from step two was run through a cation exchange resin (Chelex 100 in acid form). The resin was eluted with 1 M HCl and the eluate was evaporated to dryness. The residue was oxidized with nitric acid and made to volume with 0.1 M HCl.

All three fractions were then analyzed for their trace metal content by atomic absorption spectrophotometry preceded by APDC complexation and MIBK extraction. The standard deviation of replicate samples varied between 1.0 and 2.2%.

The discharge of water and suspended solids by the Susquehanna River over a 17 month period in 1965-66 is depicted in Figure 2. The most striking feature of the data is the large variation of both discharges during the year. As we would expect, both discharges are greatest during early spring. In fact 5.1% of the total annual discharge of solids occurred on one day, February 17; and 30% of the total annual solids discharge

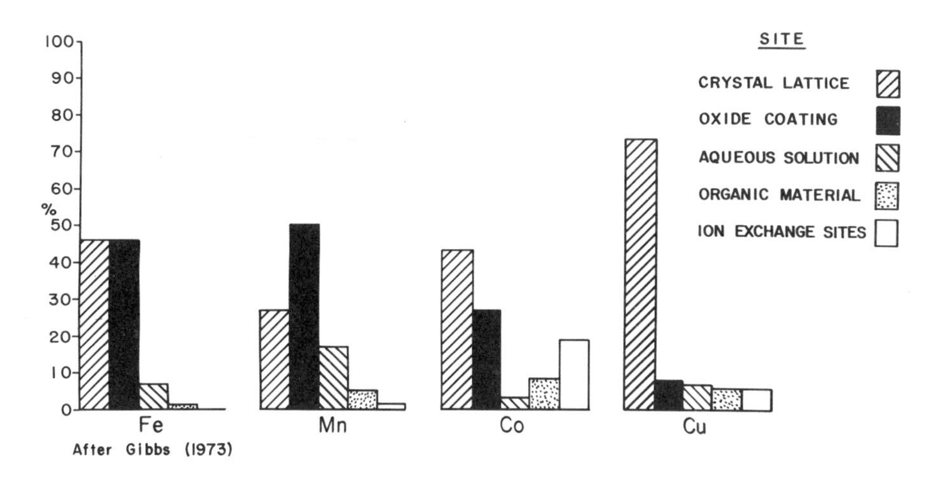


*Figure 1. Mechanisms of metal transport in the Amazon River*

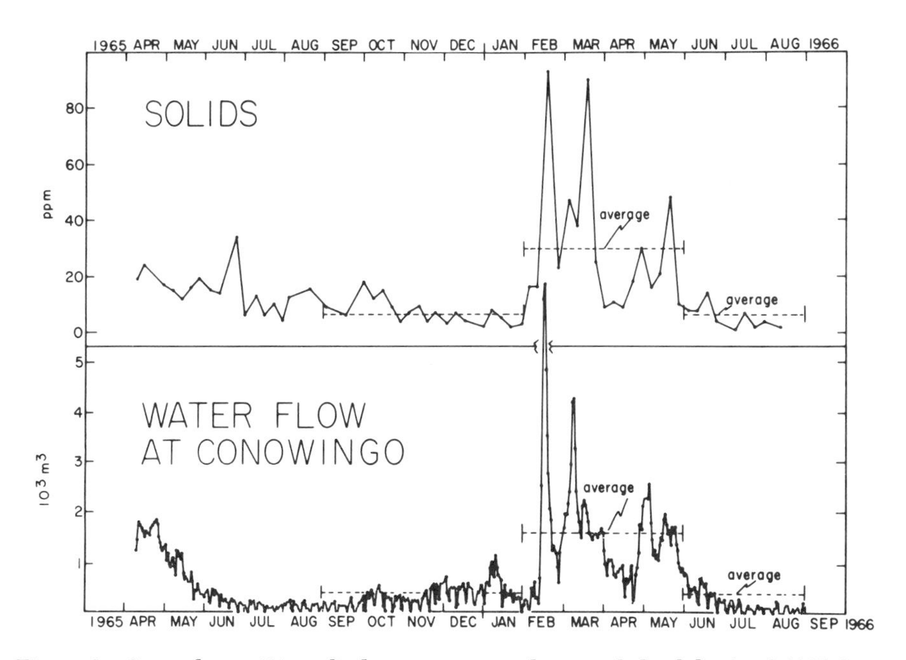


*Figure 2. Susquehanna River discharge, water, and suspended solids, April 1965–September 1966 (after Carpenter* et al. (11))

occurred in one week, February 14-20. Because of these large fluctuations in discharge rates, we conclude that estimates of total solids and total trace element discharge of rivers computed from concentrations and flow rates determined during periods of low or moderate discharge are not representative of the true total annual discharge.

As a first approximation the trace metal concentration (μg/kg water) data correlates well with the solids discharge (Figures 3, 4 and 5). The concentrations are highest in the spring and lowest in the summer and fall. Closer inspection, however, reveals differences in the patterns of particular elements and total solids. Mn, Ni, Zn and Co exhibit large concentrations in January, and Cu, Cr and Mn have concentration maxima in late spring or early summer. The temporal variability of the solids discharge can be eliminated from the trace metal data by computing the weight concentrations of the metals in the solids fraction (mg/kg solids). These data are shown in Figures 6 and 7. Generally there are peaks in concentration for all the metals during December and January and secondary peaks for Co, Cr, Ni, Cu and Mn in July. Since decaying organic matter (particulate and dissolved) is abundant in the Susquehanna during these two periods, the high concentrations may be the result of binding of the metals to particulate organisms. Figures 8 and 9 illustrate the distributions of Fe, Mn, Cu, Ni and Zn among the three fractions during the 17 month sampling period. Most of the Fe and Zn are associated with the solids throughout the year except during January when a substantial portion of the two metals is in the dissolved state. The partitioning of Mn between solution and solids varies throughout the year. There is a substantial amount of soluble manganese present during January and early spring, but the rest of the year manganese is associated predominantly with the solids fractions. Large amounts of Ni and Cu are in the dissolved form throughout the year, but the ratio of the soluble to that associated with the solids fraction increases in December and January. We believe that the high concentrations of dissolved trace metals during early winter can be associated with the increased amounts of decaying organic matter in the river during this period. Further investigation is needed to confirm this conclusion.

The importance of frequent sampling in the estimation of trace element transport is pointed out in Table 1. Turekian's estimates are based on one sampling period during the summer of 1966, and those of Carpenter and Grant are time averages over a 17 month period. The discrepancies between the two sets of data are greater than an order of magnitude for many of the metals. The higher estimates of Turekian may be attributed to the high concentrations of several trace metals in the solids fractions noted by Carpenter during July, 1966 (Figures 6 and 7).

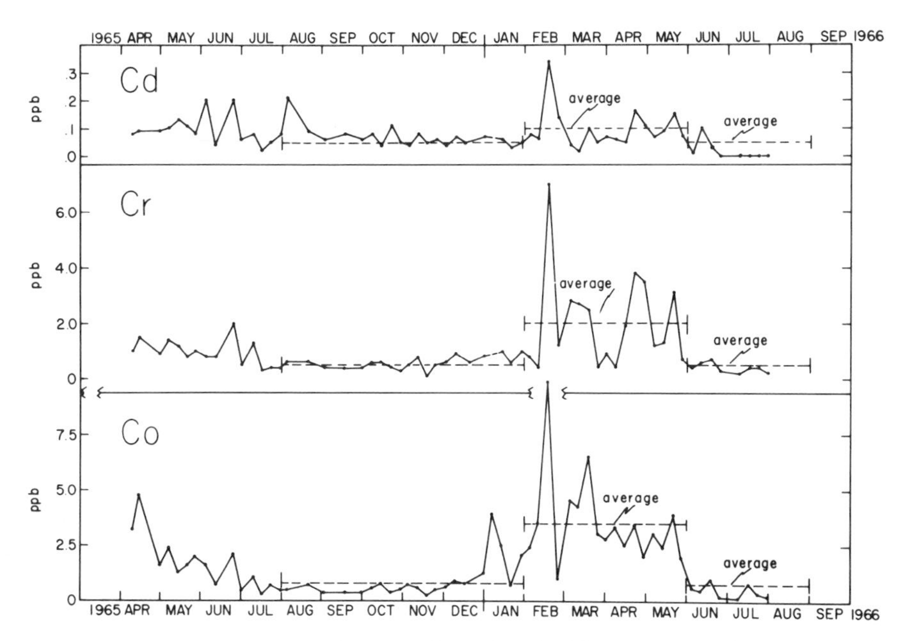


*Figure 3. Concentrations of Cd, Cr, and Co (mg/kg $H_2O$) in Susquehanna River discharge, April 1965–September 1966 (after Carpenter* et al. (11))

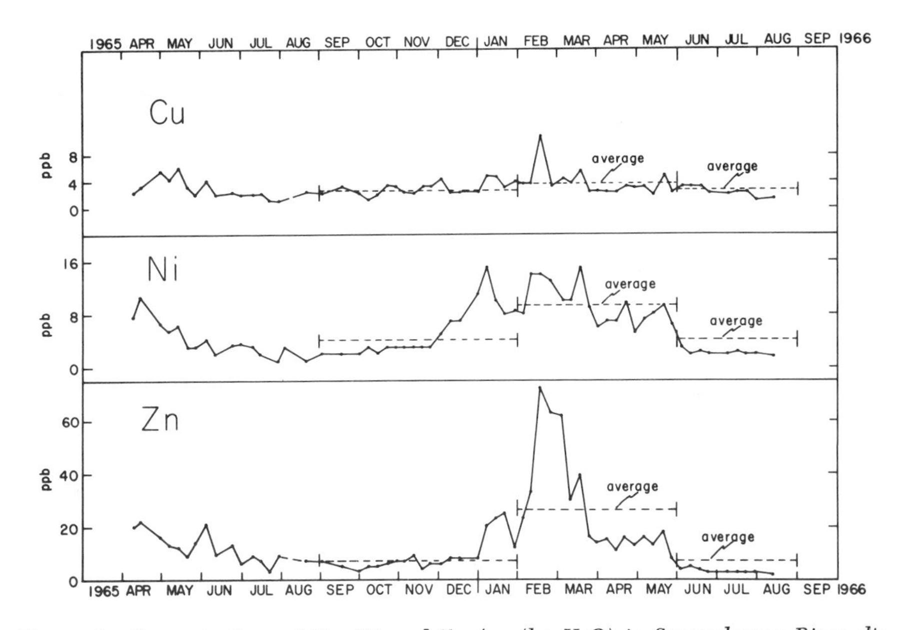


*Figure 4. Concentrations of Cu, Ni, and Zn (mg/kg $H_2O$) in Susquehanna River discharge, April 1965–September 1966 (after Carpenter* et al. (11))

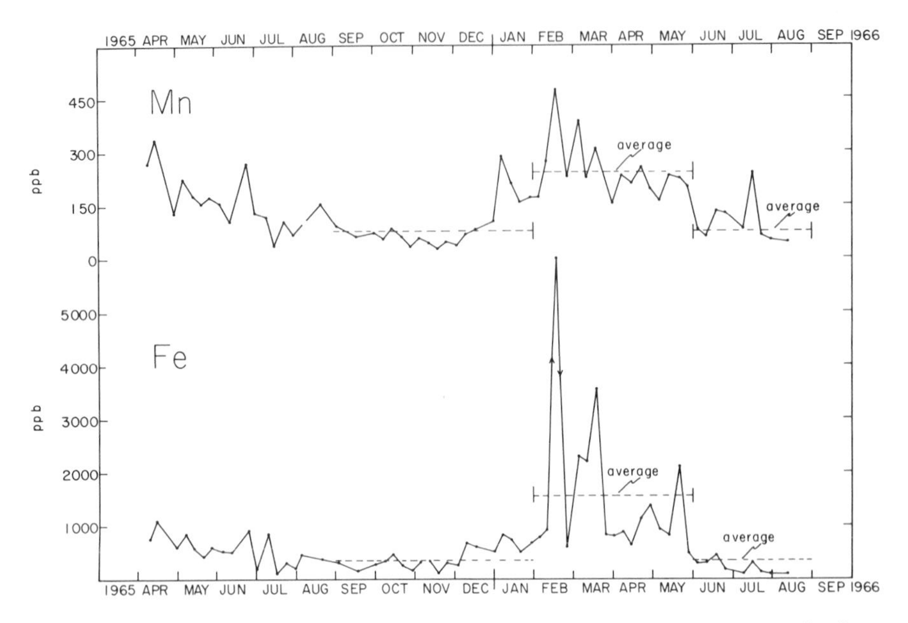


*Figure 5. Concentrations of Fe and Mn (mg/kg $H_2O$) in Susquehanna River discharge, April 1965 to September 1966 (after Carpenter* et al. (11))

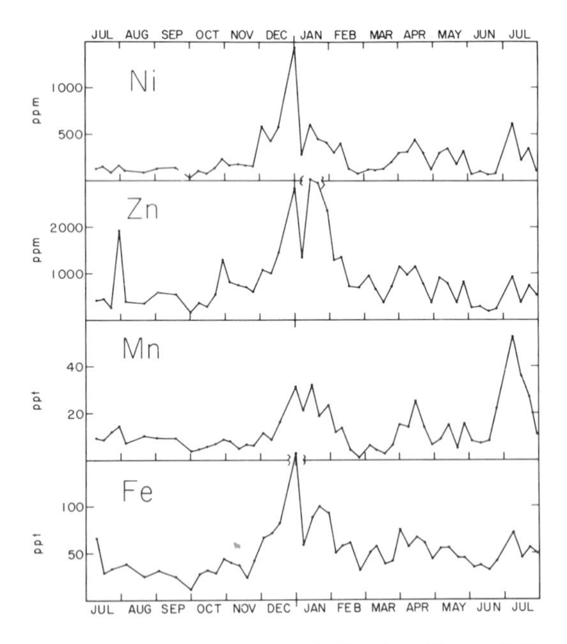


*Figure 6. Concentrations of Ni, Zn, Mn, and Fe (mg/kg solids) in Susquehanna River discharge July 1965–July 1966 (after Carpenter* et al. (11))

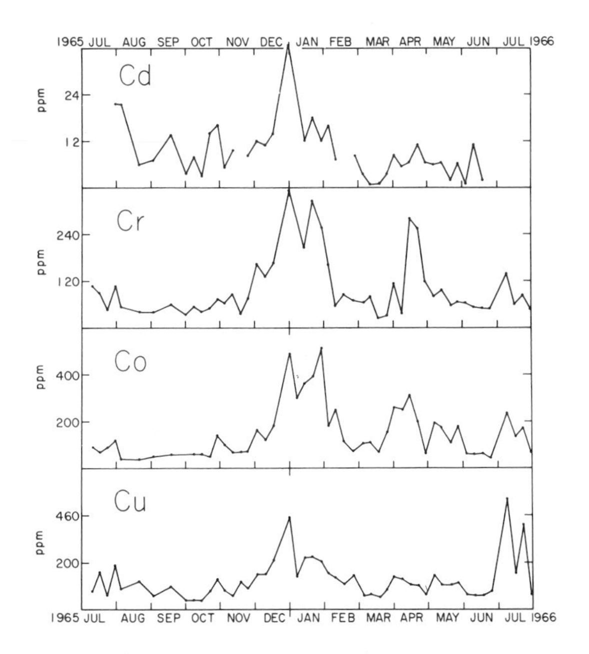


*Figure 7. Concentrations of Cd, Cr, Co, and Cu (mg/kg solids) in Susquehanna River discharge July 1965–July 1966 (after Carpenter* et al. (11))

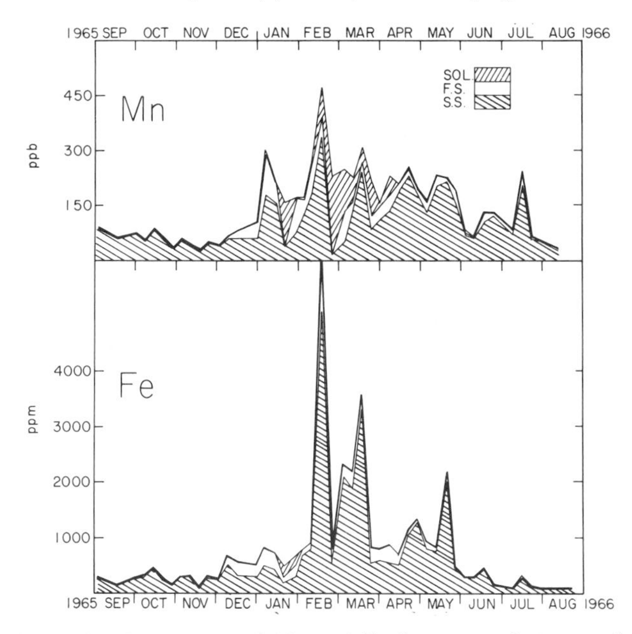


*Figure 8. Partitioning of Mn and Fe between solutions and solids in Susquehanna River discharge, April 1965–September 1966 (after Carpenter* et al. (11))

Table 1. Trace metal transport of the Susquehanna River (Tons)

| | Carpenter et al. (11) | Turekian & Scott (7) |
|---|---|---|
| Suspended solids | 767,000 | 300,000 |
| Fe | 43,000 | --- |
| Mn | 5,300 | 120,000 |
| Zn | 650 | --- |
| Ni | 215 | 3,000 |
| Cu | 105 | --- |
| Co | 88 | 1,500 |
| Cr | 53 | 870 |
| Cd | 2 | --- |
| Ag | -- | 45 |
| Mo | -- | 97 |

Although rivers are the dominant source of trace metals entering estuaries, rainfall and particulate fallout may be important sources in specific locales, particularly for volatile trace metals, such as Se, Hg and Pb. For example Patterson (12) has reported that dry aerosol deposition of Pb in the Southern California Bight equals more than 75% of the amount of Pb transported by rivers and waste waters. Similarly, in the Delaware estuary, Biggs (13) has found that the concentration of Cd and Pb in rainfall is higher than the concentrations observed in unpolluted streams discharging into the estuary.

## Biogeochemical Processes in Estuaries

Once the rivers have transported trace elements to an estuary, the elements can undergo several kinds of physical, chemical and biological reactions. It is important to identify these reactions since estuarine processes affect the amount and the form of the trace elements transported to the ocean. On the basis of their behavior in estuaries, elements can be classified as either conservative or non-conservative. Conservative elements display a linear relationship between concentration and chlorinity along the length of an estuary which reflects a simple, physical mixing process between river and sea water. Conservative behavior is typically exhibited by the major ions and some minor constituents, such as silica in the Merrimac estuary (14, and Figure 10). Most trace elements are nonconservative, like iron (Figure 10), and physical mixing processes alone cannot account for their distributions in an estuary.

There are five major categories of reactions which influence trace metal transport in estuaries:

1. Flocculation and sedimentation - these processes are encouraged by the increased salinity of estuarine waters. As a result of the salt-wedge type of circulation commonly observed (Figure 11), sedimentation is concentrated in the upper reaches of estuaries. Suspended sediment discharged by rivers is carried

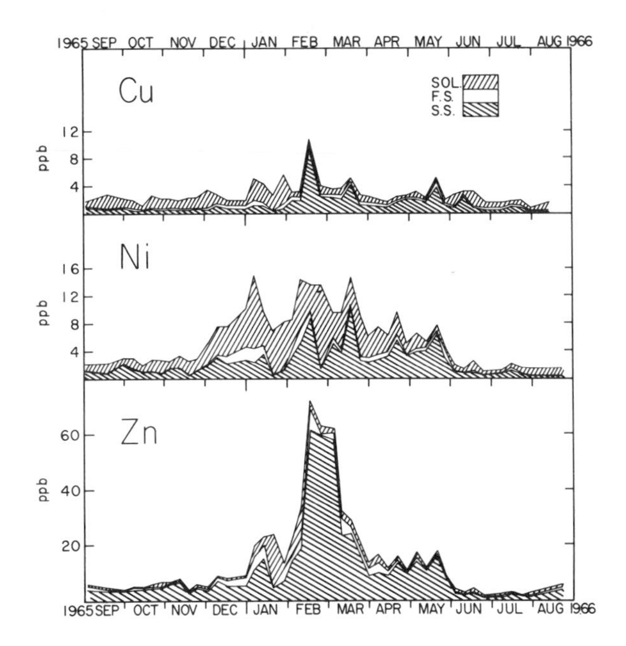


*Figure 9. Partitioning of Cu, Ni, and Zn between solution and solids in Susquehanna River discharge, April 1965–September 1966 (after Carpenter* et al. (11))

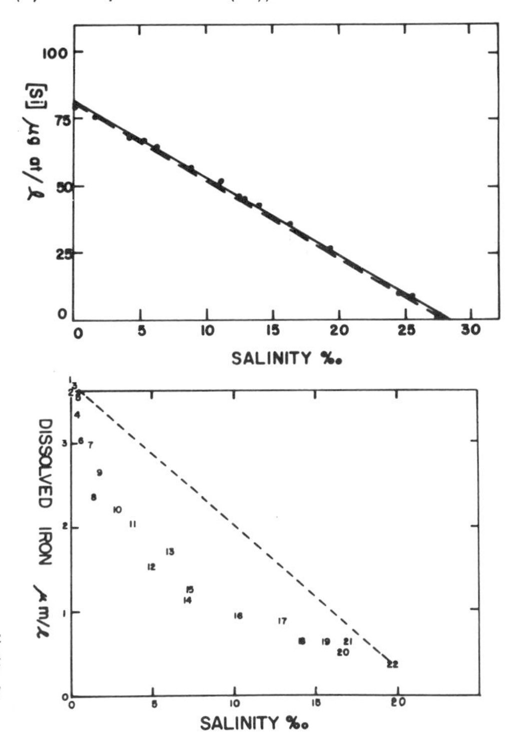


*Figure 10. Conservative behavior of silica vs. nonconservative behavior of Fe in the Merrimac River estuary (after Boyle* et al. (14))

toward the ocean by the fresh water at the surface of the estuary. As the solid particles settle, they fall into the salt wedge layer moving toward the mouth of the river. The particles continue to settle and sedimentation is concentrated in the low velocity region near the tip of the salt wedge. Schubel (pers. comm.) estimates that, in the Chesapeake Bay 90% of the solids discharge is deposited in the northern third of the estuarine system. His argument is substantiated by the distribution of Fe in sediments along the length of the bay. The time-averaged concentration of Fe in suspended solids discharged by the Susquehanna River is about 5-10% (11). The concentration of Fe in northern Chesapeake Bay sediments averages 5%, while the concentration in southern bay sediments averages 3%. These data suggest that most of the sediments of the southern bay originate from shoreline erosion rather than from river discharge.

2. Mineral-water interaction - the rapid changes in salinity, pH and oxygen content of estuarine waters can affect the precipitation or dissolution of minerals (for instance, the dissolution of the hydrous oxide coatings under low $O_2$ conditions in water or sediments).

3. Adsorption/desorption - due to the high concentration of Na and Mg in seawater, ion exchange reactions occur rapidly in estuarine waters as sediment particles adsorb the seawater cations and desorb trace elements.

4. Diagenesis and remobilization of trace metals in sediments - the decreased pH and strong pE gradient in sediment interstitial waters brought about by the decomposition of organic matter significantly affects the solubility of trace element solid phases and ion exchange equilibria. Trace metal concentrations in sediment pore waters are typically orders of magnitude greater than in the overlying water, and the flux of trace elements across the sediment-water interface may constitute an important mass transfer process in estuaries.

5. Biological processes - interactions with aquatic organisms may strongly influence the behavior of trace elements during their passage through estuaries. The generally recognized mechanisms are: ingestion of particulate suspended matter containing trace metals, trace metal complexation with organic chelates produced by organisms, trace element uptake in metabolic processes, and ion exchange and sorption on the surfaces of organisms (15).

Most trace elements undoubtedly participate in all these major categories of reactions to some extent, and it is important to identify which reactions most strongly influence the transport of each element under study. To this end, our group at Johns Hopkins has concentrated on the effect of sediment-water exchange processes on trace element cycling.

## Desorption of Zn from Estuarine Sediments

Bradford (16) investigated the influence of sediments on zinc concentrations in the overlying water in Chesapeake Bay. Using anodic stripping voltammetry, he observed large Zn concentrations just above the sediment-water interface in the northern bay during May. The distribution of total zinc (UV oxidized) and inorganic zinc (raw sample) just above the sediment-water interface at station 914S is shown in Figure 12. During May there is a large concentration of zinc at this station. The zinc is totally in the inorganic form because there is no difference in concentration between the oxidized and raw samples. The only other major change occurring during May at 914S was a rapid increase in the chlorinity from less than 0.2 ‰ to greater than 3 ‰ (Figure 13). The large zinc concentration cannot be the result of mixing of bay water and river water since the concentration of zinc never exceeded 6 ppb in either of these waters at this station. Thus, the large increase in Zn must have come from sediments. This conclusion is substantiated by the concentration gradients of Zn at station 914S during April, May and June (Figure 14). During April Zn varies from 4.5 ppb at 4 m to 5.5 ppb at 1 m above the interface. In May there is a large vertical concentration gradient which implies a flux of Zn from the interface. In June the Zn concentration is uniform at this station. We conclude from these data that zinc was desorbed from freshly deposited river sediments at this station as Mg and Ca from the salt wedge were adsorbed. During March and April the Susquehanna discharges a large quantity of suspended solids to the upper reaches of the Chesapeake Bay. These solids form a layer 1-2 cm thick in the northern bay. The ion exchange positions of these solids are in equilibrium with the river water ratios of the major cations. Once the river discharge subsides the salt wedge begins to move landward again. During May the salt wedge moved across the sediments at 914S, and in response to the ion exchange equilibria, the sediments adsorbed the major sea water cations and desorbed Zn to the overlying water. This conclusion confirms earlier observations suggesting that Zn rapidly desorbs from surface sediments as the chlorinity increases above 0.4‰ (17, 18).

## Remobilization of Trace Metals in Sediment Pore Waters

In the highly oxygenated conditions of normal river beds the hydrous Fe and Mn oxide layers coating sediment particles are stable. When the detrital particles reach the marine environment and become incorporated into the organic reach bottom sediments of an estuary or marine delta, they encounter very different conditions. Oxygen is usually depleted within the upper few centimeters of sediment by bacterially mediated oxidation of organic material. In brackish and marine waters, after oxygen is

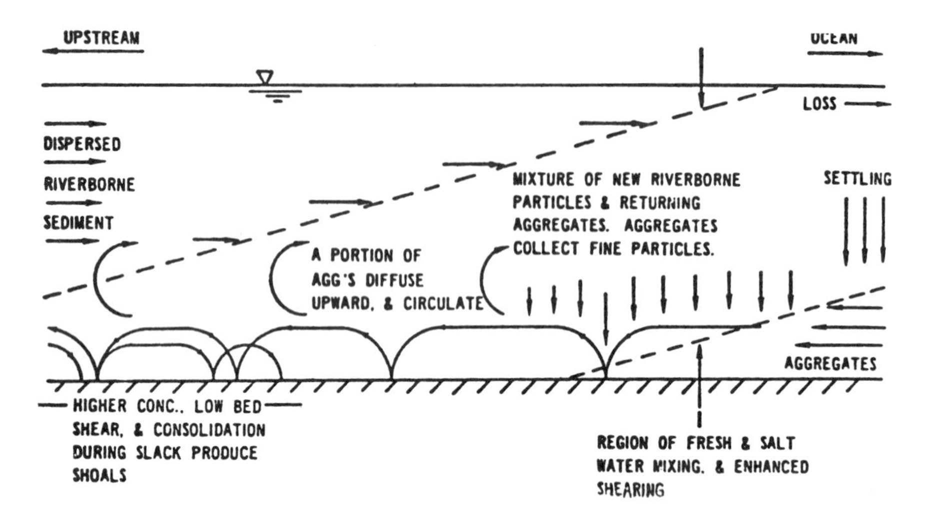


*Figure 11. Diagram of an estuarine salt wedge (after Neiheisel (22))*

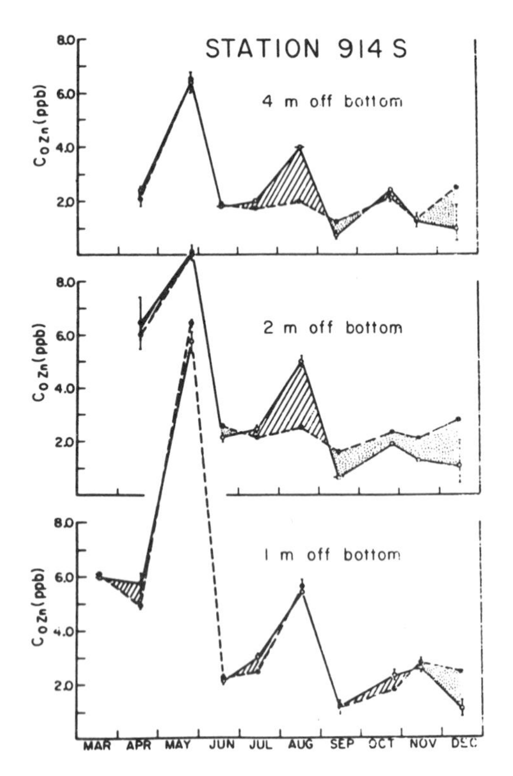


*Figure 12. Near bottom Zn concentrations with time at Station 914S in the upper Chesapeake Bay (after Bradford (16))*

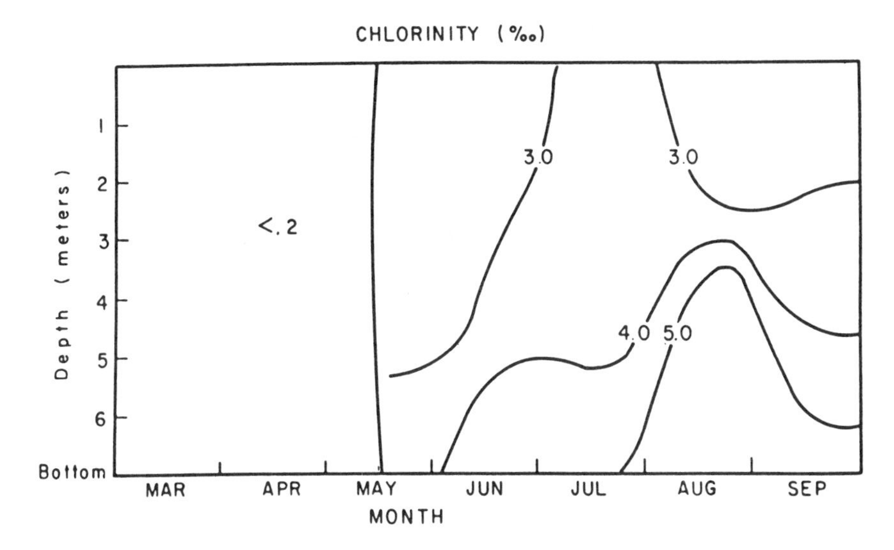


*Figure 13. The chlorinity with depth and time at Station 914S in the upper Chesapeake Bay (after Bradford* (16))

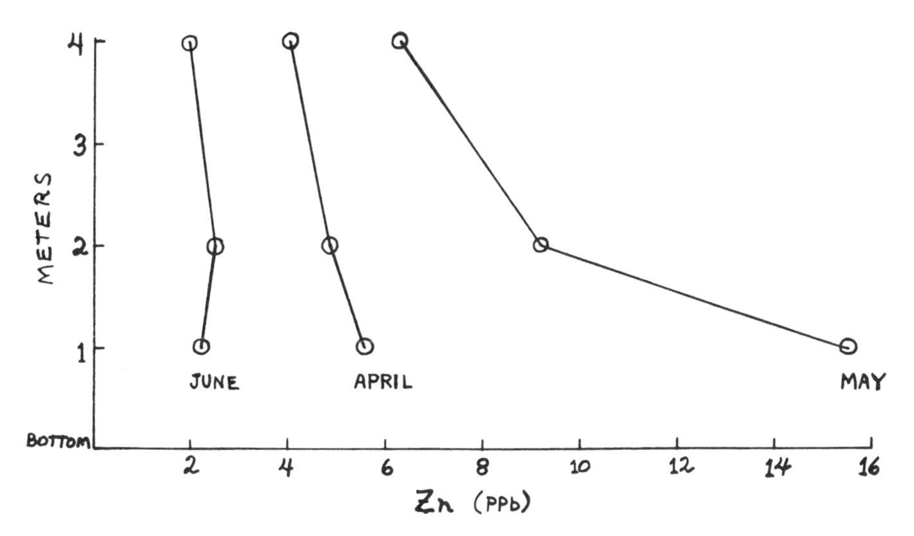


*Figure 14. Near bottom Zn concentrations in April, May, and June at station 914S in the upper Chesapeake Bay (after Bradford* (16))

depleted, bacteria use sulfate as an electron acceptor to continue the oxidation of organic matter (equation 1):

$$\text{organic matter} + SO_4^{2-} = HCO_3^- + HPO_4^- + HS^- + H^+ \quad (1)$$

These reactions decrease the sulfate concentration and pH and increase the concentrations of bicarbonate, phosphate and sulfide. Additionally, strongly reducing conditions are produced in which the oxide coatings on detrital particles are chemically unstable. Solution of the oxide coatings produces large concentrations of dissolved iron, manganese and other trace metals in the interstitial waters. The vertical distribution of dissolved Fe(II) and Mn in the interstitial waters of the upper meter of sediments at a station in the northern Chesapeake Bay is shown in Figure 15. The concentrations of both Fe and Mn increase rapidly 1 to 2 cm below the sediment water interface as a result of the dissolution of the oxide coatings. Below 10 cm, the concentrations decrease rapidly and become nearly constant below 40 cm. We attribute the rapid decreases of the concentrations to the precipitation of authigenic trace metal phases in the interstitial water as phosphate and carbonate increase rapidly with depth. Siderite and vivianite control the dissolved iron concentrations below 10 cm (19), while rhodocrosite controls the Mn distribution (20).

The concentrations of other trace metals which had coprecipitated with Fe and Mn during weathering also increase as the oxide coatings dissolve. In northern Chesapeake Bay pore waters, Cu and Pb mimic the distributions of Fe and Mn (Figure 16). Thus, just below the sediment-water interface there is a concentrated reservoir of dissolved trace metals which can be transferred across the interface by molecular diffusion, sediment resuspension and bioturbation (21). Because of the magnitude of the trace metal concentrations in sediment pore waters, very little transfer would have to occur to significantly alter the overlying water concentrations, and consequently, to increase trace metal transport to the oceans.

## Conclusions

At the present time we can draw a fairly complete picture of the transport of major ions between continents and oceans, due to their nearly conservative chemical behavior. On the other hand, trace elements are non-conservative species in estuaries, and extensive spatial and temporal surveys of their distributions are required to delineate the modes of transport of individual elements. Specifically, we need the following kinds of data:

1. River transport data for all important trace metal fractions to determine season-to-season and year-to-year variations.
2. Data on the flux of trace elements through biological systems

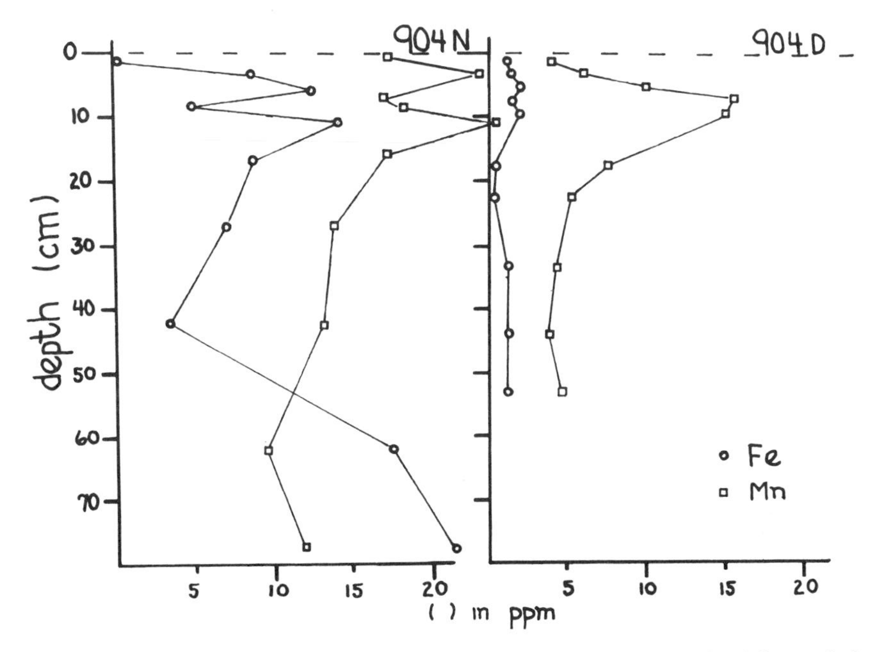


*Figure 15. Concentrations of dissolved Fe and Mn as a function of depth beneath the sediment–water interface*

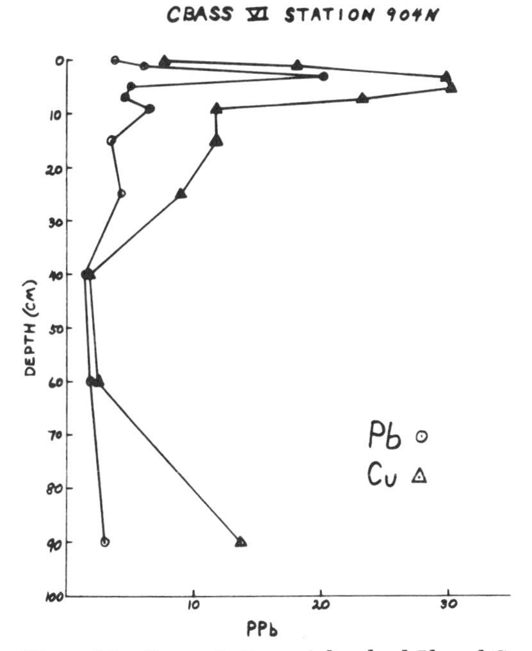


*Figure 16. Concentrations of dissolved Pb and Cu as a function of depth beneath the sediment–water interface*

3. Quantitative assessment of the role of sediments in trace metal transport.

4. Data on oceanic input of trace elements into estuaries, so that estimates of net transport to the ocean can be calculated.

In general we know the kinds of reactions which are important in trace metal transport. However, in many cases we do not know which reaction is most important for a given element, and in almost all cases we do not know the amount of each element participating in the reaction or the rate at which the reaction is occurring. The data that has been collected and that we are now collecting, will help to remedy this situation.

## Acknowledgements

We would like to thank Dr. James H. Carpenter for allowing us to use some data from his Susquehanna River study and Peter Kaerk for providing unpublished data on lead and copper in Chesapeake Bay sediments. Mr. Gerry Matisoff kindly assisted with the preparation of the figures. This work was supported by AEC Contract AT(11-1)-3292.

## Literature Cited

1. Drever, J. I. Science (1971) 172: 1334-1336.
2. Bischoff, J. L. and T. Ku, Jour. Sedimentary Petrology (1970) 40: 960-972.
3. Bischoff, J. L., R. E. Greer, and A. O. Luistro, Science (1970) 167: 1295-1246.
4. Carroll, D., Geochimica et Cosmochimica Acta (1957) 14: 1-28.
5. Tiller, K. G. and J. G. Pickering, Clays and Clay Minerals (1974) 22: 409-416.
6. Gibbs, R. J., Science (1973) 180: 71-73.
7. Turekian, K. K. and M. R. Scott, Environmental Science and Technology (1967) 1: 940-942.
8. Kharkar, D. P., K. K. Turekian and K. K. Bertine, Geochimica et Cosmochimica Acta (1968) 32: 285-298.
9. Konovalov, G. S. and A. A. Ivanova, Okeanologua (1970) 10: 628-636 (in Russian).
10. Windom, H. L., K. C. Beck and R. Smith, Southeastern Geology (1971) 1971: 1109-1181.
11. Carpenter, J. H., W. L. Bradford and V. E. Grant, "Processes Affecting the Composition of Estuarine Waters ($HCO_3$, Fe, Mn, Zn, Cu, Ni, Cr, Co, Cd)" In Proceedings of the Second International Estuarine Conference, J. H. Carpenter (ed.) (in press).
12. Patterson, C. C. In "Pollutant Transfer to the Marine Environment", R. A. Duce, P. L. Parker, and C. S. Giam (eds.), P. 12, Deliberations and recommendations of the National Science Foundation, International Decade of Ocean Exploration Pollutant Transfer Workshop, January 11-12, 1974.

13. Bopp, Frederick III, F. K. Lepple and R. B. Biggs, Trace Metal baseline studies on the Murderkill and St. Jones Rivers, Delaware Coastal Plain. College of Marine Studies, Univ. of Delaware Publication DEL-SG-10-72, 31 pp. (1972).
14. Boyle, E., R. Collier, A. T. Dengler, J. M. Edmond, A. C. Ng and R. F. Stallard, Geochimica et Cosmochimica Acta (1974) 38: 1719-1728.
15. Martin, D. F., "Marine Chemistry", Vol. 2 (New York, Marcel Dekker, 1970).
16. Bradford, W. L., "A Study on the chemical Behavior of Zinc in Chesapeake Bay Water using Anodic stripping Voltammetry", Chesapeake Bay Institute, Johns Hopkins Univ., Tech. Rept. 76 (1972) Reference 72-7, 103 p., Baltimore, Md. 21218
17. Bachman, R. W., Zn-65 in studies of the water zinc cycle. In Radioecology. V. Schultz and A. W. Klement (eds.), Proceedings of the First National Symposium on Radioecology, Colorado State University (1961), pp. 485-496.
18. O'Connor, J. T., Civil Engineering Studies, Sanitary Eng. Ser. No. 49 (1968), Dept. of Civil Eng., Univ. of Ill., Urbana.
19. Troup, B.N, O.P. Bricker, J.T. Bray, Nature (1974) 249: 237-39.
20. Holdren, G. R., Ph.D. Dissertation. Johns Hopkins University, Baltimore, Md. 21218 (in preparation).
21. Bricker, O. P. and B. N. Troup, "Sediment-Water Exchange in Chesapeake Bay" In Proceedings of the Second International Estuarine Conference, J. H. Carpenter (ed.) (in press).
22. Neiheisel, J., "Significance of clay minerals in shoaling problems". Comm. Tidal Hydraulics Tech. Bull. No. 10, Corps of Engineers, U. S. Army, Vicksburg, Miss. (1966).

# 9

# A Field Study of Chemical Budgets for a Small Tidal Creek–Charleston Harbor, S.C.

J. L. SETTLEMYRE and L. R. GARDNER

Department of Geology and Belle W. Baruch Coastal Research Institute, University of South Carolina, Columbia, S.C. 29208

The intertidal marsh is perhaps one of the most reactive geochemical environments on the surface of the earth. This is due to the combination of intense biological activity, sulfate reduction and $H_2S$ production and the presence of reactive fine grained sediment. Marshes are known to be extremely productive environments and to serve as nurseries for many marien organisms. As many chemical pollutants tend to become concentrated in the marine biosphere, it is important to understand the mechanisms by which, and the extent to which various substances accumulate and/or degrade in marshes. Also as a result of their areal extent the accumulation or export of various substances by marshes may affect the quality and productivity of nearby estuaries and coastal waters. Despite these considerations, little attention has been devoted to the interaction of marshlands and estuarine waters with respect to chemical quality, or to exploring the possible uptake or degradation of pollutants by marshes. The present study, therefore, was undertaken to identify the major water quality changes resulting from the interaction of estuarine water with a tidal marsh.

The area chosen for this study is the drainage basin of Dill's Creek, a moderate size tidal creek (Figure 1) on the south side of Charleston Harbor, S.C. Dill's Creek was chosen because it has a relatively well defined drainage basin which is isolated from inputs of fresh water and because the adjacent harbor waters are moderately polluted. Chemical and fertilizer industries on the Ashley and Cooper Rivers supply heavy metals and organophosphates to the harbor. Nearby sewerage systems on the Ashley River included the Charleston Municipal Yacht Basin and the outfall of the City of Charleston's Pine Island sewerage treatment plant (Figure 1). This situation thus offered the opportunity of detecting interactions between the marsh and the polluted waters of Charleston Harbor.

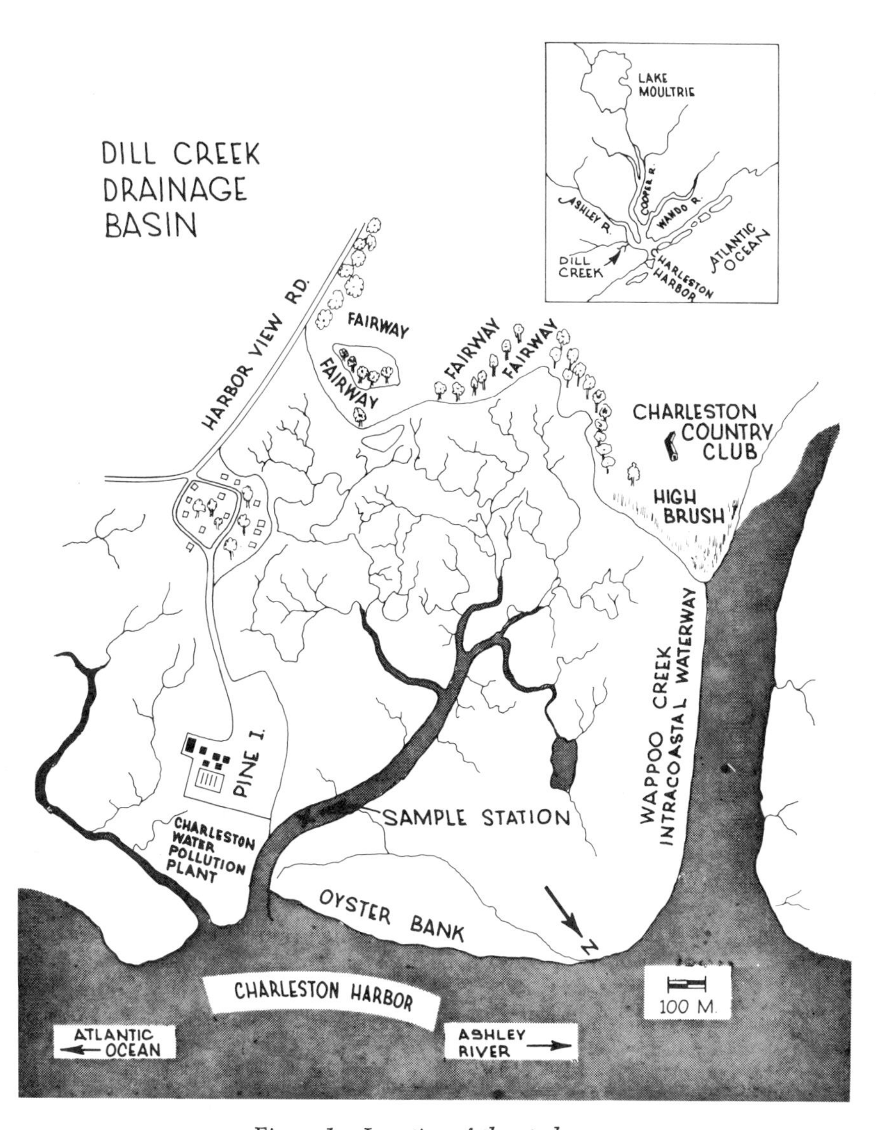


*Figure 1. Location of the study area*

Methods Of Study. This study is based on the comparison of the chemical quality of discharge weighted composite samples of flood versus ebb flow for twenty-five different tidal cycles monitored from August 1973 to August 1974. With each cycle, sampling began at slack low water through high tide to the following low tide. Thus, for each cycle the composite ebb sample was sampled from the same mass of seawater as the preceding composite flood sample. Because of logistics, the samples could not be discharge weighted by conventional current meter surveys. Instead we estimated flow into and out of the marsh on the basis of a hypsometric model and tide gage measurements in a manner similar to that outlined by Ragotskie and Bryson (1). The procedure employs aerial photographs to estimate the areas of the marsh and the channel network at bank full stage and at low water. It is assumed that the volume of water that enters (or leaves) between low water and bank full stage is equal to the difference in stage times the average of the bank full and low water channel areas. Between low water and bank full stage it is further assumed that the volume increments as a linear function of stage. The same procedure and assumptions are employed to estimate the volume that accumulates between bank full and marsh full (i.e. the stage at which water reaches the perimeter of the basin). The volume that enters between marsh full and high tide is simply the marsh area times the stage difference. The rating curve obtained by this procedure for Dill's Creek is shown in Figure 2. With such a rating curve it is possible to select a series of tidal stages which mark off equal increments of water volume. Between low tide and bank full stage we sampled so that each 250 ml sample represented 1.25 x $10^7$ liters of water according to the tide gage schedule shown in Figure 2. From bank full to high tide,sample volumes and volume increments were selected as shown in Figure 2.

During each cycle water was continuously pumped from the creek through 100 meters of plastic pipe to the edge of the sewerage treatment plant embankment where electricity was available for light and instrumentation. The sample intake was located in mid-channel (Figure 1) 1.3 meters above the bottom. Thus during each cycle the sample intake ranged from about 0.6 meters below the water surface at low tide to about 2.5 meters below the surface at high tide. The pump effluent was monitored at close intervals for salinity, temperature, and turbidity. A temperature compensated optical refractometer was used for salinity measurements, a calibrated thermistor probe for temperature, and a colorimeter for turbidity.

After collection, three aliquots of each composite sample were filtered through pre-weighed Millipore 0.45 micron filters. The filters were washed with distilled water, dried at 105°C and reweighed in order to determine total suspended sediment (TSS). The organic suspended sediment (OSS) was estimated from the igni-

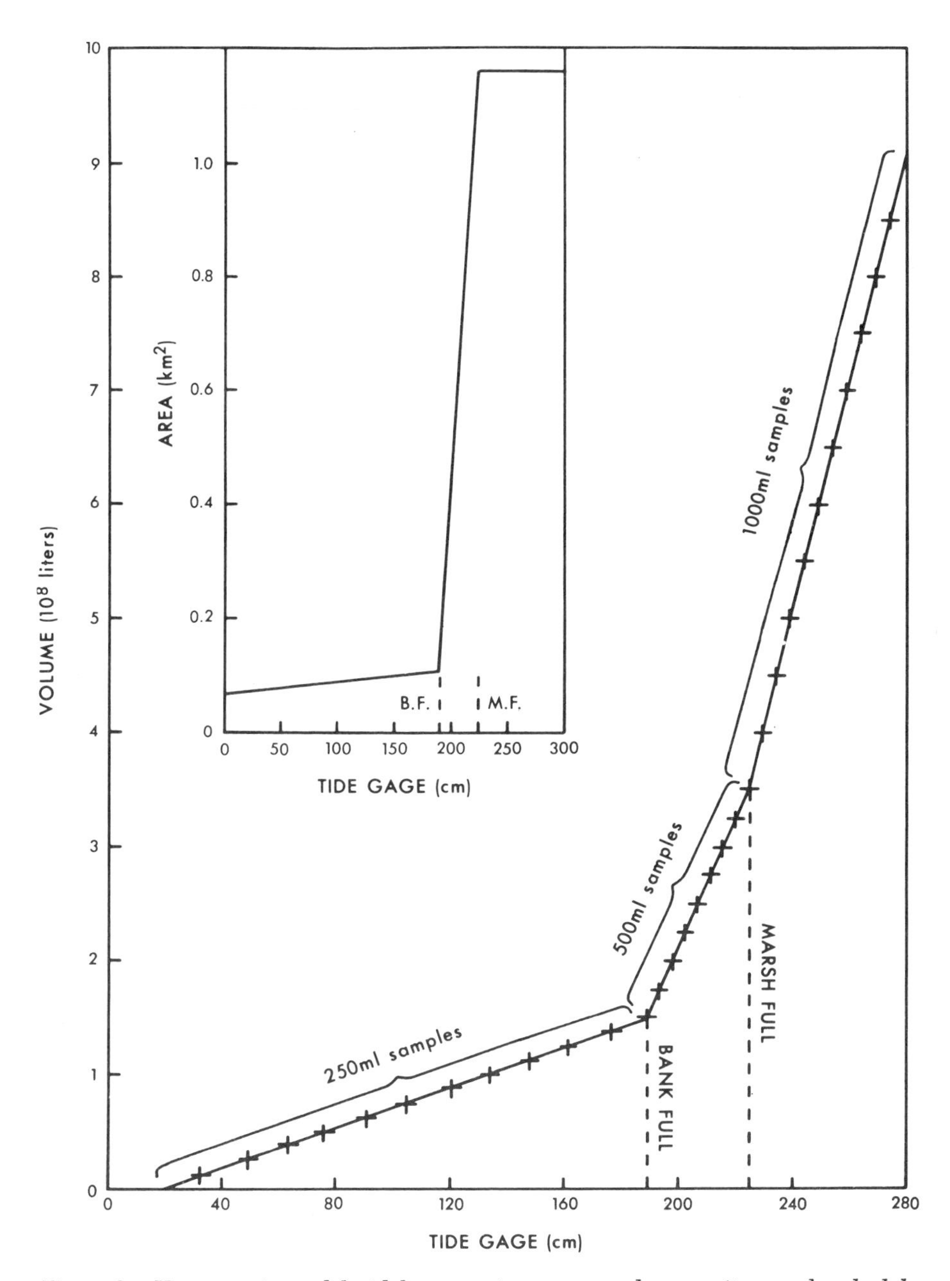


*Figure 2. Hypsometric model, tidal gage rating curve, and composite sample schedule for Dill's Creek*

TABLE I Characteristics of Dill's Creek Composite Samples

| DATE | TEMP °C | SAL ppt | VOL $10^8$ l | TSS ppm | ISS ppm | COLI $10^3$/l | $SiO_2$ ppm | $PO_4$ ppm | $BOD_7$ ppm | SURF ppm | Fe ppb | Cu ppb | Zn ppb | Pb ppb |
|---|---|---|---|---|---|---|---|---|---|---|---|---|---|---|
| 08-03-73F | 22.5 | 13.2 | 3.5 | 15.8 | 8.4 | -- | 4.80 | 0.64 | -- | 0.56 | 4.0 | 5.0 | 35.0 | 2.0 |
| E | 25.4 | 13.3 | | 5.4 | 2.6 | -- | 4.90 | 0.77 | -- | 0.27 | 10.0 | 4.3 | 30.0 | 1.0 |
| 08-06-73F | 26.3 | 13.9 | 3.7 | 75.0 | 42.0 | -- | 5.00 | 0.32 | -- | 0.62 | -- | -- | -- | -- |
| E | 25.1 | 12.0 | | 116.0 | 65.0 | -- | 5.65 | 0.57 | -- | 0.45 | -- | -- | -- | -- |
| 08-08-73F | 25.8 | 14.4 | 4.7 | 103.0 | 67.5 | 33 | 4.50 | 0.28 | -- | 0.53 | 0.5 | 2.1 | 4.0 | 1.0 |
| E | 25.4 | 14.0 | | 133.0 | 74.5 | 9 | 5.20 | 0.50 | -- | 0.24 | 0.5 | 0.1 | 5.0 | 1.0 |
| 08-20-73F | 27.7 | 16.5 | 5.0 | 14.2 | 9.0 | 22 | 4.45 | 0.41 | 4.0 | 0.55 | -- | -- | -- | -- |
| E | 29.6 | 17.0 | | 21.0 | 13.0 | 7 | 4.95 | 0.62 | 1.7 | 0.45 | -- | -- | -- | -- |
| 08-22-73F | 27.5 | 16.5 | 6.9 | 20.5 | 17.1 | 32 | 4.40 | 0.54 | 1.8 | 0.68 | -- | -- | -- | -- |
| E | 27.3 | 16.5 | | 37.0 | 30.3 | 7 | 4.90 | 0.84 | 2.0 | 0.55 | -- | -- | -- | -- |
| 09-30-73F | 27.0 | 14.7 | 6.4 | 1.0 | 0.6 | 26 | 5.00 | 0.26 | -- | 0.40 | 0.5 | 11.0 | 50.0 | 13.0 |
| E | 28.2 | 14.8 | | 1.8 | 0.9 | 7 | 5.00 | 0.37 | -- | 0.31 | 2.0 | 7.0 | 27.0 | 11.0 |
| 11-02-73F | 19.0 | 22.5 | 3.4 | 2.4 | 1.1 | 10 | 2.97 | 0.20 | -- | 0.74 | 0.5 | 14.0 | 20.0 | 15.0 |
| E | 20.0 | 23.0 | | 2.5 | 1.2 | 8 | 3.00 | 0.28 | -- | 0.71 | 4.0 | 14.0 | 48.0 | 15.0 |
| 11-16-73F | 17.1 | 22.0 | 3.5 | 28.0 | 20.0 | 3 | 3.40 | 0.16 | -- | 0.60 | -- | -- | -- | -- |
| E | 18.1 | 21.5 | | 20.5 | 10.1 | 1 | 3.05 | 0.17 | -- | 0.67 | -- | -- | -- | -- |
| 12-19-73F | 11.0 | 10.4 | 3.0 | 4.2 | 3.7 | 27 | 4.75 | 0.14 | 0.8 | 0.49 | 2.0 | 3.0 | 45.0 | 13.0 |
| E | 10.8 | 10.2 | | 5.2 | 4.2 | 10 | 4.00 | 0.24 | 0.6 | 0.42 | 44.0 | 4.0 | 15.0 | 10.0 |
| 01-08-74F | 14.5 | 12.5 | 5.4 | 80.7 | 70.0 | 33 | 3.65 | 0.31 | 2.6 | 0.26 | -- | -- | -- | -- |
| E | 13.7 | 13.5 | | 36.2 | 30.4 | 32 | 3.40 | 0.26 | 2.2 | 0.30 | -- | -- | -- | -- |
| 01-18-74F | 14.9 | 12.2 | 1.8 | 11.9 | 8.2 | 57 | 4.45 | 0.17 | 1.7 | 0.70 | -- | -- | -- | -- |
| E | 14.8 | 12.0 | | 10.2 | 7.9 | 63 | 4.35 | 0.24 | 1.8 | 0.56 | -- | -- | -- | -- |
| 01-31-74F | 16.3 | 12.4 | 2.9 | 14.3 | 10.1 | 2 | 4.10 | 0.15 | 2.6 | 0.42 | 4.0 | 8.0 | 45.0 | 0.5 |
| E | 16.0 | 11.7 | | 12.7 | 10.7 | 3 | 4.10 | 0.23 | 2.0 | 0.44 | 6.0 | 6.0 | 42.0 | 10.0 |
| 02-01-74F | 17.0 | 13.8 | 2.1 | 12.4 | 8.7 | 14 | 4.40 | 0.13 | 2.3 | 0.40 | 4.0 | 8.0 | 67.0 | 11.3 |
| E | 17.3 | 13.4 | | 15.2 | 11.9 | 1 | 3.75 | 0.14 | 1.8 | 0.48 | 4.0 | 8.0 | 53.0 | 3.0 |
| 02-21-74F | 15.0 | 10.9 | 3.5 | 21.1 | 17.1 | 45 | 4.85 | 0.15 | 1.6 | 0.36 | 2.0 | 4.0 | 31.0 | 2.0 |
| E | 14.8 | 10.6 | | 21.8 | 17.2 | 42 | 4.65 | 0.17 | 1.3 | 0.40 | 4.0 | 6.0 | 42.0 | 1.0 |

TABLE 1 (Cont.)

| DATE | TEMP °C | SAL ppt | VOL $10^8$1 | TSS ppm | ISS ppm | COLI $10^3$/1 | $SiO_2$ ppm | $PO_4$ ppm | $BOD_7$ ppm | SURF ppm | Fe ppb | Cu ppb | Zn ppb | Pb ppb |
|---|---|---|---|---|---|---|---|---|---|---|---|---|---|---|
| 03-17-74F | 16.8 | 13.0 | 1.0 | 18.2 | 12.6 | 14 | 3.35 | 0.33 | 2.7 | 0.36 | -- | -- | -- | -- |
| E | 16.9 | 12.9 | | 23.4 | 17.7 | 6 | 3.15 | 0.34 | 2.7 | 0.35 | -- | -- | -- | -- |
| 03-31-74F | 17.8 | 16.7 | 1.2 | 21.4 | 17.6 | 21 | 2.75 | 0.25 | 1.9 | 0.64 | 7.2 | 14.0 | 51.6 | 12.6 |
| E | 18.4 | 17.0 | | 21.0 | 17.6 | 7 | 2.90 | 0.28 | 2.3 | 0.60 | 4.9 | 6.6 | 46.0 | 10.0 |
| 05-04-74F | 23.7 | 16.0 | 7.1 | 77.4 | 46.0 | 22 | 2.75 | 0.58 | 3.7 | 0.54 | 4.0 | 1.2 | 33.3 | 0.7 |
| E | 22.7 | 15.5 | | 63.0 | 36.0 | 1 | 2.80 | 0.57 | 3.3 | 0.47 | 4.9 | 4.9 | 42.2 | 1.6 |
| 05-05-74F | 21.4 | 15.5 | 4.3 | 40.0 | 28.0 | 42 | 2.75 | 0.33 | 2.9 | 0.50 | 4.1 | 7.2 | 43.7 | 3.2 |
| E | 21.8 | 16.0 | | 41.0 | 18.0 | 16 | 2.85 | 0.42 | 2.9 | 0.45 | 4.2 | 9.0 | 66.6 | 4.0 |
| 05-29-74F | 25.0 | 16.6 | 3.8 | 23.8 | 19.0 | 2 | 3.10 | 0.15 | 3.0 | 0.49 | 1.0 | 7.0 | 54.0 | 6.7 |
| E | 25.7 | 16.2 | | 37.4 | 30.0 | 3 | 3.10 | 0.15 | 2.9 | 0.47 | 37.0 | 3.8 | 45.0 | 5.1 |
| 06-10-74F | 26.3 | 17.7 | 2.1 | 20.6 | 16.0 | 21 | 2.35 | 0.16 | 2.6 | 0.48 | 5.1 | 2.6 | 103.0 | 4.2 |
| E | 27.5 | 17.3 | | 28.3 | 13.0 | 17 | 2.85 | 0.25 | 3.4 | 0.45 | 1.0 | 0.2 | 46.7 | 2.6 |
| 06-17-74F | 28.8 | 21.0 | 5.0 | 41.0 | 30.0 | 8 | 1.85 | 0.25 | 3.6 | 0.24 | 2.8 | 3.7 | 65.0 | 17.0 |
| E | 27.3 | 19.5 | | 59.0 | 39.0 | 13 | 2.65 | 0.28 | 2.9 | 0.15 | 5.6 | 7.6 | 67.0 | 5.0 |
| 06-19-74F | 28.3 | 19.5 | 7.8 | 66.0 | 54.0 | 1 | 1.60 | 0.18 | 3.5 | 0.17 | 3.1 | 3.9 | 36.0 | 3.5 |
| E | 27.1 | 21.0 | | 73.0 | 61.0 | 23 | 2.25 | 0.23 | 3.0 | 0.16 | 5.1 | 6.6 | 83.0 | 2.0 |
| 07-09-74F | 27.2 | 17.5 | 4.7 | 37.0 | 32.0 | 13 | 2.80 | 0.18 | 1.7 | 0.27 | 9.2 | 8.5 | 58.0 | 6.6 |
| E | 28.5 | 17.2 | | 31.0 | 23.4 | 100 | 2.85 | 0.22 | 1.8 | 0.13 | 2.5 | 5.9 | 53.0 | 6.9 |
| 07-24-74F | 28.0 | 11.2 | 5.3 | 35.0 | 30.0 | 14 | 3.00 | 0.16 | 2.0 | 0.27 | 4.3 | 9.0 | 55.7 | 3.8 |
| E | 29.5 | 11.0 | | 44.0 | 34.0 | 6 | 3.10 | 0.16 | 2.4 | 0.21 | 3.2 | 9.6 | 108.9 | 5.2 |
| 07-25-74F | 28.0 | 10.2 | 3.6 | 29.0 | 24.0 | 0 | 3.00 | 0.15 | 1.6 | 0.27 | 3.9 | 8.7 | 58.0 | 1.9 |
| E | 28.6 | 10.0 | | 36.0 | 26.0 | 1 | 3.00 | 0.16 | 2.0 | 0.25 | 3.0 | 9.8 | 94.4 | 4.1 |

TSS = total suspended sediment
ISS = inorganic suspended sediment
OSS = organic suspended sediment = TSS - ISS
COLI= coliform bacteria

tion loss at 550°C (2).

A separate aliquot of each composite sample was filtered through Millipore 0.45 micron filters and a portion of each filtrate was sent to Orlando Laboratories, Inc., Orlando, Florida for atomic absorption analysis of dissolved Fe, Cu, Zn and Pb on a contract basis. This laboratory uses the APDC-MIBK extraction procedure (3) and employs high intensity replaceable lamp filaments for high sensitivity (4). The rest of the filtrate was used for analysis of dissolved silica, phosphate and surfactants. For these measurements Hach Chemical Company reagents and a calibrated Delta Scientific colorimeter were used. Silica was determined by amino acid reduction-molybdate blue formation, phosphate by stannous chloride reduction-molybdate blue formation, and surfactants by melthylene blue-chloroform extraction (2).

For some cycles BOD was determined on unfiltered samples by the Pine Island sewerage treatment plant laboratory on a fee basis. Millipore Coli Count water testers were used to obtain estimates of coliform bacteria concentrations in unfiltered samples.

Results. Physical data along with the concentrations of various constituents for ebb and flood composite samples for each cycle are shown in Table I. Budgets for various parameters for each cycle were calculated by multiplying the concentration difference between flood and ebb composite samples times the volume of the tidal prism and dividing by the area of the basin (1.08 $km^2$). The results of these calculations are shown in Table II.

To evaluate the significance of the derived budgets and to identify factors which may influence water quality parameters of the system the data were analyzed by a variety of statistical techniques. The paired "t" test (5) was employed in order to ascertain whether the mean deviations between flood and ebb concentrations of various water quality parameters are significantly different from zero (Table III). The significance of the number of cycles for which flood concentration exceeded ebb was tested by means of binomial probability theory, the null hypothesis being that the chance of flood concentration exceeding ebb on a particular cycle is 0.50 (Table IV). Because water temperatures change dramatically between September and November and between March and May the budget results have been classified into those of the winter season (November-March) versus those of summer (May-September). The "t" test (6) was used to ascertain the significance of differences between the mean budgets for the two seasons. The results of the analysis of seasonal effects are shown in Table V along with the simple and seasonally adjusted mean yearly budgets. Seasonal adjustment is necessary because the number of summer cycles, 15, substantially exceeds the number of winter cycles, 10. Finally, an attempt was made to ascertain whether day versus night conditions have an impact on any of the budgets. Cycles were classified as day or night cycles according to the time of high tide. If high tide occurred at night or within one hour of sunset or

Table II. Budgets for Dill's Creek Water Quality Parameters

| DATE | SALT $mg/m^2$ | TSS $g/m^2$ | ISS $g/m^2$ | COLI $10^6/m^2$ | $SiO_2$ $mg/m^2$ | $PO_4$ $mg/m^2$ | SURF $mg/m^2$ | $BOD_7$ $mg/m^2$ | Fe $mg/m^2$ | Cu $mg/m^2$ | Zn $mg/m^2$ | Pb $mg/m^2$ |
|---|---|---|---|---|---|---|---|---|---|---|---|---|
| D08-03-73 | -33 | 3.37 | 1.88 | -- | -32 | -42 | 93 | -- | -1.90 | 0.23 | 1.62 | 0.32 |
| D08-06-73 | 650 | -14.04 | -7.88 | -- | -220 | -85 | 57 | -- | -- | -- | -- | -- |
| D08-08-73 | 180 | -13.03 | -3.05 | 10.4 | -304 | -130 | 126 | -- | 0.00 | 0.87 | -0.44 | 0.00 |
| D08-20-73 | -230 | -3.17 | -1.85 | 6.9 | -232 | -97 | 46 | 1250 | -- | -- | -- | -- |
| D08-22-73 | 0 | -10.55 | -8.43 | 16.0 | -319 | -192 | 83 | -128 | -- | -- | -- | -- |
| D09-30-73 | -55 | -0.44 | -0.18 | 11.2 | 0 | -65 | 54 | -- | -0.88 | 2.37 | 13.20 | 1.18 |
| D11-02-73 | -160 | 0.00 | 0.00 | 0.6 | -10 | -25 | 9 | -- | -1.10 | 0.00 | -8.85 | 0.00 |
| D11-16-73 | 160 | 2.42 | 3.20 | 0.7 | 113 | -4 | -22 | -- | -- | -- | -- | -- |
| N12-19-73 | 55 | -0.28 | -0.14 | 4.7 | 208 | -28 | 19 | 55 | -11.65 | -0.28 | 8.32 | 0.83 |
| N01-08-74 | -500 | 22.20 | 19.80 | 0.5 | 125 | 25 | -20 | 200 | -- | -- | -- | -- |
| N01-18-74 | 38 | 0.28 | 0.05 | -1.0 | 17 | -12 | 23 | -17 | -- | -- | -- | -- |
| N01-31-74 | 185 | 0.44 | -0.16 | -0.3 | 0 | -21 | -6 | 161 | -0.54 | 0.54 | 0.81 | -2.54 |
| D02-01-74 | 74 | -0.55 | -0.70 | 2.5 | 126 | -2 | -16 | 97 | 0.00 | 0.00 | 2.72 | 1.55 |
| N02-21-74 | 93 | -0.24 | -0.04 | 1.0 | 65 | -6 | -13 | 97 | -0.65 | -0.65 | -3.56 | 0.32 |
| D03-17-74 | 9 | -0.48 | -0.47 | 0.7 | 18 | -1 | 1 | 0 | -- | -- | -- | -- |
| D03-31-74 | -37 | 0.05 | 0.00 | 1.5 | -17 | -3 | 7 | -44 | 0.26 | 0.82 | 0.62 | 0.29 |
| N05-04-74 | 324 | 9.45 | 6.57 | 13.4 | -32 | 6 | 45 | 263 | -0.59 | -2.42 | -5.85 | -0.59 |
| N05-05-74 | -195 | -0.40 | 3.98 | 10.3 | -40 | -36 | 20 | 0 | -0.04 | -0.65 | -9.15 | -0.31 |
| D05-29-74 | 139 | -4.77 | -3.85 | -0.4 | 0 | 0 | 7 | 35 | -13.00 | 1.16 | 3.26 | 0.56 |
| D06-10-74 | 74 | -1.50 | 0.58 | 0.8 | -97 | -18 | 6 | -150 | 0.80 | 0.46 | 10.95 | 0.31 |
| D06-17-74 | 695 | -8.33 | -4.16 | -2.3 | -370 | -14 | 42 | 342 | -1.30 | -1.34 | -0.92 | 5.56 |
| N06-19-74 | -1080 | -5.05 | -5.05 | -15.9 | -469 | -36 | 7 | 324 | -1.44 | -1.95 | -33.90 | 1.08 |
| D07-09-74 | 120 | 2.61 | 3.74 | -38.0 | -21 | -17 | 61 | -20 | 2.96 | 1.13 | 2.18 | -0.13 |
| D07-24-74 | 93 | -4.40 | -1.96 | 3.9 | -49 | 0 | 30 | -185 | 0.54 | -0.30 | -26.00 | -0.69 |
| D07-25-74 | 65 | -2.33 | -0.67 | -0.3 | 0 | -4 | 6 | -124 | 0.30 | -0.38 | -12.00 | -0.73 |

D= day cycle. N= night cycle OSS= TSS - ISS

Table III. Paired "t" test results on concentration deviations.

| | Sal | TSS | ISS | OSS | COLI | $SiO_2$ | $PO_4$ |
|---|---|---|---|---|---|---|---|
| $\overline{D}$ | 0.15 | -1.29 | -0.12 | -3.17 | 3.48 | -0.09 | -0.07 |
| t | 1.07 | 0.36 | 0.05 | 2.35 | 0.73 | 1.19 | 4.27 |

| | BOD | SURF | Fe | Cu | Zn | Pb |
|---|---|---|---|---|---|---|
| $\overline{D}$ | 0.19 | 0.06 | -4.65 | 0.42 | -3.28 | 1.08 |
| t | 1.27 | 3.40 | 1.53 | 0.61 | 0.51 | 1.06 |

$\overline{D}$ = mean deviation. Units for deviation are same as in Tables I & II. t(0.05) = 2.07 (23 d.f) = 2.13 (15d.f.).

Table IV. Binomial probabilities.

| | Sal | TSS | ISS | OSS | COLI | $SiO_2$ | $PO_4$ |
|---|---|---|---|---|---|---|---|
| F<E | 8 | 16 | 15 | 16 | 7 | 14 | 21 |
| F>E | 16 | 8 | 8 | 7 | 16 | 7 | 2 |
| p | 0.08 | 0.08 | 0.11 | 0.05 | 0.05 | 0.09 | 0.01 |

| | BOD | SURF | Fe | Cu | Zn | Pb |
|---|---|---|---|---|---|---|
| F<E | 6 | 5 | 11 | 8 | 9 | 6 |
| F>E | 10 | 20 | 5 | 8 | 9 | 10 |
| p | 0.23 | 0.01 | 0.11 | 0.60 | 0.60 | 0.23 |

Table V. Students' "t" results for winter versus summer budgets.

| Parameter | A | B | C | D | E | F |
|---|---|---|---|---|---|---|
| Salinity | 26.5 | -8.3 | 50.0 | 0.42 | 17.0 | 11.9 |
| TSS | -1.16 | 2.38 | -3.50 | 2.20 | -0.56 | -393 |
| ISS | 0.05 | 2.15 | -1.35 | 1.71 | 0.40 | 280 |
| OSS | -1.21 | 0.23 | -2.15 | 2.23 | -0.96 | -673 |
| Coliform | 1.20 | 1.09 | 1.07 | 0.03 | 1.08 | 760 |
| $SiO_2$ | -62.0 | 64.5 | -146 | 3.89 | -41 | -29.0 |
| $PO_4$ | -32 | -8 | -49 | 2.23 | -28 | -19.6 |
| $BOD_7$ | 124 | 94 | 146 | 0.52 | 120 | 84.5 |
| SURF | 27 | -2 | 46 | 3.94 | 22 | 15.4 |
| Fe | -1.57 | -2.28 | -1.21 | 0.51 | -1.75 | -1.23 |
| Cu | -0.02 | 0.07 | -0.07 | 0.23 | 0.00 | 0.00 |
| Zn | -3.18 | 0.01 | -4.78 | 0.80 | -2.39 | -1.69 |
| Pb | 0.39 | 0.07 | 0.55 | 0.59 | 0.31 | 0.22 |

A -- simple average, units as given in Table II.
B -- winter average, units as given in Table II.
C -- summer average, units as given in Table II.
D -- "t" statistic for difference of winter versus summer mean
E -- seasonally adjusted average, units as given in Table II.
F -- yearly budget based on E in $gm^{-2}$ $yr^{-1}$ or $10^6$ cell $m^{-2}$ $yr^{-1}$.

sunrise the cycle was considered to be a night or subdued light cycle; otherwise, it was considered a day or full light cycle. A "t" test was used to ascertain the significance of differences between the mean budgets for night versus day (Table VI). Also the day-night effect was analyzed non-parametrically by means of 2 x 2 hypergeometric contingency tables (7). The hypergeometric probabilities of the observed contingency tables are shown in Table VII.

Table VI. Student's "t" test results on day versus night budgets.

| | Sal | TSS | ISS | OSS | COLI | $SiO_2$ | $PO_4$ |
|---|---|---|---|---|---|---|---|
| t | 1.72 | 2.34 | 2.81 | 1.82 | 0.17 | 0.94 | 1.35 |

| | BOD | SURF | Fe | Cu | Zn | Pb |
|---|---|---|---|---|---|---|
| t | 0.25 | 1.64 | 0.51 | 2.65 | 1.02 | 1.13 |

t(0.05) = 2.07 (23 d.f.) = 2.13 (15 d.f.)

Table VII. Hypergeometric 2 x 2 contingency probabilities (7) for day versus night effect on budgets.

| | Sal | TSS | ISS | OSS | COLI | $SiO_2$ | $PO_4$ |
|---|---|---|---|---|---|---|---|
| p | 0.17 | 0.16 | 0.30 | 0.17 | 0.31 | 0.11 | 0.09 |

| | BOD | SURF | Fe | Cu | Zn | Pb |
|---|---|---|---|---|---|---|
| p | 0.20 | 0.14 | 0.08 | 0.06 | 0.24 | 0.32 |

Discussion. In this section we will attempt to analyze the results presented in Tables I-VII in the light of presently understood physical, chemical and biological processes relevant to the salt marsh environment. This then will be an attempt to evaluate the compatibility of our results with previous related research and to suggest mechanisms which may govern the import and/or export of substances from the marsh. In addition, we wish also to assess the suitability and limitations of the method of investigation employed in this study.

Salinity. The primary reason for measuring salinity is that it provides a check on the method of composite sample collection. Salinity in Dill's Creek commonly varies by five to six parts per thousand between low and high tide. On rainless days unbiased composite samples of flood and ebb flow should show identical salinities. Consistently higher composite flood salinities would suggest that the sampling method results in sampling more saline waters on flood flow whereas the opposite result would indicate over sampling of fresher water on flood flow. The paired "t" test (Table III) indicates that the mean deviation between flood and ebb salinities is not significant. Binomial probability theory, however, indicates that the probability of flood salinity exceeding ebb on 16 out of 24 cycles is only 0.08 (Table IV). It should be noted, however, that on four cycles (8-3, 8-6, 8-22, 12-19) rainfall occurred during ebb flow and this might explain the lower ebb salinities observed on these days. On a seasonal basis salinity budgets are negative in winter and positive in summer but the two seasons are not significantly different (Table V). No day-night effects are present (Tables VI and VII). These considerations, then, together with the fact that there are no obvious correlations with the budgets of other constituents, suggest that the failure to obtain balanced salinities on most samples is due to random errors in sampling and measurement rather than to systematic errors associated with the sampling procedure. Thus, we have confidence that persistent patterns shown by certain constituents (most notably phosphate) are not artifacts of the sampling procedure but are the results of marsh processes.

Suspended Sediment. In general the average total suspended sediment (TSS) concentration in Dill's Creek is about 31 ppm. Of this, the volatile fraction averages about 29 percent. These figures compare closely with those of the Federal Water Pollution Control Administration (8) which reported an average of 32 ppm and 28 percent volatile matter for a surface water station in the Ashley River near the mouth of Dill's Creek. The paired "t" test (Table 3) indicates that the mean deviations for total and inorganic suspended sediment (ISS) are not significant whereas the deviation for organic suspended sediment (OSS) is significant. Binomial probabilities (Table IV) suggest that cycles showing export of suspended sediment tend to occur more frequently than cycles showing import. However, it should be noted that the average budgets for winter versus summer cycles are significantly different at the 0.05 level for TSS and OSS and at the 0.10 level for ISS (Table V). Thus export of suspended sediment probably occurs preferentially in summer whereas in winter there is import (TSS and ISS) or balance (OSS). On a seasonally adjusted basis the yearly budget for TSS, ISS and OSS are -393 g $m^{-2}$ $yr^{-1}$, +280 g $m^{-2}$ $yr^{-1}$ and -673 g $m^{-2}$ $yr^{-1}$ respectively.

The significance of these budgets will be discussed in detail in a future report. For the sake of brevity we will simply note that the theoretical budget for ISS required for vertical marsh growth to keep pace with rising sea level, 0.15 cm $yr^{-1}$ (9,10), can be shown to be about 570 gm $m^{-2}$ $yr^{-1}$, which is about twice the observed seasonally adjusted budget. With regard to the budget for OSS it is reasonable to assume that the maximum export of OSS should not exceed the yearly above ground production of *Spartina* which for this region probably falls between 660 and 970 g $m^{-2}$ $yr^{-1}$ (11,12). By comparison our value for OSS, -670 g $m^{-2}$ $yr^{-1}$, may be somewhat excessive since allowance must be made for the fact that a portion of the annual *Spartina* production is either incorporated into the sediment or solubilized during microbial decomposition. These figures are set forth here because they have a bearing on the budgets of some of the substances discussed below. As for seasonal effects it is of interest to note that Boon (13) in a study of a small tidal creek in Virginia also found major export of TSS in summer with minor import in winter.

Silica and Phosphate. The budgets for dissolved silica and phosphate indicate that the Dill's Creek marsh is releasing these substances. Since both are involved in the growth of phytoplankton their export by marshes could play a role in the ecology of coastal waters. Release of phosphate occurred on twenty-one out of twenty-five cycles. Both the binomial and paired "t" test results indicate that the export budget for phosphate is statistically significant at the 0.05 level (Tables III and IV). On a seasonally adjusted yearly basis the release of phosphate amounts to about 19.6 gm $m^{-2}$ (Table V). Release of silica occurred on fourteen cycles, uptake on seven cycles and no difference on four cycles. It should be noted, however, that all of the uptake cycles occurred during the winter months (November through March) and that persistent release occurred during summer. Thus there is a statistically significant seasonal pattern to silica uptake and release (Table V). On a seasonally adjusted yearly basis the net silica release averages about 29.0 g $m^{-2}$.

The rate of silica release based on the data of this study compares favorably with an earlier study of silica export associated with low tide runoff (14). In that study the discharge and silica concentration of a number of small tidal creeks near North Inlet, S.C. and also in the Dill's Creek basin were monitored during low tide exposure. The results indicate that dissolved silica in the interstitial water system diffuses from the sediment into a thin layer of water left on the marsh surface by the receding tide. Drainage of this water during low tide exposure results in the export of dissolved silica at a rate of about 1.9 ug $m^{-2}$ $sec^{-1}$. Discharge weighted composite sampling of a small creek near North Inlet and of a small creek in the Dill's Creek basin over seven separate cycles gave an average export rate of 2.3 ug $m^{-2}$ $sec^{-1}$ as compared with an export rate of 1.0 ug $m^{-2}$ $sec^{-1}$ for the Dill's Creek basin. The general concurrence of these three different scale studies indicates that diffusion of silica from marsh sediment is a continuous process (at least during summer), that the average yearly flux of silica from the sediment is on the order of 50 g $m^{-2}$ and that the method of composite sample collection used in Dill's Creek yields acceptable estimates for budget studies of certain dissolved substances.

The export rate for phosphate can be evaluated by similar comparisons. For low tide runoff with an average discharge of 62.2 liters $sec^{-1}$ $km^{-2}$ and an equivalent phosphate concentration of 0.7 ppm (14) the export rate is 0.044 ug $m^{-2}$ $sec^{-1}$. This rate, however, is based mainly on small creeks in the North Inlet area. One small creek in the Dill's Creek basin gave a low tide export rate of 0.31 ug $m^{-2}$ $sec^{-1}$. Discharge weighted composite sampling of the same creek over three different cycles gave an average export rate of 0.47 ug $m^{-2}$ $sec^{-1}$ as compared with an average rate of 0.62 ug $m^{-2}$ $sec^{-1}$ for the entire Dill's Creek basin. Thus there is better correspondence between the results for Dill's Creek at large and one of its small creeks than between Dill's Creek at large and the low tide runoff results for the small creeks in the North Inlet area. The discrepancy between the phosphate export rates for the low tide runoff from the pristine marshes of the North Inlet area as compared with the Dill's Creek basin may be partly due to the fact that the low tide studies do not measure the effect of phosphate pumping by Spartina from the interstitial water reservoir (15). From Reimold's data (15) it appears that the pumping effect could account at most for 0.13 ug $PO_4$ $m^{-2}$ $sec^{-1}$ or twenty percent of the observed Dill's Creek rate. Nevertheless the difference between the two areas can not be entirely due to the pumping effect because the low tide export rate in the Dill's Creek basin is almost an order of magnitude larger than the low tide export rate for the North Inlet area and in fact compares more favorably with the Dill's Creek rate based on composite sampling than with the North Inlet low tide export rate. This may indicate that there is a greater amount of decomposing organic detritus in the Dill's Creek marsh as compared to the North Inlet area because of greater primary production and/or because of the accumulation and decomposition of sewage.

The minimum Spartina production that would have to be solubilized in order to maintain the Dill's Creek rate of phosphate export can be estimated as follows. Assuming that the phosphorous content of dry Spartina is 0.1 percent (12) then the decomposition of 100 gm of Spartina should yield about 0.1 gram of P or 0.3 grams of $PO_4$. Thus to maintain a yearly export rate of 19.6 g $PO_4$ $m^{-2}$ would require yearly production and subsequent complete decomposition of about 6500 g Spartina per square meter. This figure is well beyond the range of Spartina production estimates even if the production of roots and rhizomes is taken into account. Because only a portion of the Spartina is solubilized the total production would have to be substantially greater than 6500 grams. Thus the phosphate export rate may be due chiefly to decomposition of sewerage and/or to dissolution of phosphatic sediment derived from fertilizer plants and/or to solubilization of ferric phosphate in the anoxic sediment followed by diffusion into the surface water.

As indicated in an earlier report (14) the magnitude of the export rates for $SiO_2$ and $PO_4$ indicates that the total amounts of these substances supplied to coastal waters by South Carolina marshes may equal or exceed the total amounts supplied by the combined runoff from the Pee Dee, Santee, and Savannah River basins. Marshes, therefore, may play an important role in determining the concentrations of these and perhaps other substances in nearby estuarine waters.

Surfactants. Uptake of surfactants occurred on 20 out of 25 cycles. All of the cycles that showed release occurred during the winter season (Table II). Thus, as in the case of silica, and to some extent with phosphate, there appears to be a seasonal pattern to the surfactant budget. Strong uptake (degradation) occurs in summer with balance or slight release in winter. On the average, yearly uptake appears to be about 15.4 g $m^{-2}$ (Table V).

This uptake rate does not necessarily mean that the marsh actually sequesters surfactants by adsorption on clays or chemical precipitation. More likely it simply means that surfactants undergo bacterial degradation while the water resides in the marsh. The question then is whether the rate of degradation in the marsh is greater than in the harbor. An approach to an answer to this question is to estimate the lifetime of surfactants in the marsh environment. The average ratio of ebb surfactant concentration to flood concentration for the 28 cycles available (including three cycles for a small creek) is 0.87. Thus there is, on the average, a 13 percent decrease in surfactant concentration during residence of the water in the marsh. On the basis of tide gage-time curves and tide gage-volume curves (Fig. 2) we estimate that the average residence time of water in the marsh is 4.5 hours. If it is assumed that surfactants are degraded in an exponential fashion in accordance with first order reaction kinetics and that there is a 13 percent reduction every 4.5 hours, then the time required to reduce an initial concentration by 95% (i.e. the lifetime) is 4.1 days. Patton (16) states that sodium p-m-decyl benzene sulfonate is biologically decomposed after four days in river water whereas alkyl benzene sulfonates have half-lives of about fifteen days. Thus the marsh degradation rate only represents significant acceleration if the surfactants present are dominantly of the ABS type. This is unlikely because most commercial detergents now contain readily degradable surfactants of the linear alkylate sulfonate type.

Trace Metals. Analyses of Fe, Cu, Zn and Pb were obtained for 18 cycles. Nickel and cobalt were measured on several cycles but the results were generally below the limit of detection (0.5 ppb) and study of these elements was abandoned. None of the trace metals show significant mean deviations (Table III) nor are their binomial probabilities (Table IV) significant. Of the metals, Fe shows the greatest and most significant mean deviation and the lowest binomial probability. An ordinary "t" test indicates that the mean flood and ebb concentrations for Fe are significantly different (t = 2.05, d.f. = 48). Thus there is some suggestion that Fe may be exported at a rate of about 1.0 g $m^{-2}$ $yr^{-1}$ (Table V). None of the metals show seasonal effects (Table V) but there is an indication (Tables VI and VII) that day-night conditions may affect the budgets of Fe and Cu.

Assessment of these results is difficult because the processes and factors which control the concentrations and migration of trace metals in aquatic environments are poorly understood. Some of the factors and processes that have been discussed in recent literature are: interactions between trace metals and hydrous iron and manganese oxides (17); precipitation of metal sulfides (18); uptake and movement of trace metals through biota (19, 20, 21, 22, 23); cation exchange between water and sediment (24, 25); complexing of trace metals by dissolved organic matter (21, 26, 27, 28, 29, 30, 31, 32, 33, 34); diffusion and advection under potential gradients (18, 35, 36). In the following paragraphs we will attempt to evaluate several processes which might play a role in controlling the budgets for trace metals in the Dill's Creek basin.

Accumulation of trace metals could occur if the incoming tide is supersaturated with respect to a metal solid phase such as ferric hydroxide. Then during residence of the tidal prism precipitation might occur resulting in a positive budget. This possibility can be evaluated by estimating the state of saturation of trace metals in the flood composite sample with respect to possible solid precipitates. In Table VIII we present estimates of the solubility of various metal hydroxides, carbonates and phosphates in water with a pH of 8 and 14 ppt salinity. Metal sulfides are not considered here because sulfide cannot be detected in Dill's Creek surface water and thus metal sulfide precipitation can only occur after diffusion into the sediment (discussed below). Equilibrium free metal ion activities, F, are based on solubility products given by Krauskopf (37) and Nriagu (38, 39, 40, 41) and assumed $CO_3^{-2}$ and $PO_4^{-3}$ maximum activities of $10^{-5}$ and $10^{-10}$ respectively. Equilibrium total activities are based on ion pair formation constants given by Krauskopf (37) and were calculated by means of the equations at the bottom of Table VIII. The equilibrium total activities are compared (Table VIII) with the maximum observed flood trace metal concentrations which represent the most favorable conditions for precipitation and positive budgets. As can be seen Cu and Zn in Dill's Creek water are undersaturated with respect to possible known precipitates. Lead may be supersaturated with respect to $Pb_5(PO_4)_3Cl$ but only on those rare occasions when both lead and phosphate reach unusually high concentrations. Iron is probably supersaturated with respect to ferric hydroxide unless it exists predominantly in the ferrous state or is highly complexed by dissolved organic matter. However, iron appears to have a negative (i.e. export) budget, as indicated above, and thus its tendency to precipitate must be offset by other processes.

As indicated above the precipitation of trace metals from Dill's Creek surface water as metal sulfides probably depends on the extent to which the dissolved metals can diffuse from the water column into the sulfidic sediment during residence of the water in the basin. The rate of diffusion in turn depends primarily

Table VIII. State of saturation of trace metals in flood tide composite sample.

| Solid | pKsp | Ref. | pF | pT | Eqt. |
|---|---|---|---|---|---|
| $Fe(OH)_2$ | 15.1 | (37) | 3.1 | * | * |
| $Fe(OH)_3$ | 39.1 | (37) | 21.1 | 10.8 | 1 |
| $FeCO_3$ | 10.5 | (37) | 5.5 | * | * |
| $Fe_3(PO_4)_2 8H_2O$ | 36.0 | (38) | 5.3 | * | * |
| $FePO_4 \cdot 2H_2O$ | 26.4 | (39) | 16.4 | 6.1 | 1 |
| $Cu(OH)_2$ | 20.5 | (37) | 8.5 | 6.3 | 2 |
| $Cu_2(OH)_2CO_3$ | 33.8 | (37) | 8.4 | 6.2 | 2 |
| $Cu_3(PO_4)_2$ | 36.9 | (37) | 5.4 | * | * |
| $Zn(OH)_2$ | 16.9 | (37) | 4.8 | * | * |
| $ZnCO_3$ | 10.8 | (37) | 5.8 | * | * |
| $Zn_3(PO_4)_2 \cdot 4H_2O$ | 35.3 | (40) | 5.1 | * | * |
| $Pb(OH)_2$ | 15.3 | (37) | 4.8 | * | * |
| $PbCO_3$ | 13.1 | (37) | 8.1 | 5.8 | 3 |
| $Pb\ HPO_4$ | 9.9 | (37) | 4.9 | * | * |
| $Pb_3(PO_4)_2$ | 43.5 | (37) | 7.7 | 5.4 | 3 |
| $Pb_5(PO_4)_3Cl$ | 81.0 | (41) | 10.0 | 7.7 | 3 |

Max. obs. conc. (-log moles /l)

| Fe | Cu | Zn | Pb |
|---|---|---|---|
| 6.8 | 6.7 | 5.8 | 7.1 |

pF= equilibrium free ion activity (-log)
pT= equilibrium total activity (-log)
* No calculation necessary

1- $Fe_T = [Fe^{+3}]\left(1 + 10^{11.7}[OH^-] + 10^{21.1}[OH]^2 + 10^{34.2}[OH]^4 + 10^{1.5}[Cl^-]\right)$

2- $Cu_T = [Cu^{++}]\left(1 + 10^{6.7}[OH^-] + 10^{1}[Cl^-] + 10^{6.8}[CO_3^{-2}] + 10^{10}[CO_3^{-2}]^2\right)$

3- $Pb_T = [Pb^{++}]\left(1 + 10^{6.2}[OH^-] + 10^{10.9}[OH]^2 + 10^{13.9}[OH]^3 + 10^{7.3}[CO_3^{-2}] + 10^{10.6}[CO^{-2}]^2 + 10^{1.6}[Cl^-] + 10^{1.8}[Cl^-]^2\right)$

on the difference in trace metal concentration between the surface water and the sulfidic interstitial water and the depth to the oxidized-reduced boundary (i.e. the gradient). Probably the most outstanding feature of the work that has been done on the occurrence of dissolved trace metals in sulfidic marine waters is the fact that, except for Mn and perhaps Fe, the concentrations observed far exceed those predicted by simple metal sulfide equilibria (18,33,34,42). Reasons suggested for the enhanced solubility of trace metals in sulfidic waters are equilibration with more soluble solid phases, organometallic complexes and bisulfide or polysulfide complexes (26). A summary of average trace metal concentrations found by various workers is presented in Table IX. Our values are for four grab samples analyzed on contract by Orlando Laboratories.

Table IX. Average concentrations of trace metals in sulfidic marine waters (ppb).

| Fe | Cu | Zn | Pb | Reference |
|---|---|---|---|---|
| 6 | 3 | 7 | - | (34) |
| 22 | 6 | 28 | - | (34) |
| 40 | 4 | 30 | - | (32) |
| 186 | 112 | 105 | 35 | (42) |
| 5 | 0.5 | 1 | - | (18) |
| 54 | 4 | 33 | 3.8 | This study |

In the case of Cu and Zn, the concentrations of these metals in sediment pore water and Dill's Creek surface waters are approximately equal if we assume pore water values of about 5 ppb Cu and 30 ppb Zn. Thus strong gradients probably do not exist to drive these elements into or out of the sediment by diffusion alone and this is perhaps the reason for our failure to observe significant uptake or export of these two elements. The concentration of Fe in sulfidic interstitial waters, on the other hand, typically falls in the range from 20-200 ppb as compared to an average 3.6 ppb for Dill's Creek flood composite samples. Furthermore, observations in the Black Sea (18) and in other anoxic water bodies indicate that dissolved iron and manganese both show concentration maxima just below the oxidized-reduced boundary whereas zinc and copper do not. While such concentration maxima have not been clearly detected for sulfidic sediments, many of the pore water profiles described by the workers cited in Table IX show maximum iron and manganese concentrations in the uppermost samples even though the concentrations in the overlying water are much lower. We have measured iron concentrations in several pore water profiles from Baruch Marsh sediments using the newly developed Ferrozine reagent (44) and found that maximum iron concentrations (200-300 ppb) tend to occur in the upper 10cm of sediment. Thus for diffusion of iron from marsh sediments the most favorable and at the same time realistic circumstances that can be envisioned is a maximum iron concentration of about 300 ppb located just below the oxidized-reduced boundary which generally lies 5 to 10 mm below the sediment surface. Location of an iron maximum near this boundary is consistent with the fact that a pH minimum is commonly found here (unpublished work in progress). Assuming then the most favorable configuration (300 ppb Fe at 0.5 cm depth) and assuming a linear concentration gradient to the surface where the concentration is less than 10 ppb, the flux calculated in accordance with Ficks first law of diffusion would be about $1.4 \times 10^{-6}$ ug $cm^{-2}$ $sec^{-1}$ using a value of $3.0 \times 10^{-6}$ $cm^2$ $sec^{-1}$ for the diffusion coefficient and a value of 0.80 for the porosity. If more or less steady state conditions prevail during tidal submergence and if, as assumed in the surfactant calculations above, the average residence time of the tidal prism is 4.5 hours ($1.6 \times 10^4$ sec), the total transfer of iron from the sediment to the surface water during submergence is $2.3 \times 10^{-2}$ ug $cm^{-2}$. For an average tidal prism with a volume of $5 \times 10^8$ liters this would produce a change in iron concentration of +0.5 ug liter $^{-1}$ which is about one seventh of the average change. Thus, if the conditions outlined above actually prevail in marsh sediments, then diffusion of iron from the sediment pore water could account for a significant portion of the observed export. Additional transfer of iron could also result from bioturbation.

Another possible mode of trace metal migration that can be roughly evaluated is release from decomposing surface Spartina. Dead Spartina is reported to contain 5,000 ug Fe and 22 ug Zn per gram dry weight (11) whereas the living mature plant contains about 300 ug Fe, 11 ug Zn and 2.5 ug Cu (12). These authors indicate that Spartina probably accumulates trace metals mainly from the interstitial water of deep sediments but that some absorption from surface water may also occur, especially after death. The dramatic difference in iron between living and dead Spartina may be due to infiltration of iron rich sediment into the plant after death as after death the ash content increases from 14 to 28 percent (11). Using the trace metal values for the living plant it can be shown that complete decomposition and solubilization on the marsh surface of the annual Spartina production (660 g dry wt $m^{-2}$ $yr^{-1}$) would yield about 210 mg Fe $m^{-2}$ $yr^{-1}$ and 7.0 mg Zn $m^{-2}$ $yr^{-1}$. Copper and zinc budgets based solely on this mechanism probably could not be detected whereas for iron the calculated yield is one fifth of the observed export and is probably the maximum to be expected from this process because a large proportion of the annual Spartina production either is exported as suspended organic sediment or accumulates in the marsh sediment.

The rate of iron release from decomposing Spartina can be further evaluated by assuming that most of the suspended organic sediment is composed of Spartina detritus and by estimating the release of iron from suspended organic sediment during residence of the tidal prism in the marsh. The classical approach to the microbial decomposition of organic substrates assumes that the oxidation process follows first-order reaction kinetics (45) such that

$$-\frac{dS}{dt} = k\ S \tag{1}$$

where S is the substrate concentration at time t and k is the first-order rate constant. This rate law is routinely applied by sanitary engineers in water quality studies and was used by Berner (46) to model the distribution of dissolved sulfate in anoxic marine sediments. For soluble substrates such as sewerage effluent the magnitude of k is about $2.7 \times 10^{-6}$ $sec^{-1}$ (46). *Spartina* detritus should probably be characterized as refractory since it is rich in cellulose and since most of the labile constituents are probably released shortly after death (47). Thus the magnitude of k for *Spartina* detritus is probably not greater than $10^{-9}$ $sec^{-1}$. Integration of equation 1 yields

$$S = S_o\ e^{-kt} \tag{2}$$

If t is assumed to be the average residence time of the tidal prism, 4.5 hrs or $1.6 \times 10^4$ sec, and if an initial suspended organic sediment concentration, $S_o$, of 10 ppm is assumed then, with k equal to $10^{-9}$ $sec^{-1}$ equation 2 yields an average ebb suspended organic sediment concentration of 9.9995 ppm. Decomposition and solubilization of $5 \times 10^{-4}$ ppm of *Spartina* with even 5000 ppm Fe in turn would increase the dissolved iron concentration by only $2.5 \times 10^{-3}$ ug $liter^{-1}$. To produce the average observed change in iron concentration, 4 ug $liter^{-1}$, by this process the value of k would have to be about $2 \times 10^{-6}$ $sec^{-1}$. Thus release of readily soluble iron from *Spartina* probably occurs shortly after death by cell wall rupture and spillage of cell contents. The proportion of iron in the cell contents to total iron in the living plant is unknown. Release of iron from suspended organic matter may be due chiefly to respiration and secretion by living microorganisms and to remineralization of dead microbes in the water column rather than to decomposition of *Spartina* detritus. It is also possible that enzymes and organic acids secreted by microbes in their efforts to digest suspended organic matter may inadvertently solubilize iron and other metals bound to nearby inorganic particles.

In summary, the biogeochemical cycle of iron in a salt marsh probably involves initial import of inorganic iron in the form of iron oxide coatings on clays. Upon burial most of this iron is converted into iron sulfide but a portion is released into the interstitial water. A portion of the dissolved iron is taken up by the roots of growing *Spartina* whereas the remainder tends to diffuse towards the surface under the influence of concentration gradients and bioturbation. Remineralization of a portion of the Spartina together with the diffusive exchange from the sediment results in a net export of dissolved iron from the marsh. Possibly superimposed on this cycle is the day-night effect which tends to retard the export of iron during daylight because of uptake by growing phytoplankton and/or because of photosynthetically induced increases in pH and dissolved oxygen which tend to promote precipitation. The latter factors are suggested because photosynthetic uptake of only 1.0 ug liter$^{-1}$ of a trace metal implies an increase in phytoplankton biomass of at least 1.0 mg liter$^{-1}$ for Fe and perhaps as much as 100 mg liter$^{-1}$ for Cu and Zn based on the trace metal composition of phytoplankton given by Martin and Knauer (48). For the other metals (Cu,Zn,Pb), concentration gradients probably are not strong enough to promote significant diffusive exchange between the sediment and overlying water. Also these metals do not accumulate in Spartina to the same extent as iron and thus remineralization cannot result in measurable export budgets. For copper the main cycle seems to consist of a more or less balance between uptake during daylight and release at night. No explanation can be offered at present for the seemingly erratic budget of lead. This scheme neglects the possible effects of purely physical adsorption and desorption from mineral and organic surfaces because the data at hand are inadequate for this purpose. These effects could be important if the concentrations of dissolved metals are not close to the appropriate adsorbent-adsorbate equilibrium values. Another possible mechanism that we are unable to evaluate is the oxidation and solubilization of iron sulfide brought to the surface by burrowing organisms.

Coliform Bacteria and BOD. Neither of these parameters show any obvious patterns that are amenable to conclusive interpretation. Except for the fact that coliform shows a binomial probability of 0.05 (Table 5), suggesting uptake, its budget does not appear to be statistically significant nor does it show seasonal or diurnal effects. As regards coliform, one would expect lower counts in ebb flow due to the fact that seawater is thought to be toxic to these organisms (49). On the other hand one might expect the BOD of ebb flow to be greater than flood because there is generally a greater concentration of organic suspended sediment in ebb flow. If in fact oxygen demanding substances do tend to accumulate in marshes this could have a favorable effect on the oxygen balance of estuarine waters providing that the reduction in BOD is not offset by a reduction in dissolved oxygen due to marsh respiration. Further investigation of this possibility is warranted by the fact that the proposed rediversion of the Santee River is expected to lead to lower dissolved oxygen concentrations in the upper portions of Charleston Harbor as a result of reduced flushing (8).

Concluding Remarks. In view of the relatively small number of cycles studied in this endeavor and in view of the less than ideal sampling design that was employed, the results reported herein should be considered as tentative and approximate. Nevertheless, as we have attempted to show above, the results generally can be interpreted in light of currently available theory and data and thereby provide insight into the magnitude and importance of various processes involved in regulating marsh chemistry. With further refinement, particularly with regard to depth averaged sampling and the installation of continuously recording instrumentation, we believe that study of tidal creeks in the fashion described above will yield not only more precise estimates of the budgets for various parameters but will reveal the subtle effects and interplay of factors such as salinity, temperature, sunlight and tidal range on the day to day variations in the magnitude of different budgets.

Abstract
Chemical and sediment budgets based on discharge weighted composite samples of flood and ebb flow over 25 tidal cycles from August, 1973 through July, 1974 have been determined for a salt marsh basin adjacent to Charleston Harbor, S.C. The results indicate that the basin exports suspended organic sediment and dissolved $SiO_2$, $PO_4$ and perhaps Fe. Inorganic sediment, surfactants and possibly coliform bacteria are imported and/or degraded. No significant budgets were detected for Cu, Zn or Pb. Budgets for Fe and Cu seem to be affected by light conditions as uptake occurs more frequently in daylight and release at night. Degradation of surfactants and export of organic sediment, $SiO_2$ and $PO_4$ are greater in summer than winter.

Acknowledgements
This study was supported by funds provided by the United States Department of the Interior, Office of Water Resources Research, as authorized by the Water Resources Research Act of 1964 and the South Carolina Water Resources Commission under OWRR Project No. S-044-SC. We wish to thank Dr. Bjorn Kjerfve and Dr. W. E. Sharp for their helpful review of this manuscript. A special expression of gratitude is extended to Mr. Andy Young, Superintendent of the Pine Island Treatment Plant, and his men for their assistance and numerous courtesies.

Literature Cited

1. Ragotzkie, R.A. and Bryson, R.A., Bull. Mar. Sci., Gulf & Caribbean, (1955) 5, pp. 297-314.
2. American Public Health Assoc., Inc., "Standard Methods for the Examination of Water and Waste Water, 13th ed., 874 pp., American Public Health Assoc., Washington, D.C., (1971).

3. Brooks, R.R., Presley, B.J., and Kaplan, I.R., 1967, Talanta, (1967), 14, pp. 809-820.
4. Hobbs, J., personal communication from Orlando Labs., Inc., Orlando, Florida (1973). Glomax hollow cathode lamps.
5. Ostle, B., "Statistics in Research," Iowa State University Press, Davenport, (1963).
6. Brownlee, K.D., "Statistical Theory and Methodology in Science and Engineering," John Wiley and Sons, Inc., New York, (1965).
7. Langley, R., "Practical Statistics Simply Explained," p. 292, Dover Publications, Inc., New York, (1970).
8. Federal Water Pollution Control Administration," Charleston Harbor Water Quality Study, Appendix F," U.S. Dept. of the Interior, Washington, D.C., 88 pp., (1966).
9. Shepard, F.P., Science, (1964), 143, pp. 574-576.
10. Stuiver, M. and Daddario, J.J., Science, (1963), 142, p. 951.
11. Williams, R.B. and Murdoch, M.B., in Symposium on Radioecology (D. Nelson and F. Evans, eds.), USAEC Conf-670503 Oak Ridge, Tenn., pp. 431-439, (1969).
12. Broome, S.W., Woodhouse, W.W., and Seneca, E.D., Sea Grant Pub. UNC-SG-73-14, North Carolina State University, Raleigh, N.C. (1973).
13. Boon, J.D., in International Symposium on Interrelationships of Estuarine and Continental Shelf Sedimentation, Inst. Geol. du Bassin d' Aquitaine, Bordeaux, pp. 67-74, (1974).
14. Gardner, L.R., Limnol. Oceanogr., (1975), 20, pp. 81-89.
15. Reimold, R.J., (1972), Limnol. Oceanogr., (1972), 17, pp. 606-611.
16. Patton, J., in "Water Quality Criteria" (McKee, J.E. and Wolfe, H.W., eds.), Calif. State Water Resources Board, 2nd. ed., pp. 392-404, (1963).
17. Jenne, E.A., in "Trace Inorganics in Water" (R.F. Gould, ed.) Adv. in Chem. Ser., No. 73, ACS, Washington, D.C. pp. 337-389, (1968).
18. Spencer, D.W. and Brewer, P.G., J. Geophys. Res., (1971) 76, pp. 5877-5892.
19. Wolfe, D.A., Cross, F.A., and Jennings, C.D., in "Radioactive Contamination of the Marine Environment" IAEA, Vienna, pp. 159-175, (1973).
20. Wolfe, D.A., and Rice, T.R., Fisheries Bull., (1972) 70, pp. 959-972.
21. Cross, F.A., Duke, T.W., and Willis, J.N., Chesapeake Sci., (1970) 11, (4), pp. 221-234.
22. Khaylov, K.M., and Finendo, Z.Z., Oceanology, (1968) 8, (6), pp. 980-991.
23. Rice, T.R., (1963), in "Radio-ecology" (V. Schutz and A.W. Klement, eds.), Proc. First National Symp. on Radioecology, Reinhold Publ. Corp., Washington, D.C., pp. 619-631, (1963).

24. Evans, D.W., and Cutshall, N.H., in "Symposium on Interaction of Radioactive Contaminants with Constituents of the Marine Environment", IAEA, Seattle, Wash., (1972).
25. Forster, W.O., Wolfe, D.A., and Lowman, F.G., in "Proceedings of Third National Symposium on Radioecology", AEC CONF 710501-32, Oak Ridge, Tenn., (1971).
26. Gardner, L.R., Geochim. Cosmochim. Acta, (1974) 38, pp. 1297-1302.
27. Bradford, W.L. Tech. Rpt. No. 76, Ref. 72-7, Chesapeake Bay Institute, The John Hopkins University (1972).
28. Baker, W.E., Geochim. Cosmochim. Acta, (1973), 37, pp. 269-281.
29. Rashid, M.A., Chem. Geol., (1972) 9, pp. 241-248.
30. Evans, W.D., in "Advances in Organic Geochemistry" (eds. U. Colombo and S.D. Hobson), pp. 263-270. MacMillan and Com., N. Y., (1964).
31. Zirino, A., and Healy, M.L., Limnol. Oceanogr., (1971) 16, pp. 773-778.
32. Slowey, J.F. and Hood, D.W., Geochim. Cosmochim. Acta., (1971) 35, pp. 121-138.
33. Presley, B.J., Kolodny, Y., Nissenbaum, A., and Kaplan, I.R., Geochim. Cosmochim. Acta., (1972) 36, pp. 1073-1090.
34. Brooks, R.R., Presley, B.J., and Kaplan, I.R., Geochim. Cosmochim. Acta., (1968), 32, pp. 397-414.
35. Piper, O.Z., Geochim. Cosmochim. Acta., (1971), 35, pp. 531-550.
36. Gross, M.G., in Estuaries (G. Lauff, ed.) A.A.A.S. pub. No. 83, pp. 273-282, (1967).
37. Krauskopf, K.B., "Introduction to Geochemistry," McGraw-Hill Book Co., N.Y., 721, (1967).
38. Nriagu, J.O., Geochim. Cosmochim. Acta., (1972) 36, pp. 459-470.
39. Nriagu, J.O., Am. Jour. Sci., (1972) 272, pp. 476-484.
40. Nriagu, J.O., Geochim. Cosmochim. Acta., (1973) 37, pp. 2357-2361.
41. Nriagu, J.O., Geochim. Cosmochim. Acta., (1974) 38, 887-898.
42. Duchart, P., Calvert, S.W. and Price, N.B., Limnol. Oceanogr. (1973) 18, (4), pp. 605-610.
43. Horne, R.A. and Woernle, C.H., Chem. Geol., (1972) 9, pp. 299-304.
44. Stookey, L.L., Analytical Chemistry, (1970) 42, pp. 779-781.
45. Barrett, M.J., in "Mathematical and Hydrologic Modeling of Estuarine Pollution," (Gameson, A.L.H., ed.) Water Pollution Technical Paper No. 13, Dept. of the Envir., London, pp. 39-48, (1972).
46. Berner, R.A., Geochim. Cosmochim. Acta., (1964) 28, pp. 1497-1503.

47. Stevenson, L.H., Personal Communication from Biology Department, University of South Carolina, (1973).

48. Martin, J.H. and Knauer, G.A., Geochim. Cosmochim. Acta., (1973) 37, pp. 1639-1653.

49. Savage, H.P., and Hanes, N.B., Jour. Water Pollution Control Federation, (1971) 43, pp. 854-860.

# 10

# Processes Affecting the Vertical Distribution of Trace Components in the Chesapeake Bay

RICHARD L. HARRIS and GEORGE R. HELZ
University of Maryland, Department of Chemistry, College Park, Md. 20742
ROBERT L. CORY
U.S. Geological Survey, Edgewater, Md.

The study of trace components is of considerable interest since certain ones are essential to life while others are highly toxic. It is within the scope of human endeavor to alter the rates at which some of these components are delivered to estuaries and coastal waters (1 - 3) and, consequently, the question of the fate of these substances is important. To answer this question, we must first understand the processes which control their transport, and ultimately their deposition. These transport processes are conveniently separated into horizontal and vertical components. In most estuaries, horizontal transport is dominated by simple advective flow. However vertical transport, especially of trace components, is influenced by a number of physical, biological and chemical processes.

This paper, which explores the relative importance of some of the vertical transport processes in estuaries, is based upon observations made during an ongoing study in the central Chesapeake Bay, Maryland. To date, the vertical distribution of phosphate and five trace metals (Cu, Fe, Mn, Pb, and Zn) have been investigated over the period of a year. Samples were collected in the deep waters off Bloody Point in the Chesapeake Bay at latitude 38°50', longitude 76°24'. This station lies in the northernmost part of a trough having depths to 60 m, that extends about thirty miles south. During late summer, the bottom waters in this trough frequently become anoxic and sulfide-bearing. Water samples were collected through polyethylene tubing connected to a gasoline powered peristaltic pump. We also obtained surface sediment samples. The iodometric method (4) was used to determine sulfide, while phosphate was found colorimetrically (5). Total trace metals were analyzed by atomic absorption, and nitric acid digestions of the particulates by flameless atomization with a carbon rod. A detailed discussion of the sampling and analytical procedures and a full presentation of the data will be given elsewhere (6).

## The Conservative Model

The simplest distribution pattern that might be expected in an estuary would be one established by conservative mixing. A conservative component varies linearly with salinity in both the horizontal and the vertical directions. Such a distribution is indeed observed for some trace components. For example Warner (7) showed that $F^-$ in the Chesapeake Bay was distributed conservatively. On the other hand, it is well-known that conservative mixing can not account for the distribution of many trace metals. As an example, in Figure 1, we show total Fe (μg/l) plotted versus salinity (°/₀₀). For reference, we have plotted the linear distributions which would be expected under conservative conditions assuming that river water entering the Bay contained, respectively, 100, 300 or 1000 μg/l Fe. Schubel (8) cites Carpenter as finding the average annual concentration in the Susquehanna River, the principal tributary to the northern Bay, to be 300 μg/l. Note that total iron observed at our station differs from conservative dilution models in two ways: (a) the iron concentrations are systematically lower than would be expected if iron were conservative; (b) the trend is wrong; salinity consistently increases downward, but on some occasions Fe was observed to pass through a minimum at intermediate depth. A similar phenomenon is sometimes observed in the vertical distribution of phosphate. In Figure 2, we show a case in which total phosphate passes through a maximum at intermediate depth while soluble phosphate, that is, the fraction which can go through a 0.45 μm Millipore filter, passes through a minimum. No conservative mixing model could account for this type of vertical distribution. Using average abundances of Mn in river and ocean waters (9), similar considerations reveal that manganese also is usually present at our station in quantities less than would be expected if it were conserved during mixing. On the other hand, Cu, Zn, and Pb are typically present in excess of the conservative values, although the data vary considerably.

## Biological Processes

Most oceanographers and marine biologists will not be surprised that the trace components we have mentioned so far are not conserved during mixing in estuaries. These substances are known to be taken into living organisms, so their distribution in an estuary might be expected to be influenced by biological processes. For example, using the gross primary productivity data of Flemer (10) and the empirical formula of Richards (11) for marine organic matter, we estimate that typically 0.6 μm/l/day of phosphate is fixed in the euphotic zone of the central Chesapeake Bay in the summer. This represents about twice the total amount present in surface waters, indicating that much of this phosphate is rapidly recycled, an observation which is consistent with other published reports (12,13). It seems likely that the large particulate phosphate concentration shown in Figure 2 at intermediate depths is due to settling plankton debris which has become trapped by buoyancy at the top of the higher density bottom water.

Although biological processes such as those described above can play a major role in the distribution of phosphorus in estuaries, they are probably much less effective for trace metals. In this regard estuaries may differ from the deep ocean where longer periods of time are available. Estimates of the time needed for phytoplankton to strip 100% of the phosphorus and the five trace metals, assuming no regeneration, are compared in Table I (bottom line). These estimates are based on a plankton summertime gross productivity rate of 1.6 g C/m$^2$/day (10) and on composition data for marine phytoplankton (14). We feel that the use of marine phytoplankton data in this environment is justified because trace element concentrations are not greatly different from those in the oceans. Furthermore, Riley and Roth (15) showed that trace metal concentrations in phytoplankton do not depend critically on plankton species. Metal rates are much longer than radioactive isotope exchange rates observed by a number of authors (16,17). The removal time is quite short for phosphorus but long for the trace metals. These data, although subject to some uncertainty, cast doubt on the widely held view that biological processes are primarily responsible for the non-conservative behavior of trace elements in estuaries.

## Chemical Processes

Chemical processes usually considered as likely

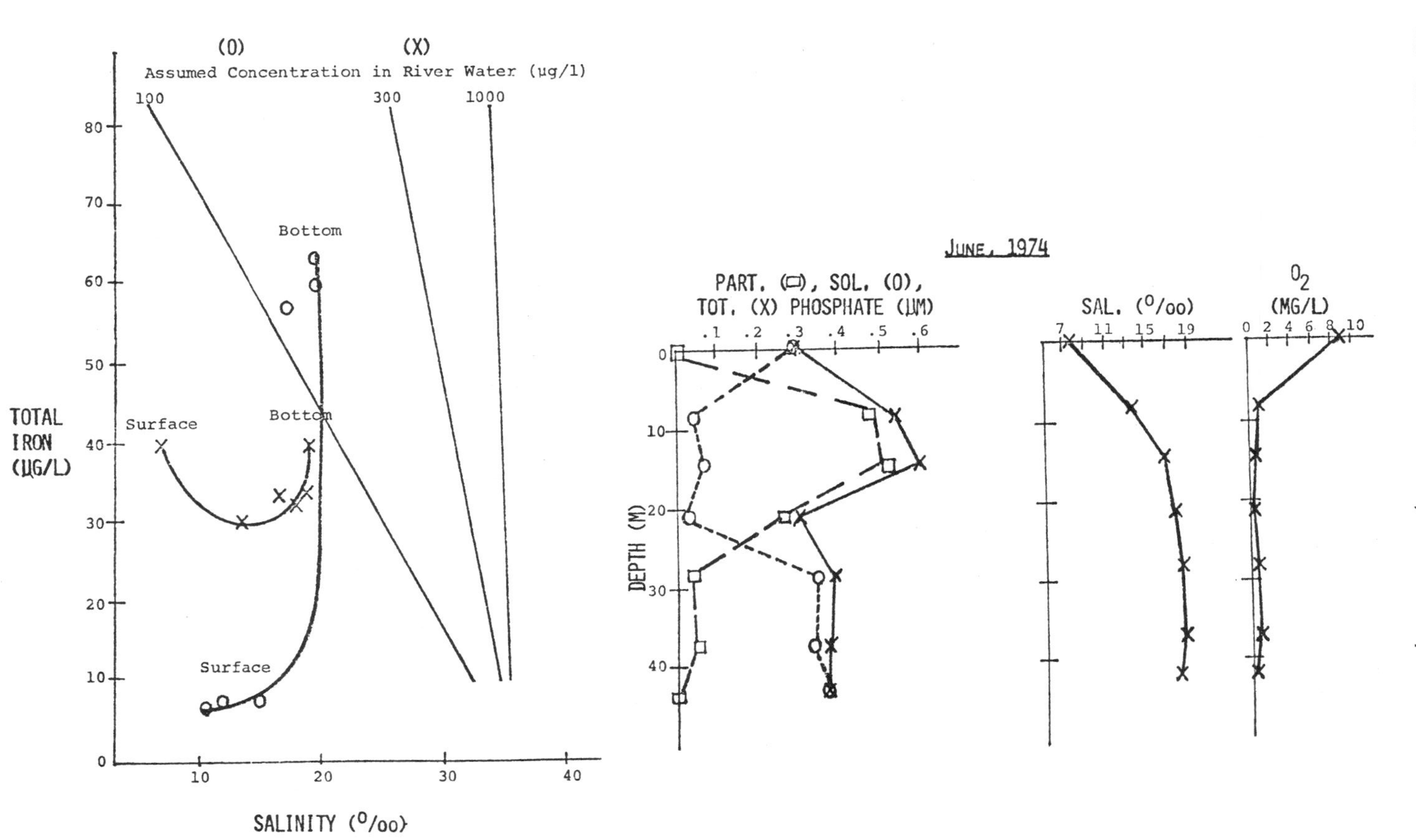


Figure 1. Total iron (µg/l) vs. salinity (‰) for March and June, 1974

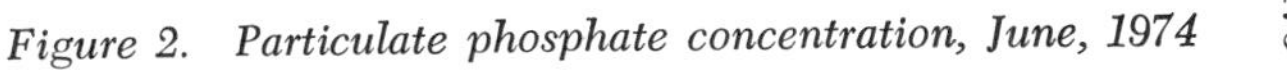

Figure 2. Particulate phosphate concentration, June, 1974

to affect the distribution of trace components in estuaries and in the oceans are precipitation and adsorption. Krauskopf (18) investigated these mechanisms and concluded that in the oceans, local precipitation of sulfides constituted a possible control for some metals. Subsequently, however, a number of authors (19,20) have found that sulfide-rich marine waters apparently fail to precipitate trace metal sulfides. We have calculated the degree of saturation of the trace metals in our samples with respect to sulfides, hydroxides, phosphates and carbonates using computational procedures similar to those described in the literature (21). In general, the waters of the central Chesapeake Bay are undersaturated with respect to these phases. Where the calculations suggest that the waters are supersaturated, there is no evidence that dissolved metal concentrations in the water are responding. For example, the highest dissolved Pb concentrations which we observed in the bottom waters during our year-long sampling period were observed in June 1974 when the bottom water contained 0.3 mg/l sulfide. Thus precipitation appears to play no role in trace metal cycling in the Chesapeake Bay. Possible exceptions to this rule, based on solubility calculations, are the oxyhydroxides of Fe and Mn. Because both biological assimilation and precipitation appear to be insignificant as vertical transport mechanisms for trace metals other than Fe and Mn, we feel that adsorption-desorption processes must be of major importance. Other workers (22,23) have discussed these processes placing emphasis on the desorption which occurs as river-borne particles encounter the increasing salinity of estuarine water. However the work of Schubel (8) and Biggs (24) reveal that most of the particles brought into Chesapeake Bay by the Susquehanna River, the major tributary to the northern section, are deposited near the head of the Bay. Consequently adsorption-desorption reactions involving river-borne particles can not account for the non-conservative vertical profiles which we observe in the central part of the Bay.

Based upon the sediment mass balance data which Biggs (24) compiled and which we show in Figure 3, the principal sources of particulate matter in the central part of the Chesapeake Bay are shore erosion and biological production. Data on the chemical and mineral composition of source material is not available, but we have estimated its composition using data obtained from the deposited sediments. Table II shows the average concentrations of clay minerals, iron and

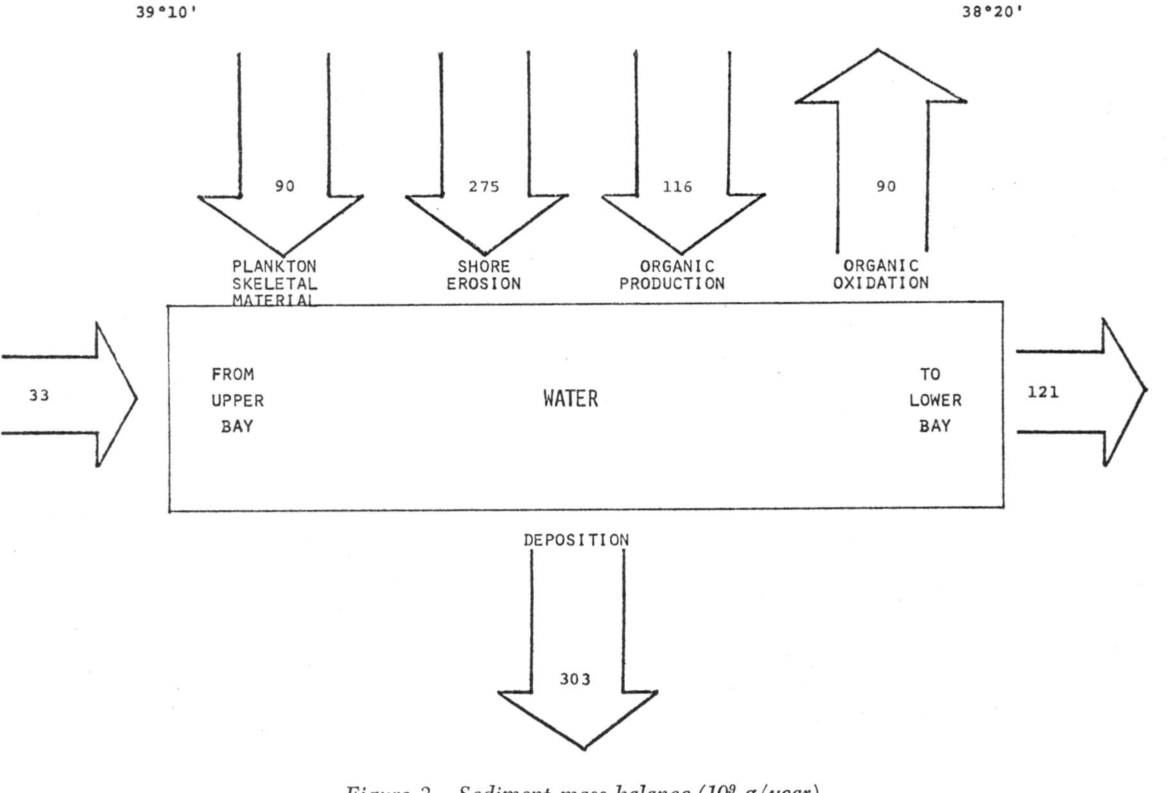


*Figure 3. Sediment mass balance ($10^9$ g/year)*

TABLE I - ESTIMATED BIOLOGICAL ASSIMILATION RATES IN SURFACE WATER

| | $PO_4$ | Mn | Fe | Cu | Zn | Pb |
|---|---|---|---|---|---|---|
| Typical Concentration in Euphotic Zone of Chesapeake Bay (Summer) (µg/L) | 30 | 11 | 20 | 3 | 6 | 7 |
| Typical Concentration in Marine Phytoplankton (µg/g dry weight) | $2.7 \times 10^4$ | 5 | 200 | 4 | 20 | 5 |
| Phytoplankton Fixation Rate in Chesapeake Bay (µg/L/day) | 60 | 0.02 | 0.8 | 0.01 | 0.06 | 0.01 |
| Turnover Rate (Days) | 0.5 | 600 | 30 | 300 | 100 | 700 |

TABLE II

MINIMUM CONCENTRATIONS AND DEPOSITION RATES FOR POTENTIAL ADSORBING AGENTS IN CHESAPEAKE BAY SEDIMENTS

| Component | Average Concentration (%) | Deposition Rate ($10^9$ g/yr) |
|---|---|---|
| Chlorite | 4 | 12 |
| Illite | 10 | 30 |
| Kaolinite | 7 | 21 |
| Montmorillonite | 6 | 18 |
| Iron | 4 | 21 |
| Organic Matter | 5 | 15 |
| | 36 | 117 |

$$\frac{117 \times 10^9 \text{ (g/yr) Total Deposition}}{9 \times 10^{13} \text{ (L/yr) Total Water Flow}} = 1.3 \text{ mg/L}$$

organic matter in sediments based on the work of Ryan (25), Owens, et al (26), and our own work. The percentage figures in the middle column are given on a dry weight basis. Only the clay minerals found in the clay sized fraction, which makes up 27% of the average sediment, are included. In the right hand column, the annual deposition rate for each of these components is shown. These are based on the mass balance data of Figure 3. At the bottom of Table II we have computed the average concentration of these components in Bay water by distributing the total amount deposited annually over the total amount of water which flushes through this section of the Bay. This figure is a minimum value because the computation assumes that sediment passes once through the section of the Bay under study. Actually, because of the small size of the clay minerals and the low density of the organic matter, these particles have small settling velocities and thus may be recycled by turbulence several times before finally being deposited. Thus the trace concentration of potential absorbing agents in Bay water might be several times higher than the 1.3 mg/l shown in Table II. Biggs (24) has found total suspended matter up to 20 mg/l in the Chesapeake Bay. If we assume that the adsorptive capacity of this material is on the order of 0.1 - 1%, as is usually observed (27, 28), then adsorption-desorption processes could be important for those trace metals, such as Cu, Zn and Pb, which are typically present in dissolved form at concentrations below 10 μg/l. This will be true, however, only if the trace metals are able to dominate the adsorption sites in competition with the much more abundant Na, Mg and Ca.

## Conclusions

Vertical distributions of phosphate, Fe, Mn, Cu, Zn, and Pb in central Chesapeake Bay are not explainable by conservative mixing, so active vertical transport processes must be operating. For phosphate, biological processes are undoubtedly involved, but for the metals, biological assimilation appears to be too slow. Solubility reactions also appear to be unimportant, except possibly for Fe and Mn. Generally the waters are undersaturated with respect to pure mineral phases of Cu, Zn and Pb, but in summer when sulfide appears in the bottom waters and supersaturation is reached with respect to CuS, ZnS and PbS, there is still no evidence that these metals are being removed. Because biological assimilation and solubility reactions do not appear capable of explaining the non-conservative vertical distribution of Cu, Zn and Pb, we feel that adsorption-desorption processes must control vertical transport of these metals. Estimates of the adsorption capacity available on the kinds of particulate matter which must be present in the central part of Chesapeake Bay suggest that this mechanism is feasible provided adsorption is selective for trace metals.

## Abstract

A one-year, single station study has been completed in the central Chesapeake Bay, Maryland to investigate seasonal changes and vertical transport mechanisms of phosphate and five trace metals (Cu, Fe, Mn, Pb, Zn). Vertical distributions are not explainable by conservative mixing. Biological processes can account for phosphate distributions, but appear to be too slow to affect vertical metal transport. Mid-Bay waters never reach saturation with respect to carbonate, phosphate, or sulfide minerals. Precipitation of oxyhydroxides may be the controlling process for Fe and Mn, while adsorption appears to be the only reasonable mechanism for Cu, Pb, and Zn.

## Literature Cited

(1) Bertine, K.K. and Goldberg, E.D., Science (1971) 173, p. 233-235.
(2) Young, D.R., Young, C.S., and Hlavka, G.E., in Curry, M.G. and Gigliotti, G.M. (Ed.), "Cycling and Control of Metals; Proceedings of an Environmental Resources Conference", p. 21-39, U.S. Environmental Protection Agency, 1973.
(3) Helz, G.R. (1975), in press.
(4) Brown, E., Skougstad, M.W., and Fishman, M.J., "Techniques of Water Resources Investigations of the United States Geological Survey", p. 154, Department of the Interior, Washington, D.C., 1970.
(5) Strickland, J.D.H. and Parsons, T.R., "A Practical Handbook of Seawater Analysis", p. 49, Fisheries Research Board of Canada, Ottawa, 1972.
(6) Harris, R.L. (1975), Ph.D. Thesis, Univ. Md.
(7) Warner, T.B., Deep-Sea Research (1971) 18, p. 1255-1263.
(8) Schubel, J.R., "The Physical and Chemical Conditions of Chesapeake Bay; An Evaluation", p. 59, Chesapeake Bay Institute, John Hopkins University, Baltimore, 1972.

(9) Riley, J.P. and Chester, R., "Introduction to Marine Chemistry", p. 65, Academic Press, New York, 1971.
(10) Flemer, D.A., Chesapeake Science (1970) 11 (2), p. 117-129.
(11) Richards, F.A., in Riley, J.P. and Skirrow, G. (Ed.), "Chemical Oceanography, Vol. 1", p. 624, Academic Press, New York, 1965.
(12) Pomeroy, L.R., Mathews, H.M., and Min, H.S., Limnology and Oceanography (1963) 8, p. 50-55.
(13) Rigler, F.H., Ecology (1956) 37 (3), p. 550-562.
(14) Martin, J.H. and Knauer, G.A., Geochimica et Cosmochimica Acta (1973) 37, p. 1639-1653.
(15) Riley, J.P. and Roth, I., J. Mar. Biol. Ass. (1971) 51, p. 63-72.
(16) Duke, T.W., J. Water Pollution Control Federation (1967) 39 (4), p. 536-542.
(17) Wolfe, D.A., Second International Estuarine Research Conference Preprint (1973), 49 pp.
(18) Krauskopf, K.B., Geochimica et Cosmochimica Acta (1956) 9, p. 1-32.
(19) Presley, B.J., Kolodny, Y., Nissenbaum, A., and Kaplan, I.R., Geochimica et Cosmochimica Acta (1972) 36, p. 1073-1090.
(20) Piper, D.Z., Geochimica et Cosmochimica Acta (1971) 35, p. 531-550.
(21) Zirino, A. and Yamamoto, S., Limnology and Oceanography (1972) 17 (5), p. 661-671.
(22) Kharkar, D.P., Turekian, K.K., and Bertine, K.K., Geochimica et Cosmochimica Acta (1968) 32, p. 285-298.
(23) Martin, J.M., Kulbicki, G., and DeGroot, A.J., in "Proceedings of Symposium on Hydrogeochemistry and Biogeochemistry, Vol. 1", p. 463, The Clarke Company, Washington, D.C., 1973.
(24) Biggs, R.B., Marine Geology (1970) 9, p. 187-201.
(25) Ryan, J.D., "The Sediments of Chesapeake Bay", 120 pp., Md. Dept. Geol., Mines and Water Resources, Bull. 12, 1953.
(26) Owens, J.P., Stefansson, K., and Sirkin, L.A., J. Sedimentary Petrology (1974) 44 (2), p. 390-408.
(27) Grim, R.E., "Clay Mineralogy", p. 189, McGraw-Hill Book Company, N.Y., 1968.
(28) Stumm, W. and Morgan, J.J., "Aquatic Chemistry", p. 543, Wiley-Interscience, N.Y., 1970.

# 11

# An ESCA Study of Sorbed Metal Ions on Clay Minerals

MITCHELL H. KOPPELMAN and JOHN G. DILLARD

Department of Chemistry, Virginia Polytechnic Institute and State University, Blacksburg, Va. 24061

In natural water systems, clay minerals comprise a large part of the suspended loads of rivers which empty into estuarine environments. A portion of this load ultimately enters the ocean domain. Numerous investigators have demonstrated that the clay mineral portion of this suspended load can act as a sponge for trace metals and serve as a taxi for the transport of the metals into the estuarine and ocean systems.[1,2] It is well known that the mode of interaction between the aqueous metal ion and the clay mineral is adsorption at the Stern layer.[3,4] The interaction is believed to occur as a result of the negative surface potential generated on the clay mineral by isomorphous substitution in either tetrahedral or octahedral sites, broken surface bonds or hydrolysis of exposed surface hydroxyl groups in the clay mineral.[5]

Specifically it has been shown that various clay minerals adsorb iron species from solution.[6-12] Although the degree and conditions under which iron is adsorbed were carefully documented in these studies, little information was presented regarding the bonding interaction between the clay mineral and the metal ion. Information on the nature of the bonding between the metal ion and the clay mineral is important in providing insight into various ion exchange phenomena which are important to the maintenance of the chemical balance in the ocean system.

A study of the bonding phenomena between the clay mineral and adsorbed metal ions Fe(III) and Cr(III), has been initiated using X-ray photoelectron spectroscopy (XPS or ESCA). ESCA would seem to be an ideal method for this study since it has been postulated that the adsorbed ions are located on the clay mineral surfaces. In X-ray photoelectron spectroscopy the binding energy of core and valence electrons can be measured from a sampling depth of up to about 50Å.[13] The measured binding energy can be related to the electron density or charge on an element of interest and can thus be used to examine the oxidation state (bonding nature) of elements in clay minerals and metal ions adsorbed at clay mineral surfaces. To compliment the ESCA measurements the bonding nature of iron in clay mineral systems has also been examined using Mössbauer spectroscopy. In this paper the results of a study of the bonding between the clay minerals, chlorite, illite, and kaolinite and the metal ions Fe(III) and Cr(III) will be reported.

## Experimental

The clay minerals and compounds investigated were chlorite (from Ishpeming, Michigan obtained from Ward's Natural Science Establishment), illite (Fithian, Illinois, A.P.I. standard #35 obtained from Illinois Geological Survey), kaolinite (Hydrite RT, obtained from Georgia Kaolin, Inc.), nontronite (Washington, obtained from Dr. C. I. Rich, VPI & SU Agronomy Department), $Fe_2O_3$ (Fisher Chemical Company, reagent grade), and air dried amorphous ferric hydroxide (precipitated from 100 ppm $Fe(NO_3)_3$ solution at pH 3). These clays were selected because of their non-expansive nature in an effort to decrease interlayer substitution and enhance surface adsorption. The clays were ground with an agate mortar and pestle and a ball mill (using tungsten carbide balls) such that all particles were <5 μm.

The adsorption studies of Fe(III) and Cr(III) onto chlorite, kaolinite, and illite were initially performed in untreated glass vessels to test the feasibility of using ESCA to detect surface iron and chromium species. The vessels were open to the atmosphere, and were maintained at room temperature which varied from 21-26°C. Data for the amount of Fe(III) and Cr(III) adsorbed reported herein refer to experiments identical to those described above, but under controlled atmospheric conditions. Glass vessels were coated with an aqueous dispersion of Teflon (Teflon 30).[14] To insure a pin hole free coating, the vessels were then treated with methyl triethoxysilane.[15]

Iron(III) and chromium(III) solutions of 100 and 1000 ppm were prepared from $Fe(NO_3)_3 \cdot 9H_2O$ (Fisher reagent grade chemical) and $Cr(NO_3)_3 \cdot 9H_2O$ (Fisher reagent grade chemical), respectively. These solutions were degassed by heating to 50°C and stirred while purging with argon. The pH was adjusted to 1.0 using NaOH and $HNO_3$ to prevent the precipitation of the metal hydroxides. Each clay sample, 3.25 gms, and a magnetic stirring bar were placed in coated vessels, evacuated to $10^{-3}$ torr, and then filled with 870 torr of argon. The 100 and 1000 ppm Fe(III) and Cr(III) solutions (650.0 mls) were placed in the appropriate vessels and sealed under a positive pressure (870 mm) of argon with a rubber septum seal. The vessels were then attached to a vacuum manifold so that argon could be replenished after removal of each solution sample aliquot. The vessels remained in a constant temperature bath (25°C $\pm$ 0.5°C) for periods of 7-9 days, with 50 ml samples removed periodically using a 50 ml syringe and needle to pierce the septum. About 5-7 solution sample aliquots were removed during the duration of each experiment. After each sample of solution was removed, the vessel was repressurized to a positive pressure of argon (870 mm). The solutions remaining in the vessels were stirred twice daily to suspend the clay particles. Upon termination of the adsorption experiments, the supernatant liquids were removed using a water aspirator and the clay residue collected by centrifugation @ 10,000 rpm for 20 minutes in 50 ml tubes (5°C). The supernatant liquid was then discarded and the clay samples were air dried and reground in an agate mortar and pestle for ESCA studies.

Solution samples were analyzed for pH immediately upon removal. pH was measured with a Corning Model 707 pH meter and glass electrodes under an Ar atmosphere. Dissolved Fe, Cr, K, Mg, and Al were measured using atomic absorption spectrophotometry. A Perkin Elmer Model 503 atomic adsorption spectrometer with hollow cathode lamps was employed. When concentrations exceeded the normal linear working range, the burner head was turned approximately 75° relative to the beam to increase the range of absorbance linearity (Beer's law).

Dissolved $SiO_2$ was determined spectrophotometrically using the silico-molybdate method. p-methyl aminophenol sulfate (metol) was employed as the reducing agent and $Na_2SiF_6$ was used to prepare $SiO_2$ standards.[16,17]

The core electron binding energies were measured using an AEI ES 100 photoelectron spectrometer. Al K$\alpha$ radiation (E = 1486.6 eV) was the photon source. Typical sample chamber pressures were less than $10^{-7}$ torr. Data acquisition and spectrometer control were accomplished using the AEI DS-100 data system and a

Digital PDP 8/e computer. Spectra were plotted using the MADCAP IV plotting routine on a Digital PDP 8/I computer. The Fe $2p_{3/2}$ peaks were deconvoluted using the GASCAP IV[18] program in conjunction with a PDP 8/I computer. The program requires input for the number of peaks and the height, FWHM, and position for each peak suspected of comprising the multiplet. Based on these parameters, a spectrum is calculated which can be compared to the experimental spectrum. The program contains a routine which performs a point by point subtraction of the calculated spectrum from the experimental spectrum. In all of the deconvoluted spectra presented in this paper, subtractions yielded nearly straight lines, thus indicating a good fit between calculated and experimental spectra.

Clay mineral samples were all ground in an agate mortar and pestle and then dusted onto double stick cellophane tape which in turn was mounted on a brass probe. Since all the minerals examined in this study are non-conductors, it was necessary to correct the measured core electron binding energies for charging.

To account for charging of the clay samples in the evaluation of the spectrometer work function, the binding energy of the Si $2p_{1/2,3/2}$ core level was used as an internal standard. The absolute binding energy of the silicon level was measured in two separate experiments. In one measurement a thin gold film was vapor deposited onto the sample surface and the binding energy of the Au $4f_{7/2}$ (BE = 83.4 eV)[19,20] level used to calibrate the energy scale for the Si $2p_{1/2,3/2}$ level. In another determination a thin film of background spectrometer carbon was allowed to build up on the sample, and the C $1s_{1/2}$ (BE = 284.6 eV) level used in the calibration of the binding energy scale. The Si $2p_{1/2,3/2}$ binding energy determined by the two methods agreed to within the experimental precision ($\pm$ 0.1 eV).

The measured binding energies for all samples (Tables I, II, III) represent the average of no less than three separate determinations. The deviations in binding energies (one standard deviation) are all $\pm$ 0.1 or less.

Mössbauer spectra were investigated on a Northern Scientific model 605-512 multichannel analyzer spectrometer. A 3 m Curie $^{57}Co$ (in a Pt matrix) source was used (2½ years old). Isomer shifts reported are relative to Fe foil. A linear least squares regression program written by G. W. Dulaney and Calcomp plotting routine in conjunction with an IBM 370/158A computer were used for data processing. Percentages

Table I - Core Electron Binding Energies and FWHM For Mineral Lattice Elements. (in eV ± 0.1)

| | CHLORITE | ILLITE | KAOLINITE |
|---|---|---|---|
| Si $2p_{1/2,3/2}$ | 102.1<br>2.4 | 102.5<br>2.5 | 102.7<br>2.2 |
| Al $2p_{1/2,3/2}$ | 74.2<br>2.5 | 74.3<br>2.5 | 74.4<br>2.1 |
| O $1s_{1/2}$ | 531.3<br>2.8 | 531.7<br>2.9 | 531.9<br>2.4 |
| K $2p_{3/2}$ | 292.9<br>3.2 | 293.2<br>2.2 | ----- |
| Ca $2p_{1/2,3/2}$ | 350.7<br>3.5 | 350.7<br>3.6 | ----- |
| Mg $2p_{1/2,3/2}$ | 50.0<br>2.1 | 49.7<br>3.4 | ----- |

Table II - Binding Energy for Fe $2p_{3/2}$ Electrons

| | | Assignment | Binding Energy (eV) | FWHM (eV) |
|---|---|---|---|---|
| Chlorite | ($Fe^{+2}$) | native: | | |
| | | composite experimental peak | 710.6 | 5.2 |
| | | lattice $Fe^{+2}$ | 710.3 | 3.8 |
| | | with adsorbed $Fe^{+3}$: | | |
| | | composite experimental peak | 711.3 | 6.3 |
| | | lattice $Fe^{+2}$ | 710.3 | 3.7 |
| | | adsorbed $Fe^{+3}$ | 711.4 | 3.9 |
| Illite | ($Fe^{+2}$,$Fe^{+3}$) | native: | | |
| | | composite experimental peak | 712.6 | 6.4 |
| | | lattice $Fe^{+2}$ | 710.4 | 3.9 |
| | | lattice $Fe^{+3}$ | 712.6 | 4.0 |
| | | with adsorbed $Fe^{+3}$: | | |
| | | composite experimental peak | 712.5 | 6.5 |
| | | lattice $Fe^{+2}$ | 710.4 | 3.9 |
| | | lattice $Fe^{+3}$ | 712.6 | 3.8 |
| | | adsorbed $Fe^{+3}$ | 711.5 | 3.7 |
| Kaolinite | | native: No Fe detected | | |
| | | with adsorbed $Fe^{+3}$: | | |
| | | adsorbed $Fe^{+3}$ | 711.4 | 3.8 |
| Nontronite | | native: | | |
| | | lattice $Fe^{+3}$ | 712.5 | 4.9 |
| $Fe_2O_3$ | | lattice $Fe^{+3}$ | 711.1 | 4.9 |
| Amorphous Ferric hydroxide | | "lattice" $Fe^{+3}$ | 711.9 | 4.3 |

Table III - Chromium Adsorption Data

| | Binding Energies ($2p_{3/2}$) | FWHM | ppm Adsorbed |
|---|---|---|---|
| chlorite | 577.2 | 3.2 | 38 |
| illite | 577.3 | 3.5 | 22 |
| kaolinite | 577.5 | 3.1 | 19 |
| uvarovite | 578.3 | 3.1 | -- |

of Fe(III) and Fe(II) in the illite sample were determined from ratios of peak areas of Lorentzian peaks representative of the specific ferric and ferrous iron sites.

## Results and Discussion

The core electron binding energies for the lattice elements in chlorite, kaolinite, and illite are presented in Table I. The small differences in binding energies of Si $2p_{1/2,3/2}$, Al $2p_{1/2,3/2}$, and O 1s electrons between these three minerals are not unexpected, since there are only minor bonding differences between the minerals. The range of binding energy values of the Si $2p_{1/2,3/2}$ electron observed here (102.1 eV - 102.7 eV) is in good agreement with those found by Adams et.al.[21] (101.8 - 102.4 for various silicates), Huntress et.al.[22] (103.0 for lunar materials) and Anderson et.al.[23] (102.8 - 103.2 for a series of aluminosilicates). The Al $2p_{1/2,3/2}$ and O $1s_{1/2}$ binding energies reported here are in good agreement with those published in the literature.[21-24] Schultz et.al.[25] presented binding energy data for the Si $2p_{1/2,3/2}$ electrons in kaolinite (107.4 eV) and illite (109.3 eV). Differences between their values and the ones presented in this report probably arise from failure[25] to completely correct for the sample charging phenomena.

The investigation of the binding energy of the Fe $2p_{3/2}$ electrons in chlorite, illite, and kaolinite revealed rather interesting results. These binding energies are tabulated in Table II. Adams et.al.[21] had previously reported difficulty in distinguishing between $Fe^{+2}$ and $Fe^{+3}$ species using ESCA. However, when comparing the binding energy for the Fe $2p_{3/2}$ level in nontronite where iron is in the +3 (ferric) state (substituted for some $Al^{+3}$ in the octahedral layers) and chlorite where the iron is in the +2 (ferrous) oxidation state (substituted for $Mg^{+2}$ in the octahedral layers), a distinct difference (1.9 eV) is observed. The XPS spectrum of Fe $2p_{3/2}$ electrons in illite showed a rather broad peak with a binding energy of 712.6 eV. Fe $2p_{3/2}$ levels always exhibit broad peaks in the ESCA spectrum due to multiplet splitting phenomena.[26-28] The full widths at half maxima for the Fe $2p_{3/2}$ peak for nontronite and chlorite (each containing only one type of iron) are 4.9 eV and 5.2 w respectively, while illite is 6.4 eV. This increase in peak width is suggestive that illite contained more than one type of iron.

Mössbauer data (Table IV) confirms that nontronite contains only Fe(III) in one environment and that chlorite contains only Fe(II) in one environment. Illite, on the other hand, was shown by Mössbauer spectra to contain both Fe(II) and Fe(III), each in only one environment.

Since the Mössbauer results indicated that Fe(II) and Fe(III) in illite were similar to Fe(II) and Fe(III) in chlorite and nontronite, respectively, it was of interest to detemine the binding energies of Fe(II) and Fe(III) in illite using deconvolution. The input data used for the deconvolution were the binding energies of Fe(II) in chlorite and Fe(III) in nontronite. The input data also included satellite peak structure determined for chlorite, (Fe(II)) and nontronite (Fe(III)), as shown in Figure 1. By small alterations of the relative intensities of the peaks and by slight changes in peak half widths and positions, a good fit of the calculated and experimental peaks was obtained (Fig. 2). Furthermore it is of interest to note that the percentage of Fe(III) in illite obtained from XPS peak areas is 71%, which is in good agreement with the percentage (∿80%) obtained from Mössbauer measurements. Binding energy data for both experimental and deconvoluted Fe $2p_{3/2}$ levels are given in Table II.

Although Mössbauer data revealed that kaolinite contained Fe(III) (a weak peak was observed, with long (>24 hours) counting times needed for resolution), no ESCA Fe $2p_{3/2}$ peak was observed (12 hour scanning time; typical of all these studies). This probably indicates that the small amount of iron in kaolinite is present in the lattice at depths greater than the escape depth for the photoejected electrons. Since Mössbauer is a bulk technique compared to the ESCA surface technique, it is reasonable that small amounts of lattice iron present in the bulk would be detected. When kaolinite is stirred with 100 ppm $Fe(NO_3)_3$ solution (pH remained low enough to prevent precipitation of any iron hydroxide species), the Fe $2p_{3/2}$ peak was observed in the XPS spectrum (Fig. III). The Mössbauer spectrum indicated approximately a three fold increase in the amount of Fe(III) present. The binding energy of this adsorbed Fe(III) species was found to be 711.4 eV, 1.2 eV lower than that for Fe(III) in a lattice site in the octahedral layer of a silicate mineral. It is suggested that the lowering of the Fe $2p_{3/2}$ binding energy is due to the interaction of Fe(III) with the clay surface upon adsorption. Cations in the clay surface region (Stern layer) are attracted to

Table IV - Mössbauer Spectroscopy Data

| | Iron Species | Quadrupole Splitting (mm/sec) | Isomer Shift (mm/sec) |
|---|---|---|---|
| chlorite | | | |
| pure | $Fe^{+2}$ | 2.54 ± .02 | .836 |
| w/100 ppm $Fe^{+3}$ | $Fe^{+2}$ | 2.59 ± .02 | .826 |
| illite | | | |
| pure | $Fe^{+2}$ | 2.57 ± .03 | .948 |
| | $Fe^{+3}$ | 0.60 ± .04 | -.034 |
| w/100 ppm $Fe^{+3}$ | $Fe^{+2}$ | 2.48 ± .05 | .882 |
| | $Fe^{+3}$ | 0.67 ± .02 | -.024 |
| kaolinite | | | |
| pure | $Fe^{+3}$ | --------- | ~.15 |
| w/100 ppm $Fe^{+3}$ | $Fe^{+3}$ | --------- | ~.15 |
| nontronite | | | |
| pure | $Fe^{+3}$ | 0.40 ± .04 | .005 |
| amorphous ferric hydroxide | $Fe^{+3}$ | 0.62 ± .01 | .015 |

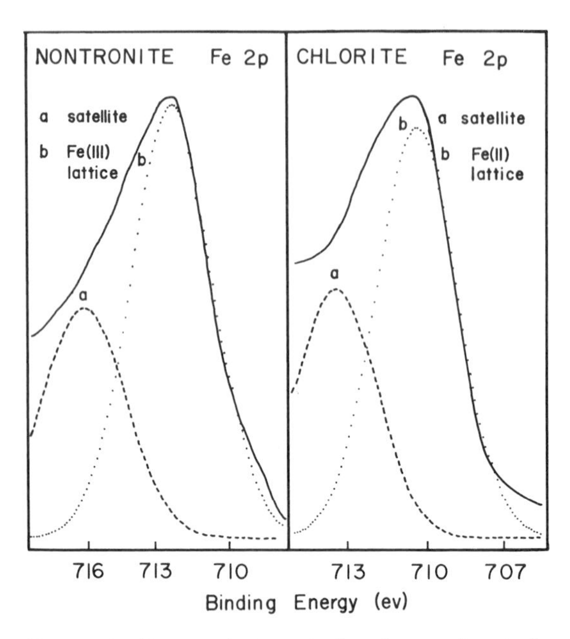


*Figure 1. Deconvolution of the photopeaks of the Fe $2p_{3/2}$ level in nontronite and chlorite*

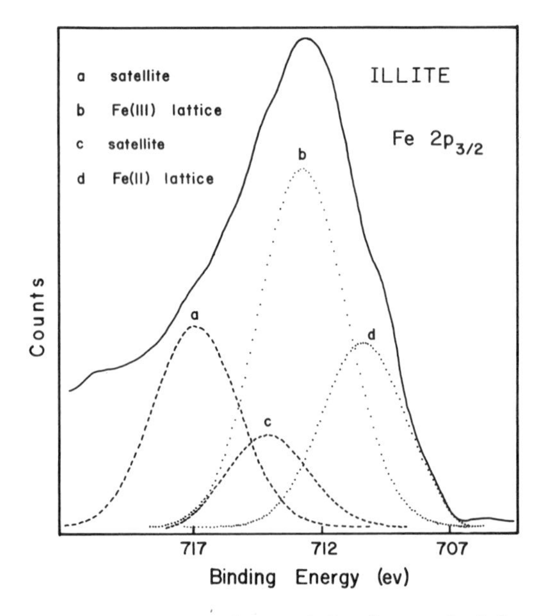


*Figure 2. Deconvolution of the photopeak of the Fe $2p_{3/2}$ level in pure untreated illite*

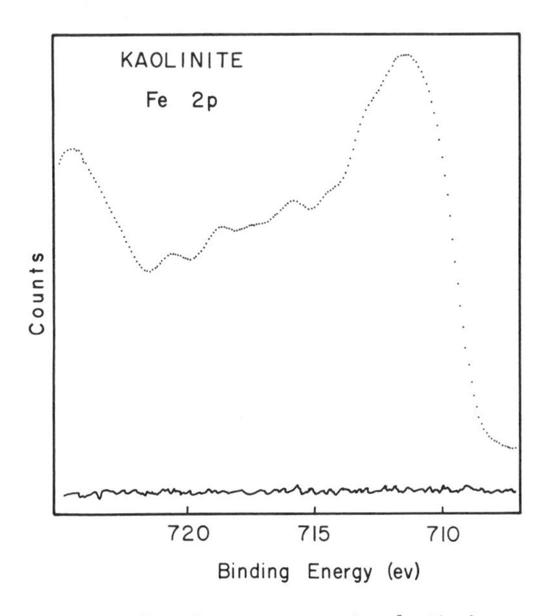


*Figure 3. Photoelectron spectra for the Fe $2p_{3/2}$ level in untreated kaolinite and kaolinite treated with 100 ppm $Fe(NO_3)_3$*

negatively charged surface bonding sites where they experience a negative site potential.[3] The effect of a negative bonding site potential on iron would be to lower the binding energy of the adsorbed cation. In similar adsorption studies of Fe(III) on chlorite, the binding energy measured for iron from the experimental photopeak is 711.4 (see Table II). This value is greater than that measured for iron in native chlorite and similar to that determined for Fe(III) adsorbed on kaolinite. It was determined that the composite experimental photopeak could be deconvoluted into two iron peaks (including satellite structure for each peak) with binding energies of 711.4 and 710.3 eV. The iron species represented by the peak at a binding energy of 711.4 is attributed to adsorbed Fe(III) while that at 710.3 corresponds to Fe(II) in the chlorite lattice.

Similarly, the composite experimental Fe $2p_{3/2}$ peak for illite, obtained when the clay mineral was stirred at pH 1 in a 100 ppm $Fe(NO_3)_3$ solution, could be deconvoluted into three iron peaks (including satellite structure for each peak) with binding energies of 712.6 eV, 711.5 eV, and 710.4 eV. The iron species represented by the peak at 712.6 eV is attributed to lattice Fe(III), that at 711.5 eV to adsorbed Fe(III), and the peak at 710.4 eV is characteristic of lattice Fe(II).

In a further effort to investigate the effect of the negative surface potential upon adsorbed cations, Cr(III) adsorption experiments onto kaolinite, chlorite and illite were performed. Uvarovite,[29] in which Cr(III) is in an octahedral layer site, was investigated with XPS to determine the binding energy for Cr $2p_{3/2}$ electrons for the lattice substituent Cr(III). XPS investigation of the three untreated clays gave no evidence for a Cr $2p_{3/2}$ peak. Upon adsorption of Cr(III) from 100 ppm $Cr(NO_3)_3$ (pH = 1.0) solutions, the XPS spectrum of each of these clays revealed a Cr $2p_{3/2}$ peak, indicating that chromium had indeed been adsorbed. The binding energy data for the Cr $2p_{3/2}$ electrons for a lattice substituted Cr(III) as well as the data for the adsorbed Cr species can be found in Table III.

The binding energy results for the adsorbed Cr(III) species reflected a similar decrease in binding energy for the adsorbed cation, as was found in the Fe(III) adsorption study. The negative surface potential at the adsorption site manifests itself by increasing the electron density on the adsorbed metal ion, thus lowering the binding energy of the metal.

The differences in binding energy for Cr $2p_{3/2}$ electrons for a lattice substituted Cr(III) as compared to an adsorbed Cr(III) ranged from 0.8 eV (Cr(III)) on kaolinite) to 1.1 eV (Cr(III) on chlorite).

## Conclusions

The technique of X-ray photoelectron spectroscopy has been shown to be a useful tool in the investigation of adsorbed metal ions on solid surfaces. Not only may it prove to be an analytical tool for detecting small (ppm level) quantities of adsorbed cations, but may be instrumental in determining the bonding nature of an adsorbed cations. Through further investigations of the adsorption of cations of different oxidation states, it is hoped that a site-bonding model for cations adsorbed on mineral surfaces can be developed.

## Acknowledgements

This work has been supported by the NSF (Grant GP-29178 ESCA Spectrometer) and by the VPI Small Projects Fund. We would also like to thank Mrs. N. Crews for assistance with the compilation and interpretation of the Mössbauer data and Drs. J. Craig and C. I. Rich for help in obtaining the minerals used in this study.

## Abstract

The bonding nature of metal ions sorbed from aqueous solutions by the marine clay minerals kaolinite, illite, and chlorite has been examined using X-ray Photoelectron Spectroscopy (XPS). Binding energies for lattice elements Si, Al, O, Mg, K, and Ca are reported, and are in good agreement with published values. Significant differences (1.9 eV) in binding energies for the Fe $2p_{3/2}$ level for iron in nontronite and chlorite were observed. This difference in binding energy was attributed to the two different oxidation states of iron, Fe(II) in chlorite and Fe(III) in nontronite. Illite, which contains both Fe(II) and Fe(III) in lattice positions was found to have an unusually broad Fe $2p_{3/2}$ photopeak. This broad peak could be deconvoluted into two peaks whose binding energies were indicative of both Fe(II) and Fe(III) lattice constituents. Fe(III) adsorbed onto kaolinite had an Fe $2p_{3/2}$ level binding energy 1.2 eV lower than lattice Fe(III). This lowering of binding energy for the adsorbed iron species as compared to lattice Fe(III) may arise from the negative potential of the electrical double layer at the mineral surface when the cation is adsorbed. Comparison of binding energies of sorbed ions ($Fe^{+3}$, $Cr^{+3}$) and their stoichiometric or mineral lattice counterparts indicate a decrease in binding energy for the sorbed ions. Extent of sorption was determined by monitoring sorbed metal ion species concentrations as well as examining changes in solution pH, dissolved $SiO_2$ and concentration of stoichiometric lattice elements ($Al^{+3}$, $K^+$, $Mg^{+2}$, $Fe^{+2,+3}$).

## Literature Cited

1. Krauskopf, K. B., Geochim. Cosmochim. Acta, (1956) 9, 1.
2. Kharkar, D. P., K. K. Turekian, and K. K. Bertine, Geochim. Cosmochim. Acta, (1968) 32, 285.
3. Stumm, W., and Morgan, J. J., Aquatic Chemistry, Wiley-Interscience, New York, 1970.
4. Van Olphen, H., An Introduction to Clay Colloid Chemistry, Wiley-Interscience, New York, 1963.
5. Grim, R. E., Clay Mineralogy, 2nd ed., McGraw-Hill, New York, 1968.
6. Whittig, L. D., Page, A. L., Soil Sci. Soc. Amer. Proc., (1961), 278.
7. Follett, E. A. C., J. of Soil Science, (1965) 16, 334
8. Fordham, A. W., Aust. J. Of Soil. Res., (1969) 7 185.
9. Fordham, A. W., Aust. J. of Soil Res., (1969) 7, 199.
10. Fordham, A. W., Aust. J. of Soil Res., (1969) 8, 107.
11. Blackmore, A. V., Aust. J. of Soil Res., (1973) 11, 75.
12. Fordham, A. W., Clays and Clay Minerals, (1973) 21, 175.
13. Lindau, I., Spicer, W. E., J. Of Electron Spectroscopy and Related Phenomena, (1974) 3, 409.
14. Singleton, J. H., Vac. Symp. Trans., (1963) 15,267.
15. Personal communication with Dr. Peter Orenski of Union Carbide Corporation.
16. Mullin, J. B., Riley, J. P., Analytica Chimica Acta (1955) 12, 162-176.
17. Fanning, K. A., Pilson, M.E.A., Anal. Chem., (1973) 45, 136.
18. Written by G. W. Dulaney, VPI & SU, 1969 (Presently at Digital Equipment, Corp., Maynard, Mass.)

19. Seals, R. D., Alexander, R., Taylor, L. T., Dillard, J. G., Inorg. Chem., (1973) 12, 2485.
20. Burness, J. H., Dillard, J. G., Taylor, L. T., Inorg. Nucl. Chem. Lett., (1974) 10, 387.
21. Adams, I., Thomas, J. M., Bancroft, G. M., Earth and Planetary Science Letters, (1972) 16, 429.
22. Huntress, W. T. Jr., Wilson, L., Earth and Planetary Science Letters, (1972) 15, 59.
23. Anderson, P. R., Swartz, W. E. Jr., Inorg. Chem. (1974) 13, 2293.
24. Yin, L. T., Ghose, S., Adler, I., Science, (1971) 173, 633-635.
25. Schultz, H. D., Vesely, C. J., Langer, D. W., Appl Spect., (1974) 28, 374.
26. Carver, J. C., Schweitzer, G. K., Carlson, T.A., J. of Chem. Phys., (1972) 57, 973.
27. Fadley, C. S., Shirley, D. A., Phys. Rev., A. (1970) 2, 1109.
28. Fadley, C. S., Shirley, D. A., Freeman, A. J., Bagus, P. S., Mallow, J. V., Phys. Rev. Lett., (1969) 23, 1397.
29. Uvarovite - $Ca_3Cr_2(SiO_4)_3$ (garnet group) - from Outokumpo, Finland, sample provided by Dr. J. Craig, Dept. of Geological Sciences, VPI & SU, Chemical composition data may be found in Rock Forming Minerals by W. A. Deer, R. A. Howie, and J. Zussman, Longmans Press, London, 1962.

# 12

# The Use of Natural Pb-210 as a Heavy Metal Tracer in the River–Estuarine System

LARRY K. BENNINGER, DALE M. LEWIS, and KARL K. TUREKIAN

Department of Geology and Geophysics, Yale University, New Haven, Conn. 06520

The naturally occurring radioactive isotope of lead, Pb-210, provides a valuable tracer for the behavior of heavy metals in the soil-stream-estuary system. Since it is continuously produced only as a member of the U-238 decay series, it is free from the problems of environmental or analytical contamination so often encountered in stable heavy metal studies. In addition, because of its half-life of about 22 years it is useful not only as a tracer but also as a dating tool to monitor events of the past 100 years in various repositories.

Lead-210 can be supplied to the soil-stream-estuary system through two pathways: (1) atmospheric--Rn-222 released from soils to the atmosphere decays to Pb-210 which then follows the fate of aerosols, ultimately to be returned to the Earth's surface via atmospheric precipitation; (2) terrigenous--Ra-226 in soils, rocks, streams and groundwater generates Pb-210 which is in some degree subject to mobilization. By following the pathways of Pb-210 from both these sources we can make predictions about the expected behavior of common lead at the Earth's surface. Further, by analogy with lead we can get an idea of the behavior of other trace metals in the Earth's aqueous reservoirs.

## The Fate of the Atmospherically-Supplied Pb-210

Precipitation, as rain, snow or dry fallout, continually strips the atmosphere of aerosols, including the Pb-210 produced there from Rn-222 decay. It has been estimated that tropospheric aerosols associated with Pb-210 have a mean residence time of a week or less (1). As Pb-210 is deposited it can fall either on standing bodies of water, the ocean or lakes, or on land. There is sufficient evidence now to indicate that in lakes it is rapidly removed from the dissolved state to the lake sediments (2, 3, 4). It is also being removed from the ocean surface. We need to know the fate of Pb-210 (and presumably other metals) when it encounters soil. One of the ways of determining this is to see how the atmospherically-derived Pb-210 is distributed in the soil profile and how well the total inventory of atmospherically-derived Pb-210 in the soil profile reflects the expected flux. If the soil contains all the Pb-210 that is delivered to it by precipitation, we can be sure that the properties of the soils make them efficient scavengers for lead and other trace metals similar to lead in chemical behavior.

Continuous collection of precipitation and dry fallout and monthly analysis for Pb-210 content have been carried out at the Kline Geology Laboratory at Yale for the past year-and-a-half. These indicate a flux of 1 dpm Pb-210/$cm^2$/y in the New Haven area. If we assume that this is an adequate measure of the Pb-210 flux for the northeastern United States, then by assaying soil inventories we can arrive at the extent of mobilization of Pb-210 from the soil.

If a soil is to be used in such a study it must be in steady state with respect to the input of Pb-210 from atmospheric sources; this requires that the soil profile be undisturbed by man's activities for 5 half-lives of Pb-210 or around 100 years. Then the flux of atmospherically-derived Pb-210 to the soil is given by the standing crop of excess Pb-210 divided by the mean life of Pb-210, if there is no loss from the soil profile by leaching.

Several soil profiles from the Northeast have now been analysed to test this. We have studied the Cook Forest State Park in north central Pennsylvania, where there is a virgin stand of timber up to 300 years old, and two sites in coastal Connecticut, the Farm River salt marsh and an associated upland island. In addition to our data, Fisenne (5) has reported Ra-226 and Pb-210 data for an undisturbed soil profile from a tobacco-growing region of Maryland. Table I shows the integrated flux to these sample sites compared to the rainfall data. Within the uncertainties, soils and salt marshes retain virtually all of the Pb-210 supplied from the atmosphere.

Table I

Flux of Pb-210 to the Eastern U.S.A.

| location | flux dpm/$cm^2$/yr | type of measurement |
|---|---|---|
| New Haven, Conn. | 1 | precip. & dry fallout |
| East Haven, Conn. | 1 | salt marsh profile (6) |
| East Haven, Conn. | 0.8 | soil profile (6) |
| Cook Forest St. Park, Pa. | 1 | soil profile |
| Maryland | 1.2 | soil profile (5) |

The Pb-210 is associated with the organic fraction of the soil. This material has the capacity to sequester other metals as well, as the data from a soil profile of the Hubbard Brook Experimental Forest in New Hampshire clearly show (Table II). Thus we conclude that metals supplied to a soil profile have a low mobility in the dissolved phase probably because of the strong sequestering properties of soil organics.

Table II

Hubbard Brook, N.H. soil profile (7)

| soil horizon | Pb-210 excess dpm/gm | Cu ppm | Zn ppm | Cd ppm | Pb ppm | organic content % |
|---|---|---|---|---|---|---|
| $A_1$ | 280 | 56 | 88 | 11 | 802 | 73 |
| $A_2$ | 0 | 1.5 | 0.9 | 1.1 | 2.1 | 0 |

## Lead-210 in Ground Waters

There are three possible sources of Pb-210 in ground water: (1) Atmospherically-derived Pb-210 which infiltrates the soil and enters the ground water reservoirs. We know from the soil data that this cannot be a large fraction of the amount supplied to the soil. In most of the eastern United States the mean annual precipitation is ∿100 cm. However the process of evapotranspiration returns about half this water to the atmosphere, resulting in a net addition of 50 cm of water per year or about 0.05 $\ell/cm^2/y$. Since the flux of Pb-210 is 1 $dpm/cm^2/yr$, the concentration of Pb-210 in annually charged aquifers would then be about 20 dpm/ℓ if no removal occurred during infiltration. (2) Ra-226 and Rn-222 in solution in the ground water, derived from leaching or recoil from rocks. Ra-226 in ground water ranges from 0.2 dpm/ℓ to 40 dpm/ℓ but in general is around 10 dpm/ℓ (8).
(3) Pb-210 directly leached from rock and soil during weathering.

Holtzman's (8) study of well waters in Illinois shows that the activity of Pb-210 is always lower than the Ra-226 activity. The shallow wells contain low amounts of Ra-226 from rock weathering and the highest Pb-210 concentrations of the ground waters analysed. Consequently, they most directly reflect the annual precipitation. However, the highest value in these shallow aquifers, 0.4 dpm/ℓ, is still much lower than the value expected in the Northeast from precipitation (20 dpm/ℓ). Thus, less than 2% of the Pb-210 supplied to a terrain from the rain is transmitted to ground water.

If we consider the deep wells analyzed by Holtzman (8) as tapping aquifers in which the water has a residence time of about 100 years, we note that the activity ratios of Pb-210/Ra-226 are so low as to predict a mean residence time for Pb-210 of about one month (with deviations by a factor of ten around this value). Such a short residence time in dominantly limestone and sandstone aquifers implies that any heavy metals injected into the ground water system will be adsorbed on the rocks close to the point of injection, and very little can be supplied to the streams draining the adjacent ground water reservoir.

## The Fate of Pb-210 in Streams

The principal sources of dissolved Pb-210 to streams are direct precipitation onto the water surface and surface runoff, although some terrigenous Pb-210 is supplied by ground water discharge. In mining areas acid mine waters supply Pb-210 as well. What has been learned about the behavior of Pb-210 in normal soil profiles and in ground water indicates that very little Pb-210 can reach the streams in the dissolved state under normal weathering conditions. However, the injection of Pb-210 into streams by acid mine waters does provide a test of the behavior of soluble Pb-210 in a normal stream system.

The first study of Pb-210 in river waters impacted by mine drainage was carried out on the Colorado River system (9). The source waters were as high as 14 dpm Pb-210/ℓ because of leaching of tailings from uranium mining operations in this semi-arid region. As the river continued downstream this high value was found to be reduced to 0.29 dpm/ℓ. This indicates that processes within the Colorado River remove Pb-210 onto particles.

A detailed Pb-210 study of the Susquehanna River system in the northeastern U.S. has been done. The West Branch of the Susquehanna River is massively affected by mine drainage from coal mining and is high in dissolved (<0.4 μm) Pb-210, Fe and Mn, and low in pH, due to the addition of sulfuric acid.

Along the course iron hydroxide is precipitating at a pH of between 4 and 4.5, and the Pb-210 supplied by the acid mine water is diminished by about a factor of two. As the West Branch of the Susquehanna River enters the valley and ridge province of the Appalachians it has a Pb-210 concentration of 0.17 dpm/ℓ (a value lower than the lowest reported value in the Colorado River system). At this juncture it receives a considerable influx of alkalinity from tributaries draining carbonate terranes. This results in neutralization of the sulfuric acid and increase of the river pH to around 6.5 to 7. This pH adjustment is accompanied by the precipitation of Mn. Due to the slow rate of Mn removal from solution, the Mn precipitation extends a considerable distance downriver from the point at which acid neutralization occurs. Analyses for Pb-210 in the river,at points in or below the region of Mn precipitation show that Pb-210 is completely scavenged from solution onto suspended particles. At the mouth of the Susquehanna River, which integrates the waters from the varied terrains, including those influenced by acid mine drainage, there is no measurable dissolved Pb-210.

The stream-borne suspended particles with their associated Pb-210 are derived mainly from soil. We can assess how much Pb-210 derived from atmospheric precipitation on a terrain leaves the system via streams if the drainage area, the atmospheric flux of Pb-210, the sediment flux of the stream, and the Pb-210 concentration of the sediment are known. This has been done for the Susquehanna River drainage basin, and the results are presented in Figure 1. Stream-borne particles carry away no more than 0.8% of the annual atmospheric flux of Pb-210 to the soil, or 0.02% of the soil standing crop. Our soil profile data show a Pb-210/organic matter ratio which is nearly constant with depth in the soil. This implies that the Pb-210 added to the soil surface is homogenized within the organic-rich layers rapidly. Other heavy metals added at the surface, whether by atmospheric precipitation or infall of dead vegetation, should be similarly homogenized. Moreover, the same Pb-210/organic matter ratio is observed in the suspended matter of streams (Table III). This

Table III

| Material | Excess Pb-210 dpm/gm organic* |
|---|---|
| Susquehanna River sed. | $\sim$57 |
| Cook Forest soil | |
| 0-2.9 cm | 48 |
| 2.9-5.9 cm | 45 |
| 5.9 cm | 40 |

*organic - weight loss on ignition

indicates that stream-borne particles directly reflect the metal content of soils, including that added by human activity. On the basis of the material balance calculation using Pb-210, this means that only 0.02%/yr. of the metals in soils are transported out of the system on stream particles under normal weathering and erosion conditions. Therefore, heavy metals with behavior similar to Pb-210 will have a mean life in the organic-rich soil horizons of 5000 years.

## Lead-210 in Estuaries: An Example from the Long Island Sound

Metals carried in the particle load of rivers may be solubilized on first mixing with seawater (see e.g. 10; 11; 12). If this indeed occurs as a common feature of fresh water-seawater encounters, the subsequent fate of the released metals as well as any metals added from other sources must be understood to assess the impact of streams in the marine cycle. A study of Pb-210 in Long Island Sound provides such an insight.

A station in central Long Island Sound was occupied 15 times in 16 months in 1973 and 1974 for collection of water and plankton samples. In Figure 2 total Pb-210 in unfiltered water samples is shown to be a linear function of suspended sediment concentrations determined in duplicate aliquots. The intercept is indistinguishable from zero, which implies that there is virtually no soluble Pb-210 in Long Island Sound. Since the direct atmospheric supply of Pb-210 is effectively added in solution, the absence of soluble Pb-210 in the water requires a very short residence time for soluble lead in Long Island Sound.

The observed behavior of thorium isotopes in the open ocean (Th-234: 13, 14), the New York Bight (Th-228: 15) and Long Island Sound (Th-234: R. Aller and J. K. Cochran, Yale University personal communication) indicates that the residence time for thorium decreases as the total suspended matter in the water column increases. In Long Island Sound it is a few days. Probably Pb-210 behaves in a fashion similar to thorium, and other trace metals may be expected to do likewise, so that the mean residence time of many dissolved heavy metals must be less than a week in Long Island Sound.

Table IV is an estimate of the overall balance for Pb-210 in the Long Island Sound system. From the results of Thomson et al. (16), supply of excess Pb-210 to the silty sediments of central

Table IV

Long Island Sound Pb-210 balance

| Pb-210 source | $dpm/cm^2/y$ |
|---|---|
| *In situ* Ra-226 decay | 0.01 |
| Rainfall | 1.6 |
| Rivers | $\leq$0.3 |
| Total | $\leq$1.9 |
| Pb-210 excess in sediments | 1.6 |

Long Island Sound is around 1.6 $dpm/cm^2/y$. *In situ* production from Ra-226 is trivial. Rainfall contributes 1 $dpm/cm^2/y$, as noted above, but this is probably concentrated by deposition only on the silty bottom, which covers 64% of the area of Long Island Sound (Bokuniewicz and Gordon, Yale University, personal communication). Thus the atmospheric flux to the Sound's surface is equivalent to $\sim$1.6 $dpm/cm^2/y$ in silty sediments. The riverine contribution has been calculated for 1973 on the basis of four samples from the Connecticut River, the dominant river draining into Long Island Sound. Total Pb-210 from this source does not exceed 0.3 $dpm/cm^2/y$ when distributed over the silty bottom.

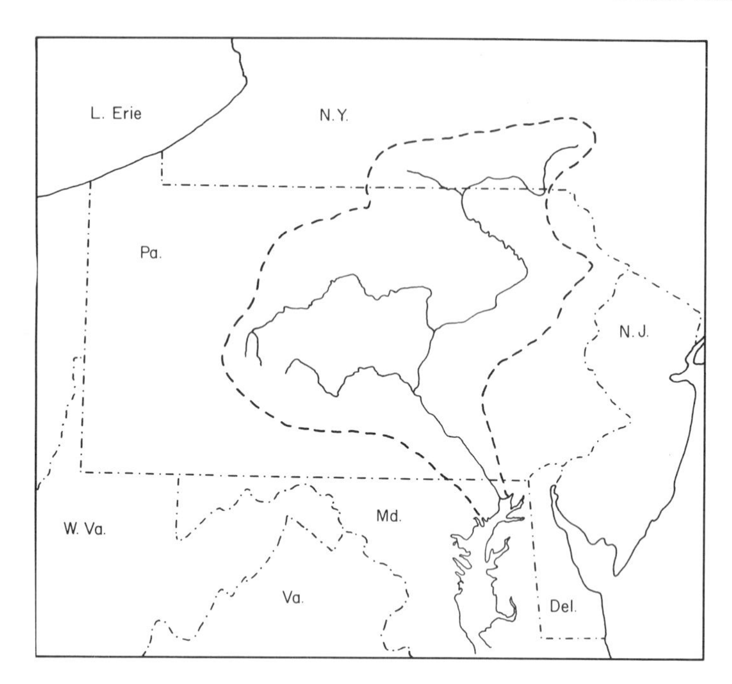


*Figure 1. Susquehanna River Basin. Area = 24,100 mi²; sediment flux = 2 × $10^{-3}$ gm/cm²/yr; particulate $^{210}Pb_{excess}$ flux = 8 × $10^{-3}$ dpm/cm²/yr; atmospheric $^{210}Pb$ flux = 1 dpm/cm²/yr; efficiency of $^{210}Pb$ removal = ≤ 0.8%; soil standing crop = 30 dpm $^{210}Pb/cm^2$; soil metal removal efficiency = 0.02%/yr.*

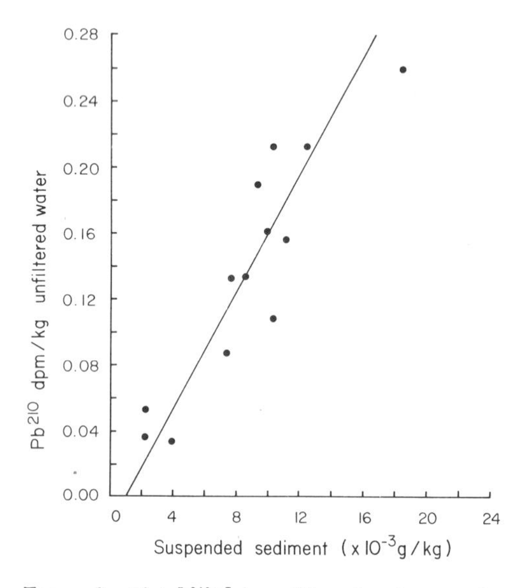


*Figure 2. Total $^{210}Pb$ in unfiltered water samples*

This is less than 20% of the atmospheric input. Thus, given the uncertainties, supply and deposition appear to be roughly in balance, strongly suggesting that Pb-210 is trapped in Long Island Sound. It is likely that other heavy metals are also efficiently retained in the sediments of Long Island Sound.

## Conclusions

Our most important conclusion must be that metals resembling Pb-210 in behavior will tend to be removed on particles at virtually every step along their journey from the atmosphere to the sea.

The atmosphere is scavenged of its aerosols as a result of normal processes of precipitation such as rain. Metals transferred to the soil by atmospheric precipitation or rock weathering will tend to accumulate and be homogenized in the organic-rich fraction of the soil, where they have a mean life of 5000 years. Ground waters tend to lose their heavy metals to the surrounding soil and rock fairly rapidly if the disequilibrium of Pb-210 relative to Ra-226 in ground water is an applicable index of the process.

In streams, any dissolved metals behaving like Pb-210 will be scavenged by manganese or iron oxides or organic matter. By the time a large stream system like the Susquehanna encounters the estuarine zone virtually all the dissolved Pb-210 has been extracted on the particulate load.

If metals are released from particles at the fresh water-seawater boundary, the evidence from Pb-210 and short-lived natural thorium isotopes is that in the estuary the metals are rapidly removed to the sediments. Thus, the flux of dissolved heavy metals from the estuary to the open ocean must be quite small. Where normal estuarine circulation pushes bottom sediments towards the continent it is obvious that very little leakage of particulate heavy metals to the deep sea will occur. Only low density or extremely fine-grained particles can escape this process and may provide a mechanism of supply of metals to the open ocean.

Alternatively, it appears that transport of metal-rich topsoil via the atmospheric route as indicated by the data of Duce et al. (17) may provide another efficient way of supplying metals to the open ocean.

## Acknowlegements

This study was supported by Atomic Energy Commision grant AT(11-1)-3573 and NOAA, Department of Commerce grant 04-4-022-35.

## Literature Cited

1. Poet, S. E., Moore, H. E. and Martell, E. A., Jour. of Geophys. Res. (1972) 77.
2. Krishnaswami, S., Lal, D., Martin, J. M. and Meybeck, M., Earth Planet. Sci. Letters (1971) 11.
3. Koide, M., Bruland, K. W. and Goldberg, E. D., Geochim. Cosmochim. Acta (1973) 37.
4. Robbins, J. A. and Edgington, D. N., Geochim. Cosmochim. Acta (1975) 39.
5. Fisenne, I. M., U.S. Atomic Energy Commission Report No. UCRL-18140 (1968).
6. McCaffrey, R. J., Yale University, personal communication.
7. Kharkar, D. P., Thomson, J., Turekian, K. K. and McCaffrey, R. J., U.S. Atomic Energy Commission Report No. COO-3573-8 (1974).
8. Holtzman, R. B., The Natural Radiation Environment (ed. Adams, J. and Lowder, M.) (1964), Univ. of Chicago Press.
9. Rama, Coide, M. and Goldberg, E. D., Science, (1961) 134.
10. DeGroot, A. J., DeGoeij, J. J. M. and Zegers, C., Geologie en Mijnbouw (1971) 50.
11. Martin, J. M., Ph.D. Thesis (1971) Univ. of Paris.
12. Turekian, K. K., Impingement of Man on the Oceans (ed. D. W. Hood) (1971), J. Wiley and Sons, Inc.
13. Bhat, S. G., Krishnaswami, S., Lal, D. Rama and Moore, W. S., Earth Planet. Sci. Letters (1969) 5.
14. Matsumoto, E., Geochim. Cosmochim. Acta (1975) 39.
15. Feely, H. W., 7th Annual Long Island Sound Conference, Jan. 11, 1975, New York, New York (1975).
16. Thomson, J., Turekian, K. K. and McCaffrdy, R. J., Estuarine Research: Recent Advances (ed. L. E. Cronin) Academic Press, in press (1975).
17. Duce, R. A., Hoffman, G. L. and Zoller, W. H., Science, (1975), 187.

# 13

# Origin and Distributions of Low Molecular Weight Hydrocarbons in Gulf of Mexico Coastal Waters

WILLIAM M. SACKETT and JAMES M. BROOKS

Department of Oceanography, Texas A&M University, College Station, Tex. 77843

The dissolved low-molecular-weight hydrocarbons, methane through the pentanes, in oceanic waters are derived from natural phenomena and/or human activities. Natural inputs have two sources; (1) biological, with methane overwhelmingly predominant and having methane to ethane plus propane ratios [$CH_4$/ ($C_2H_6$ + $C_3H_8$)] over 100 and (2) petrogenic, with ratios generally in the range of 1 to 30 (18). Man derived inputs, which are almost entirely petrogenic, result from offshore and onshore petroleum production, transportation and manufacturing operations. They are particularly significant in coastal waters. This paper attempts to define contemporary sources and describe the distributions and fates of the light hydrocarbons in the ocean with particular emphasis on coastal waters of the Gulf of Mexico.

Most early work by oil industry scientists was stimulated by the potential use of light hydrocarbon concentrations in locating offshore gas seeps. Hydrocarbon seepage had been used since the beginning of oil exploration efforts to locate commercial deposits. Link (16) reported that about one-half of the oil re-reserves of the world were discovered by drilling in the vicinity of seeps. The detection of offshore hydrocarbon seepage by means of concentration anomalies was described as early as 1960 by Dunlap *et al.* (10). During the sixties, periodic reports appeared in the commerical literature about offshore seep detection, but almost none of the large amount of work done in this subject area by the major oil companies has been reported in the open scientific literature. Thus, the usefulness to the oil industry of hydrocarbon concentration anomalies as an exploration tool is unknown to the general scientific community.

The development of a new gas chromatographic method by Swinnerton and Linnenbom (23) led to a new surge of activity in fundamental ocean studies (2, 11, 24, 26) and in oil exploration applications (20). Recent reports by Brooks et al. (6), Brooks and Sackett (7), Lamontagne et al. (13), Lamontagne et al. (14) and Swinnerton and Lamontagne (25) provide hydrocarbon baseline data which give a partial understanding of the contemporary marine geochemistry of dissolved light hydrocarbons. An attempt is made in this paper to outline the general distribution of low-molecular-weight hydrocarbons in the ocean and point out significant trends and problems.

## Procedures

Our hydrocarbon analyses are performed using three different procedures. For discrete samples, the methods described by Swinnerton and Linnenbom (23) and McAullife (17) are applied. The Swinnerton and Linnenbom method is used for the analysis of natural levels of light hydrocarbons in the open ocean. The procedure consists of helium-purging one liter samples and trapping the ethane plus higher hydrocarbons on an activated alumina column and methane on an activated charcoal column, both traps being in series and cooled to dry ice temperatures. The traps are isolated after purging and warmed to 90°C to desorb the hydrocarbons. Each trap is then separately injected into a gas chromatograph equipped with a flame ionization detector for separation and determination of hydrocarbon concentrations. McAuliffe's method consists of equilibrating a water sample with pure helium in a glass syringe. When a 1:1 mixture of sea water and helium is equilibrated, more than 95% of the dissolved hydrocarbons partition into the helium phase. A sample loop is used to inject the helium phase into the gas chromatographic stream. In our laboratory, discrete samples are now being analyzed using McAuliffe's method for the determination of methane and a modification of the method of Swinnerton and Linnenbom for the determination of the higher hydrocarbons. The modification consists of purging a 1-liter sea water sample at 200 cc/minute and using a single trap of activated alumina cooled to liquid nitrogen temperature for quantitatively trapping all the hydrocarbons except methane. The average deviation determined by replicate analyses of samples containing open ocean concentrations of methane, ethane and propane is less than ten percent.

Continuous determinations of light hydrocarbons in surface water (3 meters below the sea surface) is performed at sea using a hydrocarbon "sniffer". The procedure, described by Brooks and Sackett (7), involves using a vacuum produced by the inflow restriction of a 12-stage booster pump to continuously outgas the stream of intake water. The bubbles from the outgassing are collected and passed through a sample loop where they are injected every five minutes into the chromatographic stream of a Beckman Industrial Process Gas Chromatograph. Because the vacuum extraction procedure only outgasses approximately 25 percent of the total dissolved gases in the intake stream, relative values are obtained by comparison with open ocean concentrations and calibration with the quantitive methods described above. This procedure, no doubt, involves more uncertainity than discrete sample analysis, but considering the six orders of magnitude range in concentrations that have been observed in Gulf of Mexico surface water, the uncertainities involved in the calibration are insignificant. Hydrocarbon "sniffer" values are normalized relative to open ocean equilibrium values. Thus relative values of one for methane, ethane plus ethene and propane are equal to open ocean surface concentrations of 45, 3 and 1 nannoliters per liter, respectively.

## Results and Discussion

Natural Sources and Distributions. The primary processes for the production of light hydrocarbons in nature are: (1) bacterial catalysis, involving the reduction of $CO_2$ or fermentation of acetic acid or methanol in anoxic environments and yielding principally methane and (2) cracking, either thermal or catalytic, yielding a large spectrum of saturated and unsaturated products. Since neither of these processes are normally operative in aerobic environments the light hydrocarbons in the ocean must enter chiefly from man-related sources or across the sea-air or sea-sediment interface. Sediment generation is also responsible for the biologically reactive gases such as methane, produced in riverine and estuarine environments and found in high concentrations in runoff.

The concentration of a dissolved non-reactive gas; for example, argon, in sea water is determined, according to Henry's Law, by its partial pressure in the atmosphere and its solubility coefficient at the temperature and salinity during water mass formation. As a water mass sinks and spreads from polar regions it becomes a few degrees warmer due to geothermal heating. This warming results in a slight supersaturation of the inert dissolved gases. Nitrogen behaves in a similar manner except that processes such as nitrogen fixation and denitrification may result in slight variations of less than one percent from the concentrations expected by solution of air (4). On the other hand, dissolved oxygen is utilized by oxidative processes during water mass movement until in the north equatorial Pacific it is almost completely depleted. Methane and other dissolved gaseous hydrocarbons exhibit a behavior similar to oxygen in that equilibrium values are observed in surface water and there is a depletion in concentration with age of the water mass. To calculate the equilibrium concentration of a hydrocarbon in sea water, it is necessary to know the atmospheric partial pressure and the solubility coefficient. However, except for methane, little information is available about atmospheric hydrocarbon concentrations or solubility coefficients. For methane, almost all atmopheric concentrations are in the range of $1.4 \pm 0.1$ ppmv. Values have been obtained for near surface atmosphere samples over Greenland and Norway (15), tropical Atlantic and Pacific Oceans (14) and the Antarctic (3). Also, thirteen separate samples collected in May 1974 over the northern Gulf of Mexico along the edge of the continental shelf had values of about 1.4 ppmv. Thus, methane appears to have a nearly constant global atmospheric concentration. Assuming a partial pressure of 1.4 ppmv and the solubilities reported by Atkinson and Richards (2), calculated equilibrium concentrations in surface water are 65 and 35 nannoliters per liter at $-2^{o}C$ and $25^{o}C$, respectively. These levels are approximately 10 nannoliters per liter lower than measured surface concentrations at similar temperatures in the Ross Sea, Antarctica and the Gulf of Mexico. Profiles in these two areas are shown in Figure 1. The values higher than equilibrium concentrations at these stations and reported elsewhere (14) are apparently due to the ubiquitous biological activity in the surface layer. The deep water in the Ross Sea had concentrations very near calculated atmospheric equilibrium values, suggesting that high surface values may be a seasonal phenomenon. Although methane is not known to be formed in aerobic environments, it is found in the digestive tracts of fish and other marine organisms and therefore must be added to near surface water by biological activity.

Cooling of surface water and convection in regions of deep water formation result in isothermal water columns as shown in Figure 1 for the Ross Sea. Similar processes in the North Atlantic produce dense water

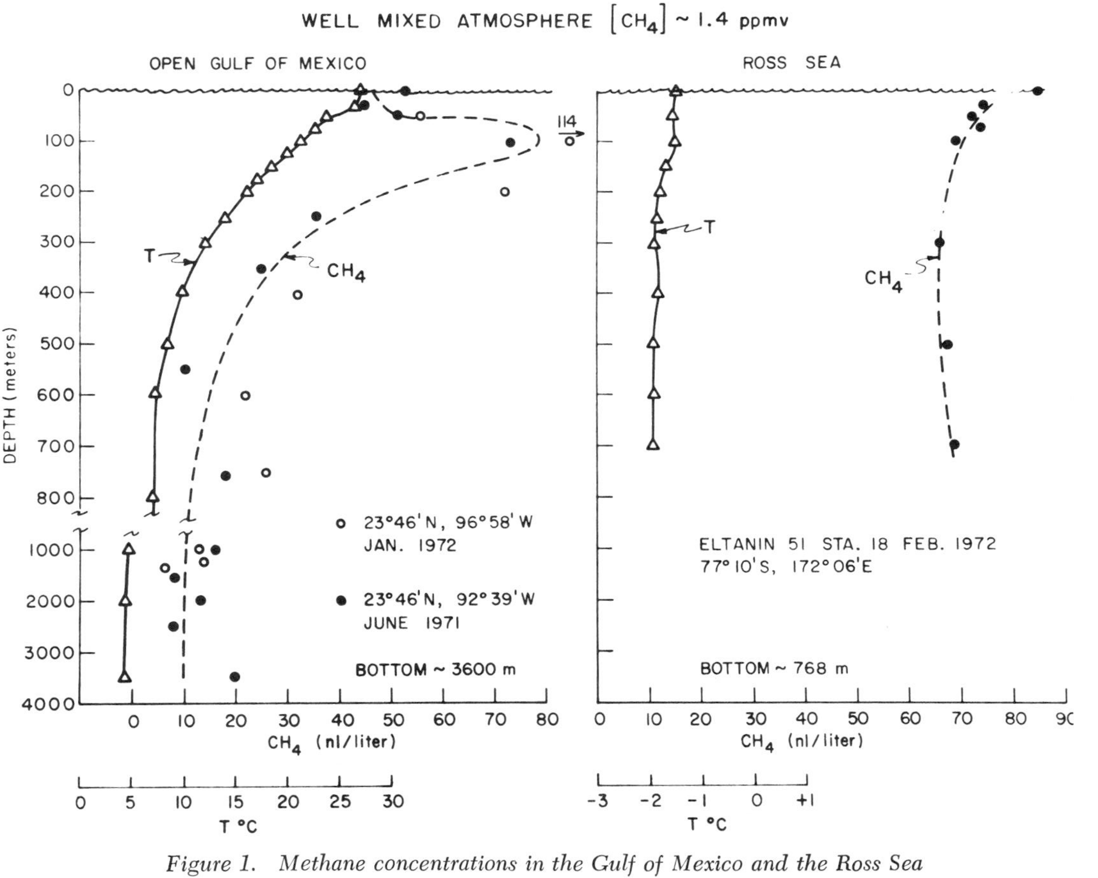


*Figure 1. Methane concentrations in the Gulf of Mexico and the Ross Sea*

which spreads laterally at depth and eventually finds its way into the Gulf of Mexico. This cold dense water begins its journey with methane levels of about 70 nannoliters per liter and when found in the Gulf, the concentrations have decreased to about 10 nannoliters per liter. This remarkable depletion apparently continues to the point where only 3 nannoliters per liter are found in 4200 meter water in the Pacific (13). The explanation for this decrease in methane concentrations is unknown but may be due to bacterial utilization, partitioning into the lipids of marine organisms and/or some other process.

The second major natural input of hydrocarbons to the ocean is across the sea-sediment interface. Again, there are two major sources of sedimentary generated hydrocarbons; biological activity and petroleum.

Methane is, by far, the preponderant biogenic hydrocarbon although unsaturated types such as ethene and propene are produced in the euphotic zone of the ocean (7, 25). However, these hydrocarbons have a shorter lifetime in sea-water than methane before degrading to non-hydrocarbon products. They are found only in trace amounts below the euphotic zone.

Claypool and Kaplan (9) have shown that methane is formed in anaerobic marine sediments during bacterial respiration processes principally below the zone of sulfate reduction where methane formation is thermodynamically unfavored (19). Also, Claypool and Kaplan (9) suggest that $H_2S$ is toxic to methane bacteria. Thus, methane is formed deep in the sulfate-free sediment column (>5 meters) and eventually may diffuse and bubble upwards into the overlying water column. Intuitively, the magnitude of this process should be proportional to the amount of organic substrate available to methane producing bacteria.

Anoxic basins such as the Cariaco Trench are environments which have high amounts of organic carbon (∿5% in sediments). The minimal exchange of these basin waters with the open ocean allows for a buildup of high concentrations of $H_2S$, produced in the sulfate reducing zone near the sediment-water interface, and $CH_4$, produced in the sediments below the sulfate reducing zone. Thus, extremely high methane levels (up to 0.2 $ml_{STP}$/liter) are observed in Cariaco Trench Basin waters (2).

The flux of methane from the few anoxic basins to the open ocean is seemingly small compared to the addition from riverine, estuarine and continental shelf sediments around the world. Coastal and estuarine sediments usually have only moderate amounts of organic matter and probably become sulfate-free within a few meters of the sediment-water interface. It is highly likely that all of these widely distributed sediments are sources of methane which manifest themselves as widespread band of diffusive seeps. These may be either small bubbling seeps or non-bubbling with methane being exchanged by diffusive processes in the sediments. In many regions of the continental shelf small bubbles will dissolve in the overlying water column before reaching the surface. Regardless of whether the seep is bubbling or non-bubbling the greatest deposition of hydrocarbons will take place in near bottom water.

Along the northern continental shelf of the Gulf of Mexico, over one hundred actual seeps have been located by sonar techniques (1, 12, 26). These seeps are associated with four types of geological settings: (1) shallow nonstructured sedimentary beds, (2) faulting, (3) domes, and (4) mud mounds (26). The only two gas samples which have been analyzed and reported in the literature have molecular (>99.9% methane) and isotopic ($\delta^{13}C_{CH_4} \sim -60$ vs PDB) compositions indicative of a biogenic origin (8). However, both of these seep gases were associated with banks that are not known to be associated with salt domes or deep faulting. A concentration versus depth profile at one seep location (Figure 2) shows high dissolved methane concentrations at depth, indicating that appreciable amounts of the seep gases dissolve in the overlying water column. The relative lack of higher concentrations of ethane and propane at depth than observed at the surface is also characteristic of a biogenic seep.

It is highly likely, although not substantiated at the present time to the authors' satisfaction, that geological faults and salt domes may give rise to petrogenic gas seeps. As previously suggested petrogenic seepage would be characterized by low $C_1/(C_2+C_3)$ ratios and $\delta^{13}C_{CH_4} \sim -45$ $^o/oo$ vs PBD compositions Brooks *et al.* (8). In the future, compositional studies of seep gas should determine if acoustically located bubble plumes are valid indicators of petrogenic gas seepage.

Anthropogenic Sources and Distributions. Over the past six years, the hydrocarbon "sniffer" system, has been used to determine the dissolved hydrocarbon concentrations in surface water at thousands of locations in the Gulf of Mexico, contiguous rivers and estuaries and the Caribbean Sea. Approximately 40 discrete sample profiles have also been analyzed. These

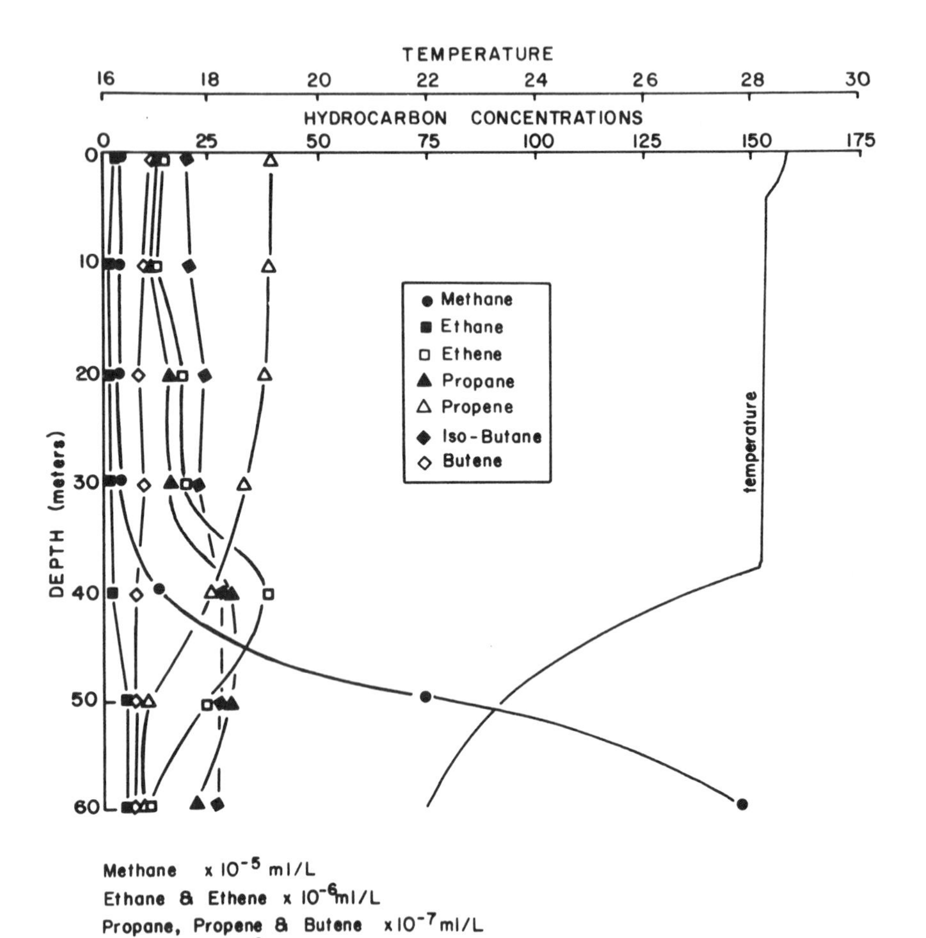


*Figure 2. Vertical hydrocarbon profile in the vicinity of a bubbling gas seep on the Louisiana Shelf*

abundant data were used to establish the general methane distribution in surface water shown in Figure 3. High dissolved low-molecular-weight hydrocarbon concentrations (up to six orders of magnitude higher than open ocean values for some of the butanes) are often found in areas along the continental shelf of the northern Gulf of Mexico in close proximity to offshore petroleum production, transportation, and refining operations. Anthropogenic sources and distributions in the Gulf of Mexico are discussed separately for each of three-types of geographical settings: (1) the open ocean; (2) the continental shelf; and (3) rivers and estuaries.

Open ocean. Most surface waters in the open ocean, including most of the Gulf of Mexico, have near equilibrium hydrocarbon concentrations. The most likely anthropogenic sources of light hydrocarbon contamination on the high seas are tankers. In the case of catastropic accidents with spillage of thousands of tons of liquid petroleum, the relatively small amounts of light hydrocarbons are unimpressive compared to the visible and physical manifestations of the high molecular weight components. On the other hand, one standard operating procedure used on many tankers is the Load-On-Top (LOT) procedure which involves combining all the tank washings (tank cleaning is usually accomplished with high velocity water jets) into one tank, allowing the oil to separate from the water, pumping out the underlying "clean" water and retaining the oil and the water near the oil-water interface for processing onshore or incorporation into the next shipment. Essentially, the tanker becomes a gigantic separatory funnel in which a solvent extraction takes place with the relatively soluble components of petroleum such as the low-molecule-weight aliphatic and aromatic hydrocarbons being substantially extracted into the water phase which is subsequently discharged. Thus tankers using this standard operating procedure will leave a trail of the water soluble fraction of petroleum.

On one cruise in October 1971, we happened to approach a tanker anchored on Misteriosa Bank in the northwestern Caribbean. As was learned sometime later, pumping of the water underlying an oil layer had just been completed. In fact, the pumping had gone somewhat too far, in that an oil emulsion and slick were clearly visible. About 1 kilometer from this tanker the concentrations of ethane, propane, the butanes and the pentanes suddenly increased by one to two orders

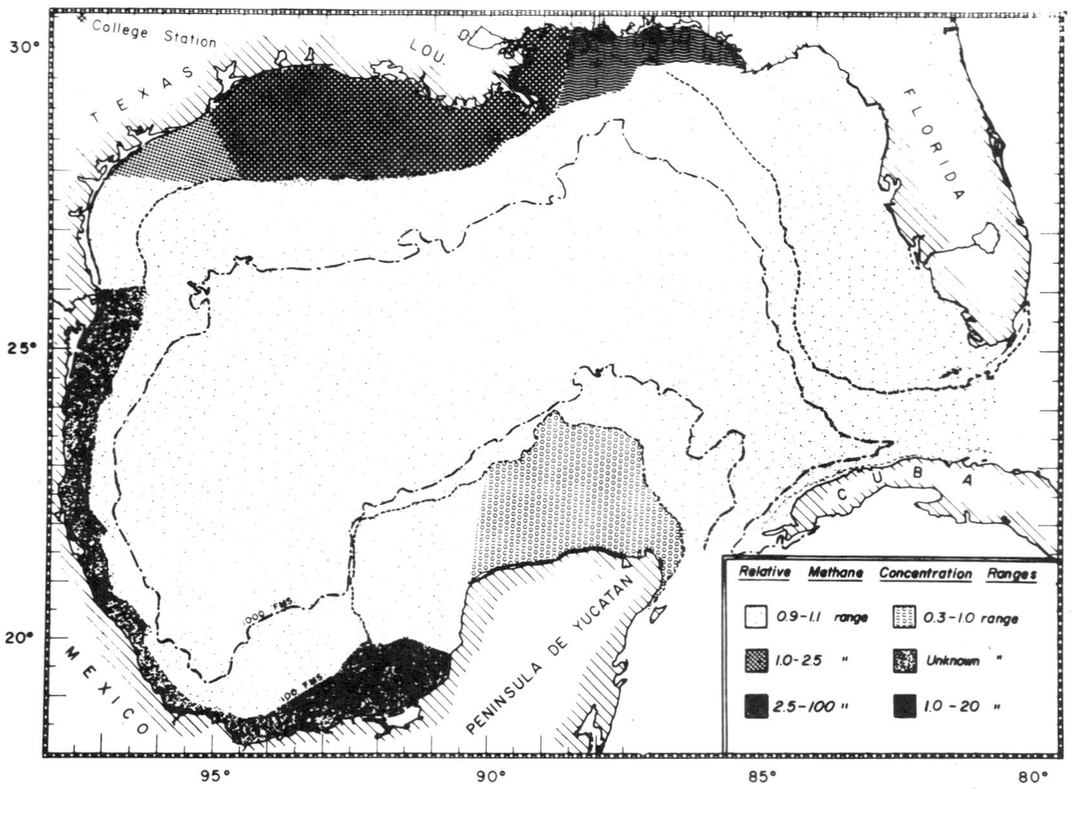


*Figure 3. The distribution of methane in surface waters of the Gulf of Mexico where a relative value of one is equal to 45 nl./l.*

of magnitude over normal levels. A chromatogram obtained at this time is given in Figure 4B. This situation dramatized the fact that a standard internationally approved procedure is nevertheless resulting in significant pollution of open ocean surface waters by the water soluble fraction of petroleum. This fraction includes some of the most toxic components of petroleum such as the light aromatics and various nitrogen, sulfur and oxygen containing compounds.

Continental shelf. Several thousand oil and gas wells are being produced from hundreds of offshore platforms along the northern continental shelf of the Gulf of Mexico. Minor accidents and normal handling procedures result in rather low day to day losses of liquid hydrocarbons. On the other hand, two standard operational procedures are introducing very large quantities of the volatile and water soluble components of petroleum into the areas near production platforms. These procedures that are sanctioned with some limitations by the Federal Government at the present time are (1) underwater venting of "waste" gases and (2) the discharging of co-produced brines.

Offshore production platforms provide a working facility to separate water, liquid and gaseous hydrocarbons. Separation is required offshore because of the extremely large pressure drops which are experienced during two-phase flow of liquids and gases in a pipeline. Separation then is made offshore to allow single-phase transmission of crude oil and gas through separate pipelines. Depending on the economics of transportation, the separated gas may either be transported for sale or disposed of in one of several ways. The gas may be disposed of by venting the gas either below or above water, by flaring the gas, or by compressing the gas and pumping it back into the reservoir to provide gas lift. The first two options account for the majority of the natural gas disposal in the Gulf of Mexico. Although some gas is flared, it appears that most of the produced gases are vented underwater. According to figures obtained from the USGS, $1.7 \times 10^{11}$ liters of gas were vented and/or flared in the Gulf of Mexico during September 1974 (21). They estimated that roughly 70% of this amount was vented underwater. The petroleum industry considers underwater venting preferable to above water flaring because it eliminates possible ship-pipe collisions or hurricane damage associated with the latter procedure. Since platform operations involve the separation of volatile and flammable gases and liquids, underwater venting also eliminates

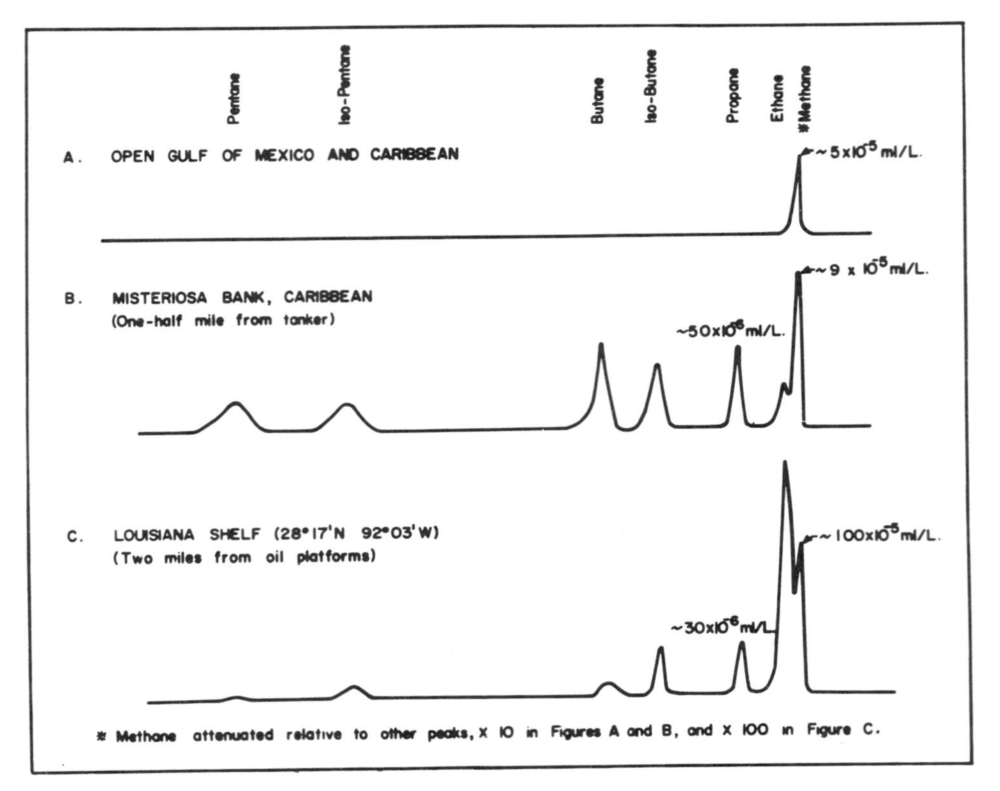


*Figure 4. Gas chromatograms for three typical situations. Retention times in minutes are as follows:* $C_1$ *—0.4,* $C_2$ *—0.6,* $C_3$ *—1.0, i-*$C_4$ *—2.0, n-*$C_4$ *—2.6, i-*$C_5$ *—5.8, and n-*$C_5$ *—7.2.*

the danger associated with burning flares in the vicinity of producing platforms.

At the sea surface underwater vents are visible as a boiling turbulent area of "white water covering an area of several hundred square meters. Table 1 shows the analyses of the water overlying some vents sampled during cruises aboard the R/V Gyre during 1974. They all have the same basic ratio of hydrocarbon components although the absolute concentrations are dependent on the proximity and depth of the underwater vent.

Figure 5 shows the distributions of relative methane, ethane, propane and isobutane concentrations in surface water around a platform group in South Marsh Island Block 6. All saturated $C_2$ to $C_4$ hydrocarbons, which are certainly petrogenic, as well as methane, show approximately the same relative concentrations. The highest values were found within a few meters of the surface manifestation of an underwater vent. It appears that this particular vent was responsible for the concentrations three orders of magnitude higher than equilibrium found some 8 kilometers down current from the vent.

The sum total of the hydrocarbon inputs from these vents seem to be responsible for the estimated average relative hydrocarbon levels of about 30 in coastal waters. Using the USGS estimates for September 1974 of 1.7 X $10^{11}$ liters of gas vented, with 70% being vented underwater, and assuming 5% solution, would give a methane flux into coastal waters of the northern Gulf of Mexico of about 800 nannoliters/liter/month. This type of calculation gives an indication of the contribution of anthropogenic methane and sup-

Table 1. Gaseous Hydrocarbon Concentrations in Surface Water near Underwater Vents

| Vent | $CH_4^*$ | $C_2H_6^*$ | $C_3H_8^*$ | $i\text{-}C_4H_{10}^*$ | $N\text{-}C_4H_{10}^*$ |
|---|---|---|---|---|---|
| Exxon SMI 6-A | 1500 | 40 | 15 | 3.6 | 3.6 |
| G.A.T.C. GI 40B | 1400 | 40 | 14 | 3.7 | 3.0 |
| G.A.C.C. GI 47C | 840 | 20 | 7.6 | 2.0 | 1.8 |
| G.A.T.C. WD 96R | 200 | 5 | 2.2 | 0.59 | 0.48 |

*Concentrations expressed as $10^{-6}$ liters/liter

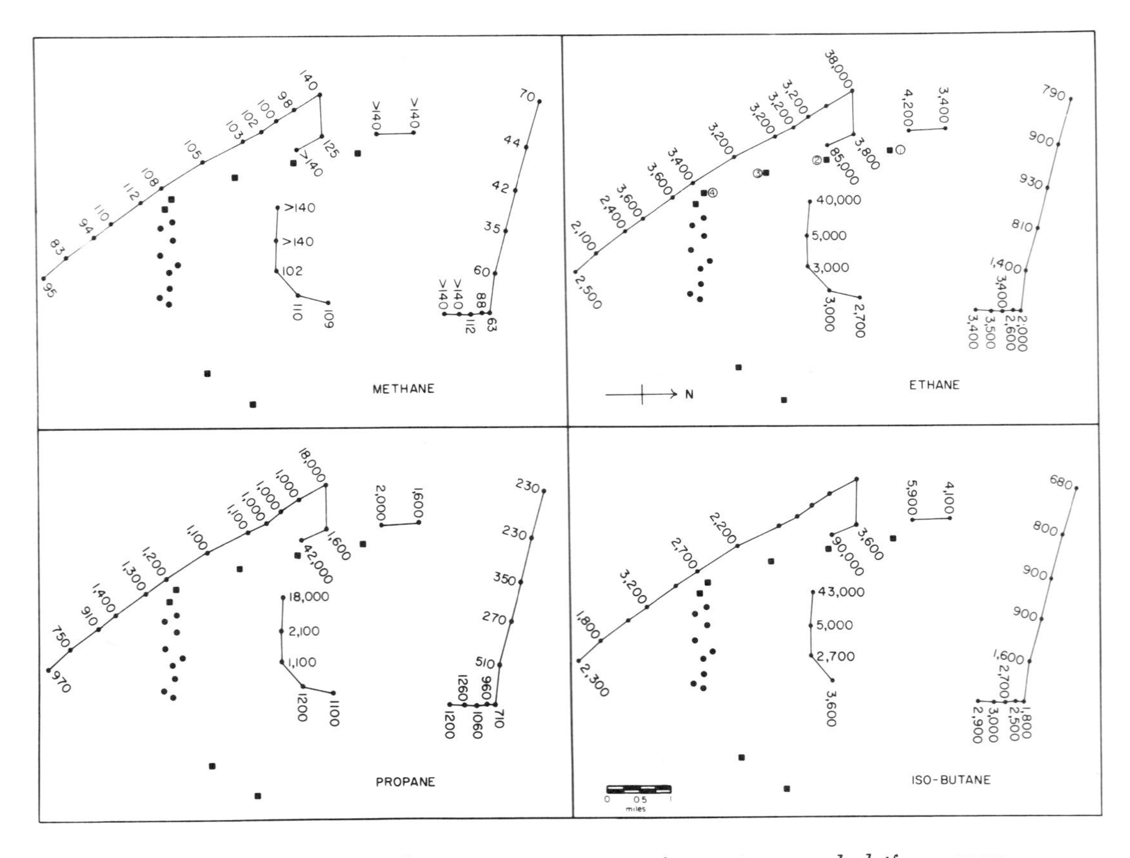


*Figure 5. Relative hydrocarbon concentrations in surface water around platform group at 28° 58'N and 92° 00'W*

port the contention that underwater vents are responsible for the high hydrocarbon levels in Gulf of Mexico coastal waters.

Brines which are co-produced with oil and gas are also sources of the light and relatively water soluble hydrocarbons in coastal areas. In many fields the brine production exceeds oil production. Thus in 1969, the 700 wells in South Pass Blocks 24 and 27 fields produced 1.4 X $10^7$ and 3.1 X $10^7$ liters per day of oil and brines, respectively (22). These brines are usually discharged to the sea after passing through an oil/water separator. Although government regulations require that brines be treated to reduce oil levels down to less than 50 ppm, this concentration represents for some constituents a significant input from the 4.6 X $10^7$ liters of brine per day discharged into Outer Continental Shelf areas in 1973. Table 2 shows some light hydrocarbon concentrations in produced brines. The concentrations in these brines represent lower limits since only one of six brines was properly collected and stored. Its methane concentration of 13 x $10^6$ nannoliters/liter and benzene concentration of 10 mg/liter are both within government regulations. In addition to brines disposed of in offshore areas, there is another 15 X $10^7$ liter/day discharged in coastal waters within the three mile limit. If all of the 20 x $10^7$ liters of brine per day had the methane concentrations listed above, the daily methane input would represent only one percent of the normal open ocean equilibrium levels predicted in these coastal waters. However, the calculated input of 730 metric tons of benzene per year could be significant in discharge areas in so far as toxicity to marine organisms is concerned.

Rivers and estuaries. There are numerous rivers entering the Gulf of Mexico from the United States and Mexico. Many of the associated estuaries are ringed with petroleum production, refinery and petrochemical facilities. These activities are concentrated along the banks of the Southwest Pass of the Mississippi River, the Port Arthur-Beaumont-Orange areas along the Sabine and Neches Rivers in Texas, the Houston ship channel leading into Galveston Bay, the Brazos River estuary near Freeport, Texas, and the estuary of the Santa Maria River near Tampico, Mexico. Intuitively the input of hydrocarbons to the Gulf of Mexico is related to the size of the facility and to the amounts and types of effluents from a particular operation rather than the flow of fresh water to the Gulf. However, the Mississippi, draining 2/3 the total area of

Table 2. Light Hydrocarbon Concentrations in Produced Brines

| Sample[1] | $C_1$ | $C_2$ | $C_3$ | $\sim C_4$ | $n-C_4$ | $\sim C_5$ | $n-C_5$ | Benzene[2] | Toluene | $C_1/(C_2+C_3)$ |
|---|---|---|---|---|---|---|---|---|---|---|
| Brine 1[3] | 13,000 | 970 | 300 | 75 | 85 | 31 | 17 | 9.8 | 3.3 | 10 |
| Brine 2 | 160 | 30 | 24 | 9.1 | 4.0 | 4.2 | -[4] | 3.3 | 0.29 | 3 |
| Brine 3 | 171 | 8 | 0.3 | 0.02 | <0.01 | - | - | - | - | 21 |
| Brine 4 | 966 | 48 | 5.8 | 0.14 | <0.01 | - | - | - | - | 18 |
| Brine 5 | 105 | 7.9 | 0.4 | <0.01 | <0.01 | - | - | - | - | 13 |
| Brine 6 | 36 | 2.0 | 0.1 | <0.01 | <0.01 | - | - | - | - | 17 |

[1] $C_1$ to $C_5$ hydrocarbon concentrations expressed as $10^{-6}$ liters of gas/liter of brine

[2] Benzene & Toluene concentrations expressed as mg/L by weight (ppm).

[3] Brine 1 is the only sample stored in glass, other brine samples were weathered for 2 to 3 weeks by storage in 1-liter plastic containers with an air space above the brine.

[4] (-) indicates the component was not determined.

the United States and contributing about 62 percent of the total runoff to the Gulf, may be considered a special case. Low-molecular-weight petrogenic hydrocarbons added to the upper portion of the Mississippi are probably lost during flow to the Gulf. Water soluble but relatively non-volitile components such as benzene may be retained and ultimately carried into the Gulf. An example of relative hydrocarbon concentrations for one of several trips up the Southwest Pass and down the South Pass of the Mississippi River Delta is shown in Figure 6. For this particular cruise relative $C_1$ to $C_4$ hydrocarbon concentration ratios ranged from $10^2$ to $10^4$. Methane increased dramatically at the mouth of each pass, reflecting the acknowledged high productivity in the water column and sediments in the transition zone between river and ocean environments. Methane to ethane plus propane ratios were 20 (petrogenic) in the river and increased sharply due to the increase in methane levels, discussed above. The entire pattern is complex but each of four cruises in the delta showed similar trends.

The overall effect of the industrial activities along the northern Gulf coast is to produce a band of high hydrocarbon levels which decrease seaward due to mixing with cleaner offshore water and loss to the atmosphere.

In summary, low-molecular-weight anthropogenic hydrocarbons, particularly from underwater vents, have increased mean concentrations in the Gulf of Mexico coastal waters by an estimated one to two orders of magnitude over equilibrium open ocean levels. This non-equilibrium situation results in a flux of these hydrocarbons from the sea to the atmosphere. Using the classical stagnant layer model discussed in several recent papers; eg, Broecker and Peng (5), the piston velocity or exchange rate is on the order of one to two meters per day for methane with slower rates for higher molecular weight and/or more hydrophilic species. By eliminating all anthropogenic sources, it would take about one-month for coastal waters with an isopycnal water column structure, such as found for winter conditions, to return to equilibrium with respect to the atmosphere. This estimate is based on a mean depth of about 50 meters. For summer conditions when a thermocline develops, this exchange would only take place readily within the mixed layer. High hydrocarbon levels in the deeper water such as shown on Figure 2. might well be laterally advected out into the open Gulf, and thereby provide a tag for studying the mixing dynamics between coastal and open Gulf waters.

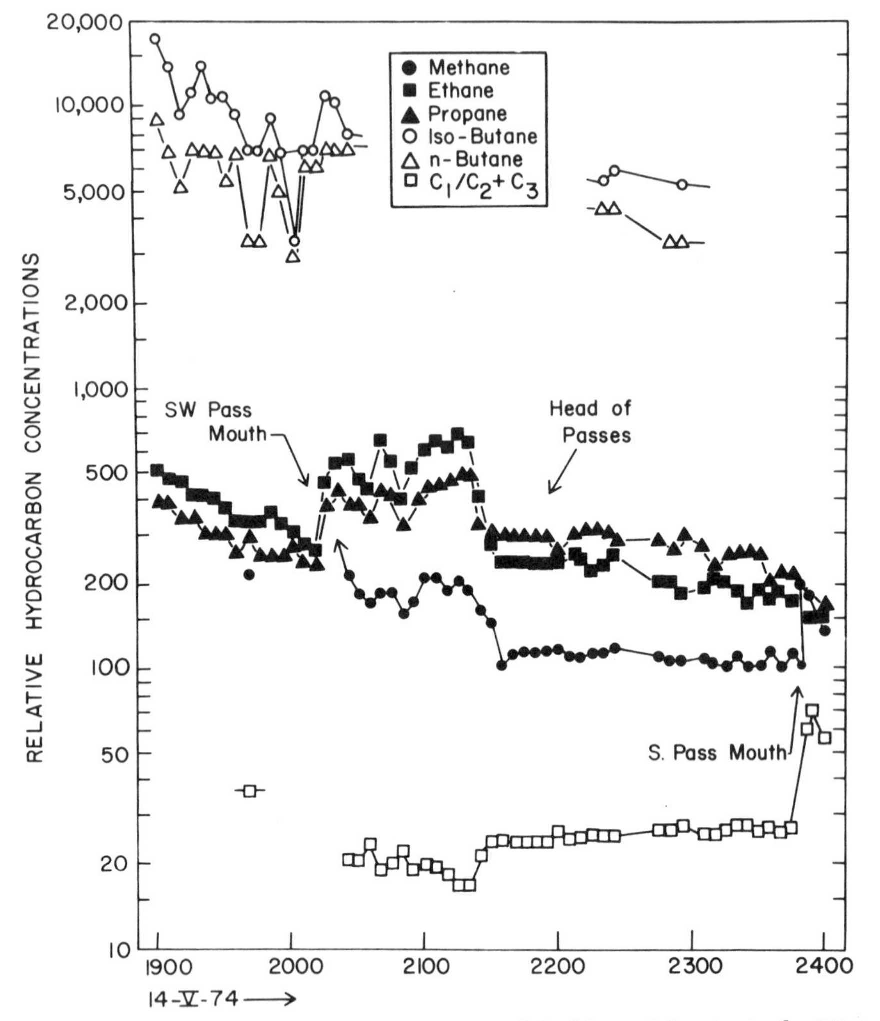


*Figure 6. Relative hydrocarbon concentrations and $C_1/(C_2 + C_3)$ ratios in the Mississippi River at the delta for 1900–2400 LMT on 14 May 1974*

Summary

Thousands of $C_1$ to $C_5$ dissolved hydrocarbon determinations have been made on water samples from the Gulf of Mexico in order to locate and/or characterize (1) natural seeps and (2) man-derived sources of petroleum contamination. Rising bubbles from over 100 natural gas seeps have been located by various groups using conventional sonar techniques, but the only bubbles analyzed thus far have hydrocarbon (>99% $CH_4$) and isotopic ($\delta^{13}C_{CH_4} \sim -60^o/oo$ vs PBD) compositions, indicative of a biological rather than a petroleum origin. Extremely high dissolved hydrocarbon concentrations were found in areas receiving effluents from petroleum production, transportation and refining operations. Two oil industry practices are especially important from a pollution point of view; the underwater venting of non-commercial amounts of gases issuing from gas/liquid separators and the discharging of brines with high concentrations of the relatively water soluble components of petroleum. The hydrocarbon inputs into coastal waters along the Gulf coast due to natural seeps are apparently negligible compared to man-derived inputs. Loss to the atmosphere is the primary fate of dissolved hydrocarbons in coastal waters.

Acknowledgment

Financial support for these studies was provided by National Science Foundation Grant Nos. GX-37344 and GA-41077.

Literature Cited

1. Albright, J. L., Offshore (1973) 33, 123-128.
2. Atkinson, L. P. and Richards, F. A., Deep Sea Res. (1967) 14, 673-684.
3. Behar, J. V., Zafonte, L., Cameron, R. E. and Morelli, F. A., Antarctic Journal (1972) 7, 94.
4. Benson, B. B., and Parker, P. D. M., Deep Sea Res. (1961) 7, 237-253.
5. Broecker, W. S. and Peng, T. H., Tellus (1974) 26, 21-35.
6. Brooks, J. M., Fredericks, A. D., Sackett, W. M. and Swinnerton, J. W., Environ. Sci. and Technol. (1973) 7, 639-642.
7. Brooks, J. M. and Sackett, W. M., J. Geophys. Res. (1973) 78, 5248-5258.
8. Brooks, J. M., Gormly, J. R. and Sackett, W. M., Geophys. Res. Let. (1974) 1, 213-216.
9. Claypool, G. E. and Kaplan, J. R., "Natural Gases

in Marine Sediments," 99-140, Plenum Press, New York, 1974.

10. Dunlap, H. F., Bradley, J. S. and Moore, T. F., Geophys. (1960) 25, 275-282.
11. Frank, D. J., Sackett, W. M., Hall, R. and Fredericks, A. D., Am. Assoc, Pet. Geol. Bull. (1970) 54, 1933-1938.
12. Geyer, R. A. and Sweet, W. M., Trans. Gulf Coast Assoc. Geol. Soc. (1973) 23, 158-169
13. Lamontagne, R. A., Swinnerton, J. W., Linnenbom, V. J. and Smith, W. D., J. Geophys. Res. (1973) 78, 5317-5324.
14. Lamontagne, R. A., Swinnerton, J. W. and Linnenbom, V. J., Tellus (1974) 26, 71-77.
15. Larson, R. E., Lamontagne, R. A., Wilkniss, P. E. and Wittman, W. J., Nature (1972) 240, 345-347.
16. Link, W. K., Am. Assoc. Pet. Geol. Bull. (1952) 36, 1506-1514.
17. McAullife, C., Chem. Technol. (1971) 1, 46-51.
18. Moore, B. J., Miller, R. D. and Shrewsbury, R. D., Inform. Circ. 8302, 1-142, U. S. Bureau of Mines, Pittsburgh, 1966.
19. Richards, F. A., "Chemical Oceanography," 1, 611-645, Academic Press, New York, 1965.
20. Schink, D. R., Guinasso, N. L., Sigalove, J. J. and Cima, N. E., Offshore Technol. Conf. (1971) Paper No. 1339, 131-142.
21. Solanas, D. W., private communication (U.S.E.S., New Orleans) 1974.
22. Sport, M. C., Proc. Offshore Technol. Conf. (1969) Paper No. 1015, 145-152.
23. Swinnerton, J. W. and Linnenbom, V. J., J. Gas Chrom. (1967) 5, 570-573.
24. Swinnerton, J. W. and Linnenbom, V. J., Science (1967) 156, 1119-1120.
25. Swinnerton, J. W. and Lamontagne, R. A., Environ. Sci. and Technol. (1974) 8, 657-663.
26. Tinkle, R. A., Antoine, J. W. and Kuzela, R., Ocean Ind. (1973) 8, 139-142.
27. Wilson, D. F., Swinnerton, J. W. and Lamontagne, R. A., Science (1970) 168, 1577-1579.

14

# Genesis and Degradation of Petroleum Hydrocarbons in Marine Environments

T. F. YEN

Departments of Chemical Engineering, Medicine (Biochemistry) and Program of Environmental Engineering, University of Southern California, Los Angeles, Calif. 90007

Petroleum and its related organic matter comprise the following entities: (a) gas, a gaseous mixture of methane to butane; (b) petroleum, a liquid mixture of mainly hydrocarbons; (c) asphalt, a semi-solid mixture of complex nonhydrocarbons and (d) kerogen, a solid, cross-linked, insoluble multipolymer. Gas, petroleum, and asphalt can be found in reservoirs, but kerogen is located only in source rocks. Usually, gas, asphalt and kerogen are associated with petroleum production.

The elemental distribution and composition of a typical oil, asphalt, and kerogen are illustrated in Table I. In going from gas (alkanes), to oil (petroleum), asphalt and kerogen, there is a decrease in the ratio of hydrogen to carbon in the material.

Table I <u>Chemical Composition of Gas, Oil, Asphalt and Kerogen</u>

| | Gas | Oil | Asphalt | Kerogen |
|---|---|---|---|---|
| Carbon | 76 | 84 | 83 | 79 |
| Hydrogen | 23 | 13 | 10 | 6 |
| Sulphur | 0.2 | 2 | 4 | 5 |
| Nitrogen | 0.2 | 0.5 | 1 | 2 |
| Oxygen | 0.3 | 0.5 | 2 | 8 |

Composition of a typical crude oil consists of 30% gasoline ($C_4$-$C_{10}$), 10% kerosene ($C_{11}$-$C_{15}$), 15% gas oil ($C_{13}$-$C_{20}$), 20% lube oil ($C_{20}$-$C_{40}$) and 25% asphaltic bitumen ( $C_{40}$). In terms of molecular types, an oil contains 30% paraffins, 50% naphthenics, 15% aromatics and 5% asphaltics. The relationship of different petroleum products together with the boiling ranges of representative hydrocarbons are illustrated in Figure 1.

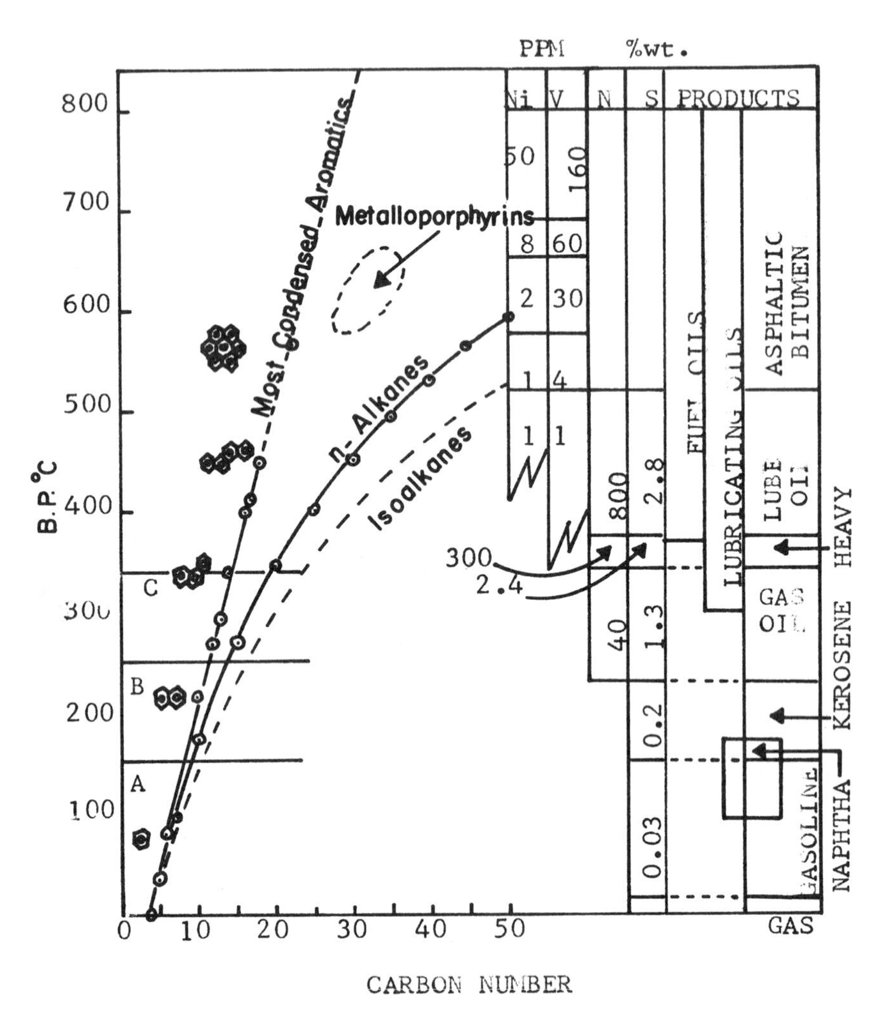


*Figure 1. (Right) Composition and boiling points of major petroleum fractions. (Left) Summary of the boiling points of pure hydrocarbons (based on Ref. 34).*

Asphalts are derived either from straight run residues of distilled crude, or from the bottom of the pot with the "visbreaker" thermal cracking process. They can also be formed by oxidation of crude residues by the "blown" air oxidation process. Typical asphalt is a composite material consisting of asphaltene, wax, and resins. Asphaltene is responsible for many properties of the asphalt system. Furthermore, asphalts also may be converted to asphaltites or asphaltoids by weathering and metamorphism (1). One of the properties of the asphaltic bitumen or "resid" fraction is its concentration of metals such as vanadium and nickel (2), as indicated by Figure 1.

Kerogen is a solid, insoluble, organic material derived through biological origin and laid down with the sediments. Structurally, it is a cross-linked multipolymer (3) composed of subunits similar to asphaltic bitumens. Kerogen is the major organic component in source rock and shale.

In terms of carbon distribution on earth, kerogen in shales and sandstone constitutes at least 1000 times more than the biomass of living organisms (Table II) (4).

Table II Carbon on Earth

| | Carbon g/cm$^{2}$* |
|---|---|
| Carbonates | 2340 |
| Shales and Sandstones | 633 |
| Ocean ($HCO_3^{-1}$+$CO_3^{-2}$) | 7.5 |
| Coal and Petroleum | 1.1 |
| Living matter and dissolved Organic Carbon | 0.6 |
| Atmosphere | 0.1 |

*Gram per square centimeter earth surface based on Skirrow (Ref. 8).

Actually, kerogen is the most abundant organic material in the ecosystem; surveys indicate (5) the dilute reserves are 3200 trillion metric tons. This amount is at least 5000 times the volume of known petroleum reserves (Fig. 2).

Gas ($C_1$-$C_4$) usually admixes with petroleum in producing reservoirs to a small extent. It is evolved from source rock through the terminal maturation sequence as shown in Figure 3, as well as product of metamorphism of all fossil fuels during diagenesis. The amount in subsurface water is considerable.

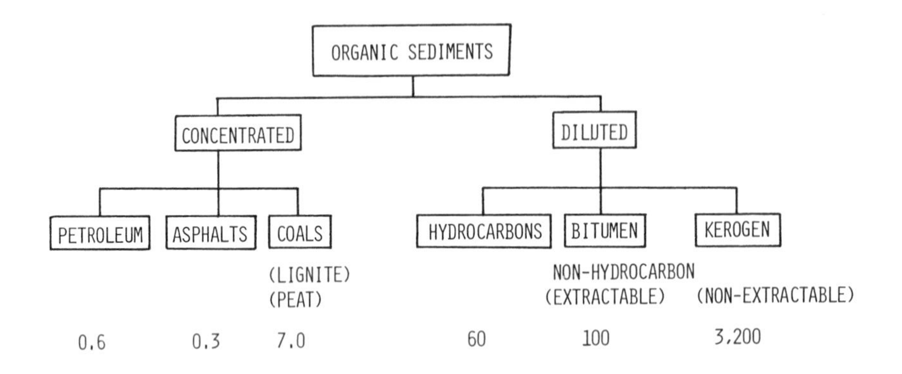


*Figure 2. Classification of organic sediments. The abundance of various fossil fuels in the ecosphere is presented in trillion tons.*

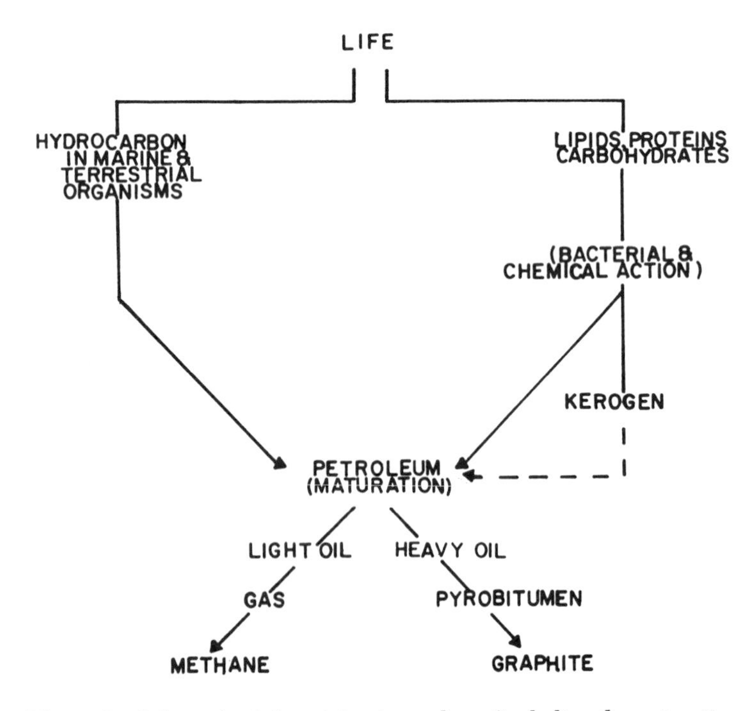


*Figure 3. Schematic of the origin of petroleum (including the maturation process)*

## Biogenesis

At the 8th World Petroleum Congress for the Panel Discussion on "Recent Advances in Understanding the Origin, Migration and Accumulation of Oil and Gas and the Resulting Methods of Evaluating Existing Petroleum Prospects" in 1971, scientists working in this field have agreed to rule out any theories on inorganic origin. The general acceptance of a biotic origin is due to the overwhelming evidence of biogenic origins supplied by recent research. However, even recently, a few people still advocate the Fischer-Tropsch formation from the "Petroleum Rain" (6).

It is not the intention of the present paper to trace the biogenesis of all the components in petroleum since the required lengthy discussion is out of the scope of this presentation, and an excellent review has been compiled by Hodgson (7). Rather, the emphasis here is to point out the important evidence for organic origin.

The general similarity of petroleum composition and the nature of its constituents to the basic structural subunits of components occurring in organisms is a major reason for believing the biotic origin of petroleum. For example, in living matter, both bacteria and spores contain major concentrations of lipids, as shown in Table III, and simple decarboxylation of those lipids will yield hydrocarbons. Long-chain hydrocarbons derived from organisms do form portions of petroleum, but the amount is less than 10%. The bulk of the petroleum constituents are derived from the fossilized biomass through bacterial and chemical action. As indicated in Figure 3, kerogen is the controlling factor for such biostratinomy and taphonomy processes.

Table III Composition of Living Matter

| | Weight % of Major Constituents | | |
|---|---|---|---|
| Substance | Lipids | Proteins | Carbohydrates |
| Green plants | 2 | 7 | 75 |
| Humus | 6 | 10 | 77 |
| Phytoplankton | 11 | 15 | 66 |
| Zooplankton | 15 | 53 | 5 |
| Bacteria (veg.) | 20 | 60 | 20 |
| Spores | 50 | 8 | 42 |

As indicated in Table IV, long-chain hydrocarbons are derived from microorganisms such as algae and fungi. These include n- and iso-alkanes, alkenes, alkadienes and isoprenoids (9). There is no need for the more advanced organism to yield steroids and terpenoids since steroids have been found in prokaryotic organisms. Furthermore, the algae are responsible for most $C_{17}$ hydrocarbons (Table V). The same n-$C_{17}$ hydrocarbon has been found in 2.5 billion-year Sudan Shale. Non-photosynthetic bacteria yield hydrocarbons in the general range of $C_{15}$-$C_{25}$. For the photosynthetic bacteria, the contribution is shifted to the higher molecular weight range of cyclics and unsaturates (Table VI) (10). The hydrocarbon distribution in $C_{25}$-$C_{35}$ range is mostly derived from plant and insect material (Fig. 4).

In most anaerobic and aerobic bacteria, there are unsaturated long-chain carboxylic acids. These unsaturated acids contain multiple unconjugated olefinic double bonds (Table VII) (11). Usually, these isolated double bond sites may crosslink to form high molecular weight kerogen analogs or precursors.

One of the direct evidences of biotic origin of petroleum is the remains and debris found in kerogens through peleobiology. These are spores and even specimens of _Botryococus brannii_ in shale samples. The kerogen may originate from lacustrine basins, shallow seas, continental platforms and shelves, or in small lakes, bogs, lagoons associated with coal-forming swamps. The types of microorganisms in each set of geological environments are different.

Active enzymes have been isolated from kerogen-bearing shales or rocks. Typical analyses by Bergmeyer's method for some shales and lignites are summarized in Table VIII (12).

## Biological Markers

In Eglinton's definition (13), a biological marker is "a compound, the structure of which can be interpreted in terms of previous biological origin". It is essential that this particular compound is a stable one that can survive through a number of environmental stimuli (14). Hitchon (15) has discussed detailed classes of biological markers. This review is only limited to certain basic principles of their nature.

The best example is porphyrin. Petroporphyrin is widespread in all petroleum and related substances (16). These porphyrins chelate either with vanadium or nickel; other

Table IV Long-Chain Hydrocarbons Produced by Microorganisms

| Hydrocarbons | Range and predominant carbon no. | Type of Micro-organisms |
|---|---|---|
| n-Alkanes | $C_{16}$-$C_{33}$,$C_{17}$ | *Anacystis cyanea* |
| | | *Chroococcus turgidus* |
| | | *Lyngbya aestuarii* (A)* |
| | $C_{13}$-$C_{19}$ $C_{25}$-$C_{29}$,$C_{29}$ | *Sclerotinia sclerotiorum* (F)** |
| i-Alkanes | $C_{25}$,$C_{27}$,$C_{29}$,$C_{31}$ | *Sphacelothica reiliana* |
| | $C_{27}$,$C_{29}$,$C_{31}$ | *Ustilago maydis* |
| | | *Urocystis agropyii* |
| | 4-$M_e$ $C_{17}$ | *Chlorogloea fritschii* (A) |
| | 7-$M_e$ $C_{17}$ | *Nostoc muscorum* |
| | 8-M $C_{17}$ | *Anacystis cyanea* (A) |
| Alkenes | n-$C_{17}$-$\alpha$ | *Chlorella pyrenoidosa* |
| | | *Anacystis nidulans* |
| | n-$C_{23}$-$\alpha$,n-$C_{25}$-$\alpha$ | *Scendesmus quadricauda* (A) |
| | n-$C_{26}$-$\alpha$,n-$C_{29}$-$\alpha$ | *Anacystis montana* (A) |
| | n-$C_{25}$-$\alpha$,n-$C_{27}$-$\alpha$ | *Chorella vulgaris* (A) |
| Alkadiene | n-$C_{27}$-$\Delta$ 1,18 | |
| | n-$C_{29}$-$\Delta$ 1,20 | *Botryococcus braunii* |
| | n-$C_{31}$-$\Delta$ 1,22 | |
| | botryococcene | |
| | isobotryococcene (tetra-$M_e$ acyclic triterpene) | (A) |
| Isoprenoids | squalene | *Methylococcus capsulatas* |
| | hopene-22 [29] | *Methylococcus capsulatas* |
| | hopene-17 [21] | *Bacillus acidocaldarius* |

*Algae (A).
**Fungi (F).

Table V Hydrocarbons from Algae (percent only of total listed)

| | Nostoc muscorum (blue-green) | Anacystis nidulans (blue-green) | Phormidium luridum (blue-green) | Chorogloea fritschii (blue-green) | Chlorella pyrenoidosa (green algae) |
|---|---|---|---|---|---|
| n-$C_{15}$ | 0.35 | 20.60 | --- | --- | 0.13 |
| n-$C_{16}$ | 0.35 | 2.50 | --- | 0.26 | 0.073 |
| $\Delta$-$C_{17}$ | --- | 2.95 | --- | --- | 81.5 |
| n-$C_{17}$ | 82.75 | 73.75 | 96.00 | 87.30 | 18.5 |
| 7- and 8-methyl heptadecane | 16.10 | 0.15 | 4.00 | 0.09 | --- |
| 4-methyl-heptadecane | --- | --- | --- | 12.20 | --- |
| n-$C_{18}$ | 0.41 | --- | --- | 0.09 | 0.055 |

Table VI Hydrocarbons from Photosynthetic and Nonphotosynthetic Bacteria
(percent only of total listed)

| | E. coli (aerobic) | P. shermanii (anaerobic) | Clostridium acidiurici | Rhodospirillum rubrum | Chlorobrium (sulfurbacteria) |
|---|---|---|---|---|---|
| n-$C_{15}$ | 0.5 | 2.1 | 1.0 | 0.01 | 1.5 |
| n-$C_{16}$ | 1.7 | 2.6 | 14.4 | 0.06 | 0.75 |
| Pristane | --- | 46.5 | 2.1 | 0.10 | 0.5 |
| n-$C_{17}$ | 5.5 | 13.3 | 50.0 | --- | --- |
| Phytane | --- | 1.0 | 1.3 | 3.50 | 50.0 |
| n-$C_{18}$ | 27.6 | 3.6 | 4.5 | --- | 0.5 |
| n-$C_{19}$ | 12.0 | 3.8 | 4.9 | 0.35 | 1.3 |
| n-$C_{20}$ | 10.0 | 3.8 | 3.1 | 0.45 | 1.3 |
| n-$C_{21}$ | 5.5 | 4.2 | 1.9 | 0.32 | 1.0 |
| n-$C_{22}$ | 6.0 | 4.1 | 1.7 | 0.24 | 1.5 |
| n-$C_{23}$ | 8.3 | 3.1 | 1.0 | --- | 3.0 |
| n-$C_{24}$ | 7.4 | 1.5 | 1.0 | --- | 4.1 |
| n-$C_{25}$ | 6.0 | 1.0 | 0.7 | --- | 6.9 |
| n-$C_{26}$ | 3.3 | 0.5 | 0.5 | --- | 10.8 |
| n-$C_{27}$ | 3.3 | 0.5 | --- | --- | 13.1 |
| n-$C_{28}$ | 0.5 | --- | --- | --- | 2.1 |
| Squalene & high MW cyclics | --- | --- | --- | 94.7 | --- |

Table VII The Nature of the Unsaturated Acids Found in Some Species of Bacteria (Mainly from Scheuerbrandt & Bloch)

| Bacterial species | Number of carbon atoms in the acids | | |
|---|---|---|---|
| | 14 | 16 | 18 |
| Aerobic bacteria: | | | |
| *Mycobacterium phlei* | | 9 (3%) | 9 (7%) |
| *Corynebacterium ovis* | | 9 (14%) | 9 (6%) |
| Anaerobic bacteria: | | | |
| *Lactobacillus arabinosus* | | (4%) | 11 (35%) |
| *Pseudomonas fluorescens* | | 9 (31%) | 11 (12%) |
| *Rhodopseudomonas spheroides* | | | |
| anaerobic culture (light) | 7 (6%) | 9 (3%) | 11 (69%) |
| aerobic culture (dark) | 7 (3%) | 9 (3%) | 11 (78%) |
| *Clostridium butyricum* | 7 (0.3%) | 9 (8%) | 11 (2%) |
| *Clostridium pasteurianum* | 7 (1%) | 9 (3%) | 11 (2%) |

types of metals are rarely found (2).

The predominant types of petroporphyrin, both the DPEP (Desoxyphilloerythroetioporphyrin) and the etio (Etioporphyrin), are unsymmetrically substituted. They are quite different from the non- or symmetrically substituted porphins synthesized from electric discharge or chemical condensation of pyrroles and aldehydes. Further the sequential evolution scheme of the porphyrin structure can only be explained and traced stepwise by biodiagenesis. Fossil porphyrins from chlorophyll a to DPEP, to Etio and formally to arylporphyrin can be illustrated by Figure 5. A detailed account of the biological diagenesis of petroporphyrin has been reviewed by Yen (2).

The predominance of phytane and pristane in petroleum is also closely related to chlorophyll pigments. The sequence can be depicted in Figure 6. In recent marine sediments, biological markers such as pheophorbides and chlorins are detected. The similarity of the composition in organisms and in recent sediments has enhanced greatly the biotic origin claim (see Figure 4) (17).

Carbon isotope ratio studies indicate that organic compounds of petroleum and of the associated shales have similarities. Silverman (18) has indicated the difference between marine and nonmarine lipids and their relationship to crude oil (Fig. 7). Silverman (19) has analyzed narrow petroleum distillation fractions and revealed that there is minimum in the 425-450°C range (Fig. 8). Since fractionation will result in higher isotopic ratios by splitting of the low-isotopic ratio methane molecule, compounds located in minimum must be unaltered, naturally occurring molecules. Actually, it was found that the 425-450°C fraction consisted of steriods and triterpenoids. Furthermore, there is a maximum in optical activity corresponding to the minimum isotopic positions.

For a Rio Zulia crude oil, there is more than one minimum (Fig. 9 ). The minimum at 150°C indicates isoprenoids of $C_{10}$ type which are present in petroleum in considerable amounts (Table IX).

## Geological Fence

Classical Cox's "posts" theory (20) is still useful for the limiting factors under which organic material could transform into petroleum. Some of the "posts" are as follows: Temperature not exceeding 200°C, pressure not exceeding 500 psi, and at least

RECENT SEDIMENTS
STEVEN
BRAY
KVENVOLDEN
PLANT & INSECT
WAXES CHIBNALL
% OF TOTAL
0
10
20
30
40
21
23
25
27
29
31
33
35
37
CARBON NUMBER

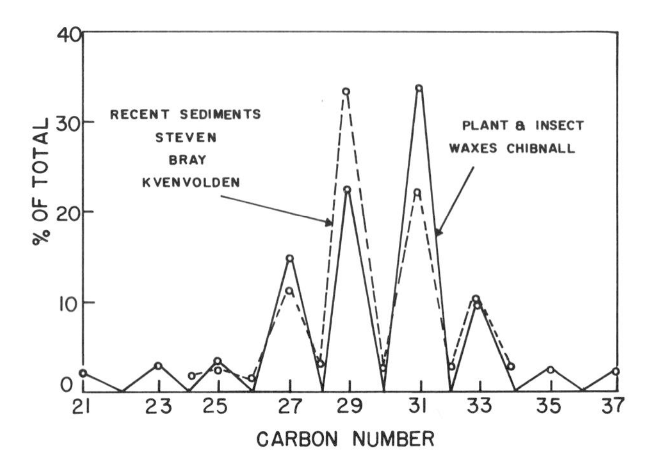

*Figure 4. n-Paraffin distribution in plant and insect waxes and recent sediments (based on Ref. 17)*

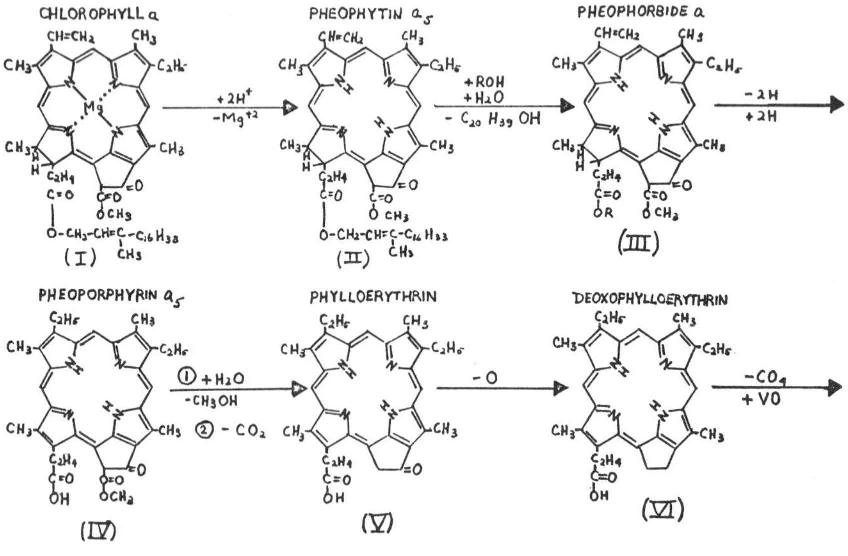

CHLOROPHYLL a
(I)
+2H+
-Mg+2
PHEOPHYTIN a5
(II)
+ROH
+H2O
- C20 H39 OH
PHEOPHORBIDE a
(III)
-2H
+2H
PHEOPORPHYRIN a5
(IV)
① +H2O
-CH3OH
② - CO2
PHYLLOERYTHRIN
(V)
-O
DEOXOPHYLLOERYTHRIN
(VI)
-CO2
+VO

*Figure 5. Diagenesis of petroporphyrins from chlorophyll* a (I) *to DPEP* (VII)

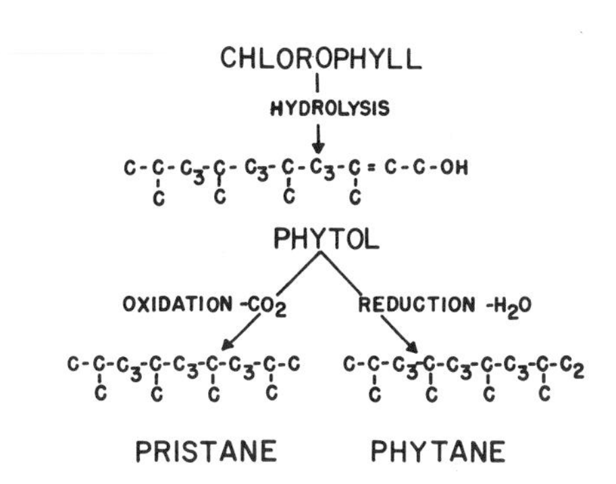


*Figure 6. Fate of pristane and phytane in petroleum*

DPEP (VII)
ETIO (ETIOPORPNYRIN III) (VIII)
(IX)
RHODO (BENZO PORPHYRIN) (X)
(XI)
ms-α- NAPHTHYL PORPHYRIN (XII)

*and Etio (VIII). The end-product of MS-α-naphthyl-porphyrin (XII) is also shown.*

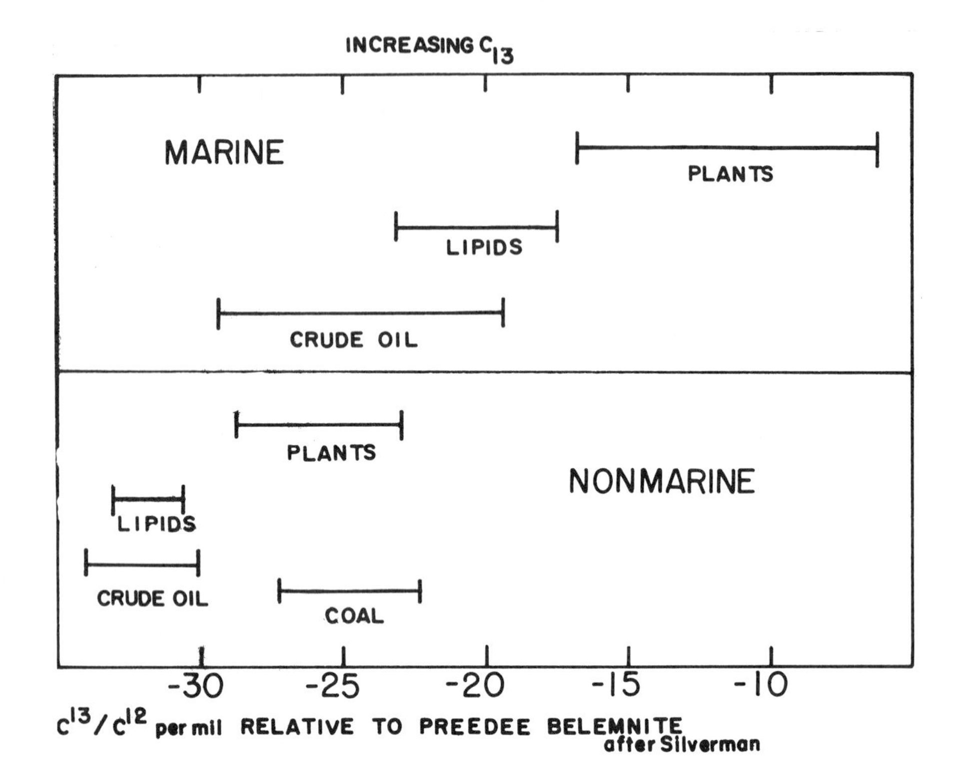


*Figure 7. Carbon isotope range of natural materials (based on Ref. 19)*

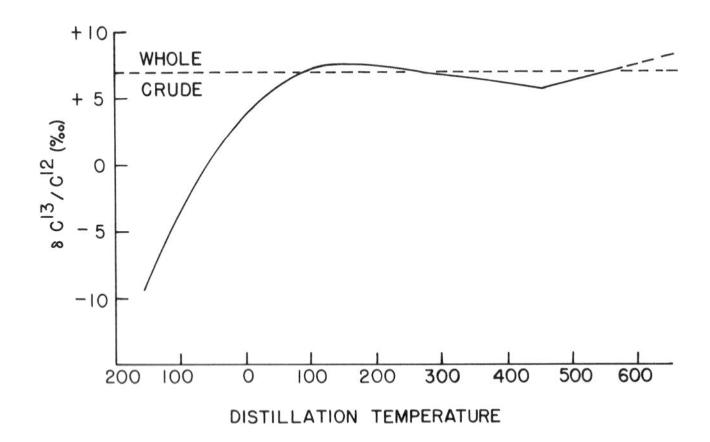


*Figure 8. Generalized curve relating carbon isotope ratios and boiling temperature of petroleum distillation fractions (based on Ref. 18)*

Table VIII Enzymes Isolated from Shales and Lignites

| Enzymes | IU (Bergmeyer's Method) |
|---|---|
| Aldolase | 3.2-25.3 |
| α-Amylase | 111.5-701.4 |
| Creatine phosphokinase | 0.0-10.2 |
| Glutamate dehydrogenase | 40.3-150.3 |
| Glutamic-oxalacetic transaminase | 130.1-800.2 |
| Glucose-6-phosphate dehydrogenase | 14.3-96.3 |
| Isocitrate dehydrogenase | 7.6-30.4 |
| Malate dehydrogenase | 14.2-66.2 |
| Leucine aminopeptidase | 60.4-260.5 |
| Trypsin | 43.5-240.3 |

Table IX $C_{10}$-Isoprenoid Fraction from Ponca City Crude Oil

| Compound | B.P. (°C) | Conc. (vol. %) |
|---|---|---|
| 2,6 Dimethyloctane | 160.4 | 0.50 |
| 2-Methyl, 3-ethylheptane | 160.9 | 0.64 |
| All other isomers (49 possible) | 155-165 | 0.44 |

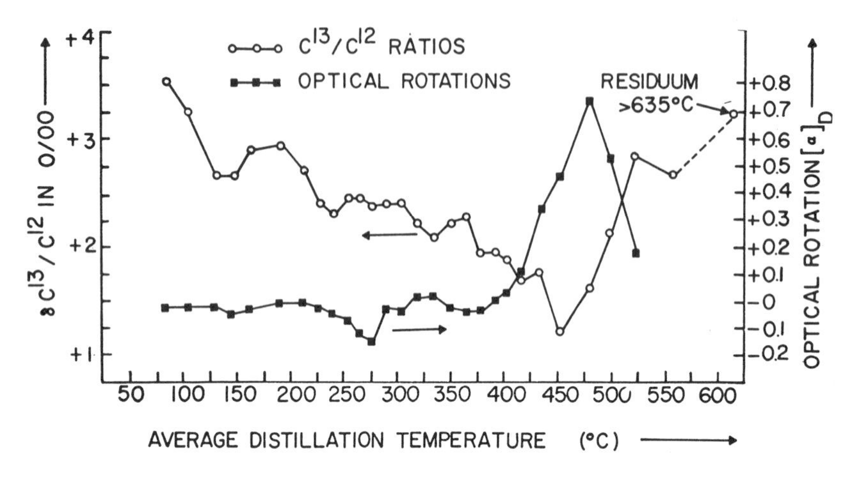


*Figure 9. Relation between $C_{13}/C_{12}$ ratio and optical rotation in Rio Zulia crude oil distillation fractions (based on Ref. 19)*

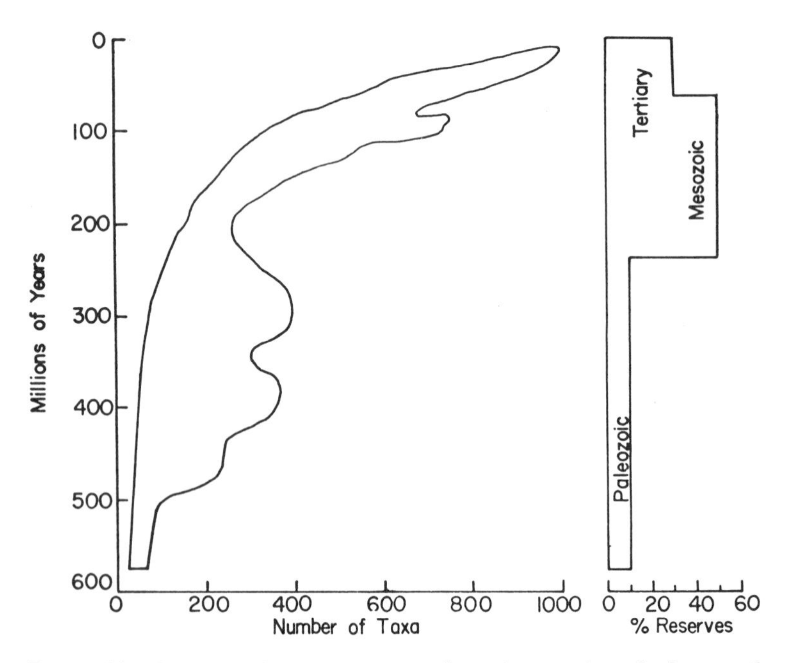


*Figure 11. Reaction of variation in number of taxa of total plants and animals over geological time. The right band rotation is percent reserves of total free world (based on Ref. 21).*

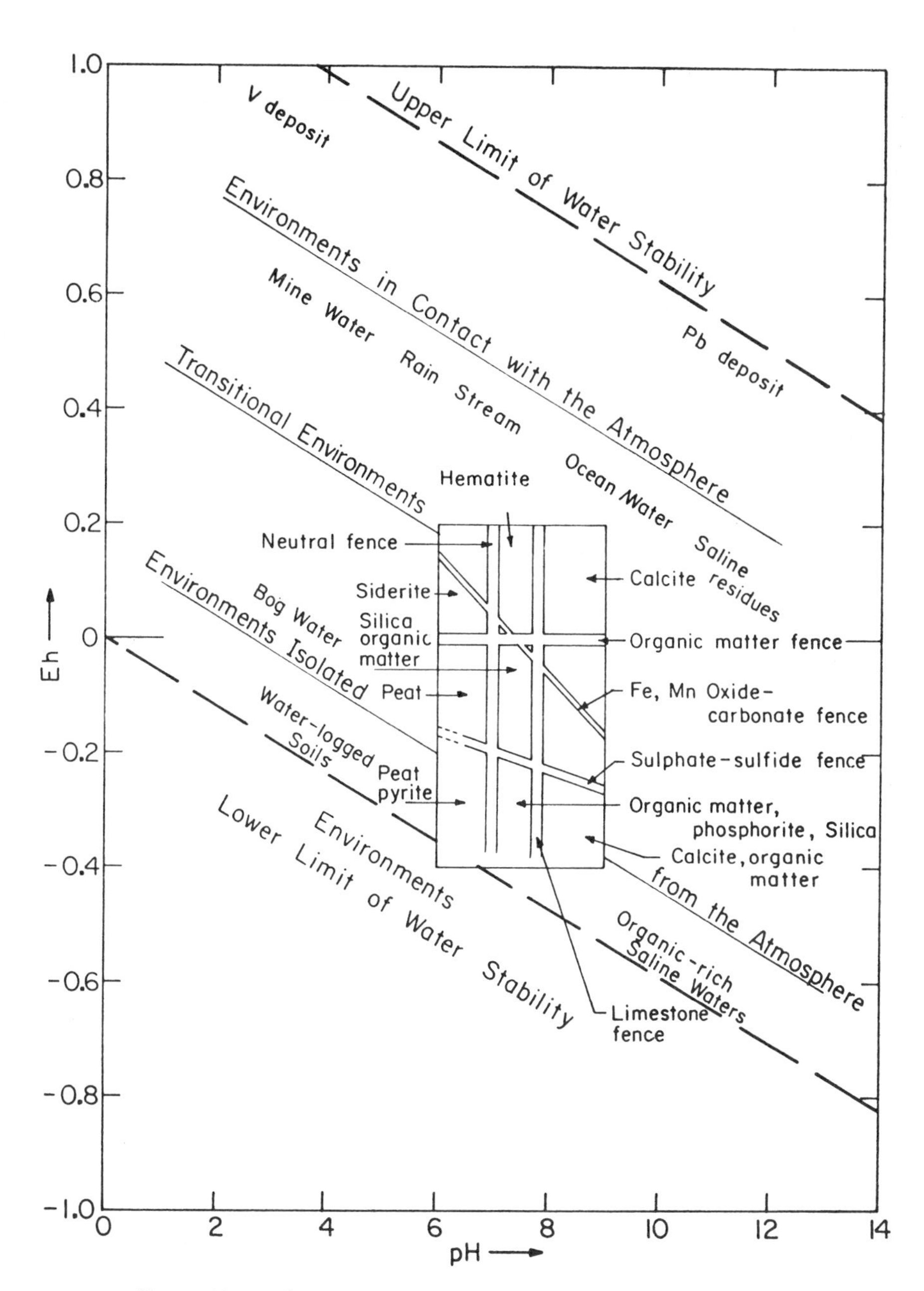


*Figure 10. Redox potential vs. pH for different types of deposits*

one million years to allow the transformation. Refinement of these geological constraints has included the elevation of temperature and pressure and the reduction of maturating time, since Pleistocene oil has been discovered. In order to preserve the organics, a reducing environment is much favored. Under marine conditions, a situation of high production of organics and a rapid sedimentation is often required.

Petroleum deposits which behave as any other mineral ore concentrate are formed at a given set of pH-Eh boundary regions (Fig. 10). Again, the condition of preservation is enhanced either in the absence of bacteria or in the presence of fine-grained sediments such as silts or clays, and carbonate reefs. In studies of the habitat of oil through geological age, the amount of reserves is proportional to the extant taxa (Fig. 11) (21).

The condition of petroleum production is well exemplified in the hinge-belt of major downwarps, for example, the east-west Tethyan Belt of Eurasia and the eastern Circum-Pacific Belt (Fig. 12). These temperate zones have the conditions favorable to the overproduction of food. The major oil fields of the world are included in those two major belts.

Major kerogen deposits are limited by the conditions listed in Table X. In this case, the microorganisms' source material is different for different set of geological conditions.

## Molecular Diagenesis

Major reactions under geochemical conditions are in dynamic equilibrium. Major reaction types can be exemplified by these types outlined in Figure 13. Decomposition (spallation) and condensation (polymerization) still explain many petroleum transformations. Using expressions in Figure 13, P is a large complex molecule when complexed to a. Examples of this could be the fragmentation of methane from pentacyclic triterpenoids, e.g., aromatization of betulinic acid to 2,9-dimethylpicene (Fig. 14). Condensation reactions can be illustrated by the melanoidin formation from the interaction of amino acids and simple sugars. This condensate can be further polymerized to melanoidin structure (22).

The most frequent conversion for fossil fuels is dehydrogenation which is illustrated by the second type. Depicted by a triangular plot, the end-product is concentrated in the shaded region (Fig. 15) from the starting material such as cellulose or carbon

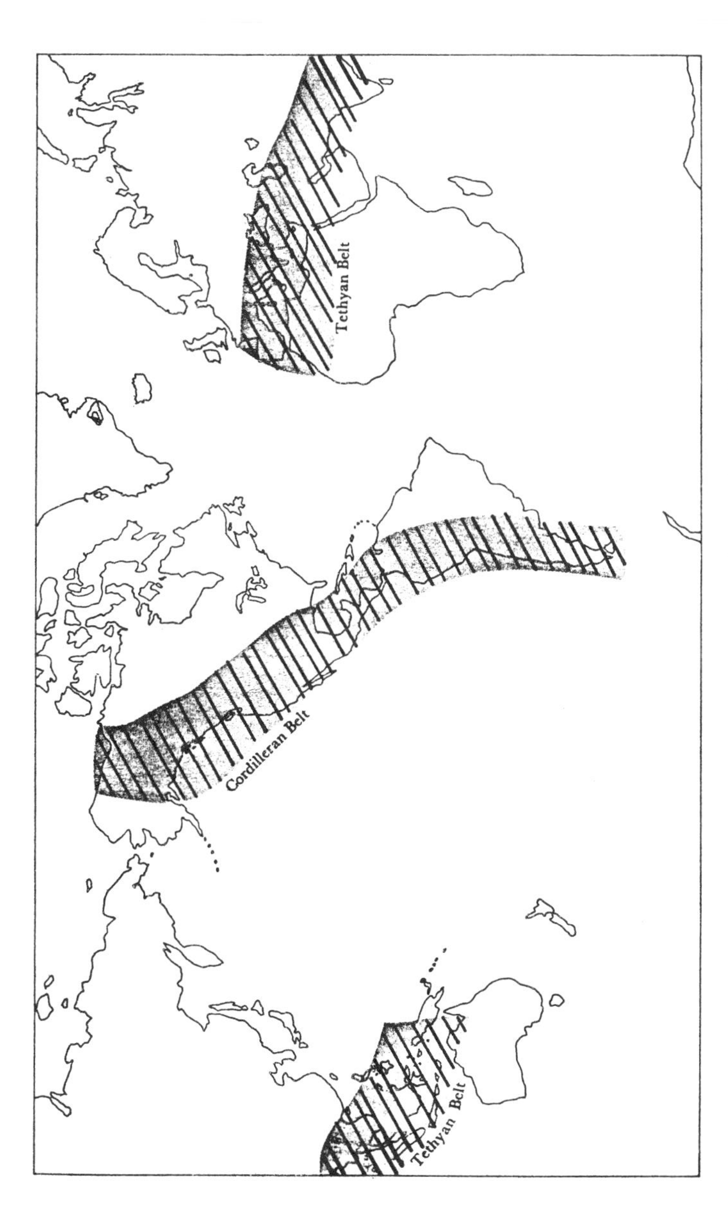


*Figure 12. Major world petroleum belts*

Table X Origin of Major Kerogen Environments of the World

| Location | Type | Source |
|---|---|---|
| Large lake basins | Green River Formation (Eocene)<br>Stanleyville Basin--Congo (Triassic)<br>Albert Shale, New Brunswick (Mississippian) | Cyanophyceae |
| Shallow seas on continental platforms and shelves | Alaskan Tasmanite, Brooks Range (Mississippian)<br>Phosphoria Formation (Permian)<br>Monterey Formation (Miocene)<br>Irati Shale, Brazil (Late Permian) | Unknown--probably red algae |
| Small lakes, bogs, lagoons, associated with coal-forming swamps | NSW Torbanite (Devonian ?)<br>Fusan, Manchuria (Tertiary) | Xanthophyceae<br>Chlorophyceae<br>(*Botryococcus braunii*) |

Decomposition and Condensation

$$a + P \rightleftharpoons aP$$

Naphthenic – Aromatic interconversion

$$N \rightleftharpoons A + H_2$$

$$N \rightleftharpoons HA + nH_2 \rightleftharpoons A + mH_2$$
$$m > n$$

Transalkylation

$$R_mA \rightleftharpoons R_nA \rightleftharpoons \cdots \rightleftharpoons R_pA \rightleftharpoons A$$
$$m > n > \cdots > p$$

*Figure 13. Major processes of petroleum diagenesis*

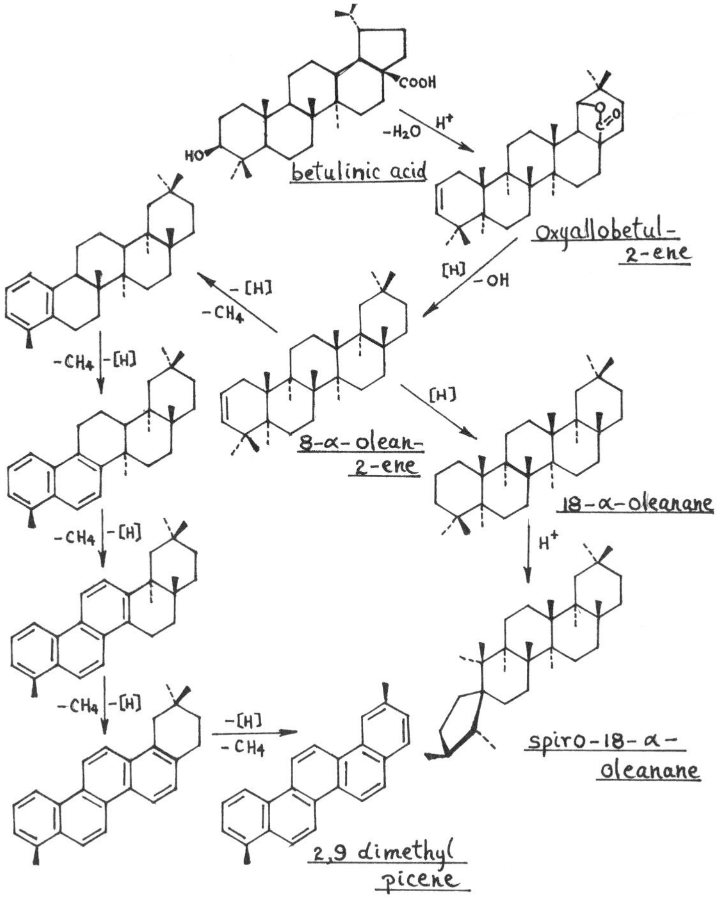

*Figure 14. Dehydrogenation and the accompanying methane formation from betulinic acid to 2,9-dimethylpicene*

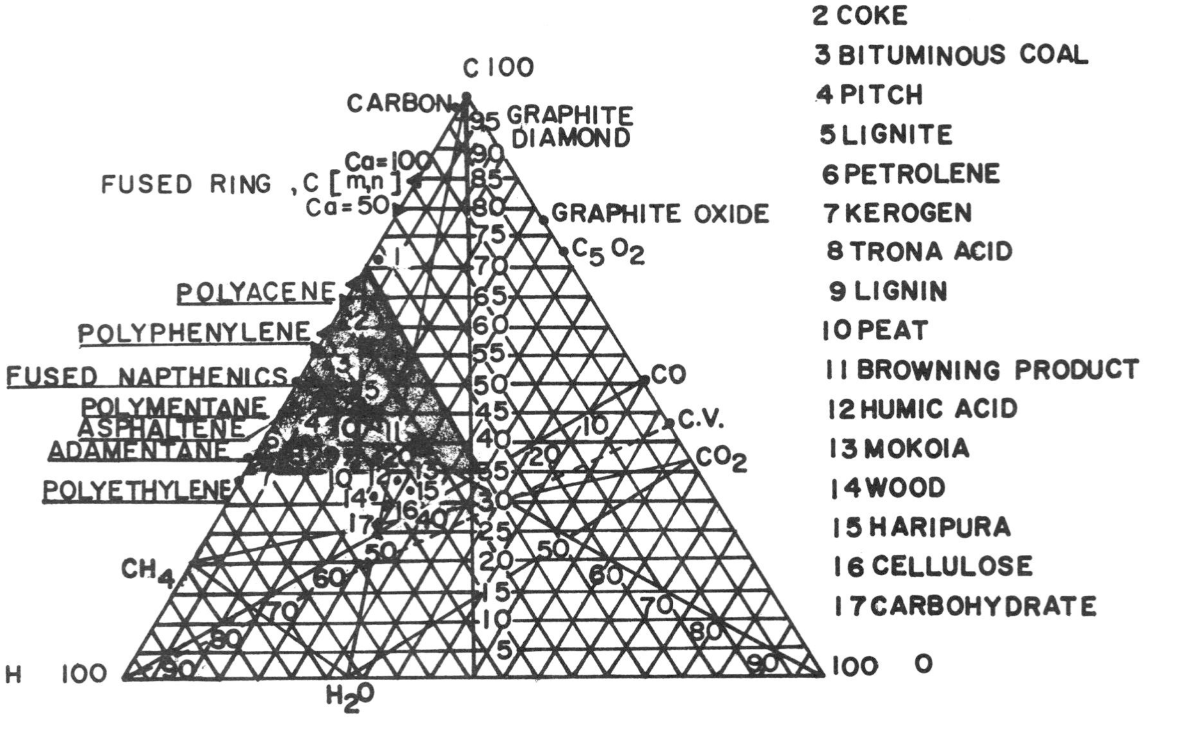


*Figure 15. Ternary diagram relating gaseous components of prebiological environment to the formation of fossil fuels. The shaded area indicates the region where the matured kerogen belongs.*

monoxide and water. Aromatization is also illustrated by the change of aromaticity for the more aged sample. In many cases, cracking of aliphatics, or the opening of the strained naphthenics, with respect to aging are found. For petroporphyrins, the ratio of DPEP/Etio is related to the depth of burial of petroleum (Fig. 16) (23).

The last type of transformation, transalkylation is the major control for maturation process. For example, the partial mass spectrum of petroporphyrin in old samples exhibit wider distribution of masses when paired with those of the recent sediments (14). This important principle forms the basis of carbon preference indexes (CPI). For recent samples, there is odd preference; as samples get old, this sharp difference of odd and even carbon numbers of the hydrocarbons in samples becomes almost equal. This CPI concept can be illustrated in Figure 17.

Other miscellaneous changes will involve weathering, or mild oxidation of petroleum components. Allomerization may illustrate the above as one of such processes (25). Actually, chlorin and purpurin have been isolated (Fig. 18). A number of decarboxylation, reduction and other reactions are occurring. For a complete survey of other reactions, readers should consult the paper of Breger (26).

## Microbial Modification

Maturation signifies, in a broad sense, the transformation from an uneven, distorted distribution of homologs of compounds to an even, symmetrically Gaussian-type, bell-shaped distribution. During the primary migration of petroleum from source to reservoir and the secondary migration which involves the uplift by buoying, shift by faults or alternative paths by capillary activities, constant contact of subsurface water to petroleum is required. Types of water in close contact are summarized in Table XI.

Table XI Subsurface Water During Petroleum Migration

| No. | Source | Type | Nature |
|---|---|---|---|
| 1 | Vadose | surface meteoric | sulfate |
| 2 | Connate | trapped | alkaline carbonate |
| 3 | Juvenile | magmatic | chloride |

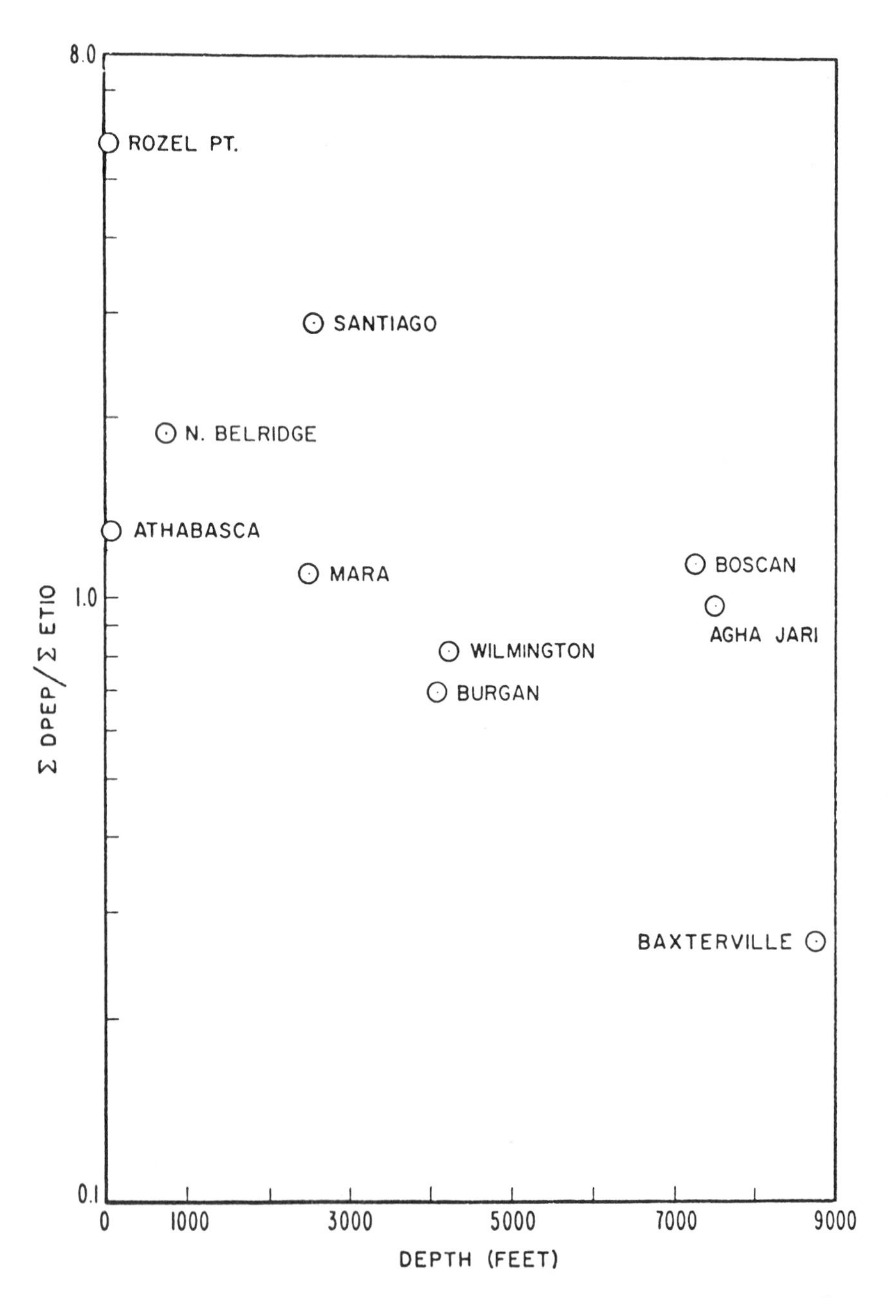


*Figure 16. A plot of the conversion of DPEP to Etio petroporphyrin vs. depth of burial in various petroleum*

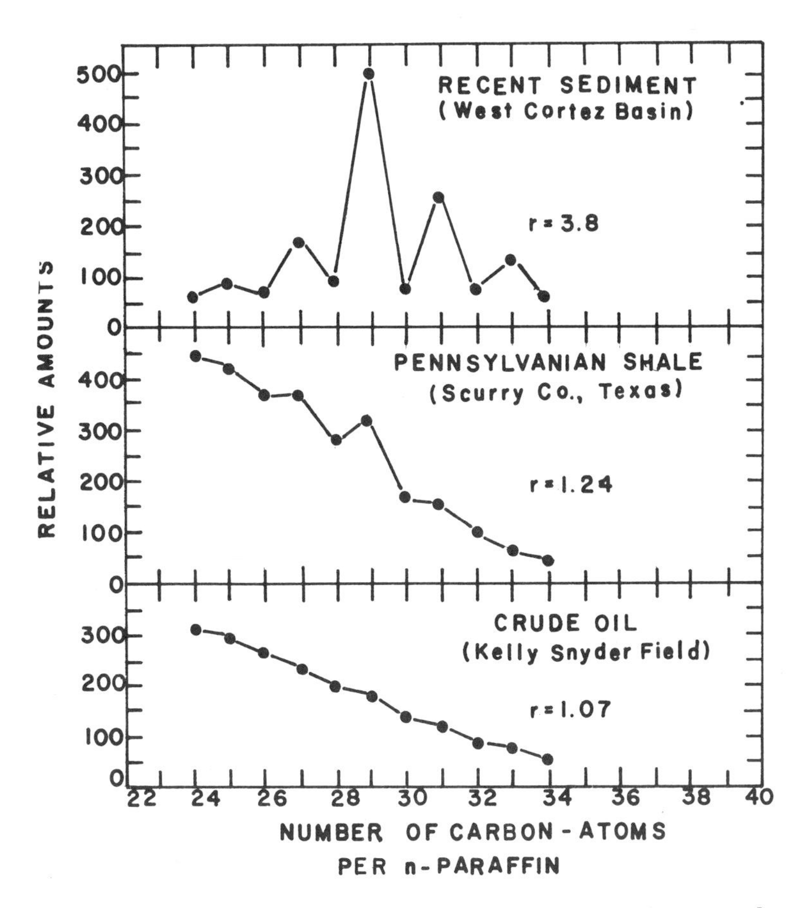


*Figure 17. n-Paraffin distributions for a recent sediment, an ancient sediment, and a crude oil (based on Ref. 35). The CPI numbers are indicated.*

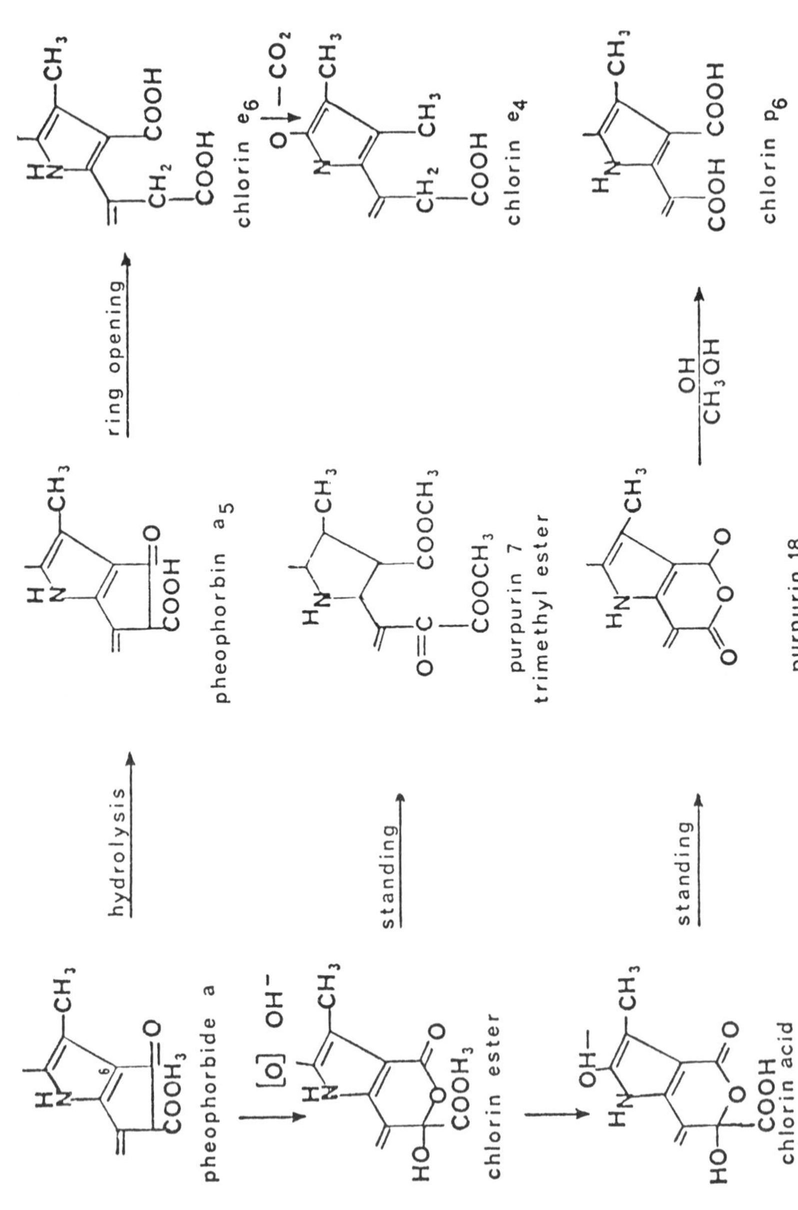

*Figure 18. Allomerization process and the formation of chlorin from pheophoride a*

Studies (27) for the microbiological alternative of crude oil in reservoirs, have indicated the following: (a) the aerobic microbes will selectively oxidize n-paraffins; (b) the altered oils are usually located in relatively shallow reservoirs which would be more easily reached by percolating surface waters with oxygen. The resultant crude oil will lose the paraffin wax fraction and increase the naphthenic content. Laboratory conditions indicate such modification is completed in one day or so by acclimated mixed culture of sewage-oxidizing bacteria (Fig. 19) (28). Bacteria which will attack petroleum hydrocarbons in saline medium include Corynebacterium, Arthrobacter and Achrobacter (29).

Kuznetsov (30) advocates that if the stratal water is rich in sulfate, the subsurface water will be active with sulfate reducers. In the presence of Desulfovibrio desulfuricans organics such as methane will be consumed:

$$Na_2SO_4 + CH_4 \longrightarrow Na_2CO_3 + H_2S + H_2O$$

Sulfide will be formed and, furthermore, deposits of secondary calcite will be accumulated, in many cases, even to seal the oil deposits for further bacterial decomposition.

$$CaCl_2 + Na_2CO_3 \longrightarrow CaCO_3 + 2\ NaCl$$

Sulfur-oxidizing bacteria such as Thiobacillus thioparus were found to be in mixed waters and subsurface waters containing hydrogen sulfide. In deep formation waters of the upper Paleozoic era in the Volga region, various group of bacteria were found. These include protein-utilizing, methane-oxidizing, glucose-fermenting, denitrifying, sulfate-reducing, methane-synthesizing and sulfur-oxidizing types (30).

In general, if the sulfate is absent and the water-transfer is low, bacteria will cause the anaerobic decomposition of oil into methane and nitrogen. In case the water-exchange is enhanced, and, especially, during the exploitation of the oil deposit, the oxygen-containing waters penetrate into the stratum and the oil is aerobically decomposed by hydrogen-oxidizing bacteria. In many cases, the oil becomes heavy and subsequently, high sulfur, or high vanadium asphaltic fractions will be formed. The possibility of cell material as a source for the vanadium and porphyrin during bituminization has been considered (31).

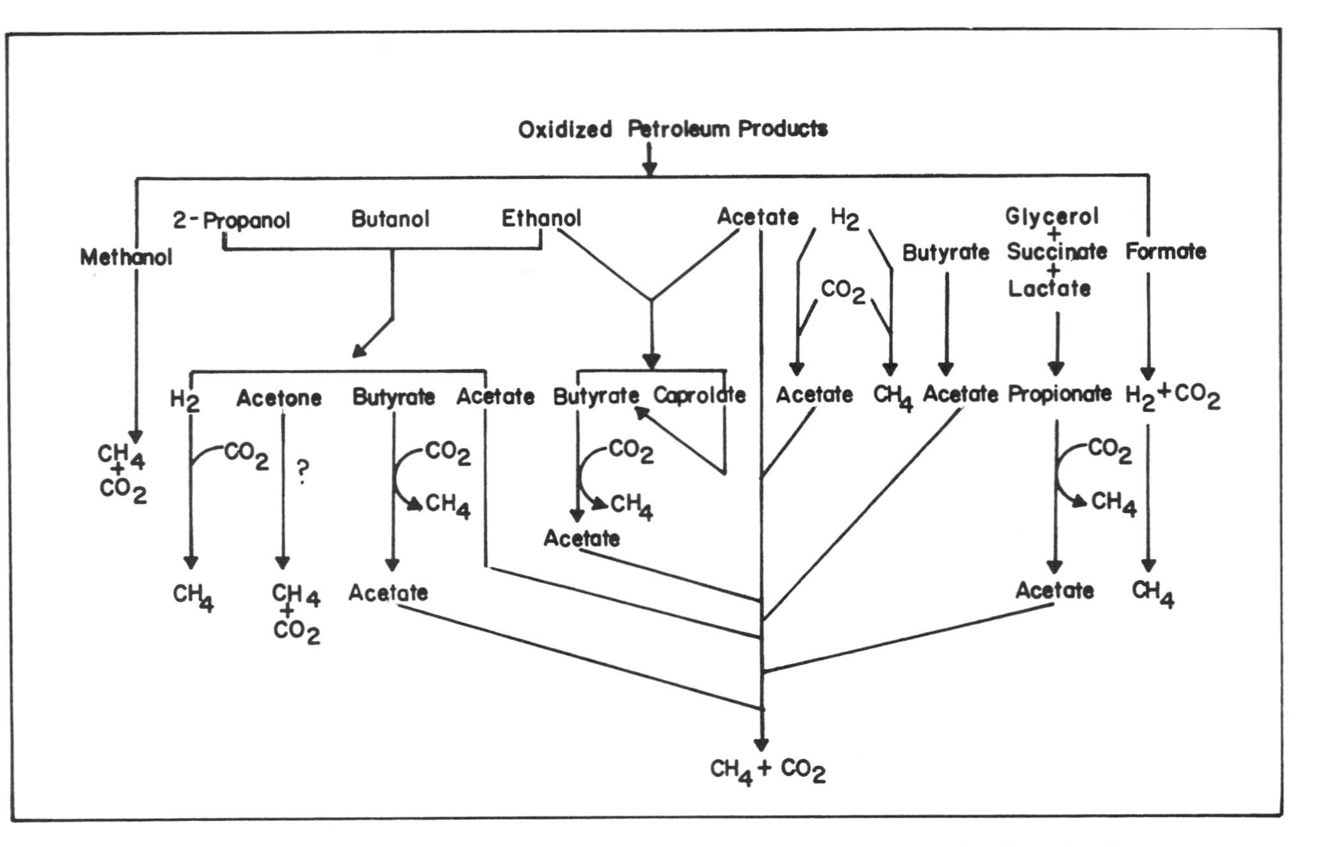


*Figure 19. Fermentation schemes representing the formation of methane and carbon dioxide*

## Biochemical Degradation

The mechanism of fatty acids degradation by hydrocarbon utilizers is well understood through bio--oxidation. For example, palmitic acid can be first, transformed into its coenzyme A ester through ATP, coenzyme A and thiokinas, then undergo de-hydrogenation (which was catalyzed by flavoprotein), hydration (catalyzed by enoylhydrase) followed by oxidation (catalyzed by dehydrogenase) and finally bond cleavage of a 2-carbon fragment resulting in a $C_{14}$ acid (catalyzed by thiolase). The alkanes can be degraded by both aerobic and anaerobic conditions. One of such routes can be expressed by the scheme shown in Figure 20. The end-products are usually sample acids or alcohols. In the case of obligate anaerobic bacterium, e.g., *Desulfovibrio desulfuricans*, the oxidation of alkanes is always favorable energetically (Table XII).

Table XII Free Energy of Alkane Oxidation with Sulfate at 25°C

| Hydrocarbon | Reaction | ΔF (kcal) | ΔF/$SO_4^{=}$ reduced (kcal) |
|---|---|---|---|
| Methane | $CH_4+SO_4+2H \rightarrow H_2S+CO_2+2H_2O$ | -22.8 | -22.8 |
| Propane | $2C_3H_8+5SO_4+COH \rightarrow 5H_2S+6CO_2+2H_2O$ | -141 | -28.2 |
| n-Octane | $4C_8H_{18}+25SO_4+50H \rightarrow 25H_2S+32CO_2+36H_2O$ | -744 | -29.8 |
| n-Undecane | $2C_{11}H_{24}+17SO_4+30H \rightarrow 17H_2S+22CO_2+24H_2O$ | -507 | -29.8 |

Other microorganisms that dehydrogenate alkanes under aerobic conditions are as in Table XIII (32).

Table XIII Anaerobic Dehydrogenation of Alkanes

| Organisms | Alkane | Products |
|---|---|---|
| *Achromobacter* sp. | n-Decane | n-Decanol, decyl aldehyde, decanoic acid |
| *Candida rugosa* | n-Decane | n-Decene, n-decanol |
| *C. tropicalis* | n-Decane | n-Decane, n-decanol |
| *D. desulfuricans* | n-Hexadecane | n-Hexadecane |
| *Nocardia* sp. | n-Decane | ----- |
| *Pseudomonas aeroginosa* | n-Heptane | n-Heptene |

Table XIV Degradation by Microorganisms

| Substrate | Organism | Product |
|---|---|---|
| ALKANES | | |
| Acetylene | *Mycobacterium lacticola* | Acetadehyde |
| Propane | *Mycobacterium smegman's* | Acetone |
| Butane | *Mycobacterium smegman's* | 2-Butanone |
| Hexane | *Pseudomonas aeruginosa* | Hexanoic acid |
| Hexadecane | *Arthrobacter* sp. | Hexadecanol |
| Pristane (2,6,10,14-tetra-methylpentadecane) | *Corynebacterium* sp. | 4,8,12-trimethyl-tridecanoic acid |
| AROMATIC HYDROCARBONS | | |
| Benzene | *Mycobacterium rhodochrous* | Succinic acid |
| | *Pseudomonas aeruginose* | Succinic acid |
| | *Micrococcus sphaeroides* | Phenol |
| | *Pseudomonas putida* | Catechol |
| Toluene | *Pseudomonas aeruginosa* | Benzoic acid |
| | *Achromobacter* sp. | Pyruvic acid |
| p-Xylene | *Pseudomonas* sp. | p-Toluic acid |
| Diphenyl methane | *Hydrogenomonas* sp. | Benzoic acid |
| Naphthalene | *Pseudomonas* sp. | α-Hydroxy-muconic semialdehyde (proposed) |
| | | Catechol |
| | | Salicyclic acid |

Table XIV Degradation by Microorganisms (cont.)

| Substrate | Organism | Product |
|---|---|---|
| Anthracene | *Corynebacterium* sp. | 2-Hydroxy-3-naphthoic acid |
| Phenanthrene | *Pseudomonas aeruginosa* | 1-Hydroxy-2-naphthoic acid |
| | CYCLOPARAFFINS | |
| Cyclohexane | *Pseudomonas aeruginosa* | Valeric acid |
| Decalin | *Flavobacterium* sp. | Pimelic acid |
| Tetralin | *Aspergillus niger* | α-Tetralol |
| | TERPENE HYDROCARBONS | |
| α-Pinene | *Aspergillus niger* | d-trans-Sorbrerol |
| Camphene | *Aspergillus niger* | 2-Nonene-2,3-dicarboxylic acid anhydride |
| β-Santalene | *Aspergillus niger* | |
| ρ-Cymene | *Pseudomonas* sp. | α-p-Tolyl-propionic acid |

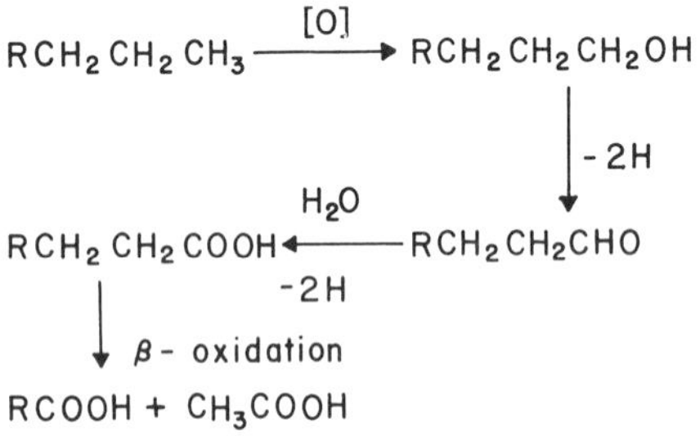

$$RCH_2CH_2CH_3 \xrightarrow{[O]} RCH_2CH_2CH_2OH$$

$$RCH_2CH_2CH_2OH \xrightarrow{-2H} RCH_2CH_2CHO$$

$$RCH_2CH_2CHO \xrightarrow[-2H]{H_2O} RCH_2CH_2COOH$$

$$RCH_2CH_2COOH \xrightarrow{\beta\text{- oxidation}} RCOOH + CH_3COOH$$

*Figure 20. Biodegradation of n-paraffins*

In a not only the alkanes, but the arenes, naphthenics as well as terpenoids also undergo degradation (Table XIV).

In many cases, simple acids and alcohols isolated from the degradation intermediates can be further fermented into methane and carbon dioxide by methane-producing bacteria. For example, acetic acid can be essentially converted to methane by acetate-fermenting species such as *Methanosarcina barberi* and *Methanococcus mazei*.

$$CH_3COOH \rightarrow CH_4 + CO_2$$

In this equation, it seems that methane is derived from the methyl radical, yet, actually, there are, simultaneously, redox reactions occurring. The methane is derived from carbon dioxide reduction:

$$2\ CH_3COOH + 2\ H_2O \rightarrow 4\ HCOOH + 8\ H \text{ (oxidation)}$$
$$CO_2 + 8\ H \rightarrow CH_4 + 2\ H_2O \text{ (reduction)}$$

which is

$$2\ CH_3COOH + CO_2 \rightarrow CH_4 + 4\ HCOOH + 2\ H_2O$$

It can be generalized that many intermediates can be directly converted into carbon dioxide and water (Fig. 19).

Other components in petroleum, such as aromatic hydrocarbons and heterocyclic compounds even with multiple rings can be degraded by microorganisms. A detailed discussion is out of the scope of the present paper, but is available elsewhere (33).

As discussed in the last section, bituminous materials are the end-products and are resistant to further biodegradation. But under a set of special conditions, miscellaneous bituminous materials can be degraded (Table XV).

Table XV Microbiological Attack of Miscellaneous Bituminous Materials

| Bitumens | Microorganism |
|---|---|
| Coal tar | *Candida* |
| Crude oil | *Cladosporia* |
| Humic acid | *Steptomycetes* |
| Mellitic acid | *Mycobacteria* |

Table XV Microbiological Attack of Miscellaneous Bituminous Materials (cont.)

| Bitumens | Microorganism |
|---|---|
| Graphitic acid | *Cephalosporium ascrimonium* |
| Graphitic acid | *Aspergillus penicilloides* |
| Asphalt | *Mycobacteria* |
| Asphalt | *Norcardia* |
| Lignin | *Pseudomonas* |
| Petroleum waxes | *Corynebacteria* |
| Tar sands | *Desulfovibrio* |

Strictly speaking, every substance synthesized by living matter also can be degraded by microorganisms. Probable reasons for resistance or refractoriness to biological degradation can be listed in Table XVI.

Table XVI Resistance to Biodegradation

1. The material is inaccessible to organisms
2. An essential factor for the growth of organisms is absent
3. The environment is toxic
4. The damaging agent is inactivated
5. The chemical or physical composition of the material is unsuitable for biotic attack
   No organism exists which might attack the material in question.

In conclusion, petroleum can be formed and also can undergo degradation according to the particular set of conditions and environments. The purpose here, for the environmental scientist, is to familiarize himself with the fundamental principles governing both types of transformations. Study of a given site should take into account all parameters underlying those dynamic conversions.

## Acknowledgment

Partial support from A.G.A. Grant No. GR-48-12, PRF Grant No. 6272-AC2 and NSF Grant No. AER-74-23797 is acknowledged.

## Literature Cited

(1) Yen, T. F. and Sprang, S. S., ACS, Div. Petrol. Chem. Preprints, 15(3), A65-A76 (1970).

(2) Yen, T. F., "Chemical Aspects of Metals in Native Petroleum," in The Role of Trace Metals in Petroleum (T. F. Yen, ed.), Ann Arbor Sci., 1975, pp. 1-30.

(3) Yen, T. F., Energy Sources, 1(4), 447-463 (1974).

(4) Hunt, J. M., Geochim. Cosmochim. Acta, 22, 37-49 (1961).

(5) Weeks, L. G., "Habitat of Oil and Some Factors that Control It," Habitat of Oil, American Association of Petroleum Geologists, 1958, pp. 58-59.

(6) Gaucher, L. P., Chem. Technol., 2(8), 471-5 (1972).

(7) Hodgson, G. W., Adv. Chem. Ser., 103, 1-29 (1971).

(8) Skirrow, G., "The Dissolved Gases--Carbon Dioxide," Chemical Oceanography, (J. P. Riley and G. Skirrow, eds.), Academic Press, Chap. 7, 1965.

(9) Yen, T. F., "Structural Aspects of Organic Components in Oil Shale," Oil Shale, (T. F. Yen and G. V. Chilingar, eds.), Elsevier, Amsterdam, 1975, Chap. 7.

(10) Han, J. and Calvin, M., Proc. Nat. Acad. Sci., USA, 64, 436-443 (1969).

(11) Asselineau, J., "The Bacterial Lipids," Holden-Day Inc., San Francisco, 1966, p. 145.

(12) Yen, T. F., "Facts Leading to the Biochemical Method of Oil Shale Recovery, in Analytical Chemistry Pertaining to Oil Shale and Shale Oil (S. Siggia and P. C. Uden, eds.), Univ. of Mass., 1975, pp. 59-79.

(13) Eglinton, G., Advances in Organic Geochemistry, Pergamon, Oxford, 1969, pp. 1-24.

(14) Yen, T. F., "Terrestrial and Extraterrestrial Stable Organic Molecules," in Chemistry in Space Research, (R. F. Landel and A. Rembaum, eds.),

(15) Hitchon, B., Adv. Chem. Ser., 103, 30-66 (1971).

(16) Baker, E. W., Yen, T. F., Dickie, J. P., Rhodes, R. E., and Clark, L. F., J. Am. Chem. Soc., 89, 3631 (1967).

(17) Knovolden, K. A., in Advances in Organic Geochemistry, Pergamon, Oxford, 1969, pp. 335-366.

(18) Silverman, S. R., J. Am. Oil Chem. Soc., 44, 691-695 (1967).

(19) Silverman, S. R., Proc. 8th World Petrol. Congr., PD 1(5), Moscow, 1971.

(20) Cox, B. B., Bull. Assoc. Petrol. Geologists, 30, 645-659 (1946).

(21) Cutbill, J. C. and Funnell, B. M., "Numerical Analysis of the Fossil Record," in The Fossil Record, Geol. Society of London, 1967, pp. 791-820.
(22) Young, D. K., Sprang, S. R., and Yen, T. F., ACS, Div. Petrol. Chem. Preprints, 19(4), 769-775 (1974).
(23) Yen, T. F. and Silverman, S. R., ACS, Div. Petrol. Chem. Preprints, 14(3), E32-E39 (1969).
(24) Ellison, Jr., S. P., "Petroleum (Origin)," in Kirk-Othmer Encyclopedia Chem. Tech., 14, 838-845 (1967).
(25) Baker, E. W. and Smith, G. D., ACS Div. Petrol. Chem. Preprints, 19(4), 744-768 (1974).
(26) Breger, I. A., Geochim. Cosmochim. Acta, 19, 297 (1960).
(27) Winters, J. C. and Williams, J. A., ACS, Div. Petrol. Chem. Preprints, 14(4), E22-E33 (1969).
(28) Mechalas, B. J., Meyers, T. J., and Kolpack, R. L., Center for Wetland Resources, LSU-59-73-01, pp. 67-79 (1973).
(29) Soli, G. and Bens, E. M., Biotech. Bioeng., 14, 319 (1972).
(30) Kuznetsov, S. I., Proc. 7th World Petroleum Congr., pp. 171-182 (1967).
(31) Radchenko, O. A. and Sheshina, L. S., Geol. Sh., 33, 274 (1955).
(32) Mahadevan, A., J. Scient. Ind. Res., 33, 39 (1974).
(33) Yen, T. F., Application of Microbiology to Energy and Environment, Ann Arbor Sci., 1976.
(34) Erskine, B. L. and Whitehead, E. V., J.S.T., 3(4), 221-243 (1975).
(35) Cooper, J. E. and Bray, E. E., Geochim. Cosmochim. Acta, 27, 1113 (1963).

# 15

# A Comparison of Analysis Methods for Hydrocarbons in Surface Sediments

J. W. FARRINGTON and B. W. TRIPP

Chemistry Department, Woods Hole Oceanographic Institution, Woods Hole, Mass. 02543

Concern about the inputs, effects, and fate of oil in the marine environment has instigated an increased level of research activity in all areas of oil pollution studies. An integral part of most studies is the analysis of sediments, water, or organisms to detect petroleum and estimate the concentration of petroleum compounds. A variety of analytical methods have been employed in this task (1-3). Our laboratory has investigated some aspects of the problem of estimating the concentrations of petroleum in marine samples and the problems inherent in trying to compare data obtained in different laboratories where different methods of analysis are used. We have reported results obtained with our colleagues concerning the intercalibration of analyses for hydrocarbons in marine tissue samples and lipid extracts of marine organisms (4-6). We have also reported some of our results comparing the efficiency and accuracy of three extraction procedures for isolating hydrocarbons from shellfish and the application of such procedures and further analytical methodology such as column chromatography and gas chromatography for estimating petroleum hydrocarbon concentrations in shellfish (7). We report here the initial results of investigations of the efficiency of extraction and precision and accuracy of some procedures for the estimation of petroleum hydrocarbons in sediment samples.

There have been several studies of the fate of oil spilled or leaked into the marine environment (1, 2, 8). It is clear from several of these studies that one fate of oil inputs is the deposition to and/or incorporation into surface sediments. Once incorporated, oil may persist for years resulting in some degree of exposure of the benthic ecosystem and thus the marine environment as a whole for that period of time. For this reason, the estimation of petroleum hydrocarbons in marine sediments is an integral part of oil pollution studies and is of concern when considering studies of the environmental impact of offshore drilling and production, offshore tanker terminals, refinery siting in the coastal zone, loading and offloading areas in harbors, and other sources of input resulting from man's activities.

The estimation of petroleum hydrocarbon composition and concentration in marine surface sediments is complicated for the following reasons:

1) There are very complex mixtures of compounds with a wide range of molecular weight and polarity in petroleum. Consequently, a complete analysis of a crude oil using one method of analysis is not possible at this time. In fact, it is often stated, and we concur, that there has yet to be a report of a complete analysis of a single crude oil and it is unlikely that there will be in the near future.

2) There are a large number of sources of petroleum hydrocarbons in the marine environment introduced by man's activities (2). The composition of these petroleum hydrocarbons varies and in some locations in the marine environment the inputs are mixed together complicating the mixture to an even greater extent.

3) Hydrocarbons are formed by biosynthetic processes of both marine and land organisms and then released to the marine environment. Our knowledge of the biosynthetic production of certain classes of hydrocarbons (n-alkanes, branched alkanes, and n-alkenes) is reasonably complete but definitive studies of the production or lack of biosynthetic production of other classes of petroleum compounds such as cycloalkanes, cycloalkenes, aromatic hydrocarbons, naphthenoaromatic hydrocarbons, and heteroatom ring compounds are scarce. Furthermore, only a limited number of species have been investigated and only a few studies of geographical and temporal variability have been reported (2, 3).

4) There are a number of geochemical processes introducing hydrocarbons into the marine environment. The natural seepage of oil is an obvious example of this category. Weathering of ancient sediments and the resulting transport of the sediment and associated ancient hydrocarbons to the marine environment by fluvial or eolian processes can result in the introduction of an assemblage of hydrocarbons and other compounds similar to the composition of petroleum. Two other processes - forest fires and early diagenesis of organic matter deposited to surface sediments also must be considered. The contribution of these geochemical processes are not well known either quantitatively or qualitatively at this time with the exception of the oil seepage rate which has been estimated with poor confidence (2).

The natural biosynthetic and geochemical processes contribute to the natural base load of hydrocarbons and other compounds on which is superimposed an ever increasing discharge of petroleum hydrocarbons and other oil components due to man's activities. The general analytical approach for isolating hydrocarbons in surface sediments is an organic solvent extraction which yields a lipid extract, removal of elemental sulfur, followed by isolation of the hydrocarbons from other lipids by column or thin layer adsorption chromatography (2, 3). The isolated hydrocarbons contain a mixture from both natural processes, and inputs from man's activities. From analysis of this hydrocarbon mixture an estimate of the concentration of petroleum hydrocarbons from man's activities must be deduced. In addition, an oil spill may occur in an area such as a harbor where there has been a chronic input from small spills and effluent discharges for several years prior to the spill under consideration. The estimation of the geographical extent and severity of the impacts of the new spill will depend to some degree on the ability of the analyst to detect the freshly spilled oil in the presence of the chronic oil inputs. We will discuss this point later in this paper and present some data pertinent to this problem. There are at present no definitive criteria for separating those hydrocarbons introduced from some of the natural geochemical processes and those from man's activities. Some criteria of assistance in distinguishing between recently biosynthesized hydrocarbons and those contributed by man's activities have been set forth elsewhere (1-3).

In this report we will concentrate primarily on saturated hydrocarbons (cycloalkanes and alkanes) and leave a discussion of the aromatic hydrocarbons and heteroatom components of petroleum to later papers. For those interested in aromatic hydrocarbons in marine surface sediments, two recent papers (4, 10) are suggested. We realize that there are several extraction, isolation, and instrumental methods which we have not yet explored in our studies. The methods we have investigated are among those most extensively applied to date in studies of oil pollution in the marine environment and the evaluation of these methods is important.

## Methods

We have examined the hydrocarbons obtained from three methods of extraction applied to a Narragansett Bay sediment. The sediment sample used was a large composite grab sample, taken from a location in West Passage that is relatively uncontaminated. Precautions to avoid contamination must be taken at the sampling site as well as during analysis (1-3). During sampling, care is taken to avoid contact with all oils, lubricants, and plastics and the sample is allowed to contact only solvent rinsed stainless steel or glass. Samples of the ship's fuel and lubricating oils are routinely taken for comparison to sediments that indicate possible contamination (1-3).

In general, the extraction techniques used will separate a complex lipophilic mixture from the environmental sample. The preliminary extraction is followed by the separation of hydrocarbons from the other lipids and then the hydrocarbon mixture can be analyzed by a variety of techniques in an attempt to deduce the relative amounts of native and fossil hydrocarbons. A blank of the complete extraction procedure is regularly run along with samples to determine if interfering compounds are introduced during the analysis.

The Soxhlet extraction procedure with benzene:methanol has been used by several investigators (1-3) and has been modified to fit our circumstances. About 100 g, wet weight, of spiked or unspiked sediment is weighed into a single thickness paper extraction thimble and Soxhlet extracted for 24 hours with 1:1 benzene: methanol. At the end of this period, the extract is transferred to a storage flask and the sample is further extracted with fresh solvent. The extraction is continued for a third day, again using fresh solvent after each twenty-four hour period.

The combined extract is transferred to a separatory funnel and shaken with acidic, saturated NaCl solution. The benzene phase is separated from the aqueous methanol phase. The latter is then extracted two more times with pentane. The combined pentane-benzene extract is back-washed with fresh saturated NaCl solution and dried over $Na_2SO_4$ overnight. All of the aqueous portion of the original extract is discarded.

The dried solution of total extractable lipid is evaporated at reduced pressure at or below room temperature. The sample is removed as soon as the pressure drops to $\sim$5 Torr. Extracts are weighed directly on an analytical balance. The total extractable lipid is saponified under reflux for two hours with 0.5 N KOH in methanol:benzene 1:1 with 25% $H_2O$ added. This alkaline hydrolysis removes fatty acids, wax esters and triglycerides which are a part of the total lipid extract. It is important that 25% $H_2O$ is added to this hydrolysis to prevent transesterification of esters which may be present. The hydrolyzed sample is cooled and extracted as above with salt solution and pentane. The pentane extract is evaporated as above and the residue weighed again.

An activated copper column is used to remove interfering elemental sulfur from the non-saponifiable lipid extract (1). The sample is charged to the column and eluted with three column volumes of pentane-benzene. The solvent is again evaporated and the sample weighed as above.

The sulfur-free, non-saponifiable lipid extract is further separated by chromatography through a column of (1:1) $Al_2O_3$ (alumina) packed over $SiO_2$ (silica). The $Al_2O_3$ and $SiO_2$ are activated overnight at 250°C and 120°C respectively and then both are deactivated with 5% water. The sample of sulfur-free, non-saponifiable lipids is charged to the column with a small volume of pentane. The ratio of column material to non-saponifiable lipids is 100:1 or greater. The following elution scheme is employed:

Fraction 1 f1 - 1 column volume of pentane
Fraction 2 f2 - 1 column volume of 10% benzene in pentane
1 column volume of 20% benzene in pentane
Fraction 3 f3 - 1 column volume of benzene

The three fractions are evaporated to dryness on a rotary

evaporator at room temperature or slightly below with care taken to remove the sample vial as the pressure drops to ∿5 Torr. The hydrocarbon fractions are then taken up in 50 to 200 μl of $CS_2$ and an aliquot of 5 to 10 μl is weighed on the Cahn Electrobalance. An aliquot of 1-5 μl is injected into the gas chromatographic column.

The second extraction technique is the direct alkaline hydrolysis of a sediment sample. Following the procedure of Farrington and Quinn (11), 50 g of the estuarine sediment is refluxed for three hours with benzene and 0.5 N methanolic KOH. After cooling, the solvent is filtered with suction and the remaining sediment stirred with methanol and 1 N HCl. This mixture is also filtered and the residue is washed on the filter with methanol followed by pentane. The combined filtrate is extracted two times with pentane, the combined pentane extract is back-washed with distilled water and dried over $Na_2SO_4$. After evaporation of the solvent, the sample is weighed and then carried through the sulfur removal and column chromatography steps as described in the preceding discussion.

An extraction procedure using an acid wash prior to three days of Soxhlet extraction with benzene-methanol has also been tested. A sediment sample was washed two times with 1 N HCl and once with distilled $H_2O$, centrifuging between washings and saving the supernate. The sample is then weighed into an extraction thimble and extracted as described above. The sample is treated exactly as in the original Soxhlet extraction procedure except that the combined acid washings are added to the benzene-methanol extract during the re-extraction with pentane plus salt solution.

Each extraction technique used results in three separate fractions of hydrocarbons at the end of column chromatography. The data presented here are our interpretations of gravimetric analyses of the three fractions and gas chromatographic analysis of the pentane fraction from the column chromatography of non-saponifiable lipids. A packed OV-101 (5 ft., 1/8" stainless steel, 15% on 100/120 HP Chrom G, helium carrier gas at 15-20 ml/min) column and an OV-101 SCOT (Perkin-Elmer, 50 ft., helium carrier gas at 4-6 ml/min) column installed in Varian model 1440 and 2740 gas chromatographs were used. The columns were routinely temperature programmed from 75°C to 275°C at 6°/min., using a chart speed of 0.5 in/min.

The SCOT column is preferred because of its capability for higher resolution, but we have experienced some difficulty with the resolution of SCOT columns at the higher end of the molecular weight range that we encounter. Analysis of quantities of greater than 10 nanograms of each hydrocarbon produces sharp peaks with good resolution, but smaller amounts on the order of 1 nanogram or less result in poorer resolution and significant losses of the alkanes above n-$C_{24}$. Some of the probable causes of this effect such as split ratio, injector and/or detector temperature and flow rate have been investigated and we have concluded that these were not the cause. Some of the SCOT columns could be kept operating by periodic injection of a concentrated solution of crude oil. Evidently, the loss of resolution is due to an interaction between "hydrophobic" sites in the column and the sample material. Periodic coating of these sites with high molecular weight components of crude oil is usually sufficient to restore the operating capacity of the column.

A standard n-alkane mixture of known concentration is used to measure detector response per unit weight of alkane. The areas of peaks plus unresolved complex mixture are measured using planimetry and compared to the standard alkane mixture.

## Results and Discussion

Comparison of Efficiencies of Extraction Procedures. The result of comparison of three methods of extraction are given in Table I. Based on our present data we conclude that there are no substantial differences in the efficiencies of the extraction procedures. Chemical treatment of the sediment by acid or base does not significantly increase the yield of hydrocarbons as compared to organic solvent extraction alone. This is in contrast to the extraction of fatty acids from sediments where KOH digestion or saponification extraction yielded a significantly greater amount of fatty acids (11). We do not have sufficient data to explain this difference and can only speculate that it may be the result of the differences in early diagenesis of the two classes of compounds.

The precision of the measurements of hydrocarbons in sediments is not as good as the precision for measurements of hydrocarbons in clams *Mercenaria mercenaria* sampled at the same locations (7), as is indicated in Table II. We think the poorer precision of the data for sediments reflects the greater difficulty of obtaining a homogeneous mixture of the wet sandy silt sediment and reproducibly subsampling in comparison to the clam homogenate. The fact that the analysis of the hydrocarbons extracted from the clams yields a better precision indicates that the column chromatography and weighing of isolated hydrocarbons is not the source of the poorer precision for the sediment hydrocarbon data. This contention is also supported by analyses of hydrocarbons in marine lipids and marine tissues conducted as part of an intercalibration program (4-6). The poorer precision of the sediment hydrocarbon analyses may also be the result of differences in the molecular weight range of hydrocarbons being quantified in each case. Gas chromatographic analyses show that the heavier molecular weight hydrocarbons are more predominant in the sediment hydrocarbon mixture compared to the clam hydrocarbon mixture. It may be more difficult to obtain precise measurements of heavier molecular weight hydrocarbons due to difficulties associated with extraction or column chromatography.

TABLE I: COMPARISON OF EFFICIENCES OF SOME METHODS FOR EXTRACTING HYDROCARBONS FROM SURFACE SEDIMENTS [a]

| EXTRACTION METHOD | NO. OF ANALYSES | uG/G. DRY WEIGHT OF SEDIMENT[b] (AVERAGE ± STANDARD DEVIATION) | | | |
|---|---|---|---|---|---|
| | | $f_1$ | $f_2$ | $f_3$ | $\Sigma f_1 f_2 f_3$ |
| SOXHLET W/BENZENE: METHANOL, 72 HRS. | (5) | 147±54 | 25±7 | 83±30 | 265±56 |
| DIGESTION W/KOH:MeOH | (4) | 136±58 | 29±6 | 45±9 | 202±54 |
| HCl PRE-WASH FOLLOWED[c] BY SOXHLET W/BENZENE: METHANOL, 72 HRS. | (2) | 190±55 | 43±13 | 115±20 | 347±65 |
| 10 PPM SPIKE, NO. 2[c] FUEL OIL FOLLOWED BY SOXHLET EXTRACTION | (2) | 161±19 | 40±6 | 77±9 | 279±34 |

a. ANALYSES OF SUBSAMPLES OF A SURFACE SEDIMENT SAMPLE. SEE TEXT.

b. $f_1$ - ALKANES, CYCLOALKANES, ALKENES, CYCLOALKENES
$f_2$ - ALKENES, CYCLOALKENES, 2 AND 3 RING AROMATICS
$f_3$ - 4-6 RING AROMATIC HYDROCARBONS AND TRACES OF METHYL KETONES

c. ESTIMATED STANDARD DEVIATION FROM TWO ANALYSES.

TABLE II: COMPARISON OF PRECISION OF SOXHLET AND DIGESTION EXTRACTION AND HYDROCARBON ANALYSIS FROM CLAMS (M.mercenaria) AND SURFACE SEDIMENTS SAMPLES FROM THE SAME LOCATION [a].

| EXTRACTION METHOD | NO. OF ANALYSES | %RELATIVE STANDARD DEVIATION: SEDIMENT $f_1$ | SEDIMENT $f_2$ | NO. OF ANALYSES | CLAM $f_1$ | CLAM $f_2$ |
|---|---|---|---|---|---|---|
| SOXHLET W/BENZENE: METHANOL | (5) | 37% | 28% | (6) | 11% | 8% |
| DIGESTION W/KOH:MeOH | (4) | 43% | 22% | (6) | 13% | 14% |

a. DATA FROM REFERENCE NO. 7

This certainly warrants further investigation.

One argument favoring use of the digestion procedure is that it simultaneously extracts and saponifies the samples whereas an additional step is required for the Soxhlet extraction. It may be feasible to conduct the saponification in the Soxhlet pot thereby eliminating this disadvantage. However, if the analyst wishes to simultaneously analyze the sediment sample for chlorinated hydrocarbons such as DDT and for nonchlorinated hydrocarbons, then saponification must not occur before splitting the lipid extract in order to avoid chemical transformation of the DDT during saponification.

The results of Soxhlet extraction and subsequent hydrocarbon analyses for samples of surface sediments from three different depositional regimes are given in Table III. The agreement of duplicate analysis for surface sediments of the continental shelf sediment station G187DG is well within the precision for the Narragansett Bay Station C subsamples. However, agreement of the duplicate analyses of the deep sea abyssal plain surface sediment is not as good. Our investigation of the influence of different sediment type and varying concentrations of hydrocarbons in the sediment on the precision of measurement is not yet complete. Also, we have not completed the analyses of the fraction 2 and fraction 3 hydrocarbons and we offer the data in Table 3 as initial results.

The concentration of hydrocarbons in the surface sediments at Station C in Narragansett Bay are close to that reported three years ago - 130 to 160 ug/g dry weight for fraction 1 and fraction 2 hydrocarbons (12) as compared to 202 and 265 ug/ gram dry weight for the latest analysis. We cannot be certain at this time whether this difference is a result of the slightly different methods of analysis or is a real increase in hydrocarbon concentration. This illustrates the need for the type of investigation of analytical methods we are pursuing.

Analyses of Sediment Samples Spiked with 10 PPM API No. 2 fuel oil. We have applied the Soxhlet extraction procedure to the analysis of Station C surface sediments spiked with 10 ug. API No. 2 Fuel Oil (13) per gram dry weight of sediment. As expected and shown in Table I, gravimetric quantification of the column chromatography fractions by weighing on the Cahn electrobalance did not detect the added No. 2 fuel oil in the presence of the base load of hydrocarbons.

Gas Chromatographic Analyses. The reproducibility of the gas chromatographic analyses of duplicates of fraction 1 hydrocarbons is demonstrated in figures 1 and 2. The reproducibility of the gas chromatogram for the saturated fraction of the abyssal plain sediment hydrocarbons was similar. The main features of the gas chromatograms are:

TABLE III: REPLICATE ANALYSES OF SEDIMENT SUBSAMPLES FROM 3 DEPOSITIONAL ENVIRONMENTS.

| LOCATION | WATER DEPTH (METERS) | SEDIMENT TYPE | μG HYDROCARBON/G. DRY WEIGHT SEDIMENT $f_1$ (SATURATES) |
|---|---|---|---|
| NARRAGANSETT BAY (STATION C) 41°33.4'N 71°25.3'W | 3.1 | SANDY-SILT | 147±54 (5 ANALYSES) |
| HUDSON TROUGH (G 187DG) 40°04'N 73°28'W | 78.0 | SILT CLAY | 58.3<br>48.0 |
| ABYSSAL PLAIN (K 19-4-9) 30°01'N 60°02'W | 5252 | CLAY | 1.2<br>2.9 |

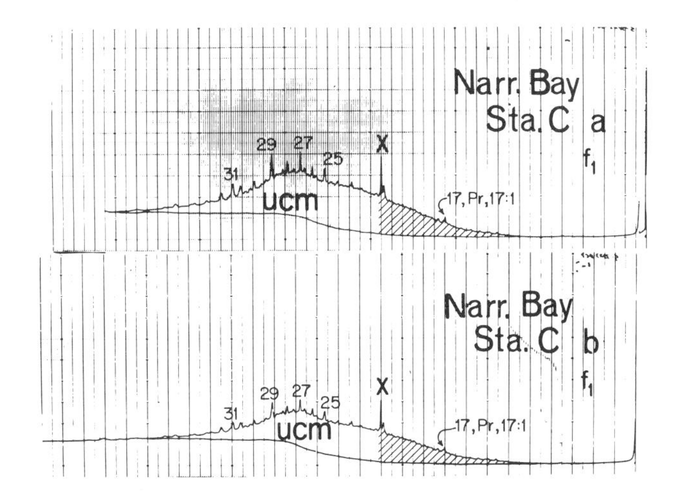


*Figure 1. Gas chromatograms of fraction 1 hydrocarbons from surface sediments of Narragansett Bay Station C subsamples a and b. Pr—pristane; 17:1 n-heptadecene; 17, 25, 27, etc. indicated number of carbon atoms in n-alkanes eluting at the position indicated by the peak. UCM is unresolved complex mixture. Bottom recorder trace is column bleed signal. Hatched area was measured to obtain values given in Table IV. × designates cycloalkenes and alkenes of unknown structure. G.C. column was SCOT OV 101.*

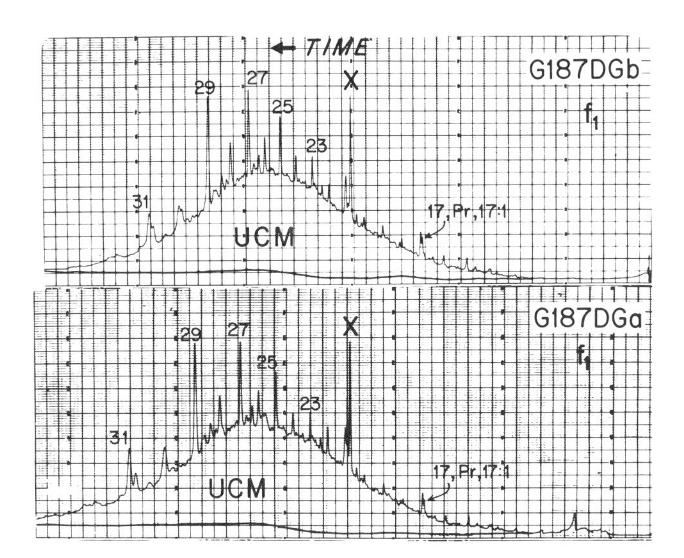


*Figure 2. Gas chromatograms of fraction 1 hydrocarbons from surface sediments of the Hudson Trough Station G187DG. Subsamples a and b. Explanation of notations given in legend of Figure 1.*

i) The predominance of an unresolved complex mixture of alkanes and cycloalkanes with a wide molecular weight range. This is similar to that found in petroleum and is not predominant, if present at all, in recently biosynthesized hydrocarbons. The sources of these hydrocarbons are most likely petroleum discharged to the marine environment via a combination of man's activities. The various factors bearing on this contention have been previously discussed (2, 3, 4, 12). We think that for concentrations of hydrocarbons about an order of magnitude lower than that reported for station G187DG, the source of the unresolved complex mixture may be weathering of ancient sediments and transport and deposition of the sediments and their associated petroleum-like hydrocarbon assemblages to the surface sediments of the continental shelf and estuarine areas. In addition, at low concentration levels of hydrocarbons there may be some as yet undefined biosynthetic or biogeochemical diagenetic processes in surface sediments which produce such a hydrocarbon assemblage.

ii) The presence of the partially resolved n-$C_{17}$, pristane, $C_{17}$ mono-olefin. These hydrocarbons are most likely of recent biosynthetic origin (2,3).

iii) The presence of the heavier molecular weight n-$C_{23}$ to n- $C_{31}$ n-alkanes with an odd carbon predominance. Land plants and marsh grasses are probable sources for these n-alkanes (2,3).

iv) The presence of fully or partially resolved components in the n-$C_{27}$ to n-$C_{31}$ molecular weight range for Station G187DG sediments which have been tentatively identified as triterpanes and steranes based on gas chromatographic retention data and gas chromatography - mass spectrometry. These compounds are probably present in Station C sediment but are not well resolved in the presence of the higher concentration of unresolved complex mixture of cycloalkanes and alkanes. The origin of these compounds is probably a combination of biosynthesis and diagenesis (2, 3, 14).

v) The presence of partially resolved hydrocarbons indicated by the X in the gas chromatograms. The structures of these hydrocarbons are not known. Gas chromatograph - mass spectrometer analyses of the compounds after trapping from gas chromatography and before and after hydrogenation indicates that the major component of the mixture is a 25 carbon cycloalkene with an isoprenoid side chain (15).

Regardless of the origin of the hydrocarbons in the surface sediments at Station C and Station G187DG, they represent present baseload concentrations and compositions. Station C is located in Narragansett Bay, an estuary which has substantial tanker traffic and effluent discharges (12). Station G187DG is on the Continental Shelf in the Hudson Trough east-southeast of the New York Bight. It is adjacent to the proposed Outer Continental Shelf Drilling and Production sites. It is also near sewage, sludge and dredge spoil dump sites, tanker routes, and could receive petroleum contaminated municipal and industrial effluent from the adjacent coastal area. Any determination of the concentrations and composition of hydrocarbons incorporated into these sediments as a result of an accidental spill or new nearby chronic releases of oil would have to take these present hydrocarbon concentrations and compositions into account.

A simple but necessary experiment was conducted to determine the feasibility of detecting a low concentration of 10 ppm freshly spilled fuel oil incorporated into the Station C sediments. The results of the gas chromatographic analyses are given in figure 4 along with the gas chromatogram of the fraction 1 hydrocarbons of the API No. 2 Fuel Oil. A comparison of the gas chromatograms of figures 1 and 3 shows that the presence of the 10 ppm spike of No. 2 fuel oil in the sediment sample is clearly detected. Obviously, higher concentrations of the fuel oil would be detected using similar procedures. However, the detection of lower concentrations of spiked fuel oil hydrocarbons i.e. at the 1 ppm level will be more difficult. The detection of the presence of the No. 2 fuel oil in sediments by gas chromatography as was accomplished with the West Falmouth oil spill studies (16, 17, 18) does not provide a quantitative measurement of low concentrations of the No. 2 fuel oil hydrocarbons. In the case of the West Falmouth oil spill it was evident for the sediments with the heaviest concentration of oil (1,000 ug/ g dry weight sediment) that gravimetric determination of the hydrocarbons using a Cahn electrobalance or analytical balance provided a fair assessment of the No. 2 fuel oil concentration. However, the results previously reported in this paper clearly demonstrate that at concentrations of about 100 ug total hydrocarbons/ gram dry weight of sediment the gravimetric determination cannot detect the presence of the No. 2 fuel oil at the 10 ppm concentration level. This applies to many of the sediment samples analyzed in the West Falmouth oil spill as well as at other oil spill sites (19). A good method of measurement is needed if correlations of acute and chronic biological effects at low concentration levels of No. 2 fuel oil are to be determined. Likewise, a method of measurement is necessary to provide a true assessment of the geographical extent of the spread of petroleum hydrocarbons from an oil spill site and the persistence of a given low concentration of oil in sediments over a period of time.

One method for measurement of petroleum hydrocarbons spiked to a marine lipid extract which has been used with some success (4) is calibration of the flame ionization detector of the gas chromatograph with a standard mixture of n-alkanes followed by quantification of the gas chromatograph detector signal for the hydrocarbons from the sediments for the spiked and unspiked samples. We have conducted an initial test of this approach by quantifying the detector signal for the n-$C_{12}$ to n-$C_{20}$ molecular weight range of the fraction 1 hydrocarbons of the No. 2 fuel oil using planimetry. The gas chromatograms used were those shown in

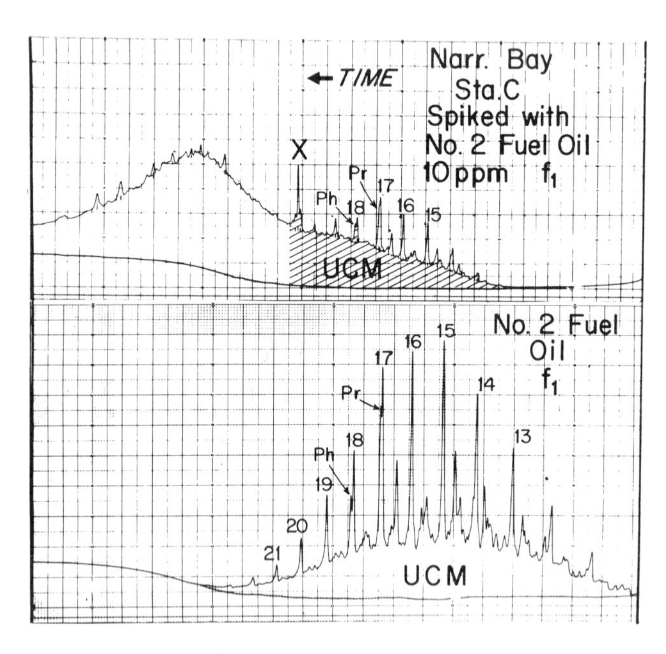


*Figure 3. Gas chromatograms of fraction 1 hydrocarbons from surface sediment of Narragansett Bay Station C spiked with No. 2 fuel oil and of fraction 1 hydrocarbons of the No. 2 fuel oil. Ph-phytane; all other notations given in legend of Figure 1.*

figure 1 and the duplicate of the one at the top of figure 3.

Agreement of the duplicate analyses is fair as is shown in Table IV. Recovery of the fraction 1 hydrocarbons from column chromatography of the API No. 2 fuel oil and evaporation to concentrate the sample are also given in Table IV. The lower boiling hydrocarbons in the fuel oil are lost during this procedure (7). The measurement of the No. 2 fuel oil hydrocarbons in the spiked sediment samples gives 2.9 ug hydrocarbons/ g dry weight of sediment which is lower than the 4.9 ug/ g dry weight predicted from the recovery experiments with the No. 2 fuel oil. This lower value is probably due to a more extensive loss of lower boiling hydrocarbons in the fuel oil spike due to the extensive handling during several concentration steps in the sediment analysis procedure. A comparison of the gas chromatograms in figure 3 suggests that this is the case. Incorporation of an internal standard into the sediment sample prior to extraction would probably improve accuracy (4). This involves splitting the sample and analyzing subsamples with and without an internal standard in order to avoid masking information on the indigenous hydrocarbons and to correct for the interference of indigenous hydrocarbons with measurement of the internal standard.

We do not have gas chromatographic analysis of the fraction 2 and fraction 3 hydrocarbons to present at this time. However, we are applying procedures similar to those described above to analyses of these fractions.

The limited data we have presented suggests that the methods as described above have a good potential for providing reliable estimates of fuel oil hydrocarbons in surface sediment samples. We realize that the experiments performed do not take into account alteration of the oil by weathering or degradation during and after incorporation into the surface sediments. However, if the methodology fails the simple tests we are applying then it has little possibility of working for the more difficult field test. Application of these methods to the field situation requires a surface sediment from the area prior to the oil pollution incident or a control station nearby which has a similar hydrocarbon composition and concentration as the oil spill area prior to the pollution incident.

We are certainly not yet at the point where we can formulate a standard method of analysis for petroleum hydrocarbons in sediments at the 10 ppm level. Our studies are ongoing and further tests with various sediment types, various baseload concentrations of hydrocarbons, and spiking with several different types of oils over a range of concentrations are contemplated and needed. The analytical methodology to be tested is by no means limited to the methods we have discussed in this paper and we urge others in this field of research to pursue testing procedures such as these to investigate the efficiency, precision, applicability, and limitations of other methods of analysis. Without such testing, interpretation of the data collected during the studies of the input,

TABLE IV: GAS CHROMATOGRAPHIC QUANTIFICATION OF THE n-$C_{12}$ TO n-$C_{20}$ MOLECULAR WEIGHT RANGE OF HYDROCARBONS IN UNSPIKED AND FUEL OIL SPIKED SURFACE SEDIMENT.

| | | μG $f_1$ HYDROCARBON/G DRY WEIGHT SEDIMENT | |
|---|---|---|---|
| SEDIMENT, NOT SPIKED | | 1.6 | |
| | | 1.7 | |
| | $\overline{AV}$. | 1.6 | |
| SEDIMENT, SPIKED WITH 10 μG FUEL OIL/G. WET WEIGHT SAMPLE | | 5.4 | |
| | | 3.4 | |
| | $\overline{AV}$. | 4.4 | |
| SPIKED - UNSPIKED | | 2.9 | |
| RECOVERY OF API NO. 2 FUEL OIL FOR COLUMN CHROMATOGRAPHY | | 4.3 | (43%) |

fate, and effect of petroleum pollution in the marine environment will be unnecessarily restricted.

## Abstract

Three extraction procedures - Methanol:benzene Soxhlet extraction with and without pretreatment of sediment with 1N HCl and alkaline hydrolysis extraction - were applied to subsamples of a surface sediment sample. The three procedures yielded similar amounts of hydrocarbons as determined by gravimetric analysis. Gas chromatographs of the alkane-cycloalkane portion of the hydrocarbons exhibited a very complex mixture of hydrocarbons with a wide molecular weight range extending from n-$C_{14}$ to beyond n-$C_{31}$. Biogenic hydrocarbons from land plants and possibly from marine organisms were present.

Subsamples of the surface sediment were spiked with 10 ppm dry weight of API No. 2 fuel oil. The spike was not detected by gravimetric analysis. However, the spike was detected by gas chromatography. An estimate of 2.9 ppm out of a possible 10 ppm No. 2 fuel oil was obtained by quantification of the g.c. detector signal.

## Acknowledgments

This is Contribution No. 3553 of the Woods Hole Oceanographic Institution and was supported by Grant R802724 of the U. S. Environmental Protection Agency.

## Literature Cited

1. National Academy of Sciences "Background Papers for A Workshop on Inputs, Fates and Effects of Petroleum in the Marine Environment", 824 pp., NAS, Washington, D. C., 1973.
2. National Academy of Sciences "Petroleum in the Marine Environment", 107 pp., NAS, Washington, D. C., 1975.
3. Farrington, J. W. and P. A. Meyers in: "Environmental Chemistry, Vol. 1" ed. G. Eglinton, 184 pp., The Chemical Society, Burlington House, London, 1975.
4. Farrington, J. W. et al, Bull. Environ. Contam. and Tox. (1973) 10(3): 129.
5. Farrington, J. W., et al in: "Marine Pollution Monitoring (Petroleum)", Nat. Bur. of Standards (U.S.) Spec. Pub. No. 409, 293 pp., Washington, D. C., 1974.
6. Medeiros, G. C. and J. W. Farrington in: "Marine Pollution Monitoring (Petroleum)", Nat. Bur. of Standards (U.S.) Spec. Pub. No. 409, 293 pp., Washington, D. C., 1974.
7. Farrington, J. W. and G. C. Medeiros in: "1975 Conference on Prevention and Control of Oil Pollution, Proceedings", American Petroleum Institute - U. S. Environmental

Protection Agency - U. S. Coast Guard, 612 pp., Washington, D. C., 1975.

8. American Petroleum Institute - U. S. Environmental Protection Agency - U. S. Coast Guard, "Proceedings of Conference on Prevention and Control of Oil Pollution", Washington, D. C., 1969, 1971, 1973, 1975.
9. Giger, W. and M. Blumer. Anal. Chem. (1974) 46(p): 1663.
10. Blumer, M. and Youngblood, W. W. Science (1975) 188(4183): 53.
11. Farrington, J. W. and J. G. Quinn. Geochim et Cosmochim Acta (1973) 37:259.
12. Farrington, J. W. and J. G. Quinn. Estuarine and Coastal Marine Science (1973) 1:71.
13. API No. 2 fuel oil available from J. W. Anderson, Biology Department, Texas A&M University, College Station 77843.
14. Maxwell, J. R., et al, Quarterly Reviews (1971) 25(4):571.
15. Gschwend, P. and J. W. Farrington. Unpublished data.
16. Blumer, M. and J. Sass. Marine Pollution Bull. (1972) 3(6):92.
17. Blumer. M. and J. Sass. Woods Hole Oceanographic Institution Tech. Report No. 72-19, part II (1972).
18. Michael, A. D., et al in: "1975 Conference on Prevention and Control of Oil Pollution, Proceedings", American Petroleum Institute - U. S. Environmental Protection Agency - U. S. Coast Guard, 612 pp., Washington, D. C., 1975.
19. McAuliffe, C. D., et al in: "1975 Conference on Prevention and Control of Oil Pollution, Proceedings", American Petroleum Institute - U. S. Environmental Protection Agency - U. S. Coast Guard, 612 pp., Washington, D. C., 1975.

# 16

# Lipid Geochemistry of Recent Sediments from the Great Marsh, Lewes, Delaware

PAUL J. SWETLAND* and JOHN F. WEHMILLER

Department of Geology, University of Delaware, Newark, Del. 19711

Understanding the significance of ancient sedimentary organic matter requires knowledge of the organic distributions common to modern environments. Organic geochemical studies of recent sediments seek to determine: 1) the biological sources from which organic compounds in modern sediments were derived, 2) the organic distributions in different modern depositional environments for comparison with ancient analogs, and 3) the early diagenetic alterations of recent sedimentary organic matter.

This study was designed to investigate the composition and early diagenesis of organic matter deposited in a modern salt marsh environment. Salt marshes are generally highly productive environments that contribute large amounts of herbaceous organic matter to sediments. Herbaceous, or leafy, land plants are characterized by long chain hydrocarbons and fatty acids which serve as a waxy protective coating (1). Because of the abundance of this waxy material in higher land plants, high-wax petroleums are thought to be derived from coastal environments where inputs of land plant organic matter are great (2,3). Since sediments of salt marshes contain about ten times more lipid material than those of marine-lagoon or open-shelf environments, salt marshes are considered probable source environments for high-wax petroleum (4).

The main emphasis of this study was placed on the identification of sediment lipid (n-alkane and n-fatty acid) distribution patterns as primary source material or as products of diagenesis. Correlations of sediment lipid distributions to specific biological sources are suggested. The nature and possible rates of various diagenetic reactions responsible for alteration of the

*Pres. Address: U. S. Geological Survey, Federal Center, Denver, CO 80225

lipid distributions are also discussed.

## Study Area and Sample Collection

The Great Marsh occupies an area of approximately five square miles along Delaware Bay northwest of the town of Lewes, DE (Figure 1). It is subjected to a tidal range of about 30 cm with salinities varying from 5 to 30 o/oo. Present marsh vegetation is dominated by the halophytic grasses *Spartina alterniflora*, *S. patens*, and *Distichlis spicata*. Other plant species include *Salicornia* sp., *Iva frutescens*, *Baccharis halimifolia*, and *Phragmites communis*, as well as a variety of marine algae.

The combination of high organic productivity and rapid sedimentation provides a strongly reducing environment favorable for the preservation of organic matter. Typical Eh and pH values for subsurface samples in this area are -200 mv and 7.1, respectively (5). Average sedimentation rates are on the order of 10-15 cm/100 years (6) based upon radiocarbon dates on a variety of peat and shell samples found at various depths in the general area.

The geological history of the Great Marsh is closely associated with the late Holocene marine transgression (7). A large part of the present Great Marsh area was once a shallow lagoon which began filling about 3000 yrs. B. P. (7). Fringing marshes adjacent to headland areas gradually accreted upward while abundant fine-grained material accumulated in near-shore lagoonal environments. Later stages in this history have involved the development of a broad marsh over both the fringing marsh and lagoonal deposits (7).

The samples of interest in this work come from one core and one surface sample taken for comparative purposes at a nearby location (see Figure 1). Both locations are within an area that is dominated by *Spartina alterniflora* (8). A core of approximately four meters length was taken by pounding a 6.35 cm i.d. plastic pipe into the marsh sediment using the procedures described by Kraft (9). Samples were immediately returned to the laboratory and were frozen at -20° C. until analyzed.

Core pipes were split lengthwise using a hand-held rotary saw and descriptions were made on wet samples. Two distinct lithologic zones are recognized in the core (Figure 2) and can be correlated to other stratigraphic sections in the immediate area that have radiocarbon age control (6). Coarse sands occupy the lower 1.5 meters of the core; overlying these are 2.5

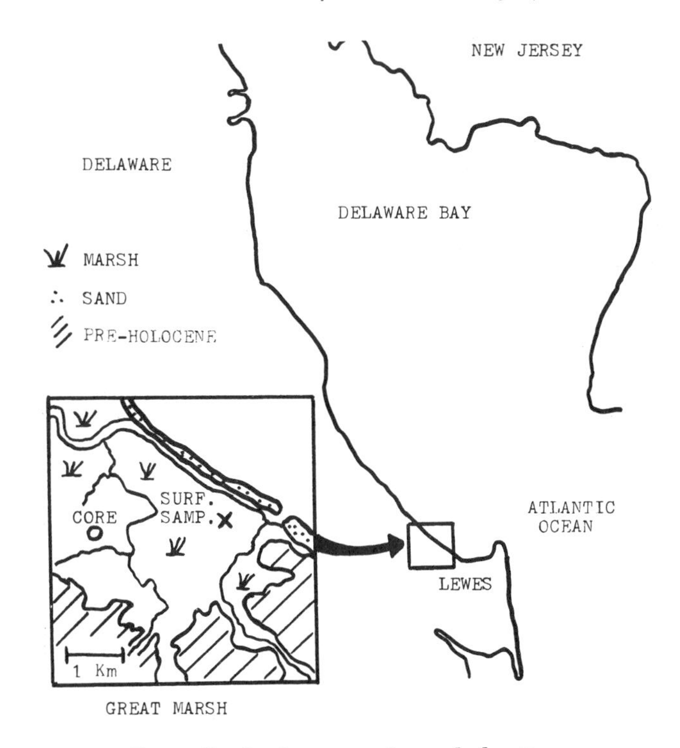


*Figure 1. Study area and sample locations*

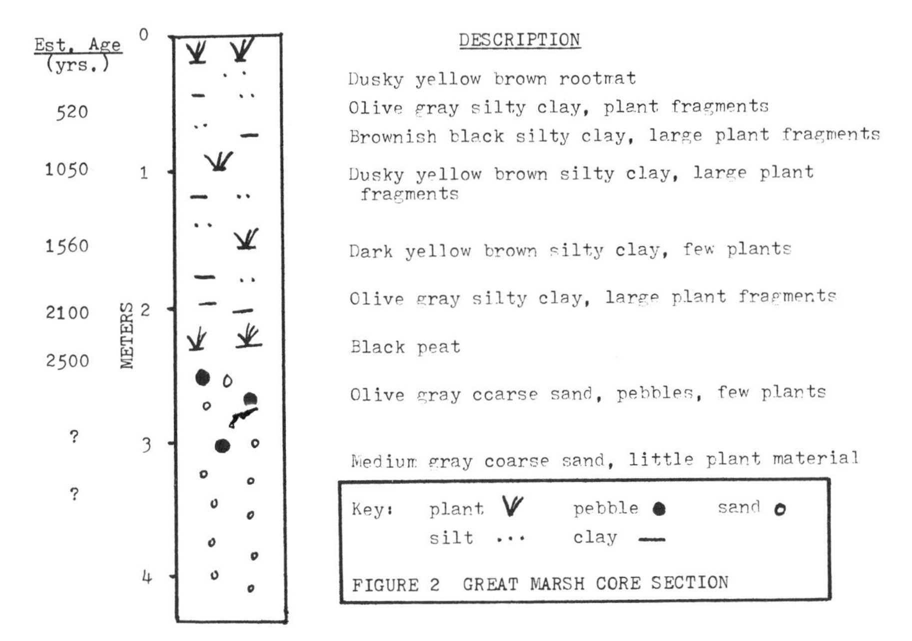


*Figure 2. Great Marsh core section*

meters of organic-rich silty clay, with a basal peat unit being the lowermost component of this organic rich zone. The organic rich zone is interpreted to record continuous marsh sedimentation (fringing marsh evolving to broad marsh) for the past 2500 years, the approximate radiocarbon age for other fringing marsh deposits at comparable depths in the area (6). This interpretation is supported by the presence of continuous interlocking network of roots and rhizomes throughout the upper 2.5 meters. The age of the lower core sandy sediments is uncertain. They are most likely pre-transgression in age, though it has been suggested that they represent channel deposits that accumulated during the transgression prior to establishment of a fringing marsh (7). Such an interpretation would imply that the sediments were in contact with estuarine waters, and, as noted below, the distribution of n-alkanes in the sediments implies a marine source for at least some of the organic matter in these lower core sands.

## Analytical Procedures

Selected 20 cm core intervals were freeze-dried and divided into three parts for 1) mechanical analysis, 2) ignition loss determination, and 3) extraction of fatty acids and hydrocarbons. The percent sand, silt, and clay of each sample were determined by pipette analysis according to the method of Folk (10). An approximate measure of the organic carbon content was obtained by determining the weight loss on ignition at 375° for 16 hours (11).

The freeze-dried core samples were extracted with a benzene-methanol azeotropic mixture followed by chloroform for one hour each using ultrasonic vibrations. Extraction mixtures were centrifuged and filtered through coarse filter paper. The solvent extracts were then combined and taken to dryness by rotary evaporation.

An 18 cm by 20 mm diameter column of activated (110° C.) silica gel in hexane was prepared to separate the crude extract residue into three fractions. The residue was rinsed, transferred to the column and successively eluted with 1) hexane, 2) benzene, and 3) benzene/dioxane/acetic acid. Two column volumes (150 mls) of each eluate were collected.

The hexane eluate containing saturated hydrocarbons was reduced in volume and transferred to an activated copper column to remove elemental sulfur (12). The sulfur-free extract was then reduced in volume by rotary evaporation, transferred to a pre-weighed vial and evaporated under a stream of nitrogen.

The benzene/dioxane/acetic acid fraction containing free fatty acids was reduced to near dryness by rotary evaporation. Fatty acids were then converted to their methyl esters with $BF_3$/MeOH by the procedure of Metcalfe and Schmitz (13). The methyl ester residue was transferred in benzene to a silica gel column and eluted with two column volumes each of benzene and benzene-ethyl acetate (9:1). The benzene fraction containing the fatty acid methyl esters, was dried under nitrogen in a pre-weighed vial. Since no saponification of the total sediment or the sediment extract was performed, these fatty acids constitute what must be defined as the free fatty acid fraction of the sediments (14).

Gas chromatographic analyses were performed on a F & M Model 1609 gas chromatograph equipped with a flame ionization detector. Saturated hydrocarbons and fatty acid methyl esters were chromatographed on a 1.52 meter by 3.2 mm O.D. copper column of 1 % Apiezon L on Chromosorb W. A flow rate of 30 mls/min. was used for the helium carrier gas with the oven temperature programmed from 45°C to 285°C at a rate of 4°C/minute. Identification of peaks was made by comparison of retention times with standards, co-injection of standards and mass spectrometry.

Urea adduction followed by gas chromatography was used to help characterize normal (straight chain saturated) alkanes and normal fatty acids. The adduction procedures of Sever (15) were followed. By this technique straight-chained compounds are separated from highly branched and cyclic compounds. In most cases core sediment fatty acids and hydrocarbons were relatively simple mixtures of normal compounds and the adduction separation was not required.

To minimize contamination all reagent grade solvents were redistilled. Glassware was sterilized by heating in an oven at 300°C for 24 hours. Procedural blanks were run periodically to check contamination levels of solvents and glassware. Working at the same concentration and sensitivity used for sediment lipids, no gas chromatographic peaks from the blanks could be detected in the lipid distribution range.

## Results and Discussion

Table I presents basic data on grain size distribution, organic carbon content, total saturated hydrocarbons and total free fatty acids. Table II presents the carbon preference indices for both n-alkanes (16)

Table I

| Core Depth, cm | %Sand | %Silt | %Clay | %Org. C | Sat. Hydro-carbons ppm[c] | Free Fatty Acids ppm[c] |
|---|---|---|---|---|---|---|
| 0[a] | 2.3 | 50.3 | 47.4 | 3.2 | 67 | 62 |
| 0-20[b] | 0.4 | 18.7 | 80.9 | 14.6 | n.d. | n.d. |
| 20-40 | 0.8 | 21.4 | 77.8 | 13.1 | 101 | 418 |
| 40-60 | 0.8 | 16.9 | 82.3 | 10.3 | 72 | 148 |
| 60-80 | 0.8 | 42.1 | 57.1 | 8.9 | 29 | 318 |
| 80-100 | 0.7 | 40.8 | 58.5 | 6.7 | 35 | 1387 |
| 100-120 | 1.3 | 31.3 | 67.4 | 6.7 | 55 | n.d. |
| 140-160 | 1.1 | 34.8 | 64.1 | 5.4 | 41 | 466 |
| 180-200 | 1.0 | 65.0 | 34.0 | 4.5 | 82 | 175 |
| 220-240 | 0.6 | 19.3 | 80.1 | 22.2 | 35 | n.d. |
| 260-280 | 64.3 | 26.8 | 8.9 | 0.1 | 3 | 18 |
| 320-340 | 73.0 | 10.8 | 16.4 | 0.2 | 2 | 3 |
| 360-380 | 73.8 | 14.8 | 11.4 | 0.1 | 7 | 4 |

Notes:

n.d. = not determined

a) Surface sample, number DC-2

b) Extract lost in preparation

c) Weights of total residues of each fraction per unit weight of sediment

Table II

| Core Depth, cm | Est. Age, yrs# | $CPI_H$* | $CPI_{FA}$** | C16:1/C16:0 | C18:1/C18:0 |
|---|---|---|---|---|---|
| 0 | 0 | 3.9 | 8.2 | 0.26 | 0.91 |
| 20-40 | 310 | 4.4 | 8.2 | 0.66 | 0.71 |
| 40-60 | 520 | 3.2 | 20.8 | 0.21 | 0.28 |
| 60-80 | 730 | 3.0 | 16.4 | 0.10 | 0.25 |
| 80-100 | 940 | 3.7 | 12.3 | 0.19 | 0.17 |
| 100-120 | 1150 | 3.4 | n.d. | n.d. | n.d. |
| 140-160 | 1560 | 5.1 | 17.0 | 0.07 | 0.08 |
| 180-200 | 1980 | 3.4 | 15.1 | 0.04 | 0.03 |
| 220-240 | 2400 | 2.3 | n.d. | n.d. | n.d. |
| 260-280 | ? | 2.2 | 17.5 | 0.00 | 0.00 |
| 320-340 | ? | n.d. | 10.8 | 0.00 | 0.00 |
| 360-380 | ? | n.d. | n.d. | n.d. | n.d. |

Notes:

n.d. = not determined

\# Ages estimated assuming constant sedimentation rate from 240 cm = 2500 yrs.

* Determined according to procedure of Cooper and Bray (16):

$$CPI_H = \frac{2 \times \Sigma \text{ odd C21-C31}}{\Sigma \text{ even C20-C30} + \Sigma \text{ even C22-C32}}$$

** Determined according to the procedure of Kvenvolden (17):

$$CPI_{FA} = \frac{2 \times \Sigma \text{ even C16-C28}}{\Sigma \text{ odd C15-C27} + \Sigma \text{ odd C17-C29}}$$

and n-fatty acids (17) vs. depth in the core, and the ratio of unsaturated to saturated fatty acids (C18:1/C 18:0 and C16:1/C16:0) at selected depths. Estimated ages for various depths in the core are also given in Table II.

The results outlined in these tables can be interpreted as being the combined effect of varying sources of organic matter during sedimentation and of varying diagenetic reactions occuring after deposition (Figure 3). A dramatic difference in total organic carbon is seen in comparison of the lower, sandy unit with the upper, organic-rich unit. Within this upper unit there is a steady decrease in the organic carbon content from about 15% to about 5%, except for the basal peat unit (220-240 cm) that contains large amounts of visible plant material and has an organic carbon content of 22%. Concentrations of saturated hydrocarbons and of free fatty acids also reflect the lithologic break at 240 cm. In the upper, organic-rich unit, total saturated hydrocarbon concentrations when normalized to total organic carbon, are variable with depth while free fatty acids show an initial increase in concentration to a depth of about 100 cm (ca. 1000 yrs) followed by a decrease in concentration at greater depths within both the upper and lower units. There does not appear to be any relation between the decreasing abundance of any particular fatty acid and the increasing abundance of the corresponding normal alkane. In the case of both fatty acids and saturated hydrocarbons, the normal acids and alkanes collectively account for a large percentage of the total fraction, as determined by relative peak area measurements by gas chromatograms. Figure 4 shows the normalized percent composition of the normal alkanes and normal fatty acids for samples from selected depths in the core. Each of these groups of compounds is discussed in detail in the sections to follow.

Saturated Hydrocarbons. There is little evidence for major diagenetic alteration of this group of compounds in the 2500 years represented by the samples from the upper core. The alkane fractions from all depths in the core show little trend toward increasing complexity (branching or cyclization) as might be expected from diagenetic alteration of a comparatively simple mixture (18, 19). The carbon preference index ($CPI_H$) varies from 3.0 to 5.1 in the upper core and is somewhat lower in the basal peat and in the underlying sands. $CPI_H$ values provide a qualitative index of source variation and diagenetic alteration. $CPI_H$

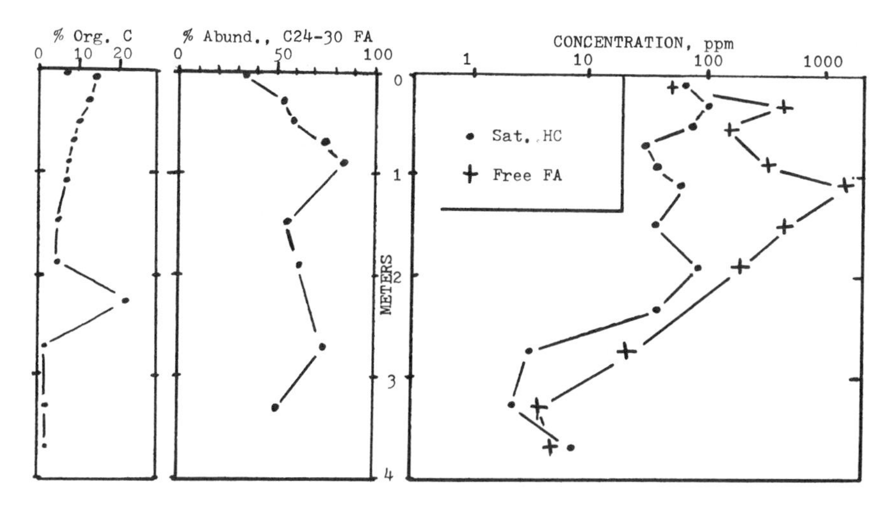


*Figure 3. Abundances of total carbon, saturated hydrocarbons, free fatty acids, and long chain (C24–30) fatty acids in Great Marsh sedimentary sequence*

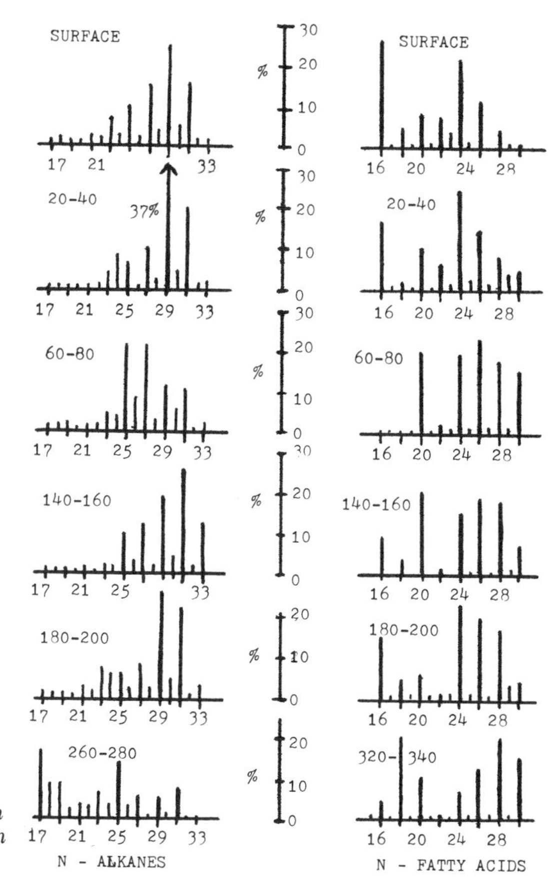


*Figure 4. Relative abundances of N-alkanes and N-fatty acids in Great Marsh sediments (depths in cm)*

values for marsh plant alkanes are generally greater than 5.0 (15) while micro-organisms generally have $CPI_H$ values much less than 5.0 and often as low as 1.0 (20). The lower $CPI_H$ values for the lower core sandy unit may result from the more rapid rates of diagenetic alteration that have been observed in sandy marsh sediments (18). However, extrapolation of these estimated rates to the probable age of the sandy core sediments of this study would predict a much greater degree of alteration than in fact is observed. This conclusion, combined with the observed large relative abundance of n-heptadecane and the smaller total amount of saturated hydrocarbon material in these sandy sediments implies that the alkane distribution in these sediments is primarily a record of the source material rather than of diagenetic alteration. As mentioned previously, the exact nature of the depositional environment of these sands is uncertain, but in terms of total organic carbon, total saturated hydrocarbons, and distribution of n-alkanes, these sandy sediments look quite similar to modern sandy sediments in lower Delaware Bay (21). The abundance of n-heptadecane suggests a marine algal source (22). Preservation of carbohydrate and chlorophyll-derived residues in these pre-transgression sands has also been recognized (23), though a specific source for these compounds was not identified.

Saturated hydrocarbon distributions in the upper core sediments are characterized by an abundance of long chain odd-carbon normal alkanes (Figure 4). The predominant n-alkane varies among n-C25, n-C27, n-C29, and n-C31. These long chain normal alkanes are common in the waxes of higher plants (24). Essentially three different patterns are seen: a pattern with n-C25 and n-C27 each constituting about 20% of the normal alkanes is most common and occurs at all intervals not shown in Figure 4, while a pattern with n-C29 dominant occurs at three levels in the upper core (surface, 20-40 cm, and 180-200 cm) and a pattern with n-C31 dominant occurs at one interval, 140-160 cm. Though there are many possible sources that might be invoked to explain the variations in these alkane distributions, it is interesting to note that two of the major plants that are common to the Great Marsh area, *Spartina alterniflora* and *Distichlis spicata*, have n-alkane peak abundances at n-C29 and n-C31, respectively (15,18,21). Sediment normal alkane distributions may therefore reflect the original marsh plant contributors of organic matter and provide a record of the changing vegetation during evolution of the marsh. The interpretation of the n-alkane sedimentary record would then be as follows: *S. alterniflora* is, and has been, the dominant grass during the period of accumulation of the upper 20-40 cm of sediment, and was also the dominant grass during the period of accumulation of sediments at the 180-200 cm interval. Distichlis spicata would have been the dominant alkane source during the time of deposition of sediments in the 140-160 cm interval. The source of the alkane distribution most commonly seen through the core, that with n-C25 and n-C27 dominant, is not known, but since this pattern is seen in the basal peat it might be associated with the Spartina patens community that presently occupies the headland regions of these marshes (8).

Free Fatty Acids. Typical upper core free fatty acid distributions (Figure 4) show a predominance of long chain even carbon normal fatty acids from C-20 to C-30. The presence of these long chain fatty acids in sediments has been attributed to hydrolysis of plant wax esters (25). Analyses of marsh plant fatty acids, however, show them to be composed primarily of the saturated n-C16 and unsaturated C-18 fatty acids (26,27). With the exception of Salicornia (26) and S. alterniflora (18), only trace quantities of normal fatty acids longer than C-19 have been reported in marsh plants. Thus there is not a clear correlation between the abundant free fatty acid material in the sediments and the most probable biological sources. A number of possible explanations exist to account for the predominance of long chain free fatty acids in sediments: 1) an unknown biologic source with a predominance of these acids, 2) preferential destruction or metabolism of the shorter-chain free fatty acids, and 3) microbial resynthesis of long chain fatty acids from precursor material. Long chain fatty acids have been observed in the saponified extracts of Florida salt marsh sediments (28) but their abundance relative to shorter-chain fatty acids was observed to decrease with depth in a sedimentary sequence with ages comparable to those in consideration here.

The abundances of total and of individual free fatty acids in the core sediments are best interpreted in light of the conceptual model for fatty acid diagenesis of Farrington and Quinn (29), and because of the time framework provided by the sedimentary sequence of the Great Marsh it is possible to estimate apparent rate constants for some of the possible reactions involved. Total free fatty acids rise to a peak concentration in sediments approximately 1000 years in age

(Figure 3) and then decrease in concentration at greater depths. The relative abundance of the long chain fatty acids (total of n-C24, n-C26, n-C28, and n-C30) follows a similar history, reaching a maximum value in sediments of the same age (Figure 3). Both of these observations can be interpreted to be the result of reactions that liberate large amounts of free, long-chain normal fatty acids in early diagenesis, followed by reactions that remove these fatty acids from the free fatty acid pool by either incorporation into humic material or degradative consumption (29,30). The liberation reactions are defined as those which produce solvent-extractable fatty acids (14), hence they could be reactions that involve hydrolysis of either plant wax esters or humic material, or reactions involving microbial synthesis or production of free fatty acids. The long-chain fatty acids increase in relative abundance as the abundance of free fatty acids increases, implying either that there is preferential liberation of long-chain acids or preferential recombination (or destruction) of the shorter-chain acids.

If the free fatty acids can be considered as non-diffusable material, then the following simple first-order kinetic model can be applied to a smooth curve drawn through the total free fatty acid data of Figure 3:

$$\underset{\text{(BFA)}}{\text{Bound Fatty Acids}} \xrightarrow{k_L} \underset{\text{(FFA)}}{\text{Free Fatty Acids}} \xrightarrow{k_R}$$

where Bound Fatty Acids (BFA) refers to those fatty acids that can only be obtained by saponification of the sediment and free fatty acids (FFA) refers to those that are recovered by solvent extraction without saponification (14). $k_L$ refers to the effective rate constant for the liberation of free fatty acids and $k_R$ refers to the effective rate constant for the removal of free fatty acids. The general equation for the concentration of total free fatty acids as a function of time will be:

$$FFA = \left[k_L/(k_R + k_L)\right] BFA_o(e^{-k_L t} - e^{-k_R t}) + FFA_o\ e^{-k_R t}$$

where $BFA_o$ and $FFA_o$ are the abundances of bound and free fatty acids, respectively, at time zero (i.e., the marsh surface).

Analysis of the results presented in Figure 3 suggests that values for $k_L$ and $k_R$ are on the order of $3.7 \times 10^{-4}$ yrs$^{-1}$ and $9.5 \times 10^{-4}$ yrs$^{-1}$, respectively. Obviously the absolute values, and the interpretation, of these rate constants require careful scrutiny in light of the assumptions inherent to the model outlined above, but these estimates provide a working model for future studies of diagenetic kinetics. Similar values for $k_R$ could be inferred from the data on free fatty acids in sediments of comparable age from Saanich Inlet, British Columbia (30). Clearly individual fatty acids would each have their own effective rate constants for the reactions presented above.

There is some evidence to suggest that the diagenetic reactions outlined above have not completely erased the record of some specific source. The three levels in the core that showed high n-C29 alkane abundance (surface, 20-40cm, and 180-200cm) also show a fatty acid pattern that is unique to these three intervals: n-C24 > n-C26 > n-C28 > n-C30, with variable amounts of n-C16 and n-C20 fatty acids. Since the dominant marsh plants do not exhibit these high abundances of long chain fatty acids (26,27), the interpretation of this unique fatty acid distribution is obviously difficult, but its apparent correlation to the normal alkane distributions that are tentatively related to specific sources warrants further investigation.

Of special interest is the normal C-20 fatty acid, arachidic acid, which ranges from six to thirty-one percent of the total free fatty acid mixture, and is in some cases the dominant fatty acid. Its identity has been confirmed by mass spectrometry. Sediment fatty acid distributions are usually characterized by high relative concentrations in the n-C14 to n-C18 and n-C24 to n-C28 ranges and low concentrations for fatty acids in the intermediate range (30). Similar amounts of arachidic acid have been reported in the Florida salt marsh study of Miller (28), and his study convincingly demonstrated that arachidic acid (as well as the longer chain fatty acids) was uniquely associated with salt marsh sediments in the area of study. Specific source material for these long chain acids remains uncertain, though reduction of long chain polyunsaturated fatty acids derived from planktonic and benthic marine organisms could be the simplest explanation (29,31).

The free fatty acids in the lower core sandy sediments are two orders of magnitude less abundant than in the overlying marsh sediments. This observation could be interpreted as a reflection of a greater age (hence more extensive removal of free fatty acids) for the sandy sediments, or as a reflection of a lower amount of material originally deposited with the sediments. The steep concentration gradient of free fatty acids across the lithologic break implies that diffusional mixing of these polar organic compounds might be affecting the composition of the lower sands, and in fact the 260-280 cm interval (not shown in Figure 4) has a fatty acid distribution almost identical to that seen in the 60-80 cm interval. The distribution of fatty acids at 320-340 cm (Figure 4 is quite similar to that seen in modern Delaware Bay sandy sediments (21), though the n-C16 and n-C18 abundances are reversed from what is seen in the modern sediments. Modern Delaware Bay sediments contain about one order of magnitude more free fatty acid material than is found in the lower core sands (21). Since it has previously been concluded from the distribution pattern of n-alkanes in these lower core sands that the sands were deposited in contact with estuarine waters, we would conclude that the fatty acid distribution represents preferential removal of n-C16 fatty acid (30) from a free fatty acid mixture much like that of modern Delaware Bay sediments.

Mono- and poly-unsaturated fatty acids are abundant components of living organisms (18,27,32,33) and these compounds could be the sources, via hydrogenation or metabolic alteration, of many of the saturated long chain fatty acids identified in the marsh sediments. The disappearance of unsaturated fatty acids with time is perhaps the most unequivocal evidence for diagenetic alteration: Table II shows that the two most readily identified and most abundant unsaturated fatty acids, C16:1 and C18:1, both disappear within the time represented by the upper core marsh sediments, i.e., about 2000 yrs. A loss of unsaturated fatty acids with depth has also been reported for algal mats (34), estuarine sediments (30,35), hypersaline lakes (36) and salt marsh sediments (18). Rhead and co-workers (37) have shown that labeled oleic acid (C18:1) is rapidly consumed (11 days) when injected in estuarine sediments, and that only 12% of the label could be accounted for among different fatty acid and fatty alcohol products. Normal fatty acids from C12 to C20 were among the products of alteration of C18:1, with C16:0 being one of the most common of the saturated fatty acid products.

In examining the rates of disappearance of unsaturated fatty acids, it is useful to consider the ratio of the abundance of an unsaturated fatty acid to the abundance of its saturated counterpart. With a few exceptions, the C16:1/C16:0 and C18:1/C18:0 ratios (Table II) show parallel decreasing trends with depth in the core. The simplest model for this overall diagenetic reaction would assume that unsaturated fatty acids are hydrogenated to their saturated analogs, and though some of the evidence cited above (37) does not support this model, its validity is suggested by the data presented here, especially for the reaction C18:1 ⟶ C18:0. The first-order model for this reaction would be:

$$\underset{(A)}{C{=}C} \xrightarrow{k_H} \underset{(B)}{C{-}C} \qquad \text{and } A/(A+B) = e^{-k_H t}$$

where $k_H$ is the first-order rate constant for hydrogenation of the double bond of the particular fatty acid in question, and A and B are the general notation for the unsaturated and saturated fatty acids involved in the reaction. An exponential decrease in the ratio A/(A+B) with time would imply that the first-order model was applicable.

Figure 5 shows that the model does seem to apply to the results for the C18:1 ⟶ C18:0 reaction, and a rate constant for hydrogenation that would be inferred from this observation would be on the order of $1.4 \times 10^{-3}$ $yrs^{-1}$. This represents a minimum value since other evidence (37) implies that C18:0 is not the only "daughter product" of diagenesis of C18:1. Since C16:0 is one of the products of C18:1 alteration (37), the erratic behavior of the C16:1/C16:0 ratio as a function of time can perhaps be explained as being the result of more complex diagenetic pathways for the C16:1-C16:0 pair than for the C18:1-C18:0 pair.

Using the value for $k_H$ derived above, and an estimate of the total amount of "unsaturation" in surface sediments, we estimate that the total steady-state amount of hydrogen consumption via the "saturation" reaction is on the order of $2 \times 10^{-2}$ μmoles $H_2/cm^2/yr$. Since this is about three orders of magnitude less than measured methane evolution rates in the Great Marsh (5) via the reaction:

$$CO_2 + 4H_2 \longrightarrow CH_4 + 2H_2O \quad (38)$$

it appears that hydrogen production is not the rate-determining step in the reduction of unsaturated fatty acids. Hence the assumption of first-order kinetics appears valid.

## Summary

Stratigraphic and chronologic control on a continuous sequence of marsh sediments provides a framework for the evaluation of early diagenetic reactions of fatty acids and alkanes derived from native salt marsh organic matter. Little evidence is seen for alteration of alkane material; variations with depth are not mono-

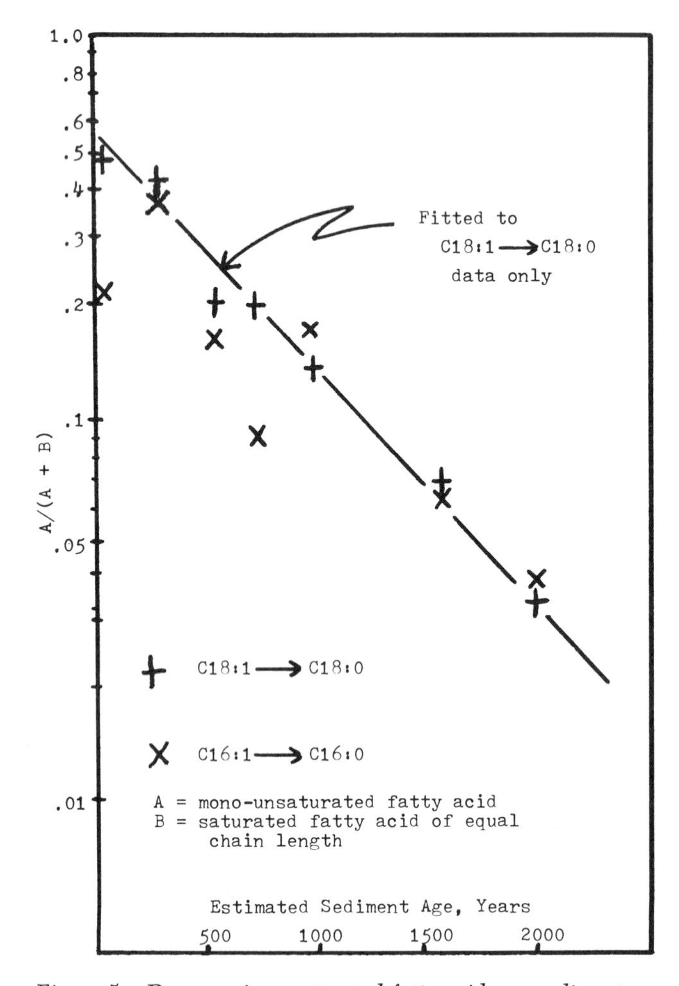


*Figure 5. Decrease in unsaturated fatty acids* vs. *sediment age*

tonic and appear to reflect variations in source material. Therefore, normal alkane distributions provide a record of the types of organic matter that have been supplied to the depositional environments of the Great Marsh as represented by one particular core.

The fatty acids determined in this study represent the free fatty acid fraction in the sediments, and their distribution in sediments of different ages is consistent with conceptual models for hydrolytic liberation, reduction, and recombination of free fatty acids during early diagenesis. It appears that some source material or some diagenetic pathway that is unique to salt marshes is responsible for the production of large relative abundances of long-chain (C20, C24, C26, C28, C30) normal free fatty acids. Consumption of abundant mono-unsaturated fatty acids is complete within 2000 years in the environment under study. The first-order kinetics observed for the disappearance of oleic acid imply that simple hydrogenation reactions, rather than complex metabolic ones, may be dominant in the early diagenesis of this particular fatty acid.

## Acknowledgements

This research has been sponsored in part by the University of Delaware Research Foundation and in part by R.A.N.N. program of the National Science Foundation. We are especially grateful to Dr. T. C. Hoering, Carnegie Institution of Washington, for his help with analytical procedures and interpretation of results. Drs. F. M. Swain and R. B. Biggs offered critical advice regarding various aspects of this research. Mass spectral analyses were provided by Drs. Swain and Hoering.

## Literature Cited

1. Kolattukudy, P. E., Lipids, (1969) v. 5, p. 259-275.
2. Hedberg, H. D., American Assoc. Petrol. Geol. Bull. (1968), v. 52, p. 736-750
3. Reed, K. J., Bull. AAPG, (1969), v. 53, p. 1502-1506
4. Biederman, E. W., Jr., Bull. AAPG, (1969), v. 53, p. 1500-1502
5. Swain, F. M., in "Adv. Organic Geochemistry, 1973," pp. 673-687, Pergamon, Oxford, 1974
6. Kraft, J. C., Geol. Soc. America Bull., (1971), v. 82, p. 2131-2158

7. Elliott, G. K., The Great Marsh, Lewes, Delaware: the physiography, classification and geologic history of a coastal marsh: College of Marine Studies, Univ. of Delaware, (1972), 139 p.
8. Klemas, V., F. C. Daiber, D. S. Bartlett, O. W. Critchon, and A. O. Fornes, "Coastal Vegetation of Delaware, the Mapping of Delaware's Coastal Marshes College of Marine Studies, Univ. of Delaware, March (1973), 29 p.
9. Kraft, J. C., "A guide to the geology of Delaware's coastal environments": College of Marine Studies, Univ. of Delaware, (1971), 220 p.
10. Folk, R. L., "Petrology of sedimentary rocks", Hemphill's, Austin, Texas, (1968), 170 p.
11. Ball, D. F., Jour. Soil Sci, (1964), v. 15, p. 84-92.
12. Blumer, M., Anal. Chem., (1957), v. 29, p. 1039-1041.
13. Metcalfe, L. D., and Schmitz, A. A., Anal. Chem., (1961), v. 33, p. 363-364
14. Farrington, J. W., and Quinn, J. G., Geochim. et Cosmochim. Acta, (1971), V. 35, p. 735-741
15. Sever, J. R., "Organic geochemistry of hydrocarbons in coastal environments", Ph.D Dissertation, Univ. of Texas at Austin, 147 p.
16. Cooper, J. E., and Bray, E. E., Geochim. et Cosmochim. Acta, (1963), v. 27, p. 1113-1127.
17. Kvenvolden, K. A., Nature, (1966) v. 209, p. 573-577
18. Johnson, R. W., and Calder, J. A., Geochim. et Cosmochim. Acta, (1973) v. 37, p. 1943-1955.
19. Zafiriou, O., Blumer, M., and Myers, J., "Correlation of oils and oil products by gas chromatography", Woods Hole Oceanogr. Inst. Ref. No. 72-55, (1972), 110 p.
20. Han, J. and Calvin, M., Proc. Nat. Acad. Sci. (1969) v. 64, p. 436-443.
21. Swetland, P. J., "Lipid geochemistry of Delaware salt marsh environments", M.S. Thesis, Univ. of Delaware, (1975), 97 p.
22. Blumer, M., Guillar, R. R., and Chase, T., Mar. Biol., (1971), v. 8, p. 183-189.
23. Swain, F. M., Jour. Sed. Petrol., (1971), v. 41, p. 549-556.
24. Eglinton, G., and Hamilton, R. J., in Swain, T., (ed.), "Chemical Plant Taxonomy": Academic Press, New York, (1963), p. 187-217.
25. Meinschein, W. G., and Kenny, G. S., Anal. Chem., (1957) v. 29, p. 1153-1161

26. Maurer, L. G., and Parker, P. L., Contr. Mar. Sci., (1967), v. 12, p. 113-119.
27. Jeffries, H. P., (1972), v. 17. p. 433-440.
28. Miller, R. E., Geol. Survey Prof. Paper, 724-B, (1972), 11 p.
29. Farrington, J. W., and Quinn, J. G., Geochim. et Cosmochim. Acta, (1973), v. 37, p. 259-268.
30. Brown, F. S., Baedecker, M. J., Nissenbaum, A., and Kaplan, I. R., Geochim. et Cosmochim. Acta, (1972), v. 36, p. 1185-1203.
31. Sassen, R., Geol. Soc. America, Abs. with Programs (Northeastern Section), (1973), v. 5, no. 2, p. 215-216.
32. Ackman, R. G., Tochter, G. S., and McLachlan, J., J. Fish. Res. Bd. Canada, (1968), v. 25, p. 1603-1620
33. Eglinton, G. in Schenck, P. A., and Havenaar, I., (eds.), "Advances in Organic Geochemistry", Perga-Press, New York, (1969), p. 1-24.
34. Parker, P. L., and Leo, R. F., Science, (1965), v. 148, p. 373-374.
35. Farrington, J. W., and Quinn, J. G., Nature Phys. Sci., (1971), v. 230, p. 67-69.
36. Nissenbaum, A., Baedecker, M. J., and Kaplan I. R. Geochim. et Cosmochim. Acta, (1972), v. 36, p. 709-729.
37. Rhead, M. M., Eglinton, G., and England, P. J. in Adv. Organic Geochemistry, H. R. V. Gaertner and H. Wehner, eds., Pergamon Press, Oxford, (1971), p. 323-333.
38. Martens, C. S. and Berner, R. A., Science, (1974), v. 185, p. 1167-1169.

# 17

# Chemical Factors Influencing Metal Alkylation in Water

K. L. JEWETT and F. E. BRINCKMAN
National Bureau of Standards, Inorganic Chemistry Section, Washington, D.C. 20234
J. M. BELLAMA
Chemistry Department, University of Maryland, College Park, Md. 20742

Because of their toxic properties, the role that heavy metals play in the environment has recently been the subject of increased study. It is observed that heavy metals in their inorganic forms exhibit relatively low toxicity toward biota when compared to these metals with certain carbofunctional groups attached. For example, alkylmercurials are from 10-100 times more toxic than inorganic mercury compounds (1,2), and similar effects are noted for organotins (2).

Many heavy metals, however, are released into the aquatic environment in relatively nontoxic form. They subsequently acquire this enhanced toxicity as organometals through environmental interactions involving both biological and non-biological processes (3). It is therefore fundamental to environmental concerns to investigate the chemical factors which bring about this transformation from inorganic to organometal compounds in aqueous media.

Two important factors responsible for the conversion of heavy metals into organometals are ligand interactions and photochemical processes. In this paper an initial effort is made to evaluate the potential importance of these factors in bringing about metal transformations.

Examples have now appeared in the literature which show the occurrence of metal transformations in aqueous solution (3), including a wide range of transmethylation reactions between heavy metal ions (4). On the one hand, the known cellular metabolite methylcobalamin has been shown to transfer with ease its methyl group to mercuric ion in aqueous solution to form the highly neurotoxic methylmercury ion (5). In another study (6), Jernelöv *et al.*, have implicated anthropogenic alkyllead discharges in formation of high methylmercury concentrations in St. Clair River sediments. Hence we have an example of methylation also occurring environmentally in apparently a straightforward metathetical reaction.

Work has also been advanced in the area of photochemistry. For example, a methylchromium bond has been formed in the photolysis of *tertiary*-butoxy radicals and chromium(II) in aqueous solution (7). It was found that in the presence of acetate ion or acetic acid in aqueous solution, $Hg^{2+}$ was photolyzed in sunlight to form a methylmercury bond (8). We have examined this latter reaction and similar ones. These last reactions are consistent with early findings (9) that $(RCO_2)_2Hg$ compounds decompose in protic organic solvents under ultraviolet irradiation to yield alkylmercurials, RHg(OCOR).

## Experimental

Metal Coordination *vs.* Rate of Reaction. Two of the organotin compounds, $Me_3SnNO_3$ and $Me_3SnClO_4$ (Me = $CH_3$) were prepared (10) by reacting $Me_3SnBr$ with an equivalent amount of the appropriate silver salt in methanol. The solutions were then filtered and the solvent was removed from the organotin product by vacuum separation techniques. Both desired tin compounds were then sublimed twice *in vacuo* and found to be suitable for use by spectrometric techniques. Other compounds used were found not to require further purification. All solutions were prepared using distilled water; the presence of dissolved air had no detectable effect on reaction rates observed.

In order to investigate the abiotic methylation of $Hg^{2+}$ in water, the effect of ligands on reaction rate was studied. It had been previously determined (4) that $Me_3Sn^+$ undergoes facile loss of one methyl group to quantitatively form $MeHg^+$ from aquo-$Hg^{2+}$ as shown by the following equation:

$$Me_3Sn^+ + Hg^{2+} \xrightarrow{H_2O} Me_2Sn^{2+} + MeHg^+ \qquad (1)$$

Reactions were monitored by proton magnetic resonance (pmr) employing procedures described elsewhere (4,11). Figure 1 shows a characteristic pmr spectrum. Areas under these curves were measured by planimetry and normalized to adjust for intensities due to the varying amounts of protons. At any time, extent of reaction (X) is therefore equal to $3/2 \cdot (\text{Area } Me_2Sn^{2+})/(\text{Area } Me_3Sn^+) \cdot [Me_3Sn^+]_{initial}$, or $3 \cdot (\text{Area } MeHg^+)/(\text{Area } Me_3Sn^+) \cdot [Me_3Sn^+]_{initial}$. These data were then subjected to regression analysis and the results are summarized in Table I.

Methylation of mercury was found to obey the second order rate law,

$$k_2t = \frac{1}{A-B} \ln \frac{B(A-x)}{A(B-x)} \qquad (2)$$

where A and B represent the initial concentrations of reactants and x is the extent of reaction at any given time (t). Figure 2 shows a plot of data typically obtained. Within experimental error, *e.g.*, ± *ca.* one standard deviation, both the formation of methylmercury and dimethyltin ions have the same slope as predicted in equation 1. The lines shown through these data points in

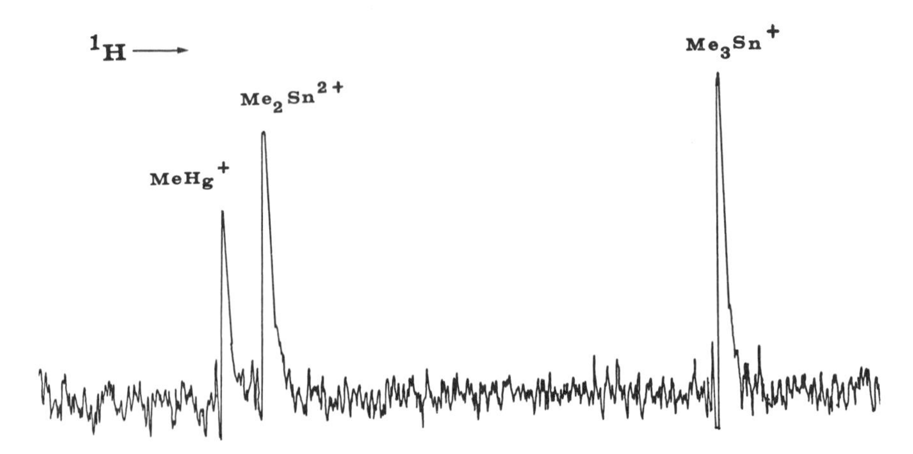


*Figure 1. Characteristic nuclear magnetic spectrum of all protonated metal species indicated by Equation 1*

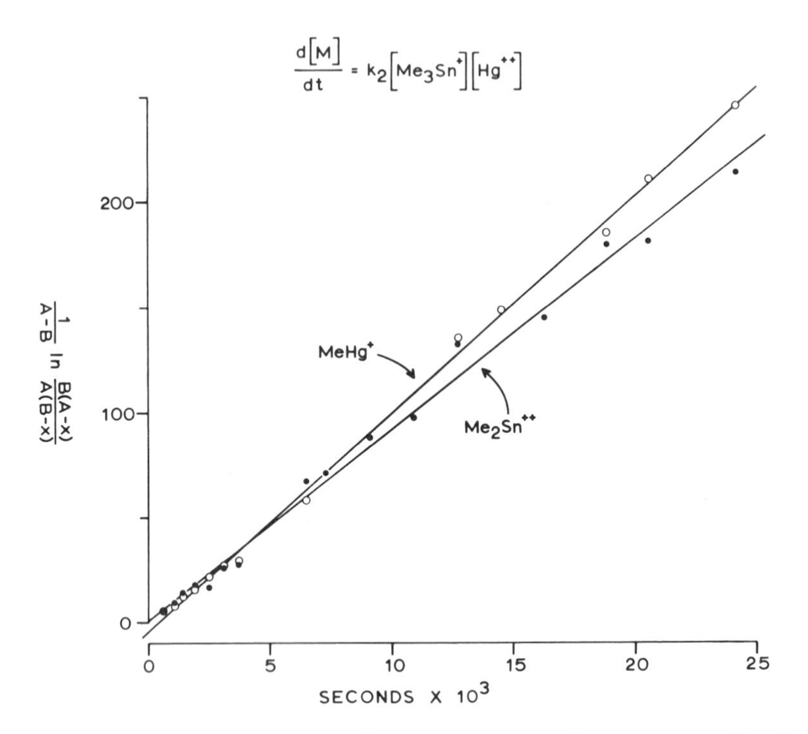


*Figure 2. When production of either $MeHg^+$ or $Me_2Sn^{2+}$ is incorporated in the 2nd order rate expression shown on the ordinate, slopes ($k_2$) for either species are in reasonable agreement. For all $Cl^-$ dependent reactions discussed here, $d[M]/dt = d[Me_2Sn^{2+}]/dt$.*

Table I

Relationship Between Total Chloride and Mercuric Ion Concentrations of Transmethylation

| Total Cl / Total Hg | $(k_{2obsd} \pm Std.Dev.) \times 10^3$ $\ell\ mol^{-1}sec^{-1}$ | Data Points (#) | Correlation Coefficient (R) | Extent of Reaction (%) |
|---|---|---|---|---|
| 1.25 | 3.56 ± .35 | 16 | 0.938 | 82.2 |
| 2.00 | 7.77 ± .38 | 13 | 0.987 | 73.5 |
| 2.92 | 9.04 ± .36 | 18 | 0.987 | 84.9 |
| 4.00 | 11.29 ± .77 | 18 | 0.965 | 90.3 |
| 5.00 | 9.56 ± .67 | 18 | 0.963 | 88.5 |
| 6.07 | 7.80 ± .58 | 17 | 0.962 | 88.5 |
| 10.07 | 7.77 ± .38 | 17 | 0.983 | 83.1 |
| 22.17 | 4.40 ± .25 | 18 | 0.975 | 72.1 |

Figure 2 represent the best statistical fit for each data set.

By varying the ratio of total chloride ion concentration to total mercury(II) ion concentration (Cl/Hg), the effect of chloride ion strength on the rate of reaction could be determined. Approximately 0.05 M solutions of each reactant in distilled water were prepared. Compounds were selected to yield appropriate *total* chloride concentrations. For example, in this scheme, for

$$Me_3SnX + HgY_2 \rightarrow Me_2Sn^{2+} + MeHg^+ + X^- + 2Y_2^- \quad (3)$$

Cl/Hg = 0, X = Y = $ClO_4^-$; for Cl/Hg = 1, X = $Cl^-$ and Y = $ClO_4^-$; and for Cl/Hg = 2, X = $ClO_4^-$ and Y = $Cl^-$. In all other cases X = Y = $Cl^-$.

Because of its propensity to remain uncoordinated with metal ions (12), perchlorate was chosen as the gegenion for situations where Cl/Hg = 0, 1 or 2. Sodium chloride was used to fortify solutions when Cl/Hg > 3. For each of these reactions, approximately 0.05 M solutions of $HgCl_2$ were prepared by dissolving that compound into saline solution of appropriate concentration. The other reactant $(CH_3)_3SnCl$ was prepared in distilled water. There was no evidence that use of $Na^+$ as a counter ion affected kinetic results.

A combination electrode was used to make pH measurements of reactant solutions. For the specific case Cl/Hg = 3, the pH of a thermostated reaction mixture was measured as a function of time, and found to follow the simple relationship, $kt = [H^+]^2$.

Photolysis of Heavy Metals Compounds. Approximately 0.05 M distilled water solutions of mercuric acetate and methylmercuric acetate were prepared in the dark. Some of each solution was placed into thin-wall fused silica nmr tubes, which were maintained in direct contact with a "pencil-probe" ultraviolet light (253.7 nm) source. While control tubes were maintained in the dark, the reaction tubes were irradiated in ten-minute increments up to a total of 100 minutes. After each exposure to radiation, both control and reaction tubes were examined by pmr spectrometry. Extent of reaction for the latter was determined relative to the concentration measured in the control tubes, utilizing methylmetal peak areas as before, as well as the methyl peak area in acetate ion. A solution of 0.05 M mercuric acetate in deuterium oxide ($D_2O$) was also irradiated and volatile decomposition products were examined.

Distilled water solutions (0.05 M) of sodium acetate and of thallium(I) acetate were also irradiated as a function of time. Experimental conditions were the same as those used in photolysis of mercury compounds.

Gas evolution and metal precipitation were characteristic for the irradiated acetate solutions containing Hg(II) or Tl(I) but not sodium acetate alone. Volatile decomposition products were analyzed by gas chromatography and mass spectrometry. For elemental mercury and trace organomercury analyses, a combination gas chromatograph-mercury specific atomic absorption apparatus was used (13).

## Results and Discussion

The abiotic transmethylation of $Hg^{2+}$ in water was examined by measuring the effect on the rate constant of varying the ratio of total chloride to total mercury (Cl/Hg) in equation 3. The results are shown in Figure 3. In the all-perchlorate system, *i.e.*, Cl/Hg = 0, no reaction was found to occur. With increasing Cl/Hg ratios the rate of reaction reaches a maximum at Cl/Hg = 4 and decreases thereafter. When nitrate ion was substituted for perchlorate in the specific case of Cl/Hg = 2, rate constants for this and the perchlorate reaction were not significantly different.

In Figure 3 it is apparent that a significant change in the rate of reaction occurs when the total chloride ion concentration introduced into equation 2 is altered with respect to the concentration of mercury present in this system. Since chloride is a very strong coordinating ligand and mercury is known to form a number of complex species in aqueous solution (14), we sought to

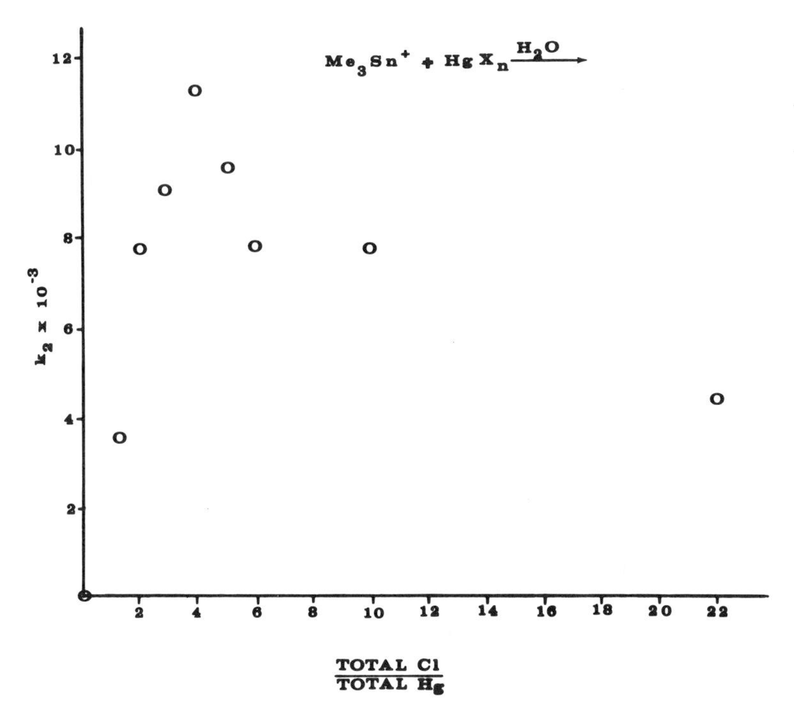


*Figure 3. Results of chloride ion dependence study on reaction rate. Regression analyses of these data yielded relative standard errors in the $k_2$ values (slopes) from 3–6%.*

identify which chlorohydroxymercury complexes were present under the varying $[Cl^-]/[Hg^{2+}]$ ratios used in the above rate studies, and to ascertain which species are actually necessary for the transmethylation reaction. It is postulated that the most reactive intermediate mercury complex will show a variation in relative abundance which will parallel the changes in the reaction rate shown in Figure 3.

It is also useful to note here that previous work illustrated in Figure 4 predicts that a large pH range has little effect on the nature of $[HgCl_n]^{2-n}$ species, whereas small changes in chloride ion concentration produce significant changes in the relative abundances of chloromercury complexes. Thus, the shaded ellipse represents pH and pCl values found by us in the estuarine environment (3), whereas the conditions approximately covered in our present laboratory investigation are shown by the dotted ellipse.

During the course of each reaction in the current study, the pH of each solution characteristically changed by approximately one pH unit. Attempts were therefore made to quantitatively relate the change in $[H^+]$ with the second order rate equation. These tests include $S \cdot [H^+]$ *vs.* time, and $S/[H^+]$ *vs.* time, and $S \cdot [H^+]^{1/2}$ *vs.* time where $S$ represents the second order rate expression shown in equation 2. Least squares regression analysis for all of the above cases incorporated data for any one run in three ways: total data points *vs.* time, first half of the data points *vs.* time and second half of data points *vs.* time. Since our rate data typically included 13-18 nmr spectra taken over an eight hour period, the latter two analyses were deemed necessary in the event that there was a significant change in kinetics during such extended periods. In any of the calculations incorporating $[H^+]$, $[H^+]^{1/2}$, or $1/[H^+]$, we found significantly poorer fits than for those runs not involving these hydrogen ion concentration terms. We therefore conclude that within the precision of our methods (*cf.* Table I), the transfer of methyl from tin(IV) to mercury(II) is not directly pH dependent in saline media.

A computer program, modified from one originally developed by Swedish workers (15) to calculate the most abundant mercury species at given pH and pCl values, indicated that several of fourteen possible $[Hg(OH)_mCl_n]^{2-m-n}$ complexes exist under conditions of environmental interest. The three complexes shown in Figure 5 predominate under such conditions, and account for over 99% of all mercury(II) present in the total system. While $HgCl_2$ decreases with increasing chloride ion concentration, $HgCl_4^{2-}$ shows the reverse trend. The only chloromercury species to reach a maximum concentration over this environmental salinity range is $HgCl_3^-$.

It was previously determined that the rate of appearance of either of the products formed in equation 1 is equal to the following expression:

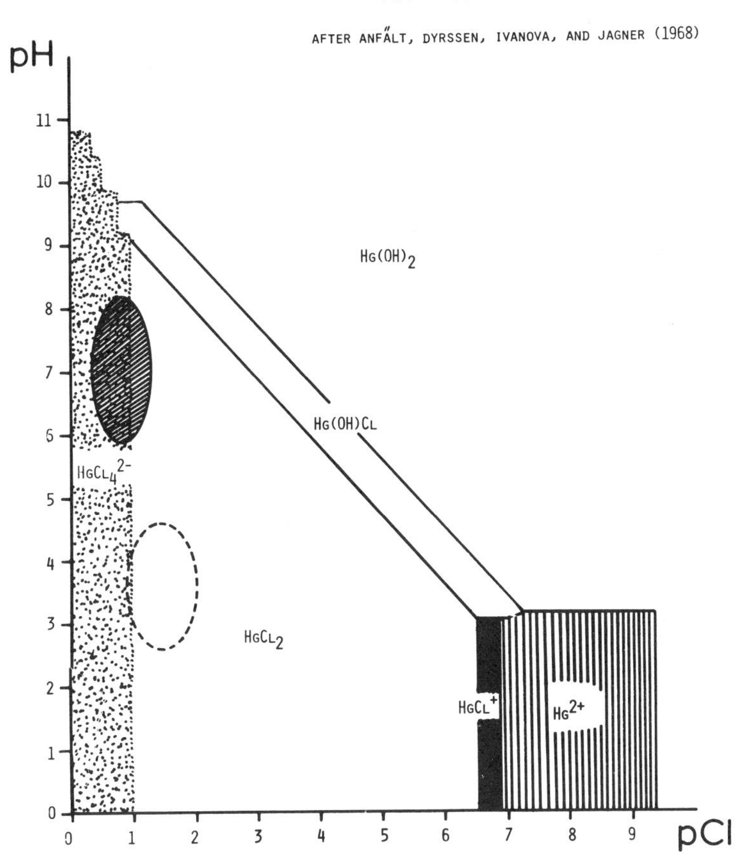


*Figure 4. Stability fields for $Hg(OH)_mCl_n$ species determined from equilibria for mercury complexes as a function of pH and pCl*

$$\frac{d[MeHg^+]}{dt} = k_{2(obsd)} \text{ [Total Mercury][Organotin]} \quad (4)$$

It is assumed that organotin species are invariant with changing Cl/Hg ratios in these aqueous solutions, and this view is consistent with the low stability constants reported for $[(CH_3)_3SnCl_n]^{1-n}$ complexes in aqueous solution (16). If the rate of product formation is dependent only (or chiefly dependent) on the presence of a specific chloromercury species, $[HgCl_n]^{2-n}$, a new rate constant $k_2'$, can be defined:

$$\frac{d[MeHg^+]}{dt} = k_2'[(HgCl_n)^{2-n}]\text{[Organotin]} \quad (5)$$

We further expect $k_2'$ to be greater than $k_{2(obsd)}$ since the reactive mercury species is present at lower concentrations under these conditions.

Moreover, the energy of activation for the reaction shown in equation 1 has been previously determined (4,11). A relatively low value of 14.2 kcal/mole is consistent with a reaction involving $Me_3Sn^+$ and $HgCl_3^-$, in that large coulombic repulsions are not expected in formation of the activated complex (17).

Although ionic interferences have hampered attempts to directly determine pCl values in laboratory experiments, refined pCl approximations have led to several conclusions. When using the modified computer program along with the estimated pCl values for the laboratory reaction conditions, it was again found that only $HgCl_3^-$ reaches a maximum concentration in the Cl/Hg range for which $k_{2(obsd)}$ reaches a maximum. Inasmuch as the reaction shown in equations 1 and 3 is first order in mercury, and the rate constant reaches a maximum when the Cl/Hg ratio is varied, we would expect a chloromercury complex to show such similar behavior. Since there is only one mercury species exhibiting the expected behavior under laboratory conditions, we conclude that $HgCl_3^-$ is the only or most important rate-determining mercury complex.

One case for usefully extending the present findings becomes apparent. In our survey (4), we have previously examined the following reaction:

$$CdCl_2 + (CH_3)_3SnCl \xrightarrow{H_2O} \text{No Reaction} \quad (6)$$

Chlorohydroxycadmium ions analogous to the present Hg cases are mainly dissociated under these conditions (18). Consequently, reaction may be mainly hindered by the anticipated large energy of activation required for the two positively charged aquated species, *e.g.*, $Cd^{2+}$ and $Me_3Sn^+$, to undergo transmethylation. In addition, preliminary results suggest that some halide ion ($Cl^-$ or $Br^-$) must also be present along with oxo-anions ($NO_3^-$ or $ClO_4^-$) to insure that methyl transfer occurs. That is, the presence of other possible negatively charged $HgL_3^-$ (*i.e.*, L = $NO_3^-$, $Ac^-$, *etc.*) complexes may not be sufficient for transmethylation. We are currently reexamining this prospect for reaction at a suitable pH where coordination of cadmium is shifted by excess $Cl^-$ to an abundance of $[CdCl_3]^-$. Good aqueous methylating agents such as trimethyltin ion or trimethylead ion (4,11) may then become effective in methyl transfer to cadmium, although the fate of any intermediate methylcadmium species in saline waters is presently unknown.

Through these studies one clearly sees that the potential effects of coordination and ionic equilibria on organometal transformations may indeed be very significant. Photolysis, however, may have a similar or an even more widespread impact on environmental transformations of metals. We have therefore initiated studies seeking to assess the scope and extent of this influence on metal transformations.

When an aqueous solution of mercuric acetate is irradiated, the results as a function of time are shown in Figure 6. Note that acetate ion ($Ac^-$) decreases in an almost exponential manner. Methylmercury ion is rapidly formed and increases to a maximum concentration, but this species finally diminishes in concentration upon further irradiation. Dimethylmercury only forms after methylmercury has reached its maximum concentration. It appears that dimethylmercury reaches some point where its rate of formation approximates the rate of decomposition, but finally this species also decreases in concentration.

During the course of reaction, gas evolution and metal precipitation are observed. These products have also been characterized. By using gas chromatography and mass spectrometry, ethane and carbon dioxide were identified. Mercury metal was determined using a gas chromatograph in tandem with mercury-specific atomic absorption spectrophotometer (GC-AA) which is described elsewhere (13).

When mercuric acetate (0.05 M) in $D_2O$ was irradiated and volatile products subsequently examined by mass spectrometry, no incorporation of deuterium was observed. Only ethane and carbon dioxide were again detected. Decomposition of reactive methyl-species must therefore involve some concerted bimolecular process in which solvent ($H_2O$ or $D_2O$) protolysis is not actively involved.

At the point where methylmercury reaches maximum concentration both methylmercury and acetate ions are observed to be of comparable concentrations. Consequently, similar behavior for the irradiation of methylmercuric acetate is anticipated. A similar pattern is seen for product formation after methylmercuric ion reaches its maximum concentration (Figure 6) when compared with the results found on irradiating methylmercuric acetate as shown in Figure 7. The slopes of each of the species represented in Figure 7 show much the same behavior as their analogues in Figure 6. The final products for this reaction are identical to those observed for the photolysis of mercuric acetate.

Photolysis of systems involving mercury(II) in the presence

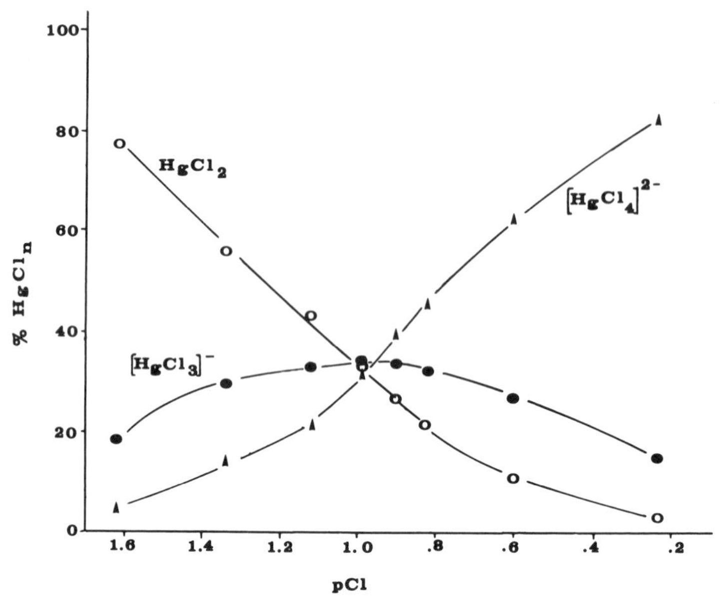


*Figure 5. At the chloride ion concentrations (pCl) shown, the above mercury species are expected to occur*

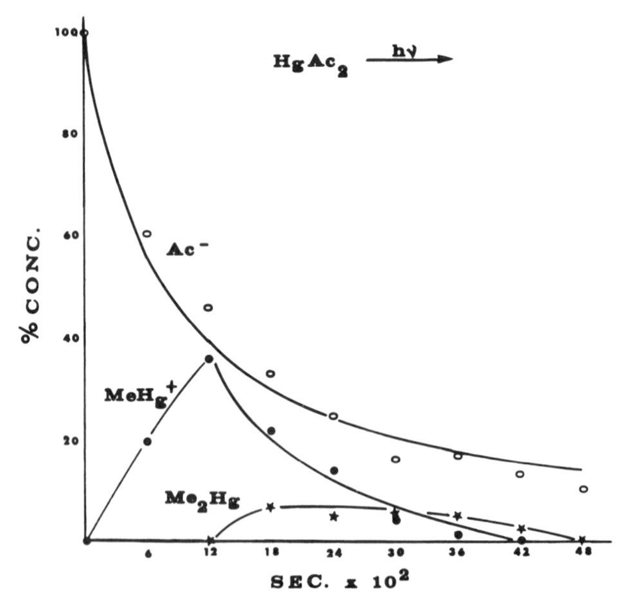


*Figure 6. Relative concentrations for the various protonated reactants and products were determined during photolysis relative to a control sample which was maintained in the dark*

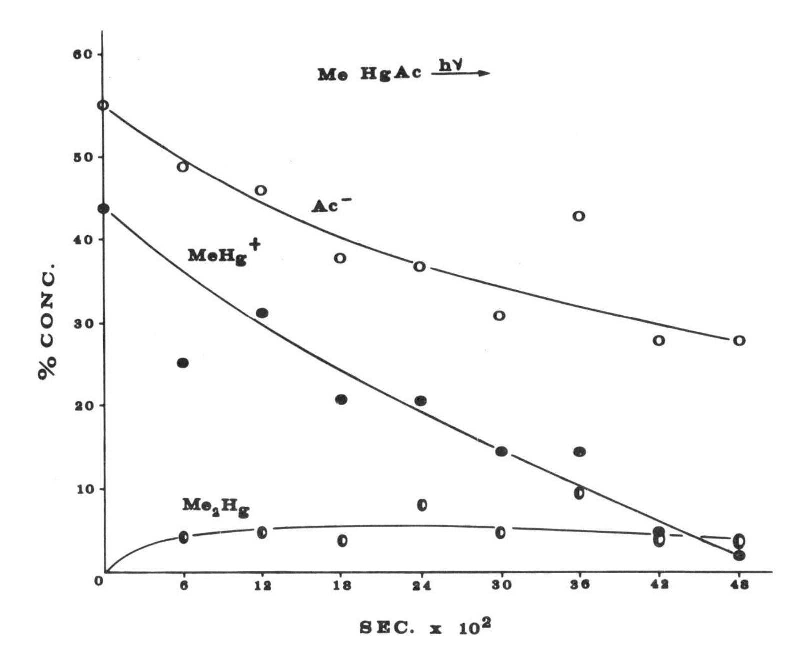


*Figure 7. The photolysis of methylmercuric acetate is seen to follow a reaction sequence similar to that observed for mercuric acetate after half-reaction*

of a number of other known bacterial metabolic products in aqueous solution has also been attempted. Thus, stoichiometric amounts of acetone, ethanol or propionic acid, when irradiated in the presence of aqueous mercuric chloride, have also been found to produce the same organomercury species, albeit at different rates.

Photolysis of mercury compounds has also been found to occur in very dilute solutions even under normal laboratory fluorescent lighting. A solution of 20 ppm (μg/g) mercury (as $HgCl_2$) and 40 ppm acetate (as NaAc), placed in a quartz cuvette, was found to produce trace amounts of methylmercury ion, dimethylmercury, and mercury metal. The amounts of these same species were found to be insignificant in the control solution which was maintained in the dark. Again, the presence of each of the products was verified by GC-AA analysis. This study implies that some mercury compounds normally used in the laboratory may be undergoing transformations by just sitting on the shelf.

A survey of the photolysis of acetate ion in the presence of other metal ions has also been initiated. Among other metals, thallium was selected for initial study because of the recent demonstration of its bacterial utilization and transport (19). Thallium(I) shows evidence of involvement in a photolytic process similar to that seen for Hg(II). Thus, when thallium(I) acetate is irradiated a decrease in acetate concentration occurs; however, if a solution of sodium acetate is similarly irradiated, no diminution of acetate concentration is observed over the same period of time. Also, as with the case for Hg, continued irradiation of Tl(I) acetate finally produces free metal, along with gaseous products.

Two important factors emerge from these results: (1) presence of a metal cation capable of appropriate (possibly bidentate) coordination of the acetate moiety may be necessary for effective photoalkylation to occur; and (2) possible formation of a transient methylthallium(I) species, similar to those reaction intermediates reported for Pd(II) and Pt(II) (4), forms but rapidly decomposes to observed products. That is, as soon as the organothallium species is produced it undergoes further reaction, presumably disproportionation to Tl(III) and the metal (20). This prospect is currently under investigation, and efforts are being made to identify elusive intermediate species, and any possible stable methylthallium(III) by-products.

These studies clearly indicate that salinity and photolysis may have important roles in bringing about organometal transformations in the environment, and should be considered in further studies of the role of metals in the environment. Salinity has been demonstrated to regulate the rate of one kind of transmethylation reaction to an important extent, and may be a key factor in other cases. In addition, photolysis has been demonstrated to be a process desirable for further study. Even in the ppm range methylation of mercury has been demonstrated as a facile process occurring under typical laboratory lighting conditions. This last point will bear future considerations on common analytical methods employed for both environmental or laboratory samples.

The aforementioned results represent model studies into the effects that ligand interactions and photolysis may play in the environment. It is through our understandings attained in these studies that we may more capably investigate appropriate reactions occuring under environmental conditions.

## Literature Cited

1. Hartung, R. and Dinman, B. D., Ed., "Environmental Mercury Contamination", Ann Arbor Sci. Publ. (1972), pp. 341-345.
2. Barnes, J. M. and Magos, L., Organomet. Chem. Rev. (1968), 3, 137-150.
3. Brinckman, F. E. and Iverson, W. P., ACS Symposium on Marine Chemistry in the Coastal Environment (T.Church,Ed.) (1975), paper 22.
4. Jewett, K. L. and Brinckman, F. E., Preprints of Papers, Div. Environ. Chem. Amer. Chem. Soc. (1974), 14, 218-225.
5. Agnes, G., Bendle, S., Hill, H. O. A., Williams, F. R., and Williams, R. J. P., Chem. Comm. (1971), 850-851.
6. Jernelöv, A., Lann, H., Wennergren, G., Fagerström, T., Asell, P., and Andersson, R., "Analyses of Methylmercury Concentration in Sediment From the St. Clair System", unpublished report of the Swedish Water and Air Pollution Research Laboratory (1972).
7. Ardon, M., Woolmington, K. and Pernick, A., Inorg. Chem. (1971) 10, 2812.
8. Agaki, H. and Takabatake, E., Chemosphere (1973), 3, 131-133.
9. Razuvaev, G. A. and Ol'deleop, Y. A., Doklady Akad. Nauk S.S. S.R., (1955), 105, 738-740.
10. Clark, H. C., O'Brien, R. J., Inorg. Chem. (1963), 2 (4) 740-4.
11. Jewett, K. L. and Brinckman, F. E. (1975), submitted for publication.
12. Johansson, L., Coord. Chem. Rev. (1974), 12, 241-261, and references cited therein.
13. Blair, W., Iverson, W. P. and Brinckman, F. E., Chemosphere (1974), 3, 167-174.
14. Gilmore, J. T., Environ. Letters (1971), 2, 143-152.
15. Anfält, T., Dyrssen, D., Ivanova, E., and Jagner, D., Svensk Kemisk Tidskrift (1968), 80, 340-342.
16. Cassol, A., Magon, L., and Barbieri, R., Inorg. Nucl. Chem. Letters (1967), 3, 25-29.
17. Abraham, M. H. and Spalding, T. R., J. Chem. Soc. (A)(1968), 2530-35.
18. Weber, Jr., W. J. and Possalt, H. S., in "Aqueous Environmental Chemistry of Metals", Ann Arbor Sci. Publ. (1974), pp. 255-264.
19. Schneiderman, G. S., Garland, T. R., Wildung, R. E., and Drucher, H., Abstract 74th Ann. Mtg. Amer. Soc. Microbiol., Chicago, Ill. (May, 1974), p. 2.
20. Kurosawa, H., and Okawara, R., Inorg. Nucl. Chem. Letters (1967), 3, 21-23.

## Acknowledgements

The authors are grateful for financial support of these studies by the NBS Office of Air and Water Measurements, Environmental Protection Agency and Department of the Navy. We thank Dr. Rolf Johannesen for his valuable assistance in nmr and computer calculations. We also thank Mr. Richard Thompson (pmr measurements) and American University Research Participation Program students, Lee Silberman and Steven Wagner (data collection) for their aid.

# 18

# Chemical and Bacterial Cycling of Heavy Metals in the Estuarine System

F. E. BRINCKMAN and W. P. IVERSON

Institute for Materials Research, National Bureau of Standards, Washington, D.C. 20234

A central problem faced by environmentalists is determining pathways by which heavy metals are transported through the aquatic media, and the rates by which these events occur. This rests on establishing those important reservoirs or "sinks" wherein such heavy metals accumulate, and the key metal-containing species affording transport between compartments (*i.e.*, sediments, water, and primary trophic levels). In practical terms, once having determined the presence of a metal of interest in the field (one is dealing here principally with potentially toxic pollutant metals), the researcher is confronted with evaluating its local distribution or concentration gradient. Also one must contend with the larger problem of characterizing the abiotic and biotic transformations of the key species, and the matrix effect upon rates of these transformations in each compartment.

Over the past several years our collaborative field studies in the Chesapeake Bay (1) have provided a consistent picture of substantial partitioning for heavy metals between major compartments. The case is presently most developed for a well-known anthropogenic element, mercury. It is found (Figure 1) that substantial accumulations of mercury, determined as total metal, appear in sediments and associated plankton, although local waters support very much lower concentrations of the metal. In general, we have found (2) that mercury concentrations in plankton reflect the extent of mercury loading in surface sediments for that locale, with so-called bioaccumulations of total metal in plankton relative to seawater in the range of >7,000 for heavily impacted sites to <1,000 for pristine locales. Metal loadings in suspended inorganic matter have not been considered in the present study.

It is apparent that in order to effectively engage in studies on transport or cycling of heavy metals in the estuary, we must deal with both geographical, or macroscopic, concentrations of metals as well as the localized, or microscopic systems. Chiefly, our approach has been to investigate the metal transformations and mobilization of metals by biological agents, primarily estuarine bacteria, and to couple these findings with our concurrent efforts in characterizing the underlying chemistry of the relevant aquated metal species.

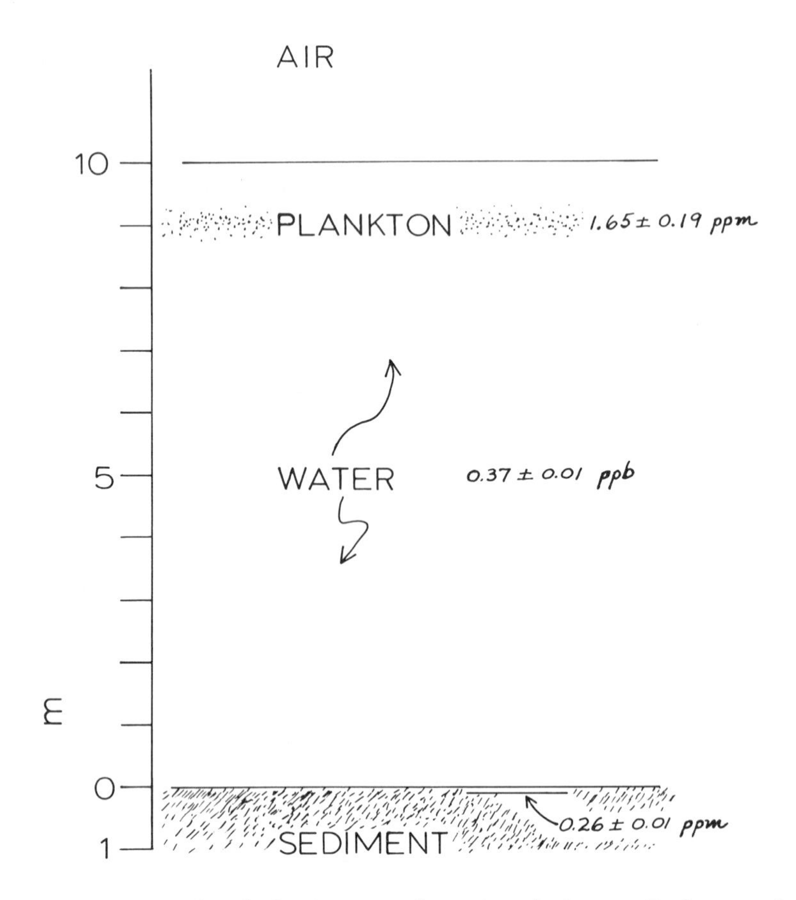


*Figure 1. The idealized water column for a Baltimore Harbor sampling site typically demonstrates substantial partitioning for total mercury concentrations (μg/g, dry weight) in the surface sediment, ambient water, and plankton compartments. The vertical scale illustrates depths in meters.*

In some instances, as has been described elsewhere (3), there is potential for a combination of both biological and non-biological pathways which might serve as means for metal transport, particularly across sediment-water-organism interfaces. Some of the results to be discussed in this paper will touch on these prospects and means for their elucidation.

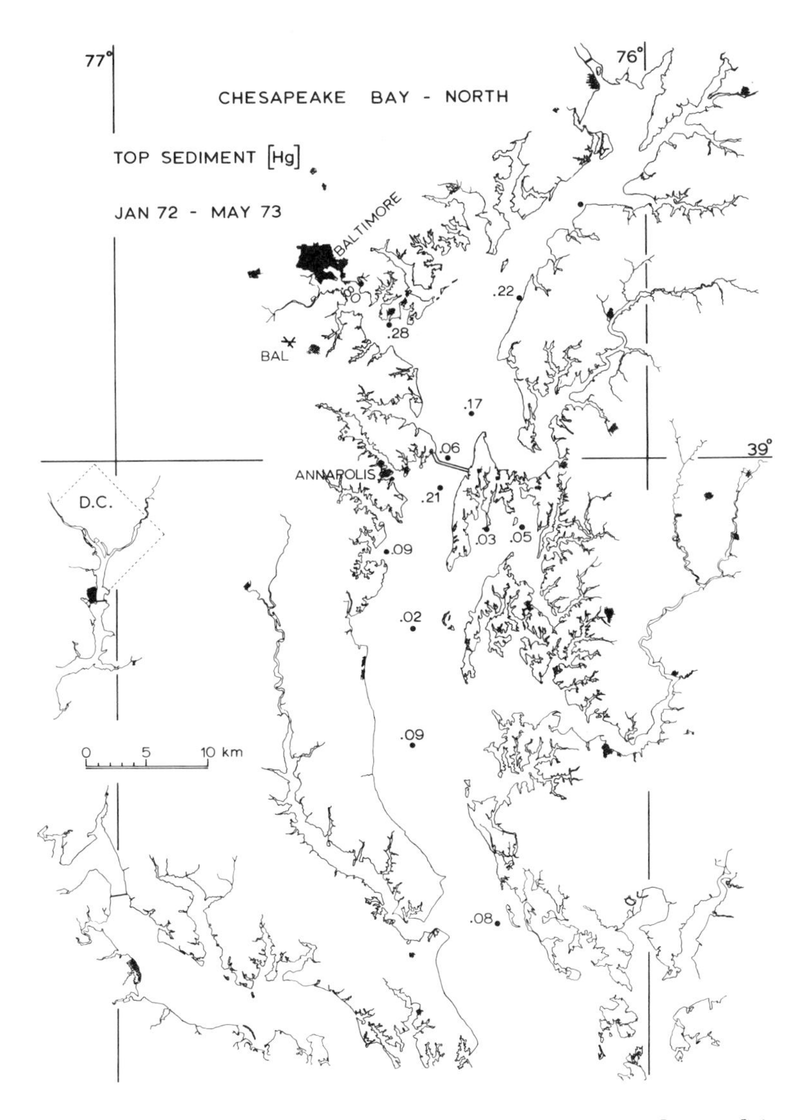


*Figure 2. Distribution of total mercury concentrations (μg/g, dry weight) in surface (0–5 cm) sediments of the northern reach of the Chesapeake Bay collected over an annual period. In general, those locales subject to heavy commercial traffic and industry show a 10 to 50-fold increase in the mercury loading as compared with "background" levels (0.01–0.02 ppm) observed at pristine sites.*

Distribution and Speciation of Heavy Metals in the Estuarine System

Not unexpectedly, our survey of total mercury distribution in surface sediments in the Chesapeake Bay reveals a pattern largely reflective of recent use of its waters and corresponding anthropogenic metal inputs (Figure 2). Comparable surveys of the entire range of the Bay for surface sediment distributions of other heavy metals are not available, although several important studies in local areas have appeared (4, 5). On the basis of our survey we have concentrated our studies in several sites, characteristic of three principal settings in mid-estuary: a highly urban-industrial, polluted harbor stream (Colgate Creek); a deep channel ship lane (off Matapeak Point); and a pristine, shallow sub-estuary exploited for shellfishing in Eastern Bay (Parson Island). In each of these sites we have attempted to develop additional information with respect to vertical concentration gradients of mercury as well. The present discussion will mainly focus on some current results available for the Colgate Creek site inasmuch as our information on metals and other contaminents present in the sediment column is best developed for that site.

Figure 3B illustrates a typical concentration gradient for total mercury in sediment columns obtained at Colgate Creek. In general, we find that total mercury concentrations increase with depth for heavily contaminated areas while the reverse is true for non-polluted locales, such as Parson or Kent Islands. The immediate question occurs as to whether or not such a mercury concentration gradient reflects differences in the form of sedimentary mercury, and whether or not there are present other materials which cause preferential accumulation of the metal (or are responsible for its active transport) within the sediment column.

Reimers *et al.* (6) have observed that mercury strongly adsorbs on almost all suspended material, particularly clays and organic complexes (*e.g.*, humates). Also, it is generally held that in sediments supporting high concentrations of sulfide ion, certainly in those anaerobic benthos subjected to organic outfalls, any mercury depositions will be immobilized as the highly insoluble sulfide, HgS (7). In fact, presumed insolubility of metal sulfides in such natural systems is not as simple as that described by laboratory solubility product data, since it has been well demonstrated that a number of "insoluble" sulfides (including those of copper, nickel and iron) are resolubilized by aerobic bacterial activity (8, 9). Even the case for mercuric sulfide is shown to be in doubt since Fagerström and Jernelöv (10) have demonstrated that the resolubilization of HgS may occur as a result of aerobic microbiological processes.

In Figure 3A we show the concentration gradient for free elemental sulfur in the Colgate Creek sediment core described above. Again, as with mercury, sulfur concentration increases with sediment depth, but the two contaminents bear no statistical relationship ($R<0.37$). Since large amounts of free sulfur are

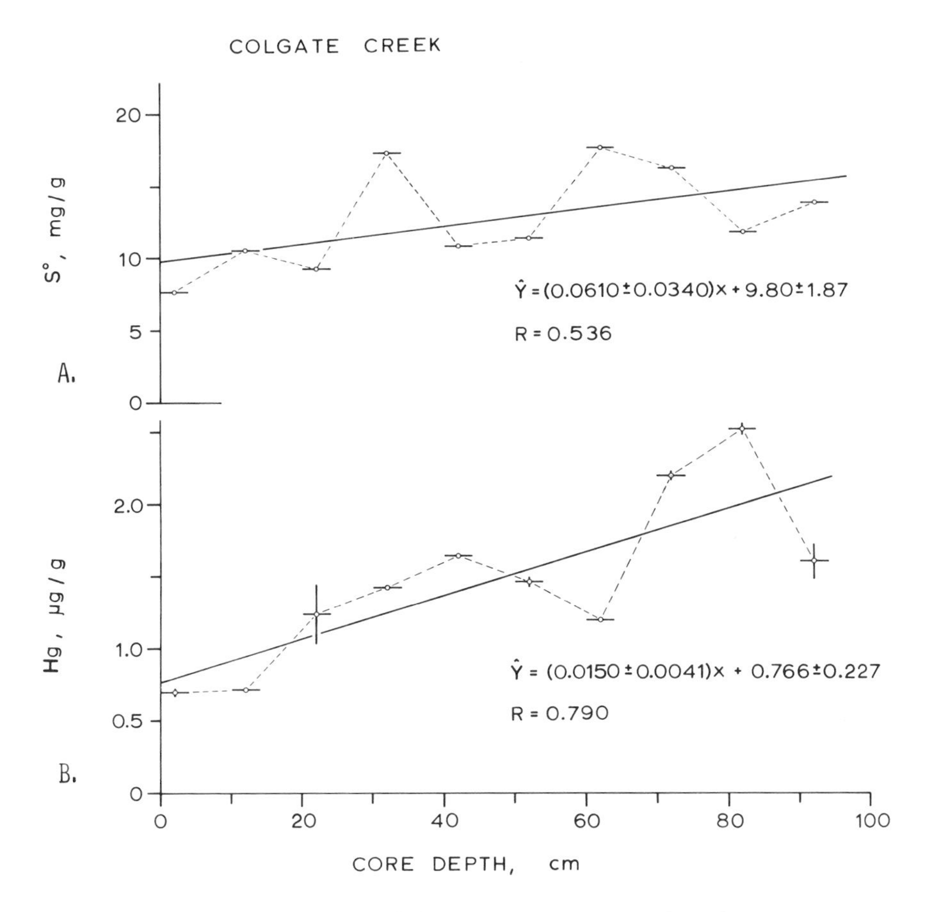


*Figure 3. Concentrations (dry weight) of elemental sulfur and total mercury as functions of depth in a core sample taken at Colgate Creek, Baltimore Harbor. Regression analyses of either the element concentration* vs. *depth or the relative concentration of both elements at given depth indicates no significant correlations (R). Generally, as indicated here, the concentration of total mercury increases with depth (to at least a meter) at more highly polluted Bay stations.*

present in this locale, this can be diagnostic of a substantial bacterial sulfur cycle (11) suggested in Figure 4.

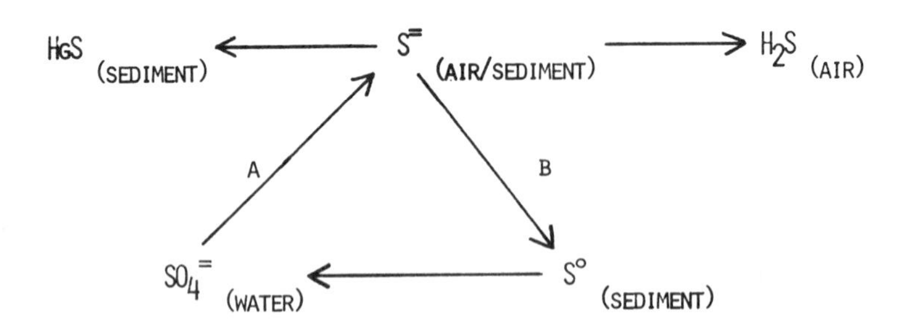


*Figure 4. Simple representation of estuarine sulfur cycle showing removal of sulfide as volatile hydrogen sulfide or insoluble mercuric sulfide*

The biogenic character of the free sulfur found by determination of the sulfur isotope ratios is suggested for this site and a number of others. The $^{32}S/^{34}S$ ratios for a number of sites are summarized in Table 1. Here, the extent of change, $\delta$, in isotope ratios for the environmental samples is compared with a reference meteoritic sulfur standard by the relationship (12)

$$\delta^{34} = \frac{(^{32}S/^{34}S)_{STD} - (^{32}S/^{34}S)_{SAMPLE}}{(^{32}S/^{34}S)_{STD}} \times 1,000$$

Thus, where $^{34}S$ increases, $(^{32}S/^{34}S)_{sample}$ decreases, and $\delta^{34}$ is positive.

It is evident that there is considerable fractionation of the sulfur isotopes which would indicate biogenic activity (12). It should be noted that while very low sulfur concentrations are found in pristine Eastern Bay (Parson Island), the depletion of $^{32}S$ has almost reached its theoretical maximum (13). A large number of reports (12, 13, 14, 15) substantiate the idea that biogeochemical cycling results in the depletion of the heavier $^{34}S$ isotope relative to the lighter one in free sulfur deposits. The assumption is generally made that the sulfur is formed under anaerobic conditions from the oxidation of biologically produced $H_2S$ (Figure 4, pathway A) which originated from aqueous sulfate. The mechanism of such an oxidation reaction in step B is not clear

TABLE 1

Comparison of Extractable Mercury *versus* Total Mercury Concentrations at Several Sites Bearing Variable Amounts of Free Biogenic Sulfur of Known $^{32}S/^{34}S$ Ratios

| CORE SITE | CORE DEPTH, cm | TOTAL Hg μg/g[a] | EXTRACTABLE Hg μg/g[b] | EXTRACTABLE Hg %[c] | FREE SULFUR, %[d] | $\delta 34$[e] |
|---|---|---|---|---|---|---|
| Colgate Cr | 0-35 | 1.017±0.075 | 0.004 | 0.4 | 1.19 ±0.08 | + 10.2 |
| Colgate Cr | 70-110 | 1.300±0.288 | 0.013 | 1.0 | 0.61 ±0.02 | 0.0 |
| Parson Is | 0-22 | 0.010±0.004 | 0.005 | 45.7 | 0.008±0.001 | + 46.0 |
| Matapeake (858C) | 0-27 | 0.077±0.001 | 0.013 | 17.0 | 0.550±0.001 | + 12.2 |
| Station 922Y | 0-30 | 0.053±0.004 | 0.015 | 28.3 | 0.065±0.004 | + 6.5 |

[a] Mean ± average deviation for three runs

[b] Reported as dry weight for original sample

[c] Percentage of total mercury

[d] Corrected for Hg blank in acetone extraction procedure, mean ± average deviation for three aliquots

[e] Determined by Teledyne Isotopes, Inc., Westwood, N. J. 07675

(16). The positive δ-values in the table, however, indicate low $^{32}S/^{34}S$ ratios, or an enrichment of the $^{34}S$ isotope. This might be expected to happen if $^{32}S$-enriched $H_2S$ were preferentially removed with respect to $SO_4^{=}$, leaving residually $^{34}S$-enriched sulfate trapped in the sediment. Since there is an active flow of water due to tidal action at these sites, a removal mechanism does exist. Moreover, it has been experimentally observed that such enrichment of bacterially produced $H_2S$ in the heavy isotope can occur with an increase in time (13).

Thus we conclude that while substantial biochemical sulfur cycling occurs in such mercury-laden sediments, no simple relationship exists for the data depicted in Figure 3 which might suggest extensive sequestering of the total metal by such large excesses of transient sulfide ion in continuous production. Consequently, we have examined the question of whether or not mercury found in these sediment columns, or those at different sites, is entirely bound to sediment (presumably as sulfide) or present in another more labile form which is not immobilized by excess sulfide. The organic extracts used for determination of free sulfur and corresponding sulfur isotope ratios have therefore also been analyzed for "extractable mercury". These results are compared with total sedimentary mercury for several stations in Table 1, and in Table 2 compared for a higher vertical resolution study of the Colgate Creek cores.

From the viewpoint of lateral distributions, the site-to-site comparison in Table 1 indicates a substantial decrease in available "extractable mercury" with increased total sulfur loading. That is, with an increase of about 100X in free sulfur we observe a corresponding decrease of 50X in extractable mercury. In all cases, the total sulfur to total mercury ratio remains at about $10^4$. Nonetheless, closer examination of the vertical relationship between sulfur and forms of mercury within a more highly polluted sediment column indicates a more complex situation. In Table 2, it is seen that substantial extractable mercury is available even in the presence of high sulfur concentrations in the surface sediment; however, no significant correlation ($R < 0.49$) results for the entire column with respect to sulfur versus extractable mercury, just as was the case for total mercury.

Clearly, "labile mercury" can exist in some locales or at certain depths within the sediment matrix which is not insolubilized by large and continuous excesses of sulfide ion. The molecular form of labile mercury found in organic solvent extracts (here, acetone or a chlorofluoroethane) is not yet known. We know that several estuarine facultative or anaerobic microorganisms isolated from this or other Chesapeake Bay sites are capable of forming methylmercurials from inorganic Hg(II) (17). In addition, evidence for prevalent, but low amounts (*ca.* 0.03%) of methylmercury relative to total sedimentary mercury in aquatic sediments is available (18). Also significant is the report that many-fold (10-30X) enrichment of interstitial mercury over that found in associated surface waters, far in excess of that predicted for HgS solubility, may be due to formation of organic and polysulfide complexes of the metal (19). Presence of only small amounts of volatile methylmercurials in the sediments under study is reflected in our evaluation of total volatile mercurials recovered from wet sediments by pumping under hard vacuum. In those experiments (see Appendix), it is shown that either slow oven-drying or pumping off water and other volatiles to constant sample weight does not alter total mercury assays significantly, nor does the volatilized water contain more mercury than that suggested by reported methylmercury levels.

TABLE 2

Comparison of Total Mercury *versus* Extractable Mercury with the Vertical Distribution of Free Sulfur in a Core Taken at the Colgate Creek Site.

| CORE DEPTH, cm[a] | TOTAL Hg µg/g[b] | EXTRACTABLE Hg µg/g[c] | %[d] | FREE SULFUR, %[e] |
|---|---|---|---|---|
| 0-4 | 0.700±0.025 | 0.316 | 45.1 | 0.77±0.00 |
| 10-14 | 0.712±0.001 | 0.187 | 26.3 | 1.06±0.02 |
| 20-24 | 1.240±0.208 | 0.114 | 9.2 | 0.93±0.01 |
| 30-34 | 1.424±0.015 | 0.124 | 8.7 | 1.73±0.00 |
| 40-44 | 1.644±0.005 | 0.196 | 11.7 | 1.09±0.00 |
| 50-54 | 1.465±0.032 | 0.189 | 12.9 | 1.14±0.01 |
| 60-64 | 1.201±0.021 | 0.238 | 19.8 | 1.77±0.01 |
| 70-74 | 2.197±0.030 | 0.073 | 3.3 | 1.63±0.01 |
| 80-84 | 2.521±0.042 | 0.135 | 5.4 | 1.18±0.01 |
| 90-94 | 1.606±0.120 | ----- | ---- | 1.38±0.01 |

[a] Effective depth taken as mean of homogenized 4 cm core sections indicated

[b] Mean ± average deviation for three determinations

[c] Reported as dry weight for original sample

[d] Percentage of total mercury

[e] Corrected for Hg blank in acetone extraction procedure; mean ± average deviation for three aliquots

Examination of the oily residues obtained in the sulfur analysis procedure (20) is underway employing a combination of gas chromatography coupled with both flame ionization and atomic absorption detectors (17). We have not identified any small mercury-containing molecules, such as alkymercurials, although our detection system can speciate such organometallic compounds at the nanogram level. Nonetheless, as is illustrated in Figure 5, we can detect mercury-containing compounds in the oils eluting at retention times characteristic for either $>C_{20}$ petroleum components or arylmercurials. Presently, it can be presumed that such extractable mercury fractions are ligated with fairly large solubilizing organic groups, and, additionally, that these species are quite resistent to precipitation as inorganic mercuric sulfide in the presence of a copious excess of active sulfur.

That the mercury, or other metals, might be closely associated with sediment "sinks" extractable by organic solvents is a very important point. In cooperation with R. Colwell's group we have determined that such petroleum-rich fractions in Colgate Creek can contain very substantial quantities of mercury - perhaps 4,000 times the ambient sediment concentration. Also significant are Walker and Colwell's additional findings (21) that several other toxic heavy metals accumulate in the oily compartment. Based on available data (4, 21) for metal concentrations in Colgate Creek sediments and oil depositions, some of these metals are also greatly concentrated in oils as compared with the ambient sediment matrix. Principal among these are chromium, copper, nickel, lead, and zinc which show partition coefficients [*e.g.*, (μg/g metal in oil)/(μg/g metal in sediment)] of 7, 1, 1, 0.7, and 2.5, respectively. Since a large mercury concentration was also found in water-borne oil isolates, we can also perceive a means for this metal's lateral transport in the active sub-estuarine flow system.

Colgate Creek is a highly polluted site within Baltimore Harbor, containing, along with high metal loadings, a large amount of petroleum pollutants. However, we should not conclude that such anthropogenic petroleum sinks are entirely responsible for concentrating or solubilizing metals. As shown in Table 1, the petroleum-free Parson Island site also demonstrates both a very high level of sulfur cycling and yet supports a large amount of extractable mercury in much less polluted circumstances. It should also be noted that these metal-containing petroleum fractions are also subject to extensive biodegradation by microorganisms shown to be capable of metal transformations as well (2, 21, 22). The process occurring in natural waters provides a pathway for substantial rerelease of the metals with potential cycling through the primary trophic level.

Thus, aside from the direct undesirable aesthetic and biological effects of petroleum, a vehicle for the mobilization and transport of heavy metals is provided. Heavy metals trapped or sequested after transport may later be released in concentra-

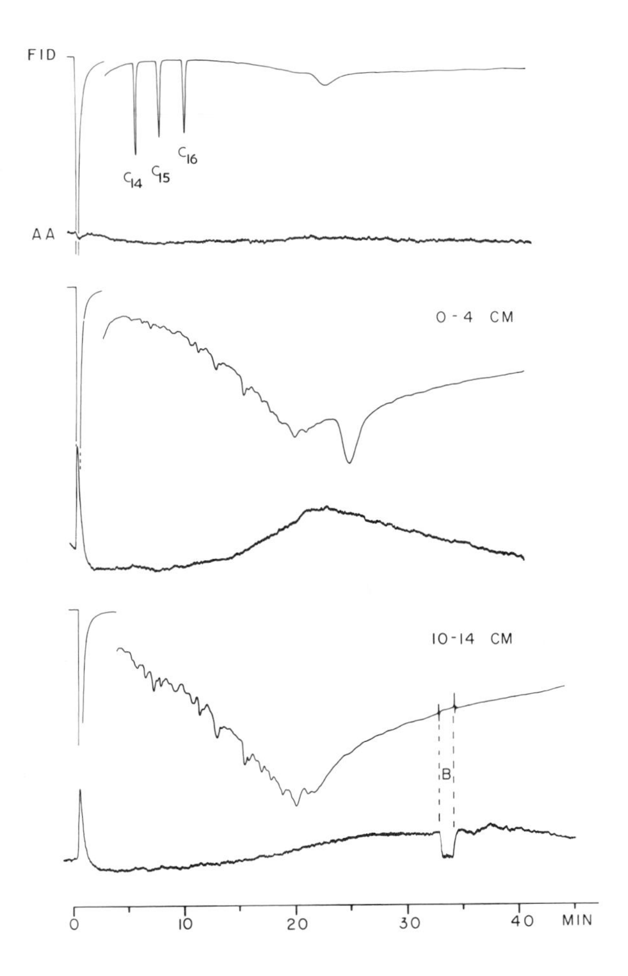


*Figure 5. Simultaneous chromatograms obtained by a flame ionization detector (FID) and a mercury-specific flameless atomic absorption detector (AA) are compared for a known, Hg-free calibration mixture of $C_{14}$–$C_{16}$ paraffins (top) and oil extracts from two sections of a Colgate Creek core. The 0–4 cm sample contained total Hg = 0.7 μg/g with 45% as extractable Hg, plus 0.8% free S°; the 10–14 cm sample contained total Hg = 0.7 μg/g with 26% as extractable Hg, plus 1.1% free S°. Significantly different retention times of* ca. *22 and 36 min were noted for the ligated mercury effluents, respectively. The parameter jump seen at "B" in the lowest set of spectra results from a Hg baseline check made by momentarily replacing the GC effluent stream with a reference air stream in one of the AA dual-beams.*

tions that could be immediately toxic to the surrounding biota and later toxic to higher biota and man as a result of accumulation mechanisms.

Further evidence indicating that mercury is not completely sequestered in sediments as the sulfide, but may also exist or be transported in the elemental state, has been found. Earlier work (22) and our later studies (17) on principal mercury processing bacteria from the Chesapeake Bay implied that substantial elimination of sediment mercury as volatile elemental gas might occur in natural settings. Direct examination of freshly obtained sediment samples has been conducted aboard the R/V Ridgely Warfield employing a dual-beam flameless atomic absorption spectrophotometer adapted to determination of gaseous mercury contained in atmospheres over benthic samples (22). Figure 6 illustrates presence of elemental mercury in sediments collected at the pristine Eastern Bay site at Parson Island. The concentration levels found here are not inconsistent with those observed for laboratory demonstrations with pure cultures on synthetic marine matrices containing comparable amounts of inorganic mercury.

The metal species found in biota are largely unknown; however, in the case of mercury, the resident form found in fish and man is chiefly the highly toxic methylmercury. The mechanism through which this form appears in these higher biota, whether it is through bioaccumulation, *via* the food chain, or direct formation on or within the body of the fish itself, is not clear (23, 24). Nevertheless, it appears that bacteria may be involved in both of these mechanisms (24, 25).

## Biotic and Abiotic Transformations of Metals

Since Jensen and Jernelöv (26) discovered that biological methylation of mercury occurs in nature, having reported its formation from $HgCl_2$ in lakes and aquarium sediments, a number of bacteria including *Clostridium cochlearium* (27), *Pseudomonas fluorescens, Mycobacterium phlei, Escherichia coli; Aerobacter aerogenes, Bacillus megaterium* (28), and bacteria isolated from the Chesapeake Bay, have been demonstrated to produce extracellular methylmercury. Intracellular methylmercury formation has also been demonstrated by several fungi (28, 29). Methylmercury may be bacterially degraded with the formation of elemental Hg (30, 31). In addition to the microbial formation of methylmercury from inorganic mercury ($Hg^{2+}$) bacteria have been reported to produce the elemental form (17, 32, 33, 34). Using a gas chromatograph coupled with a double-beam atomic absorption detector, Nelson, *et al.* (22) have demonstrated >50% conversion of phenylmercuric acetate to elemental mercury by a prevalent *Pseudomonas* strain in the Chesapeake Bay. An index of such mercury metabolizing bacteria at a given sampling site in the natural environment may be provided by determining the total number of mercury

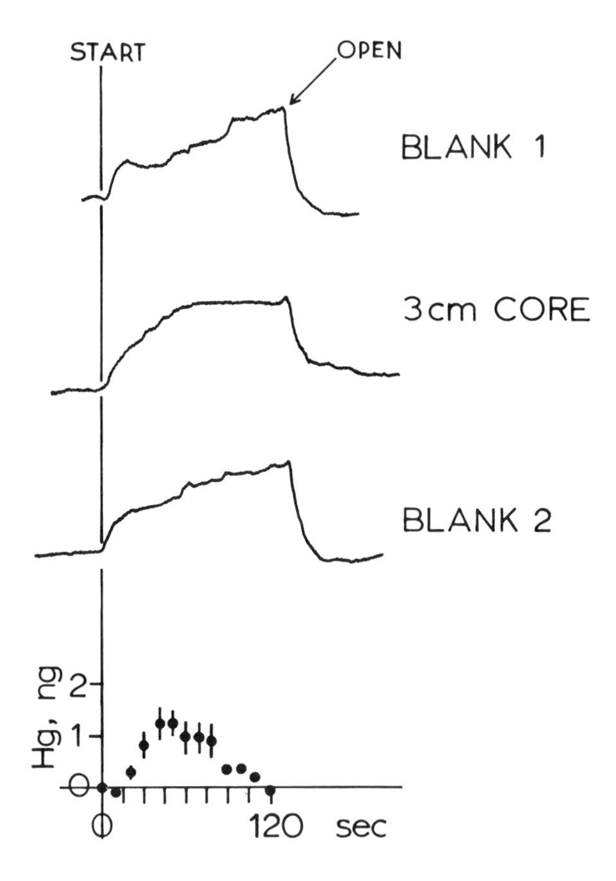


*Figure 6. The captive atmosphere above a freshly acquired surface sediment core sample (0–3 cm) taken at Parson Island is compared with that above two local bottom water samples by means of a dual-beam flameless atomic absorption spectrophotometer coupled to a circulating pump and a special sediment sample cell. The core signal illustrated here was compared against the average of two blanks and a Hg in seawater calibration curve prepared on the site. The bottom differential curve thus obtained indicates the net gaseous mercury detected for the entire sediment sample (*ca. *20 g) within the measurement volume (400 cc) over the flow period of about 2 min. Subsequent laboratory analysis indicated a dry weight concentration of 0.04 μg/g total mercury in the core sample.*

tolerant bacteria (25).

It had been known, prior to the discovery of microbial mercury methylation, that a number of other heavy metals, *viz:* arsenic, selenium and tellurium, could also be biologically methylated. Again, as in the case of mercury, the methylated species are highly toxic.

Trimethylarsine production has been demonstrated by a number of fungi and bacteria grown in media containing this element (35, 36). Arsenic volatilization has been also observed to occur in soils (37) and in lakes, wells, and ponds where as high as 68% of the total arsenic present in the form of alkylarsenicals (38). The reduction of arsenate to arsenite occurs in cultures of the yeast, *Pichia guillermondi* (44), *Chlorella* (45), and marine bacteria (46), while the oxidation of arsenite to arsenate is brought about by *Pseudomonas sp.* (47). Dimethylselenium has been reported to be formed from soils rich in selenium (39), probably as a result of microbial action. Several fungal cultures have also been shown to evolve dimethylselenium (40, 41, 42). Tellurium, likewise, may be methylated by a number of pure fungal cultures (41, 42, 43) but it has not been determined whether biological methylation occurs in natural ecosystems (39).

The most recently demonstrated metal to be biologically methylated is tin (3). The same strain of *Pseudomonas* isolated from the Chesapeake Bay, and which was quite active in forming elemental mercury from phenylmercuric acetate or $Hg^{2+}$, produced a methylated tin species, as yet unidentified, from $Sn^{4+}$. A summary of all these microbial transformations is shown in Figure 7. Of the five metals which have been demonstrated to be biologically methylated, three (*viz.:* selenium, tellurium, and mercury) have also shown to be also biologically reduced to their elemental form (32, 44).

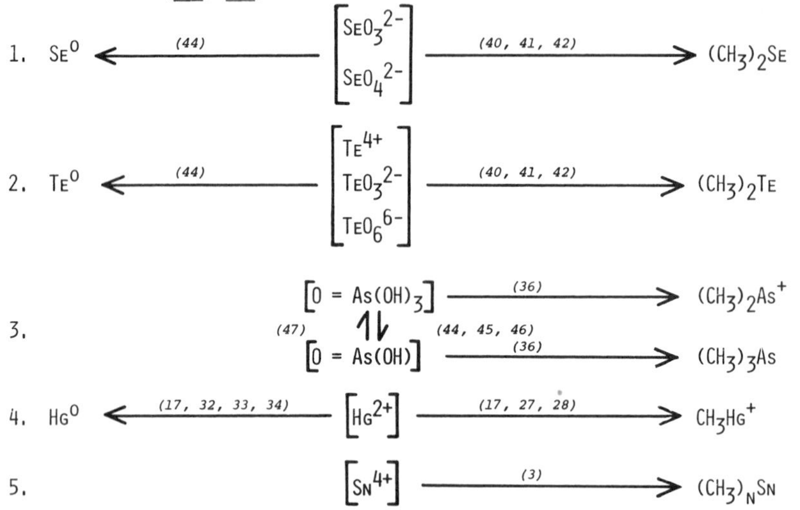

1. $Se^0 \xleftarrow{(44)} [SeO_3^{2-}, SeO_4^{2-}] \xrightarrow{(40,\ 41,\ 42)} (CH_3)_2Se$

2. $Te^0 \xleftarrow{(44)} [Te^{4+}, TeO_3^{2-}, TeO_6^{6-}] \xrightarrow{(40,\ 41,\ 42)} (CH_3)_2Te$

3. $[O = As(OH)_3] \xrightarrow{(36)} (CH_3)_2As^+$; $[O = As(OH)_3] \rightleftharpoons [O = As(OH)]$ (47) / (44, 45, 46); $[O = As(OH)] \xrightarrow{(36)} (CH_3)_3As$

4. $Hg^0 \xleftarrow{(17,\ 32,\ 33,\ 34)} [Hg^{2+}] \xrightarrow{(17,\ 27,\ 28)} CH_3Hg^+$

5. $[Sn^{4+}] \xrightarrow{(3)} (CH_3)_NSn$

*Figure 7. Biotransformations of inorganic substrates can be accomplished by either bacterial or fungal agents cited by references given in the text (italics). Such metabolites may be found to occur as labile, volatile methyl derivatives or in reduced form as the element.*

It should not be presumed, however, that metal transformations or transport in natural settings result only from biological forces. Jernelöv and his co-workers have demonstrated (48) an abiotic transalkylation process between anthropogenic alkyllead compounds and inorganic mercury in river sediments. Japanese workers have recently shown formation of methylmercury in a caustic soda factory wherein only inorganic source materials are available, including metallic mercury and calcium carbide (49). Further, there is now evidence that photoalkylation of mercury may occur in sunlight, *e.g.*, photolysis of acetate ion in the presence of aqueous Hg(II) leads to the formation of methylmercury ion (50, 51). Additional experimental results suggest that the free-radical pathway available for forming methylmercury bonds may also exist for other heavy metals, including Sn(IV) and Pb(IV) (52). Such a demonstration is reported for formation of long-lived methylpentaaquochromium(III) ion by photolysis of an organoxy substrate in the presence of aquo-$Cr^{2+}$ (53).

The underlying chemistry supporting these metathetical or photolytic alkylation processes for aqueous metal ions is not well understood (54), but a substantial increase in interest in this field has developed within the past several years, particularly with respect to structural features of organometallic ions solvated by water (55). In our laboratory, we have undertaken a survey of potential transmethylation reactions which might occur between aqueous metal ions under abiotic conditions (56). Several of the more important of these reactions are summarized in the following equations:

$$(CH_3)_3Sn^+ + Hg^{2+} \longrightarrow (CH_3)_2Sn^{2+} + CH_3Hg^+$$

$$(CH_3)_3Pb^+ + Hg^{2+} \longrightarrow (CH_3)_2Pb^{2+} + CH_3Hg^+$$

$$(CH_3)_3Sn^+ + PtCl_4{}^{2-} \longrightarrow [CH_3PtCl_3{}^-] \longrightarrow C_2H_6 + Pt^\circ + (CH_3)_2Sn^{2+}.$$

Trimethyllead is a far more rapid methylator of Hg(II) than is the analogous tin ion, being comparable to methylcobalamin (57) in its ability to transfer $CH_3$ to mercury. Noble metal ions such as Pd(II), Au(III), or Pt(II) are methylated by trimethyltin cation at about the same rate as Hg(II), but these lead to significantly different reaction pathways. Evidence for unstable aquo-methylmetal intermediates was obtained for the Pd and Pt cases, and these rapidly decompose to ethane, or methyl chloride in excess salinity, thereby providing a means for irreversibly ejecting "active methyl" from the reaction system.

For those cases where we have determined kinetics of transmethylation, we find that second-order rate laws are obeyed, *i.e.*, the reactions are first order with respect to each reactant metal ion. These reactions do not appear to be pH-dependent. These findings are important with respect to the more complicated kinetics observed for the cellular metabolite, methylcobalamin, which can similarly methylate several aqueous metal ions (57), depending sometimes on availability of the metal ions in several oxidation states as well as pH.

A number of environmentally important parameters do affect the rate of transmethylation between simple metal ions, however. In the paper by Jewett et. al. (in this Symposium) we demonstrate the substantial influence of salinity or more accuately chloride ion concentration, on the absolute rate of methyl transfer between tin(IV) and mercury(II) in water (51). We similarly describe the rich array of methylated products which can also result from mild photolysis of Hg(II) in the presence of a number of well-known bacterial metabolites which, in effect, can act as methyl precursors. These substrates include ethanol, acetone, and acetate ion.

An interesting and provocative question arises from this inquiry. Does there exist a pathway by which *both* non-biological *and* biological events may be important to transformations and transport of a metal or metals? From a purely empirical standpoint there already appears ample evidence that certain processes occurring in the environment result in selective co-concentration of remarkable metal pairs. Such data are provided for both sediment systems as well as uptake or accumulations in lower organisms.

A study of the distribution of heavy metals at sixteen stations in the Sörfjord estuary of West Norway reveals positive linear correlations (typically $R>0.95$) for ratios of certain metal concentrations, such as those of Cu or Cd with Zn (58), and more remarkably Ba/In, Ag/Cd, or Pb/In. Since this estuary has been subjected to substantial industrial effluents for many years, care was taken to evaluate the relationship between such element ratios for effluents as well as those in sediments. It was concluded that while Zn/Pb ratios of effluents (0.75) and sediments (0.77) are very similar, other metal ratios in sediments are quite different from discharge inputs, implying widespread selective fixation of certain metals like Cd and Cu by either chemical or biological processes in the aquatic system itself.

From the chemist's viewpoint several such metal pairings given by the examples of Cu-Zn or Cd-Zn do not appear unlikely on the basis of simplistic relationships of chemical similarity. Nonetheless, it is important to note that transport of such a metal pair from the sediment matrix into biota constitutes a diagnostic tool for ultimately characterizing specific transport routes for each metal, and the relative rates by which these cycles occur. Thus, Huggett, *et al.* (59) propose that "natural" metal concentration ratios exist for Cu-Zn and Cd-Zn in oysters (*Crassostrea virginica*) taken from uncontaminated portions of the Chesapeake Bay. These relationships, however, are not necessarily statistically related to the sediment concentrations for the respective metals. Indeed, in those locations where known anthropogenic metal stress occurs, the relative metal levels in oysters taken from these sites differs significantly from the so-called "natural" level.

Apparently some metals can accumulate in sediments over many sites in "fixed" ratios, but even for cases where this does not occur, the metals may appear in biota associated with the benthos or local waters as fixed concentration ratios. Evidence for this transport process is indicated by the correlation matrix which we have derived from the data of Schramel, *et al.* in their studies on heavy metal distributions in waters, plants, sediments, and fishes of Bavarian rivers (60). From the correlation coefficients shown in Table 3, it is readily seen that no significant metal pair relationships exist for metals in sediments from a number of stations reflecting both polluted and pristines sites. A singular, large positive correlation appears for the tin-mercury pair in (unspecified) water plants.

Table 3

Correlation Matrix for 5 Heavy Metals (Bavaria)

| | Hg | As | Se | Sb | Sn |
|---|---|---|---|---|---|
| Hg | | 0.2266 | 0.4296 | 0.5774 | 0.9563 |
| As | 0.6975 | | 0.3531 | 0.7819 | 0.3074 |
| Se | -0.2647 | -0.5658 | | 0.7725 | 0.6595 |
| Sb | 0.7670 | 0.3569 | 0.3630 | | 0.7315 |
| Sn | -0.2098 | -0.4055 | 0.6053 | 0.3418 | |

WATER PLANTS (6 ORIGINS) (above diagonal)
SEDIMENT SAMPLES (5 ORIGINS) (below diagonal)

In view of the known (61) propensity for absorption and growth modification of plant cells in the presence of aquated organometals, it seems useful to conclude this discussion with an appraisal of our recent discovery. An aquatic microorganism can be participant in a process which both releases soluble mercury from sediments and also produces a methylated tin species capable of subsequently methylating available Hg(II) by an abiotic mechanism (3). This organism, a mercury-and tin-tolerant *Pseudomonas* species was isolated from the Chesapeake Bay, where it is one of the predominent organisms. In the presence of Hg(II), this organism forms Hg°, but in the presence of Sn(IV) a volatile methylated tin species is produced. However, in the presence of both Hg(II) and Sn(IV), methylmercury is formed.

Since methylated tin (trimethyltin) has been shown to methylate Hg(II) abiotically, a bimetallic model for the formation of methylmercury has been demonstrated. Such a model is shown in Figure 8 where some indication of known metal concentration and estimated half-lives ($t_{1/2}$) for various transport processes are given.

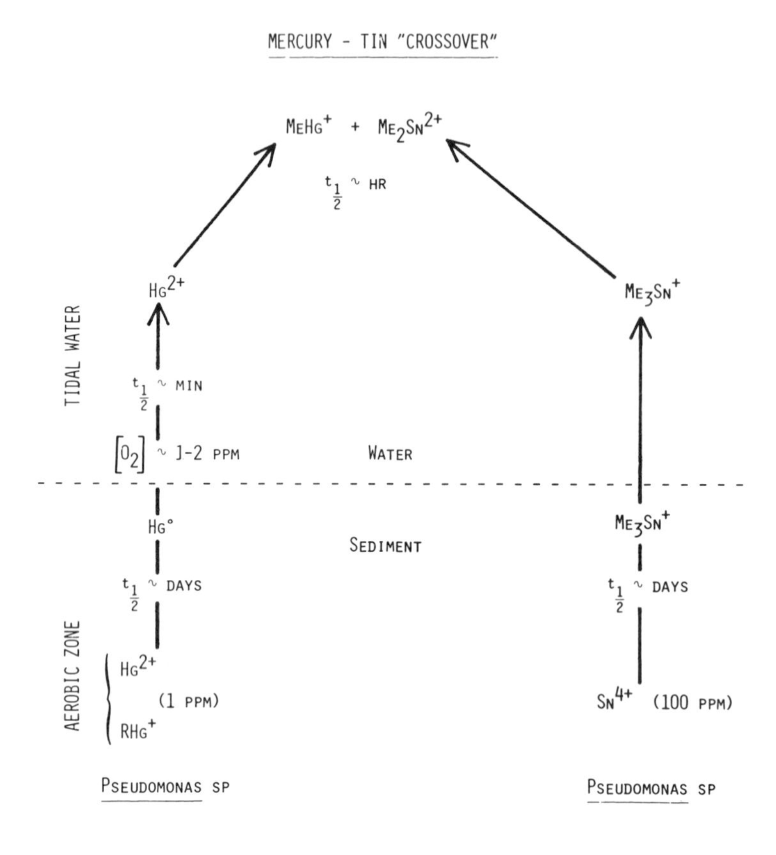


*Figure 8. Involvement of* Pseudomonas sp *in biotic release of both mercury and tin from the sediment matrix into the aqueous phase are postulated on the basis of field and laboratory experiments. Following release, the bio-methylated tin species "$Me_3Sn^+$" can abiotically methylate available $Hg^{2+}$.*

The relative importance of such a model in natural environments has not been determined. In view of the presence of tin, along with mercury pollution in heavily impacted locales in

Europe, and the presence of tin with mercury in water plants previously mentioned (60), investigation of this "crossover" phenomenon appears justified. Evaluation of other biometallic mechanisms (*viz.:* Pb-Hg, Cd-Hg) also appears warranted in view of the indictment of an organolead contaminant as a contributor to methylmercury production in the Saint Clair Waterway (48).

In summary, the chemical and bacterial cycling of heavy metals in natural environments appears to be a highly complex system involving many pathways which are determined by the nature (metal species) of the input, the physical properties and chemical composition of the aquatic environment, the metabolic properties of the various types of biota, and the ecological relationships among the various types of biota inhabiting the environment. A complete understanding of all these relationships probably may never be realized, but it seems likely that a continued strong coupling of biological and chemical approaches to environmental questions can yield increasingly useful insights regarding critical pathways and rates. A vital link in our progress will be the concurrent broad development of tools for characterization which permit speciation of metal-containing biotic or abiotic agents in the aquatic system at trace concentrations applicable to the natural setting.

## ACKNOWLEDGEMENT

We are very pleased to indicate the valuable and critical contributions of our colleagues and co-workers: Drs. R. B. Johannesen, J. Nelson, J. Walker, and Prof. R. Colwell, as well as Mssrs. K. L. Jewett and C. Huey. We are also indebted to Mssrs. W. Blair and R. Thompson, and to L. Silberman and S. Wagner for continued measurements of highest quality. We thank the National Science Foundation and the Chesapeake Bay Institute for sustained support with their excellent facilities made available with the R/V Ridgely Warfield. Financial support and esteemed guidance from the Environmental Protection Agency, the NBS Office of Air and Water Measurement, and the U. S. Naval Ship Research and Development Laboratory (Annapolis) are greatfully acknowledged.

# APPENDIX

## Experimental Procedures

Details of specific instrumental and manipulative procedures for chemicals and microbiological experiments have been described in previous papers (1, 2, 3, 17, 21, 22, 51, 56). Appropriate water, sediment, and (when possible) plankton (>20 nm tow net) samples were taken at each Chesapeake Bay station. Site locations were fixed by visual, radar, and sounding techniques. Standard water parameters, including salinity, temperature, pH, and dissolved oxygen, at depths of interest in water, were determined.

A gravity corer was used to obtain sediment samples. These were 6 cm in diameter and variable in length. Cores were immediately sectioned into 4 and 20-30 cm increments and separately frozen (-80°C) on board the R/V Warfield. Typically, alternate 4 cm sections of the fresh core were immediately measured for pH, $E_h$, and total sulfide by selective electrode. Frozen samples for later laboratory analyses were subsequently thawed at room temperature and dried to constant weight over several weeks in a still oven maintained at 50°C.

Dried sediment samples were finely ground in an acid-washed porcelain mortar and thoroughly blended. Total mercury analyses performed in accordance with EPA methods for mercury in sediments (61) used control spikes of methylmercuric chloride as an indication of the reliability of the procedure. Typically, spike recoveries were in the range of 85-110%. Free sulfur and extractable mercury were determined on samples obtained from 24 hr continuous acetone extraction of 20-100 g sediment samples in a grease-free Sohxlet apparatus. The total free sulfur was determined in aliquots of the extract solution by colorimetric titration with cyanide to a bromthymol blue endpoint (20); precision here was better than ±0.3%. Extractable mercury was determined by evaporating aliquots of the extract solution and performing the total mercury analysis by the EPA sediment method.

Possible vaporization of mercury-containing components of high vapor pressure from sediments treated in above manners was quantitatively assessed by comparison with equivalent wet sediment samples subjected to exhaustive pumping under high-vacuum. Thus, a grease-free glass apparatus, open to a Hg-free pumping station, containing 46.20 g of wet sediment sample, was pumped for 48 hrs. to a steady ultimate pressure of $7.5 \times 10^{-5}$ Torr. All condensable volatiles removed from the sample were collected in a trap maintained at -196°C; this was 27.39 g which on total Hg analysis indicated a total of 9.1 ng of Hg had been transferred. The total Hg found in the residual vacuum dried sediment was 2.17 ± 0.02 μg/g. Another sample of the same sediment was oven dried at 50°C for one week to constant weight. Total Hg determined in this material was 2.09 ± 0.05 μg/g, indicating that no significant difference in Hg assay resulted from the two different methods, and that volatized Hg was <0.02% of total Hg. Subsequent analyses indicated the sediment to contain about 9% extractable Hg along with 1.1% elemental sulfur.

Water samples from the field were collected usually at surface, bottom, and mid-depth by a Van-Dorn sampler. Shipboard samples were stored in acid washed, Hg-free 1-liter glass bottles, and stabilized with $HNO_3$ at 1% final concentration. Maintained at 2-4°C in the field and laboratory, water samples were found to yield consistent results for total Hg by the EPA method (62) for periods to 20 days.

A number of sulfur extracts were analyzed for the $^{32}S$ and $^{34}S$ isotope composition. Here, extract solutions from selected sites or sediment core depths were evaporated and the free sulfur crystals thereby precipitated and isolated. In some cases, the sulfur thus obtained was heavily contaminated with pollutant petroleum, requiring several repeated fractional recrystallizations of the sulfur to achieve pure samples. Sulfur samples so obtained were examined by a dual collector mass spectrometer for the $^{32}S/^{34}S$ ratio as compared with a geochemical meteoritic standard (12).

The gas chromatograph coupled with the flameless atomic absorption detector has been previously applied to speciation and determination of volatile methylmercurials (17). For use with the characterization of Hg-containing oil fractions extracted from sediments, only the thermal program was altered to optimize column performance with both the high-molecular weight organic fractions and metal containing eluents. The following conditions proved optimal: $N_2$ carrier at 20 ml/min with dual glass columns, 1/8-in X 2 m employing 80/100 mesh Supelcon AW DMCS support co-packed with 5% SP-2100 plus 3% SP-2401 silicone stationary phases. The chromatograph FID detector was maintained at 250°, with the injection port at 200°. The column oven was programmed at $T_1$ = 120° for 0 min (injection) $\Delta T$ = 4°/min to $T_2$ = 200° (8 min). For extraction of oils from sediments to be used in GC-AA analyses, a chlorofluorocarbon, $CCl_2FCF_2Cl$ (b.p. 47.6°) was found to be superior both in extraction efficiency and removal of excess sulfur crystals, as well as yielding highly reproducible chromatograms of low column bleed and highly reproducible retention times in both FID and AA modes (better than 0.5%). Typical calibration sample injections were 2.0 μl, while oil extracts evaporated to an estimated 1:1 oil-solvent ratio were injected in 4.0 μl amounts.

## LITERATURE CITED


(1) Nelson, Jr., J.D., Colwell, R.R., Blair, W., Brinckman, F.E., and Iverson, W.P., (to be published).

(2) Brinckman, F.E., Jewett, K.L., Blair, W.R., Iverson, W.P., and Huey, C., Progr. Water Technol., (1975) 7 (in press).

(3) Huey, C., Brinckman, F.E., Grim, S., and Iverson, W.P., "Proceedings of the International Conference on Transport of Persistent Chemicals in Aquatic Ecosystems," (Q.N. LaHam, ed.), II-73 to II-78, National Research Council of Canada, Ottawa, 1974.

(4) Villa, O. and Johnson, P.G., "Distribution of Metals in Baltimore Harbor Sediments," Environ. Protection Agency Report #EPA-903/9-74-012, Jan. 1974.

(5) Bender, M.E., Hugget, R.J., and Johnson, P.G., J. Wash. Acad. Sci., (1972) 62, 144-153.

(6) Reimers, R.S., Krenkel, P.A., and Englands, Jr., A.J., "Proceedings of the International Conference on Transport of Persistent Chemicals in Aquatic Ecosystems," (Q.N. LaHam, ed.), II-79 to II-92, National Research Council of Canada, Ottawa, 1974.

(7) Jernelöv, A., "Chemical Fallout," 68-74, Thomas, C.C., Springfield, 1969.

(8) Silverman, M.P. and Ehrlich, H.L., Adv. Appl. Microbiol., (1964) 6, 153-206.

(9) Duncan, D.W. and Trussell, P.C., Can. Metall. Qtrly., (1964) 3, 43-55.

(10) Fagerström, T. and Jernelöv, A., Water Res., (1971) 5, (3), 121-122.

(11) Meinschein, W.G., Bull. Amer. Assoc. Petrol. Geologists, (1951) 43, 925-943.

(12) Ault, W.V. and Kulp, O.L., Geochim. Cosmochim. Act., (1959) 16, 201-235.

(13) Eremenko, N.A. and Mekhtieva, V.L., Geokhimiya, (1961), 174-180.

(14) Thode, H.G., Kleerekoper, H., and McElcheran, D., Research (London) (1951) 4, 581-582.

(15) Jones, G.E. and Starkey, R.L., Appl. Microbiol., (1957) 5, 111-118.

(16) Davis, J.D., "Petroleum Microbiology," 79, Elsevier, New York, 1967.

(17) Blair, W., Iverson, W.P., and Brinckman, F.E., Chemosphere, (1974) 3, 167-174.

(18) Andren, A.W. and Harriss, R.C., Nature, (1973) 245, 256-7; also see Batti, R., Magnaval, R., and Lanzola, E., Chemosphere (1975) 4, 13-14 and references cited therein.

(19) Lindberg, S.E. and Harriss, R.C., Environ. Sci. Technol., (1974) 8, 459-462.

(20) Skoog, D.A. and Bartlett, J.K., Anal. Chem., (1955) 27, 369-371.

(21) Walker, J.D. and Colwell, R.R., Appl. Microbiol., (1974) 27 (1), 285-287.
(22) Nelson, J.D., Blair, W., Brinckman, F.E., Colwell, R.R., and Iverson, W.P., Appl. Microbiol., (1973) 26 (3), 321-326.
(23) Jernelöv, A., The Natural Conversion of Mercury and some Comments on its Importance to Ecological and Toxicologic Effects. Presented at the Swedish-Finnish Mercury Symposium, Helsinki, 1969.
(24) Gavis, J. and Ferguson, J.F., Water Res., (1972) 6, 989-1008.
(25) Colwell, R.R., and Nelson, Jr., J.D., "Proceedings of the International Conference on Transport of Persistent Chemicals in Aquatic Ecosystems," (Q.N. LaHam, ed.), III-1 to III-10, National Research Council of Canada, Ottawa, 1974.
(26) Jensen, S. and Jernelöv, A., Nature, (1969) 223, 753-754.
(27) Yamada, M. and Tonomura, K., J. Ferment. Technol., (1972) 50, 159-166.
(28) Vonk, J.W. and Sijpesteijn, A.K., Antonie van Leeuwenhoek, (1973) 39, 505-513.
(29) Landner, L., Nature, (1971) 230, 452-454.
(30) Spangler, W.J., Spigarelli, J.L., Rose, J.M., and Miller, H.H., Science, (1973) 180, 192-193.
(31) Spangler, W.J., Spigarelli, J.L., Rose, J.M., Flippin, R.S., and Miller, H.H., Appl. Microbiol., (1973) 25 (4), 488-493.
(32) Magos, L., Tuffrey, A.A., and Clarkson, T.W., Brit. J. Ind. Med., (1964) 21, 294-298.
(33) Summers, A.O. and Lewis, E., J. Bacteriol., (1973) 113, 1070-1072.
(34) Furukawa, K., Suzuki, T., and Tonomura, K., Agr. Biol. Chem., (1969) 33, 128-130.
(35) Challenger, R., Adv. Enzymol., (1951) 12, 429-491.
(36) Cox, D.P. and Alexander, M., Appl. Microbiol., (1973) 25 (3), 408-413.
(37) Epps, E.A. and Sturgis, M.B., Soil Sci. Soc. Amer. Proc., (1939) 4, 215-218.
(38) Braman, R.S. and Foreback, C.C., Science, (1973) 182, 1247-1249.
(39) Alexander, M., Adv. Appli. Microbiol., (1974) 18, 1-73.
(40) Challenger, F., Lisle, D.B., and Dransfield, P.B., J. Chem. Soc. (London) (1954) 1760-1771.
(41) Challenger, F. and Charlton, P.T., J. Chem. Soc. (London) (1947) 424-429.
(42) Fleming, J.F. and Alexander, L.T., Appl. Microbiol., (1972) 24 (3) 424-429.
(43) Bird, M.L. and Challenger, F., J. Chem. Soc. (London), (1939) 163-168.
(44) Bautista, E.M. and Alexander, M., Soil Sci. Soc. Amer. Proc., (1972) 36, 918-920.

(45) Blasco, F., Robert, J.J., and Guadin, C., C.R. Acad. Sci. Ser. D., (1972) 275, 1223-1226.
(46) Johnson, D.L., Nature, (1972) 240, 44-45.
(47) Turner, A.W., Nature, (1949) 164, 76-77.
(48) Jernelöv, A., Lann, H., Wennegren, G., Fagerström, T., Åsell, B., and Andersson, R., "Analysis of Methylmercury Concentrations in Sediment From the St. Clair System," unpublished report of the Swedish Water and Air Pollution Research Laboratory, Stockholm, (in English) (1972) 17 pp.
(49) Yamaguchi, S., Matsumoto, H., Hoshide, M., Matsuo, S., and Kaku, S., Arch. Environ. Health, (1971) 23, 196-201.
(50) Agaki, H. and Takabatake, E., Chemosphere (1973) 1, 131-133.
(51) Jewett, K.L., Brinckman, F.E., and Bellama, J.M., Amer. Chem. Soc. Symp. on Marine Chemistry in the Coastal Environment (T. Church, ed.), (1975), paper no. 21, this volume.
(52) Janzen, E.G. and Blackburn, B.J., J. Amer. Chem. Soc., (1969) 91, 4481-4490.
(53) Ardon, M., Woolmington, K., and Pernick, A., Inorg. Chem., (1971) 10, 2812.
(54) Saxby, J.D., Rev. Pure Appl. Chem., (1969) 19, 131-150.
(55) Tobias, R.S., Organometal. Chem. Rev., (1966) 1, 93-129.
(56) Jewett, K.L. and Brinckman, F.E., Preprints of Papers, Div. Environ. Chem., Amer. Chem. Soc., (1974) 14, 218-225.
(57) Agnes, G., Bendle, S., Hill, H.O.A., Williams, F.R., and Williams, R.J.P., Chem. Comm., (1971) 850-851.
(58) Skei, J.M., Price, N.B., Calvert, S.E., and Holtedahl, H., Water, Air, Soil Pollution, (1972) 1, 452-461.
(59) Huggett, R.J., Bender, M.E., and Sloane, H.D., Water Res., (1973) 7, 451-460.
(60) Schramel, P., Samsahl, K., and Pavlu, J., Internat. J. Environ. Studies, (1973) 5, 37-40.
(61) Thayer, J.S., Abstr. Papers, 8th Great Lakes Regional Mtg., Amer. Chem. Soc., West Lafayette, Indiana, (1974) INOR 131.
(62) Environmental Protection Agency, "Manual of Methods for Chemical Analysis of Water and Wastes," (1974).

# 19

# Spatial and Temporal Variations in the Interstitial Water Chemistry of Chesapeake Bay Sediments

GERALD MATISOFF, OWEN P. BRICKER III, GEORGE R. HOLDREN JR., and PETER KAERK

Department of Earth and Planetary Sciences, The Johns Hopkins University, Baltimore, Md. 21218

The calculation of chemical mass balance relations in an estuarine environment requires a careful evaluation of the material fluxes within the sediment and across the sediment-water interface. The rapid response of the estuarine environment to variations in temperature, salinity and sediment deposition rates makes this type of assessment very difficult. One approach is to examine the integrated results of these effects in terms of the spatial and temporal variability of the concentrations of dissolved species in the sediment. Spatial variations define the limits which may be placed upon the instantaneous concentrations of chemical species as a function of location. Temporal variations interpreted within the framework of the spatial limits may be used to assess the long-term effects of temperature, salinity and sediment deposition rates, thus enabling the more accurate calculations of chemical fluxes.

## Nature of the Study

The validity of a temporal study hinges upon the ability to accurately relocate the same sampling sites throughout the study period. A realistic evaluation of the navigational capabilities during the period of the study indicates that relocation within a circle of 500' diameter was possible in open waters. Where stations were close to buoys or other fixed points, accuracy of relocation was substantially improved.

Temporal data were collected at station 856-C and 856-E for the period June, 1971 through September, 1973. Most of the data were collected prior to the observations of Bray et al. (2) on the effects of sample oxidation and consequently, the data are not reliable for any metals, phosphate or Eh. The chloride, silicate, sulfate and ammonia concentrations as well as carbonate alkalinity are unaffected by oxidation during squeezing of the sample and these data are presented below.

A spatial study was conducted in Chesapeake Bay on April 14-16, 1974 to evaluate the variability in interstitial water chemistry over a small spatial range. The mid-bay region with a salinity approximating half that of sea water was chosen for this study. Three stations forming a transect across the bay at 38° 53'N (Station 853) and a fourth station at 39°03'52"N, 76°19'20"W (Station 904N) were sampled (Fig. 1).

Either four or five sets of gravity cores were obtained in a "T"-shaped pattern at each station, with approximately 100' spacing between each coring location. Two cores were taken at each sampling site, one for physical description and one for chemical analysis. The core used for chemical analysis was extruded and squeezed at room temperature with a Reeburgh-type squeezer (1) in a nitrogen atmosphere to prevent oxidation of metals (2). Measurements of pH, $pS^{=}$ and Eh were made by electrode methods in a closed cell in the nitrogen-filled glove box and samples for iron, phosphate and other oxygen-sensitive species were prepared in the box.

## Analytical Techniques

In general, ten interstitial water samples were squeezed from sediment sections of each of the cores. The first five sample sections were two centimeters in thickness, and together comprised the top ten centimeters of the core. Three centimeter sections were obtained at about 15-18 cm, and 25-28 cm. The remaining sample sections were 5 cm each, and their locations were dependent upon the length of the core. All data are plotted as the midpoint of each section. Bottom water values were determined from the water which was trapped in the core tube above the top of the sediment.

Since only 20-50 ml of sample are obtained from each squeezed section, analytical methods which require only small aliquots of sample must be used. Some of the analytical methods have been taken directly from the literature, while others are modified versions of existing methods. In all cases, the analytical methods were tested for accuracy, precision, and interferences. When found necessary appropriate corrections were made. Table I summarizes the analytical methods which we used.

## Results and Discussion

Space limitations preclude presentation of all the data from each of the four stations sampled in the spatial variation study. We show here the data for station 853-C which is representative of the quality for all the stations. Also included are additional data from a shallow water station, 853-G. The best agreement between each of the chemical species and between the physical core descriptions is found in the data from station 853-E (not shown). Although we present only a part of our total results in this communication, the estimated limits of spatial variability have been determined taking into account all four stations.

Table I. Analytical techniques used in this study

| Chemical Parameter | Method | Reference |
|---|---|---|
| Eh | Electrode | |
| pH | Electrode | |
| pS | Electrode | |
| Chloride | $AgNO_3$ Titration | (3) |
| Carbonate Alkalinity | NaOH Back Titration | |
| Sulfate | Sulfate-Phosphate | (4) |
| Reactive Phosphate | Colorimetric | (5),(6),(7) |
| Ferrous Iron | Colorimetric | (8) |
| Ammonia | Colorimetric | (1) |
| Reactive Silicate | Colorimetric | (10) |
| Manganese | Atomic Adsorption | |

A temporal study was performed at two stations. Station 856-C was sampled 16 times during a two-year period and the data, numbered sequentially, are presented here with only half plotted for the sake of simplicity.

Core Descriptions. Visual descriptions of leaf layers, shell debris, or sand layers which result from singular events can provide qualitative information which is useful in determining the degree of stratigraphic correlation between the cores. Color banding is another possible feature which may be used, although the causes of the banding are not well known. Examination of Chesapeake Bay cores shows, however, that none of these features may be used independently since they are often discontinuous from two cores obtained at the same time from the same location. The only features which appear to be reliably consistent over large distances (in fact, over most of the Bay) are 1) a thin brown layer at the sediment-water interface due to oxidation, 2) a greenish-black, water-rich, somewhat silty layer of mud (commonly containing many color bands) beneath the oxic layer, and 3) an olive green, often sandy layer beneath layer 2, which continues to the maximum depth that we have been able to sample by piston coring. (This layer is not always penetrated by gravity coring.) Contacts separating the layers range from very sharp at many places to gradational at others. We hypothesize that layer 2 is the result of large quantities of organic debris being brought into the Bay in the last 100-200 years due to the rapid rise in human population in the area. We are currently obtaining $Pb^{210}$ data to determine the validity of this hypothesis.

"Clinkers" are pieces of slag produced from coal burning ships (see Figure 2). They are particularly abundant in the sediment just south of Baltimore Harbor. Core descriptions from five cores at station 853-E show that a single "clinker" layer exists in each core at the following depths: 43 cm, 43 cm, 47 cm, 43 cm, 45 cm. In all of the cores these clinkers were found at, or just above the contact between layers 2 and 3. This spatial agreement is also offered as evidence in support of the above hypothesis.

Chloride. Chloride may be treated as a conservative tracer during mixing of natural waters. When investigating the exchange of water across the sediment-water interface, it is therefore important to examine the chloride ion distribution.

In Figure 3, spatial data for chloride concentrations at station 853-C are presented. It is quite obvious that there is excellent agreement among the five cores obtained at this deep water station. The agreement is not nearly as good at the shallow water station 853-G (Figure 4). Extremely low chloride values at depth suggest dilution by discharge from a fresh-water aquifer into the Bay environment.

Comparison of chloride concentrations between these two stations, both at the sediment surface and at depth in the sediment, reveals that the values are much higher at the deep water station. This reflects the fact that the salt-water wedge travels farther up the bay in the main channel of the estuary than it does along the shallower sides. Data for interstitial chloride values which are uncontaminated by ground-water discharge appear to be good to $\pm$ 1.5% and reflect the mean salinity of the local environment.

Temporal data for chloride is presented in Figure 5. Near the sediment surface there is considerable systematic variation with time. Late spring values are lowest due to the high volume of fresh water discharged into the Bay, while late fall values are the highest reflecting decreased river discharge and increased evapotranspiration. It appears that the upper 20 cm or so of the sediment responds rapidly to salinity changes at the sediment-water interface, while the sediment below this reflects a longer-term average of the salinity in the overlying water. This is further supported by examining cores 9, 11, 12, and 13. Core 10, obtained May 25, 1972 (not shown), plots between cores 9 and 11, and reflects a wet spring. Core 11 was obtained 5 days after Hurricane Agnes passed through the Chesapeake Bay area, and the upper part of the core is even more depleted in chloride than usual for cores collected during a normal spring. A month later (core 12), chloride is strikingly absent in the upper part of the core. Two months after Agnes, the uppermost part of the core had responded to an increase in chloride resulting from the re-establishment of the salinity wedge, although the effects of the storm are still apparent (core 13). Subsequent data (not

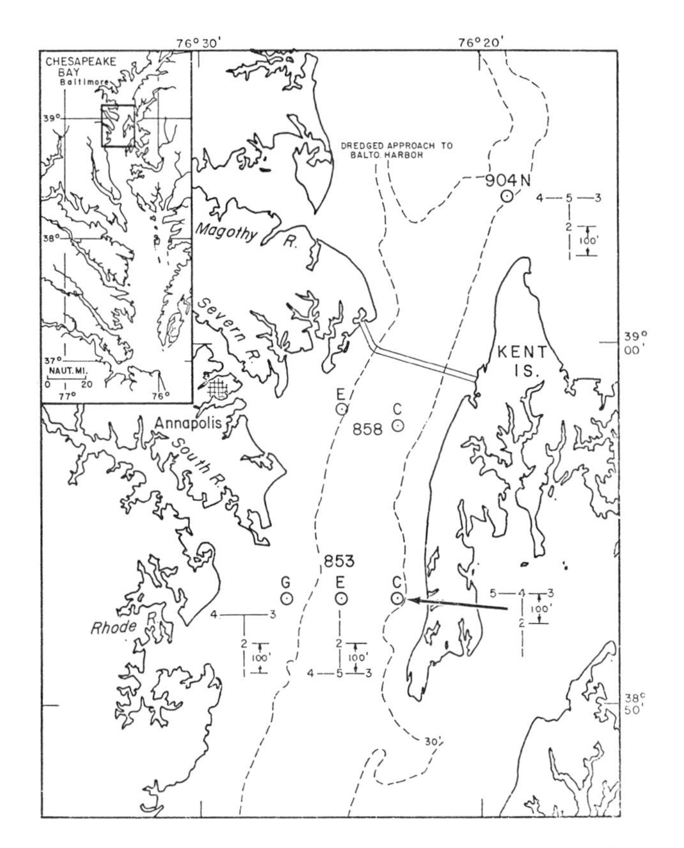


*Figure 1. Location of sampling sites in the mid Chesapeake Bay area. Stations 856-C and 856-E were used for the temporal study. The spatial study was conducted at Stations 853-C, 853-E, 853-G, and 904-N.*

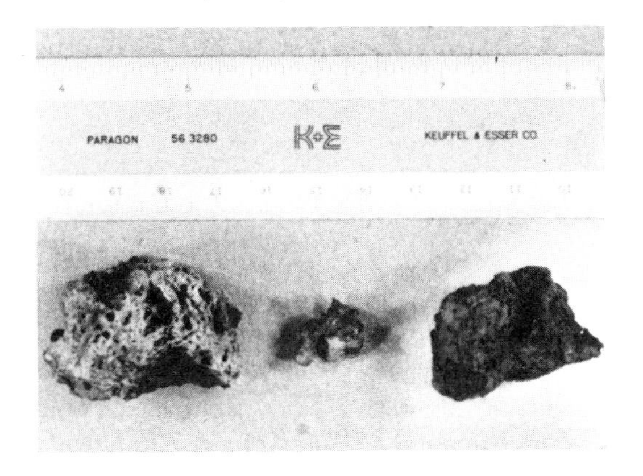


*Figure 2. Photograph of some "clinkers" obtained at Station 853-E. Upper scale in inches, lower scale in centimeters.*

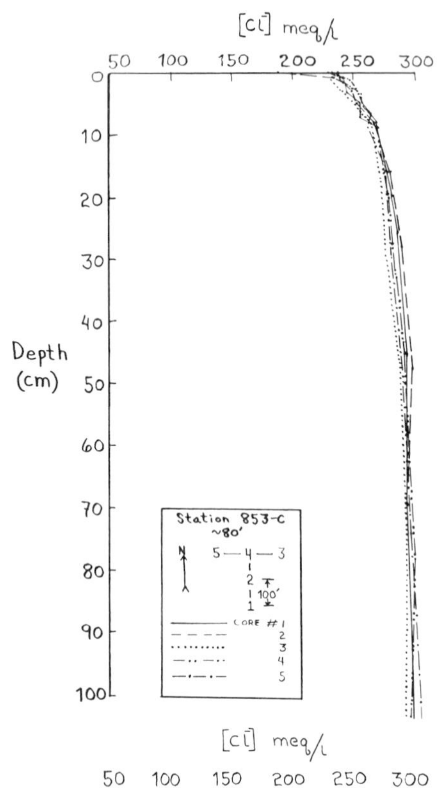


*Figure 3. Spatial variation of chloride ion concentration at Station 853-C. Excellent spatial agreement is evident. Water depth at this station is about 80 ft.*

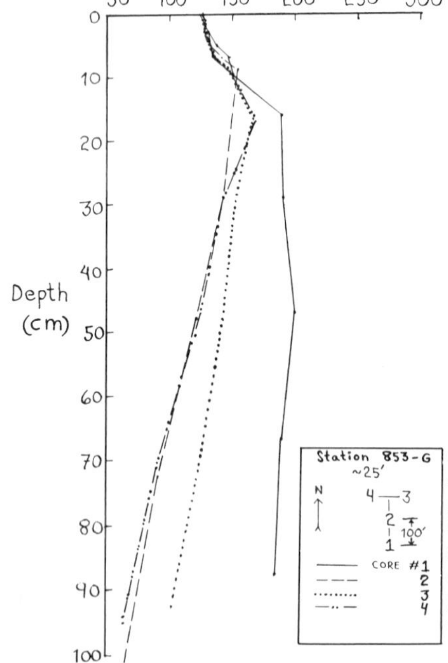


*Figure 4. Spatial variation of chloride ion concentration at Station 853-G. Poor spatial agreement and decreasing values at depth suggest the infiltration of low salinity groundwater from a source at depth. Note higher concentration of chloride at all depths at the deeper water station (853-C). This reflects the fact that the salinity wedge infiltrates farther up the bay in the main channel than it does along the shallower sides.*

presented) show complete recovery from the effects of Hurricane Agnes. This sequence of data indicates the rapidity of response of the upper part of the sediment column to fluctuations in the salinity of the overlying water. Further discussion and a mathematical model for the chloride distribution are presented in the next paper of this symposium by Holdren, et al. (11).

Silicate. Interstitial silicate concentrations are typically an order of magnitude greater than the overlying water values. Diatom dissolution is an important source of silica (13,14) and clay mineral reactions can either release or take up silica (12). Both diatom dissolution and clay mineral reactions are complicated functions of temperature, water composition, pH, sediment composition, and particle size (15 and 16), so that a simple explanation of the interstitial silica profiles is impossible.

Figure 6 shows the spatial data for silica at station 853-C. The data for this station and the three other stations not shown suggest that silica values are good to $\pm$ 10%. Examining figure 7, one can see that the data varies by as much as 450%. Note that data obtained during the winter months exhibit low concentrations of dissolved silica, and increase with depth, whereas summer data show very high values, and decrease with depth. This observed seasonal variation suggests a strong temperature control on the "in situ" dissolved silica concentrations. Since temperature gradients exist within the cores themselves (18), squeezing samples at "in situ" temperatures would be enormously cumbersome. The error introduced by our squeezing at room temperature (17) would result in an increase in winter values over the true values, and hence, would minimize the temporal variation in this data. Thus, this observed temporal variation is real, and may, in fact, be even more pronounced than what we have reported here. We are currently working on a mathematical model which will describe this seasonal variation in terms of temperature fluctuations.

Sulfate. Chesapeake Bay sediments are rich in organic material, and its decomposition plays a major role in the chemistry of the interstitial waters. The anaerobic decomposition of organic matter by sulfate reducing bacteria may be written as (19 and 7)

$$(CH_2O)_{106}(NH_3)_{16}H_3PO_4 + 53SO_4^{=} = \\ 106HCO_3^{-} + 53HS^{-} + 16NH_4^{+} + HPO_4^{=} + 39H^{+} \quad (1)$$

Thus, an interdependency exists among sulfate, carbonate alkalinity, ammonia, phosphate, pH, and Eh. In addition, the presence of large quantities of iron and manganese in the sediment results in other dependencies due to precipitation of iron and manganese sulfides, phosphates, and carbonates. The system is still affected, however, by the same physical processes that control

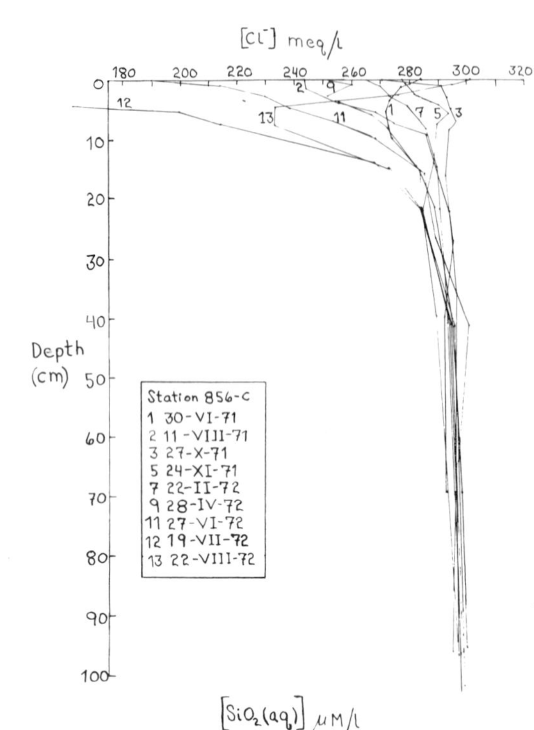


*Figure 5. Temporal variation of chloride ion concentration at Station 856-C. Dates are given as day–month (in Roman numerals)–year. Note excellent agreement at depth, but seasonal variation in the upper 20 cm.*

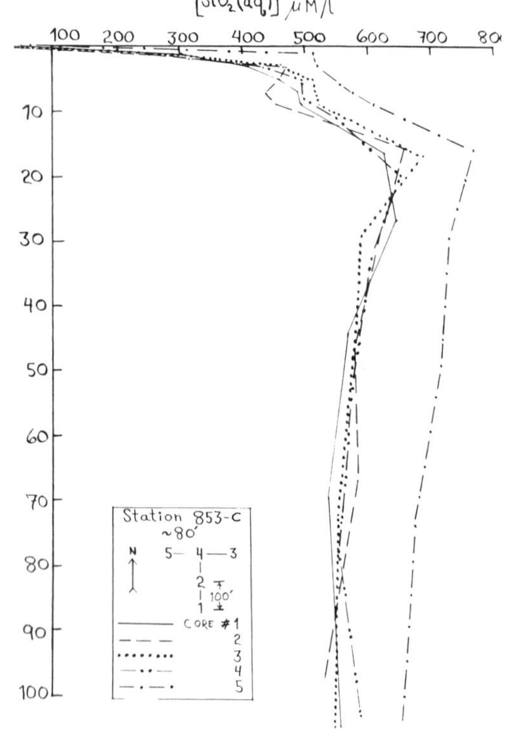


*Figure 6. Spatial variation of dissolved silica at Station 853-C*

the chloride distribution. A changing water composition at the sediment-water interface will, therefore, have an important influence on the sulfate distribution. Temperature fluctuations affect the rate of organic decomposition and, therefore, the rate of reduction of sulfate and production of the species in equation 1. One would thus expect a rough seasonal correlation among these species, but as an added complexity the composition of the interstitial waters may also be constrained by mineral equilibrium reactions.

Although the spatial data for sulfate is unavailable, it seems reasonable that its reproducibility should be good to at least ± 10%. Temporal data is presented in Figure 8. It is quite clear that the temporal variation for sulfate far exceeds these estimated spatial limits. This variation shows a seasonal trend. During the warmer months when biological processes are most active, almost all of the sulfate is reduced to sulfide, but during the colder months when these processes slow down, not all of the sulfate that has diffused into the sediment is reduced.

The effect of Hurricane Agnes can be seen by examining core 12. Sulfate has been substantially reduced in the upper 15-20 cm of the core. This is in agreement with chloride data. A month later, essentially all of this sulfate is gone.

Carbonate Alkalinity. At the pH of interstitial waters, the primary component of the carbonate alkalinity is the bicarbonate ion. It is produced not only from the oxidation of organic matter by sulfate reducing bacteria as described by equation (1), but also from any "reverse weathering" reactions (20) which might be taking place. If reverse weathering is taking place, its effect on the carbonate alkalinity is probably quite small with respect to bacterial activity. If the alkalinity is produced primarily by bacterially mediated oxidation of organic matter by sulfate, an inverse relationship should exist between sulfate and carbonate (21). Troup (22) and Bray (7) have shown that such a relationship exists for these species in the Chesapeake Bay. Figures 9 and 10 give the spatial data for 853-C and 853-G, respectively. The data appear to be accurate to ± 10%. The shallow-water station exhibits considerably more variation and lower values than the deeper-water station. Comparison of the carbonate alkalinity data with the chloride data for station 853-G suggests that the variation in the carbonate alkalinity reflects a variation in chloride, and hence, a variation in the amount of sulfate available for organic decomposition. Similarly, the magnitude of the carbonate alkalinity at stations 853-G and 853-C reflects the chloride concentration at the two stations. Thus, by equation (1), the same effects should be seen in the ammonia and phosphate profiles at the two stations. Examination of Figures 3 & 4, 9 & 10, 12 & 13, and 15 & 16 for these effects shows excellent agreement with prediction.

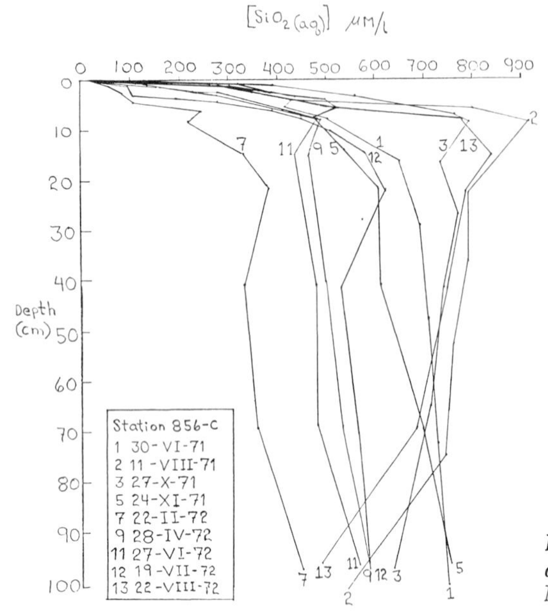


*Figure 7. Temporal variation of dissolved silica at Station 856-C. Note strong seasonal variation.*

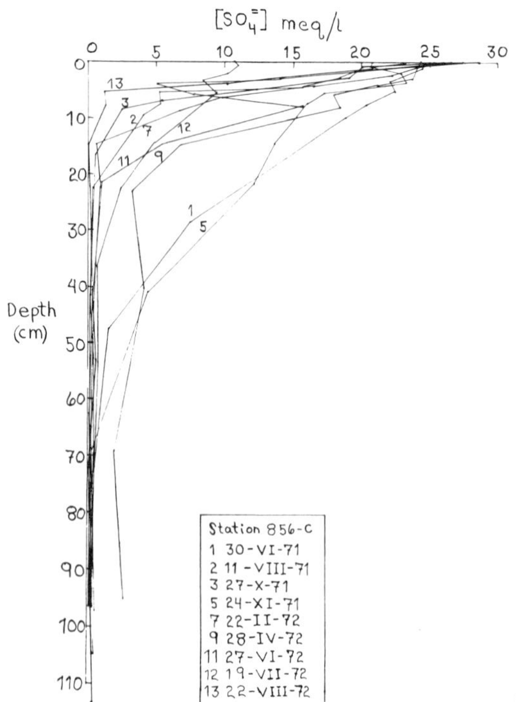


*Figure 8. Temporal variation of sulfate concentration at Station 856-C*

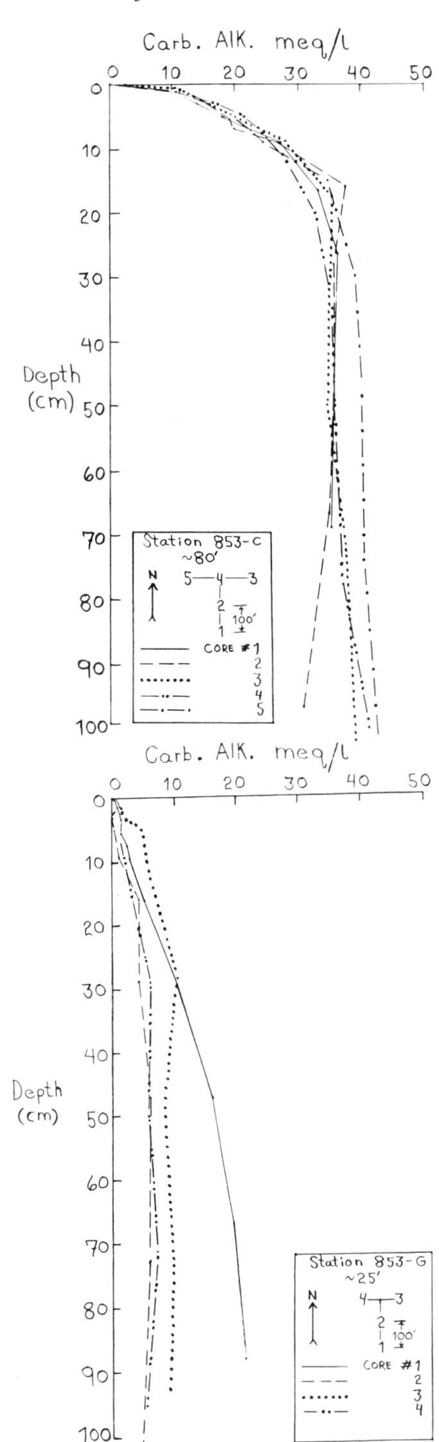


*Figure 9. Spatial variation of carbonate alkalinity at Station 853-C*

*Figure 10. Spatial variation of carbonate alkalinity at Station 853-G. Note lower values than Station 853-C and agreement with chloride at the same station, Figure 4.*

The temporal data in Figure 11 shows greater variation than can be attributed to sampling location. The rough seasonal correlation observed in the sulfate values is less well displayed by the carbonate alkalinity. In addition, the mirror image relationship between sulfate and carbonate alkalinity which exists for a single core is missing when the temporal data are compared as a group (see Figures 8 and 11).

Ammonia. Ammonia is supplied to the interstitial waters by the decomposition of organic matter. Consequently, it should vary inversely with sulfate and directly with carbonate alkalinity. Spatially, ammonia values appear to be reproducible to ± 15% as seen in Figures 12 and 13. As predicted, the shallow-water station exhibits a lower ammonia value and greater variation than the deeper-water station. Temporally, ammonia behaves very much like carbonate alkalinity as would be predicted from equation (1) (compare Figures 11 and 14). Like carbonate alkalinity, ammonia also exhibits an excellent inverse correlation with sulfate on a core by core basis, but the correlation is poor when the two sets of temporal data are compared as groups.

Phosphate. Interstitial phosphate is derived primarily from the oxidation of organic matter, although unlike carbonate alkalinity and ammonia, its concentration is dependent not only upon the extent of oxidation of organic matter, but also upon an equilibirum with vivianite ($Fe_3(PO_4)_2 \cdot 8H_2O$) (2). Therefore, to explain the phosphate profiles, one must look for superimposed inverse relations with both sulfate and iron. In addition, it has been shown (2) that both iron and phosphate values are substantially lower in pore waters which were not protected against oxidation during squeezing than in those which were. Since the majority of the temporal data were collected prior to the development of oxidation protection procedures, these iron and phosphate data are suspect and are not presented.

Figures 15 and 16 give the results of the spatial study. There is wide variation of spatial agreement for each of the four sampling stations, probably as a result of local equilibrium with vivianite. In agreement with the theory that the major source of phosphate is from the oxidation of organic matter, station 853-G is considerably lower and exhibits greater variation than station 853-C. The profiles differ from those of carbonate alkalinity and ammonia in that they often decrease with depth. It can be seen in Figure 18 that this is due to an inverse response to iron, implying that phosphate is controlled by an equilibrium with vivianite.

pH. Hydrogen ion activity is predominantly regulated by the bacterial destruction of organic material, with perhaps a minor contribution from "inverse weathering" reactions and from reduction of iron and manganese hydroxides. Thus, a slight

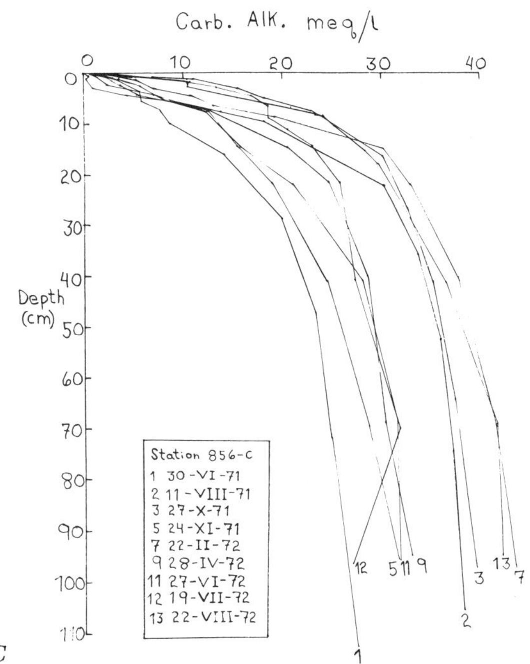


*Figure 11. Temporal variation of carbonate alkalinity at Station 856-C*

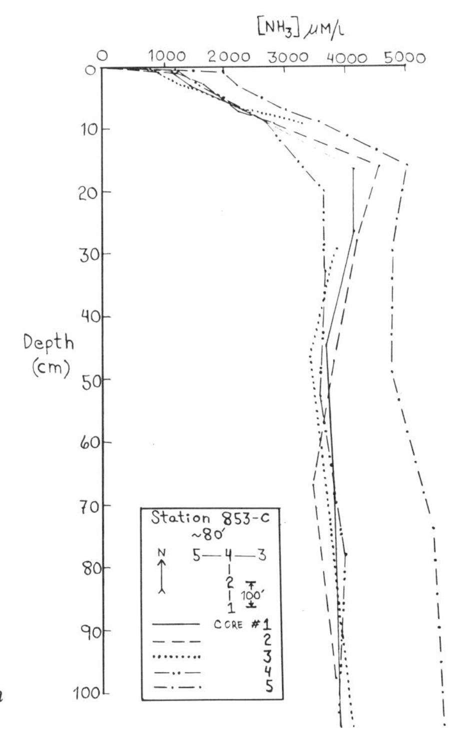


*Figure 12. Spatial variation of ammonium ion concentration at Station 853-C*

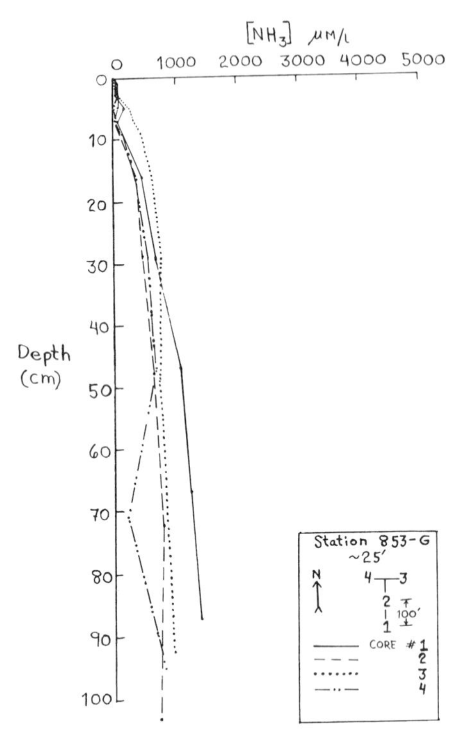


*Figure 13. Spatial variation of ammonium ion concentration at Station 853-G. Note lower values than Station 853-C and agreement with chloride and carbonate alkalinity at the same station, Figures 4 and 10, respectively.*

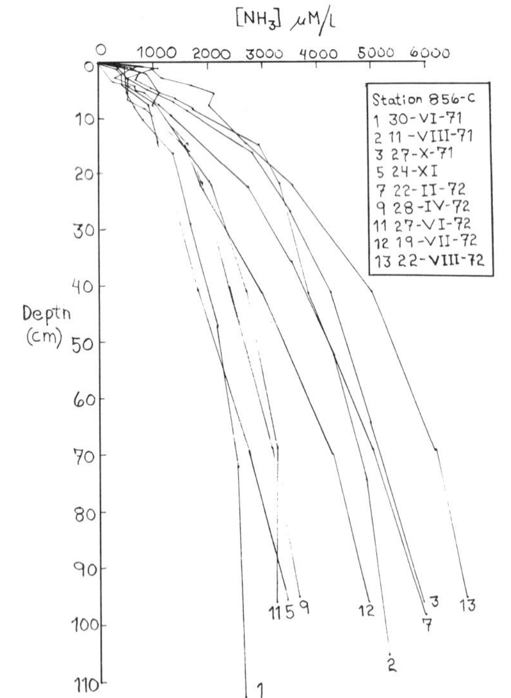


*Figure 14. Temporal variation of ammonium ion concentration at Station 856-C*

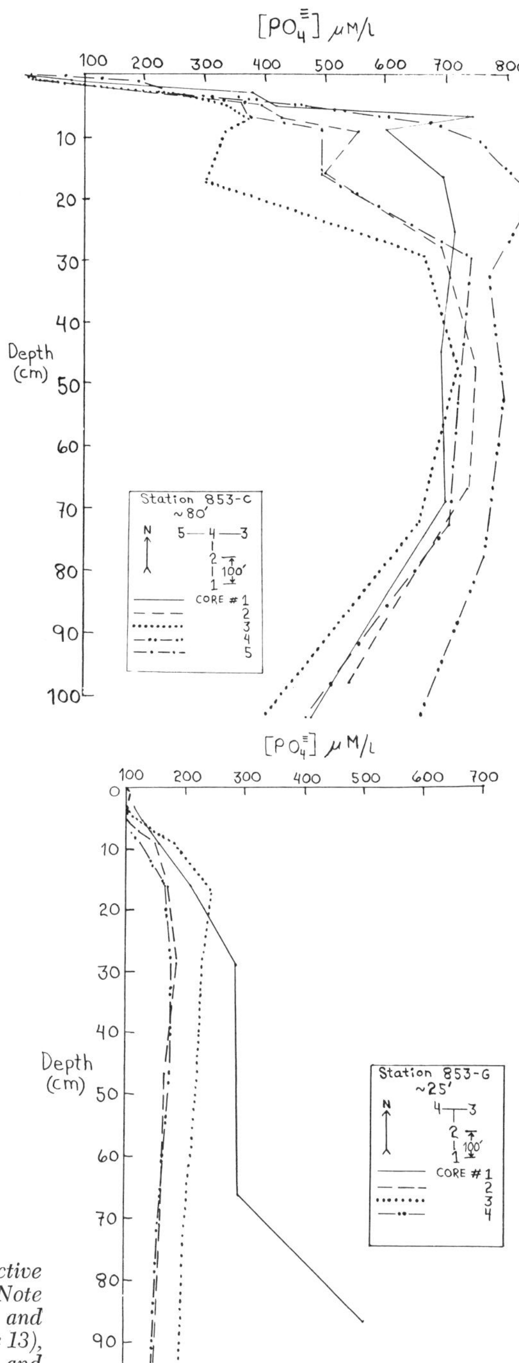


*Figure 15. Spatial variation of reactive ortho-phosphate at Station 853-C*

*Figure 16. Spatial variation of reactive ortho-phosphate at Station 853-G. Note lower values than Station 853-C and agreement with ammonium (Figure 13), carbonate alkalinity (Figure 10), and chloride (Figure 4) from the same station.*

decrease in pH with depth might be expected. Figure 17 shows some of the results of the spatial study. The pH values are good to approximately ± .2 pH units, with a mean pH of about 7.4. The temporal data are incomplete, and therefore not presented, but they show no systematic variations, all having values of approximately 7.4 ± .4.

Eh. The platinum ion electrode potentials are interesting even though the specific reactions taking place at the electrode surface cannot be determined. The redox potential (platinum electrode potential) drops dramatically from about +100 millivolts to 0 or to negative values in the first few centimeters of the sediment, and usually remains relatively constant below 20 cm. Aside from this similar behavior, the profiles from all four stations are very different in shape. It is possible that this is a result of a variation in the intensity of reduction of sulfate. The reason that the oxidation-reduction potential appears to stabilize at the value it does is not yet understood.

Values of the potential appear to be reproducible within about ± 20 millivolts, as seen in Figure 18. Troup (22) has demonstrated that the absolute value of Eh is unimportant in the reduction of ferric oxides and hydroxides to the more soluble ferrous ion as long as the Eh is no greater than about 0 mv. Holdren (23) has shown that this is also true for manganese. Thus, it can be seen that except for the very top of the core, the sediment is anoxic, and hence, iron and manganese species are mobile. A slight increase in the oxidation potential with depth may be due to a subsequent depletion of sulfide by the precipitation of iron sulfides (24). Obviously, any sample to be used for the measurement of the redox potential must be protected against oxidation by air during sampling. Because Eh data collected before 1973 are not reliable, temporal data for Eh are not given.

Iron and Manganese. Weathering processes produce fine grained iron and manganese oxides and hydroxides which are brought into the Bay as colloids or as adsorbed coatings on detrital particles. Because of reducing conditions beneath the sediment-water interface, the metals are reduced to lower oxidation states and become more soluble. Sediment sampling procedures that minimize the possibility of oxidation have resulted in significantly higher observed concentrations for dissolved metals than did prior sampling techniques. These higher metal concentrations are compatable for saturation with phases such as iron carbonates and phosphates. The presence of vivianite in the sediment has already been shown (2) and is discussed above in connection with phosphate. Concentrations of trace metals in interstitial waters thus becomes a function not only of pH and Eh, but possibly of various metal-anion mineral equilibria.

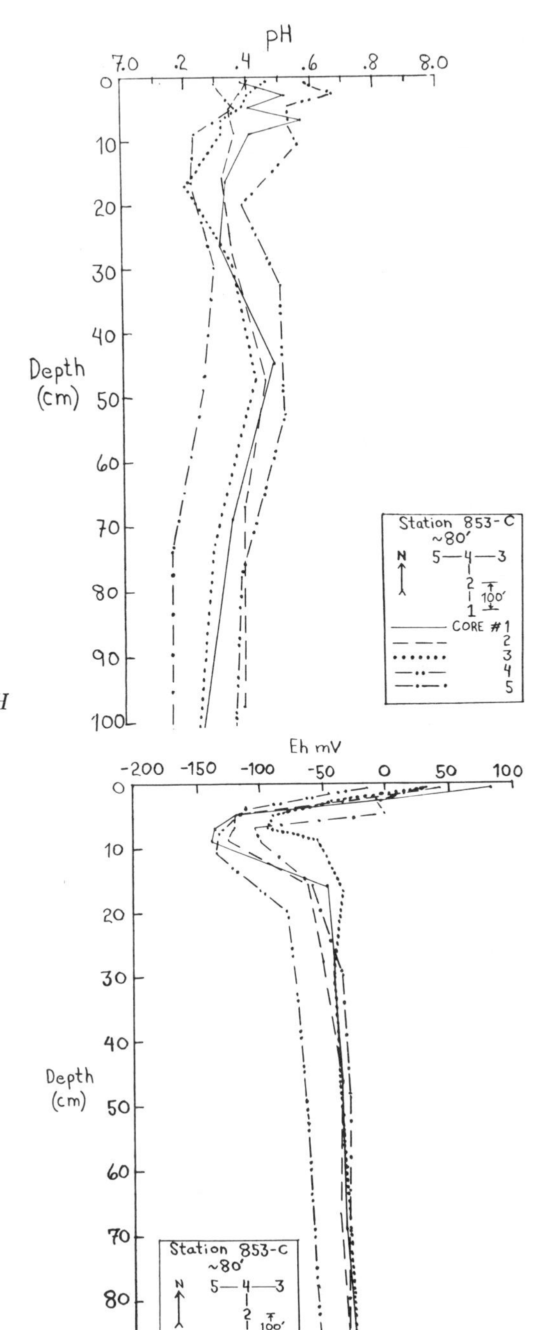


*Figure 17. Spatial variation of pH at Station 853-C*

*Figure 18. Spatial variation of Eh at Station 853-C*

Spatial variation of iron is depicted in Figure 19. This station exhibits the poorest agreement among the cores of the four stations sampled. The concentrations measured at 853-E agree to within about ± 10%. Nevertheless, even at this station quantitative comparison of the cores is difficult, although qualitatively, there is excellent agreement. The large variations observed are probably related to local equilibrium with vivianite (compare the inverse relationship of Figures 15 and 19).

Soluble manganese distribution shown in Figure 20 for station 853-C also exhibits the largest variation among the four stations. Quantitatively, the data are probably good to about ± 25%. The manganese distribution is significantly different from iron, suggesting that its concentration is probably controlled by a different mechanism than that of iron. Holdren (23) has suggested an equilibrium with rhodochrosite ($MnCO_3$), and gives a mathematical model for the manganese distribution in the next paper of this symposium (11).

## Conclusions

Acquisition of a core from a given station is subject to two fundamental problems which affect the reliability of the data: 1) navigational accuracy which limits relocation of sample sites to 100'-500', and 2) variations in temperature, salinity, and sediment deposition rates on seasonal and on other time-scale cycles which can influence both the biological and inorganic processes taking place within the cores. Qualitatively, the reliability of spatial data is excellent for most chemical species studied. In general, spatial variations were only slightly greater than the limits set by the analytical techniques. Some important exceptions exist. Ground water infiltration can severely affect the interstitial water chemistry and may be an important influence in some areas of the bay. Also, the distribution of some chemical species is controlled by local mineral equilibria, and this decreases the reproducibility of spatial data.

Temporal variations far exceed the limits of spatial variation for each chemical species investigated. For parameters which are conservative and/or influenced predominantly by inorganic activity, seasonal changes in salinity and temperature control the interstitial water profiles. Those species which are involved in the decomposition of organic matter also show a gross seasonal correlation, but other processes must be taking place. Additional work is needed to fully understand these temporal variations and their importance in governing the mass flux across the sediment-water interface.

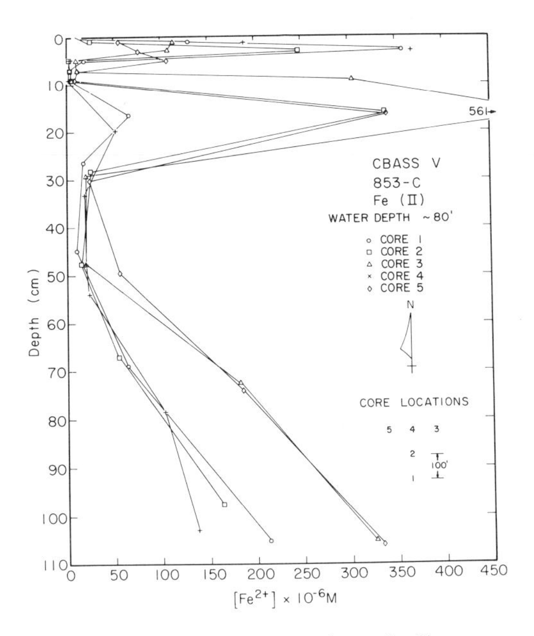


*Figure 19. Spatial variation of iron (+2) concentrations at Station 853-C*

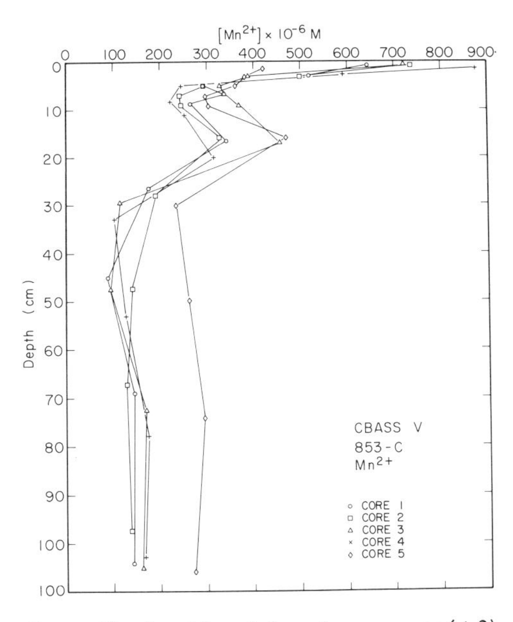


*Figure 20. Spatial variation of manganese (+2) concentration at Station 853-C*

## Acknowledgements

Most of the temporal data were collected by John Bray, Bruce Troup, Mimi Uhlfelder, and Virginia Grant. The authors thank John Ferguson, David Given, Ruth Braun, and Betsy Daniel for their help in acquisition of the spatial data. Financial support was provided by AEC grant #AT(11-1)-3292. G. Matisoff is the recipient of a Baltimore Gas and Electric Company Fellowship.

## Literature Cited

1. Reeburgh, W. S., Limnol. Oceanogr. (1967) 12, 163-165.
2. Bray, J. T., Bricker, O. P., and Troup, B. N., Science (1973) 180, 1362-1364.
3. Knudsen, M., "Hydrographic tables" 63 pp., G.E.C. Gad., Copenhagen, 1901.
4. Dollman, G. W., Environ. Sci. Technol. (1968) 2, 1027-1029.
5. Murphy, J. and Riley, J. P., Anal. Chim. Acta (1962) 27, 31-36.
6. Harwood, J. E., van Steenderen, R. A., and Kuhn, A. L., Water Res. (1969) 3, 417-423.
7. Bray, J. T., Ph.D. dissertation, 149 pp., The Johns Hopkins University, Baltimore, 1973.
8. Sandell, E. B., "Colorimetric determination of traces of metals" 1032 pp., Interscience, New York, 1959.
9. Solórzano, L., Limnol. Oceanogr. (1969) 14, 799-801.
10. Strickland, J. D. H. and Parsons, T. R. "A practical handbook of seawater analysis", 311 pp., Bull. Fisheries Res. Board of Can. no. 167, 1968.
11. Holdren, G. R. Jr., Bricker, O. P. III, Elliott, A. J. and Matisoff, G. (this volume).
12. Mackenzie, F. T., Garrels, R. M., Bricker, O. P. and Bickley, F., Science (1967) 155, 1404-1405.
13. Hurd, D. C., Earth Planet. Sci. Lett. (1972) 15, 411-417.
14. Lewin, J. C., Geochim. Cosmochim. Acta (1961) 21, 182-198.
15. Siever, R. and Woodford, N., Geochim. Cosmochim. Acta (1973) 37, 1851-1880.
16. Hurd, D. C., Geochim. Cosmochim. Acta (1973) 37, 2257-2282.
17. Fanning, K. A. and Pilson, M. E. Q., Science (1971) 173, 1228-1231.
18. Reeburgh, W. S., Ph.D. dissertation, 94 pp., The Johns Hopkins University, Baltimore, 1967.
19. Richards, F. A., In "Chemical oceanography", vol. 1, J. P. Riley and G. Skirrow (eds.), 611-645, Academic, New York, 1965.
20. Mackenzie, F. T. and Garrels, R. M., Amer. Jour. Sci. (1966) 264, 507-252.
21. Berner, R. A., Scott, M. R. and Thomlinson, C., Limnol. Oceanogr. (1970) 15, 544-549.

22. Troup, B. N., Ph.D. dissertation, 114 pp., The Johns Hopkins University, Baltimore, 1974.
23. Holdren, G. R. Jr., Ph.D. dissertation (in preparation), The Johns Hopkins University, Baltimore.
24. Nissenbaum, A., Presley, B. J. and Kaplan, I. R., Geochim. Cosmochim. Acta (1972) 36, 1007-1028.

# 20

# A Model for the Control of Dissolved Manganese in the Interstitial Waters of Chesapeake Bay

G. R. HOLDREN, JR., O. P. BRICKER, III, and G. MATISOFF

Department of Earth and Planetary Sciences, The Johns Hopkins University, Baltimore, Md. 21218

Recent interest in the origin of marine and fresh water ferro-manganese deposits has resulted in a number of investigations of the distribution of dissolved manganese in recent sediments (1-6). Attempts to model the observed manganese distribution have been made by several investigators.

Michard (7) devised a model to describe the concentration of manganese as a function of depth in the sediment by dividing the sediment into three chemically distinct zones and by using a partition coefficient, $\alpha$, to describe the distribution of manganese between the solid and solution phases. The differential equations were developed independently for each zone, and, then, they were coupled at the boundaries between the different zones in order to maintain continuity in the calculated profiles. Calvert and Price (8) developed a qualitative model to describe the profiles of manganese in the sediments of Loch Fyne, Scotland. Unfortunately, their sample spacing was too large to delineate the fine structure of the dissolved manganese profile typically found in the top few centimeters of an estuarine sediment. Also, analyses of chemical parameters other than the trace metals were not done, and so, they could only presume that the chemical controls in Loch Fyne were the same as have been found in other systems. Robbins and Callendar (9) developed a model, purportedly continuous over the depth of the sediment column, to describe the diagenesis of manganese in Lake Michigan sediments. This model, however, requires that a point source of manganese exist at some arbitrary depth in the sediment and that the concentration of dissolved manganese slowly approach equilibrium with some elusive detrital manganese phase at greater depths. While a nice qualitative fit is obtained for their field data, very little is actually known about the chemical nature of these interstitial waters, and, thus, the operative equilibrium controls in this system are again left to conjecture.

Other models have been developed to explain the frequently observed enrichment of manganese in the solid phases of surface sediments. Bender (10), Anikonchine (11), and Lynn and Bonatti

(12) have developed diffusion models to describe this phenomona. The emphasis of these works has been to investigate if upward diffusion of manganese could supply the metal required to explain the observed enrichment. In light of this goal, no attempts were made to describe the specific diagenetic reactions involved in the control of manganese in these sediment systems.

In this paper, we explore possible chemical and physical mechanisms that may control the distribution of manganese in the Chesapeake Bay estuarine sediments. Interstitial waters of the bay sediments contain greater concentrations of dissolved manganese than have been reported in any other marine or brackish water sediment system (1-4). It is not uncommon to find manganese concentrations that exceed 400 μM (∿20 ppm) and concentrations as high as 950 μM (52.5 ppm) have been observed. Based on these observations and the general chemical composition of the interstitial waters, we develop a model to describe profiles of manganous ion in the Chesapeake Bay sediments.

## Field Study and Methods

A two phase field program was initiated to investigate the spatial and temporal variability in the pore water composition of Chesapeake Bay sediments. Figure 1 shows the location of some of our standard sampling stations. The problems involved in relocating at any particular stations in the bay and the attendant sampling errors have been discussed (13).

Temporal changes were investigated by sampling monthly at a mid-bay station for the period June 1971 to August 1972. This station, 858-8, is located at 38°58'20"N x 76°23'W, east of the mouth of the Severn River and is located in about 33 m of water. Three gravity cores were collected each month using a Benthos gravity corer. The sediment was held in cellulose-acetate-butyrate plastic coreliners. A plastic butterfly valve was used to retain the sediment during retrieval. The water trapped above the sediment in the core liners during this operation was siphoned off, filtered and saved for chemical analysis. Predetermined sections of the sediment were extruded directly into Reeburgh-type sediment squeezers (14). The pore waters to be used for chemical analysis were expressed through Whatman filter paper and 0.22 μm Millipore membrane filters by 150 psi pressure exerted by nitrogen gas against a rubber diaphram in the squeezer.

Aliquots of these samples were analyzed for carbonate alkalinity, chloride, ammonia, reactive phosphate, ferrous iron, pH, $pS^{=}$ and Eh onboard ship. The remainder of the sample was returned to the lab for analysis of silica and sulfate. A complete description of both the analytical techniques and the sample handling procedures are found elsewhere in this symposium (13).

In the second phase of the field program, the spatial variability of the pore water composition in the bay was investigated.

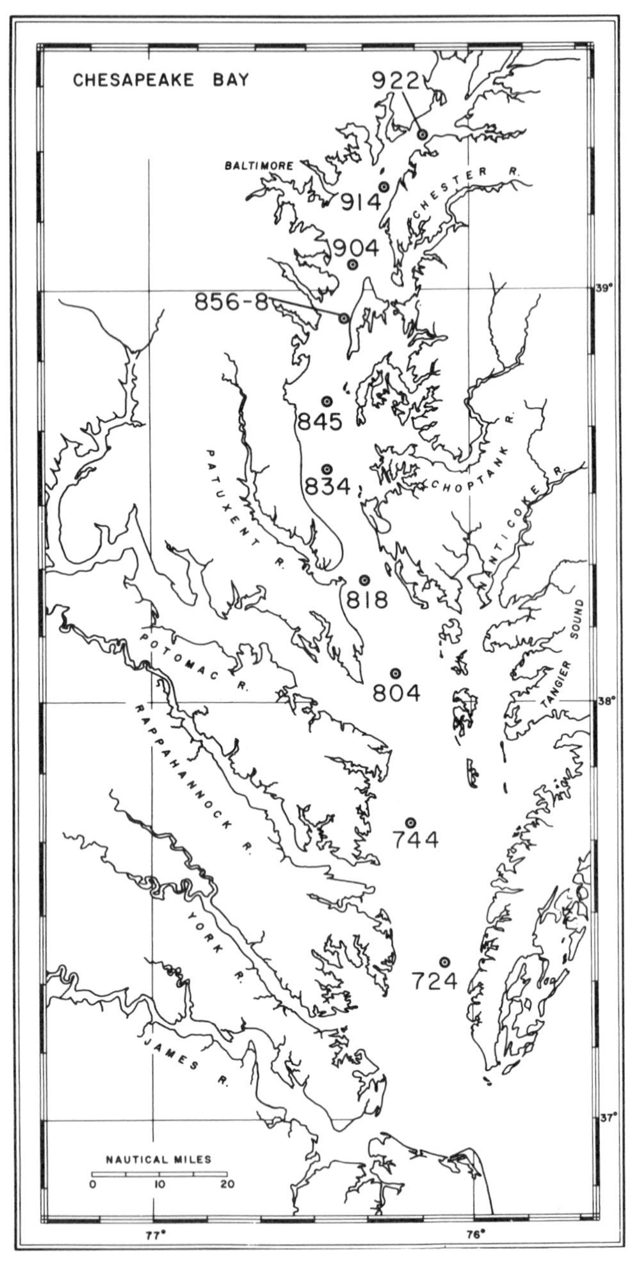


*Figure 1. The Chesapeake Bay and the locations of our sampling stations along its central axis. Stations occupied during this study, but not shown on the map, are cross-bay transects at the latitude of the stations shown.*

Six cruises conducted between August 1972 and December 1974 allowed us to collect over 700 individual interstitial water samples along with the associated sediment. Sampling locations ranged from station 935, located at the mouth of the Susquehanna River near Havre de Grace, Md., to station 724R, located between the York and Rappahannock Rivers in Virginia. During this phase of the program, all sample handling operations were done in a glove box under an inert nitrogen atmosphere to avoid the loss of trace metals and phosphate (15). Otherwise, all onboard and laboratory analytical techniques remained unchanged. In addition to the above analyses, dissolved manganese was determined on acidified subsamples of the pore water collected during this phase of the study by direct aspiration of the sample into an atomic absorption spectrophotometer. Typically, samples were diluted by a factor of between 5 and 100 with 0.01 N HCl to lower the concentration of manganese into the linear range of detection for our instrument. A detailed description of the techniques is found elsewhere in this symposium (13).

## Results and Discussion

Physical Influences on Transport: Chloride Data. The Chesapeake Bay estuary is a very productive area, biologically. This is reflected in the organic content of the sediments in the estuary which is typically 2 to 3% on a dry weight basis. A large infaunal benthic community is supported by these organics. The resulting activity mixes the upper portion of the sediment and enhances the exchange of material between the sediments and the overlying water. To investigate the magnitude of this mixing effect, along with other physical processes such as diffusion, we have studied the time dependent changes that occur in pore water chloride concentration with depth beneath the sediment/water interface.

Chloride is an ideal tracer to study these effects in an estuary such as the bay. It is essentially inert in terms of chemical reactivity in the estuarine environment. Thus, only the changing physical environment affects its distribution. Because of the seasonal variations in the fresh water input to the bay, the chloride distribution in the bottom waters is constantly changing. This produces a continually varying concentration gradient between bottom waters and interstitial waters. By following the response of the chloride profile in the sediment to changes in the chlorinity of the overlying waters, an estimate of the net rate of transport in the sediment can be made.

Figure 2 shows the results of our study at station 858-8. Easily measurable changes occur in the chloride profile on a month-to-month basis. The surface sediments respond most quickly and, with increasing depth, the magnitude of the changes decreases until at a depth of about 20 cm, variations are essentially within the analytical limits of the measurements. The mean concentration of chloride in the upper 20 cm of the profile is considerably more dilute than the concentrations deeper in the sediment. This is a result of the year-to-year fluctuations in the mean discharge of the Susquehanna River which supplies between 90 and 97% of the fresh water to this portion of the bay.

If the primary mechanism for the transport of chloride is diffusional in nature, the diffusion equation should adequately describe the shapes of the measured profiles. Lateral concentration gradients of chloride in the sediments are small compared to the vertical gradients, and so the situation is reduced to a one dimensional diffusion problem. Several models for diffusion in modern sediments have appeared in the recent literature (16-19). Only one of these was designed specifically to deal with a boundary condition which is oscillatory in nature (17). However, for our purposes in this paper, the numerical approach used by Scholl et al., is too involved.

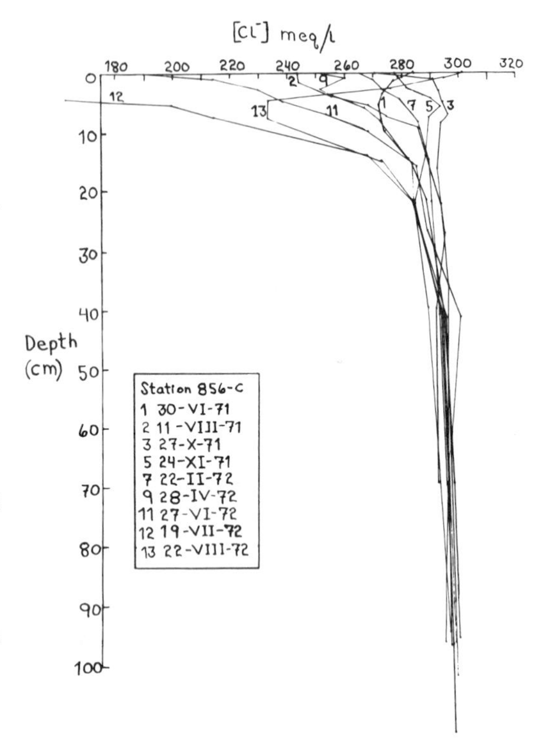


*Figure 2. Interstitial water chloride profiles collected at station 856-8 (C) over a one-year period. Only about half of the profiles are shown here. Sampling date for each profile is given in the legend.*

One other may be important in setting up this diffusion model for chloride. Sediment deposition in the Chesapeake Bay is on the order of 1 cm/yr (20). The sediment is derived from both shoreline erosion and suspended sediment discharge from the inflowing rivers and streams. While this probably does not significantly alter the chloride distribution in the sediment over the period of one year, provisions should be made in the model to account for any long term effects resulting from the sediment accumulation on the chloride distribution.

The equation, incorporating a sedimentation term, is:

$$\frac{dCl}{dt} = D\frac{d^2Cl}{dX^2} - U\frac{dCl}{dX} \quad (1)$$

With the boundary conditions:

$$Cl(0,t) = Cl_0 + Cl_1 \cos(\omega_1 t) + Cl_2 \cos(\omega_2 t) \quad (2)$$

$$\left.\frac{dCl}{dt}\right|_{X \to \infty} = 0 \quad (3)$$

where $Cl(X,t)$ = depth and time dependent chlorinity
$Cl_0$ = long term mean chlorinity
$Cl_1$, $Cl_2$ = short and long term chlorinity variations, respectively
$\omega_1$, $\omega_2$ = frequency of the short and long term variations, respectively
D = the molecular diffusion coefficient
X = depth in cm below the sediment/water interface
and U = sedimentation rate

Equation (1) is the standard one-dimensional form of the diffusion equation with an advective term ($-U\, dCl/dX$) to describe the effects of sediment deposition. The boundary conditions are based on the physical observation made for the system.

The first boundary condition describes the chloride concentration in the overlying water as a function of time. The first term on the right hand side of the equation is the long-term mean chlorinity. The second term accounts for the seasonal fluctuations in the chloride concentration. These are the variations observed in the month-to-month changes in the chloride profiles. The last term describes the long term changes in the mean annual chlorinity. It is this term which accounts for the skewing of the upper portion of the profile toward more dilute concentrations relative to the deeper pore waters.

The second boundary condition simply states that there is no net diffusional flux of chloride at great depth in the sediment. This is equivalent to saying that the estuary is "lined" by impermeable bedrock beneath the sediment.

For constant D and U the solution to this equation is found in Carslaw and Jaeger (21):

$$Cl(X,t) = Cl_0 + Cl_1 \cdot \cos\{\omega_1 t - X a_1^{1/2} \sin(\tfrac{1}{2}\phi_1)\} \cdot \exp\{\frac{UX}{2D} - X a_1^{1/2} \cos(\tfrac{1}{2}\phi_1)\} + Cl_2 \cdot \cos\{\omega_2 t - X a_2^{1/2} \sin(\tfrac{1}{2}\phi_2)\} \cdot \exp\{\frac{UX}{2D} - X a_2^{1/2} \cos(\tfrac{1}{2}\phi_2)\} \quad (4)$$

where $a_1 = \{\frac{U^4}{16D^4} + \frac{\omega_1^2}{D^2}\}^{1/2}$, $a_2 = \{\frac{U^4}{16D^4} + \frac{\omega_2^2}{D^2}\}^{1/2}$

$\phi_1 = \tan^{-1}(-4D\omega_1/U^2)$, $\phi_2 = \tan^{-1}(-4D\omega_2/U^2)$

and all other terms as defined above.

By picking values for U, D, and $\omega_2$ ($\omega_1 = 2\Pi$/year), theoretical, time dependent chloride profiles can be calculated. The range of the values for U and $\omega_2$ are available from independent sources (20, 22). Therefore, an extimate of the diffusion coefficient typical of bay sediments can be made by matching calculated profiles to the field data.

The results of some representative calculations are shown in figure 3a-c. All three parameters, D, U and $\omega_2$, were varied in the calculations to determine the net effect of each on the profiles. Results indicate that reasonable rates of sedimentation has very little effect on the chloride profiles. It makes little difference whether U is $3x10^{-5}$ cm/yr or 3 cm/yr in the final results. Changes in $\omega_2$ has only a slightly greater effect.

Changes in the diffusion coefficient, D, had the greatest effect of these three parameters. Comparison of the calculated profiles to the field data indicate that the best value for a constant D is $5x10^{-6}$ $cm^2$/sec. This is in good agreement with values that have been reported in other sediment systems (23,24). Results of the model indicate that the diffusion coefficient is not strictly a constant with depth. We have a numerical model which calculates chloride concentration profiles through time for any arbitrary functional form of D. However, the purpose here is not to generate exact replicates of the observed chloride profiles in the bay sediments, but rather is to obtain a feeling for the magnitude of the combined effects of diffusion, bioturbation and sedimentation on the distribution of any dissolved component of the interstitial waters. The simple model described above accomplishes this goal.

Manganese: Field Data. The next step in determining the overall diagenetic behavior of dissolved manganese in anoxic pore waters is to identify which, if any, specific reactions or apparent equilibria may be involved in controlling the manganese cycle. To do this it first helps to examine the concentration profiles of manganese. Figure 4 shows some typical profiles of dissolved manganese for stations located along the axis of the bay. These samples were collected in the summer of 1973.

Several features of these profiles should be noted. The concentration of dissolved manganese in the overlying waters never exceeded 6 μM on this cruise. These values, which are relatively concentrated by open water standards, are probably the result of the resuspension and subsequent mixing of the top few millimeters of sediment which occurred during the coring operation. Within the sediment, concentrations of dissolved manganese increase quickly below the sediment/water interface. Commonly, the concentration in the top two centimeters of the sediment column is the highest in the core. Concentration of dissolved manganese usually decreases with depth. Samples collected at other times of the year exhibit the same gross features. However, during colder periods, the maximum concentration is reached five or ten centimeters below the sediment/water interface.

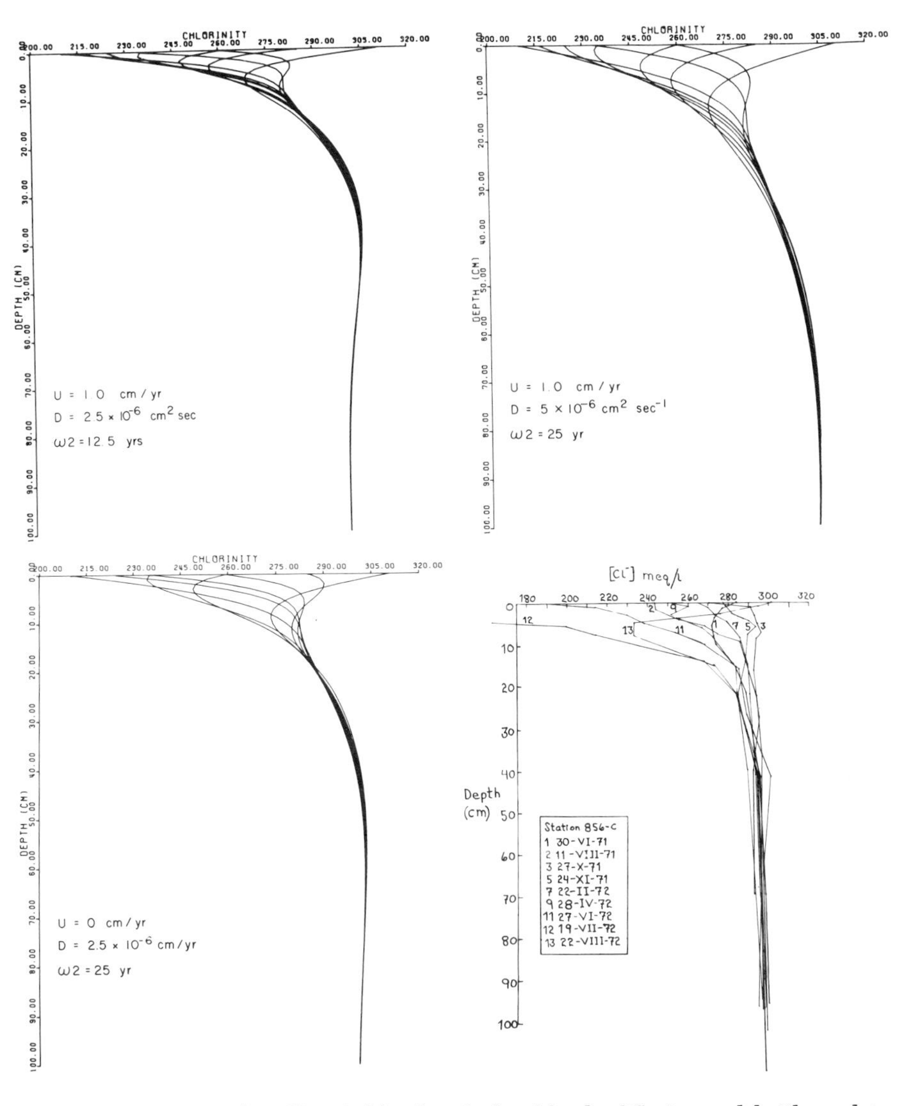


*Figure 3. a-c. Vertical profiles of chloride calculated by the diffusion model. These plots show the effect of varying D, U, and ω2 on the profiles. d. For comparison to the model profiles, the field data is replotted.*

To determine whether any heterogeneous equilibrium constraints are being imposed on the concentration of dissolved manganese by the pore water composition, activity calculations were made on each sample. These calculations were checked for possible saturation of a number of common sedimentary manganese minerals including rhodochrosite ($MnCO_3$), reddingite ($Mn_3(PO_4)_2 \cdot 3H_2O$) and albandite (MnS). The calculation used a modified form of the Garrels and Thompson model for sea water (25) to describe the ionic medium and determine ionic strengths. Activity coefficients were estimated from the extended form of the Debye-Hückel equation. An ion pairing model was then used to calculate activities of manganous ion from the composition of the pore waters. Free energy data used in calculations were obtained from several sources (26, 27).

The results of these calculations indicate that rhodochrosite is the only mineral for which the pore waters exceed saturation. This supersaturation exists at all stations and for most levels within the sediments of the bay. In the northern bay, the pore waters are between 1.5 and 2.5 orders of magnitude supersaturated and in the southern bay, the pore waters are generally in the range of 0.5 to 1.5 orders of magnitude supersaturated with respect to rhodochrosite.

Alabandite is the only other mineral that even approaches saturation in the pore water system. This situation occurs in the southern portion of the bay where pore water sulfide values are generally higher because of the greater sulfate concentrations in the overlying water.

To describe manganese profiles in the bay, the interaction between manganese and carbonate must be further investigated. To this end it is helpful to understand the behavior and genesis of bicarbonate in the pore waters of the bay.

Carbonate. At pore water pH's, bicarbonate ion concentration is essentially equal to the carbonate alkalinity. Figure 5 shows some profiles of carbonate alkalinity measured at stations located along the axis of the bay. Concentrations of bicarbonate ion in surface sediments and overlying waters rarely exceed 1.5 meq/l. The concentration generally increases with depth; however, individual profiles can be quite complex.

Bicarbonate ion is a byproduct of bacterial oxidation of organic matter in the sediment. In an anoxic marine environment, sulfate is used by the bacteria as an oxygen source. The general equation for this oxidation is written:

$$(CH_2O)_{106}(NH_3)_{16}(H_3PO_4) + 53\ SO_4^{=} = 106\ HCO_3^{-} + 53\ HS^{-} + 16\ NH_4^{+} + HPO_4^{2-} + 39\ H^{+} \quad (5)$$

For every equivalent of sulfate reduced, one equivalent of bicarbonate is generated, along with lesser amounts of ammonia,

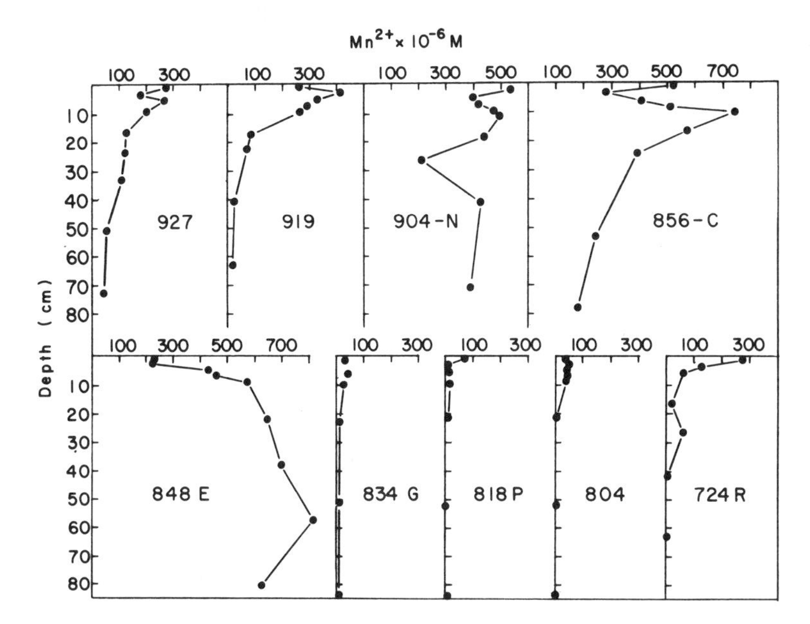


*Figure 4. Profiles of dissolved manganese in the interstitial waters of the sediments obtained along the central axis of the bay. These are typical summer profiles.*

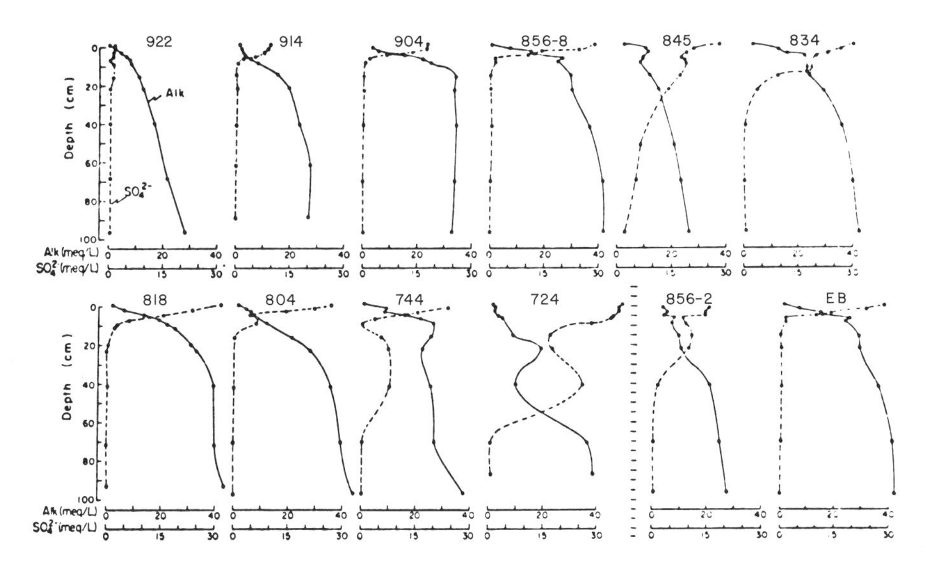


*Figure 5. Sulfate and alkalinity data for the interstitial waters along the axis of the bay. These are typical profiles.*

phosphate and sulfide. Measured profiles of sulfate in the bay sediments are shown in figure 5. The one-to-one correspondence between the amount of sulfate reduced and the amount of bicarbonate generated as predicted by (5) does not hold true. However, as can be seen in figure 5, a nearly linear relation does exist.

Bicarbonate Ion Control of Manganese. The reaction of bicarbonate ion with manganous ion can be expressed as:

$$Mn^{2+} + HCO_3^- = MnCO_{3(s)} + H^+ \qquad (6)$$

As indicated by the activity calculations, this is the reaction which controls the concentration of manganese in the sediment. The ion activity product (IAP) of the reaction components of (6) calculated from pore water compositions usually exceeds the thermodynamically derived solubility product of rhodochrosite. This occurs for two reasons.

First, one of the assumptions made in setting up the ion pairing model to calculate manganous ion activities was that only inorganic ion pairs need be considered. No attempt was made to account for organic complexes of manganese. The interstitial waters of the Chesapeake Bay contain up to 70 ppm dissolved organic carbon. Others (28) have shown that manganese can complex strongly with naturally occurring organics. Because the association constants for reactions of this type are not well known, they cannot be included in the equilibrium calculations. This exclusion results in calculated activities which are larger than actually occur.

The second reason apparent supersaturation exists in the bay sediments is that the IAP we calculate is compared to the solubility of pure rhodochrosite. It is highly likely that the rhodochrosite in the sediments is not a pure phase, but a solid solution with an enhanced solubility compared to that of pure rhodochrosite.

To incorporate the effects of these two factors, we calculated an "apparent stability constant", $K'_{sp}$, to describe the reaction between the dissolved components and the solid sedimentary carbonate phase. pH, alkalinity and manganese data from the deepest sample from each core was used for this purpose. We felt that these samples had had the greatest opportunity to attain equilibrium with the solid phase. This calculation yields a Gibbs free energy for the manganese carbonate phase of -193.1 kcal/mole. These combined effects reduce the apparent stability of the sediment phase by about 2 kcal compared to the free energy of pure rhodochrosite which lies in the range -195.05 (29) to -195.7 kcal/mol (30).

If the pH and bicarbonate concentration are known, the concentration of manganese can be determined from the mass action relation for (6).

$$Mn^{2+} = \frac{K'_{sp} \cdot [H^+]}{[HCO_3^-]} \tag{7}$$

Sample by sample adjustments of pH could be made in applying (7) to the calculated manganese profiles. However, because variations in pH with depth in any one core are generally small, and in order to maintain continuity, the mean value of the measured pH within a core is used for all depths in that core.

Similarly, bicarbonate concentrations on a point by point basis could be used in the calculation. However, to expedite the computational process and again for sake of continuity in the profiles, we chose to use a simple model to describe the generation of bicarbonate. Because of the observed relation between sulfate and bicarbonate, a modification of Berner's model for sulphur diagenesis was used (31) to calculate the bicarbonate distribution in the sediment. Bicarbonate concentration is described as a function of depth by the equation:

$$HCO_3^- = HCO_3^-{}_{(I)} + \frac{U^2 G_O}{U^2 + K_1 D} [1 - \exp(-K_1 X/U)] \tag{8}$$

where $HCO_3^-{}_{(I)}$ = bicarbonate in the overlying water
$U$ = the sedimentation rate
$D$ = the diffusion coefficient for bicarbonate
$G_O$ = the organic content of surface sediments
$K_1$ = the first order rate constant for bicarbonate generation
and $X$ = distance below sediment/water interface (cm)

There are several assumptions in this model. U and D must be constants through time and space, respectively. Generation of bicarbonate is assumed to be a first order reaction with respect to the amount of available organic material in the sediment. Finally the bicarbonate profile is assumed to have reached steady state. Since we are simply fitting this model to the bicarbonate data, these assumptions are of little concern to us.

Estimates for U and D are obtained by independent means. By adjusting the values of $K_1$ and $G_O$ the model can be fit to the data. The results of this method of calculating the bicarbonate concentrations are shown in figure 7 for several of our stations. The values of pH, U, $K_1$, and $G_O$ used for each station are listed in table I.

By using (8), concentrations of dissolved manganese can be calculated for most of the sediment column. However, the results of the calculation in the top few centimeters of the sediment are inconsistent with the field data. This portion of the profile must be controlled by some other process.

Oxidation and Diffusion of Manganese. The concentration of dissolved manganese in the waters immediately overlying the sediment are generally small. The concentration jumps to as high

TABLE I

| Station | U | pH | $K_1$ | $G_o$ | $K_2$ | $(MnO_2)_0$ |
|---|---|---|---|---|---|---|
| | cm/yr | | $year^{-1}$ | mmoles $l^{-1}$ | $year^{-1}$ | moles $l^{-1}$ |
| 904D | 1.0 | 7.5 | 0.0133 | 0.091 | 0.0173 | 0.010 |
| 834G | 0.5 | 7.65 | 0.0385 | 0.265 | 0.0173 | 0.0025 |
| 914Q | 1.0 | 7.2 | 0.0345 | 0.082 | 0.0173 | 0.012 |
| 848F | 0.5 | 7.7 | 0.0198 | 0.179 | 0.0173 | 0.005 |
| 919T | 1.0 | 7.0 | 0.0277 | 0.105 | 0.0173 | 0.015 |

$D = 3x10^{-6}$ $cm^2/sec.$ for all stations.

as 857 μM within the top two centimeters of the sediment. To maintain such a large concentration gradient over a small distance for any length of time, a sink for dissolved manganese must exist at the sediment/water interface. Manganese is sensitive to the oxidation potential of the environment. Upon diffusing from the anoxic mud into a zone containing free molecular oxygen, manganese would precipitate as a hydrous-oxide phase. Then, upon burial, this metal would be available for remobilization.

The concentration of dissolved manganese in the zone immediately beneath the oxic layer is dependent on two factors: 1) how fast the metal is released from the solid phase, and 2) how quickly it diffuses away from its source. The rate a material is released from a solid depends on many parameters. The surface area of the solid is one of the major factors (32, 33). The hydrous oxide phase is present essentially as a two dimensional coating on clay particles. For this reason, the amount of solid manganese is roughly proportional to the surface area of the solid available for dissolution. If we assume that the rate of dissolution of manganese is first order relative to the amount of available solid phase, the rate of production will be expressed as

$$\frac{d\ Mn^{2+}}{dt} = -K_2(MnO_2) \tag{9}$$

To apply this expression to bay sediments, we must assume that for any station the supply of solid manganese to the surface sediments is constant with time. Since this surface zone rarely extends more than about 5 cm into the sediment, representing a maximum period of about 10 years, this is a reasonable assumption.

Finally, the balance between the rate of dissolution and subsequent upward diffusion of the manganous ion must be established. If we assume the system is in steady state, this balance can be written

$$D\frac{d^2Mn^{2+}}{dX^2} - U\frac{d\ Mn^{2+}}{dX} + K_2(MnO_2) = 0 \tag{10}$$

with the boundary conditions

$$Mn^{2+}(0,t) = 0$$

and $$Mn^{2+}(\infty,t) = Mn^{2+}{}_f$$

The solution to this equation is

$$Mn^{2+} = \frac{U^2 \ (MnO_2)_0}{U^2 + K_2D} [1 - \exp(-K_2X/ \ )] \qquad (11)$$

where $U$ = the sedimentation rate
$D$ = the diffusion coefficient for $Mn^{2+}$
$K_2$ = the first order rate constant for the dissolution of the hydrous manganese oxide phase.
and $(MnO_2)_0$ = the amount of solid manganese in the surface sediments.
$X$ = distance below sediment/water interface (cm)

This solution is similar to (8). The values of $(MnO_2)_0$ and $K_2$ are listed in table I.

The two processes that dominate manganese chemistry have now been described. They must be integrated into a unified model in order to predict the distribution of dissolved manganese over the whole sediment column. Each reaction is limiting over that part of the sediment where it is dominant. This is to say, in the upper part of sediment, there is only a limited rate at which $Mn^{2+}$ is produced. It cannot attain concentrations large enough to become saturated with respect to any mineral because it simply diffuses to the sediment/water interface too quickly. At greater depths, the amount of dissolved $Mn^{2+}$ which can be maintained by dissolution is greater than that which is allowed by the solubility of the manganese carbonate. Precipitation of the carbonate then becomes the limiting factor in the observed concentration of the metal. Therefore, by calculation the concentration of manganese using both (7) and (11), the observed concentration will be the lesser of the two values at any particular depth.

Results of Model Calculations. Figure 6 shows the results of our calculations. The second in each set of graphs shows the concentration of dissolved manganese as a function of depth in the sediment at several northern and mid-bay stations.

The solid line is the concentration of dissolved manganese predicted by the model. Remember that this single line is the combined result of two competing processes. The portion of the curve increasing with depth is the result of the dissolution and upward diffusion of manganous ion. The lower portion of the profile is the part controlled by equilibrium with the carbonate phase.

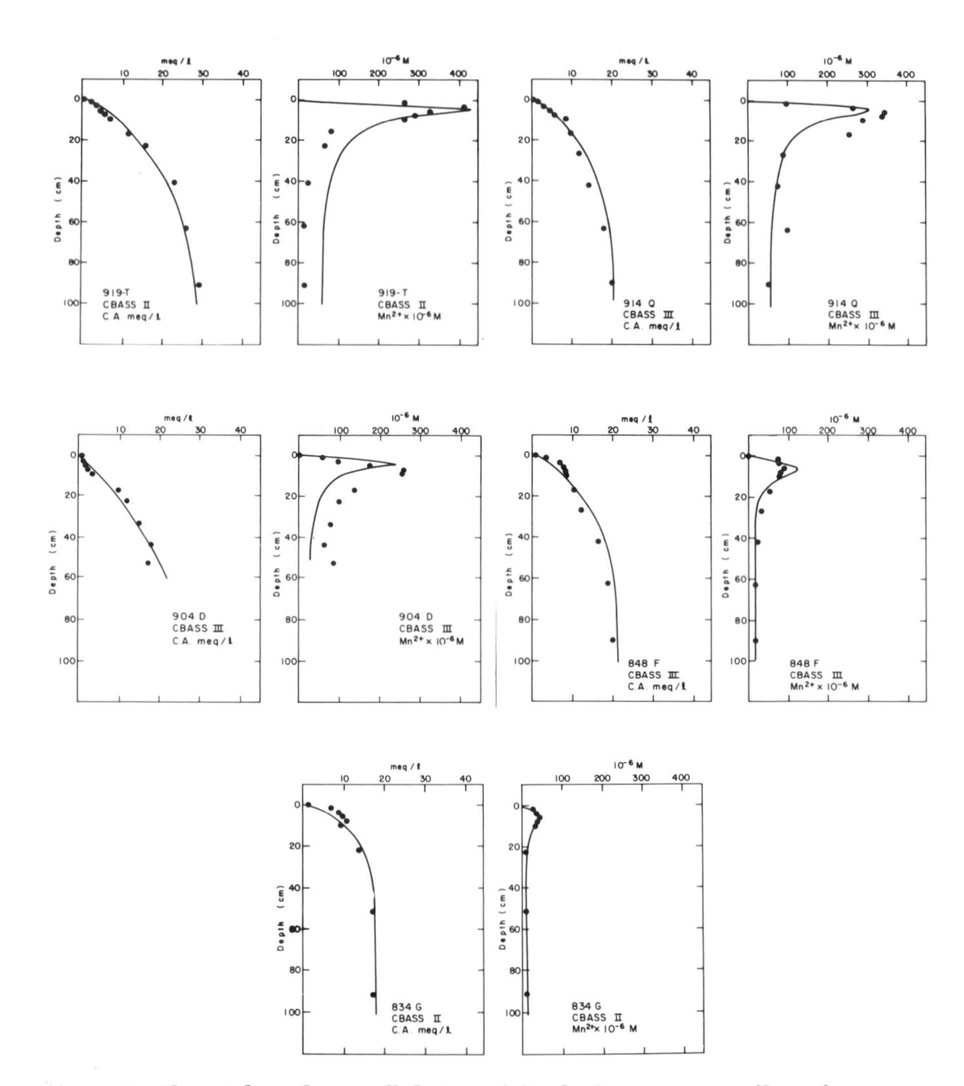


*Figure 6. Plots of the carbonate alkalinity and dissolved manganese profiles at five upper and mid-bay stations. The alkalinity plots show the field data (●) and the model representation of that data used in the calculation of the manganese profiles (——). The second graph in each set contains the dissolved manganese profile at each station (●). The concentration profiles predicted by our model using the pH and alkalinity data are shown by the line.*

## Conclusion

We have developed a model for the prediction of dissolved manganese distribution in the anoxic pore waters of the sediments of the Chesapeake Bay. The model requires knowledge of the pH of the pore waters, the distribution of bicarbonate ion with depth in the sediment, the amount of manganese oxide in the surface sediment and the rate of release of manganous ion from those solids. In the calculations presented, a modification of Berner's model for sulphur diagenesis was used to describe bicarbonate ion distribution. This model was fit to the observed profiles. Other techniques, however, such as fitting a power series to the data, could equally serve this purpose.

There were several assumptions used in the model. The diffusion coefficient and sedimentation rates were assumed to be constant through space and time, respectively. We assumed that steady state had been reached in the system, and that with depth in the core manganous ion was in equilibrium with a poorly crystalline carbonate phase.

The model was developed from observations on the pore water composition. The model describes the results of two independent competing reactions. Both reactions are continuous over the whole sediment column, and the final calculated concentration of dissolved manganese at any particular depth is dictated by the process most limiting that concentration at that depth.

Agreement between the model and the field data is generally good. This suggests that the processes controlling the distribution of dissolved manganese in the bay sediments are basically understood. The results of the model are qualitatively the same as reported profiles of dissolved manganese in other marine sediment systems (1, 3, 9). It would be most interesting to see if the model can describe these interstitial water systems with the same accuracy as was obtained in Chesapeake Bay sediments.

## Acknowledgements

The authors thank John Bray, Robert Mervine, Bruce Troup and Mary Uhlfelder who helped develop most of the field techniques used in this work and who conducted the first phase of the field program. We also thank Ruth Braun, Betsy Daniel, Jeff Elseroad, John Ferguson, Dave Given, Peter Kaerk and the many others who assisted us in both the field and lab. Special thanks goes to Mrs. Virginia Grant who did much of the lab work and to whom we went when the inevitable problems arose in the lab. This work was supported by AEC contract no. AT(11-1)3292.

## Literature Cited


1. Duchart, Patricia, S. E. Calvert and N. B. Price, Limnol. Oceanogr. (1973), 18 (4), 605-610.
2. Bischoff, James L., T-L Ku, J. Sed. Pet. (1971), 41 (4), 1008-1017.
3. Li, Yuan-Hui, J. L. Bischoff and G. Mathieu, Earth Planet. Sci. Letters (1969), 7, 265-270.
4. Presley, B. J., R. R. Brooks and I. R. Kaplan, Science (1967) 158, 906-910.
5. Hartmann, vonMartin, Meyniana (1964), 14, 3-20.
6. Spencer, Derek W. and P. G. Brewer, J. Geophys. Res. (1971), 76 (24), 5877-92.
7. Michard, Gil, J. Geophys. Res. (1971), 76 (9), 2179-86.
8. Calvert, S. E. and N. B. Price, Earth Planet. Sci. Letters (1972), 16, 245-249.
9. Robbins, John A. and E. Callendar, Amer. J. Sci. (1975), accepted for publication.
10. Bender, Michael L., J. Geophys. Res. (1971), 76 (18), 4212-4215.
11. Anikouchine, William H., J. Geophys. Res. (1967), 72 (2), 505-509.
12. Lynn, D. C. and E. Bonatti, Mar. Geol. (1965), 3, 457-474.
13. Matisoff, Gerald, O. P. Bricker, G. R. Holdren Jr. and P. Kaerk, Proc. ACS-MARM Symposium, Mar. Chem. in the Coastal Environment, Philadelphia, Pa. (1975).
14. Reeburgh, William S., Limnol. Oceanogr. (1967) 12, 163-165.
15. Troup, Bruce N., O. P. Bricker and J. T. Bray, Nature (1974), 249 (5454), 237-239.
16. Lerman, A. and B. F. Jones, Limnol. Oceanogr. (1973), 18 (1), 72-85.
17. Lerman, A. and R. R. Weiler, Earth Planet. Sci. Letters (1970), 10, 150-156.
18. Tzur, Y., J. Geophys. Res. (1971), 76 (18), 4208-4211.
19. Scholl, David W. and W. L. Johnson, 7th Int'l. Sedimentological Congr. (1967), Abstract.
20. Schubel, J. R., Beach and Shore (1968), April.
21. Carslaw, H. S. and J. C. Jaeger, "Conduction of Heat in Solids," 510 pp, Oxford Press, London, 1959.
22. U. S. Dept. of the Interior, Geological Survey, "Estimated Stream Discharge Entering Chesapeake Bay," Monthly Reports 1951- .
23. Manheim, F. T., Earth Planet. Sci. Letters (1970), 9, 307-309
24. Li, Yuan-Hui, and S. Gregory, Geochim. Cosmochim Acta (1974), 38, 703-714.
25. Garrels, R. M. and M. E. Thompson, Amer. J. Sci. (1962), 260, 57-66.
26. Garrels, R.M. and C. L. Christ, "Solutions, Minerals, and Equilibria," 450 pp., Harper and Row, New York, 1965.

27. Wagman, D. D., W. H. Evans, V. B. Parker, I. Halow, S. M. Bailey, and R. H. Schumm, "Selected Values of Chemical Thermodynamic Properties," NBS-TN-270-4, 141 pp., U. S. Gov't. Printing Office, Washington, D. C., 1969.
28. Crerar, David A., R. K. Cormick and H. L. Barnes, Acta Mineral. Petrogr. (1972), XX (2), 217-226.
29. Robie, R. A. and D. R. Waldbaum, U.S.G.S. Bull. 1259, U. S. Gov't Printing Office, Washington, D. C., 1968.
30. Bricker, O. P. Amer. Mineral. (1965), 50, 1296-1354.
31. Berner, R. A., Geochim. Cosmochim. Acta (1964), 28, 1497-1503
32. Hurd, David C., Earth Planet. Sci. Letters (1972), 15, 411-417.
33. Rickard, D. T., Amer. J. Sci. (1974), 274, 941-952.

# 21

# Release of Heavy Metals from Sediments: Preliminary Comparison of Laboratory and Field Studies

KERILYN C. BURROWS* and MATTHEW H. HULBERT†

Center for Marine and Environmental Studies and Department of Chemistry, Lehigh University, Bethlehem, Penn. 18015

The work reported in this paper is the initial portion of a study intended to clarify the rates and mechanisms of heavy metal uptake and release by sediments. Particular reference is made to estuarine sediments and the metals zinc, cadmium, and lead. The significance of these metals is, of course, their detrimental effects on a variety of animals, including man, and plants via interference with certain enzymes. At present the biologically critical levels of various chemical species of the heavy metals are imperfectly known, and these for only a few biological species. Evidence for synergistic and antagonistic effects has been reported, but data is so limited as to be merely tantalizing. Estuarine sediments are of particular interest since the estuaries serve as breeding grounds of many economically important species, and, in the anoxic state, the sediments appear to serve as traps or sinks for heavy metals. (6,7)

Anoxic sediments are to be expected in areas where large quantities of organic wastes are discharged into overlying water; heavy metals accompany the organic wastes in industrialized, urban areas. Below Philadelphia such sediment must be dredged to maintain navigation. In the past this sediment has been used to "reclaim" salt marshes, but with the growing recognition of the values of unreclaimed marshland and the increasing political muscle of conservation interests, new methods of disposal have become mandatory. A particularly apt disposal technique is the construction of artificial wildlife habitats including structures designed to form fresh water ponds. But what happens when oxygenated water percolates through these anoxic sediments? Would reversal of the metal-sequestering reactions lead to undesirable water quality in the surroundings?

## Experimental

Reagents. Stock solutions (1 mg/ml) of zinc, cadmium, copper, and lead were prepared by dissolving 100 mg high purity metal (Alfa-Ventron) in a few milliliters Ultra-Pure nitric acid

*Present address: The Wetlands Institute, Box 91, Stone Harbor, N.J. 08247

†Present address: Department of Chemistry, Connecticut College, New London, Conn. 06320

(65%, Alfa-Ventron) and diluting to 100 ml with distilled water. These solutions were stored in polypropylene bottles. Working solutions were prepared daily by serial dilutions of the stock solutions with pH 1.3 $HNO_3$.

pH 1.3 nitric acid, which was also used for rinsing glassware, was prepared by adding 750 μl Ultra-Pure $HNO_3$ to 1 ℓ distilled water, and further purified by electrolysis at -1.5 V (vs. SCE) for approximately 24 hours. One M sodium citrate solution was prepared from USP tri-sodium citrate and purified by electrolysis. Hydrofluoric acid (48%, Fisher Reagent Grade) was redistilled using a sub-boiling Teflon still (4). All other chemicals were reagent grade and used without further purification.

Sediments. Unpolluted samples were collected manually from the upper 50 cm of bottom sediments in approximately 0.5 meter of water at low tide in Jenkins Sound, Stone Harbor, New Jersey. The Sound has a salinity of 25 to 30‰ and is located about one mile from both major roads and major pleasure boat traffic. Samples were transported in covered plastic buckets, under several inches of overlying water. Portions not immediately processed were frozen at -20°C in polyethylene bags.

Polluted samples were collected in the Schuylkill River in south Philadelphia, just below the Gulf refinery and Naval Yard, on 3 November 1974 by the University of Delaware's R/V Ariadne, using a modified Foster Anchor Dredge. The upper 10 cm of the bottom sediments were taken, under 1.5 meters of water. Samples were stored in plastic bags at 0°C for about 16 hours, then frozen to -20°C.

Sediment Digestion. Sediment samples were dissolved by an adaptation of the method used by Presley.(5) Prior to analysis, the samples were dried at 70°C, ground by hand, and washed by shaking for one minute with equal volumes of distilled water, centrifuged 1.5 minutes at 4100 rpm, and dried again. This was designed to remove excess salts from the sediments.

Adsorbed fraction. Two grams of ground, dried sediment were placed in a 15 ml polypropylene centrifuge tube with 10 ml distilled water, shaken by hand for five minutes, centrifuged at 4100 rpm for five minutes, filtered into a 25 ml volumetric flask, and taken to volume with pH 1.3 $HNO_3$.

Reducible fraction. The residue remaining in the centrifuge tube and on the filter paper was washed into a 125 ml flask with 50 ml of 0.25 M $NH_2OH \cdot HCl$ in 25% acetic acid, and stirred magnetically for 4 hours at room temperature (23°C). The solution was transferred to the centrifuge tube, centrifuged at 4100 rpm for three minutes, and filtered into a 250 ml Teflon beaker. The residue was washed by shaking for one minute with five 10 ml portions of distilled water, centrifuging at 4100 rpm for 30 seconds, and combining the supernatant with the filtrate. The filtrate was heated slowly to near dryness on a hot plate. After cooling, one ml $HNO_3$ and 5 ml HCl were added. The solution was filtered into a 25 ml volumetric flask and taken to volume with pH 1.3 $HNO_3$.

Oxidizable fraction. The residue remaining in the centrifuge tube and on the filter paper was washed back into the flask with 50 ml distilled water, heated to dryness, and cooled. Twenty-five ml 30% $H_2O_2$ was added in 10 ml portions, and the solution left to stand overnight. On the following morning, it was heated for 30 minutes, cooled, centrifuged at 4100 rpm for three minutes, filtered into a Teflon beaker and prepared as for the reducible fraction.

Residual fraction. Ten ml distilled water was used to wash the residue into the 25 ml Teflon cup of a Parr bomb, and the sample was heated to dryness. After cooling, 10 ml HF and 2.4 ml $HNO_3$ were added, the solution heated in the bomb for three hours at 125°C, cooled overnight, and taken to dryness. This process was repeated using another 10 ml of HF and 2.6 ml of $HNO_3$. After cooling, 3 ml HCl and approximately 15 ml pH 1.3 $HNO_3$ were added and heated for ten minutes, then the solution was transferred to a 50 ml glass flask and boiled to dissolve the remaining solids. The solution was filtered, when cool, into a 50 ml volumetric flask and taken to volume with pH 1.3 $HNO_3$.

Total. One-half gram of ground, dried sediment was prepared as for the residual fraction.

All samples were stored in two-ounce polypropylene bottles which had been leached for a week prior to use with 6 M nitric acid.

Leaching Studies. Short term leaching studies were conducted using washed and dried samples of clean and polluted sediments in both distilled and saline waters under atmospheres of air and nitrogen. The caps of one liter wide mouth polypropylene bottles were drilled for gas inlet and outlet tubes, and leached for a week with 6 M nitric acid. Twenty-four hours before use, they were filled with the leaching liquid, which was either distilled water or fresh sea water filtered through a three micron filter (30‰ salinity). For the studies, this was discarded, replaced with 500 ml of fresh solution, and the bottles connected to a nitrogen tank or air pump and purged for one hour at a flow rate of 200 ml/min. Ten grams of polluted sediment or 20 grams of the clean were added, and the systems stirred magnetically for twenty-four hours. After settling for 45 minutes, a 50 ml portion was pipetted into a two ounce polypropylene bottle and acidified with 20μ 1 $HNO_3$.

Long-term leaching studies were conducted on clean sediment exposed to the weather. A hole was drilled in the bottom of a 20 gallon plastic trash barrel and a plastic drainage plug (11 cm inside diameter) was inserted. The cap of a one liter narrow mouth polypropylene bottle was drilled to accommodate the plug, and held in place by a rubber O ring. The cans were supported in a wooden frame. Seepage samples were collected in polypropylene bottles screwed into the cap. A similar system was used for rain- and dust-fall collection. In early June, the barrel was filled with clean wet sediment. Three bottles of seepage were collected within the next two days; afterwards, the bottles were changed at irregular intervals, generally after rain falls. Samples were preserved by the addition of 0.5 ml $HNO_3$. A surface sample of weathered sediment was taken for analysis in late January.

Instrumentation. Samples were analyzed for zinc, cadmium, lead, and copper by differential pulse anodic stripping voltammetry (DPASV) using a hanging mercury drop electrode (HMDE). The analyses were performed with a Princeton Applied Research model 174 polarographic analyzer, and data recorded on a Hewlett-Packard 7001 A XY flatbed recorder. The analytical cell consisted of 50 ml capacity borosilicate glass bottom (PAR 9301) with plastic top and fittings supplied by PAR. A metro-ohm mercury micro-feeder (model E-410) served as the HMDE; a standard colomel electrode (SCE) isolated by a pH 1.3 $HNO_3$ bridge and Pt wire were the reference and counter electrodes, respectively. The instrument was used in the differential pulse mode, with a 1.5 V scan range, 25 mV modulation amplitude, and one second drop time; the recorder's X axis was calibrated at 100 mV/inch. A setting of three divisions of the micro-feeder provided a mercury drop with a surface area of 1.8 $mm^2$. One analytical cycle consisted of purging with nitrogen and stirring at 15 rpm for one minute, plating and stirring for one minute, plating for 30 seconds, and scanning. During the analysis the nitrogen flow was directed over the solution.

Analysis. Sediment fractions were diluted 1:5 with pH 1.3 $HNO_3$ before analysis. A ten ml aliquot of the solution to be analyzed was placed in the cell and purged with nitrogen for 15 minutes. Sodium citrate solution (0.5 ml for rainwater samples, 1.0 ml for seepages, 2.0 ml for sediments and leachates) was added (2), and a preliminary analysis made. This was run from -1.4 V to +0.1 V, at 10 mV/sec from -1.4 V to -0.8 V, 5 mV/sec from -0.8 V to -0.4 V, and 10 mV/sec from -0.4 V to +0.1 V, at a current range of 10 μA. This scan was used to estimate the peak potentials and current ranges needed for the individual metals. In general -1.4 V served as the initial potential for Zn, -0.8 V for Cd, -0.7 V for Pb, and -0.4 V for Cu. Zinc and copper were scanned at 10 mV/sec and cadmium and lead at 5 mV/sec. Once the peak had been scanned, the drop was rapidly scanned to

+0.1 V; this portion was not recorded as its only purpose was to strip the remaining metals back into solution. Quantitation was accomplished by standard additions; two spikes of 20 to 100 µl and appropriate concentrations were added.

## Results and Discussion

Sediments used in these studies were selected to represent a relatively unpolluted area and an urban area subjected to industrial pollution. These will be referred to in tables and figures as "clean" and "dirty" respectively. Both sediments were dark, anoxic, and rich in organic matter; the unpolluted sample was characterized as a clayey silt, the polluted as silt.

TABLE I

Grain Size Analysis of Sediments

| Sample | %Sand | %Silt | %Clay |
|---|---|---|---|
| CLEAN | 16.7 | 63.0 | 20.3 |
| DIRTY | 9.2 | 81.1 | 9.7 |

Short-term Studies. Short-term leaching experiments were designed to test the effects of oxygen and salinity on heavy metal removed from unpolluted and polluted sediments (Figure 1).

With respect to zinc, the unpolluted sediments showed very little difference with the various treatments, losing approximately 15 µg zinc per gram of dry sediment. Loss from the polluted sediments under salt water was somewhat less, around 5 µg/g. Under fresh water the polluted sediment lost 60 µg Zn/g. Presence or absence of oxygen had no appreciable effect. These variations imply that zinc is bound in different ways in the different sediments.

Cadmium studies show a tendency for the sediments to actually remove the metal from fresh water, amounting to an increase of 0.9 µg/g for the unpolluted sample and 1.7 µg/g for the polluted one. In salt water this appears to be affected by the oxygen content. The unpolluted sample releases 3 µg Cd/g in the absence of oxygen; release from the polluted sample, 2 µg/g, occurs in the presence of oxygen. Here a synergistic effect may be taking place, with the organic content of the sea water being of greater importance than the salinity.

Under fresh water the unpolluted sample shows similar losses of lead, 1 µg/g, regardless of oxygen content. The polluted sample shows a release more than doubled in the absence of oxygen, 14 µg Pb/g as opposed to 6. Meaningful data in salt water is difficult to obtain as a consequence of analytical error, but the lead is below detection limits when the unpolluted sample is exposed to oxygen, and when the polluted sample is not. This may also result from a synergistic effect.

Sediment Analysis. Samples of both sediments, as well as

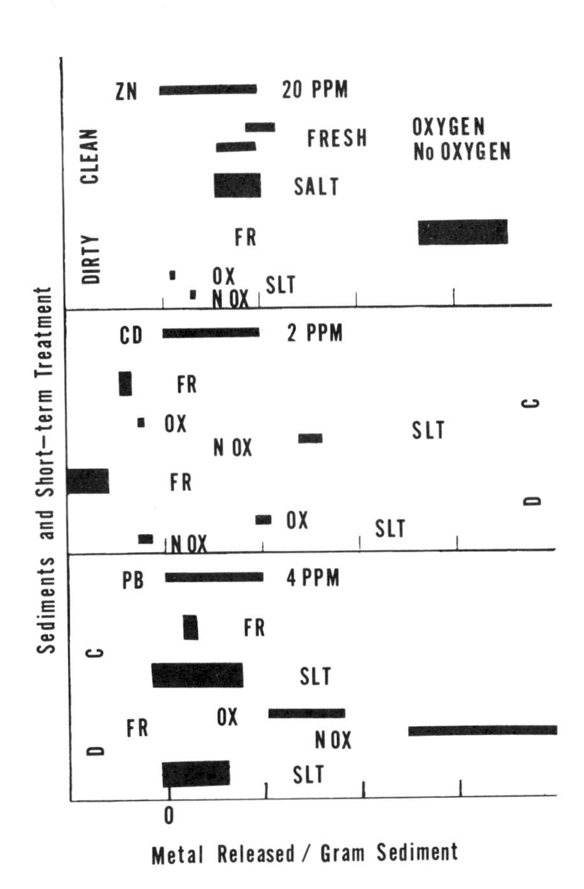


*Figure 1. Effects of oxygen and salinity on unpolluted and polluted sediments during short-term leaching*

unpolluted sediment weathered for six months, were subjected to a four-stage analysis to determine metal distribution. The first stage, a brief washing with distilled water, was designed to remove labile adsorbed species. This was followed by treatment with hydroxylamine hydrochloride in acetic acid to dissolve the reducible fraction, consisting of carbonates and iron-manganese oxides. Hydrogen peroxide was employed to dissolve oxidizable species such as sulfides and organic complexes. The final residual portion consists primarily of silicates; under normal conditions the metals in this phase are environmentally inert. The summations of the fractional metal analyses are in generally good agreement with a total dissolution of the sample.

Prior to weathering no one fraction of the unpolluted sample can be identified as metal-rich (Figure 2). Zinc is concentrated in the reducible fraction; less than 1% occurs as oxidizable. Cadmium is detectable only in the oxidizable fraction; it is below the detection limit in the total digestion. Roughly half the lead is adsorbed, and an additional third is associated with the reducible fraction. After weathering, virtually all the cadmium and lead are found in the residual fraction; about 2% of the latter remains adsorbed. Approximately three-fourths of the zinc is also residual, but 20% remains associated with the reducible fraction (note added in proof - Ref. 8).

The greatest contrast between the polluted and unpolluted sediment metal distributions is found with cadmium. The oxidizable fraction of the former contains no detectable amount, over half is associated with the reducible fraction. This portion also accounts for 40% of the zinc; all but 1% of the remainder is divided evenly between the oxidizable and residual portions. Lead is undetectable in the oxidizable fraction; it is located primarily in the residual component with less than 1% adsorbed.

Seepage Studies. During the six-month period of natural weathering, seepage and rainfall were monitored for zinc, cadmium, and lead. The rainfall averaged 40 ppb, 8 ppb, and 20 ppb in the respective metals; the variation was generally random, with two exceptions, and all three metals followed similar patterns. Two marked increases in metal content were noted, one of roughly ten-fold in the sample taken over the Fourth of July and one of three-fold in the sample taken over Labor Day. As these samples were collected in a resort community, the increases probably result from the influx of land and marine vacationers.

Metals in the seepage from unpolluted sediment over the same period exhibit a gradual but definite increase (Figure 3). Zinc levels range from undetectable to 13 ppm, cadmium from 10 to 130 ppb, and lead from 10 to 90 ppb. Thus, rainfall appears to be a minor source of metal input to the seepage. Analysis of the sediment at this stage is ambiguous, showing a slight decrease in zinc content, but increases in both cadmium and lead. However,

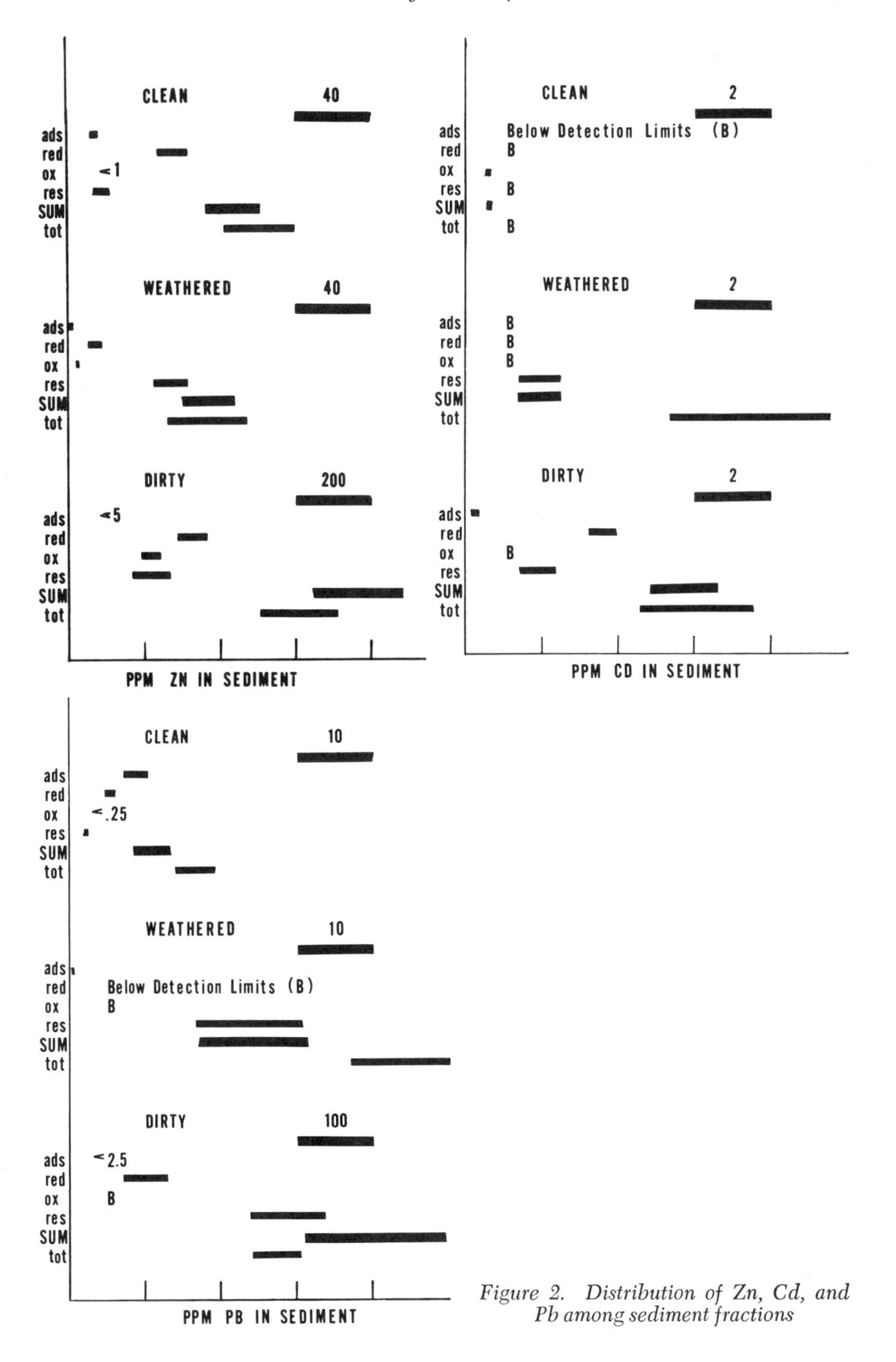


*Figure 2. Distribution of Zn, Cd, and Pb among sediment fractions*

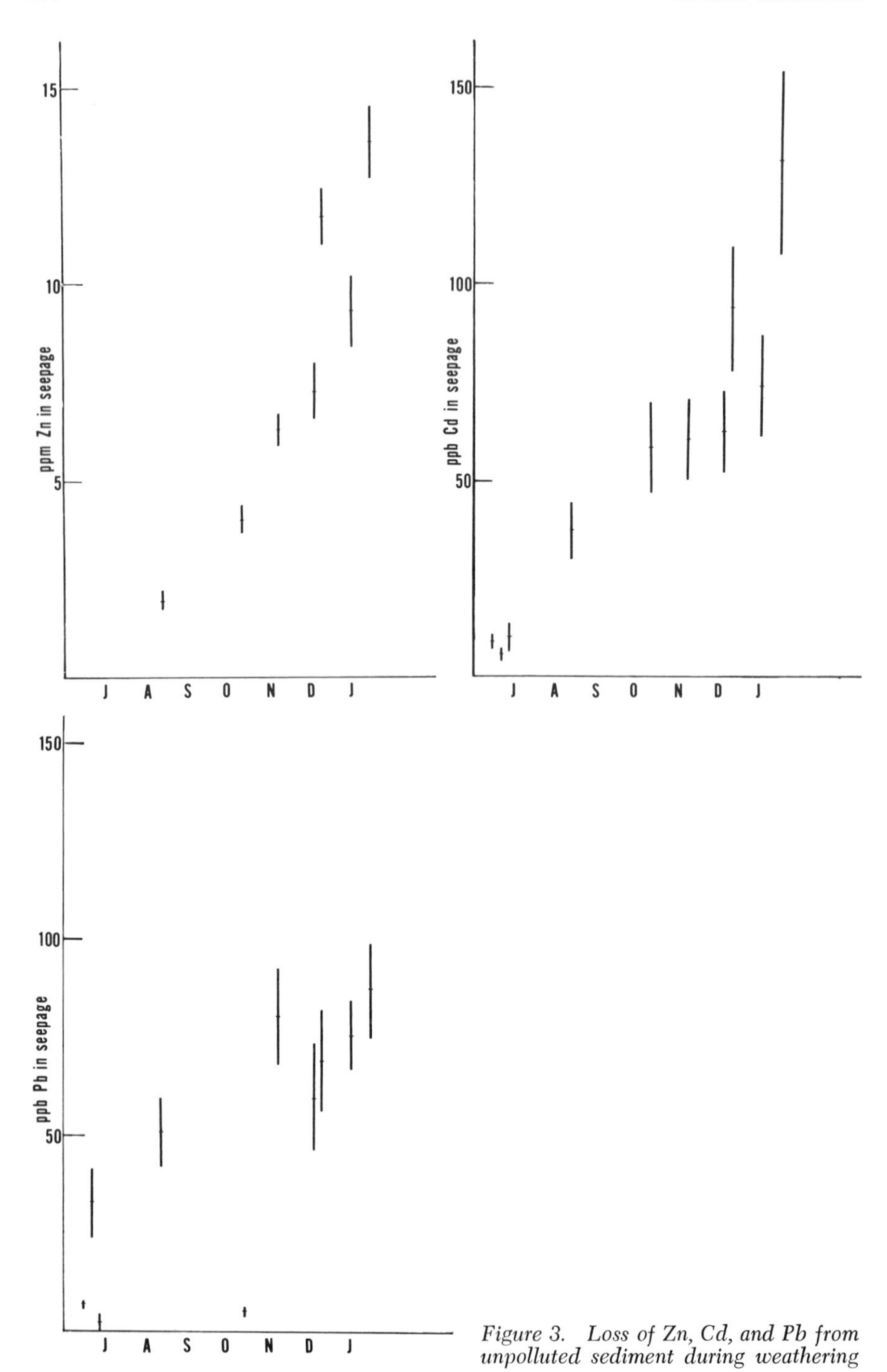


*Figure 3. Loss of Zn, Cd, and Pb from unpolluted sediment during weathering*

a distinct change in metal distribution is observed, as previously noted.

Copper. Under the experimental conditions reported here, the determination of copper was difficult. At analyte levels below 0.5 ppm Cu, the copper peak could not be effectively detected on the shoulder of the mercury oxidation wave. In order to exhibit well-defined peaks, copper levels above 1 ppm were necessary. Such levels are not desirable for simultaneous determination of zinc, due to the formation of a zinc-copper intermetallic which decreases the zinc signal and enhances the copper response (1).

## Conclusions

The three metals under consideration exhibit differing responses to short-term leachings. Zinc removal is not influenced by the presence of oxygen, but changes from fresh to saline water do appear to affect removal from the polluted sample. Cadmium responds to the presence or absence of oxygen in both sediments, but only under salt water. A similar result for lead is shown only by the polluted sample in fresh water. Correlations between metal distribution in the sediment fractions and metal removal have not become apparent.

Changes in sediment zinc content due to weathering and to short-term fresh water leaching are comparable, and the negative short-term result for cadmium may be paralleled by the slight increase in cadmium content of the weathered sediment, but no correlation is found for lead. However, the seepage shows that all three metals are, and continue to be, leached from the unpolluted sediment at appreciable levels.

As may be anticipated for natural systems, none of the relatively simple laboratory techniques used seem sufficient to reflect the heavy metal release which actually occurs as sediments are weathered. Total sediment dissolution, in addition to being time-consuming, provides no distinction between environmentally-active metals and those which may be classed as inert. Results of the various partial digestion procedures employed here are not easily correlated with metal removed. Simple short-term leaching is qualitative at best, and even here active elements may be undetected.

The key to this question may lie in specific portions of the gross sediment fractions, notably hydrous iron and manganese oxides in the reducible fraction and organics or sulfides in the oxidizable fraction (3). More selective techniques than those currently employed are called for. Natural weathering of polluted sediment samples is essential to determine the utility of short-term techniques, although it appears reasonable to expect higher metal levels than those found in the seepage from unpolluted sediments. Rates of such release may be expected to depend as well on the type of pollution; industrial, essentially metal-contaminated sediments may differ substantially from urban sources containing significant amounts of organic matter.

When several areas are being considered as sediment sources for creation of artificial habitats, a six-month period of weathering is impractical as a technique of sample preparation. Rapid, representative methods are essential, both for determination of sediment suitability and for estimation of the removal times involved. For example, a sediment with a small but constant metal loss may prove satisfactory in an open, easily flushed situation, while the construction of a fresh water pond may lead to the accumulation of potentially harmful metal levels.

This work has been confined to the effects of geochemical processes. In reality biologically induced transformations, such as soil oxidation adjacent to plant roots, or active biological transport may prove to be of equal or greater importance in heavy metal mobilization. While the construction of physically suitable wildlife habitats poses no great difficulties, the question of chemical suitability is much more subtle, complex, and vital.

Future work in this laboratory is directed toward reducing the time required to obtain an environmentally meaningful sample from raw sediments, and toward improving both the speed and sensitivity of the analytical technique, particularly with respect to copper. The first objective involves investigations of both selective sediment digestion and other parameters of short-term leaching, notably organic content. Currently we are investigating the applicability of thin-film mercury electrodes and suppression of intermetallic interferences following the gallium technique proposed by Copeland et al. (1).

Acknowledgements. The authors wish to acknowledge the financial support of this work by the Victoria Foundation, and express appreciation for the cooperation of the staff of the Wetlands Institute. Sampling assistance was generously provided by Fred Bopp of the University of Delaware and Mike Criss of the Wetlands Institute, and Joe Kelley of Lehigh University performed the grain size analysis. We are especially grateful to Mark Brindle of Lehigh for his invaluable aid during the early stages of the polarographic work.

## Abstract

Sediments taken from polluted and unpolluted areas were exposed to short-term leaching in the laboratory, and an unpolluted sample was exposed to natural weathering for six months. The sediments were subjected to fractional digestion. Zinc, cadmium, and lead levels in leachates, seepage, and sediment were determined by differential pulse anodic stripping voltammetry. The metals continue to be leached during natural weathering, a process not adequately reflected by short-term studies, and become associated primarily with the residual sediment fraction. Implications of these results must be considered in the use of dredge spoils for artificial habitat construction.

## Literature Cited


1. Copeland, T. R., R. A. Osteryoung, and R. K. Skogerbee, Anal. Chem. (1974) 46, 2093.

2. Florence, T. M., J. Electroanal. Chem. (1972) 35, 237.

3. Jenne, E. A. in "Trace Inorganics in Water," Advances in Chemistry Series No. 73, p. 337, American Chemical Society, Washington, D. C., 1968.

4. Mattinson, J. M., Anal. Chem. (1972) 44, 1715.

5. Presley, B. J., PhD Thesis (1969), UCLA, Los Angeles, California.

6. Presley, B. J., Y. Kolodny, A. Nessenbaum, and R. J. Kaplan, Geochim. Cosmochim. Acta (1972) 36, 1073.

7. Windom, H., J. Waterways, Harbors, Coastal Engineering Division, Amer. Soc. Civil Eng. (1972) 98, 475.

8. Note: Added in proof: The hydroxyl amine HCl technique is not sufficiently rigerous to dissolve free iron oxides and hydroxides; this may account for the increased amount of heavy metals found in the residual (HF soluble) fraction of the weathered sample (E.A. Jenne, personal communication).

# 22

# Trends in Waste Solid Disposal in U.S. Coastal Waters, 1968–1974

M. GRANT GROSS

Chesapeake Bay Institute, The Johns Hopkins University, Baltimore, Md. 21218

## Sources of Waste Solids

Large volumes of solids are transported to sea by barges or seagoing hopper dredges for disposal in coastal waters. These wastes (listed in approximate order of volumes dumped at sea) include sediment deposits dredged from waterways, solids in waste chemicals, sewage sludges, rubble from building construction and demolition, and fly ash (1,2). Floatable waste solids such as garbage, refuse, and rubbish as well as explosives are excluded. Although produced in large quantities in urban areas, these materials are not now dumped in coastal or estuarine waters, except from ships (3).

Riverborne sediments carried into harbors by estuarine circulation in the adjacent coastal ocean and by sands moving along beaches (4) must be dredged from harbors. Artificially deepened navigation channels and irregular shorelines formed by slips and basins used for shipping are effective sediment traps and must periodically be dredged to maintain water depths needed for navigation.

If clean gravels, sands, or coarse silts were the only materials dredged, there would be little problem in most coastal urban areas. Dredged materials have been used for landfill, especially hydraulic landfills, and for other construction purposes. In urban areas, however, dredged materials must commonly be handled as wastes because the natural riverborne sediments have been polluted by municipal and industrial wastes. Moreover, convenient landfill sites are usually unavailable at the time when wastes must be disposed of. In both cases, ocean disposal is an attractive option.

Municipal wastes, including sewage solids, are major contributors of materials dredged from navigable waterways. Untreated ("raw") sewage is often discharged to the waterways, which then provide primary treatment through gravitational settling of solids and secondary treatment through biological degradation of organic matter. The solids settle out in the waterways. Effluents from sewage treatment plants also carry suspended solids into adjacent waterways, and some of these solids doubtlessly add to the deposits that must be removed by dredging. In addition to the human wastes, many urban areas also discharge wastes and debris washed from streets by storm water discharges.

## Regulation of Ocean Disposal

In the United States, prior to 1973, regulation of dredging and disposal of dredged materials was primarily the responsibility of the U.S. Army Corps of Engineers. In most areas, these regulatory functions were conducted under the authority of the Refuse Acts (5), which prohibited waste disposal in navigable waters without prior permission of the local Corps of Engineers District. In New York, Baltimore, and Hampton Roads, Virginia, dredge spoil disposal was regulated under provision of the Supervisor of Harbor Act (6). On the Pacific coast, especially in California, regional and state agencies were also involved in regulation of waste discharges (1).

Since April 1973 waste disposal in U.S. coastal waters has been regulated through the issuance of permits by the Environmental Protection Agency (EPA), except for dredged "spoils," under provisions of PL 92-532, The Marine Protection, Research and Sanctuaries Act of 1972. Dredge spoil disposal is conducted under permits issued by the Corps of Engineers, subject to EPA review. This regulatory program is directed toward a "no harmful discharge" goal and covers all dumping seaward of the baseline from which the territorial sea is measured (7).

## Waste Disposal Sites

Most waste disposal sites are situated in open estuarine waters or on the continental shelf, relatively close to city or industry generating the wastes. Disposal sites are usually deeper than 20 meters in the United States except where the materials are used for beach replenishment or construction of artificial fishing reefs. (In the Gulf of Mexico some sites are located in waters as shallow as 4 meters.) Beneficial uses of waste solids appear to be the exception rather than the rule.

In order to avoid interference with navigation and to remove disposal sites as far as possible from beaches and other public areas, many are located more than 5 kilometers from the nearest land, often in international waters. Of thirty-one sites along the open Atlantic coast from Maine to Cape Hatteras, North Carolina, for which good location data were available in 1970, seventeen were more than 5 kilometers from the coast and eight were more than 10 kilometers from the coastline. A site used for toxic chemicals is approximately 190 kilometers from the entrance to New York harbor.

Estuarine areas have in the past received large volumes of

waste solids. Chesapeake Bay, Long Island Sound, and Puget Sound each have several waste disposal sites (Table 1). All were used primarily for disposal of locally dredged wastes except for Long Island Sound, which has received wastes from the New York Metropolitan Region as well as small volumes of waste chemicals dumped in the eastern end.

In general, more recently established waste disposal areas are farther offshore than older disposal areas and this trend continues. And use of estuarine disposal sites seems likely to diminish further, owing to local opposition from fishing, shellfishing, and conservation interests.

In the late 1960's, there were at least 140 waste disposal sites in U.S. estuarine and coastal ocean waters (Table 1). Along the U.S. Atlantic coast, there were fifty-nine active waste disposal sites (Figures 1 and 2). Puerto Rico had two active sites. At least twenty sites were actively used along the Gulf of Mexico coast, and at least sixty sites were in use along the U.S. Pacific coast, including Alaska. No data were available on waste disposal operations in the Hawaiian Islands or U.S. territories or possessions. Although this report does not include the Great Lakes, ninety-five waste disposal sites were reported to be active there in the late 1960's (8).

Despite the limitations of the data, it is apparent that the number of actively used disposal sites in U.S. waters has dropped from 140+ in the late 1960s to 110 in 1973 (7). Many of the inshore and estuarine sites have been abandoned or are scheduled to be moved because they are too close to shore or interfere with commercial fishing operations.

## Natural Sediment Sources

Except for the Gulf Coast, most continental shelf areas now receive little sediment (9) which might dilute or bury waste deposits. Therefore, it is essential to consider the tonnage of sediment brought by rivers to each of the coastal areas in comparison with the tonnage of wastes deposited there. This permits a crude assessment of potential environmental effects resulting from waste deposits remaining in contact with overlying waters or with bottom-dwelling organisms for long periods of time.

Waste discharges in the Canadian Maritime Provinces, the U.S. portion of the Gulf of Maine, Long Island Sound, and the mid-Atlantic coast areas greatly exceed the probable suspended sediment discharge by rivers in these areas (10). For example, the discharge of riverborne suspended sediment into Long Island Sound is probably about $10^5$ metric tons per year, whereas waste discharges prior to 1970, exceeded $10^6$ metric tons per year (11). Furthermore, the sediment comes primarily from the Connecticut River near the eastern end of the Sound. Waste disposal activities were concentrated in western Long Island Sound because the bulk of the wastes came from nearby densely populated urban areas.

Table I. Location and number of waste disposal sites in various coastal ocean areas with estimated tonnages of wastes dumped in each region, 1960s.

| Disposal Sites (no. of active sites) | Estimated Tonnage ($10^6$ tons/year) | |
|---|---|---|
| Atlantic Coast (61) | | |
| Gulf of Maine (13) | 1.4 | 14.3 |
| Mid-Atlantic coast (29)* | 7.8 | |
| South Atlantic (17) | 5.0 | |
| Puerto Rico (2) | 0.1 | |
| Gulf of Mexico (20+) | | |
| Eastern Gulf (9) | 3.6 | 29.3 |
| Western Gulf (5+) | 10.0 | |
| Mississippi River (6) | 15.7 | |
| Pacific Coast (59) | | |
| California (3) | 1.7 | 12.6 |
| Oregon-Washington (38) | 10.8 | |
| Total | | 56.2 |

* Includes Long Island Sound (12) and Chesapeake Bay (2)(after 11)

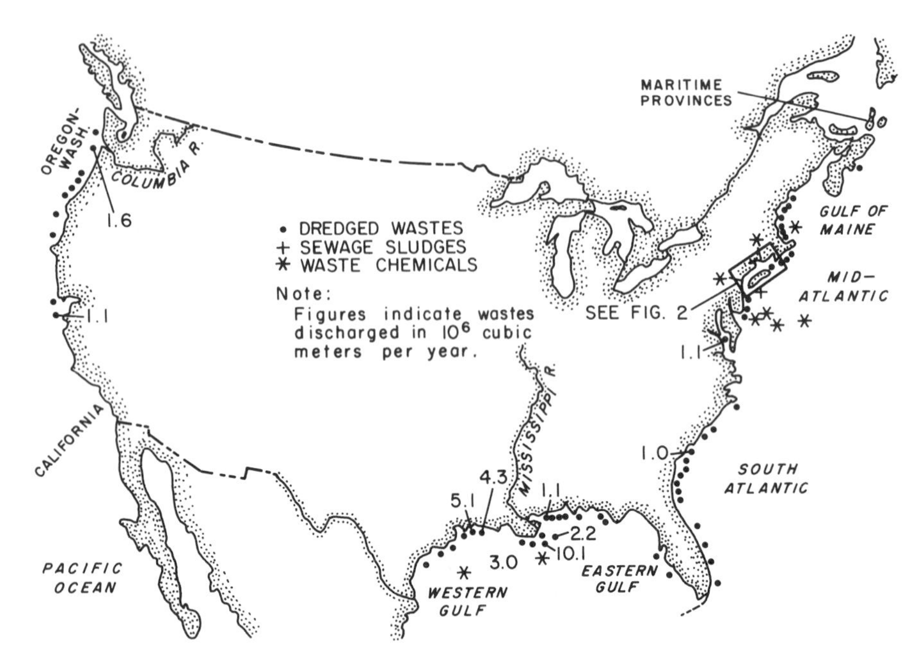


Figure 1. Location of waste disposal sites and volumes of dredged wastes ($10^6 m^3/year$) discharged in 1968

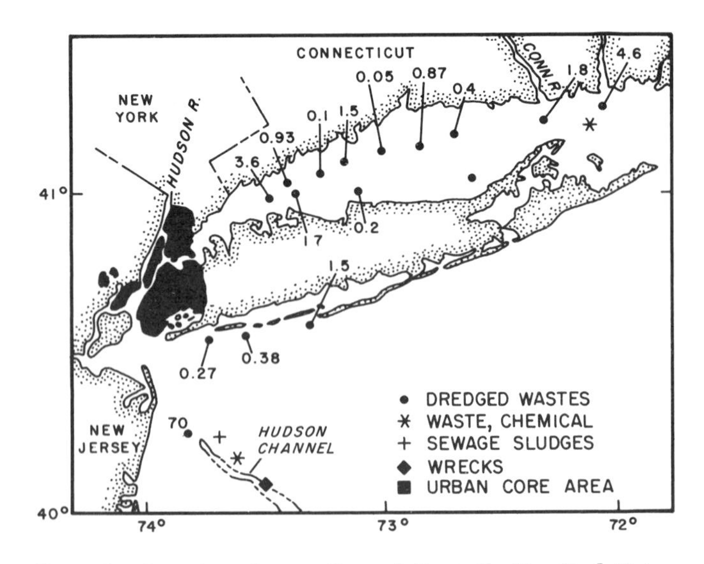


Figure 2. Locations of waste disposal sites in the New York Metropolitan Region in 1968 and volumes of dredged wastes ($10^5 m^3/year$) discharged in each

Near the mouth of the Mississippi River the river's suspended sediment load (12), about $3.1 \times 10^8$ metric tons per year, greatly exceeds the tonnage of wastes dumped there so that wastes are likely to be buried rather quickly. Similar situations probably prevail near the mouths of other rivers in the Gulf of Mexico, near the mouth of the Columbia River (13) and in the Strait of Georgia (14). Elsewhere on the continental shelf, wastes probably accumulate and remain at the water-sediment interface for long periods of time.

Based on these rather sketchy data, one can surmise that large volumes of wastes dumped at sea are likely to have the least effect in the Gulf of Mexico, and greatest effect on the Atlantic Coast.

## Impact of Dredged Waste Disposal

Waste solids pose immediate and potential long-term problems in the coastal ocean. The immediate effects arise in two ways--from dredging and from waste disposal operations. Wastes and sediments are disturbed during dredging causing increased turbidity and releasing nitrogen compounds, phosphates, and various reduced substances to the water. Increased oxygen demand arising from introduction of reduced substances (15) and decomposition of excessive phytoplankton growth (resulting from the phosphate and nitrate enrichment of the waters) deplete dissolved oxygen concentrations in near-bottom waters (16).

Potentially more troublesome is the long-term exposure of unburied waste deposits on the continental shelf or estuary bottom. Since most waste disposal sites are located in relatively shallow water, tidal currents and wave action may resuspend and move them outside the designated disposal area. Under certain conditions such as upwelling, movement of near-bottom waters is directed landward, toward the estuary or river mouth. Some resuspended wastes may move toward beaches or back into the harbor from which they were removed. There is no compelling evidence that large volumes of waste solids have moved from disposal sites to be deposited on beaches.

Studies of two waste disposal sites near New York City receiving dredged wastes and sewage sludges indicate that populations of bottom-dwelling organisms in both waste disposal areas were severely reduced over several tens of square kilometers (16). Factors thought to be responsible include low dissolved oxygen concentrations, in waters over the disposal site and presence of toxic compounds, or pathogenic organisms in the wastes. In addition, it seems likely that physical factors may also play a role. The bottom was originally hard sand or gravel, but the waste deposits formed a fine-grained soft bottom. Furthermore, rapid accumulation of wastes (ranging from 0.4 to 30 cm/year, assuming that the wastes are spread uniformly throughout the affected areas) may make it impossible for many bottom-dwelling organisms to live in waste disposal sites (17). And the changed bottom conditions inhibit (or prevent) repopulation of the disposal sites by organisms from nearby unaffected ocean areas.

Since 1968, data have become available on the environmental impact of waste solid disposal in coastal waters; much of the data come from studies of the New York Bight (18). Enrichments of several metals (including Cu, Pb, Ni, Zn, and Cr) were substantial, typically 10 to 25 times greater than in nearby uncontaminated sediment deposits (19). These metal-rich sediments were found nearly 40 km down the axis of Hudson Channel seaward of the designated disposal sites.

Studies of the effects of sewage sludge disposal in the Thames Estuary (20), the Firth of Clyde (21), and Liverpool Bay (22) have not found, in general, the obvious local effects on bottom-dwelling organisms that were observed in the New York Bight disposal sites. This may be due, in part, to better dispersal of the sludge in the coastal waters around Great Britain because of strong tidal currents there. In this case, the effects of waste disposal operations may eventually be felt over a much wider area.

## Trends in Ocean Disposal Operations

Despite changes in regulation of disposal operations, coastal ocean areas will probably receive even larger volumes of wastes over the next five to ten years. Coastal urban areas are short of land necessary for the present style of waste disposal operations (23); they find ocean disposal to be an attractive solution for existing and foreseeable waste disposal problems. At present there is no equal-cost alternative to ocean disposal for large-volume wastes. Hence, there will likely be increasing pressure for new types of wastes to be dumped in ocean waters, at least on an interim basis.

It is useful to consider changes in tonnages of wastes disposed at sea in 1968 and 1973 (Table 2). The amounts of industrial wastes, sewage sludges, and construction-demolition debris have increased, at compound rates of +2.9, +3.9 and +14.2% per year, respectively. Ocean discharges of garbage and refuse have been greatly reduced and disposal of explosives at sea has ceased.

Tonnages of dredged spoils disposed at sea have apparently decreased from an estimated 62 million tons in 1968 to 37 million tons in 1973. (Table 2.) This apparent decrease may well be the result of incomplete data.

It is also instructive to consider trends in ocean disposal since 1949 (Table 3). These data indicate a compound annual change of +7.7% between 1949 and 1968 in the amount of waste disposed at sea. (Data in Tables 2 and 3 are not directly comparable, although each indicate trends in the two periods involved.) It is too early to judge the impact of the new regulatory procedures, implemented in 1973, and the use of the ocean for disposal of industrial and municipal wastes. There has, however, been a de-

Table II. Ocean disposal of various waste solids, 1968[1] and 1973[2] (in millions of tons).

| Waste type | Atlantic Ocean | | Gulf of Mexico | | Pacific Ocean | | Total | | Change/year |
|---|---|---|---|---|---|---|---|---|---|
| | 1968 | 1973 | 1968 | 1973 | 1968 | 1973 | 1968 | 1973 | (%)[3] |
| Industrial wastes | 3.01 | 4.00 | 0.70 | 1.41 | 0.98 | 0 | 4.69 | 5.41 | + 2.9 |
| Sewage sludges | 4.48 | 5.43 | 0 | 0 | 0 | 0 | 4.48 | 5.43 | + 3.9 |
| Construction and demolition debris | 0.57 | 1.16 | 0 | 0 | 0 | 0 | 0.57 | 1.16 | +14.2 |
| Subtotal | 8.06 | 10.6 | 0.70 | 1.41 | 0.98 | 0 | 9.74 | 12.0 | + 4.2 |
| Dredged wastes[4] | 30.9 | 18.4 | 13.0 | 9.01 | 8.32 | 9.45 | 52.2 | 36.9 | - 6.9 |
| TOTAL | 39.0 | 29.0 | 13.7 | 10.4 | 9.30 | 9.45 | 61.9 | 48.9 | - 4.7 |

[1] 1968 data from (2).

[2] 1973 data from (7).

[3] Compound increase.

[4] To calculate tonnages of dredged wastes, a solid content (dry) of 0.9 metric ton per cubic meter of dredged material is assumed.

Table III. Ocean waste disposal, excluding dredged spoil, 1949-68, in millions of short tons (2).

| Coastal region | 1949-53 | 1954-58 | 1959-63 | 1964-68 |
|---|---|---|---|---|
| Atlantic | 8.0 | 16.0 | 27.3 | 31.1 |
| Gulf | 0.04 | 0.28 | 0.86 | 2.6 |
| Pacific | 0.49 | 0.85 | 0.94 | 3.4 |
| Total U.S. | 8.53 | 9.13 | 29.1 | 37.1 |
| Average/yr | 1.7 | 1.8 | 5.8 | 7.4 |

These data include the tonnage of liquids discharged as well as waste solids. They cannot be compared directly with Table 2.

crease in dredge spoil disposal in the past five years.

It seems most likely that there will be increasing volumes of dredged wastes in the future as new port facilities are prepared for deep-draft vessels, requiring 20 to 25 m of water. In the U.S., channel depths now rarely exceed 15 m; thus extensive dredging will likely be required to accommodate larger ships. The magnitude of this potential problem can be appreciated by considering that the dredging necessary for maintaining depths exceeding 20 m at the entrances to the Hook of Holland (Rotterdam-Europort, Netherlands) resulted in the annual dredged-waste discharges increasing from $5.1 \times 10^5 m^3$ in 1957 to $2.2 \times 10^7 m^3$ in 1968, a forty-fold increase in eleven years, an amount 3.9 times that dredged from New York Harbor in 1964-68 (18). These figures refer mainly to maintenance dredging. At present the amount of dredged material is substantially increased because of the deepening of the entrance channel for passage of deep-draft tankers. A comparable increase in dredged waste disposal was experienced at the port of Amsterdam during this same period (Director, Rijkswaterstaat, written communication, January 2, 1970). The total volume of material dredged from these ports was $1.3 \times 10^8 m^3$, about one-third the present annual sediment discharge of the Mississippi River (12). If comparable port developments occur in North America, the volume of dredged materials will probably increase sharply.

Because of the widespread use of ocean disposal sites, their potential long-term effects, and the present inadequate knowledge about the wastes and their effects on the ocean waters and marine organisms, marine scientists face a challenge to apply their research capabilities in this area. There is also a chance to contribute toward possible solutions for environmental problems as well as to learn more about coastal ocean processes that are otherwise difficult or impossible to study.

Far more studies with more careful experimental design will be needed to determine what is causing the environmental degradation observed in the New York Bight and to devise means of dealing with it. Present studies of such sites can best be described as inclusive rather than discriminating. Virtually every conceivable item is measured or examined. At best this is costly and is likely to be self-defeating.

The recent abandonment of disposal sites raises the issue of rehabilitation of waste deposits. On land, a garbage dump is closed by covering it with several feet of earth and landscaping. But for ocean areas we lack even this simple technology. It seems reasonable to expect that a complete ocean waste management program would include efforts to isolate waste deposits in abandoned sites from possible interactions with the overlying water column or marine life. This is probably a simple problem but one that merits attention.

Another trend to be considered is the probable increased volume of sewage sludges as the required level of treatment is raised. And the use of scrubbers to remove $SO_2$ from power plant exhausts will also generate large volumes of fine-grained watery wastes which are difficult to dispose of on land, especially in urban areas. For coastal cities and industries, ocean disposal is a convenient and, by conventional economic analysis, a relatively cheap disposal method. Such disposal operations (if permitted) pose new challenges to marine scientists to provide the data and insights on the coastal ocean and its organisms needed for proper planning and monitoring of the disposal operations.

## Abstract

Estuaries and coastal oceans receive large volumes of waste solids dredged from harbors or removed from sewage treatment plants or industrial plants and placed on the adjacent continental shelf, which often receives little sediment derived from other sources. Legislation, implemented in 1973, has caused a significant reduction in the number of sites used; from 140+ in the late 1960s to 110 in 1973. Dredged spoil disposal apparently decreased; from 56 million tons in 1968 to 37 million tons in 1973. Dumping of industrial wastes, sewage sludges and construction debris has increased at annual rates of +2.9%, +3.9% and +14.2%. The present trend is to operate fewer disposal sites located farther from shore.

## Literature Cited

1. Brown, R.P., and Smith, D.D. Marine Pollution Bulletin (1969) 18:12-16.
2. Smith, D.D. and Brown, R.P. "Ocean disposal of barge-delivered liquid and solid wastes from U.S. coastal cities," Environmental Protection Agency Publication SW-19c, Washington, D.C. (1971).
3. Cox, G.V. Environmental Science and Technology (1975) 9(2): 108-111.
4. Meade, R.H. Journal of Sedimentary Petrology (1969) 39(1): 222-234.
5. U.S. Congress, 1899. River and Harbor Act, approved March 3 1899 (30 Stat. 1152; 33 U.S.C. 407); River and Harbor Act, approved March 3, 1905 (33 Stat. 1147; 33 U.S.C. 419).
6. U.S. Congress, 1888. Supervisor of Harbor Act, approved June 29, 1888 (33 U.S.C. 441-451), amended July 12, 1952 and August 28, 1958 (Pub. L. 85-802, 72 Stat. 970).
7. EPA. Administration of the ocean dumping permit program: Second Annual Report, Environmental Protection Agency, Washington, D.C. (1974).
8. U.S. Army Corps of Engineers. "Dredging and Water Quality Problems in the Great Lakes," Buffalo, New York (1969).
9. Emery, K.O. Bulletin of the American Association of Petroleum Geologists (1968) 52:445-464.

10. Dole, R.B. and Stabler, H. U.S. Geological Survey of Water Supply Paper (1909) 234:85.
11. Gross, M.G. Waste-solid disposal in coastal waters of North America, p. 252-260. In W.H. Matthews and others (ed). Man's Impact on Terrestrial and Oceanic Ecosystems. M.I.T. Press, Cambridge, Mass. (1971).
12. Holeman, John N. Water Resources Research (1968) 4(4):737-747.
13. Gross, M.G., McManus, D.A. and Ling, Y-Y. Journal of Sedimentary Petrology (1967) 37(3):790-795.
14. Waldichuk, M.W. Fisheries Research Board of Canada, Progress Report (1953) No. 95:59-63.
15. Brown, C.L. and Clark, R. Water Resources Research (1968) 4(6):1381-1384.
16. Pearce, J.B. The effects of solid waste disposal on benthic communities in the New York Bight, p. 404-411. In M. Ruivo (ed). Marine pollution and sea life. Fishing News (Books) Ltd., London (1972).
17. Gross, M.G. Water Resources Research (1970) 6(3):927-931.
18. Gross, M.G. Geological Society America Bulletin (1972) 83:3163-3176.
19. Carmody, D.J., Pearce, J.B. and Yasso, W.E. Marine Pollution Bulletin (1973) 4(9):132-135.
20. Shelton, R.G.J. Marine Pollution Bulletin (1971) 2:24-27.
21. MacKay, D.W., Halcrow, W. and Thorton, I. Marine Pollution Bulletin (1972) 3(1):7-11.
22. Dept. of the Environment. "Out of Sight, Out of Mind: Report of a Working Party on the Disposal of Sludge in Liverpool Bay", Her Majesty's Stationery Office, London, 3 volumes (1972).
23. Regional Plan Association. Waste management, Regional Plan Association Bulletin, 107 (1968).

# 23

# Industrial Viewpoint on Ocean Disposal

LLOYD L. FALK

Engineering Service Division, Engineering Department, E. I. du Pont de Nemours & Co., Wilmington, Del. 19898

Advocates or opponents of specific waste disposal systems must recognize that systems finally adopted must adhere to natural laws. For example, the laws of thermodynamics and of the conservation of mass energy say: "The only thing you get for nothing - is nothing". Since these laws apply to environmental control problems, we see that some ideas which have formed the basis for Federal water pollution control laws are unattainable. The idea of total waste recycle or "zero" discharge has magical appeal. The concept, however, fails to consider that industrial societies draw upon reservoirs of nonrenewable resources which occur in the oceans, in the atmosphere, and in the ground.

If we use these resources, those laws of nature require those materials to be returned eventually to the oceans, to the air, or to the ground whence they came. We may change the nature of the materials and how the deposition is distributed, but we cannot alter the ultimate requirement that deposition take place. Thus, we see that the idea of total recycle of materials is a will-o'-the-wisp unless we ultimately include redeposition in the environment within our definition of total recycle. Regardless of how many times a material is repeatedly cycled through a box we might call "World Society", it must be returned to the earth from which we extracted it. Furthermore, while recycling can minimize a natural material drain, it is usually done at the cost of an energy drain.

Once we take something from nature and use it, we must eventually return it. The question then becomes: How do we return those materials we have extracted from the world around us in such a way as to be compatible with the environment we wish to maintain? If we approach this problem wisely, we will recognize the oceans can serve as a receiver of some excess by-products which have no economical means of recycle. Such excess by-products we call "wastes". The term "economical means" includes all techniques used for recycling while still maintaining economic health both within the society and in competition with other societies. We must also recognize that the act of recycling itself can have economic and environmental costs. For example, if we neutralize sulfuric acid with hydrated lime to form a potentially recyclable gypsum, we must expend energy to extract limestone from the earth, transport it, calcine it, transport it again, use it, and transport the gypsum. This requires energy, and the possibility of creating other by-products from fuel combustion. Thus, even recycling entails an environmental cost - nothing is for free.

The oceans, the ground, and the air all have assimilative capacities which our duty, as users of resources, requires us to determine and use wisely. In the case of waste disposal in the oceans, industrial societies need to regulate how they use that capacity. I want to mention here that when Congress was considering the passage of ocean dumping regulatory legislation in 1972, industry generally favored such regulations rather than see the practice abused.

Barging wastes to sea has distinctive advantages over other ways of recycling waste products back into the environment, products which often would ultimately reach the ocean anyway. A barge has a mobile discharge point. You have a wide selection of locations to disperse wastes. If wastes, even treated, are discharged to rivers, you cannot divert mouths of rivers. Sewer pipes discharging to the ocean can be moved only within narrow limits. This means that, by proper prior evaluation, barge wastes can be dispersed in an area with minimum, or even no, environmental impact.

The 1972 law Congress passed is known as the "Marine Protection, Research, and Sanctuaries Act". Essentially, that law prohibits transport of wastes for the purpose of dumping them into the ocean without an appropriate permit issued by the Environmental Protection Agency. Subsequent to the passage of the law, EPA adopted a complex set of regulations in 1973. These specify the kinds of permits that can be obtained, and establishes criteria for the kinds of wastes which may be discharged. EPA is presently in the process of revising its existing regulations and criteria. I expect Mr. T. A. Wastler, who follows later in this symposium, will discuss the law and regulations. It appears, however, that EPA may be moving in the direction of making the regulations more complex, and the obtaining of permits even more difficult than now.

But aside from the complexity of obtaining permits, and regardless of how the regulations are phrased, we in industry detect that, at least on an EPA regional basis, EPA is moving in the direction of eliminating ocean disposal of wastes transported to sea by vessels. This is being done in spite of the expressed purpose in the Federal law to regulate ocean disposal, not eliminate it. For example, even though a waste disposal operation may meet the current criteria for obtaining what is called a Special (and renewable) permit which might be issued for a period of three years, EPA regional offices so far have issued only Interim permits which can only last for a maximum of one year and are not renewable. A discharger is then required to reapply for a new permit and undergo all of the complex process of obtaining it. He must do this even though he has specifically indicated that a particular disposal operation will be phased out within two or three years.

We believe the regulations on ocean dumping should be simplified, not made more complex. For example, the drafts of the revised regulations which I have seen go into great detail as to how disposal operations are to be monitored and how studies of sites are to be conducted prior to the start of disposal operations. The requirements for such specific studies should be developed on a case-by-case basis depending upon the amounts, kinds, frequency of wastes, and where the disposal operation is to occur. Further, the EPA Region should have greater flexibility in deciding the nature of such studies.

As a result of the apparent trend toward elimination of ocean disposal, our society is being required to seek alternative ways of disposing of waste products. These alternatives may prove to be environmentally less sound overall than would be ocean disposal. We believe that any regulatory agency should not categorically foreclose any method of waste disposal. Rather, all alternatives, including ocean disposal, should be evaluated so that the least costly, considering both total environmental as well as economic needs, is selected. Instead of forcing an end to any particular waste disposal technology or technique, regulatory agencies need to educate the public that there are no categorically unusable techniques.

I would like to summarize as follows:

1. As long as we take nonrenewable resources from the ground, air, or waters of the earth, we must eventually return them. We cannot ignore the laws of thermodynamics and conservation of mass-energy.
2. The ocean can be a satisfactory place for nonrecyclable by-products or wastes. We need to assess its assimilative capacity to determine how to dispose of wastes commensurate with maintaining a satisfactory ocean environment. While we cannot have "zero" impact, we can minimize it.
3. Regulatory agencies should not categorically rule out any specific alternative for waste disposal, including ocean disposal, since that alternative may prove, in certain instances, to be the most environmentally acceptable overall. Other alternative disposal methods which appear "more acceptable" to the public may ultimately prove to be environmentally more costly when one considers the energy requirements as well as the disposal of residues from the waste disposal operations themselves. Ocean disposal, adequately controlled and monitored, can often provide the most economical, environmentally sound alternative available to us in managing waste products of industrial societies.

# 24

# Sludge Disposal and the Coastal Metropolis

NORMAN NASH

Department of Water Resources, City of New York,
Municipal Building, New York, N.Y. 10007

In August 1973 the U.S. EPA published a report on its first year of administration of the ocean dumping program. On the whole, this was an admirable work. However, early in the report it mentioned the deaths and illnesses attributed to mercury in Minamata Bay, Japan, went on to cite the fact that about 20 percent of the shellfish beds in the United States are closed because of polluted waters, and then said, "Finally, attention is drawn to the case of the New York Bight, New York City's 'dump'."

We are inured to such remarks, although we can't help wincing at each blow, but it is discouraging to hear it from the EPA, which should know better. The bight is not "New York City's dump." It also is Nassau County's dump, and Westchester County's, and especially New Jersey's. It is the dump for a multitude of industrial wastes, greater in volume than New York City's sludge, and for a veritable mountain of dredge spoil, nearly four times the volume of sludge discharged by the City.

In 1973 dredge spoil constituted 56 percent of the total volume of dumped wastes; municipal sludges from New York City, New Jersey and Nassau and Westchester Counties amounted to 26 percent, and the remainder, 18 percent, was industrial wastes, all from New Jersey (Figure 1). Dredge spoil has been described by the EPA as "34 percent polluted"; if sludge is considered to be 100 percent polluted, then dredge pollution. That is, by virtue of its large volume, inspite of its lesser percentage of pollution, the total impact of dredge spoil discharged at disposal sites cannot be ignored.

However, the volume of spoil is so huge, the necessity for dredging is usually so obvious, and alternatives to ocean disposal are so non-existent, that opposition to the practice never got off the ground. Region II of EPA is considering moving the dredge spoil site farther offshore, which is the only solution to the problem, if indeed it is a problem.

The opponents of ocean disposal concentrate their fire on the sewage sludge that is discharged at the site about 12 miles offshore. Yes, it is true that New York City is responsible for 58 percent of the total volume; New Jersey accounts for 34 percent, Nassau County 6 percent and Westchester County 2 percent. But the population of the City that is served by treatment plants, and thus by the disposal site, is just about 58 percent of the total served population of all areas. Therefore, it might be said that New York City discharges no more than its rightful share. Furthermore, our sludge is produced by secondary treatment, and so is much less dense and much more voluminous than the primary sludge discharged by Westchester County and by virtually all of the New Jersey sources. If those communities practiced secondary treatment, then the percentages would be reversed: New Jersey would be the major contributor, with about 60 percent, and New York City's share would be only about 35 percent. Better treatment of wastes, which is the heart of our business, produces more sludge; cities which practice secondary treatment should be commended for their committment to higher treatment, not scolded as ocean polluters.

The discharges of New York City and New Jersey cannot properly be compared without considering the large volume of industrial wastes, all from New Jersey, that is discharged at the dumping sites (Figure 2). The 281,600 cu ft/day of industrial wastes, plus New Jersey's municipal sludge volume of 143,800 cu ft/day, make New Jersey's share of the total volume 61 per cent New York City's share is reduced to 35 percent, and Nassau's and Westchester's to only 3 and 1 percent. If the primary vs. secondary treatment factor were introduced into this accounting, New York City's true share would be reduced even further, to only 24 percent, and New Jersey's would increase to 72 percent.

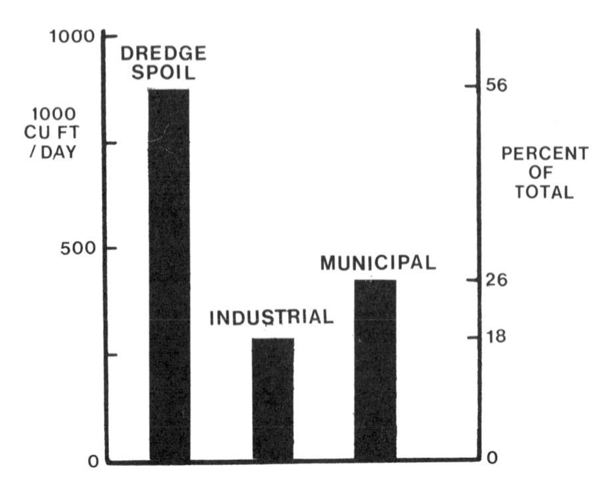


*Figure 1. Volumes of wastes discharged at New York Bight disposal sites, 1973*

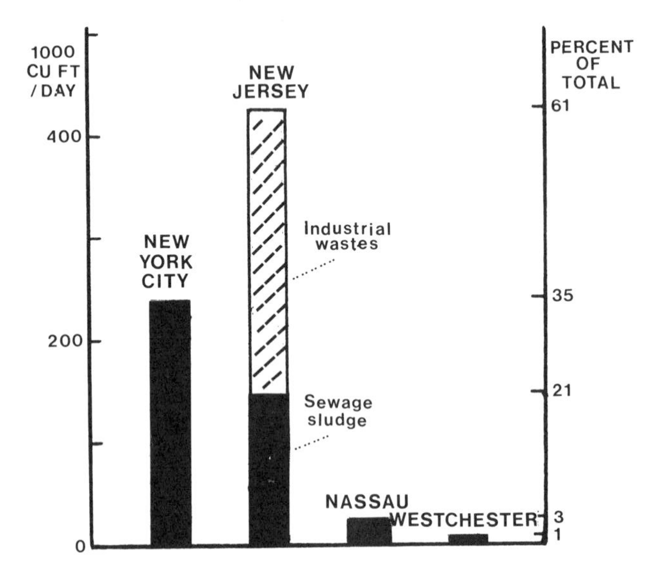


*Figure 2. Discharges from municipal and industrial sources, 1973*

So much for the relative volumes of the several sources; now for some of the constituents of the wastes. The finger has been pointed at heavy metals as grave dangers to marine and even to human life, and so in Figures 3 and 4 are presented the average daily additions of seven metals and oil at the three dump sites, including those from industrial wastes. This does not mean that we agree that metals are a paramount danger; mighty little real evidence has been presented to justify the noise and fear those claims have aroused. Furthermore, metals are ubiquitous. Substantial amounts exist even in purely residential wastes and are only partly removed in the sludge, and in storm water runoff where none is removed.

Cadmium, mercury and nickel have been grouped in Figure 3, and four other metals and oil in Figure 4, because the lbs/day discharged for those in each set are similar. It can be seen that for the first three metals, New Jersey is the principal contributor. The electroplating industry is the largest source of nickel; some idea of the number of such firms and/or their degree of attention to housekeeping possibly can be gleaned from the data: 125 lbs/day from New York City and 217 from New Jersey. (The data for Nassau and Westchester counties are not shown, except for Westchester in Figure 4, because their discharges are insignificant.)

New York City contributes more copper and lead than does New Jersey, but less chromium and zinc. Nearly 1.4 tons/day of oil are discharged, 58 percent from New York City, 41 percent from New Jersey, and 1 percent from Nassau and Westchester. Thus, about 80 barrels a day are discharged in sludge and industrial wastes.

So sin is not confined to one side of the Hudson River; but soon we'll all be in the same boat, the laggards with the leaders, all chained to 85 percent removal of BOD and year-round disinfection, regardless of whether we discharge to a potable stream or to a tidal estuary. The amount of sludge that will be produced by the present users of the site, including a large increase from New York City's new and upgraded plants, is an incredible 1-million cu ft/day. This

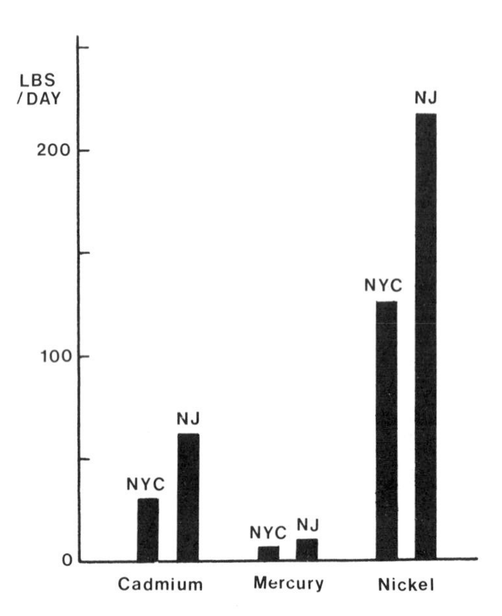


*Figure 3. Cadmium, mercury, and nickel from New York City and New Jersey sources, 1973*

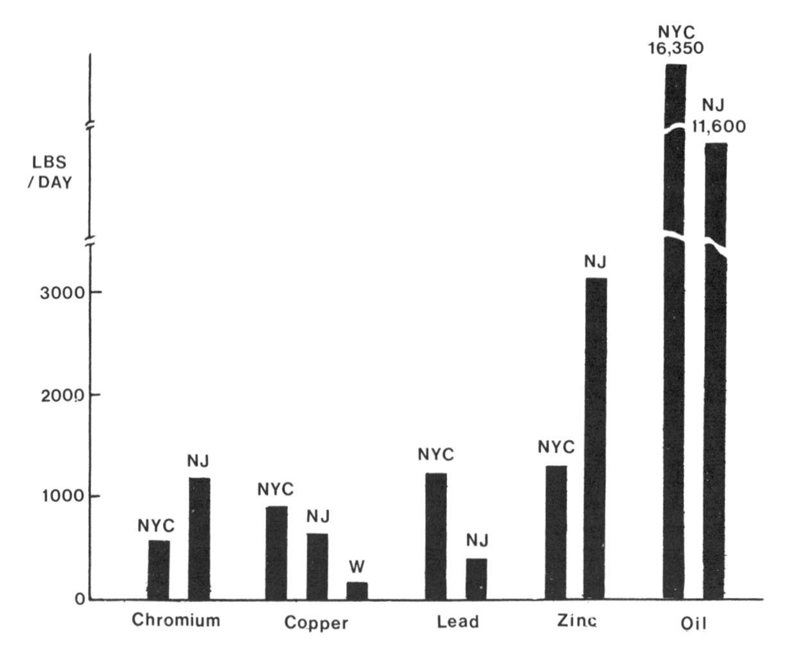


*Figure 4. Chromium, copper, lead, zinc, and oil from New York City and New Jersey sources, 1973*

would fill Yankee Stadium 10 feet deep, including the bull pen, to the last row of the lower box seats.

What is to be done with it? Very little can be placed on the land, and none in New York City, unless we are prepared to lagoon it in Central Park. It can be burned in sludge incinerators, but at a fearful capital cost and a great continuing cost in energy and dollars. Perhaps it can be consumed in incinerators designed to burn a mixture of refuse and sludge, and so solve two problems.

And there is the ocean, a resource which should be used with care. Of course, the ocean should not be the dumping ground for truly toxic wastes. It may be, too, that some limit should be placed on the volume of sludge disposed there. It may even be necessary to end sludge disposal altogether. But whatever decision is made must be based on scientific fact, not on emotion and semi-hysteria. The ocean is not a virgin to be protected against all contact with the coarseness of man; it is the once and future sink with an undoubted capacity to assimilate some volume of waste. It is folly to, without evidence, disregard that capacity and burden the land and the air with consequences possibly far more serious.

Let us divorce emotion and scare tactics from this matter. Well-meaning but ill-informed criticism must not be allowed to deter scientists and engineers from a disciplined resolution of the problem that impartially considers the total environment.

ABSTRACT

When both sludge and industrial wastes are considered, it is evident that New Jersey, not New York City, is the largest contributor to the New York bight disposal sites. New Jersey wastes contain more cadmium mercury, nickel, chromium and zinc than does New York City sludge, but New York discharges more copper, lead and oil.

Whatever decisions are made on the future of ocean disposal must be based on scientific fact, not on emotion. Without evidence, the ocean should not be ruled out as a disposal site, and the entire burden transferred to the land and the air.

# 25

# Chemical Needs for the Regulation of Ocean Disposal

T. A. WASTLER and C. K. OFFUTT

Oil and Special Materials Control Division, Environmental Protection Agency, Washington, D.C. 20460

Shortly after the passage of the Federal Water Pollution Control Act Amendments of 1972 (FWPCA), came the passage of the Marine Protection, Research, and Sanctuaries Act of 1972, often known as the Ocean Dumping Act, which prohibits the disposal in ocean waters of certain materials and which provides a permit system to regulate all other forms of ocean disposal. Also included in the Act are provisions for research to determine effects of disposal on the marine environment and authorization to establish specified areas as marine sanctuaries for recreational, conservational, and ecological purposes. The legislation was initiated by the Administration following the release by the Council on Environmental Quality in 1970 of the report "Ocean Dumping -- A National Policy" which surveyed current and past ocean disposal activities. Both the Senate and House versions of this Bill reflected the concern that pollutants, which were previously discharged into the nation's waters or air and are now restricted by the FWPCA and the Clean Air Act, not cause unreasonable degradation of the marine environment.

The Act prohibits, except by permit, disposal in any ocean waters of materials which have been transported from U.S. ports or from any port in U.S. flag vessels. It also prohibits disposal without a permit by any ship in U.S. territorial waters and contiguous zone. No permit may be issued for disposal of radiological, biological, and chemical warfare agents and high-level radioactive wastes.

The Administrator of EPA is authorized to establish regulations under which permits will be issued and criteria by which to evaluate permit applications considering the need for the proposed dumping, the effects of dumping on the marine environment, alternative methods of disposal, and alternate uses of the oceans. EPA is also authorized to designate disposal sites. These regulations have been issued and are contained in Title 40 of the Code of Federal Regulations at Sections 220-227. The regulations establish categories of permits which may be issued and also designate disposal sites in which dumping may occur.

Concurrent with the development of domestic legislation to strictly regulate ocean disposal, the International Convention on the Prevention of Marine Pollution by Dumping of Wastes and Other Matter was negotiated in 1972 at a conference attended by 92 nations. In February, 1973, the President submitted the Convention to the Senate for ratification, which was accomplished in August, 1973. The Convention will become effective after ratification by 15 nations, at which time the ocean dumping law becomes the enabling domestic legislation for the treaty. At the present time nine nations have ratified the Convention, and it is anticipated that it will come into force by the end of 1975.

EPA initiated the permit system by developing regulations for issuing permits for ocean disposal, criteria consistent with both the Act and the Convention under which to evaluate wastes, and enforcement programs to insure compliance. On April 23, 1973, the program became operational and unpermitted ocean disposal was no longer allowed. By July, 1973, 47 permits had been issued under interim regulations. Final regulations were promulgated on October 15, 1973, and to date 106 permits including renewals and reapplications have been issued under the final regulations for strictly regulated ocean disposal of wastes.

EPA is currently revising the ocean dumping regulations based on our operating experience and, in compliance with the National Environmental Policy Act (NEPA), is beginning studies for the preparation of environmental impact statements on the disposal sites.

The criteria for evaluating each permit application are in Section 227 of Title 40 of the Code of Federal Regulations. It is useful to understand that the the criteria are based on specific requirements of the International Ocean Dumping convention as well as the statutory requirements in the Ocean Dumping Act. In addition to the prohibitions on dumping biological, chemical, and radiological warfare agents and high-level radioactive wastes, the convention prohibits the dumping of wastes containing mercury, cadmium, organohalogens, persistent plastics, and petroleum oils, as other than trace contaminants. The Convention, however, does not define "trace contaminants"; and, therefore, each country must make its own interpretation of what a "trace contaminant" is, and what levels can be permitted in wastes to be ocean dumped. The EPA criteria reflect these prohibitions and go one step further by also prohibiting the disposal of materials insufficiently described to permit evaluation of their impact on the marine environment, which is one of the reasons why chemical analysis of the wastes is so important.

A technical evaluation must be made on each application, not only as to the compliance with these criteria, but also as to the need for the dumping, the availability of alternatives, and feasibility of treatment.

The basic decision to issue a permit for ocean dumping depends on the determination by the Administrator that the "dumping will not unreasonably degrade or endanger human health, welfare, or amenities, or the marine environment, ecological systems, or economic potentialities." Such decisions are difficult to make in a regulatory program without adequate criteria, and without strong technical capability for evaluating permits and for understanding the effects of pollutants on the marine environment. The technical needs for the regulation of ocean disposal involve the ability:

(1) to develop criteria which reflect accurately the impact of waste materials on the marine environment;

(2) to characterize fully waste materials proposed for ocean disposal;

(3) to accurately assess the existing conditions at a disposal site; and

(4) to accurately detect any effects of the dumping activities on the disposal site.

It must be understood clearly that what we are really concerned about in terms of the impact of pollutants on the marine environment is their ultimate impact on the living organisms of the ocean and on their interrelationships. Recognition of this basic concern frequently tends to put the chemical aspects of marine pollution control purely in the position of a technical support requirement for the larger ecological concerns rather than as the basic technique and approach through which situations of potential ecological concern can be identified and mechanisms for their correction developed.

In discussing the role of the chemist and of chemistry in securing data for supporting regulatory programs for the marine environment, we will emphasize not so much specific problems that presently exist, but rather the types of approaches we feel would be most rewarding in assessing the impact of pollutants on marine ecosystems. These can be grouped roughly into three general categories of investigation, which we would like to characterize briefly, and then relate to each of the regulatory needs already mentioned.

First, there is sheer analytical capability. This may appear to be perfectly straightforward, but consider that in the particular regulatory situation that concerns us, we are interested first in detecting a chemical constituent (an element, ion, or specific compound) in "trace" quantities in a waste material containing many similar chemical species;

then, after it is dumped, we want to be able to detect it at several orders of magnitude smaller concentrations - not in a pure solution in distilled water but in a three percent salt solution of variable composition--and with sufficient accuracy that we can determine if there is any progressive buildup in concentration attributable to the discharge of the constituent from the waste in which it was originally identified as a "trace" contaminant. We would, of course, like to be able to perform the necessary analyses on a routine basis with minimally trained technicians using fool-proof methods of analysis which require inexpensive equipment.

However, it is not enough to simply be able to measure concentrations of specific constituents under highly variable conditions. We also need to know the kinetics of their transformations in the marine environment. This is the second of the major categories we mentioned, but it leads indefinably into the third, which is the consideration of synergism and anatagonism among various waste constituents when a waste material of one composition is dumped suddenly and without preparation into a medium of somewhat different characteristics.

The point we wish to bring to your attention is simple to state, but is extremely complex when applied to the real world. To use a simple example, if you know what the kinetics of the methylation of mercury are in a pure solution, can you extrapolate this accurately to the methylation of mercury in sewage sludge or in the marine environment? In a regulatory program, we are mildly interested in the "pure solution" case, but we are vitally interested in the latter two cases and their relationship. Therefore, kinetics, synergism, anatagonism, and what happens in a three percent salt solution to affect any or all of these types of reactions of a waste constituent are the basic information upon which an environmentally responsible regulatory program must rest, and it is in this context in which we would like to discuss what we feel is needed in each of the four types of technical requirements outlined earlier.

## Criteria Development

There are certain aspects of the ocean disposal criteria which are now, and will probably remain, a matter of judgment on the part of EPA as to what is acceptable and what is not. These include the amounts of inert materials and the amounts of acidic, basic, or oxygen - consuming materials which can be dumped without unreasonable degradation in any particular case. More specific than these are the overall toxicity criteria determined by a bioassay according to specified approved procedures which sets limits on how toxic wastes can be disposed of in the marine environment. Beyond this are definite limits set on the amounts of certain specific waste constituents prohibited as other than "trace contaminants."

The problems associated with setting rational limits on such constituents may be illustrated by the problems surrounding the setting of reasonable limits on the discharge of mercury to the marine environment. The existing criteria are based on allowing, within the mixing zone, an increase of 50 percent above average ambient levels for mercury in the water column or in the sediments. Recent information sets an acute toxicity level for mercury in seawater of 50 ppb, and a chronic level of 0.10 ppb, which is in the range of normal ambient levels for mercury in seawater.

There are several problems that appear in relating the known toxic values for mercury in the environment to those which are permissible in wastes to be dumped, and we would like to relate these to the concepts of kinetics, synergism, and antagoism discussed earlier. Mercury is generally reported as total mercury, whereas the biologically active species is organic mercury. Thus, we must first determine how much of the total mercury becomes biologically active and at what rate; second, how is the rate of conversion to organic mercury related to the amount of sewage sludge or other waste present and its organic content; and, third, if one puts this complex mixture into a three percent salt solution, how does this affect the conversion of inorganic mercury to the biologically active methylated mercury? To answer such questions requires an integrated, cohesive approach toward determining the kinetics of Mercury reactions in seawater and the synergistic or antagonistic effects of mercury in the environment, and the development of final criteria for mercury discharges must depend upon the chemical research to identify these relationships.

## Waste Characterization

Complete analysis of waste materials is essential to identify the constituents, their nature and their concentration. Reference can be made to data available in the literature on toxicological effects, and a prediction can be made on the behavior of the constituents in the marine environment - whether a material will be completely soluble, concentrate in a surface slick, or adsorb onto bottom sediments.

From a regulatory standpoint, we have reasonably good capability in being able to determine the constituents of waste materials themselves. The thirteenth edition of Standard Methods for the Examination of Water and Wastewater, prepared and published jointly by the American Public Health Association (APHA), American Water Works Association, and Water Pollution Control Federation, is currently available and is revised and updated periodically. Another common reference is the Annual Book of ASTM Standards and specifically Part 23 for Water.

Section 304(g) of the Federal Water Pollution Control Act Amendments of 1972 requires that EPA promulgate guidelines establishing test procedures for the analysis of pollutants. Such test procedures are to be used by permit applicants to demonstrate that effluent discharges meet applicable pollutant discharge limitations, and by the States and other enforcement activities in monitoring of effluents to verify effectiveness of pollution control measures.

In response to that legislative requirement EPA published guidelines for test procedures in the Federal Register of October 16, 1973 (40 CFR 136). As part of that effort EPA has compiled in the recent edition of "Methods for Chemical Analysis of Water and Wastes" (1974) chemical analytical procedures used by its laboratories for the examination of ground and surface waters, domestic and industrial waste effluents, and treatment process samples. Except where noted the methods are applicable to both water and wastewaters, and both fresh and saline water samples. Although other test procedures may be used, as provided in the October 16, 1973, Federal Register issue, the methods described in the manual will be used by EPA in determining compliance with applicable water and effluent standards established by the Agency. Most of the EPA methods are direct modifications or clarifications of the APHA or ASTM methods, some of which have been adapted to instrumental methods in preferance to manual procedures because of the improved speed, accuracy, and precision.

In developing test procedures for the implementation of the Ocean Disposal Permit Program. The methods that were developed in response to Section 304(g) of the FWPCA, were adopted for analyzing wastes proposed for ocean disposal.

## Assessment of Disposal Site Conditions

In order to evaluate impact of dumping on the marine environment, an accurate representation of the existing conditions is necessary. These disposal site evaluations need to include examination of the water column and benthic regions with respect to water quality, biota, and sediments. But there are a number of complicating factors. Any sampling is occurring in a dynamic system. Even the sampling techniques themselves are not standardized. The ambient concentration of many constituents of interest are on the range of parts per million and parts per billion, and although preconcentration techniques are available, they do introduce chance of error. The very nature of the seawater causes analytical problems with saline interferences.

The objective in making baseline assessments is not only to determine the ambient concentrations of particular constituents, but also to understand the variation of levels in time and space, to provide information statistical variability and the range of values.

Here, again, we must face the problems of kinetics, synergism, and antagonism, because in the initial baseline assessment of a particular disposal site, attention must be directed toward the rates at which certain reactions occur, and the factors which tend to stimulate or suppress certain reactions. It is only through a thorough understanding of the processes which are most significant at a particular location that the fundamental rate processes which occur at a disposal site can be understood to the extent that effective control can be exerted over the site to reduce adverse effects and to enhance positive effects.

The major goal of any environmental regulatory program is to find out what forms of waste disposal have unacceptable adverse effects on the environment and to eliminate, or at least reduce the unacceptable adverse impact. In the marine environment this is very difficult to achieve for a number of reasons. First, the existing baseline data are poor simply because no agency has ever had the mandate to conduct the extensive surveys necessary to establish baselines adequate for regulatory purposes. Second, the major portion of the existing chemical research efforts during the past has been directed toward how to improve our analytical capability, which is of course an essential precursor to finding out what analytical results mean in terms of environmental protection, but which does not contribute directly toward an understanding of the effects of specific pollutants on the marine environment.

Third, the volumes of water involved in ocean disposal are so great that many very critical pollutants, such as trace metals and organohalogens become diluted very rapidly, so that the measurements that are needed to detect pollutants and their impacts require techniques that are so sensitive that contamination of samples by the sampling equipment itself becomes a critical problem in some cases.

Fourth, many ocean sediments contain potentially toxic trace metals, but in forms which are not available to the ecosystem under normal ambient conditions in the oceans. Certain materials, even in small quantities, might upset this balance and cause the release into the seawater of toxic materials from natural sediments. Conversely, toxic materials which might be a problem in a fresh water ecosystem could become non-toxic in seawater or be adsorbed onto or absorbed into some natural sediments and be permanently removed from the ecosystem.

This is merely a listing of some of the types of chemical

problems we face in implementing a regulatory program to protect the marine environment. It is not intended to be exhaustive, but just to indicate the scope of the chemical problems which need to be attacked and solved so that we can deal with pollution problems in the marine environment with the same degree of certainity that we have in the fresh water environment.

Problems associated with marine pollution have been highlighted by the national concern about the effect of ocean dumping of toxic materials on the marine environment which culminated in the passage of the Ocean Dumping Act. It is well to remember, however, that the vast majority of pollutants reaching the ocean enter the marine environment through ocean outfalls or by way of rivers and estuaries, and that the needs and problems outlined here apply equally well to pollutants from these sources.

The problem of ocean pollution is not a temporary one. In the United States the rate of both population and industrial growth is greater in the coastal area than in the interior; in addition, we are entering an era of greater exploitation of offshore marine resources and the development of deepwater ports near shore. These activities and this growth will undoubtedly result in greater direct stress on the marine environment and greater pressure for increased waste disposal into ocean waters.

It is possible that the ocean may be the most environmentally acceptable place to put some types of wastes after they have been adequately treated. Before we can accept this, however, we must know much more about the marine environment and how pollutants behave in it.

I hope that our comments here have served to focus your attention on what some of that "much more" might be as far as chemistry is concerned.

# 26

# Pollutant Inputs and Distributions off Southern California

D. R. YOUNG, D. J. McDERMOTT, T. C. HEESEN, and T. K. JAN

Southern California Coastal Water Research Project, El Segundo, Calif. 90245

The coastal plain of southern California is inhabited by approximately 11 million persons, about 5 percent of the Nation's population. A wide variety of activities in this region generate waste products that may contribute pollutants to the adjacent marine environment, which is known as the Southern California Bight.

Most of the freshwater that enters the Bight is municipal wastewater discharged from four submarine outfall systems lying off the cities of Los Angeles, Santa Ana, and San Diego (Figure 1). These systems carry approximately 95 percent of the more than 1 billion gallons of wastewater released daily off southern California. In addition, surface runoff contributes, on the average, the equivalent of 400 million gallons per day (mgd) of freshwater through 15 major channels distributed fairly evenly along the coast. However, about 80 percent of this flow results from a few storms, which generally occur during the fall and winter seasons.

Another potential source of pollution is industrial waste discharged directly into the marine environment. A great number of these discharges enter the Los Angeles/Long Beach Harbor in San Pedro Bay, the largest harbor in the Bight This harbor also receives the greatest variety of industrial wastes.

The 14 major harbors or marinas along the southern California coast provide anchorage for approximately 40,000 recreational craft, in addition to numerous commercial and naval vessels. The relatively large quantities of antifouling and other bottom paints applied annually at these anchorages represent another potential source of marine pollutants, particularly trace metals and synthetic organic compounds.

A fifth route by which man-made contaminants may

reach the coastal environment is aerial precipitation. In view of the severe air pollution problems in southern California, one would expect aerial fallout to be a significant factor in marine pollution in the area. At present, we have relatively little information on the precipitation of trace elements; however, we recently established the importance of this input route for chlorinated hydrocarbons such as DDT and polychlorinated biphenyls (PCB).

The Bight is roughly defined on the west by the eastern edge of the California Current, which generally carries large quantities of trace materials (1). Thus, it is instructive, for comparative purposes, to estimate the quantity of a given material that fluxes through the surface layers of the Bight annually via ocean current advection.

In the past 4 years, the Coastal Water Research Project has investigated the inputs and distributions of more than a dozen potentially harmful trace elements (1, 2, 3, 4, 5). Here we summarize our results on one of these elements, mercury, as an example of our findings to date.

During Water Year 1971-72, we sampled storm runoff from four major channels in the urban and agricultural regions of the southern California coastal plain and analyzed the samples for trace metals and chlorinated hydrocarbons. We often observed a correlation between flow rate and concentration of suspended sediments and trace constituents (Figure 2). During 1972-73, we repeated our survey of chlorinated hydrocarbons in Los Angeles River storm runoff and obtained flow-weighted mean concentrations very similar to those of the previous year. Therefore, we have extrapolated our data to unsampled channels and time periods to obtain Bight-wide emission estimates for Water Year 1972-73. The flow-weighted mean concentrations of total mercury found in storm runoff from the four channels ranged from 400 to 1030 ng/L.

To obtain an initial indication of the importance of direct industrial waste discharges, during 1973, we sampled effluents from a number of different kinds of industries located in Los Angeles Harbor. Our results for total mercury are summarized in Table 1. The highest concentration we observed (240 ng/L) occurred in wastewater from a fish cannery; however, this value was only eight times the average levels (30 ng/L) we have measured in seawater collected off southern California. Most of the concentrations observed were lower than the cannery waste value and do not suggest significant mercury contamination via this route.

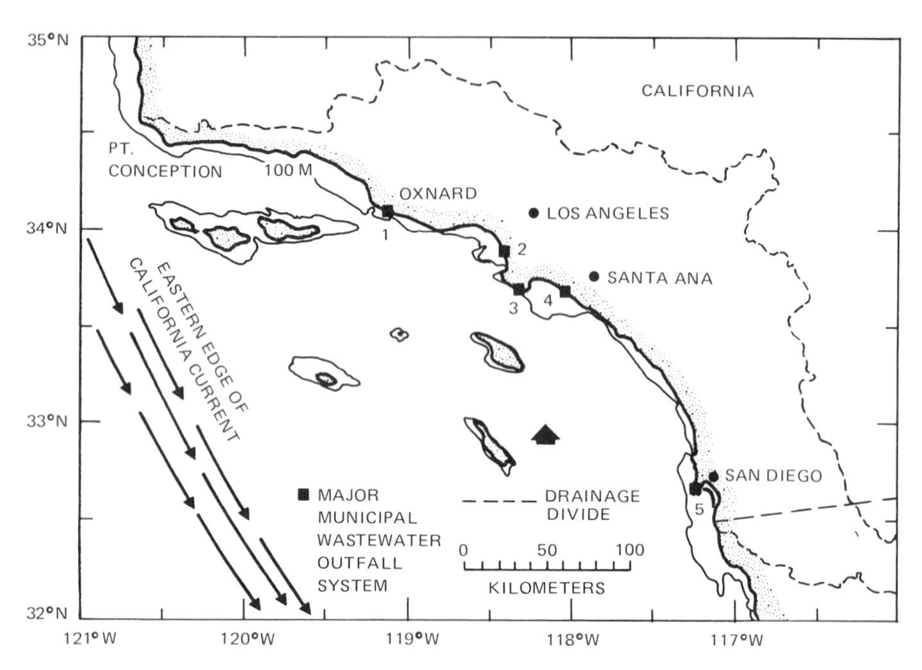


*Figure 1. The Southern California Bight. Outfall systems are: (1) Oxnard City, (2) Hyperion, Los Angeles City, (3) Whites Point, Los Angeles County, (4) Orange County, and (5) San Diego City.*

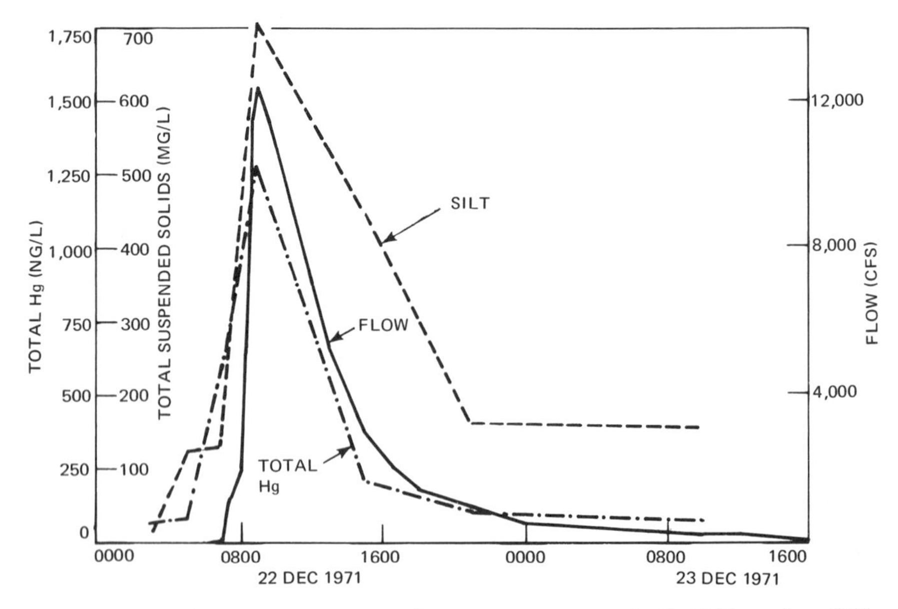


*Figure 2. Total mercury in Los Angeles River storm runoff, 22–23 December 1971*

During 1972-73, we obtained several 1-week composites of final effluent from the five major municipal wastewater treatment plants along the coast of the Bight. Several grab samples per day were collected and frozen. At the end of the week, replicate composites were obtained and filtered, and the filtrates and residues were analyzed for total mercury. Average results are listed in Table 2. These data indicate that the municipal effluents contain concentrations of mercury an order-of-magnitude greater than those found in direct industrial discharges to Los Angeles Harbor. In addition, the wastewater particulates have relatively high mercury concentrations: Typical levels are at least 100 times the average concentration (0.04 mg/dry kg) we have measured in uncontaminated coastal marine sediments.

In Table 3, we have summarized our values for annual mass emission rates of total mercury to the Bight via the three types of wastewater discussed above. The table also contains an estimate of the amount of mercury flowing through the Bight via ocean current advection, based on an average seawater concentration of 30 ng/L measured during several surveys sponsored by the Project, and on oceanographic assumptions outlined by Young (2). These results indicate that municipal wastewaters are the dominant fluvial source of mercury to the Bight.

The input of mercury via municipal wastewaters would be undetectable if it were well mixed with the advecting seawater. However, this is not always the case, as is illustrated by the data of Figure 3. This figure summarizes the results of a 1973 survey of mercury in surface sediments around the JWPCP[1] submarine outfall system off Whites Point on the Palos Verdes Peninsula. As with effluent particulate values listed in Table 2, contamination levels in the sediments are two orders of magnitude above concentrations in coastal control areas. Vertical profiles of mercury in box cores collected at Stations B18 through B21 are illustrated in Figure 4. These results indicate that the influence of the outfalls extends to approximately 30 cm below the sediment surface at the discharge depth of about 60 m. On the basis of numerous vertical profiles obtained at the monitoring stations shown in Figure 3, we estimate that roughly 4 metric tons of anthropogenic mercury are contained in these sediments,

1. Joint Water Pollution Control Plant of the County Sanitation Districts of Los Angeles County.

Table 1. Concentrations of total mercury in industrial effluents discharged directly to the Los Angeles Harbor, 1973.

| Type of Discharge | Flow (mgd) | Mercury (ng/L) |
|---|---|---|
| Fish Cannery | | |
| Waste No. 1 | 5.55 | 240 |
| Waste No. 2 | 3.20 | 140 |
| Retort Discharge | 0.12 | 81 |
| Condensor Water | 1.94 | 120 |
| Shipyard | | |
| Cooling Water No. 1 | 0.04 | 84 |
| Cooling Water No. 2 | 0.43 | 86 |
| Oil Tanker Cleandown | 0.25 | 120 |
| Ship Ballast No. 1 | 0.04 | 95 |
| Ship Ballast No. 2 | 0.29 | 150 |
| Oil Refinery | | |
| Cooling Water | 0.02 | 48 |
| Power Plant | | |
| Cooling Water | 257 | 38 |
| Chemical Plant | | |
| Combined Processes | 5.51 | 79 |
| Average Seawater | | 30 |

Table 2. Concentrations of total mercury in municipal wastewater discharged off southern California, 1972-73.

| Discharger | Flow (mgd) | Wet (mg/L) | Dry (mg/dry kg) |
|---|---|---|---|
| Oxnard City | 10 | 0.0005 | 3.3 |
| Los Angeles City | | | |
| Effluent | 330 | 0.0005 | 5.2 |
| Sludge | 5 | 0.05 | 8.4 |
| Los Angeles County | 360 | 0.001 | 4.0 |
| Orange County | 150 | 0.001 | 6.7 |
| San Diego City | 100 | 0.001 | 6.0 |

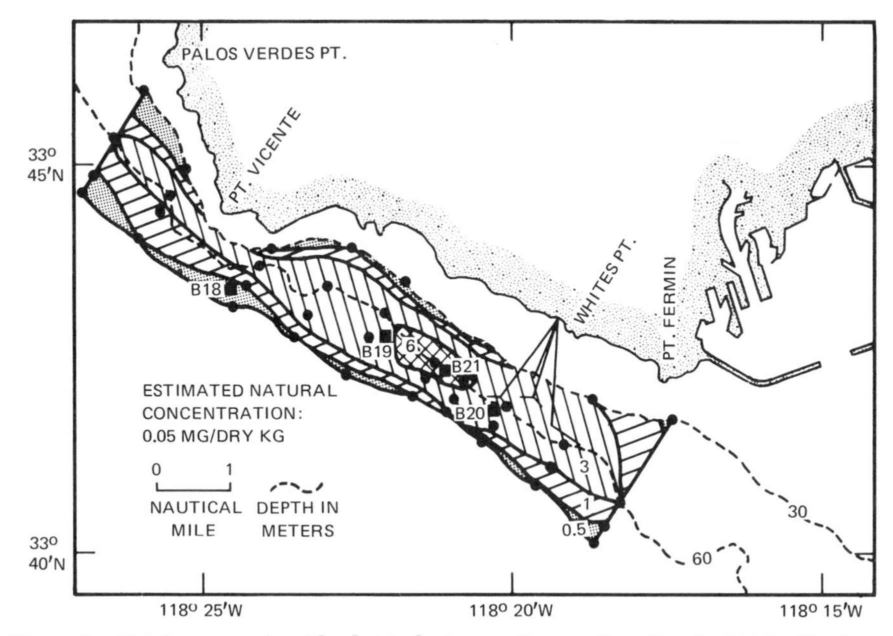


*Figure 3. Total mercury (mg/dry kg) in bottom sediments (0 to 5 cm) off Palos Verdes, 5 September 1973*

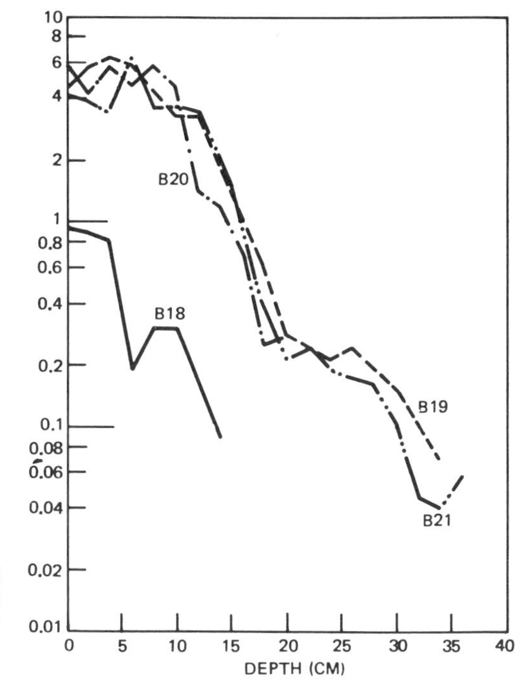


*Figure 4. Vertical profiles of total mercury concentrations in box core sediments collected northwest of the Whites Point outfall system, July 1971. Stations shown on Figure 3.*

with approximately 20 percent of this quantity located in the upper 5 cm.

A number of other trace elements also are highly concentrated above natural levels in the Whites Point discharge area. Table 4 summarizes typical sediment contamination factors observed in a bottom trawling zone between Whites Point and Point Vicente, "down-stream" of the outfall diffusers. Also listed are average concentrations of these trace elements in livers of resident flatfish, the Dover sole (Microstomus pacificus), trawled from these contaminated sediments and from an uncontaminated region off Santa Catalina Island. These investigations were conducted in collaboration with J. Galloway (University of California, San Diego) and V. Guinn and J. de Goeij (University of California, Irvine).

The most striking result from the Dover sole study is the fact that the outfall specimens (known by their high DDT concentrations and frequently eroded fins to have occupied the contaminated region) did not exhibit increased concentrations of the contaminant elements. The only statistically significant differences observed were reduced liver concentrations of arsenic, cadmium, and selenium in the fish collected near the discharges. Subsequent studies still in progress have confirmed this lack of general enhancement of elemental concentrations in this species on the Palos Verdes shelf.

With the support of the U. S. Environmental Protection Agency we have investigated inputs of chlorinated hydrocarbons to the waters of the Bight via six different routes during the past two years. The municipal, industrial, surface runoff, and advection input studies generally followed the patterns outlined above for mercury. Detailed surveys of antifouling paint usage were conducted at three major southern California harbors, and analysis of approximately thirty major brands of paint were made (7). In addition, using a technique developed by V. McClure[2], we have conducted extensive studies of dry aerial precipitation of chlorinated hydrocarbons throughout the Bight during three different seasons.

The results obtained to date from the chlorinated hydrocarbon studies are summarized in Table 5. The principal chlorinated hydrocarbons observed were DDT and PCB compounds. Major coastal sources during 1973-74 were municipal wastewater and aerial dry precipitation.

2. National Marine Fisheries Service, La Jolla, California.

Table 3. Inputs of total mercury to the Southern California Bight, 1972-73.

| Route | Estimated Flow (mgd) | Mercury MER (metric tons/yr) |
|---|---|---|
| Municipal Wastewater* | 950 | 2.5 ± 1 |
| Direct Industrial** | 1,300 | 0.04 |
| Storm Runoff | 400 | 0.4 |
| Ocean Currents | $1 \times 10^7$ | 600 |

* Estimate based on average of mercury concentrations obtained by SCCWRP and major treatment plants.

** San Pedro Harbor only, MER estimate excludes the natural mercury in seawater used for cooling.

Table 4. Trace elements in flatfish collected off Los Angeles during 1971 and 1972. Outfall specimens were trawled off Palos Verdes Peninsula; control, off Santa Catalina Island.*

| Trace Element | Sediments Outfall/Control | Flatfish Livers† (mg/wet kg) | |
|---|---|---|---|
| | | Outfall | Control |
| Antimony | 13 | 0.003 | 0.003 |
| Arsenic | 15 | 1.3 | 3.1 |
| Cadmium | 160 | 0.19 | 0.58 |
| Copper | 23 | 2.0 | 2.2 |
| Mercury | 85 | 0.11 | 0.11 |
| Selenium | 14 | 0.65 | 1.2 |
| Silver | 3 | 1.8** | 2.2** |
| Zinc | 17 | 26 | 27 |

* Sediment data after Galloway (6); flatfish values after de Goeij *et al*., (5).
** Arbitrary units.

† Arsenic, cadmium and selenium values significantly different at the 95 percent confidence level.

Surface runoff inputs were of secondary importance, and contributions from direct industrial discharges and antifouling paints currently in use appear to be insignificant. However, a few dried samples of old antifouling paint showed extremely high PCB concentrations (approximately 10 percent on a dry weight basis); this suggests that past contributions from vessel paints may have been on the order of several metric tons per year.

By far the largest input of chlorinated hydrocarbons to the Southern California Bight has been the release of hundreds of tons of DDT wastes via the JWPCP outfalls off Palos Verdes Peninsula. This input to the sewer system was traced to a large manufacturer of the pesticide in 1970 and stopped. However, contaminated sediments in the collection system continued to be carried in JWPCP effluent, and during 1971, an estimated 20 metric tons of DDT wastes were discharged off Whites Point. We recently inventoried approximately 200 metric tons of these wastes in the upper 30 cm of bottom sediments in a 50-sq-km region on the Palos Verdes shelf: Seventy percent of this total lies in the first 12 cm of sediment, the layer most available to benthic organisms (8). Figure 5 illustrates the surface sediment concentrations we measured in 1972 collections from the discharge region. A subsequent survey (summer 1973) did not reveal any decrease in surface sediment concentrations.

The input of DDT wastes via the JWPCP system has resulted in extensive contamination of the biota living off southern California (9, 10, 11). During 1971, we examined intertidal mussels (*Mytilus californianus*) from stations throughout the Bight and observed enhanced concentrations in specimens collected more than 100 km from the JWPCP outfalls (12). We also found similar gradients of total DDT in flesh of benthic fish and crabs. As shown in Figure 6, Dover sole that were collected around the outfalls in 1971 and 1972 often contained concentrations 100 times those in specimens from the island control areas. During this period, approximately two-thirds of the Dover sole collected off Palos Verdes had DDT concentrations in flesh that exceeded the 5-ppm Federal guideline for DDT residues in seafood. In 1973, specimens of two popular sportfish, black perch and kelp bass, were collected in this area; 65 percent of the perch analyzed, and 45 percent of the bass, also exceeded the Federal DDT limit. However, lower DDT values were found in yellow market crabs (*Cancer anthonyi*) taken near the outfalls in 1971 and 1972, levels ranging from

Table 5. Chlorinated hydrocarbon inputs to the Southern California Bight, 1973-74.

| Route | Metric Tons/Year | |
|---|---|---|
| | Total DDT | Total PCB |
| Municipal Wastewater | 1.6 | 6.5 |
| Direct Industrial | 0.02 | 0.05 |
| Antifouling Paint | 0.001 | 0.001 |
| Surface Runoff | 0.3 | ≤0.8 |
| Aerial Fallout | 1.3 | 1* |
| Ocean Currents | 7 | 4* |

* 1254 PCB

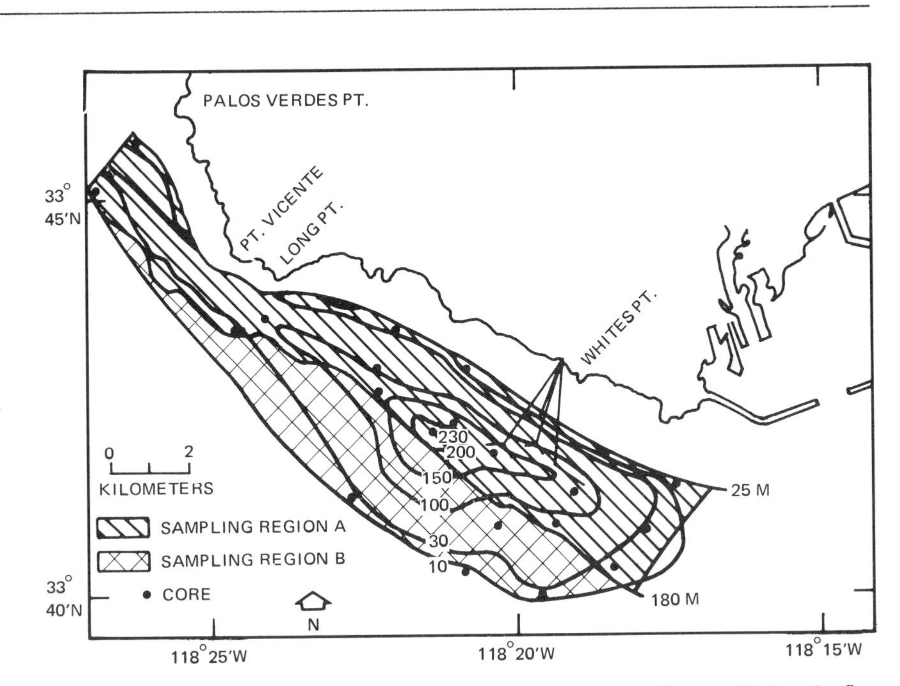


*Figure 5. Concentrations of total DDT (mg/dry kg) in surface sediments (0–2 cm) off Palos Verdes, June 1972. Samples in Region A were 3-cm-ID Phleger cores; those in Region B were 8-cm-ID barrel cores.*

0.5 to 2 ppm.

Unlike DDT wastes, inputs of PCB to the Bight have been distributed much more evenly along the southern California coast. These industrially important synthetic compounds are released from all the major municipal wastewater systems in similar amounts. During 1974, estimated input rates of total PCB for each of the four largest dischargers ranged from approximately 1 to 3 metric tons per year. This has resulted in a fairly even distribution of 1254 PCB in benthic organisms collected off major urban areas, as is illustrated in Figure 7. However, levels in flatfish and mussels from these areas were still two orders of magnitude greater than those in specimens from island control regions.

From our 1973 survey of southern California anchorages, we determined that approximately 300,000 L of antifouling paints are applied annually to recreational, commercial, and naval vessels harbored there (7). Although we measured no significant amounts of PCB in major brands of paint currently being used, the relatively high levels measured in a few samples of old paint suggest that inputs to harbor waters from this source may have been quite significant in the past. Therefore, during 1974, we surveyed levels of 1254 PCB in soft tissues of the bay mussel, *Mytilus edulis*, from three of the largest harbors in the Bight (13). In each case, the specimens from interior regions of the harbor had higher concentrations than did specimens of the same species collected from nearby coastal stations. Highest values occurred in areas of intense vessel activity (e.g., near bottom scraping and repainting facilities): The median and maximum concentrations found at 14 San Diego Harbor stations were 0.3 and 0.9 mg/wet kg, respectively, while the coastal control level was approximately 0.05 mg/wet kg. Thus, the most contaminated specimens, which were collected at the mouth of a small "commercial basin," contained nearly 20 times as much 1254 PCB as mussels growing outside the Harbor.

To determine the extent to which tidal flushing of major anchorages contribute vessel-related pollutants to the coastal environment, during fall 1974 we sampled seawater from the mouths of the three major harbors in the Bight. The collections were made over 12-hour intervals during periods of maximum tidal difference. Chlorinated hydrocarbons were concentrated from replicate 40-L samples on polyurethane, using a procedure developed by de Lappe and Risebrough (University of California, Berkeley). The results for the

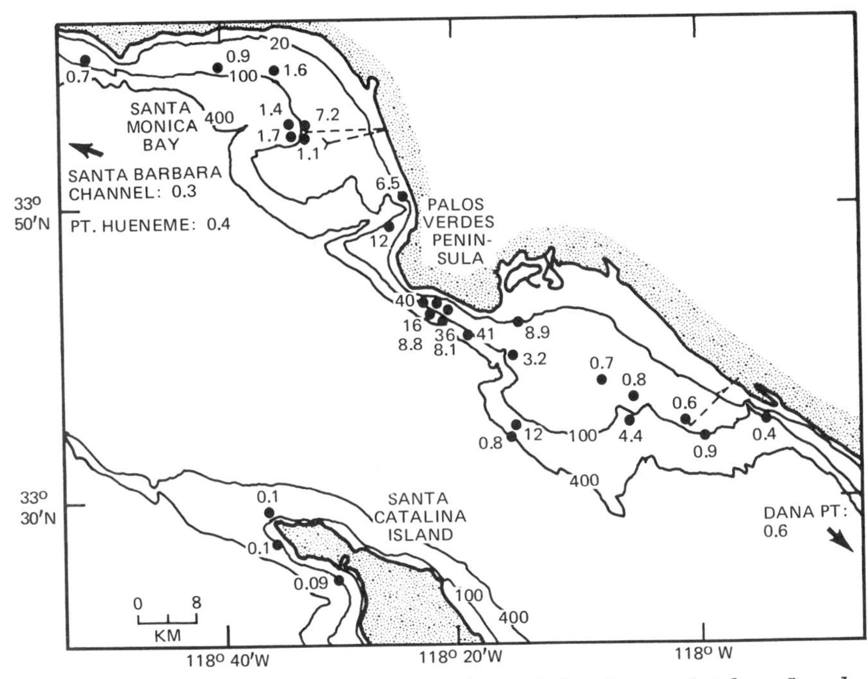


*Figure 6. Total DDT concentrations (mg/wet kg) in flesh of Dover sole taken off southern California, 1971–72*

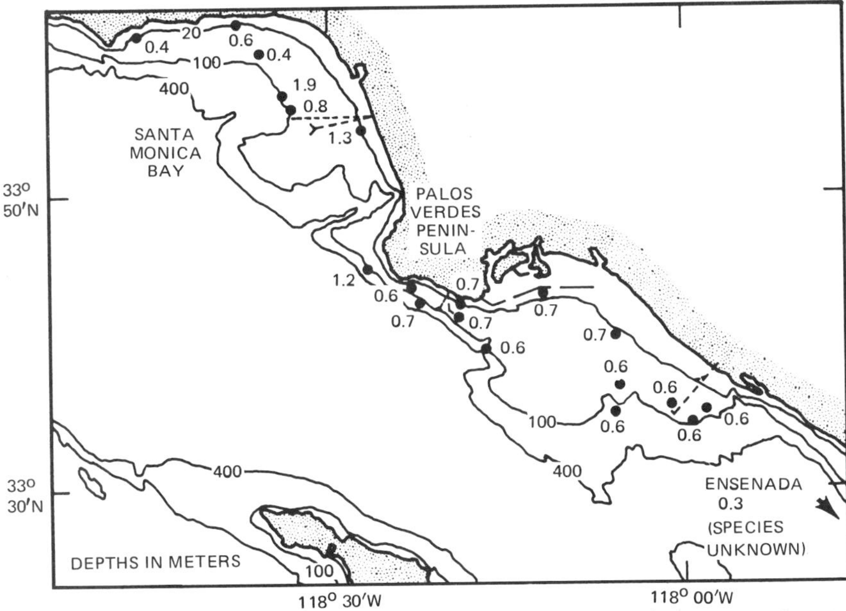


*Figure 7. 1254 PCB concentrations (mg/wet kg) in flesh of the crab,* Cancer anthonyi, *taken off southern California, 1971–72*

San Diego Harbor collections are presented in Figure 8.

These data illustrate the extremely low concentrations of 1254 PCB that we generally find off southern California; the values in this series range from 1 to 3 parts per trillion. Nevertheless, there is a distinct indication that the higher concentrations occurred near the end of outward flow, when water most characteristic of the harbor interior reached the sampling station at the harbor mouth. However, it does not appear that the net flux of 1254 PCB from this contaminated harbor to the adjacent marine ecosystem could be very large. If this limited pilot study is at all representative of present average conditions, the net annual flow of 1254 PCB through the harbor channel does not exceed 50 kg. In comparison, PCB inputs from the submarine discharge of San Diego municipal wastewater are estimated to be an order of magnitude higher.

We also investigated trace metal fluxes at the mouths of the three harbors. Triplicate samples were collected with a special acid-cleaned polycarbonate sampler, which was suspended from a polyethylene line midway between the surface and the bottom of the channel. The blank of each sample bottle was checked prior to use, and process blank samples were collected with each triplicate set. The samples were filtered through 0.4-μ nuclepore membrane filters on shipboard and adjusted to pH 1 using ultrapure acid. A subsample was spiked with approximately 0.8 ppb copper as cuprous sulphate ($CuSO_4$) and approximately 70 percent recovery by ion exchange was obtained using Chelex-100 columns. An APDC-MIBK organic solvent extraction of the filtrate (14) was also employed on a single sample per collection; internal standardization using $CuSO_4$ spikes of previously extracted seawater was employed in this case. On the average, blank values were less than the net copper concentrations measured by the three techniques employed: Ion exchange, APDC-MIBK, and particulate digestion. Results for the San Diego Harbor collections are illustrated in Figure 9.

These data indicate that most of the copper measured in these harbor water samples was in a soluble form (<0.4 μ), with particulate copper constituting only about 15 percent of the total. Although there is some indication of higher concentrations near the end of the outflowing cycle, additional studies are needed to confirm these observations. Based on these data, on the order of 20 metric tons of copper could be flowing from San Diego Harbor annually; if so, this would equal the annual mass emission rate of copper via

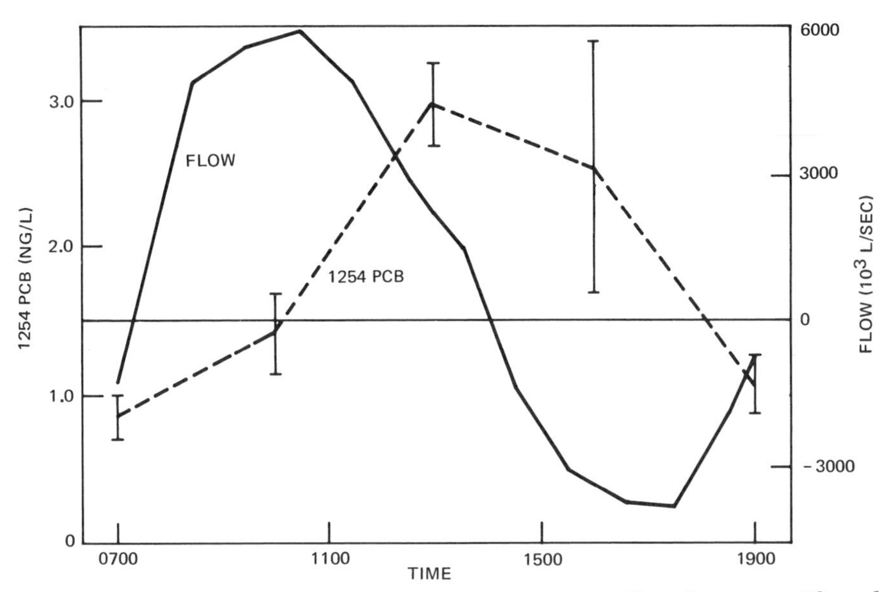


*Figure 8. Concentrations of 1254 PCB in surface seawater collected over a semidiurnal tidal cycle at the mouth of San Diego Harbor, 12 November 1974. Positive flow represents outflowing water; negative, inflowing. Vertical bars indicate individual replicate values.*

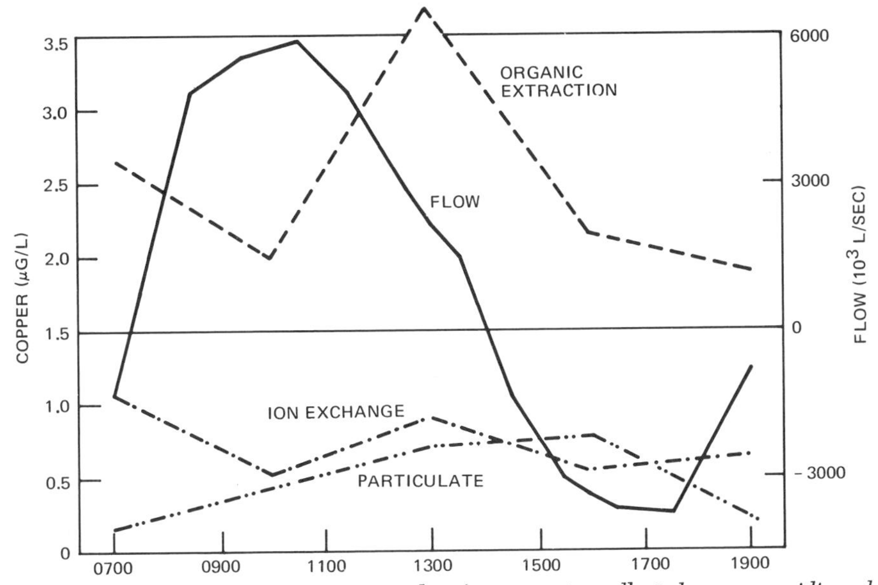


*Figure 9. Concentrations of copper in subsurface seawater collected over a semidiurnal tidal cycle at the mouth of San Diego Harbor, 12 November 1974. Positive flow represents outflowing water; negative, inflowing.*

San Diego municipal wastewater. Although such an output from the Harbor is consistent with the fact that approximately 50 metric tons of copper in antifouling paints were applied to vessel bottoms there during 1974, the data in this case are still too scattered to draw definite conclusions. We are continuing our studies in order to better establish the importance of harbors, submarine outfalls, and other routes such as aerial fallout as sources of potential pollutants to the Southern California Bight.

## Summary

Municipal wastewaters are the dominant known source of numerous trace contaminants to the Southern California Bight. Annual inputs of mercury, DDT, and PCB via this route are an order of magnitude larger than those from direct industrial discharges and storm runoff. Aerial precipitation is also a major source of chlorinated hydrocarbons. Bottom sediments around the largest outfall system are highly contaminated by various trace elements and DDT residues. Corresponding contamination of benthic organisms occurs for DDT but not the trace elements studied. Seawater collected from the mouths of three important harbors contained copper and the chlorinated hydrocarbons at the part-per-billion and part-per-trillion level, respectively.

## Acknowledgements

We thank the following individuals for contributing to the research efforts described here. Analyses for total mercury were made by Joseph Johnson (SCCWRP). Dr. Robert Risebrough and Brock de Lappe (University of California, Berkeley) provided advise in the analysis of chlorinated hydrocarbons. Sediment samples were obtained with the assistance of Douglas Hotchkiss (County Sanitation Districts, Los Angeles County) and Dr. Ronald Kolpack (University of Southern California). Personnel of the several major wastewater treatment plants listed above provided effluent samples. SCCWRP personnel assisting in sampling and analysis were Dr. Alan Mearns, Elliott Berkihiser, Ileana Szpila, Harold Stubbs, and Michael Moore. This research was supported in part by the U.S. Environmental Protection Agency (Grant number R-801153) and in part by a contract with the University of California at San Diego in connection with the State of California Agreement M-11, Marine Research Committee, Dept. of Fish & Game. Equipment used to sample chlorinated hydrocarbons in harbor water was developed with the support of the National Science Foundation International Decade of Oceanography Program, grant number ID072-06412A02.

## Literature Cited


(1) Southern California Coastal Water Research Project. "The ecology of the Southern California Bight: Implications for water quality management." Rept. TR 104, So. Calif. Coastal Water Res. Proj., El Segundo, 1973.

(2) Young, D.R. "Mercury in the environment: A summary of information pertinent to the distribution of mercury in the Southern California Bight." Rept. TM103, So. Calif. Coastal Water Res. Proj., El Segundo, 1971.

(3) Young, D.R., C.S. Young, and G.E. Hlavka. Sources of trace metals from highly-urbanized southern California to the adjacent marine ecosystem. In "Cycling and Control of Metals, Proceedings of an Environmental Resources Conference", pp.21-39, National Environmental Research Center, Cincinnati, 1973.

(4) Young, D.R., J.N. Johnson, A. Soutar, and J.D. Isaacs. Nature (1973), 244:273-375.

(5) deGoeij, J.J.M., V.P. Guinn, D.R. Young, and A.J. Mearns. Neutron activation analysis trace element studies of Dover sole liver and marine sediments. In "Comparative Studies of Food and Environmental Contamination", pp.189-200. International Atomic Energy Agency, Vienna, 1974.

(6) Galloway, J.N. "Man's alteration of the natural geochemical cycle of selected trace elements." Ph.D. dissertation, Univ. of Calif., San Diego, 1972.

(7) Young, D.R., T.C. Heesen, D.J. McDermott, and P.E. Smokler. "Marine inputs of polychlorinated biphenyls and copper from vessel antifouling paints." Rept. TM212, So. Calif. Coastal Water Res. Proj., El Segundo, 1974.

(8) McDermott, D.J., T.C. Heesen, and D.R. Young. "DDT in bottom sediments around five southern California outfall systems." Rept. TM 217, So. Calif. Coastal Water Res. Proj., 1974.

(9) Risebrough, R.W. Chlorinated hydrocarbons in marine ecosystems. In "Chemical Fallout", pp.5-23. Charles C. Thomas, Springfield, Ill., 1969.

(10) Burnett, R. Science (1971), 174:606-8.

(11) MacGregor, J.S., Fish. Bull. (1974), 72:275-93.

(12) deLappe, B.W., R.W. Risebrough, and D.R. Young. Mytilus californianus: An indicator species of DDT and PCB contamination in California coastal waters, (Submitted for Publication, 1975).

(13) Young, D.R., and T.C. Heesen. "Inputs and distributions of chlorinated hydrocarbons in three southern California harbors." Rept. TM 214, So. Calif. Coastal Water Res. Proj., El Segundo, 1974.

(14) Brooks, R.R., B.J. Presley, and I.R. Kaplan. Talanta (1967), 14:809-16.

# 27

# Use of Silver and Zinc to Trace Sewage Sludge Dispersal in Coastal Waters

F. RUTHERFORD and T. CHURCH

College of Marine Studies, University of Delaware, Newark, Del. 19711

Sewage sludge from the municipal plants of Philadelphia, Pennsylvania, and Camden, New Jersey, has been dumped some 12 miles off the mouth of the Delaware Bay from 1961 to 1972 in amounts averaging 389,340 $m^3$/year.(1) Since 1973 the disposal area has been moved some 40 miles off-shore. The City of Philadelphia commissioned the Franklin Institute Research Lab (FIRL) to investigate the dispersion of the sludge on the ocean bottom at the former dump site.(2) According to the FIRL findings, "The area has abundant life. . . well distributed number of species and that those present are healthy. Generally sediments appeared clean both *in situ*. . . and in samples brought to the surface". The only indication of potential pollution to Delaware shores that FIRL cited was that 52% of the seabed drifters recovered were found on Delaware's ocean beaches. Since the FIRL report depended on rather discriptive techniques of assessing sludge presence (3), a new method was deemed appropriate to trace the actual dispersion of the sludge material.

The secondary treated sludge is probably highly resistant to marine decomposition since it has been activated already by anaerobic digestion at 35°C for 30 days. In its barged form it forms stable flocs, has a high moisture content (80%), moderate ignition content (38%), and his heavily laden with greases, oils, and tars, having a 17% bituminous extract.(4) Also, it appears to be concentrated in several trace metals.(5) In studying the New York Bight, silver was identified as a possible tag for tracing sludge materials.(6) Also, chromium, copper, lead, nickel, and zinc were enriched in contaminated sediments by factors of 10 to 100 over uncontaminated sediments in the New York Bight.(7) In the southern California coastal zone, the anthropogenic fluxes of several metals, including silver and zinc, silver and zinc fluxes have increased, with silver originating principally from direct municipal waste, and the zinc from either atmospheric washout and/or from outfall waste.(8) Comparison between barged sewage sludge from Philadelphia and normal bottom sediments from the vicinity of the dump site showed for 26 elements that silver and zinc had enrichment factors of 100 to 200 respectively. (5) Thus, it seemed fruitful to employ assays of silver and zinc in bottom sediments surrounding the Philadelphia dump site to disclose the possible dispersion of sewage sludge.

## SAMPLING

Sediment samples had been collected from the former nearshore dump site during cruises over the past four years. Figure 1 shows the date and location of each sample. Samples obtained from FIRL (7) were collected using a La Fond Dietz Grab, while Watling et.al. ('72) and CMS 666-667 ('73-'74) used a $0.1m^2$ Van Veen Grab. The sub-samples were taken from the interior portions of the grabs by use of plastic spatulas, transferred to plastic bags and frozen. The FIRL samples were obtained in an already dry state. All precautions were taken to avoid contamination from dust and handling.

## ANALYSIS

A. Reagents and Standards: All reagents used were purified in a silica sub-boiler. Two commerical standards (Ventron, Inc. and Spex, Inc.), and one prepared in the lab using pure silver nitrate were prepared to identical titer and compared. All gave similar assay within 8%. The lab prepared silver nitrate standard was employed, stored in 30% $HNO_3$ in a linear polythylene bottle, and the bottle completely covered by aluminum foil to prevent photoxidation of the solution. No standard materials were run because of absence of a matrix similar to sludge, or lack of both Zn and Ag assays.

B. Sample Analysis: The samples were thawed and dried in an oven at 80°C for 48 hours and then stored in plastic Millipore petri dishes. Normally, two duplicate 0.5 gram subsamples were weighed out from each mother sample, homogenized after drying, and each run with replicates. The sub-samples along with 3ml conc. $HNO_3$ were placed in teflon crucibles and capped with teflon lids. These capped crucibles were placed in an aluminum bomb at about 80°C for 36-48 hours. After digestion the samples were transferred quantitatively into plastic centrifuge tubes and washed with distilled $H_2O$. The samples were then centrifuged for 10 minutes, the supernate liquor decanted, and diluted to 10ml in a glass volumetric flask for analysis. Silver was analyzed using a Varian Model 63 carbon rod flameless atomizer mated to a Varian Model 1200 atomic absorption spectrophotometer. The Model 1200, using an air-acetylene flame was employed in the zinc analysis. The instruments settings used are listed in Table I.

Table I

Instrument settings for the Varian 1200 Atomic Absorption Unit and Model 63 Carbon Rod Atomizer used for the analysis of Ag and Zn in digestates of Philadelphia sewage sludge diluted in 0.1N $HNO_3$.

| Lamp | Wave-Length (nm) | Slit (nm) | Lamp Current | Carbon Rod Settings Dry | Ash | Atomize |
|---|---|---|---|---|---|---|
| Ag | 328.1 | 0.2 | 3mA | 1 at 20 sec. | 5 at 10 sec. | 6 at 2.5 sec. |
| Zn | 213.9 | 0.5 | 5mA | Flame Analysis | | |

For each set of samples a standard curve was run, using values of 1 to 40 ppb for Ag and .5 to 5 ppm for Zn, which fit in the linear working range of the spectrophotometer and its carbon rod atomizer. If any sample was out of this linear working range, it was diluted and rerun. A blank was also determined for each set of samples and the values were corrected when necessary. The hydrogen lamp was used to correct for non-atomic absorption, which was appreciable only for Zn at its shorter wavelength.

Three separate 5μl injections from an Oxford pipette were placed in the carbon rod for each sub sample, and the peak retrieval mode was used to obtain the absorbance of the silver. For zinc, two determinations of each sub-sample were aspirated into the air-acetylene flame and the absorbance determined using the three second integration mode.

## RESULTS AND DISCUSSION

Table II lists the results of the anlaysis which are plotted in Figures 2, 3 and 4 respectively for silver, zinc and organic carbon concentration. The organic carbon was analyzed by a modified Coleman nitrogen analyzer coupled to a Beckman IR non-dispersive $CO_2$ analyzer. The concentration recorded is the average value for each sub-sample of the source material. The results indicate that silver and zinc are concentrated along shoals boardering the shipping channels entering Delaware Bay to the north and southwest of the dump site. The rectangle of the dump site and the area due west seem to have the lowest concentration of silver and zinc. The only high reading in the dump rectangle was recorded on the same day a sludge barge had discharged as "starred" in Figures 2 and 3 and noted by "during dump". In Figure 2, "dumping ceased" notes a mininution

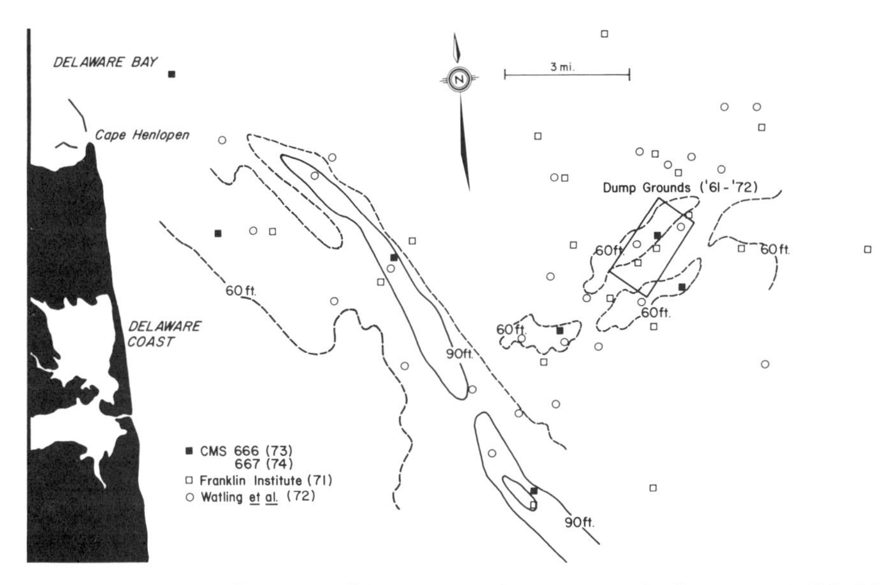


Figure 1. *Surface sediments used for trace metal survey to study dispersion of Philadelphia sewage sludge dumping ('61–'72)*

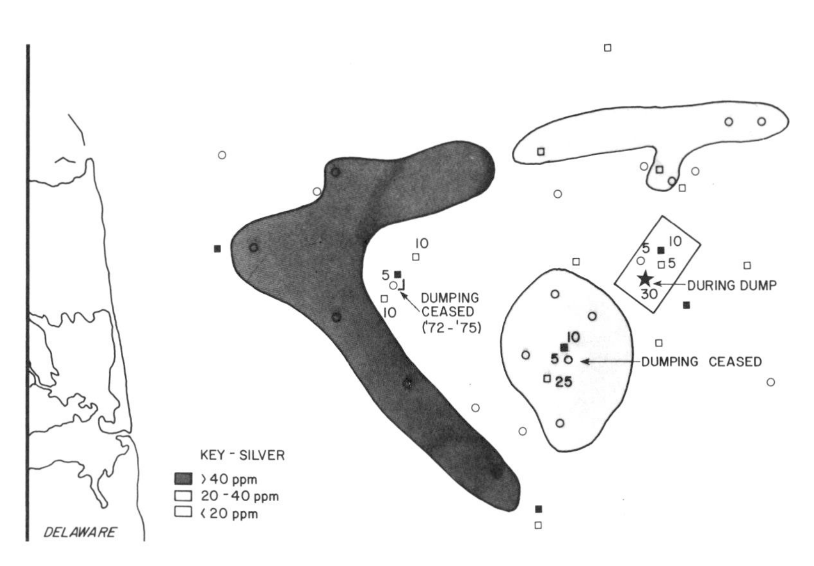


Figure 2. *Silver concentrations of sediments in vicinity of Philadelphia sewage sludge dump site (± 5 ppm)*

Table II

Bottom Sediments in the Vicinity of Philadelphia Sewage Sludge Disposal (off Delaware) which are Analyzed for Silver and Zinc

| Station | Date of Collection | Description | [Ag], ppm | [Zn], ppm | Organic Carbon (Wt%) | # of Nucula/ $.1m^2$ (9) |
|---|---|---|---|---|---|---|
| M-1 | 7-14-72 | red-brown sand particles | 21.6, 35.5 | 14.7, 14.6 | .05 | |
| M-3 | 7-14-72 | brown and red sand | 10.0 | 6.0 | .16 | |
| M-5 | 7-14-72 | very coarse sand | 30.9, 49.4 | 4.1, 3.0 | | |
| M-6 | 7-14-72 | reddish-brown coarse sand, shell material | 59.6, 6.0 | 4.7, 7.0 | .06 | |
| M-7 | 7-14-72 | gray-brown fine silt | 54.4, 85.4 | 52.3, 49.6 | .39 | 5522 |
| M-8 | 7-14-72 | gray-red brown sand | 18.5, 15.6 | 10.2, 19.5 | .13 | 293 |
| M-9 | 7-20-72 | brown coarse sand | 18.5, 32.2 | 10.3, 6.9 | .11 | 155 |
| M-10 | 7-20-72 | brown clay and large sand particles | 25.9, 20.9 | 9.9, 6.0 | .2 | |
| M-11 | 7-20-72 | large brown sand particles | 6.5 | 4.0 | .11 | |
| M-12 | 7-20-72 | gray-black sand | 14.0, 14.1 | 12.2, 12.0 | .30 | |
| M-13 | 7-20-72 | gray-white sand, some shell material | 16.0, 6.3 | 3.0, 3.1 | .38 | |

Table II (cont'd.)

| Station | Date of Collection | Description | [Ag], ppm | [Zn], ppm | Organic Carbon (Wt%) | # of *Nucula*/ $.1m^2$ (9) |
|---|---|---|---|---|---|---|
| M-14 | 7-20-72 | gray sand | 16.5, 29.6 | 4.5, 4.6 | .07 | |
| M-15 | 7-20-72 | gray sand, shell material | 38.3, 32.9 | 10.0, 10.6 | .16 | |
| M-16 | 7-20-72 | black fine silt | 20.1, 19.6 | 70.6, 68.7 | 1.54 | 1010 |
| M-17 | 7-21-72 | black fine silt with shell material | 9.1, 14.3 | 30.1, 25.5 | .15 | 3019 |
| M-18 | 7-21-72 | very fine silt, dark gray, some shells | 12.3, 13.7 | 32.9, 38.1 | .43 | 3575 |
| M-19 | 7-31-72 | light gray to gray brown coarse sand | 23.8, 25.4 | 6.4, 7.4 | .02 | 7 |
| M-20 | 7-31-72 | fine grain silt | 45.8, 47.7 | 43.8, 45.7 | .36 | |
| M-21 | 7-31-72 | gray, coarse sand, many shell fragments | 73.0, 49.5 | 11.1, 8.9 | .45 | |
| M-22 | 7-31-72 | brown sand | 5.0, 10.1 | 10.9, 8.1 | .28 | |
| M-23 | 7-31-72 | brown sand, shell fragments | 44.1, 17.2 | 6.7, 5.4 | .16 | |
| M-24 | 7-31-72 | gray-brown medium sand | 54.7, 18.4 | 10.5, 15.3 | .14 | |

Table II (cont'd.)

| Station | Date of Collection | Description | [Ag], ppm | [Zn], ppm | Organic Carbon (Wt%) | # of *Nucula*/ $.1m^2$ (9) |
|---|---|---|---|---|---|---|
| M-25 | 7-31-72 | fine silt, gray-black | 126.7,36.8 | 73.9, 65.0 | 1.05 | 12238 |
| M-26 | 7-31-72 | very fine silt, dark gray | 27.7, 49.8 | 49.3, 49.8 | 1.01 | 929 |
| M-27 | 7-31-72 | gray black silt | 60.4, 83.5 | 68.1, 43.9 | 1.12 | 23 |
| E-2 | 3-14-71 | light colored sand, several shell fragments | 20.2, 23.9 | 12.1, 17.9 | .4 | |
| E-4 | 2-21-71 | dark sand, some pebbles and shell fragments | 31.2, 24.8 | 13.5, 10.3 | .4 | |
| E-5 | 1-23-71 | brown sand, large pebbles | 5.8, 13.3 | 0.0, 0.0 | .00 | |
| E-9 | 5-21-71 | coarse sand, large pebbles | 6, 16.7 | 20.2, 6.3 | .1 | |
| E-13 | 7-16-71 | fine silt, few pebbles | 27.7, 30.7 | 49.5, 61.5 | .2 | |
| E-7 | 1-25-71 | sand, pebbles, some shell fragments | 6.4, 8.9 | 0.0, 0.0 | .1 | |
| E-11 | 7-16-71 | medium grain sand | 6.8, 10.3 | 10, 5.1 | .1 | |
| E-6 | 3-14-71 | coarse sand, small shell fragments | 8.5, 10.4 | 11.6, 6.2 | .1 | |

Table II (cont'd.)

| Station | Date of Collection | Description | [Ag], ppm | [Zn], ppm | Organic Carbon (Wt%) | # of Nucula/ .1m$^2$ (9) |
|---|---|---|---|---|---|---|
| E-1 | 2-21-71 | fine sand, few pebbles, small shell fragments | 15.2, 20.0 | 0.0, 7.7 | .1 | |
| CMS-4-3 | 6-27-74 | medium sand, few shells and pebbles | 6.3, 6.2 | 5.3, 5.5 | .1 | |
| CMS-4-4 | 6-27-74 | fine sand, several shell fragments | 10.0, 5.8 | 4.8, 10.6 | .2 | |
| CMS-4-2 | 6-27-74 | dark fine sand, many shells | 4.6 | 16.5 | .08 | |
| CMS-3-3 | 7-17-73 | medium sand, a few shell fragments | 5.7, 4.0 | 4.8, 6.9 | | |
| CMS-3-7 | 7-17-73 | fine gray sand with pebbles & shell fragments | 34.1, 30.4 | 54.1, 54.6 | | |
| CMS-3-2 | 7-17-73 | fine gray sand, many shell fragments | 8.0, 8.0 | 27.0, 26.9 | | |
| CMS-3-6 | 7-17-73 | fine gray sand, a few shell fragments | 20.0, 21.6 | 53.8, 49.1 | | |
| CMS-3-4 | 7-17-73 | medium to coarse sand, several large pebbles | 10.0, 20,8 | 5.0, 5.0 | | |

Table II (cont'd)

| Station | Date of Collection | Description | [Ag], ppm | [Zn], ppm | Organic Carbon (Wt%) | # of *Nucula*/ $.1m^2$ (9) |
|---|---|---|---|---|---|---|
| CMS-3-5 | 7-17-73 | medium to coarse sand, some shell fragments | 3.9, 4.1 | 6.9, 7.2 | 0.4 | |
| Philadelphia Sewage Sludge: | | | | | | |
| | 7-8-71 | Alliquot 1. | 7742, 9619 | 5997, 6871 | 21.5 | |
| | | Alliquot 2. | 9961, 7619 | 6114, 6286 | 19.8 | |

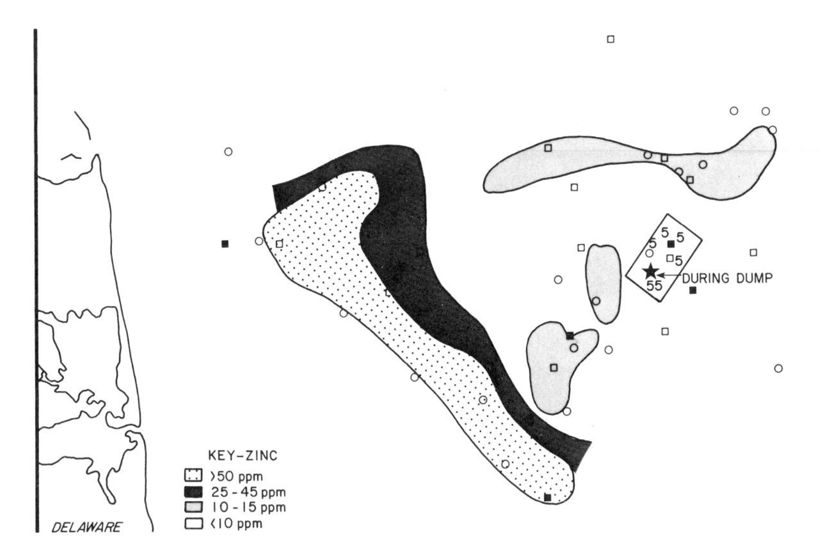


*Figure 3. Zinc concentrations of sediments in vicinity of Philadelphia sewage sludge dump site (± 5 ppm)*

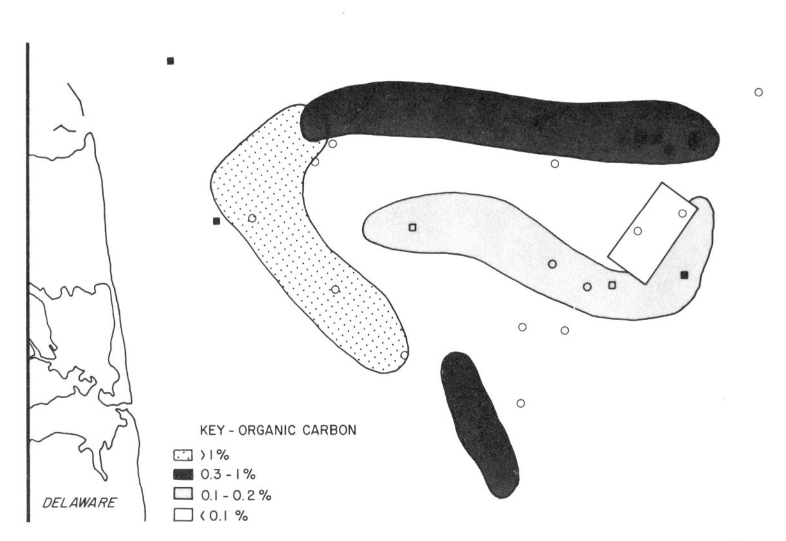


*Figure 4. Percent organic carbon of sediments in vicinity of Philadelphia sewage dump site (± 0.1%)*

in metal concentration in the years subsequent to dumping cessation. These results suggest that when the sludge settles to the bottom it is swept by the prevailing bottom drift currents out of the dump site primarily shoreward.(2) Alternatively, the strong prevailing thermocline, during the summer, may possibly be a significant trap and dispersing agent.

There is little agreement between the silver and zinc results obtained by FIRL and this report, probably due to different extraction techniques and instrumentation. FIRL reported using only 5ml of concentration HCl per 100ml of sample. This perhaps explains low metal recovery from an already resistant sludge that had received secondary treatment. The variation between sub-samples in the areas of high metal concentration is similar to that reported in the New York Bight.(7) Such heterogeneity may indicate poor mixing of sludge and sediment.

In another report on the Philadelphia dump area dealing with benthos (9), numerous _Nucula proxima_ were discovered also in the region to the south and west of the dump rectangle. This bivalve species is known to live in fine grained sediments with substantial organic content.(10) The number densities of _Nucula_ (35,750 to 122,380/$m^2$) are some of the highest ever recorded, suggesting that these deposit feeding bivalves may be either feeding directly upon the sludge utilizing its organic matter for sustenance, or indirectly utilizing the carbon and nitrogen of the associated microfauna. Both forms of behavior are suggested in the caprophagic behavior of deposit feeders.(11)

Table III lists the pair correlation coefficients between silver, zinc, organic carbon, and _Nucula_ concentrations. There is a confidence level for a relationship greater than 98% for all but one of the pairs (Zn to _Nucula_). The correlation of the numbers of _Nucula proxima_ in the area of high silver concentration support the idea that both trace metals and particular infaunal organisms are sensitive indicators that can be used to delineate the dispersal of sewage sludge. Implicated in this association is a possible link for the introduction of sludge-derived toxins and pathogens into the marine food web, the evidence for which should receive future attention.

## SUMMARY

The trace metals, silver and zinc are highly concentrated in the secondary treated sewage sludge from the Philadelphia metropolitan area, and display sensitive indications as tracers for the bottom dispersal of the sludge off the mouth of Delaware Bay. From such metal distributions, it is fair to conclude that the sludge has moved out of the former designated area of disposal and dispersed some tens of miles shoreward, toward the

Table III

Statistical Correlation Between the Concentrations of Silver, Zinc, Organic Carbon, and Number Densities of the Mollusc Nucula proxima for Sediments in the Vicinity of the Former Philadelphia Sewage Sludge Disposal Site.

| Variable Pair (x,y) | Correlation Coefficient (r) | Confidence of a Relation (%) |
|---|---|---|
| Ag to Zn | .59 | 99% exists |
| Ag to %C (org) | .44 | 98% exists |
| Ag to # Nuc. | .52 | 98% exists |
| Zn to %C (org) | .85 | 99.9% exists |
| Zn to # Nuc. | .61 | 99% exists |
| %C (org) to # Nuc. | .37 | 90% exists |

where

$$r = \frac{\Sigma xy - (\Sigma x \Sigma y/n)}{\left[\left(\Sigma x^2 - \frac{(\Sigma x)^2}{n}\right)\left(\Sigma y^2 - \frac{(\Sigma y)^2}{n}\right)\right]^{\frac{1}{2}}}$$

for n=20 degrees of freedom in 22 samples.

west. The concentrations of these two metals match closely the distribution of organic carbon and number density of the deposit feeding bivalve mollusc, Nucula proxima. A possible link between disposed sewage sludge and the food web is implicated.

Acknowledgement

Christian Wethe and John Cronin assisted with some phases of sample collection and analysis. Generous support was provided by an institutional Sea Grant to the University of Delaware under auspices of NOAA (04-3-158-30). Dr. Robert Biggs and Mr. Joel Goodman were vital instigators in this respect.

Literature Cited

1. Buelow, R. W., (1968). Ocean Disposal of Waste Material. Ocean Science and Engineering of the Atlantic Shelf. Symposium, Philadelphia, pp. 311-337.

2. Davey, C. F., (1972). Assessment of Effect of Digested Sewage Sludge Disposal Off the Mouth of Delaware Bay. Franklin Institute Research Lab. Tech. Rept. F-C2970.

3. Roth, W., (1972). Statement of Senator W. Roth: In Ocean Disposal Practices and Effects. U. S. Environment Protection Agency, Washington, D. C.

4. Kupferman, S. and Murphy, L., (1973). A Preliminary Investigation of Characteristics of Philadelphia's Digested Sewage Sludge. Technical Notes No. 1, College of Marine Studies, University of Delaware.

5. Seidel, G., (1972). Delaware State University, unpublished communication, October 6, 1972.

6. Gross, M. G., (1972). Marine Waste Deposits Near New York. Marine Pollution Bull. 3 (7): pp. 102-105.

7. Carmody, D. J., Pearce, J. B. and Yasso, W. E., (1973). Trace Metals in Sediments of New York Bight. Marine Pollution Bulletin 4 (9): pp. 132-135.

8. Bruland et. al., (1974). History of Metal Pollution in Southern California Coastal Zone. Environmental Science and Technology, 8 (5): pp. 425-432.

9. Watling, L. et. al., (1974). An Evaluation of Sewage Sludge Dumping on the Benthos off Delaware Bay. Marine Pollution Bull. 5 (3): pp. 39-42.

10. Bader, R. C., (1954). The Role of Organic Matter in Determining the Distribution of Pelecypods in Marine Sediments. J. Mar. Res. 13, pp. 33-47.

11. Frankenberg, D., (1967). Caprophagy in Marine Animals. Limnol. and Oceanog., 12, pp. 443-450.

# 28

# Alternatives to Marine Disposal of Sewage Sludge

RICHARD I. DICK

Civil Engineering Department, University of Delaware, Newark, Del. 19711

Sludge is an inevitable consequence of wastewater treatment. In current wastewater treatment practice, chemical or biological processes commonly are used to convert dissolved or colloidal substances into suspended solids of large dimension. This permits pollutants to be removed by solids separation procedures and allows discharge of the treated wastewater, but leaves behind a suspension of solid particles in water--sludge--which constitutes an appreciable problem. The sludge produced at a wastewater treatment plant typically occupies a volume in the order of one percent or less of the volume of wastewater treated. However, the cost of effectively managing the residual material is in the same order of magnitude as the cost of treatment of the wastewater. And, the potential for environmental impact and public discontent probably exceeds that associated with discharge of the treated wastewater itself.

## Sludge Characteristics

Generalizations concerning sludge characteristics are possible only when considering municipal wastewater treatment sludges or potable water treatment sludges. Even in these cases, appreciable variations from the norm occur. Sludges from industrial wastewater treatment are as varied as industry. Insofar as it is possible to make generalizations concerning sludge properties, those properties are briefly reviewed here. More comprehensive reviews of sludge properties have been prepared by Vesilind (1), the author (2), and others.

Chemical Characteristics. The solids particles in sludge are comprised of the waste constituents which the wastewater treatment plant was designed to remove, any precipitating or coagulating chemicals used in the wastewater treatment process, and materials incidentally removed in the treatment process. Initially, the liquid phase of the sludge has a chemical composition similar to the wastewater at the point in the treatment process at which the sludge was produced. However, this may change appreciably during storage or treatment of sludges.

Because the principal purpose of most wastewater treatment plants is to remove oxygen demanding materials, the principal chemical constituents of most sludges are organic compounds. In typical municipal wastewater treatment plants, these organic solids represent materials which were in suspended form in the raw wastewater and microorganisms synthesized in biological wastewater treatment processes designed to remove dissolved and colloidal organic compounds.

Sludge typically contains significant amounts of other materials which incidentally are removed during the course of wastewater treatment. To illustrate, wastewater treatment plants are not ordinarily designed specifically to remove heavy metals; yet, the majority of most heavy metals are typically removed by conventional wastewater treatment processes, and these heavy metals are incorporated into the sludge.

Most municipal wastewater treatment plants do not currently add chemicals to precipitate or coagulate waste constituents, but when such chemical addition is practiced, significant changes in chemical characteristics of sludges occur. Treatment plants which use lime, iron, or aluminum salts to precipitate phosphorus are good examples, and the same chemicals have been used to increase the removal of colloidal and suspended solids from raw wastewaters. Synthetic organic polyelectrolytes have also been used to improve clarification of raw waste or final effluents, and they have the advantage of not so significantly altering the quality and quantity of sludge produced.

Physical Characteristics. The physical properties of sludges of most concern in considering alternatives to marine disposal are those related to the fact that sludges are difficult to thicken and dewater. The solids which are suspended in water to form sludges are light, flocculent, and hydrophillic. To illustrate, it is not unusual to encounter waste sludges from biological treatment of wastewaters which cannot readily be concentrated by gravity sedimentation to beyond about one percent solids by weight. Typical values for mixed sludges from primary and secondary wastewater treatment are in the range of 4 to 5 percent solids by weight. This means that for every pound of residual solids from wastewater treatment, 20 to 25 pounds of water also typically need to be processed.

Chemical, physical, and biological techniques for conditioning sludge solids to alter the physical properties and render the sludge more amenable to dewatering processes are available, but they are expensive. Furthermore, some techniques (inorganic chemical conditioning) appreciably increase the mass of solids to be processed and alter the chemical characteristics of the sludge, while others (notably, thermal conditioning) result in appreciable impairment of the quality of the liquid phase of the sludge which is subsequently returned to the wastewater treatment plant. Often, it has been found to be more economical to leave water in the sludge than to pay the costs of taking it out by use of sludge conditioning and dewatering techniques (3).

Biological Properties. The biological property of sludge of most interest from the standpoint of sludge treatment and disposal is the content of viruses and pathogenic organisms. Simple sedimentation removes significant numbers of pathogenic organisms and some virus from municipal wastewaters. Secondary treatment removes additional pathogenic organisms and significant removal of virus occurs in the secondary stage of treatment (4).

Thus, sludges from typical municipal wastewater which potentially contains agents of enteric diseases, such as typhoid fever, cholera, amoebic and bacillary dysentery, and infectious hepatitis must be assumed to contain the organisms and virus as well. [Note, however, that Lund (5) has presented data indicating that virus associated with biological sludges may be complexed in a non-infective form.] Additionally, intestinal worm eggs may be contained in sludge (6).

## Alternative Sludge Management Systems

Coastal cities in the United States and other countries have found discharge of sludge at sea to be a convenient and economical means of sludge disposal. Most often, sludges have been anaerobically digested to reduce the amount of decomposible organic material and the numbers of pathogenic organisms prior to discharge at sea. When deep water can be reached within a reasonable distance of shore, sludge outfalls have been used. For example, at Los Angeles, a 21 inch diameter digested sludge outfall extends 7 miles into Santa Monica Bay. More often, tankering or barging have been used to transport sludge to sea. This is the case on the East Coast of the United States where shallow water overlies the continental shelf for appreciable distances out to sea. The City of Philadelphia, for example, barges digested sludge down the Delaware Bay and discharges it at sea 50 miles from the coast.

In 1972, two pieces of Federal legislation were enacted which were to significantly influence marine sludge disposal. One was the Marine Protection, Research, and Sanctuaries Act, and the other the 1972 Amendments to the Water Pollution Control Act. These laws required that the U. S. Environmental Protection Agency develop guidelines to be used as criteria for issuing permits for discharge of sludge at sea. Such criteria have been promulgated (7). They specify, for example, maximum permissable concentrations of heavy metals in sludges to be discharged at sea. As a practical matter, the quality of municipal sludges do not generally seem to meet the EPA requirements, and only interim permits ordinarily are being issued. To illustrate, Philadelphia's latest permit, issued in February, 1975 is an interim one-year, ocean disposal permit to allow discharge of 150 million gallons of sewage sludge per year. A request for permission to dump an additional 25 million gallons of sludge was denied by EPA. The permit requires that sludge dumping be reduced by 50 percent by January, 1979 and by 100 percent by January, 1981 (8).

It is not the purpose of this paper to consider the relative advantages and disadvantages of ocean disposal of sludges. Rather, given the current state of affairs with regard to Federal policy on ocean dumping, what are the alternative sludge disposal schemes available to cities such as Philadelphia?

If water is to be excluded from consideration as a receptacle for waste sludges, then only air and land remain as possible places for sludge disposal--there are no other places on earth. A third, and potentially attractive, alternative would be to avoid disposal by making productive use of sludges. A fourth option which might be proposed would be to eliminate the need for sludge treatment and disposal by avoiding the generation of sludges. In the scenario of a developed industrialized nation with desires for control of environmental quality, avoidance of the need for sludge disposal is not a viable option, for wastewater treatment inevitably will produce some type of residual material. However, it is worthwhile to introduce the idea of eliminating sludge production, for some diminuation of the problem of sludge treatment and disposal could be accomplished by selection of wastewater treatment processes, control of industrial discharges, and alteration of life style so as to improve the quality of sludge and/or to reduce its volume.

Given the facts that the accumulation of residual materials from wastewater is inevitable and that water, land, and air are the only places on earth, the only identifiable options to marine disposal of sludge would, then, seem to be: (1) reclamation of sludge constituents, (2) discharge to air, and (3) discharge to water. These options are considered in more detail in the following sections.

## Sludge Reclamation

Reclamation of chemical sludges comprised of a single type of precipitant or simple mixtures of precipitants which can be easily separated has been practiced. Sludges from water softening offer a good illustration. The calcium carbonate produced by lime softening of water can be recalcined to recover the lime for reuse (and, indeed, to "mine" lime from water!) (9). And, techniques for reclaiming the magnesium precipitants in softening sludges for possible use as coagulants also have been described (10). Similar techniques for lime reclamation have been used in conjunction with physical-chemical treatment of wastewaters (11). The required technology for recovery and recycle of aluminum from sludges produced at water coagulation plants would seem to exist (12), but few plants make use of the technique. Additional examples of recycle of comparatively homogeneous chemical sludges can be found in industrial waste treatment practice where rather "pure" sludges sometimes can be produced by segregation of wastewaters.

Recovery of the constituents of sludges from typical municipal wastewater treatment plants is far more difficult. This is true even though sludges from typical municipal wastewater treatment plants contain a wide variety of potentially valuable materials. Normal sludges contain many metals, for example, as well as proteins, vitamins, and other ingredients. It is this wide diversity of constituents in sludges and their relatively low concentration which makes reclamation difficult and expensive.

Aside from the beneficial use of municipal sewage sludges on agricultural land (see later section), few illustrations of municipal sludge reclamation can be cited. With current technology, reclamation of sludge constituents would not appear to be a viable alternative to ocean disposal of municipal sludges.

Reclamation, however, is an appealing option, and it would appear to be an approach worthy of attention. The most attractive possibility for reclamation of resources from municipal sludges would seem to be to make use of the protein in waste biological sludges. Such sludges, as from the activated sludge wastewater treatment process, contain about 40 percent protein. Activated sludge has been shown [for example, by Hurwitz (13)] to be a satisfactory animal food, and, indeed, waste activated sludge from industrial waste treatment plants has been incorporated into animal feeds. Reocvery of protein for subsequent sale would seem to be another attractive possibility. Vitamin $B_{12}$ recovery from activated sludge was, at one time, practiced in Milwaukee (14), and schemes for protein recovery could well incorporate provisions for such constituents which are less plentiful than protein in sludges but more valuable. In development of such processes, it would be necessary to take into account the content of heavy metals and of trace recalcitrant organic compounds which might be found in activated sludge.

There has been appreciable interest in the possibility of forming construction materials from sludges [see, for example, Minnich (15)]. Various proposals for producing lightweight aggregates or ceramic materials by thermal processes or by use of chemical additives have appeared from time to time. With some sludges, it would seem that more valuable utilization of sludge constituents might be envisioned than to incorporate them into building materials.

While many challenges remain in the development of schemes for reclaiming sludge constituents, an important advantage is that it is not necessary to operate the sludge reclamation facilities at a profit. Current costs for treatment and disposal of sludges by other means are typically in a range of $50 to $100 per ton of dry solids. Sludge reclamation schemes which lose even this appreciable amount of money would still be competitive.

## Disposal to the Air

If a typical chemical analysis of sludge is compared to the chemical composition of air, it becomes readily apparent that disposal of sludge to the air is not a satisfactory alternative. Most constituents of sludge are not compatible with the air environment. Constituents which comprise a significant portion of sludges, however, are "at home" in the air. The organic compounds in sludge could satisfactorily be returned to the air as carbon dioxide (by incineration or biological oxidation), the nitrogeneous compounds could satisfactorily be returned to the air as nitrogen gas (by properly controlled combustion or biological denitrification), and water could be returned to the air as water vapor (by incineration). Other sludge constituents would have to be retained for disposal. Presumably, the disposal of ash would need to be on the land because it is the only place left--note, however, that incineration on ships is being explored (16).

Typical municipal sludge solids are about 70 percent combustable. Incineration in multiple hearth furnaces, fluidized bed incinerators, and by the wet air oxidation process, is practiced in the United States. With the exception of the wet air oxidation process, the remaining ash is small in volume as compared to the original mass of sludge and is physically easy to handle. Incinerators must be designed and operated in a fashion to control air pollution, and this is a factor which potentially reduces the attractiveness of incineration. With most types of incinerators, the emission of particulates is of concern, and, with some, the discharge of volatile organic compounds is a potential problem. Farrell and Salotto (17) presented data indicating that common pesticide and polychlorinated biphenyls can be adequately destroyed by incineration. Analysis of the fate of heavy metals in incinerators by the same authors indicated that, while mercury is volatilized and contained a gaseous form in discharges from incinerators, other heavy metals are predominantly contained in the ash or the fly ash. The Environmental Protection Agency's performance standards for sludge incinerators are based on use of a venturi scrubber with monitoring of mercury discharges.

A second factor which reduces the attractiveness of sludge volume reduction by incineration is the difficulty (or expense) of removing sufficient moisture from typical municipal sludges to allow the sludge solids to support combustion. The cost and energy and resource consumption involved in dewatering sludge to the point of achieving thermally self-sustaining conditions in an incinerator are very high. Most commonly, supplemental fuel must be used. The current energy situation has caused some cities to reevaluate sludge disposal schemes based on incineration. Minneapolis-St. Paul, for example, currently incinerates sludge, but, because of increasing fuel costs and the lack of reliability of fuel supplies, they are changing their approach (18).

Other sludge treatment processes in addition to incineration also result in discharge of some sludge constituents to the air. Anaerobic digestion of sewage sludges is a well established technique which accomplishes partial removal of sludge constituents and produces a fuel (methane gas) which, ultimately, is returned to the atmosphere. Anaerobic digestion is a rather standard part of sludge disposal schemes proceeding discharge to land or water. Aerobic digestion also results in return of some organic sludge constituents to the environment. Pyrolysis has recently been evaluated as a means of fuel production from sludges (19).

## Disposal to Land

Two different types of sludge disposal on land are practiced. In some cases, land is used for stockpiling of sludges; and, in other cases, sludges are beneficially used on agricultural land.

Stockpiling operations involve the discharge of liquid sludge to lagoons and the placing of dewatered sludge in landfills. These are very common types of sludge disposal--Farrell (20) estimated that 40 percent of the municipal sludges in the United States are disposed of by landfilling. The fraction is probably higher with industrial wastewater treatment plant sludges, because barriers to beneficial use on agricultural land more commonly occur. Landfilling of sludges is not necessarily an economical solution to the problem of sludge disposal, for removal of water from sludges is an expensive proposition. The alternative method of stockpiling--the accumulation of liquid sludge in a lagoon--results in storing of excessive volumes of sludge, and lagoons ordinarily remain permanently in the liquid state.

Landfilling often is not a permanent solution to the problem of sludge disposal. This is because disposal sites inevitably become filled, and new locations must be sought, and because future land use requirements might require use of the site for a different purpose. In cases where it has become necessary to vacate "permanent" sludge lagoons, appreciable expense has been involved (21).

An additional factor which makes lagooning and landfilling of sludge unattractive is that the potential is presented for groundwater and surface water contamination by sludge constituents. With modern environmental quality standards, situations are created in which sludge stockpiling might be accompanied by a perpetual obligation for monitoring and, perhaps, treatment of leachate from the stockpiling operation.

Landfilling is not a very satisfying means of dealing with sludges, for sludges are not returned to a sector of the environment with which they are compatible or in which they can beneficially be used. The author previously has suggested that stockpiling frequently has been practiced when more desirable sludge disposal techniques would have been appropriate (22). However, it must be admitted that in many cases--particularly with industrial sludges--no other viable option currently exists. This would be true in the case of an industrial sludge containing a constituent which would be incompatible with agricultural utilization of sludge, which was not combustible, and for which no feasible means of reclamation exists.

Use of typical municipal sludges (and industrial sludges with similar characteristics) on agricultural land allow productive use to be made of sludge constituents. Sludge is applied to land, in the liquid, dewatered, and dried forms. An analysis by Riddell and Cormack (3) indicated that, for most situations, it is more economical to transport the water contained in sludges to the disposal site than to attempt to remove it by dewatering and/or drying processes.

In the order of four to five percent of the weight of solids in undigested, mixed, primary, and secondary sludges from municipal plants is nitrogen, and the sludge contains a similar fraction of phosphorus (as $P_2O_5$) but only about a tenth as much potassium (as $K_2O$). When the N, P, and K content of sludges is compared to plant requirements, sludge is seen to be potassium deficient, and supplemental addition of potassium may be necessary. Anaerobic digestion solubilizes about half of the nitrogen and phosphorus. Plants which dewater sludge prior to land application recycle these solubilized nutrients to the wastewater treatment plant. When sludge is applied in the liquid form, the nutrients are not recycled to the treatment plant. In addition to nitrogen and phosphorus, sludges contain many micronutrients, humus, and (when applied in liquid form) irrigation water.

Systems for agricultural utilization of sludges must be operated with due regard for safeguarding human health and avoiding environmental impact. Nitrogen, heavy metals and other trace constituents of sludges, pathogenic organisms, and viruses are constituents of sludges which warrant special concern.

Nitrogen is contained in sludges as the ammonium ion and in the form of organic compounds. Transformations of nitrogen in soil systems include mineralization, nitrification, denitrification, immobilization, volatilization, adsorption, and uptake by plants. As a result of these transformations, some nitrogen is lost by volatilization of ammonia, some is nitrified and then denitrified to form nitrogen gas which also is lost to atmosphere, some is taken up by plants, some is converted by soil bacteria to rather stable organic compounds which slowly release nitrogen, and some is carried away with percolating water as the nitrate ion. It is the latter fate of nitrogen which is of concern from the standpoint of protection of environmental quality and human health. The concern stems from the fact that, while the positive ammonium ion is held by soil particles, nitrate ions formed by oxidation (by autotrophic bacteria in the soil) of ammonia are mobile. Nitrate contamination of ground and surface waters is of concern because it serves as a nutrient and, of more concern, because it causes infant methemoglobnemia. The U. S. Public Health Drinking Water Standards (23) limit nitrate concentrations in drinking water to 10 mg/ℓ as N.

Heavy metals in sludge are of concern when sludge is applied to land due to their potential toxicity to plants and to animals, including humans. Similar concern exists for trace quantities of recalcitrant organic compounds which might be found in sludges. Some of the heavy metals (such as zinc, copper, and nickel) are principally of concern because of their toxicity to plants while others (like cadmium, lead, and mercury) are of interest principally because of their possible accumulation in food chains.

Although appreciable quantities of sludge have been applied to agricultural land over long periods of time [Farrell (20) estimated that 25 percent of the municipal sludge in the United States currently is used on agricultural land], much remains to be learned about the fate of heavy metals in soil systems and about appropriate guidelines for making judgments on appropriate sludge application rates. In Great Britain, regulations have been adopted which are based on the "zinc equivalent" of the sludge (24). The zinc equivalent is an approximate measure of the potential toxicity of sludge to plants due to its content of zinc, copper, and nickel. A similar technique recently has been proposed in this country (25). This approach considers not only the zinc, copper, and nickel content of sludge, but also the cation exchange capacity of the soil. Neither the British technique nor the American proposal take into account the possible problem of heavy metals in the food chain.

Heavy metals are of significant concern in all types of sludge reclamation and disposal. The obvious solution to problem would seem to be the pretreatment of industrial wastewaters to avoid heavy metal introduction into sewer systems. However, there is increasing evidence to indicate that industries may not be the principal sources of heavy metals in municipal sewage. For example, Klein (26) recently indicated that if there were zero discharge of industrial wastes into the New York City sewers, still 91 percent of the copper, 80 percent of the chromium, 94 percent of the zinc, and 84 percent of the cadmium would remain in the sewage.

It must be assumed that wastewaters from municipalities may contain bacteria, protozoans, viruses, and helminths which can cause serious human diseases. The same is true of industrial wastewaters when wastes from employees are not segregated. Conventionally, sludges have been anaerobically digested prior to application on land. This may significantly reduce the number of human pathogens, but their absence cannot be assumed (27). Thus, as conventionally practiced, land application of waste sludges must be practiced with concern for the possibility of human disease transmission. The opportunity for disease transmission could arise from production of aerosols during application of sludge, by human contact with soil to which sludge has been applied, from water containing runoff or leachate from sludged areas, or from contact with or consumption of agricultural products grown on land which received sludge.

However, in spite of the fact that sludges contain pathogenic organisms and viruses, the disease agents are not totally destroyed in conventional sludge treatment processes and application of sludge on land is a rather prevalent practice in the United States, there is little indication of actual disease transmission by the sludge route. The author is unable to cite literature which documents a case of infectious disease being attributed to application of sludge on land in the United States.

Should the possibility of disease transmission interfere with schemes for disposing of sludge on agricultural land, then, consideration may be given to inactivation of the organisms prior to disposal. Sludge pasteurization is being carried out in Germany (28), and a recent study in the United States (29) suggests that pasteurization can be provided at a cost which would not preclude its consideration in sludge treatment and disposal schemes.

## People Problems

As outlined briefly in the previous sections, very significant technical problems exist in implementing alternatives to ocean disposal of sludge. However, these technical problems seem almost trivial as compared to the "people problems" involved. Indeed, the current state of affairs which forces coastal cities to seek alternatives to ocean disposal of sludge was probably created, in some degree, by social rather than technical considerations. Proposals for sludge disposal on agricultural land are particularly vulnerable to public criticism, because the people on the receiving end of the schemes are not the same ones that generate the rather unglamorous material. Experiences of the Metropolitan Sanitary District of Greater Chicago serve to illustrate the problem (30). In the mid 1960's the District outlined a sludge disposal plan which involved transport of liquid digested sludge about 65 miles south to Kankakee County, Illinois for application on low quality agricultural land. The plan was expected to reduce significantly the District's sludge disposal cost, allow productive use to be made of sludge constituents, provide fertilizer for crops, improve the value of agricultural land, and increase the tax base of the Kankakee County. The proposal was flatly rejected by Kankakee County. [More recently, the Chicago Metropolitan Sanitary District has been able to implement a plan for transporting liquid digested sludge about 180 miles to Fulton County, Illinois where it is used to convert abandoned strip mines to productive agricultural land (31).] The City of Philadelphia has encountered similar resistance in attempting to begin developing a land-based alternative to ocean disposal. In recent weeks, it has been announced (32) that a joint Environmental Protection Agency-City of Philadelphia pilot project for studying land disposal of Philadelphia sludge in Letterkenny Township, Pennsylvania as an alternative to ocean disposal was abandoned because of public resistance.

Schemes for returning major portions of sludge to the air by incineration have been perhaps somewhat less likely to be curbed by public pressures. However, they are not immune to such pressures, as experiences of the Piscataway Plant of the Washington Suburban Sanitary Commission illustrate (33). In the early 1970's, alternatives for disposal of sludge from the Piscataway Plant were studied, and incineration was adopted as the most feasible solution. The Washington Suburban Sanitary Commission then proceeded with development of the incineration scheme, and construction was completed in 1974. After only 70 hours of operation, incineration was banned by the Prince Georges County Council and the U. S. Environmental Protection Agency because of reports of high levels of lead in the blood of children in the vicinity of the incinerator. On the basis of an extensive study of blood lead levels in children and treatment plant employees, the County Health Officer concluded that no health hazard existed and the Environmental Protection Agency retracted its restrictions. However, the County Council denied a request by the Washington Suburban Sanitary Commission to operate the incinerators to permit collection of data on environmental impact.

Experiences at Chicago and Piscataway are illustrations of the socio-political barriers which may be expected to be encountered by coastal cities seeking to discontinue disposal of sludge at sea. Cities (and large industries) simply do not have within their jurisdiction the resources needed to accomplish satisfactory disposal of sludges. Regional schemes involving the cooperation of urban producers of sludge and rural recipients would seem to be necessary to cope with the problem of managing the residues produced in treatment of wastewater generated by modern societies. Such sludge disposal programs require extensive public education efforts and, perhaps, legislative action to enable cities to be able to rely on rural areas in developing plans for sludge treatment and disposal.

## Costs

The cost of treating and disposing of residues from wastewater treatment currently approaches the cost of wastewater treatment itself. All indications are that upgrading of environmental quality standards will increase the cost of sludge management even further (22).

Available tabulations of estimated costs of alternative sludge disposal are rather consistent in ranking ocean disposal, land application, and incineration in order of increasing costs [see, for example, Riddell and Cormack (3) and Burd (34)]. A pertinent illustration of costs of alternatives to ocean disposal has been provided by Guarino and Townsend (35) who indicated that while ocean disposal at Philadelphia currently costs $17 per ton of dry solids, land application would be expected to cost $50 per ton of dry solids and incineration $62 per ton of dry solids.

## Summary

There are but two alternatives to marine disposal of sewage sludges--disposal to land or disposal to air. There are no other places on earth. An additional, but potentially attractive alternative is to avoid the need for disposal by making productive use of sludge constituents. In general, the costs of the potential alternatives to ocean disposal of sludges exceed the costs of ocean disposal.

Many of the sludge constituents which are objectionable when sludges are discharged at sea are equally objectionable if they are discharged to the air or land. Heavy metals, for example, are as out-of-place in the atmosphere or in terrestrial environments as they are in the ocean. Technical problems are also associated with attempts to recycle waste sludges into productive use to avoid the need for disposal. The heterogeneous composition of many sludges, for example, renders reclamation difficult.

But perhaps the most difficult problems in establishing acceptable sludge disposal schemes are not technical in nature. "People problems" abound in the area of sludge treatment and disposal. Cities which generate sewage sludges do not have within their jurisdiction the land or air resources required to receive the sludges. Significant attention to public relations, regional planning, and the political process will be required to satisfactorily manage waste sludges.

## Literature Cited

1. Vesilind, P. A., "Treatment and Disposal of Wastewater Sludges," 236 pp., Ann Arbor Science Publishers, Ann Arbor, Michigan (1974).
2. Dick, R. I., "Sludge Treatment," In "Physicochemical Processes for Water Quality Control," 533, John Wiley and Sons, New York (1972).
33. Riddell, M. D. R. and Cormack, J. W., Paper presented at 39th Annual Meeting of Central States Water Pollution Control Association, Eau Claire, Wisconsin (1966).
4. Clark, N. A., Stevenson, R. E., Chang, S. L., and Kabler, P. W., American Journal Public Health (1961), 51, 1118.
5. Lund, E. and Rønne, V., Water Research (1973), 7, 863.
6. Krige, P. R., Journal Institute of Sewage Purification (1964), 63, 215-220.
7. Environmental Protection Agency, Federal Register (1973) 38 (198), 28610.
8. Anon., Air/Water Pollution Report (1975), 13 (7), 68.
9. Singer, P. C., Water and Wastes Engineering (1974), 11 (12), 25.
10. Garcia-Bengochea, J. I. and Black, A. P., Water-1972, Amer. Inst. of Chem. Engr., Symposium Ser. 129 (1973), 69, 329.
11. Parker, D. S., Niles, D. G., and Zadick, F. J., Journal Water Pollution Control Federation (1974), 46, 2281.
12. Fujita, H., Water and Sewage Works (1967), 114 (3), 73.
13. Hurwitz, E., Proceedings, 12th Industrial Waste Conference, Purdue University, Extension Series (1957), 94, 395.
14. Leary, R. D., Proceedings, 9th Industrial Waste Conference, Purdue University, Extension Series (1954), 87, 173.
15. Minnick, L. J., Professional Engineer (1973), 43 (6), 36.
16. Anon., Conservation News (1974), 39 (21), 5.
17. Farrell, J. B. and Salotto, B. V., Proceedings, National Symposium on Disposal of Wastewaters and Their Residuals, Water Resources Research Institute, North Carolina State University, Raleigh (1974), 186.
18. Bergstedt, D. C., Proceedings, National Conference on Municipal Sludge Management, Pittsburgh, Pennsylvania, Information Transfer, Inc., Washington, D. C. (1974), 187.
19. Harkness, N., Oliver, A. R., Gough, A. J., Tabberer, R., and Keight, D., Water Pollution Control (1972), 71, 17.

20. Farrell, J. B., Proceedings, National Conference on Municipal Sludge Management, Pittsburgh, Pennsylvania, Information Transfer, Inc., Washington, D. C. (1974), 5.
21. Anon., Illinois Central Magazine (1970), 62 (4), 2.
22. Dick, R. I., Journal Environmental Engineering Division, Proceedings American Society Civil Engineers (1974), 100 (EE5), 1077.
23. U. S. Public Health Service, "Drinking Water Standards," 61 pp., U. S. Government Printing Office, Washington, D. C. (1962).
24. Chumbley, G. G., "Permissible Levels of Toxic Metals in Sewage Used on Agricultural Land," 12 pp., British Ministry of Agriculture, Fisheries, and Food, Pinner, Middlesex, England (1971).
25. Walker, J. M., Compost Science (1975), 16 (2), 12.
26. Klein, L. A., Lang, M., Nash, N., and Kirschner, S. L., Journal Water Pollution Control Federation (1974), 46, 2653.
27. McKinney, R. E., Langley, H. E., and Tomlinson, H. D., Sewage and Industrial Wastes (1958), 30 (12), 1469.
28. Kuhrl, H., Water Research (1972), 6, 555.
29. Stern, G., Proceedings, National Conference on Municipal Sludge Management, Pittsburgh, Pennsylvania, Information Transfer, Inc., Washington, D. C. (1974), 163.
30. Bacon, V. W. and Dalton, F. E., Public Works (1966), 97 (11), 66.
31. Halderson, J. L., Lynam, B. T., and Rimkus, R. R., Proceedings, National Conference on Municipal Sludge Management, Pittsburgh, Pennsylvania, Information Transfer, Inc., Washington, D. C. (1974), 129.
32. Gould, H., Philadelphia Inquirer (1975), 292 (87), 2-B.
33. Brough, K. J., Proceedings, National Conference on Municipal Sludge Management, Pittsburgh, Pennsylvania, Information Transfer, Inc., Washington, D. C. (1974, 187.
34. Burd, R. S., " Study of Sludge Handling and Disposal," Federal Water Pollution Control Administration, Washington, D. C. (1968).
35. Guarino, C. F. and Townsend, S., Proceedings, National Conference on Municipal Sludge Management, Pittsburgh, Pennsylvania, Information Transfer, Inc., Washington, D. C. (1974), 151.

# 29

# Effects of Baled Solid Waste Disposal in the Marine Environment—A Descriptive Model

THEODORE C. LODER

Department of Earth Sciences, University of New Hampshire, Durham, N.H. 03824

The disposal of municipal and residential solid waste materials is becoming more of a problem each year as disposal costs increase, as suitable areas for sanitary landfill and open dumping are used up and as the federal water and air pollution regulations become more strict. The concept of partial recycling has met with only limited success because of economic and political problems, beside not being capable of handling the majority of our solid wastes for a number of years (1). Even with recycling, there is still a certain amount of material unsuited for recycling or burning, which must be disposed. Disposal of this material in the ocean is one alternative that has attracted increasing interest in the past several years.

There has not been sizeable ocean dumping of refuse during the past 25 years, although prior to that time several U.S. coastal cities (New York, Oakland, and San Diego) dumped part of their refuse at sea (2). This practice was stopped in part due to wash up of materials on nearby beaches and general public disfavor. Recently the only dumping of solid wastes recorded was in the Pacific Ocean (26,000 tons in 1968, 240 tons in 1973) with most of this from military institutions (2).

The problem with the earlier dumping of solid wastes (refuse) was the post-dump dispersal of both flotables and low density sinkable materials. The movement of this material on the bottom results from a combination of wave orbital velocities (associated with passing waves) and tidal currents (2). First (3) reported that bottom current velocities of 15.2 cm/sec. are great enough to move cans on the bottom; however, at a disposal site off Boston at 55 m (180 ft) incinerator residues did not drift significantly.

In 1967, Dunlea (4) proposed that solid wastes be shredded, compacted and encapsulated prior to disposal at sea to aid in sinking and prevent scattering of materials. Devanney *et. al.* (5) later studied the economics of waste disposal at sea and found it to be economically feasible. Bostrom and Sherif (6) proposed that baled solid waste be dumped in tectonically active downwelling areas with high sedimentation rates. Others have suggested that compacted bales be used as artificial reef materials (7, 8, 9, 10).

There emerge two rather distinct concepts concerning marine disposal of wastes (5). The first is that there will be major ecological destruction of an area in the ocean and consequently the dumping and its resultant effects should be confined to as local an area as possible. This is effectively the present federal policy in choosing specific dumpsites. In line with this concept, it would be best to dump in an area of low level biological activity, below the permanent thermocline (below 1000 m) to reduce the biological impact and reduce the rate of decomposition because of the cold temperatures and increased pressures.

The second concept is that the wastes represent a resource whose degradation at the proper rates could provide nutrients and organic matter in nutrient poor areas, such as many tropical surface waters. In addition, compacted bales could provide substrate for the formation of artificial reefs (10). These second concepts require that the waste material be placed in relatively shallow shelf waters preferably below wave base, but close to the photic zone to have some effect (perhaps 30 to 100 m). Because the potential for impact is greater on the shelf, both from a biological and a public point of view, a dumpsite in this area would have to be chosen and monitored carefully. Certainly from an experimental point of view it would be easier to study a shallow dumpsite than one on the continental slope or rise. In addition, it would also be more economical with regard to transport of the waste. Because of these two different philosophical views, I have chosen to look at both a shallow and a deep dumpsite in this model.

Research providing much of the basic data for this paper has been carried on by several groups who have studied the effects of both simulated and actual compacted solid wastes placed in seawater for up to three years. Bogast (11) placed solid waste bales in enclosed saltwater lagoons in Hawaii and monitored the water chemistry. Pratt *et. al.* (12) carried out a detailed chemical and biological examination of shredded and compacted solid waste materials in the laboratory. Rowe (10) has emplaced commercial bales at several depths and monitored them since 1971. Loder *et. al.* (13) emplaced small (30-40 kg) experimental bales of shredded solid waste materials in 15 m of water off the New Hampshire coast in 1971, and several bales have been emplaced on a rack off the edge of the continental shelf in 1974 (14).

## A Commercial Dump--Sequence of Events

The following sequence of events might take place after a commercial dump has occurred and is based on the experimental work of the above-mentioned authors.

The bales will most likely consist of about one and one half tons of shredded compacted solid waste (residential trash) with a density slightly greater than seawater and an approximate composition shown in Table I.

Table I. Physical and rough chemical composition of municipal refuse from the United States East Coast (15).

| Physical | Weight %* | Rough Chemical | % | Dry % |
|---|---|---|---|---|
| Cardboard | 7 | Moisture | 28.0 | ---** |
| Newspaper | 14 | Carbon | 25.0 | 34.7 |
| Miscellaneous | 25 | Hydrogen | 3.3 | 4.6 |
| Plastic Film | 2 | Oxygen | 21.1 | 29.3 |
| Leather, molded plastics, rubber | 2 | Nitrogen | 0.5 | 0.7 |
| Garbage | 12 | Sulfur | 0.1 | 0.1 |
| Grass and dirt | 10 | Glass, ceramics, etc. | 9.3 | 12.9 |
| Textiles | 3 | Metals | 7.2 | 10.0 |
| Wood | 7 | Ash, other inerts | 5.5 | 7.6 |
| Glass, ceramics, stoves | 10 | | 100.0 | 100.0 |
| Metallics | 8 | | | |
| TOTAL | 100 | | | |

* This value will vary depending on the location and season.
** Because of the relatively high moisture content of solid waste (28%) a 3000-pound bale would actually only contain 2160 pounds of waste material. Of this approximately one-third would be carbon.

If the dumpsite is in shallow water where it might be used as an artificial reef, the type of bale covering is very critical for possible attaching organisms. Figures 1 and 2 show the difference in growth of benthic algae on experimental bales with and without a quarter-inch (6.4 mm) plastic mesh covering. Bales with a suggested alternative solid plastic covering will probably take longer to sink because the release of air will be initially slowed by the plastic. The solid plastic sheeting will also slow bottom water interchange with the bale and this would slightly reduce the decomposition rate.

Once the bale has reached the bottom the oxygen concentration within the bale will drop to nearly zero in only a few hours to several days. This initial drop is probably due to inorganic oxidation of metals (mainly iron and aluminum) present in the bale and by oxidation of food decay products produced during storage of the bales after manufacture and before emplacement (13). Once this initial drop has occurred the oxygen concentration may stay near zero or rise as oxygen diffuses into the bale. It appears

*Figure 1. A 15–20-kg waste bale after immersion for about 11 months at 15 m. This bale was not meshed, and there has been no benthic organism attachment except on straps.*

*Figure 2. A 40–45-kg waste bale after immersion for about 14 months at 15 m. This bale was meshed, and there has been substantial attachment of benthic algae to the bale. The white patches are mats of sulfur-oxidizing bacteria such as in Figure 3.*

to take 1-2 months in shallow marine waters for the bacterial populations to become established well enough to produce hydrogen sulfide in solid waste materials. This might take substantially longer at a deep dumpsite because of the reduced decomposition rate which results from a combination of lower temperatures and increased pressure (16).

Oxygen consumption during the biological degradation of the waste bales and subsequent depletion of dissolved oxygen in the bottom water will be one of the most important impacts of the dumping. The rate of decomposition of the bales will be mainly a function of the depth of water, temperature, and type of waste material (i.e. percent of easily decomposed organic material, such as food wastes). The size of the shredded material in the bale and the degree of compaction will affect the porosity of the bales, which will in turn affect the diffusion of sulfate, oxygen, and nutrients in and out of the bale and ultimately affect the decomposition rate. Porosities of shredded, baled solid waste are about 50 to 70 percent (17). Although theoretical models have been developed for these processes in sediments (18), it is uncertain if these models can be applied directly to the bale problem.

Measurements of oxygen consumption have been made by several workers on actual compacted solid waste materials, both in the laboratory and the field (Table II). Although the data have a wide range, actual values at a dumpsite where mean temperatures might be 4-10°C would show less variation, perhaps 15 to 60 ml $O_2/m^2$-$hr^{-1}$ with an estimated average of 30 ml $O_2/m^2$-$hr^{-1}$. High values would be expected in a shallow water site during late summer months while lower rates might be found at a continental rise site because of the lower degradation rates.

The oxygen content of the bottom water will also be a function of the season, location, and depth of the site. Shallow shelf areas (50 m) off the northeast coast have annual ranges of 5 to 8 ml/l, while a shelf basin (> 200 m deep) such as the Wilkinson Basin (Gulf of Maine) with somewhat restricted flushing, has lower values (4 to 5 ml/l) (19). A continental slope site might have values around 7 ml/l (20).

The pH of the bale water will initially drop, probably due to food acids, formation of metal hydroxides, and release of carbon dioxide. Loder *et.al.* (13) found an average pH drop to 6.5 while Pratt *et. al.* (12) reported a drop to pH of 6.0 within a day after immersion. The pH in our experimental bales rose to between 8 and 9 after over three months of immersion. It has remained this high for over two years for most of the bales, although some bales have pH values of between 6.5 and 7.0 (13). The pH of these systems is controlled by complex reactions involving carbon dioxide, organic acids, sulfides, and ammonia (23). The pH can have an important effect on biological activity, metals solubility, and the forms of sulfides present in the bale.

There will be some phosphorous in the waste, in a quantity

Table II. Results of oxygen consumption experiments on baled solid wastes immersed in seawater.

| Experiment description (Ref.) | Temperature (°C) | Oxygen consumption ($ml\ O_2/m^2$-$hr^{-1}$)* Ave. (#values) | Range |
|---|---|---|---|
| Laboratory Test blocks (12) | | | |
| 2 months immersion | 3.2-7.5 | 39 (6) | 23-55 |
| 5 months immersion | 8-9 | 26 (4) | 16.6-30.6 |
| 7 months immersion | 16-17 | 83 (2) | 70-90 |
| Whole, exper. bales (30-40 kg) in 15 m water at Isles of Shoals, N.H. after 12 months (13) | 8-10 | 16 (4) | 10.5-17.4 |
| Whole, exper. bales (10-20 kg) in laboratory after 12-18 months (21). | 3-5 | 25 (2) | 23-28 |
| | 15 | 22 (1) | |
| | 14-17 | 76 (2) | 46-106 |
| Commercial sized bales in 15 m water at WHOI (22) | 21.7 | | 30-126 |

* ($ml\ O_2/m^2$-$hr^{-1}$) x (0.0248) = micro-moles oxygen (O)/$m^2$ - $sec^{-1}$

of probably well less than one percent of the organic matter present. The phosphorous will be oxidized to phosphate and will initially adsorb to the metal oxy-hydroxides in the waste, or form ferric phosphate. This phosphate will be released as the bale system goes anoxic (24).

In several experiments on solid waste, phosphate reached levels of 20-30 μM and then dropped off rapidly after 3 to 4 months (12, 13). It appears that the greatest amounts of phosphate will be released during the first few months of decomposition.

Most nitrogen in the bales will be converted to ammonia which then may diffuse from the bale system. Concentrations of ammonia reached a little over 400 μM in closed solid waste-seawater systems (12), while ammonia in experimental bales at sea averaged between 12 to 35 μM with highest values around 100-150 μM (13). The ammonia diffused from a shallow waste dumpsite could have an impact on productivity in inshore waters where nitrogen is often the limiting nutrient (25). In a deep water dumpsite the impact would be small. Since the relative amount of nitrogen to carbon is very low as a result of the high cellulose content of trash (Table I), the bacteria will depend on nitrate diffused into the bales as a nitrogen source once the nitrogen within the original bale is utilized.

Dissolved organic matter concentrations within the bale will rise to 2 to 3 orders of magnitude greater than the surrounding water. Pratt *et. al.* (*12*) reported concentrations as high as 1200 mg dissolved organic carbon (DOC)/l after about three months, while I found concentrations of 50 to several 100 mg DOC/l after two years of immersion for our experimental bales. Since normal ocean concentrations of DOC range from about 0.5 to several mg DOC/l, there will be a diffusional flux of dissolved organic matter out of the bale, which could contribute to the bale decomposition or solution at a rate 1 to 2 orders of magnitude faster than sulfide reduction rates predict.

A diffusion equation (Table IV) can be used to calculate the rate of loss of organic materials where $\Phi = 0.6$; $C_{x'} = 1$ mg C/l; $C_o = 100$ mg C/l; and x' = 10 cm. Estimating the diffusion coefficient ($D_s$) is difficult, but it is probably smaller than the molecular value for simple ions ( $10^{-5}$ $cm^2/sec.$); consequently $D_s = 10^{-6}$ was used in this example. The resultant loss rate then is 5 μ-moles $DOC/m^2\text{-sec.}^{-1}$. This could be an order of magnitude slower or faster depending on how realistic the diffusion rates are and whether or not the organic matter will continue to be solubilized at the same rate after the first few years of immersion. At the above rate a 3000-pound bale would "dissolve" in about 30 years.

Trace metals in solid wastes have been determined by analysis of incinerator wastes (*3*), and reactions of some of these metals have been discussed by Pratt *et. al.* (*12*). Many of the trace metals will form relatively insoluble sulfides, precipitate, and remain within the bale. However, some of the metals may form soluble polysulfide complexes or dissolved organic complexes due to the relatively high amount of dissolved organic matter in the bales. This may mobilize the metals and allow them to diffuse out of the bale. It is not possible to estimate the proportions of each metal that will remain or diffuse out of a bale at this time. This important area of research needs more work.

Hydrogen sulfide is produced within the bales as a result of the reduction of sulfate in seawater through a complex series of microbiological reactions, which decompose the organic matter within the bale releasing carbon dioxide, ammonia, hydrogen sulfide and phosphate. Pratt (*12*) reported the release of hydrogen and methane early after solid waste materials in seawater had gone anoxic, but found only $H_2S$ was produced after about 5 months.

The rate of production of hydrogen sulfide is a function of the temperature and pressure affecting the microbiological activity with the limiting rate a function of the diffusion rate of sulfate into the bale. Once formed, the hydrogen sulfide may be bound within the system by metal sulfide precipitation, or may leave the system by volatization or diffusion. Once it reaches the surface of the bale, it may be oxidized to sulfur by sulfur-oxidizing bacteria such as the Beggiatoa, sp. and Thiothrix, sp. (Figs. 2 and 3) commonly found on the solid waste bales (10, 12, 13, 26). It may undergo chemical oxidation to thiosulfate, sulfite, or sulfate with most all of it ending up as sulfate after 30 to 40 hours in seawater (27).

The maximum rate of production of sulfides within solid waste materials appears to roughly be about 0.1 μ-moles/l-day$^{-1}$ (0.0012 μ-moles/l-sec$^{-1}$) (12, 13). This is several times higher than a value of 0.04 μ-moles/l-day estimated for some anoxic sediments (28). Every μ-mole of $H_2S$ produced will use up five μ-moles of oxygen if totally oxidized. If we assume the amount of sulfide lost from a bale equals the production rate, then for a 3000-pound bale with 60 percent void space and a 5.8 meter surface area, we can calculate an equivalent oxygen consumption rate of 0.78 μ-moles oxygen/m$^2$-sec$^{-1}$. Although this is probably about the maximum rate, it represents a high percentage of the oxygen consumption by a bale surface as measured (Table II) with the rest being due to respiration by bacteria and organisms on the outside of a bale.

Using the maximum rate of sulfide production one can calculate the maximum rate of decomposition or oxidation of the organic matter in a bale. It is generally assumed that two moles of organic carbon are oxidized to carbon dioxide per mole of sulfate reduced (28, 29).

If the maximum sulfide production rate is about 0.1 m-mole/l-day$^{-1}$ then 0.2 m-moles of organic carbon would be oxidized. Thus a 3000-pound bale of solid waste with a 60 percent porosity and 25 percent carbon will take a minimum of about 500 years to decompose. This time is substantially longer than predicted based on dissolved organic carbon diffusion loss. Since this sulfide production rate is the maximum estimated rate, the bale decomposition could take an order of magnitude longer if decomposition rates are an order of magnitude too high. This might be the case for the deep sea. It is not surprising then that newspaper removed from our submerged bales after several years is easily readable.

It is difficult to predict what will happen to a single bale or group of bales (a dump) over an extended time period because the biochemical and biological degradation rates will change. The easily solubilized and oxidized materials will be removed first, leaving more resistant materials, resulting in slower degradation rates. As a bale becomes covered with sediment, the diffusion rates will also slow. However, as the biochemical processes slow, the bales should become more attractive to encrusting organisms. Destruction by boring organisms such as various species of polychaetes or isopods has been shown (Fig. 4). Boring shipworms (Teredinids) may also attach and colonize in these bales (30). If boring organisms do colonize the bales in large numbers the degradation rate of a bale could decrease to several decades from the 500 year estimate based only on sulfide

*Figure 3. Close-up of a mat of sulfur-oxidizing bacteria on a food waste bale immersed at 15 m. Both genera* Beggiotoa *and* Thiothrix *of the order Beggiatoales are present. (For scale reference the mesh openings are about one-quarter inch (6.3 mm).)*

*Figure 4. Photograph of newspaper from the outside 10 cm of an experimental waste bale immersed for 13 months at 15 m. The burrows on the left were found in the outside 2–4 cm and were made by a burrowing isopod* Limnora ligorum. *The dark areas are due to precipitation of iron sulfides.*

rates.

Since most of the volume of a bale is paper-type material, there will be a large volume decrease of a bale as this material is removed. In the end there would remain a pile of carbon-rich relatively inert materials consisting of glass, plastics, stone, dirt and some metals. These remaining materials will be either scattered by organisms and currents or buried depending on their quantity and dump site conditions. All the above parameters will probably result in different overall rates for a continental shelf dump vs. a deep water dump. Overall rates will probably be slower at a deep water site, but whether they will be significantly slower is not known and difficult to estimate.

## Input to Dump Model

In the following section the various parameters are discussed that will affect the final size and chemical changes at a hypothesized dumpsite located either on the continental shelf or on the continental slope. Some of the input data can be stated quite accurately such as the size and density of the bales. For other data one is forced to choose a most-likely estimate and give it a range of an order of magnitude or more. Some of these estimated values are site specific, such as currents, and could be better known prior to an actual dump. Other values such as the diffusion rates are only estimates based on limited measurements and poorly known diffusion coefficients. The parameters chosen are summarized in Table III.

The weight of commercial-sized bales that have been used in experimental work and landfill has been about 3000 lbs (1360 kg) (10, 31). Generally one dimension is variable while 2 directions remained fixed as a function of the size of the final compressing chamber. Although the San Diego bales were rectangular in shape (31), it is assumed that for this model the bale shape is a cube with dimensions such as to contain a given weight at a given density. A nearly cubic bale of about 3000 lbs (1360 kg; 1.08 m on a side) is easy to handle in terms of strapping, wrapping, and shipping, and is commercially feasible to manufacture (32).

Bale density should be greater than that of seawater (about 63.9 lbs/ft.$^3$ - 1.024 g/cc) so they will sink as soon as they enter the water. It is possible to make bales of shredded waste to densities of slightly over 70 lbs/ft.$^3$ (1.121 g/cc) although a density of slightly under that would be more likely.

A bale must be strapped together prior to dumping. This may be done using No. 6 to 10 carbon steel wire (possibly vinyl coated) or polypropylene strapping with metal or plastic clips. Uncoated steel wire lasts at least 2 to 3 years in shallow seawater (22); however, plastic strapping would last substantially longer. Encapsulation may be done to prevent initial loss of floatables and shredded material, but will reduce the rate of interchange with the water. This could range from a plastic

Table III. Summary of ranges and average or most-likely values for input variables to model.

| Variable Name | Units | Symbol | Estimated Average | Estimated Range |
|---|---|---|---|---|
| 1. Bale weight | (lbs) | W | 3000 | 2800-3500 |
| 2. Bale density | ($lbs/ft^3$) | $\rho_\beta$ | 68 | 64-72 |
| 3. Dumping rate of Input | (tons/day) | I | 2000 | 1000-6000 |
| 4. Bottom coverage density | (percent) | C | 20 | 5-50 |
| 5. Bottom currents | (cm/sec) | v | | |
| (a) shallow and mid-depth | | | 10 | 5-40 |
| (b) deep | | | 2 | 0.5-5 |
| 6. Thickness of mixed layer above bottom | (m) | h | 2 | 1-5 |
| 7. Concentration of X in bottom water | (μ-moles/l) | $[X]_i$ | | |
| (a) oxygen | | | 540 | 350-700 |
| (b) sulfides | | | 0 | 0 |
| (c) carbon (DOC) | | | 80 | 40-160 |
| (d) ammonia | | | 0.5 | 0-2.0 |
| 8. Diffusion rate of X | ($\mu\text{-moles/m}^2\text{-sec}^{-1}$) | $(X)_D$ | | |
| (a) oxygen uptake | | | 0.74 | 0.37-1.5 |
| (b) sulfide release* | | | 0.1 | 0.01-0.2 |
| (c) carbon (DOC) release** | | | 5 | 1-50 |
| (d) ammonia release*** | | | 0.1 | 0.01-0.5 |

* Based on about maximum sulfide production and steady-state release.

** Based on diffusion equation and estimated $D_s$ (see text)

***Based on diffusion equation with $D_s = 3.5 \times 10^6$ $cm^2/sec$. (18) (see text)

mesh allowing total chemical interchange (13) to polyethylene sheeting (6 mil) with a few holes punched to release gases (10).

The dumping rate for this model was chosen as 2000 tons/day. A rate below 1000 tons/day would not be economically feasible, while 5000 tons/day represents almost the entire output from a fairly large city such as Boston (32).

The percent of the bottom actually covered by dumped bales would depend on an environmental decision whether to spread a low-level input over a large area or to have a higher level input in a more local area. The less dense the dumping, the less the impact on benthic fauna and bottom water quality. It would also depend on whether the deposit was in shallow water and was used as an artificial reef (hence high percent of coverage) or deep water where spreading out might be necessary to reduce impact. The amount of scatter in the typical bales would also depend on how they were dumped and the water depth at the dump site. Bale coverage might range from a few percent up to nearly 100 percent. A barge dumping bales over a 5-minute period while drifting at one knot would end up with localized coverage of about 20 percent assuming a dump path of about 33 m and a constant dump rate with little water movement. Repeated dumping in the same spot would increase the coverage, however, 20 percent coverage will be considered most likely with an approximate range of 5 to 50 percent.

Because the bales are represented as cubes with five sides exposed to the water, each bale increases the effective surface area of the bottom by 5-fold. As coverage is increased the effective area of bale surfaces would increase, but the exposure of a bale surface to water currents would start to decrease as more bales inhibited the free flow of bottom water past the sides of the bales and only the tops had access to free current flow. Consequently, the actual situation in a real dump would be far more complex than described.

Choosing average values for bottom currents at a dumpsite is very difficult because the current regimes can be very site specific. Most data available has been for short durations and, consequently, long term seasonal as well as storm driven currents variations are not well known. In addition, most current measurements have been made well above the bottom to eliminate benthic boundary layer effects which result from bottom friction and slow the currents down just along the bottom (33).

A further consideration is that the currents are often tidal in nature even on the continental slope, and thus may change direction on a diurnal basis. It is conceivable that a parcel of water could oscillate over a dumpsite several times before being carried out of the dumpsite area, however, only undirectional flow is considered in this model. The rate of dumping (I) and bottom coverage (C) should be partly dependent on the currents at the dump site, because the currents will renew oxygen and remove degradation products. It is assumed that a dumpsite would not be chosen unless it had reasonable currents.

At a shallow water site on the continental shelf (30 to 200m) current speeds tend to show greater variation of both speed and direction and could be strongly affected by storms at some shallow depths up to 50 to 100 m. Current speeds range from nearly zero to about 100 cm/sec. with average currents of 5 to 40 cm/sec. (0.1 to 0.8 knots). Bottom drifter movements give an estimate of very minimum current speeds which can range from 0.5 to 4 cm/sec. on some shelf areas (34, 35).

There is less data available for deeper water off the continental shelf. Schmitz (36) reported near-bottom currents averaging 2.7 cm/sec (range 1.5 to 3.7), 2 to 3 m off the bottom in water 943 to 993 m deep in an area just east of Block Canyon.

I have chosen to calculate the effects of a dump on the concentrations of oxygen, hydrogen sulfide, dissolved organic carbon and ammonia in the bottom water at the bale site. The initial values used for oxygen are from (19), for dissolved organic carbon from (37) while for ammonia and hydrogen sulfide they are estimated. The diffusion rates for these parameters were estimated from various sources discussed in the previous section. Using this data and the following assumptions, the concentration of each parameter $[X]_t$ is determined for bottom water that just passed through the dumpsite after one year of dumping.

Assumptions made to simplify the model are:

1. Bales are of uniform density and size and are cubic in shape.
2. The bales are placed on the bottom in a uniform grid covering a given percent area of the bottom.
3. A water parcel surrounds each bale and remains around each bale for a given time determined by the bottom current speed. During this time period a calculated amount of material is added to it or removed as in the case of oxygen.
4. The volume of a water parcel is calculated using equation (8) in which (h) is the height above bottom to which complete mixing occurs (Fig. 5). The value for h would increase downstream at a real bale site, but here is assumed to remain constant. This value (h) is a complex function of current speed (v), turbulence induced by the bales and benthic boundary layer effects, all of which affect the vertical eddy diffusivity, and all of which are difficult to determine without actual field measurements. A value of 2 m (about 1.5 x the bale height off the bottom) was estimated for h.
5. Concentrations for each component $[X]_t$ are calculated using equations (2-10) in Table IV and are given in Table Vb.

There are nine input variables and each one can vary, but to simplify I have chosen only three combinations whose inputs and results are listed in Table Vb.

Results of Dump Model

Table Va summarizes the physical parameters describing the bales and dumpsite after one year of dumping at a rate of 2000

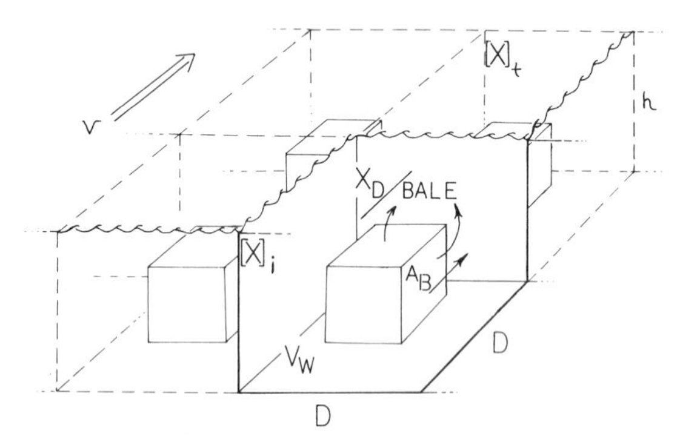


*Figure 5. Cross-sectional schematic of bale dump model for calculating the chemical effect of bale decomposition on bottom water quality just downstream of the dumpsite where: $v$ = current; $hDV_w$ = height, side, and volume at the mixed water parcel; $[X]_i$ = initial concentration at component X; $[X]_t$ = final concentration of component X at downstream edge of dumpsite; $A_B$ = area of bale side; $X_D/Bale$ = amount of X diffused from bale per unit time. Equations for calculations are listed in Table IV.*

Table IV. Equations used in model.*

$$J_x = \frac{(-D_s)\ (\Phi)\ (C_x{'} - C_o)\ (10^4)}{x'} \qquad (1)$$

$J_x$ = mass diffusing from bale in ($\mu$-moles/$m^2$-$sec^{-1}$) (38)

where $D_s$ = diffusion coefficient (molecular diffusion about $10^{-5}$ $cm^2$/sec)

$\Phi$ = porosity of the bale (assumed value 0.6)

$C_x{'}$ = concentration of component x in water outside bale ($\mu$-moles/l)

$C_o$ = concentration of component x inside bale ($\mu$-moles/l)

$x'$ = distance over which linear diffusion is assumed to occur (assumed 10 cm)

$$V_B = (W/\rho_\beta)\ (2.8317 \times 10^{-2}) = \text{Volume of bale } (m^3)* \qquad (2)$$

$$S_B = V_B^{1/3} = \text{side dimension of bale (m)} \qquad (3)$$

$$A_B = V_B^{2/3} = \text{side area of bale } (m^2) \qquad (4)$$

$$N = (I)\ (7.3 \times 10^5)/w = \text{bales per year} \qquad (5)$$

$$K = (1 \times 10^{-4})\ (N)\ (A_B)/(c) = \text{area bottom covered } (km^2/yr) \qquad (6)$$

$$D = (100\ A_B/C)^{1/2} = \text{water parcel side for one bale (m)} \qquad (7)$$

$$V_w = \left[(D^2)\ (h) - V_B\right] 1000 = \text{water parcel volume for one bale (liters)} \qquad (8)$$

$$X_D/\text{Bale} = (A_B)\ (5)\ (X_D) = \text{amount of material diffusing out of bale/sec.} \qquad (9)$$

$$[X]_t = \frac{X_i\ V_w + \left[\frac{(100)\ (D)\ (X_D/\text{bale})}{v}\right] (Nt)^{1/2}}{V_w} \qquad (10)$$

* Symbol definition and units are given in Tables 3 and 5, unless given above.

tons per day. The dumpsite covers a relatively small area of the ocean bottom at 20 percent coverage (2.8 $km^2$ or 1.09 $mi^2$).

The values of $\Delta$ X (Table Vb) give an estimate of the amount of change in the concentration of the four chemical components considered. The results are calculated for both a shallow or shelf site as well as a deep site. The impact for all parameters is greater at the deep site example, because the currents were estimated at one-fifth the velocity of those at the shallow site. The relative impact of the $\Delta$ X is a function of the concentration of the component in the bottom water. For example, since there is effectively no sulfide in oxygenated ocean water, the $[X]_t$ represents all the sulfide added. Sulfide is oxidized in the water (a ratio of 1 mole sulfide to 5 moles of oxygen) and the oxygen content reduction as a result of this sulfide oxygen demand can be easily calculated. Only for the worst case (deep site) would this reduce the oxygen to zero after one year. For all the other situations the additional sulfide oxygen demand less than doubles the $\Delta$ X for oxygen.

The amount of organic carbon released is substantial and will exert an oxygen demand on the water column. The time necessary to oxidize it all could range from weeks to years, during which time there will be substantial dilution and the end effect on the oxygen concentration should be small. The same is also true for ammonia because it takes several months for ammonia to be nitrified (39). For the worst case (deep site) the concentration of ammonia could reach levels at which it becomes toxic to certain fish types (12).

All of these values are at the end of one year. At longer periods of time, the concentration ($\Delta$ X ) increases by a factor of 1.4 for every doubling of amount of material or years. The effect of doubling other input variables is shown in Table VI. For example, it would take about 60 years of dumping at 2000 tons per day to reduce the oxygen concentration to zero at the downstream edge of the bale site for the deep site if only oxygen consumption for the average case is considered.

## Conclusions

Thus it appears on the basis of this simple model, using only short-term-based decomposition rates, that dumping of baled solid waste would not be harmful to the water quality for the parameters examined and over the short term. This suggests that an experimental dump of compacted baled solid waste could be undertaken. The physical, biological, and chemical characteristics of the site must be carefully considered prior to its selection. For example, an area with very low oxygen concentration and low currents should not be chosen.

Once a site is chosen and an experimental dump is begun, careful monitoring of the site area must be maintained. This monitoring will provide actual field data which will enable

Table Va. Calculated results for input data given in Table 3 using equations in Table 4.

| Variable name | Units | Symbol | Average |
|---|---|---|---|
| 1. Side dimension of bale | (m) | $S_B$ | 1.077 |
| 2. Area of bale side | ($m^2$) | $A_B$ | 1.159 |
| 3. Volume of a bale | ($m^3$) | $V_B$ | 1.249 |
| 4. Number of bales at a given time (1 year) | (#) | Nt | $4.87 \times 10^5$ |
| 5. Area of bottom covered | ($km^2$/yr) | K | 2.82 |
| 6. Side of water parcel around bale | (m) | D | 2.41 |
| 7. Volume of water parcel (liters) around bale | (liters) | $V_w$ | $1.04 \times 10^4$ |

Table Vb. Concentrations of X and ΔX in μ-moles/l just downstream of dump at time (t) for deep (D) and shallow (S) sites when t is one year.

| | | Average | | Worst* | | Best* | |
|---|---|---|---|---|---|---|---|
| | | $[X]_t$ | $\Delta[X]$ | $[X]_t$ | $\Delta[X]$ | $[X]_t$ | $\Delta[X]$ |
| Oxygen | S | 533 | -7 | 322 | -23 | 699 | -1 |
| | D | 505 | -35 | 68 | -282 | 693 | -7 |
| Sulfide | S | 0.94 | --- | 3.77 | --- | 1.07 | --- |
| | D | 4.7 | --- | 37.7 | --- | 0.19 | --- |
| Carbon | S | 127 | 47 | 981 | 971 | 162 | 2 |
| | D | 315 | 235 | 9,450 | 9,410 | 179 | 19 |
| Ammonia | S | 1.44 | .94 | 9.41 | 9.41 | 2.02 | 0.02 |
| | D | 5.21 | 4.71 | 94 | 94 | 0.19 | 0.19 |

* For both worst and best cases W, $\rho_\beta$, I, C, and h were held at average estimates. For $[X]_i$ and v, worst was low range value; best was high range value. For $(X)_D$ worst was high range value, best was low-range value.

Table VI. The effect on ΔX after one year by doubling variables.

| Variable | | Change ΔX by factor of: |
|---|---|---|
| Double bale weight | (W) | 0.92 |
| Double bale input/yr. | (I) | 1.4 |
| Double percent coverage | (C) | 1.6 |
| Double height of mixing | (t) | 0.5 |
| Double current speed | (v) | 0.5 |
| Double diffusion rate | ($X_D$) | 2.0 |
| Double time | (t) | 1.4 |

scientists to modify and update the dumping impact predictions. It must also be possible for the dumping scheme to be altered or stopped anytime the monitoring shows rapid deleterious effects to the area. Any dumping program should remain on an experimental basis since ocean dumping should be considered only as an interim solution to the solid waste problem. At least partial resource recovery will ultimately be of more benefit both to man and the oceans rather than ocean disposal of the entire solid waste problem, in spite of the fact that everything ultimately ends up in the oceans.

Abstract

Compacted experimental bales of municipal solid waste (trash) have been immersed in seawater for several years and monitored for chemical, biological and physical changes. This experimental data is used to develop a descriptive and simple mathematical model for a commercial-sized disposal operation on the continental shelf or deep water off the shelf, handling the needs of a medium-sized coastal city. Factors considered in the model include: bale size; quantity per unit time; bale composition; packaging; areal extent of dump; depth; biota; current regime; decomposition rates; and the effects on water quality such as oxygen consumption, dissolved organics, nutrients, and hydrogen sulfide. The effects of a dumpsite on bottom water quality at the end of one year are calculated for a shallow and deep site for the "average-most likely" dump and plausible extremes of the above variables.

## Literature Cited


1. ES & T Special Report, Environ. Sci. Technol. (1970) 4: 384.
2. Smith, D.D., Brown, R.P., Ocean disposal of barge-delivered liquid and solid wastes from U.S. coastal cities. EPA Solid Waste Management Series (SW-19c), 119 p, 1971.
3. First, M.W. (Ed.) Municipal Waste Disposal by shipboard incineration and sea disposal of residues. Final Rept. on EPA Grant #UI-00557. Harvard University School of Public Health, 1972.
4. Dunlea, J.V., Jr., U.S. Patent 3,330,088, July 11, 1967.
5. Devanney, J.W., Livanos, V., Patell, J., Massachusetts Institute of Technology Rept. No. MITSG 71-2, 124 p, 1970.
6. Bostrom, R.C., Sherif, M.A. Nature (1970) 228: 154-156.
7. Stone, R.B., R/V Challenger Cruise Rept. (1968), Copy in (5).
8. Pearce, J.B., Mar. Poll. Bull. (1972) 3: 157-159.
9. Stone, R.B., Mar. Poll. Bull. (1972) 3: 27-28.
10. Loder, T.C., Rowe, G.T., Clifford, C.H., Rept. in "Proc. Inter. Artificial Reef Conf.", Houston, March, 1974 (in press) 1975.
11. Bogost, M.S., Mar. Tech. Soc. Jour. (1973) 7: 34-37.
12. Pratt, S.D., Saila, S.B., Gaines, A.G., Krout, J.E., Mar. Tech. Rept. Series No. 9, University of Rhode Island, 53 p, 1973.
13. Loder, T.C., Anderson, F.E., Shevenell, T.C., Sea Grant Rept. UNHSG-118, University of New Hampshire, 107 p, 1973.
14. Loder, T.C., Culliney, J., Turner, R., Univ. of New Hampshire, Harvard, (1974) Unpublished reports.
15. American Chemical Society, "Cleaning Our Environment, The Chemical Basis for Action", Washington, D.C., 167 p, 1969.
16. Jannasch, H.W., Eimhjellen, K., Wirsen, C.O., Farmanfarmaian, Science (1971) 171: 672-675.
17. Torrest, R.S., Loder, T.C., Univ. of New Hampshire, (1974) Unpublished report.
18. Berner, R.A., "The Sea, Vol. 5: Marine Chemistry", p. 427-450, John Wiley and Sons, New York, 1974.
19. Colton, J.B., Jr., Marak, R.R., Nickerson, S., Stoddard, R.E., Bur. Comm. Fish., U.S.F.W.S. Data Rept. No. 23, 190 p, 1968.
20. Smith, K.L., Jr., Teal, J.M., Science (1973) 179: 282-283.
21. Loder, T.C., Eppig, P., University of New Hampshire, (1973) Unpublished report.
22. Rowe, G., WHOI, (1974) Unpublished report.
23. BenYaakov, S., Limnol. Oceanogr. (1973) 18: 88-95.
24. Stumm, W., Morgan, J.J., "Aquatic Chemistry", Wiley Interscience, New York, 583 p, 1970.
25. Ryther, J.H., Dunston, W.M., Science (1971) 171: 1008-1013.
26. Behan, R.W., Burton, S.D., Loder, T.C., Abs. in Proc. Amer. Assoc. Microbiologists Ann. Mtg., 1974.
27. Cline, J.D., Richards, F.A., Environ. Sci. Technol. (1969) 3: 838-843.

28. Orr, W.L., Gaines, A.G., "Advances in Organic Geochemistry", p. 791-812, Editions Technip, Paris, 1973.
29. Richards, F.A., In J.P. Riley and G. Skirrow, "Chemical Oceanography 1", p. 611-645, Academic Press, New York, 1965.
30. Turner, R., Harvard (1975), Pers. Comm.
31. Anonymous, Solid Wastes Management/RRJ (1971), Oct., p. 2-3.
32. Dunlea, J., Seadun, Inc. (1975), Pers. Comm.
33. Wimbush, M., Munk, W., "The Sea, Vol. 4, New concepts of sea floor evolution, Part I", p. 731-758, John Wiley and Sons, New York, 1970.
34. Gross, M.G., Morse, B.A., and Barnes, C.A., J. Geophys. Res. (1969) 74: 7044-7047.
35. Bumpus, D.F., Prog. in Oceanogr. (1973) 6: 111-157.
36. Schmitz, W.J., Jr., Jour. Mar. Res. (1974) 32: 233-251.
37. Loder, T.C. (1971), Ph.D. Thesis, University of Alaska, Rept. R71-15, 236 p.
38. Berner, R.A., "Principles of chemical sedimentology", 240 p., McGraw-Hill, New York, 1971.
39. Harvey, H.W., "The chemistry and fertility of sea water", 224 p., Cambridge University Press, London, 1955.
40. Acknowledgments. This research has been supported by the New England Regional Commission, UNH Contract No. 10230372 and the UNH Coherent Area Sea Grant Program 2-35244. I would like to thank Dr. Wendell Brown, UNH and Mr. Robert Stewart, MIT for discussions about the model and Drs. Gilbert Rowe, WHOI; Franz Anderson and Henri Gaudette, UNH; and M. Grant Gross, Johns Hopkins for discussions and review of the manuscript.

# 30

# Pioneering the New Industry of Shrimp Farming

PAUL F. BENTE, JR.

Marifarms, Inc., P.O. Box 2239, Panama City, Fla. 32401

In the late 1930's, Dr. M. Fujinaga (Hudinaga), working at Aio, Japan, studied the biology and reproduction of kuruma shrimp (Penaeus japonicus) caught commercially in the Seto Inland Sea. His classic paper (1), containing detailed drawings and descriptions of the many larval transformations and complete biological development of this shrimp, was received for publication (in the English language) one week before Pearl Harbor day. Such research, interrupted by the war, was resumed in the 1950's. Hatching and raising shrimp commercially got underway in the 1960's. The Japanese government also installed several large hatcheries with the goal of hatching baby shrimp in order to restock the Seto Inland Sea to improve commercial shrimping. Hatching and growing shrimp became the most advanced of some two-dozen mariculture operations that were under intense study in Japan.

Kuruma shrimp are now cultivated commercially in Japan in many small farms, which have growing ponds that generally are from 2 to 20 acres in size. To harvest a crop, a pond is drained, and the shrimp are collected by hand, chilled, packed in sawdust, shipped by air, truck or train, and delivered alive, to tempura restaurants. This specialty market has grown to about 700,000 lbs. per year. Its prices are several times higher than those charged for processed frozen or canned shrimp sold in the world markets.

Efforts were not made to develop the know-how into large scale farming to supply world markets, primarily because there were no suitable, large, unused coastal areas available for such operations. Water pollution and unfavorably low temperatures for much of the year were other deterring factors. Marifarms, Incorporated, was formed to undertake such a venture in the United States, starting in Florida.

With approval of the Japanese government, Marifarms acquired shrimp mariculture rights and know-how from the Taiyo Fishing Co., Ltd., of Tokyo, Japan. To begin exploratory tests in 1968, Dr. M. Miyamura, then president of Taiyo's subsidary engaged in shrimp farming at Takematsu, came to Panama City, Florida, with four experienced technicians. Americans were also employed to understudy and to help. As operations later expanded, a total of 14 Japanese were employed at one time or another, but gradually all returned to Japan, except for Dr. Miyamura who still serves on the company's Board of Directors.

In this pioneering work Marifarms encountered a formidable array of problems -- business, financial, public relations, socio-cultural, technical, etc. This paper describes a few of the many significant technical advances that the company made in wrestling with the problem of making large scale operations viable.

## Hatching

Although the Japanese team was very experienced in hatching the kuruma shrimp, of course, it had never worked with the three commercial varieties found in the Gulf of Mexico -- the brown (*Penaeus aztecus*), the white (*Penaeus setiferus*), and the pink (*Penaeus duorarum*) shrimp. Moreover, there was no available hatching experience in U.S. laboratories comparable to that in Japan, and no experience in catching large numbers of shrimp, ready to spawn. Hence, hatching became the first order of business. Hatching is first described and thereafter mother shrimping, that is, procuring shrimp for spawning to begin hatching.

Three concrete, 28,000 gallon tanks, approximately 30 ft. by 30 ft. by 6 ft. deep, the largest size that had been used successfully in Japan, were installed for the hatching tests. After a few abortive starts in the summer of 1968, the hatchery turned out 7.5 million baby shrimp of the pink variety. In 1969, about 14 million white, 5 million pink, and 70 thousand brown shrimp were hatched. Fortunately, the hatching of each variety sufficiently resembled the process for the kuruma shrimp that the technicians were able to adapt procedures to achieve good yields.

On the strength of this progress, 32 more of these tanks were constructed for commercial work. During the 1970-74 seasons it turned out a total of about 1.5 billion baby shrimp. Each year 20 million were stocked in public waters, and modest quantities were provided to several research centers.

The hatching of baby shrimp is shown by Table I.

Table I

Characteristics of Hatching Cycle

| Form | Time in Form Indicated (days) | Common Food | Size at Close of Period (mm.) |
|---|---|---|---|
| Egg | 1 | - | 0.25 |
| Larva | | | |
| Nauplius (Stages I-V) | 2 | egg yolk | 0.5 |
| Zoea (Stages I-III) | 4 | phytoplankton (5-10 μ in size) | 2.2 |
| Mysis (Stages I-III) | 3 | zooplankton, artemia, yeast | 5.0 |
| Shrimp | | | |
| Baby (suspended in water) | 6 | artemia, pellet feed | 7.5 |
| Baby (bottom dwelling) | 14 | marine life incl. benthic, pellet feed | 20.0* |

* Weight at this size is about 0.02 gms.

This is a fascinating, complex process involving 12 transformations after the spawned eggs have been fertilized. The literature is now replete with biological details of these changes. Excellent descriptions are given in two U.S. Patents (2)(3), which have been dedicated to the public.

This paper is concerned only with a few of the procedures in which modifications of commercial significance have been made. In general, the hatching water must be gently aerated. It should be at 25°-28°C. and have a salinity of at least 14 parts per thousand. If salinity is low, it can be increased by addition of rock salt rather than sea salt, a simpler, cheaper procedure which the company developed as flexibility in operating procedures was explored.

The zoea does not swim and must be kept suspended in water and fed by having its mouth collide with phytoplankton. The original process involved culturing special phytoplankton (*chlorella* and *Skeletonema costatum*) by adding nutrients, including residues obtained from the manufacture of soy sauce. One may judge what this hatching water looked like from the fact that the Japanese called the mixture "black tea." This procedure was gradually modified to culturing naturally occurring phytoplankton by adding chemical fertilizers as nutrients to enhance their multiplication. Yeast was also added as a food.

The mysis swims after its food and lives on zooplankton and other meat. The original process added various forms of supplemental meat, such as finely ground up clams. This was modified to ground fish, and that was later displaced by additions of a live food, in the form of hatched artemia, commonly called brine shrimp. These procedures simplified operations, reduced labor, and improved yields.

In this process the capacity of the hatching tanks is governed chiefly by the total weight of living shrimp (biomass) at the time the shrimp are finally moved to the growing areas of the farm. This varies for each variety of shrimp. White shrimp can tolerate a much higher density than brown or pink shrimp. Time for the full hatching cycle also varies and is another limitation on productivity. Details are given in Table II.

Table II

Productivity Limits of 28,000 Gallon Hatchery Tanks

| Kind of Shrimp | Avg. Limiting Population* | Size of Shrimp at End of Cycle (mg.) | Duration of Total Cycle (days) |
|---|---|---|---|
| Brown | 1,250,000 | 5-10 | 35-40 |
| White | 4,000,000 | 3-6 | 22 |
| Pink | 2,000,000 | 6-8 | 30 |

* Occasionally under ideal conditions densities 2.5 times this high have been obtained.

The limiting or maximum number of eggs or larvae to start the process is thus determined by these facts, taking into account mortality expected as the larvae move through their many transformations.

To increase capacity, the process has been divided into two steps, involving scale-up of an exploratory laboratory research procedure used by Cornelius Mock and coworkers of the U.S. Galveston Laboratory. The first step is conducted in a cylindrical, open top, 7000 gallon, fiber-glass tank standing on legs in one corner of the 28,000 gallon concrete tank. It is partially submerged in the water of the big tank. Water in the cylindrical tank is gently circulated -- by being removed from the bottom into an air lift pipe which discharges it at the top so as to cause a slow rotary stirring motion of the water column.

This operation accomplishes several things. The circulation prevents the shrimp larvae, food particles, and general debris, such as egg shells, from settling to the bottom, a situation which generally kills off a large proportion of a hatching if it is allowed to persist. To avoid that in the 28,000 gallon tanks, technicians occasionally had to swim about with flippers to stir up the settlings accumulating on the bottom of the tank despite vigorous air agitation. The procedure also allows the first part of the process to be done in much higher density, thus enhancing the larvae's capture of the food particles and thereby reducing the total amount of extra feed required. It turns out that this also increases the survival of the larvae as they pass through the transformations.

At the appropriate time the 7000 gallon tank is drained into the 28,000 gallon tank which contains water prepared for the second step of the hatching cycle when the shrimp are in the post larval form. While this step is being carried out, if desired, the first step of the next hatching can commence and run concurrently. Thus, once operations are underway, the number of cycles that can be turned out is essentially doubled. Since survival is increased, the new procedure is capable of essentially doubling and at times even tripling production rate, while simultaneously reducing the amount of food used per given number of shrimp hatched. Though every tank is generally operating in a different stage of the complex hatching process, all this is done by four biologists and two helpers who are deployed to other operations when the hatchery is shut down.

With this overview of the hatching process in mind consider now some of the developments in procuring gravid shrimp needed by the hatchery.

## Mother Shrimping

After hatching was demonstrated, the company made many exploratory trips, chartering commercial trawlers, to learn where in the Gulf and when during the calendar year each variety of shrimp in the gravid condition could be caught in sufficient numbers to operate commercially. Gravid shrimp are females carrying roe sufficiently developed so that spawning can be induced in a hatchery. Fishing is good when 5 percent or more of a normal catch are gravid.

Though many sites were found where mother shrimping can be done successfully, those most suited to the company's location are the waters off Mobile during February through April for brown shrimp,

Appalachicola during May and June for white shrimp, and both locations during July through September for pink shrimp. During the seasons of 1970 through 1974, the company procured a total of 154,000 mother shrimp, equivalent to about 13,000 pounds, heads on basis.

Mother shrimping in itself is a highly technical operation, for which the crew must be specially trained and the trawler specially equipped to hold the shrimp alive in aerated, chilled water, and then to package and ship them successfully to the hatchery by chartered plane or, if the run is short, by company truck.

Handling white mother shrimp poses a special problem. The sack of sperm acquired by the mother shrimp during copulation is carried outside the shell. (Pink and brown mother shrimp carry it inside the shell.) If the gravid white shrimp becomes agitated enough to snap its tail, as it does in trying to escape, it generally loses the spermatophore making fertilization of the subsequently spawned eggs impossible. Despite such odds, by careful handling, the hatching of white shrimp was quite successful. These odds were finally reduced by placing each mother shrimp, as soon as it was caught, into a perforated cylinder, capped on each end, and of a diameter that prevented each shrimp from flipping its tail. After transportation the shrimp was released safely in the hatching tank.

When a gravid shrimp is placed in the hatching tank, spawning almost invariably takes place during the next night. Since this happens in the dark, no one sees what goes on. By special techniques, the Japanese government has captured the spawning process on movie film. This shows that as the mother shrimp swims around it releases a stream of roe that is dispersed in the water by the gentle paddling action of its 10 swimmerets. The spawning also ruptures the spermatophore so that the released eggs are fertilized in the sea water.

## Feeding Shrimp in Growing Areas

In Japan little space was available for use as shrimp farms. Therefore, the pattern of development there was to cultivate what space was available as intensively as possible. Benthic foods, on which shrimp normally feed, did not grow naturally in the bottoms of the ponds that were constructed, for these were generally on a hard pan that had been used in the former practice of evaporating sea water to manufacture salt. In fact, sand usually had to be hauled in, in order to provide a bottom in which the shrimp could submerge themselves. At the time Marifarms acquired its know-how it was common practice to feed the shrimp crushed clams.

Marifarms selected sites where space permitted extensive operations. A 2500 acre embayment with tidal exchange of water was leased from the State of Florida and two 300 acre marine lakes with pumped water exchange were constructed on uplands leased from a private owner to compare farming in each type of area.

Both types of farms had bottoms which were typically soft and sandy and in which benthic organisms abound. It was intended that the shrimp feed partially on naturally occurring marine life which would provide sufficient growth stimulants and nutrients that they could be fed a lower grade supplement of ground up trash fish, since a higher quality ration of clams was not available.

Laboratory tests showed that conversion ratios for a ration of fish only were very high, and that overall growth was slow and would probably never reach full potential, regardless of how long shrimp were fed such a ration. But fish meat fed in combination with natural live food, or with hatched artemia, gave much better results.

The company installed an insulated refrigerated barge to hold a supply of fish. Pogy fishing boats delivered loads of fish every week or two, pumping as much as 90 tons at a time into the barge. The fish were pumped out as needed, passed through a grinder to make a slurry that was transported to the growing areas and discharged into the prop wash to feed the shrimp as the feed-boat toured the shrimp farm in a cross-grid pattern. During 1970, the company fed a total of 800,000 lbs. of raw fish. Because weather interferred with fishing and delivery when needed, it was difficult to maintain steady operations. Further, it was a messy, malodorous operation, conducted under the threat of being overtaken by spoilage.

Therefore, the company sought a better and more reliable supplemental feed and abandoned the feeding of raw fish in 1971 when its experimental work with dry feeds showed promise. Many dry feeds were tested, made from various combinations of ingredients, using various binders in the conventional method of feed preparation by pressing the mixture into a mold. In general the formula for pelletized shrimp feed, on a weight basis, is: 30% fish meal, 24% shrimp meal or crab meal or combinations thereof, 20% soy meal, 20% meat scraps, 2% whey, 2% vitamin mix, and 2% binder such as Durabond.

A local feed mill produced four million pounds of such pellets for operations during 1971-73, but all formulations had the disadvantage of disintegrating in sea water in several minutes. This obstacle was finally overcome by Ralston-Purina by using the cooking-extrusion process. Their proprietary pellets had much better cohesive properties; after being dropped into water they did not disintegrate for 20 to 24 hours, allowing the shrimp adequate time to find and eat the pellets. In 1974, 2.2 million pounds of these pellets were used.

Last year, in two 300 acre production units yielding harvests of 300 and 200 thousand pounds of shrimp, respectively, the conversion ratio using this pellet feed was 1.8 and 1.9 pounds (dry weight) fed per pound of whole shrimp landed. Indications are that conversion ratios as low as 1.4 will be possible in raising shrimp of a size averaging 50 per pound, heads on basis.

It is interesting to note that in Japan proprietary dry pellet feeds of a different composition were simultaneously and independently developed and adopted as standard procedure. The industry was driven to this by the growing shortage and rising prices of clams. The pellets developed in Japan were based on higher quality ingredients, since the ration had to provide all the growth stimulants in view of the fact that natural foods did not occur in the ponds used in Japan. Consequently, the cost of such feed was very much higher. Nevertheless, Marifarms purchased a large lot of this feed and evaluated it. Used under Marifarms' conditions of having a partial supply of natural food available, the results were about the same as those obtained using the feed developed at Marifarms, but, of course, the expense was far greater.

## Confining Shrimp to the Growing Areas

In farming tidal waters, the baby shrimp were initially planted in circular pens enclosed by fine mesh nylon nets having openings about like that of window screening. A chain fastened to one edge of the net sealed off the bottom, while a line of cork floats at the other edge held the net to the surface. Since small shrimp move about much as do suspended particles in riled water, in rough weather the shrimp would sometimes be washed out with waves breaking over the top of the pens. Though still on the farm, these liberated shrimp were in an area where predators had not been removed and where feed was not added. Therefore, survival was low, and growth somewhat slower at least until the shrimp grew large enough to move about in search of food.

Such difficulties were not experienced in the marine lakes. Moreover, growth of natural foods in closed systems could be stimulated by fertilizing the waters. Hence, operations in the bay were improved by constructing starting ponds on the adjacent shore. The procedures used to enrich the natural foods in these pond waters, while at the same time suppressing undesirable organisms, are those developed by Texas Parks Wildlife Department (4). In addition a supplement of powdered pellet feed is spread on the waters for the baby shrimp.

Marifarms conducted many tests to determine the size that shrimp must be to be retained by nets of various mesh sizes -- or stretch, as openings are termed by the fishing industry. Marifarms has found it most productive to retain baby shrimp behind pen nets or within diked ponds until they are at least 0.3 gm in size. Population has to be adjusted at the outset so that the shrimp do not exceed the permissible bio-density under the operating conditions that are encountered (rate of water exchange, temperature, oxygen supply, kind of shrimp, condition of the bottoms, etc.).

When the shrimp become 0.3 gm in size they are released into a much larger nursery enclosed by a coarser net designed to restrain shrimp of that size. Openings in this net are approximately 1/4" x 1/4" (or 1/2" nominal stretch). By the time the shrimp exceed 0.8 gm in size, they are released into still larger adjacent growing areas confined by a still coarser net sized to hold back these shrimp. This net has openings of 5/16" x 6/16" (or 21/32" nominal stretch). After the shrimp reach 3.0 gm in size they are released into the final growing area which is enclosed by a net with openings 7/16" x 7/16" (or 7/8" nominal stretch).

Shrimp behave very differently in large growing areas than in small experimental ponds of but a few acres. White shrimp migrate in schools; whereas brown and pink do not. Shrimp also move to find natural feeding grounds. As shrimp grow larger they seek deeper water, but they frequently move back into warm shallow waters to feed on natural foods that seem to abound where more sunlight reaches the bottoms. They generally do not seek to escape until they are young adults, at which time they would normally move out to sea with the tides. This tendency occurs just prior to and during harvest periods. At such times Marifarms attaches a specially designed supplementary collar of floats to the top of its nets in order to stop the shrimp from swimming over the top during rough weather.

Marifarms uses some 15 miles of nets. The hydrodynamics of construction may be compared to constructing a suspension cable that is subjected to fluctuating heavy loads, i.e., tides moving back and forth. To hold the nets down on the bottom, a heavy chain is lashed to the net. Various head ropes on which floats were placed have been tried to hold the net up to the surface. All the plastic type ropes have caused trouble, because they gradually elongate permanently with repeated stretching by the tides.

The result is that the net at the top is gradually stretched out, though not at the bottom, until it finally tears from top to bottom. This difficulty has now been overcome by using plastic coated galvanized steel cables in place of head ropes.

Nets often used in the fishing industry do not have a balanced construction -- that is a heavier nylon cord is used in one direction than in the other. In Marifarms' use when such nets tear, they part as if being unzippered, breaking the weaker strands progressively and rapidly. Marifarms has had its nets specially made in a balanced construction. No matter what size openings are used, the company has found, that to be strong enough, the nets must have at least a pound of nylon for every 25-30 sq.ft. of net.

Nets are dipped in a proprietary antifouling mixture to reduce their picking up barnacles, hydroids, and other marine growths. The dip is made up of 84% copper oxide, 8% tributyl tin, and 8% triphenyl tin dispersed in petroleum distillate.

In addition nets must be washed every few days by pumping many heavy jets of water, under water, onto each side of the net. Equipment for doing this is suspended from a barge on which are located two pumps. This is one of the "tractors" of the farm which had to be developed. It traverses the nets from one end to the other. Another "tractor" of the farm is equipped with a pair of large rotating nylon bristle brushes which can scrub the nets clean if more vigorous action is necessary. In appearance these brushes look like the ones you see in a car wash line, but they are much stiffer. Without periodic cleaning, the openings of the net become so plugged that the nets break under the strain of heavy tides.

Another barge tractor has been developed to deploy nets to enclose areas for shrimp farming operations and to set up outer nets to keep out fish. This tractor can put out a mile of net in one day. The same barge can remove the nets and the chain which holds the nets to the bottom. Every winter after harvesting has been completed, these removal operations are done in two separate steps, first removing floats and net, and then chain which has been cut free. The nets are cleaned, repaired, coated with antifouling agents, rebuilt with chain and float line, and put out for use again in the spring. With proper care, a net can be used for three years.

## Multiple Cropping

Marifarms has learned enough about the availability of mother shrimp and the hatching and growing patterns to make it practical to grow two consecutive crops of shrimp in the very same areas. The first crop is brown shrimp. These are hatched in February and March, planted in the starting areas in April and May, moved progressively through the nursery and into the grow-out areas, and harvested in June to August.

The hatchery meanwhile shifts to white shrimp, which are hatched in the May and June period. These move progressively through the same areas arriving in the grow-out areas after the summer harvest of brown shrimp has been completed. The white shrimp are then harvested in the October - December period.

While it is theoretically possible to run a third crop of pink shrimp through the system, hatching these in July-August-September period, holding through the winter in juvenile form while it is too cold for growth to continue, and subsequently continuing growth in the spring and harvesting them together with the brown crop during the next summer, tests show that wintering over operations should have fairly deep holding areas so that the water doesn't cool down as much as it does in shallower areas during cold snaps.

Survival in the shallow pond areas in particular is too low to justify such efforts. When water temperature drops below 10°C., survival falls off markedly. At 4°C. the effect is lethal. Operations further south ought to make a third crop feasible.

## Harvesting Procedures for Quality Product

In shrimping at sea, the catch is manipulated by hand. Sometimes as little as 10 percent of what is caught are shrimp. First of all, the trash fish and other marine life are sorted out by hand and thrown overboard. Then while trawling continues, the crew members squeeze the heads off the shrimp by hand, place the tails in wire baskets, and finally store these with ice below deck. When a load is accumulated, the trawler returns to port to unload and stock up on provisions, and then returns to fishing at sea. Ships stay out a week or more at a time.

By contrast, Marifarms uses similar trawlers, but since the farmed bay waters contain little but shrimp, there is no need to sort the catch by hand on deck. The catch is dropped on deck, spread out and layered with shaved ice. During trawling the catch is generally kept covered by a canvas. After a few hours of trawling, the boat has a load of several thousand pounds on its stern deck. Then the boat returns to a dock at shore on the farm where the catch is unloaded by a vacuum lift with the aid of a heavy stream of water.

In the procedure used in harvesting the farmed marine lakes, as they are caught, the shrimp are placed into a transfer net on the special skiff used for the trawling. Every hour or so the skiff stops at a docking point so that the transfer net can be lifted out and immediately emptied into a tank of ice water mounted on a trailer. Every few hours this is hauled to a vacuum lift for final unloading.

The shrimp are discharged from the lift into a large agitated tank of ice water from which they are removed by a metal chain link type conveyor belt. The shrimp on the conveyor pass across an inspection table where debris, and the relatively few fish, crabs, etc., are removed by hand. (The crabs are accumulated and sold as a by-product. Crab yield in 1974 was about 20,000 lbs., live weight.) At the end of the conveyor the shrimp are weighed into standard fish boxes holding 100 pounds each. They are kept chilled by placing a layer of ice on the bottom, in the middle, and on the top of each box of shrimp. These boxes are loaded into a tractor trailer, and loads of 20,000 to 30,000 pounds are hauled off to nearby processing plants within a day of being caught.

The shrimp are delivered untouched by human hands. Consequently, the bacteria count is low. The shrimp do not give off a strong fishy smell when cooked. They have a sweet taste, devoid of the iodine-like flavor often encountered in other shrimp. Thus, the quality of the cultivated shrimp produced by Marifarms is superior to the standard product of the market place.

## Outlook

Since 1970, the world catch of fish has declined. Despite greater efforts and improved equipment and larger amounts of fuel being used, the catch of shrimp seems to be reaching a plateau. In this period of mounting concern for producing high quality, nutritious foods while at the same time conserving fuel, it is important to consider where shrimp mariculture fits into the picture as food and fuel consumption are compared for producing various animals for food.

As Table III shows, cultivating shrimp uses much less feed than does raising cattle for beef. It is also considerably lower than feed consumed in raising hogs or turkeys, and in fact it falls slightly below that needed to raise chickens. The percentage of meat obtained from whole shrimp is also higher than that obtained from many animals raised for food.

It is also worth noting that in the case of raising beef it takes about 960 days from successful fertilization to final slaughter and in the case of producing pork it takes 240 days. However, in the case of cultivating shrimp, depending on variety and season, it takes only 120 to 180 days, from day of spawning to day of harvest.

The direct and indirect consumption of fuel in shrimp mariculture is presently about equal to that for beef production. However, when the mariculture operations reach normal capacity, fuel consumption will fall to half that needed to raise beef, and to less than one fourth that needed to catch an equivalent amount of shrimp at sea. Hence, development of shrimp mariculture is in keeping with the demands and concerns of the time for conservation in the use of fuel and for more efficient use of grains in producing meat for human consumption.

Table III

Consumption of Competitive Feeds in Raising Animals for Meat

| | Lbs. Feed Consumed Per Animal | Lbs. Raw Lean Meat Produced | Conversion Ratio Col 2/Col 3 |
|---|---|---|---|
| Cattle | 2,625[a] | 340[b]<br>(68)[d] | 7.75[c]<br>(38.6)[d] |
| Swine | 840 | 85[e] | 9.9 |
| Turkey | 70 | 12[f] | 5.8 |
| Chicken | 8 | 2.1[g] | 3.8 |
| Shrimp - in 1974 | 550,000[i] | 152,500[h] | 3.55 |
| Shrimp - projected | - | - | 2.8 |

(a) In addition to 10,875 lbs. of forage.
(b) Slaughter weight = 1050 lbs. Grade is "high-good" to "low-choice". "Choice" and "prime" require feeding more grain.
(c) Forage excluded from calculation. Forage above accounts for 272 lbs. of lean meat, giving a conversion ratio of 40 to 1.
(d) Grain alone accounts for 68 lbs. of lean meat, giving conversion ratio of 38.6 to 1.
(e) Slaughter weight = 240 lbs.
(f) Slaughter weight = 20 lbs.
(g) Slaughter weight = 4 lbs.
(h) Gross weight of crop = 305,000 lbs., heads on.
(i) In the case of shrimp, this feed supplied for a mariculture crop of 21 million shrimp, averaging about 70 per pound, heads on basis.

In all, Marifarms has expended over 310 man years of effort in pioneering shrimp farming. To date, the company has harvested 2 million pounds of shrimp (about 140 million shrimp), 825,000 pounds (65 million) of this last year. The goals are to more than double this in 1975. The capacity of the farm is expected to be about 5 million pounds per year, based on two crops per year raised during an 8-month period. Marifarms considers this only the beginning. In another decade shrimp farming has the prospect of increasing about 100 fold. It can spread to developing countries where a new source of food and a cash crop should be welcome. In techniques, mariculture of shrimp stands at the threshold where agriculture was one or two generations ago. But mariculture of shrimp will grow rapidly with the advantages of today's know-how in marine biology coupled with availability of essential materials, technology, and engineering.

Abstract

Commercial scale procedures are described which have been developed in pioneering the new industry of hatching, growing and harvesting shrimp in netted off tidal areas and in diked marine lakes with pumped water exchange. The technology described is a fusion of Japanese and U.S. developments in marine biology extended by a 7-year (310 man year) endeavor to commercialize shrimp farming. To date, harvests at the Panama City, Florida, farm have landed approximately 140 million shrimp, weighing nearly 2 million pounds, heads-on basis. Sequential steps in the hatching and growing operations are reviewed. Development of pelleted, dry feeds to supplement live food naturally available in the sea water is discussed, and conversion ratios are given. Unique harvesting procedures are also described which result in a product that is superior in quality to that generally available in commercial quantities.

Literature Cited

1. Hudinaga, M. Japanese Journal of Zoology (1942), (No. 2), pp. 305-393.

2. Miyamura, M. U.S. Patent No. 3,473,509, assigned to Marifarms, Incorporated, dedicated to the public. (October 21, 1969).

3. Motosaku, M. U.S. Patent No. 3,477,406, assigned to Marifarms, Incorporated, dedicated to the public. (November 11, 1969).

4. Texas Parks Wildlife Dept. Salt Water Pond Research, Project 2-169-R, Austin, Texas.

Acknowledgement

The author extends credit for the progress reported in this paper to all the employees of Marifarms. Their combined, sustained efforts in the face of many unexpected difficulties made this pioneering accomplishment possible.

# 31

# Renatured Chitin Fibrils, Films, and Filaments

C. J. BRINE and PAUL R. AUSTIN

College of Marine Studies, University of Delaware, Newark, Del. 19711

A resurgence of interest in chitin has been stimulated by the NOAA-Sea Grant marine resource development program (1, 2), and recognition of the important, though little understood, role of mucopolysaccharides in accelerating wound healing and alleviating inflamations of the skin (3, 4). Research on the problems of cell wall structure and composition in both fungi and animal tissue (5, 6) has broadened the scope of chitin studies.

Very recently, chitin itself has been shown to be an essential nutrient of the crawfish diet (7, 26), cell wall chitin microfibrils have been prepared *in vitro* with a chitin synthetase (8) and the present authors have purified and crystallized chitin from solution in fibrillar and film form for the first time in its natural or renatured state (9, 11), showing characteristic spherulitic structures.

In nature, chitin not only exists in well-recognized crystalline forms, but in tendons and other stress-bearing fibrous portions of marine animals, it is even more highly organized into oriented molecular structures showing typical fiber orientation characteristic of man-made fibers such as nylon and polyethylene terephthalate (12-16).

It is the purpose of this report to extend and amplify the earlier disclosures of techniques of chitin fibril and film preparations, and to describe an improved method for the casting of films and extrusion of filaments of chitin with high molecular organization. These renatured structures may be cold drawn to more than twice their original length, which induces fiber orientation and an increase in tensile strength equal to or surpassing that of natural chitin filaments.

It is recognized that this research is still at an early stage and that much refinement of technique and data are required to confirm and enlarge upon these findings. Nevertheless the implications for developing applications for chitin as continuous film and filaments, as well as for a better understanding of chitin properties and functions in such diverse areas as fungi, insect physiology and marine life has prompted this progress report.

## Natural Chitin

As background, several samples of natural chitin were obtained and examined by scanning electron microscopy (SEM), polarizing microscopy and x-ray diffraction techniques. Red crab chitin, used for most of our studies, occurs as both flakes and small fibrils as shown by SEM in Figures 1 and 2. The ordered layers or bundled structures are striking.

With the polarizing microscope, the presence of spherulites in natural chitin is indicated by the typical Maltese Cross patterns, Figure 3. This evidence of crystalline ordered structure is confirmed by x-ray diffractograms with their well-defined concentric Debye rings, Figure 4.

In some chitin preparations, for example, those from Dungeness and King crabs, the fibrous material balls up in a sample bottle and may contain filaments as long as 0.7 cm. Some of these filaments, apparently arising from tendons or stress-bearing portions of the crab, were found to have strong fiber orientation, as evidenced by their characteristic nodal x-ray patterns, Figures 5 and 6.

Comparison of d-spacings from these diffractograms with literature values for chitin (12, 17), Table I, showed good conformity considering the likely variations in sample history, moisture content and morphology.

## Renatured Chitin Films

Because of its intractability, chitin is normally prepared by removing calcareous and proteinaceous materials from crab shells, for example, by successive treatment with dilute acids and alkalies. The insoluble residue is chitin. Although it is soluble in concentrated mineral acids and certain salt solutions, these agents cause degradation or are inconvenient for purification and subsequent handling of the chitin.

In our prior work (9-11) it was shown that chitin could be dissolved with reasonable facility in trichloroacetic acid (TCA) systems, which permitted filtration, and then could be renatured by precipitation with acetone or other anhydrous non-solvent. In this way, amorphous and fibrillar forms and unsupported films of chitin were obtained with varying degrees of spherulitic crystallinity. The new solvent system contributed advantages of reduced rate of degradation as well as a greater tendency to reprecipitate ordered structures. These structural features were confirmed by comparisons with natural chitin: Figure 7, showing fibrils imbedded in a renatured film, using polarizing micrography; Figure 8, with spherulites in renatured film, indicated by Maltese Crosses; Figure 9, with concentric Debye rings in renatured film, which establishes its crystalline ordered chitin structure; and

*Figures 1–6. Natural chitin studies*

*Figure 1. Red Crab flake, SEM 340X*

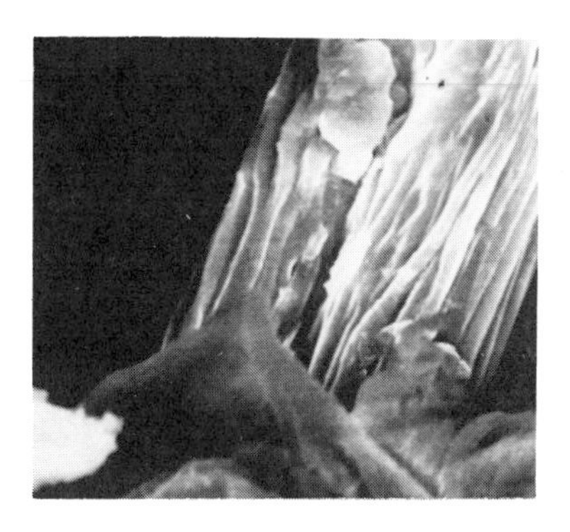

*Figure 2. Red Crab fibril, SEM 340X*

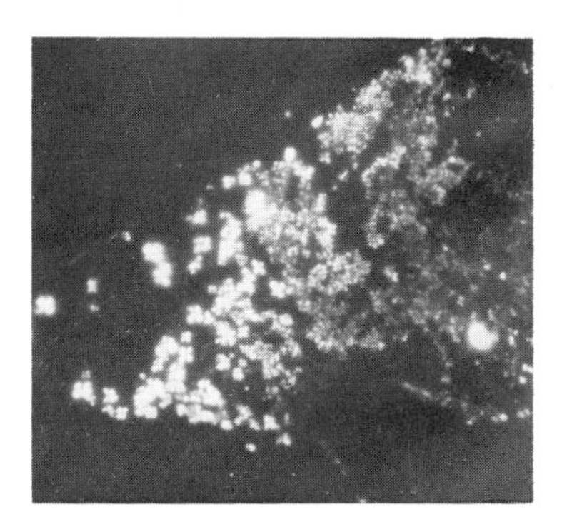

*Figure 3. Red Crab flake; spherulites; polar. microgr. 170X*

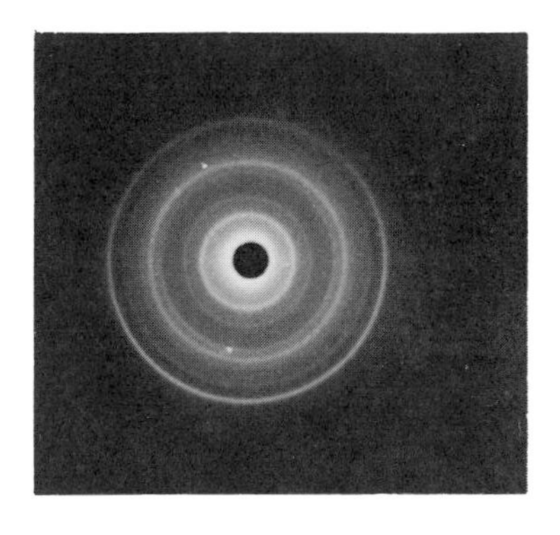

*Figure 4. Red Crab flake; x-ray shows crystalline structure*

*Figure 5. Dungeness Crab filament; x-ray shows orientation*

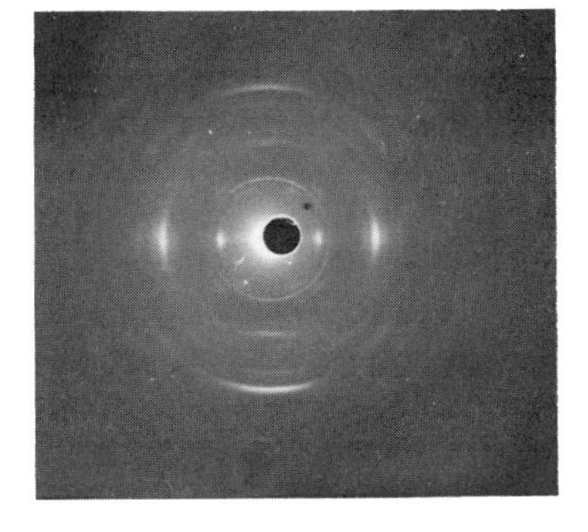

*Figure 6. King Crab filament; x-ray shows orientation*

TABLE I - NATURAL CHITINS, d-SPACINGS (Å)

| Red Crab (Rings)[a] | Shrimp (Rings)[a] | King Crab (Nodes) | Dungeness Crab (Nodes) | Lobster (Nodes)[c] (13, 17) |
|---|---|---|---|---|
| 11.08[b] | -- | -- | -- | -- |
| 11.00 | -- | 10.83 | -- | -- |
| 10.34 | 10.15 | -- | 9.84 | 9.65 |
| -- | -- | 9.16 | 9.26 | 9.56 |
| -- | -- | 8.85 | -- | -- |
| 7.86 | 7.33 | 7.82 | -- | 7.62 |
| 6.81 | -- | -- | 7.30 | 6.96 |
| 5.38 | -- | 5.68 | 5.36 | 5.65 |
| 5.08[b] | -- | 5.08 | -- | 5.13 |
| 4.87 | 4.81 | -- | 4.82 | 4.65 |
| 4.81[b] | -- | -- | -- | -- |
| 4.30 | -- | -- | 4.20 | 4.25 |
| 3.89 | -- | 3.79 | 3.96 | 3.81 |
| -- | -- | -- | 3.74 | |
| 3.51[b] | 3.54 | -- | 3.56 | 3.38 |
| -- | -- | -- | 2.71 | 2.57 |

[a]Flake, [b]spherulitic region of fibrillar samples, [c]high intensity group

Supplemental Notes to Table I. The x-ray diffraction analyses were performed throughout with the Jarell-Ash Microfocus x-ray Generator #80-000 and the Phillips Microcamera #56055 for the x-ray photographs. By referring to the geometry of the film and specimen loading in the microcamera and making the appropriate measurements of the diffraction positions on the film, the characteristic angles, $\theta$, for the reflections can be derived. The d-spacings can then be calculated using the Bragg relation, where the wavelength of the x-radiation used is 1.542Å for CuK$\alpha$ radiation and specimen to film distance is 15.0 mm. All samples were dried over night at ambient temperature in a vacuum desiccator over Drierite.

Crustacean chitins used in this study were *Opilio chionectes* - Japanese Red Crab (Eastman Kodak Co.), *Pendalis borealis* - Alaskan Pink Shrimp, *Paralithodes camtschatica* - Alaskan King Crab and *Cancer magister* - Pacific Dungeness Crab (Food, Chemical and Research Laboratories).

*Figures 7–12. Renatured Red Crab chitin products*

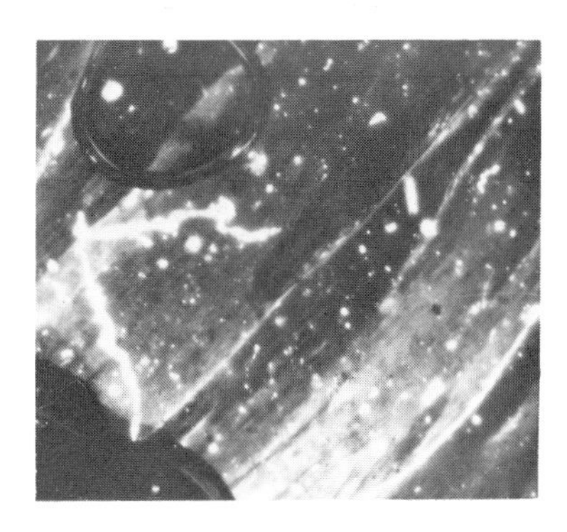

*Figure 7. Fibrils in Film BII; polar. micrgr. 170X*

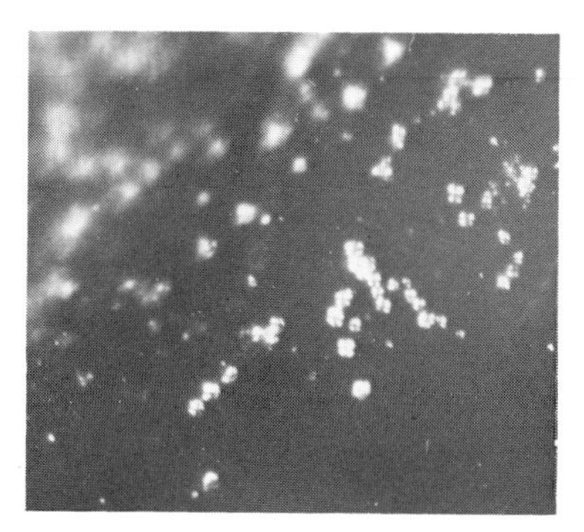

*Figure 8. Film BII; spherulites; polar. micrgr. 170X*

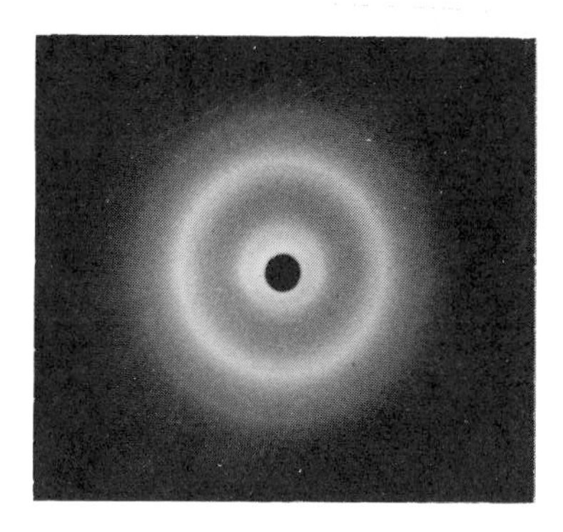

*Figure 9. Film BII; x-ray shows crystalline structure*

*Figure 10. Fibrils in Film III; SEM 475X*

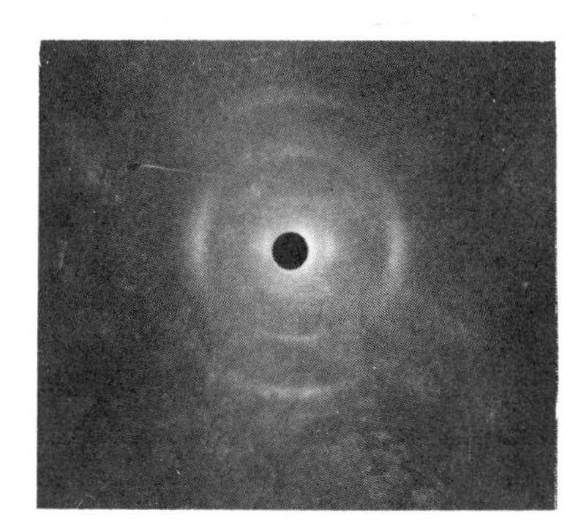

*Figure 11. Drawn Film BIIIA; x-ray shows orientation*

*Figure 12. Drawn Filament VI; x-ray shows orientation*

Figure 10, SEM, renatured chitin film showing the presence of fibrillar chitin in the matrix, responsible for the crystalline manifestations.

Oriented Chitin Films. With the stage thus set, we appeared to be at the threshold of achieving our long-range goal, that of orienting renatured chitin by cold drawing to enhance its properties and to approximate the structure of natural fibrous chitin. Indeed, we found out somewhat laboriously that to obtain chitin films of sufficiently high quality to be cast, coagulated and renatured in crystalline form, and sufficiently ordered to be cold-drawn to high tenacity products, several criteria must be met simultaneously:

1. An anhydrous solvent; trichloroacetic acid-methylene chloride was best when supplemented by chloral hydrate as a critical component.

2. High molecular weight polymer; achieved by short solution time, usually an hour or less, with the resultant solution to be as concentrated and as viscous as possible.

3. Immersion of the film in an anhydrous non-solvent, preferably acetone.

4. Neutralization with anhydrous alkali such as potassium hydroxide in 2-propanol.

5. Cold drawing of cut film from a few to over a hundred percent, and

6. Extraction and/or decomposition of residual solvents to reduce them to a low level.

Again, x-ray diffraction was used to demonstrate the fiber-type orientation realized: Figure 11 is the renatured cold-drawn chitin film (BIIIA), nodal pattern. Although not as distinct a pattern as that of the natural product, there can be no doubt that the desired chitin polymer orientation has been accomplished.

Some of the experiments that established the critical factors for the preparation of high quality films are outlined in Table II, Renatured Chitin Films.

For example, in the first line (A-105B) a reasonably good film was obtained, as judged by its tensile strength, pliability and toughness, but the chitin had been kept in acid solvent for four hours, probably too long to maintain its molecular weight, and it was neutralized in an aqueous system. The film contained very few or no spherulites and it could not be cold drawn. Note that even with alkali extraction, not all the TCA was removed.

TABLE II - RENATURED CHITIN FILMS

| Film No. | Film Preparation | Anal. N,% | Cl,% | Spherulites | Cold Draw | Path Diff.[f] | X-ray |
|---|---|---|---|---|---|---|---|
| A105B | 40% TCA in $CH_2Cl_2$, 4 hours[d] | 6.52 | 0.58 | No | No | -- | -- |
| BII | As above, 0.5 hour[e] | 6.52 | 0.58[a] | ++ | 15% | 25, 125 | Sharp rings |
| BIII | Standard (chloral) | 5.71 | 10.06[a] | +++ | 100% | 100-125 | Sharp rings |
| BIIIA | Standard (chloral) | 3.94 | 33.28 | | Good | -- | -- |
| BIIIA Drawn | Above, ext. 12 hrs. with acetone - $CH_2Cl_2$ | 5.03 | 9.45[b] | +++ | 85% | 125 | Rings; Nodal after draw |
| BIV | Standard (chloral) | 5.19 | 8.88[a,c] | +++ | 100% | -- | -- |

[a]Extracted 4-12 hrs. with $CH_2Cl_2$ before analysis

[b]Treating film further with b. 1% NaOH in 2-propanol then gave N, 6.56; Cl, 0.68

[c]Treating film further with b. 1% NaOH in 2-propanol then gave Cl, 1.16.

[d]Film coagulated with two 20 min. acetone washes; neutralized with aqueous sodium hydroxide.

[e]Film coagulated with four 15 minute acetone washes; neutralized with aqueous sodium hydroxide.

[f]Relative birefringence; path differences obtained with first order red plate and imbibing fluid with refractive index of 1.552; values in nm.

An important advance was made with the use of chloral hydrate in the TCA-methylene chloride system (BIII and BIIIA). This mixture speeded solution and dissolved more chitin, up to about 2-3 percent. It also aided the cold drawing operation, but contributed markedly to the chlorine content of the film. The retention of chloral undoubtedly affected the properties of the film, but the x-ray patterns nevertheless appear very similar to those of natural unoriented chitins as shown previously; comparisons of d-values of drawn film (Table III) bears this out.

TABLE III - RENATURED CHITIN, DRAWN FILM AND FILAMENT, d-SPACINGS (Å)

| Drawn Film BIIIA[a,c] | Drawn Fil. VI[a,c] | Lobster (Nodes)[b] (13, 17) |
|---|---|---|
| -- | 12.22 | -- |
| 11.63 | 11.00 | -- |
| 9.64 | 8.95 | 9.56 |
| 7.13 | 7.20 | 6.96 |
| 5.93 | 5.66 | 5.65 |
| 5.30 | 5.54 | 5.13 |
| 3.82 | 4.07 | 3.81 |
| -- | 3.63 | 3.38 |

[a]N, 5.03%; Cl, 9.45%; [b]high intensity group; [c]all spacings are close to calculated best fit values, orthorhombic system

The chlorine compounds in the film were very difficult to remove and confirm the strong affinity of chitin for acids and other organic compounds (18). Simple solvent extraction with methylene chloride or a mixture of it with acetone still left a substantial amount of chloral or TCA in the film. Most of it could be removed, however, by treatment with hot alcoholic alkali; it is decomposed apparently by the haloform reaction.

On the basis of crystallographic evidence, Rudall (14) indicated that approximately one in every six acetyl groups in chitin is deacetylated. If this is true, one might speculate that chloral may react with the acetamide residues (19) or the free amine groups to form an aldehyde ammonia or Schiff's base type of structure, which would account for the difficulty of removing halogenated impurities. Since chloral contains 72 percent of chlorine, a small amount has a significant effect on the composition as a whole.

It is of interest that microscopic indication of fibrillar structure and spherulites in the film correlates well with ability of the film to be cold drawn, that is, a high degree of molecular order in the film is a prerequisite for fiber-like orientation (BIII and BIIIA vs. A105B). Both cloral and water have plasticizing effects that facilitate cold drawing.

Finally, the tensile strength of the films is strikingly improved by cold drawing (A105B vs. BIIIA), as brought out with comparisons from the literature in Table IV. Here it is seen that our renatured chitin films, when they have a good degree of crystalline order or are oriented by cold drawing, are far superior to those prepared by the viscose process and equal or surpass the best values for natural, oriented chitin fibers.

TABLE IV - CHITIN FILMS

| Film | Ten. Str. kg/sq mm | Elong., % | Ref. & Comments |
|---|---|---|---|
| Natural (oriented fiber) | 58 | --- | (13) |
| Regenerated (xanthate) | 9.49 | --- | (20) |
| Renatured (A105B)[a] | 32.5 | 11 | Non-crystalline |
| Renatured (BIIIA)[a] | 52-58 | 125 | Crystalline; extracted |
| Renatured (BIIIA)[a] | 75-95 | 4 | Pre-oriented by partial cold drawing |

[a]Conditioned at 60% R. H., room temperature.

Method of Preparation. The biological source of chitin is the Japanese red crab, *Opilio chionectes*, obtained through the Eastman Kodak Company. It is pulverized dry in a blender to pass a 24 mesh screen. Two parts by weight of chitin are dissolved in 87 parts by weight of a solvent mixture comprising 40 percent chloral hydrate and 20 percent methylene chloride, with gentle warming and mechanical stirring for 30-45 minutes; methylene chloride is added to replace that lost by evaporation. The very viscous solution is filtered through wool felt and/or a glass fiber mat. It is immediately cast upon glass and doctored with a glass rod to an even thickness, usually about one-sixteenth inch. The glass plate and film is immersed in four successive washes of dry acetone, each wash lasting 15 minutes. The films become free from the glass during the acetone washing and are subsequently treated several times with 1-5 percent potassium hydroxide in 2-propanol. The films are then washed with deionized water until neutral. The dry films are clear, pliable and strong.

## Renatured Chitin Filaments

As soon as the technique for preparing good chitin films was found, it was adapted to the preparation of filaments by extrusion of the chitin solution through a syringe and hypodermic needle or by laying a ribbon filament down on glass. Solution preparation, filtration, acetone renaturing and washing were carried out as for films. In several cases the same parent solution was used for the preparation of both films and filaments. Either a sodium hydroxide-potassium hydroxide mixture or potassium hydroxide in 2-propanol was used for neutralization and the filaments were then washed with deionized water until neutral. When filaments were extruded into acetone, they were allowed to range freely; nevertheless, strain birefringence from extrusion was noted in some cases.

The filaments contained much crystalline material, as indicated by their birefringence and x-ray diffraction patterns. They could be cold drawn with necking down and the drawn filaments had good knot strength. The results of these experiments are summarized in Table V, Renatured Chitin Filaments.

TABLE V - RENATURED CHITIN FILAMENTS

| No. | Preparation | Cold Draw | Drawn Fil. | Path Diff., nm | Film Ref. No. |
|---|---|---|---|---|---|
| F-II | Standard[a], Syringe and needle | 200-300% | Knotted | 175-200 | BIII |
| V | Standard[a,b], Syringe and needle | Fair | Coherent | 175-200 | BIIIA |
| VI | Standard[a,b], Ribbon on glass | Good, Necks down | Good knot strength | 175 | BIIIA |

[a]Standard solution in TCA, chloral hydrate, $CH_2Cl_2$; coag. acetone; neut. KOH in 2-propanol, wash with water. Ext. and anal. as for Film Ref.

[b]Neut. with NaOH-KOH in 2-propanol.

With polarizing microscopy, the observations paralleled those of the films. Undrawn filaments showed the Maltese Crosses indicative of the presence of spherulites and upon drawing showed strong birefringence. Using the polarizing technique for birefringence, relative path differences were in the range of 175-200 nm for cold drawn fibers as compared with 125-150 nm for red and Dungeness crab fibrils and filaments.

In x-ray studies, unoriented filaments gave patterns of crystalline materials with fairly sharp, concentric Debye rings, whereas the cold drawn, oriented filaments gave the beautiful nodal patterns as shown in Figure 12 (VI). The close similarity to natural chitin, Figures 5 and 6, is apparent.

To characterize the nodal x-ray patterns of renatured,

oriented chitin filaments, the d-spacings were determined and compared with those cold drawn film (BIIIA) and literature values of the lobster (13, 17), Table III. Although crustacean chitins have in general the same alpha-chitin structure (14, 15, 17, 25), the x-ray diffractograms may differ slightly because of sample history, thickness, degree of order or crystallinity, impurities and polymorphism as is known with cellulose (21). In spite of these variables, it was found that the d-spacings of our oriented chitin filaments, even with residual chlorine-containing components, showed reasonable agreement with the literature and those of other oriented chitin filaments determined concurrently (Table I). The renatured filaments appear to have the established natural chitin structure.

Finally, the contribution of cold drawing and orienting to the strength chitin filaments is illustrated in Table VI, Oriented Chitin Filament Properties. In this table, data for natural chitin fibers and a regenerated product, taken from the literature, are compared with our cold drawn filaments. The value for viscose rayon is cited from the same reference.

It is seen that our renatured chitin filaments are equal or superior to natural chitin fibers, even when calculated on the original fiber dimensions. When calculated on their break dimensions, as if a cold drawn product were being measured, the higher values are indicated. The small amount of filament available precluded obtaining data directly on the oriented filaments.

TABLE VI - ORIENTED CHITIN FILAMENT PROPERTIES

| Filament | Ten. Str. kg/sq mm | Elong.[a] % | Reference |
|---|---|---|---|
| Natural (lobster) | 58 | -- | (13) |
| Regenerated from sulfuric acid | 35 | -- | (22) |
| (Viscose rayon) | 25 | -- | (22) |
| Renatured, as extruded[b] | 56 | 13 | V |
| Calcd. on dimension at break | 63 | -- | |
| Renatured, as extruded[c] | 72 | 44 | VI |
| Calcd. on dimension at break | 104 | -- | |

[a]Instron TT-CM Tensile Testing machine; conditioned at 60% R.H., room temperature.

[b]Cross section 0.08 x 0.10 mm, T at break 0.44 kg.

[c]Cross section 0.014 x 0.740 mm, T at break 0.75 kg.

## Perspective

With the establishment of the principle that chitin can be renatured, even into highly oriented form, attention is focused on the need for a superior solvent system to avoid chitin degradation, provide more concentrated solutions, avoid solvent retention in the filaments and films, and facilitate wet or dry spinning, or casting, of such structures. Studies in this direction are continuing.

Considering the modest potential supply of chitin (50-100 million pounds per year in the U. S.) and the high cost of initial manufacture of both chitin and the proposed filaments and films, applications have been hypothesized that involve high value-in-use. Examples include surgical sutures and other hospital supplies, taking advantage of body tolerance and absorbability; sewing thread and decorative fibers, capitalizing on high softening point and unusual dye receptivity (23, 24); and oven and other food wrap, based on its edibility and temperature stability.

It is noteworthy that there are only a handful of commercial fibers oriented by cold drawing; with the leads developed in this study, perhaps in time chitin can be added to that list.

## Abstract

A method has been developed for the solution and subsequent precipitation of chitin that renatures it in highly ordered, crystalline form closely similar to native chitin. Critical factors include the use of a high molecular weight, soluble chitin and an anhydrous precipitation and neutralization system. A mixture of chloral hydrate, trichloroacetic acid and methylene chloride is an effective solvent, although some solvent residue is retained by the chitin. Acetone is preferred for renaturing. Filaments and films with a good degree of order and crystallinity can be cold drawn to more than double their original length to enhance their properties; tough, pliable films and high strength filaments have been prepared in this way.

## Acknowledgements

The assistance of Dr. Jerold M. Schultz of the Department of Chemical Engineering on the x-ray studies and Dr. Peter B. Leavens of the Department of Geology on the polarizing and scanning electron microscopy is greatly appreciated. The chitin investigation was sponsored in part by NOAA, Office of Sea Grant, Department of Commerce, under Grant No. 02-3-158-30.

## Literature Cited

1. Pariser, E. R. and Bock, S., "Chitin and Chitin

Derivatives, Bibliography 1965-1971," 199 pp. M.I.T. Sea Grant Report 72-3, Cambridge, Mass. 02139 (1972).

2. Pacific Northwest Sea, Oceanographic Inst. of Washington, Seattle, WA 98109, (1973), 6-12, 6 (No. 1).

3. Prudden, John F.; Migel, Peter; Hanson, Paul; Friedrich, Louis and Balassa, Leslie, The Am. J. Surgery (1970), 119, 560-564.

4. Prudden, John F. and Balassa, L. L., Seminars in Arthritis and Rheumatism, (1974), 3, 287-321.

5. Ballou, B., Sundharadas, G. and Bach, Marilyn L., Science, (1974), 185, 531-533.

6. Brimacombe, J. S. and Webber, J. M., "Mucopolysaccharides", Elsevier, New York, (1964).

7. Meyers, S. P., Aquanotes, La. State Univ., (1974), 3, 5.

8. Ruiz-Herrera, J. and Bartnicke-Garcia, S., Science, (1974), 186, 357-359.

9. Brine, C. J. and Austin, Paul R., "Utilization of Chitin, a Cellulose Derivative from Crab and Shrimp Waste," Sea Grant Report DEL-SG-19-74, 12 pp., University of Delaware, Newark, DE 19711 (1974).

10. Austin, Paul R., 1973, U.S. Patent Application No. 418,441.

11. Brine, C. J., "The Chemistry of Renatured Chitin as a New Marine Resource," M.S. Thesis, 88 pp. (1974), University of Delaware, Newark, DE 19711.

12. Ramakrishnan, C. and Prasad, N., Biochim. Biophys. Acta, (1972), 261(1), 123-135.

13. Clark, G. L. and Smith, A. F., J. Phys. Chem., (1936), 40, 863.

14. Rudall, K. M., "Advances in Insect Physiology," (1963), 1, 257-311, Academic Press, N. Y.

15. Bittiger, H., Husemann, E. and Kuppel, A., J. Pol. Sci., (1969), Part C, p. 45-56, Polymer Symposia, Proc. 6th Cellulose Conf., Interscience, John Wiley, New York.

16. Dweltz, N. E. and Colvin, J. R., Can. J. Chem. (1968), 46 (6), 1513.

17. Carlstrom, D., J. Biophys. Biochem. Cytol., (1957), 3, 669-683.

18. Giles, C. H., Hassan, A. S. A., Laidlaw, M. and Subramanian, R. V. R., Soc. Dyers and Colorists, (1958), 74, 653.

19. Rodd, E. H., "Chem. of Carbon Compounds," p. 545, 600, Elsevier, New York, 1951.

20. Thor, C. J. B. and Henderson, W. F., Am. Dyestuff Reporter, (1940), 29, 489.

21. Atalla, R. H. and Nagel, S. C., Science, (1974), 185, 522.

22. Kunike, G., Soc. Dyers and Colorists, (1926), 42, 318.

23. Hackman, R. H. and Goldberg, M., Aust. J. Biol. Sci., (1965), 18, 935-946.

24. Knecht, E. and Hibbert, E., Soc. Dyers and Colorists,

(1926), 42, 343.
25. Okafor, N., Biochim. Biophys. Acta, (1965), 101 (2), 193-200.
26. Hood, M. A. and Meyers, S. P., Gulf and Caribbean Fisheries Institute, Proceedings, (1973), 81-92.

# 32

# Seawater Desalination by Reverse Osmosis

N. WALTER ROSENBLATT

E. I. Du Pont de Nemours & Co., Inc., Wilmington, Del. 19898

In reverse osmosis (RO), a semipermeable membrane acts as a molecular separator or filter removing the pure water ("permeate") from the saline feed stream by pressure driven transport. The process is today established in the desalination of brackish water, where bidding and selling plants of one million gpd* and more on the basis of specified water quality data is an accepted practice in the trade. The published performance record of some plants installed over the last five years is available as realistic evidence to back the warranties given by manufacturers. Research in seawater RO - supported to a large measure by the U.S. Department of the Interior (Office of Water Research and Technology)- has now reached the point where reverse osmosis is moving also in this field from the R & D phase into the commercial arena.

## Process Principles

The phenomenon of natural osmosis takes place when a dilute solution and a concentrated solution are separated by a semipermeable membrane, the solvent migrating from the dilute to the concentrated compartment (Figure 1). The process is reversible and under a hydraulic pressure in excess of the osmotic pressure, solvent (water) will flow into the dilute compartment. Considerable work of a fundamental nature was carried out in the late 19th and early 20th centuries as part of the studies on colligative properties of solutions, but the field was dormant

*gallons per day

by the late 1920's. Interest was rekindled in the 1950's with the intensified search for economical methods to desalt brackish and seawater. The thermodynamic efficiency of reverse osmosis appealed to physical chemists active in the field and the search for membranes with semipermeability began. The goals were high rejection of the solute, i.e. salts, high solvent or water flux and in addition, like with so many other good things in nature - long life. The first important lead was the discovery by Reid and Breton that cellulose acetate films have semipermeability towards salt solutions and activities in this field subsequently gathered momentum when Loeb and Sourirajan invented asymmetric cellulose acetate membranes. In these membranes a thin rejecting layer several hundred Angstroms thick is supported on a thicker but more porous substrate that has little resistance to water flow (Figure 2)(1). With asymmetry water flux could be increased by one to two orders of magnitude without losing rejection. From that time on, the pace quickened at the University of California and at other places. The first attempts to desalt seawater were more a creditable demonstration of pioneering spirit than a technical accomplishment. The reasons were in part that cellulose diacetate membranes did not have the chemical properties for prolonged stable performance in desalting seawater on top of all the other formidable challenges of the marine environment. Since then efforts were largely turned towards brackish water desalting, where RO has established a well recognized position. Several comprehensive texts on the state of the art in reverse osmosis technology have been published (2, 3, 4, 5).

Let's now turn to the problems we face with seawater RO:

TABLE I

SALT CONCENTRATION AND OSMOTIC PRESSURE IN SEAWATER RO

| | Salt Concentration | Osmotic Pressure (25°C) |
|---|---|---|
| Feed | 3.45% | 369 psia |
| Concentrate or Reject (30% recovery of product water) | 5.15% | 567 psia |
| Average | 4.3% | 460 psia |

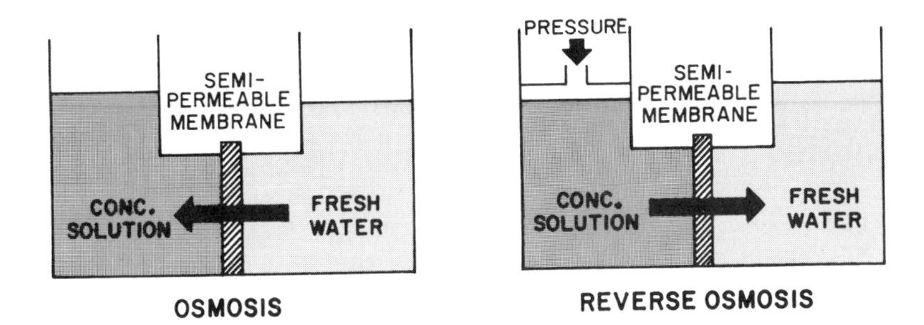


*Figure 1. Reverse osmosis (RO) principles*

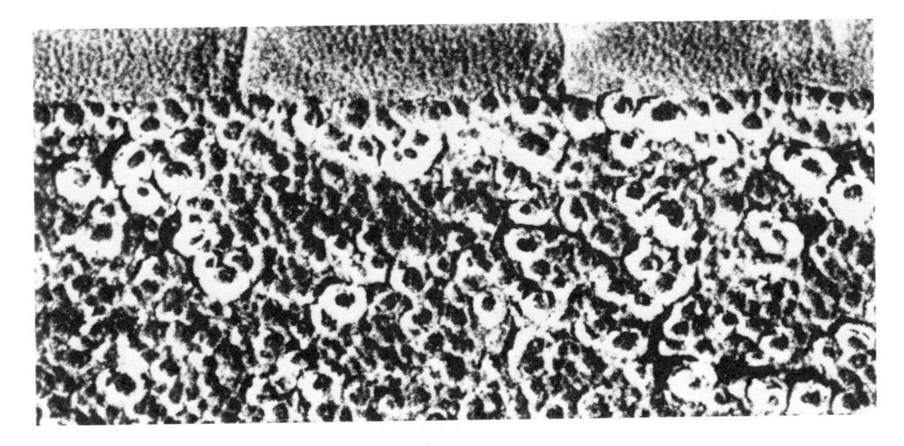

Macromolecules

*Figure 2. Skin structure of asymmetric cellulose acetate membrane* (1)

Standard seawater with a concentration of 3.45% salt has an osmotic pressure of more than 350 psi. Assuming that 30% of the incoming feed water is recovered as product water, the concentrate stream leaving the module has a concentration of 5.15% salt and more than 550 psi osmotic pressure. In reality the true salt concentration at the membrane surface is higher as a result of a phenomenon called concentration polarization (Figure 3). Salt and water migrate to the membrane surface, the water passes and the salt is rejected and accumulates. Ultimately the salt returns from the boundary to the main stream as a result of the concentration difference. The higher the membrane flux the larger the accumulation problem. Turbulent flow or a narrow gap between membranes in a laminar flow configuration will aid in depleting the boundary layer. The adverse effects of concentration polarization or accumulation of salt in the boundary layer are:

- higher osmotic pressure leading to reduced water flux,

- higher salt concentration difference across the membrane between the brine side and the product side leading to increased trans-membrane migration of salt,

- increased risk of precipitating sparingly soluble salts.

To achieve realistic flux levels and productivities in seawater RO, the applied pressures across the membrane are 800-1500 psi.

The problems of concentration polarization were recognized early and well understood from theoretical analyses of membrane permeability. Another problem encountered in RO practice was appreciated fully only when more field experience was gained. Any surface in a marine environment is rapidly covered with deposits that will interfer with water transport across the membrane and with the effective removal of the accumulated salts from the membrane surface. Sources of these deposits are 1) biological, 2) particulate matter in the seawater, or 3) corrosion products generated by the metals used in the construction of the RO plant.

Experience has shown that the decrease in membrane productivity with time from fouling and other causes can be correlated on a log/log plot. The productivity loss with time for several levels of logarithmic flux decline rate are illustrated in Figure 4 and the importance of effective fouling control is evident. We have seen cases where membrane productivity dropped at a rate corresponding to 20% residual productivity after one year. Effective fouling control can maintain a logarithmic flux decline rate better than 0.05, corresponding to 75% of the initial productivity after the first year and 70% after three years and we believe based on our experience that this is a realistic goal. Interestingly, fouling in seawater RO interferred less with salt rejection than with productivity.

The product water (permeate) from a desalting plant is expected to conform with U.S. Public Health Standards, i.e. to contain not more than 500 ppm total dissolved solids (TDS). The limit for chloride ions is 250 ppm. These water quality specifications call for a salt rejection capability above 98.5%. In practice the rejection has to be better than 99% to compensate for the increased salt concentration resulting from product water recovery and from concentration polarization.

As much as RO appeals thermodynamically, the process poses a cost efficiency problem because it is a relatively slow rate process. For example, the water generating capacity of good film membranes for seawater desalination is 10-15 gallon/sq. ft./day. In comparison a well engineered evaporator with a heat transfer coefficient of 1000 BTU/(hr.)(sq. ft.)($^{\circ}$F) generates 24 lbs. or 3 gallons of water per square foot per day per $^{\circ}$F or 30 gallons at a 10$^{\circ}$F driving force. Since seawater RO has to be conducted inside a pressurized space designed for 800 to 1500 psi operation, an obvious goal is to package as much membrane area as possible into a pressure vessel.

## Engineering of Membrane Devices

We can now compile a list of the problems that have to be considered in the development of RO devices for seawater desalting:

1. Support of the thin fragile membrane against the differential pressure of 800 to 1500 psi in the case of seawater.

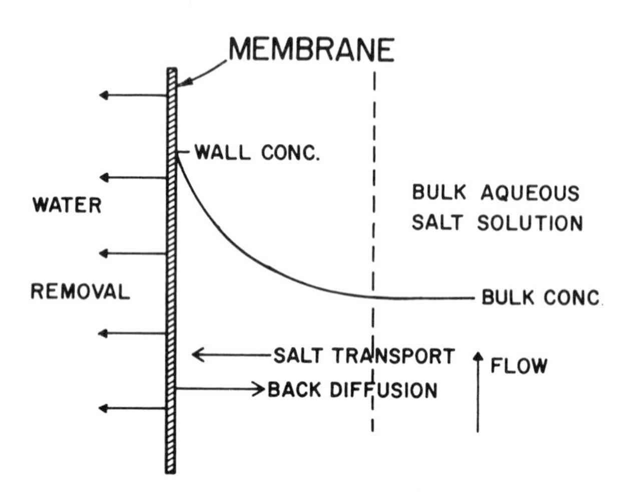


*Figure 3. Concentration polarization*

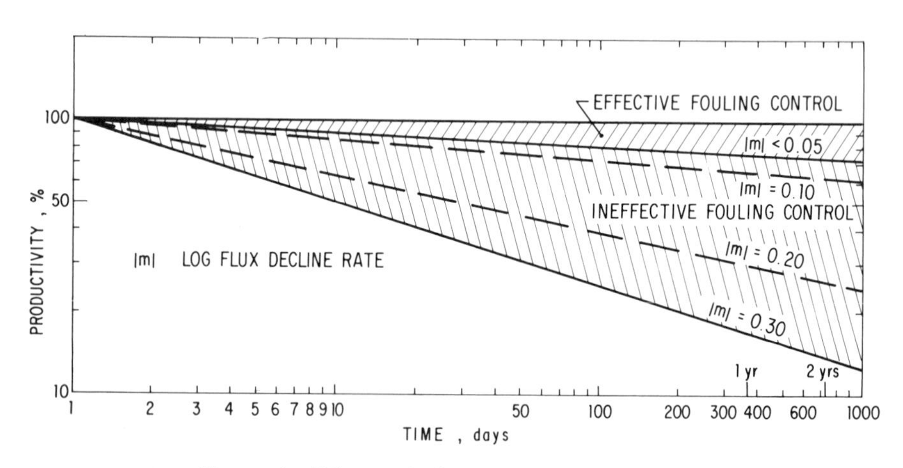


*Figure 4. Effect of fouling control on productivity*

2. Compactness - high productivity per unit volume.

3. Low concentration polarization.

4. Fouling control.
   Prevention and foulant removal by cleaning.

5. Effective sealing of the membrane to prevent by-passing of feed into product.

A compact, high pressure mass exchange device accommodating all these requirements was indeed a novel and formidable challenge to imaginative engineering. Three configurations among all the ideas proposed at one time or another remained as the principal candidates: the tubular, the spiral, and the hollow fiber concept listed here in increasing order of compactness, i.e. specific surface area per unit volume:

| | |
|---|---|
| tubular device | 100 $ft^2/ft^3$ |
| spiral device | 300 " " |
| hollow fiber device | 3300 " " |

The earliest concept was the perforated or porous tubular configuration, usually 0.5 inch ID with a membrane contained inside the tube and supported on a liner from fabric, paper or non-woven materials. Feed water under pressure flows through the bore and the product water permeates through the membrane and leaves via the perforations (4). Tubular devices, being relatively simple, saw the most diverse effort to optimize them technically and economically by:

- judicious selection of materials, e.g. sand logs, porous fiberglass tubes,
- compact packaging, e.g. helical coil configuration,
- advanced fabrication technology, e.g. laser drilling of perforations.

Another approach to achieve large specific areas is the spiral wound device, wherein a flat membrane envelope is formed around a fabric spacer and closed on three sides (Figure 5). The open side terminates at a porous or perforated product water tube. The envelope or leaf together with an external spacer for the feed water stream is rolled spirally around the product tube and is then installed in a pressure vessel. The feed stream flows axially through the channels between the spiral windings. Water permeates through the membrane and flows radially inside the leaf towards the product tube. This membrane configuration is predominantly the result of work done by the Roga Division of General Atomics (now owned by Universal Oil Products) with funds provided by the Office of Saline Water (now Office of Water Research and Technology) in the U.S. Department of the Interior.

The most compact membrane devices developed so far are hollow fiber permeators. Du Pont's hollow fiber membranes have the thickness of a human hair. The fiber is asymmetric with a rejecting skin on the outside supported on an inner porous layer (Figure 6). Pressurized saline water flows around the outside of the fiber and the water permeates through the skin and the porous support and leaves via the bore. The OD is twice the ID and the fiber has the mechanical characteristics of a thick-walled pressure vessel and the membrane is self-supporting. The fiber bundle consisting of several hundred thousand up to a few million fibers is similar to a U-tube heat exchanger. The open end is encased in an epoxy tube sheet which separates the saline water from the permeate. The feed stream enters through a central distributor pipe, flows radially outward through the bundle and is removed via the annular flow screen at the vessel wall (Figure 7). Details on the construction and history of the Permasep permeator were published elsewhere (6).

Table II summarizes the characteristics of the different designs.

More recently a new entry, the "spaghetti" module having the membrane on the outside of a polypropylene rod with longitudinal grooves was announced by the United Kingdom Atomic Energy Establishment (7); but little is known on its performance in seawater desalting.

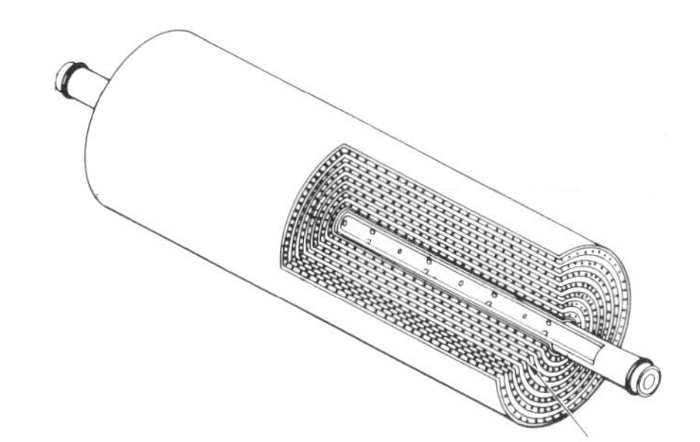


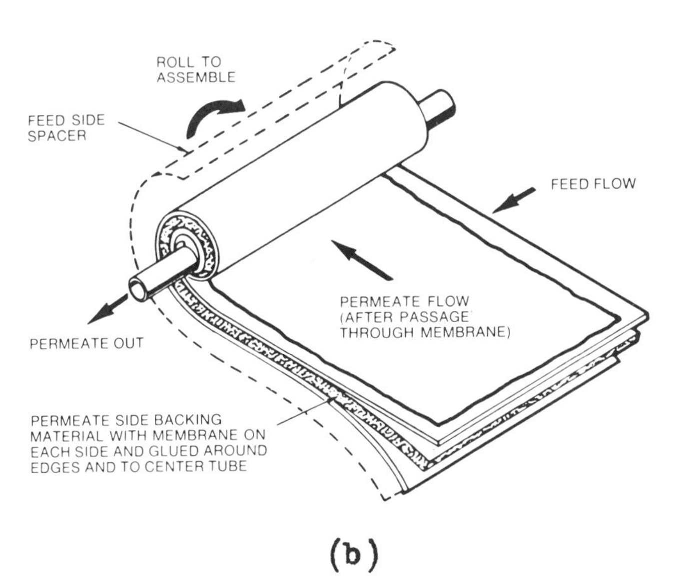


*Figure 5. Spiral wound membrane configuration. (a) spiral element, (b) partially unrolled.*

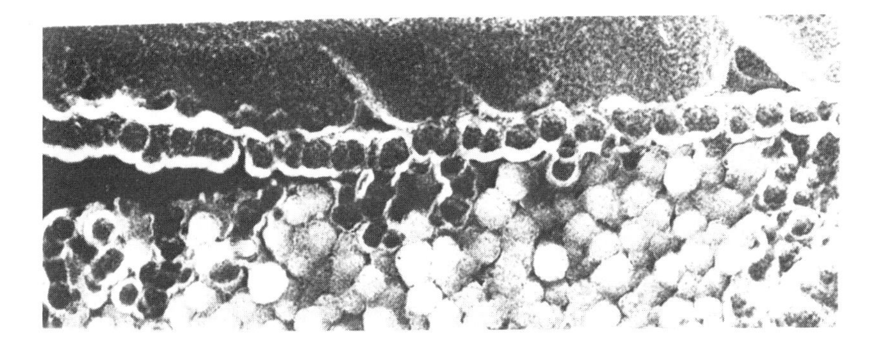

Macromolecules

*Figure 6. Skin structure of asymmetric hollow fiber membrane* (1)

TABLE II

CHARACTERISTICS OF PRINCIPAL RO DEVICES

| | Tubular | Spiral | Hollow Fiber |
|---|---|---|---|
| Compactness $ft^2$ membrane/$ft^3$ vol. | 100 | 300 | 3300 |
| Membrane Support | Supported | Supported | Self-Supporting |
| Operating Pressure, psi (seawater) | 1000-1500 | 1000-1500 | 800 |
| Particulate Matter Tolerance | Tolerant | Moderate | Sensitive |
| Cleaning | Mechanical Chemical | Chemical | |
| Cost | Expensive | Economical | |

## Plant Engineering

The essential elements of an RO plant are intake facilities, pretreatment equipment to control membrane fouling, a high pressure pump to provide the driving force and of course membrane modules (Figure 8). The two principal system concepts under consideration for seawater desalting were the one-pass and the two-pass processes illustrated in Figure 9. The two-pass approach with two stages of salt interception appealed as the technically more reliable route, but the membranes in the two-pass RO plant should have at least twice the flux of one-pass membranes to be economically competitive. Field tests with live seawater to develop know-how on plant engineering were carried out at Ocean City, New Jersey on a site operated by the Ocean City Research Corporation. The seawater was diluted with run-off water from the areas bordering the bay, a representative concentration being 2.7% TDS. The seawater contained relatively large quantities of suspended matter complicating the control of membrane fouling.

In 1968 we had lab tested with synthetic seawater an asymmetric polyamidehydrazide film membrane (DP-1) which produced 10-12 gfd* (>99% rejection) at 1000 psi (Figure 10) (8). Soon thereafter this film membrane was modified for two-pass operation to produce more than 30 gfd* at 1000 psi and rejection levels of 90%. Experiments were carried out at Ocean City with flat test cells (Figure 11) (9) to clarify the relative merits of one-pass versus two-pass desalination and to learn enough about field operation and fouling control to move towards pilot plant trials.

The two major classes of membrane foulants are particulate matter and biologically active species which form foulant layers on the membrane. The first category is best intercepted by combining coagulation with filtration through sand and/or diatomaceous earth filters. The biological foulants are conveniently rendered inactive by chlorination which is performed more effectively below the normal pH 8 of natural seawater. Since free chlorine is harmful to polyamidehydrazide and polyamide membranes it has to be removed chemically or with a carbon treatment.

*gallon/sq. ft.-day

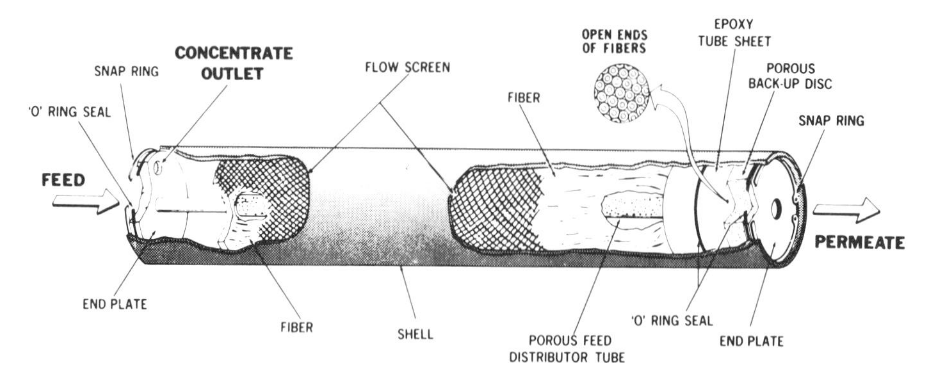


*Figure 7. Cutaway drawing of Permasep permeator*

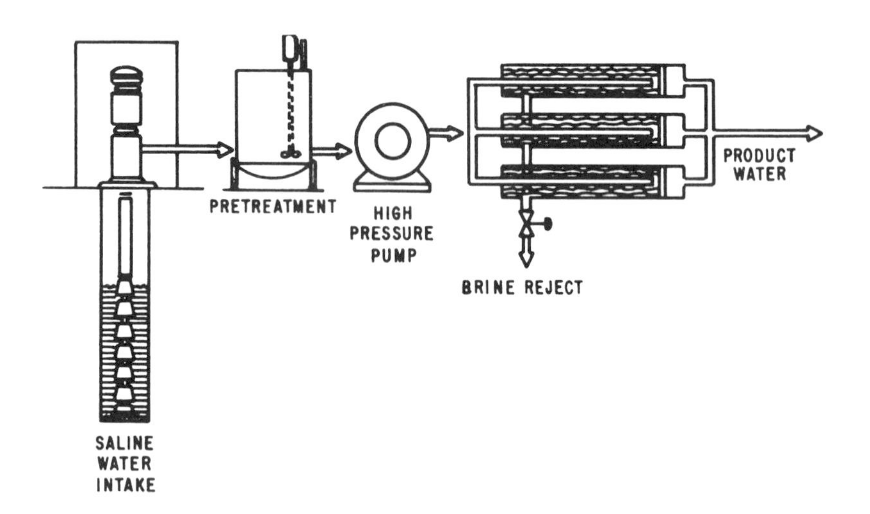


*Figure 8. Schematic of RO plant*

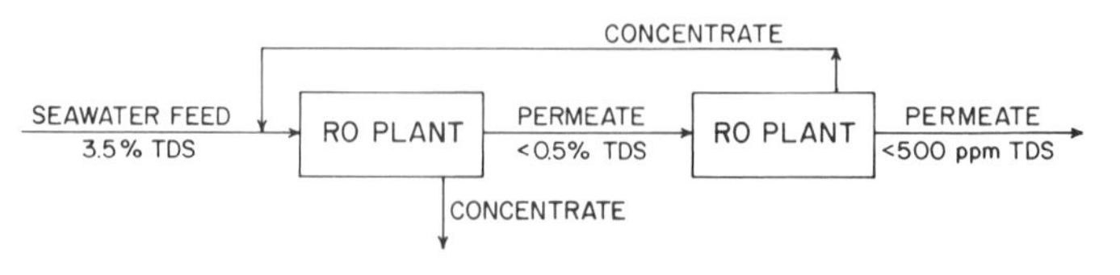


*Figure 9. Process concepts for seawater RO*

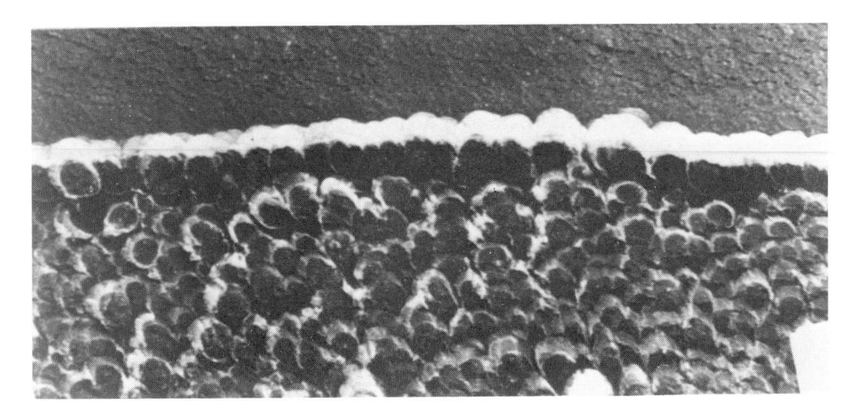

Macromolecules

*Figure 10. Skin structure of asymmetric polyamide-hydrazide membrane* (1)

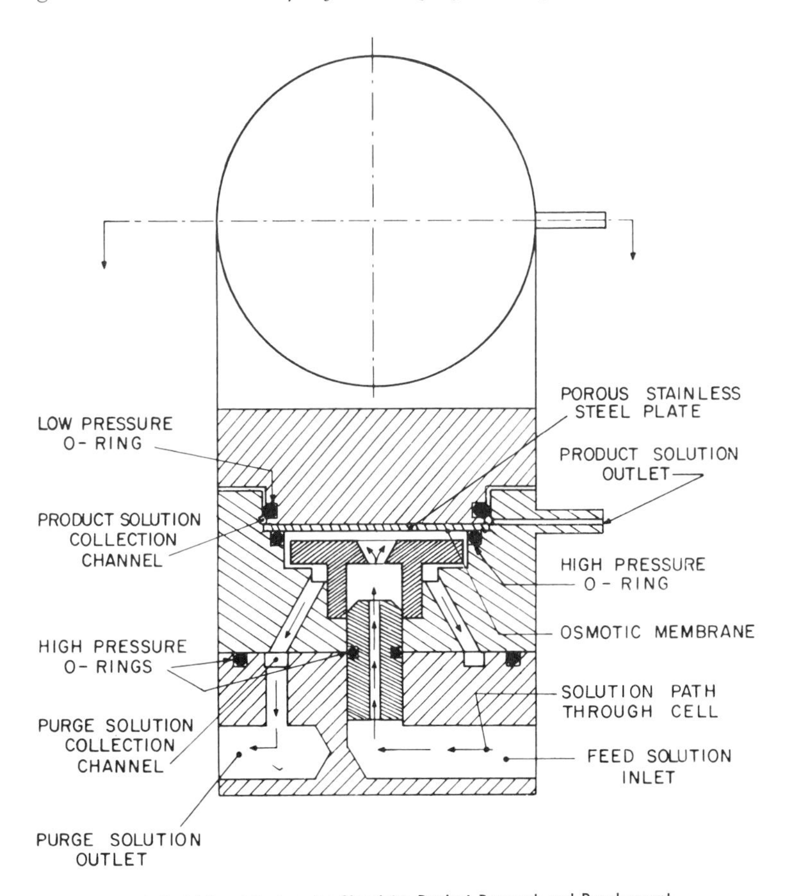


Industrial and Engineering Chemistry Product Research and Development

*Figure 11. Cross-section of radial flow cell* (9)

Foulants generated in the plant result from the corrosion of metallic components. The design of an RO plant has therefore not only to assure structural integrity but in addition reduce the adverse effects of corrosion products on membrane performance. The advantages of non-metallic materials are obvious. But Monel is satisfactory at low seawater velocities and SS 316 can be adequately engineered by maintaining velocities higher than 6 ft/sec. and eliminating stagnant zones. The interference of iron fouling can be further reduced by acidification to pH 5-6. The control of corrosion products is simplified, if only one metal is used in the plant (10).

The cell tests with the one-pass film membranes showed that fouling prevention is more effective than foulant removal by cleaning. Good performance stability was demonstrated with live seawater filtered to better than 0.5 Formazin Turbidity Units, chlorinated and then treated chemically or with carbon to remove free chlorine. The fouling control of the higher flux/lower rejection membranes for the two-pass process was more complicated and the type of pretreatment adequate for one-pass membranes was not sufficient (Figure 12). Flux declined rapidly after the first week. Cleaning restored performance briefly. The restoration, though temporary showed that the intrinsic capability of the membrane was better than indicated by the field test data and pointed to fouling and not membrane deterioration as the cause of flux decline. Tests with agitated test cells did not change the picture; evidently mixing near the membrane surface did not counteract the tendency to foul. Cost calculations did not identify sufficient incentives to solve the fouling control problem of the two-stage process and all subsequent work concentrated on the one-pass process. The elements of a pretreatment system to control fouling of one-pass membranes comprising

chlorination,

pH adjustment,

anthracite/sand and diatomaceous earth filtration,

removal of residual chlorine with carbon or chemicals,

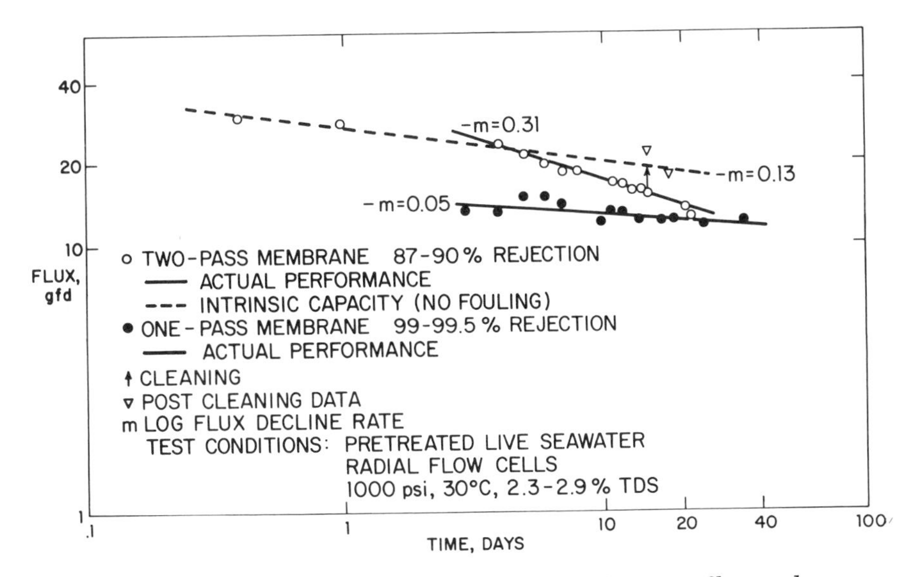


*Figure 12. Representative performance of one-pass and two-pass film membranes*

were defined and equipment to process 70,000 gal/day was installed at Ocean City to evaluate the performance of the membranes in practical modules. Provisions to heat the incoming seawater in winter were needed at Ocean City to maintain year-round operation. The only major change through the field test trials was the addition of inline coagulation. The final set-up is shown schematically in Figure 13 and the intake and pretreatment facilities can be seen in Figures 14 and 15. The two sand filters and diatomaceous earth filters were operated in parallel; the carbon filters were operated in series - except during backwash - to assure complete removal of all chlorine. The large storage tank provided sufficient residence time for effective chlorine treatment. Control instruments in the pretreatment system monitored the pH, chlorine concentration before the carbon filters as well as the absence of free chlorine between the first and the second carbon filter and finally seawater temperature before feeding the RO equipment. The other important process variables in the RO plant itself are operating pressure, flow rates and salt concentration. The one-pass membranes reject divalent ions essentially completely and therefore only the chloride ion concentration of the feed and the product stream has to be monitored. Rejection data refer to total chlorides.

## Feed Water Quality

As stated before, particulate or colloidal matter and microbiological species are the major sources of membrane foulants in seawater RO. Methods to monitor the latter category are only gradually evolving, but a meaningful measure of the first category is now available in the plug factor test. In the plug factor test the change in filtrate flow with time through an 0.45 micron filter at constant pressure is determined. Results are expressed as "percent plug factor" (P)

$$P = (1 - \frac{t_1}{t_2}) \times 100$$

wherein $t_1$ and $t_2$ are the time intervals required to collect a fixed volume (500 ml.) of filtrate. The

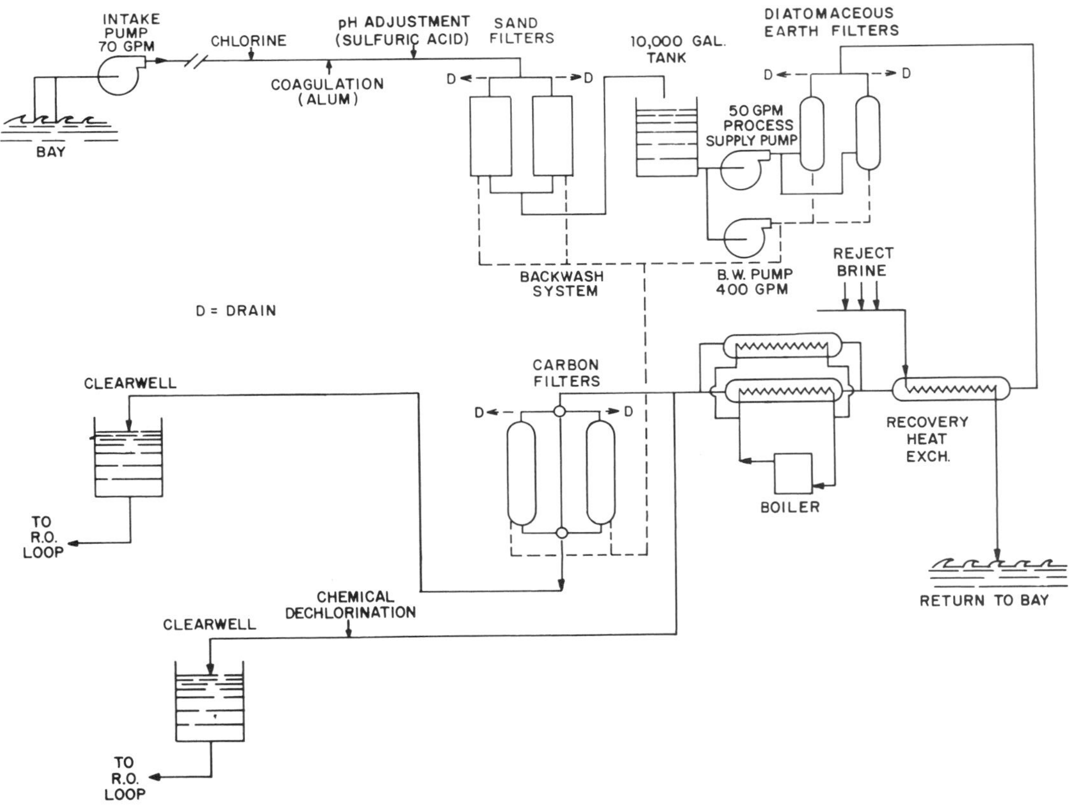


*Figure 13. Schematic of seawater pretreatment system Ocean City test site*

*Figure 14. Seawater intake at Ocean City, N.J.*

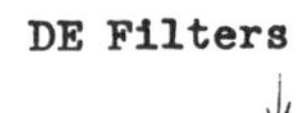


Carbon Filters →

Sand Filters ←

*Figure 15. Pretreatment facility at Ocean City, N.J.*

test has to be carried out at constant pressure - conventionally 30 psig - and the time interval between the two measurements is usually 15 minutes. The test can be performed with a Millipore Inline - 47MM Filter Holder. The ratio of percent plug to the time interval between the two measurements is termed the fouling index.

The basic operations of the manual procedure were recently automated and can be performed continuously at a frequency of three measurements per hour with the instrument shown in Figure 16.

## Field Tests of RO Devices

This section discusses the field trials which culminated in the demonstration of seawater desalting in one-pass with commercial hollow fiber permeators.

Du Pont's membranes can be fabricated in film or in hollow fiber form and are therefore adaptable to all three principal membrane configurations. Seawater RO was, in the late sixties/early seventies, a pioneering undertaking at the frontiers of desalting technology and reports on earlier attempts were less than encouraging. We adopted for this first phase of our work the tubular configuration as a matter of prudence, because

- the larger brine channels operated in turbulent flow provided flexibility in controlling the two major problem areas of concentration polarization and membrane fouling, e.g. the tubular device can be cleaned mechanically with foam balls,

- the development work to achieve operating pressures of 1000-15000 psi was less than needed for the other more complex configurations.

Desalination in one-pass was demonstrated with single perforated ss 316 tubes which contained tubular DP-1 membranes. Polymeric film membranes can lose their high salt rejection, if the support medium has an uneven surface. Continued satisfactory operation in the field was possible after smooth membrane supports from non-woven materials were provided. The performance of these early tubular devices is of more than passing technical interest because some of them have retained their salt rejection properties after more than two years in the field pointing to the inertness of the polyamide and polyamidehydrazide in natural seawater; their integrity is affected only by additives, e.g. chlorine. The DP-1 membranes were then incorporated in several multitube clusters of 18 tubes each suitable for operation up to 1500 psi. These prototypes of commercial seawater modules were fabricated by Paterson Candy International Ltd. in the United Kingdom (Figure 17). Modules were tested with two modes of fouling control, namely complete pretreatment (chlorination, pH adjustment, sand/DE filtration and carbon treatment) or mild pretreatment (sand/DE filtration) combined with frequent membrane cleanings. The following results

| Pretreated Water (1 Module) | | Sand Filtered Water (2 Modules) | |
|---|---|---|---|
| Start | 12.5 gfd/99% Rej. | Start | 11.7 gfd/99.3% Rej. |
| 77 Days | 10.4 gfd/99% Rej. | 74 Days | 10.0 gfd/98.8% Rej. (30 cleanings) |

(1200 psi, normalized to 3.5% TDS, 25°C)

pointed to realistic prospects for seawater RO. It became, however, doubtful at that stage whether the tubular configuration was a technical necessity. The experiments with test cells which had narrow brine channels of 0.020 to 0.050 inch width and some preliminary trials with hollow fiber permeators showed that the more compact membrane devices had promise. Cost calculations indicated that even with the added cost of complete pretreatment the compact configurations were economically attractive over tubular modules operated without pretreatment. Emphasis in the program shifted therefore to one-pass desalination with spiral and hollow fiber modules.

*Figure 16. Continuous plug factor monitor*

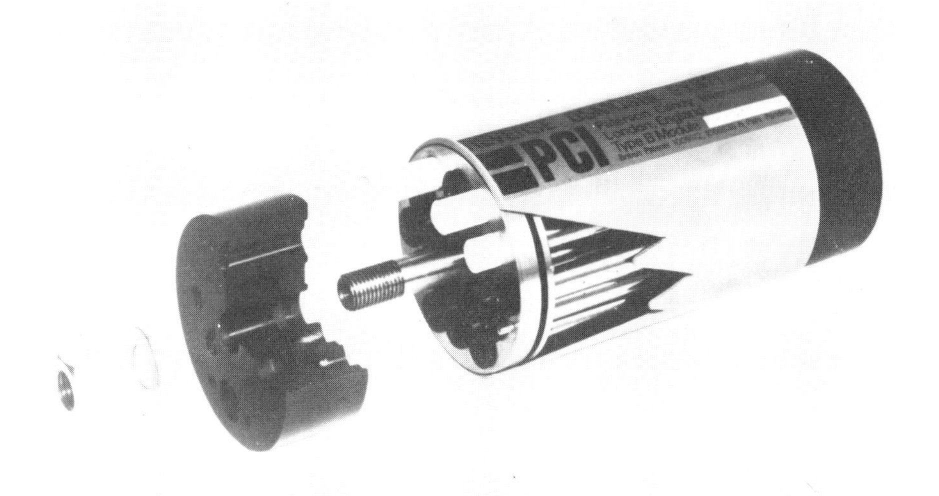


*Figure 17. Tubular RO module*

Small spiral elements (2 inch diameter - 3 sq. ft. membrane area) with DP-1 polyamidehydrazide film membranes showed good integrity in lab tests with the following results:

| | |
|---|---|
| 1000 psig | 8.4 gfd/99.3% rejection |
| 1250 psig | 11.5 gfd/99.5% rejection |

(3.5% seasalt, 25°C)

Performance stability was maintained at Ocean City in field tests lasting 3 to 4 months. Field tests on pilot plant scale are planned when larger elements (4 inch diameter) are available.

Field tests with hollow fiber permeators were carried out with 4 inch B-10 devices and with prototypes of an experimental 8 inch device. Data on the two permeators are listed in Table III. The progress made in maintaining longer and longer periods of stable operation with hollow fiber permeators is shown in Figure 18. Representative lines were drawn through actual performance data as illustrated for permeator 2787. The first permeator lost performance stability after approximately 500 hours because of inadequate removal of particulate matter. We added coagulation to the pretreatment. We then encountered slime formation and today we suspect that this slime may have been induced through our own actions by the addition of sodium thiosulfate to eliminate free chlorine. The permeators lost additional productivity- though not rejection - during a failure of the thiosulfate injection system. With carbon treatment as the more reliable dechlorination route, stable performance with moderate productivity decline from fouling was maintained for more than 4000 operation hours till November 1974 when the equipment was dismantled for transfer to the Government Test Site at Wrightsville Beach, North Carolina. The productivity loss encountered during a lengthy shutdown (permeator 3338) can be avoided, if some formaldehyde (1%) is kept in the permeator. Periodic shock treatments during RO operation with a 500-1000 ppm sodium bisulfite solution which has biocidal properties were beneficial. Sodium bisulfite is also a practical agent for chemical dechlorination, but reliable controls to intercept any residual chlorine will have to be demonstrated before it can replace the carbon.

TABLE III

B-10 SEAWATER PERMEATORS

| | | |
|---|---|---|
| Diameter | 4 inch | 8 inch |
| Membrane Type | Annealed Aromatic Polyamide | |
| Membrane Configuration | Hollow Fiber 92μ OD | 41μ ID |
| Membrane Area, sq. ft. | 1500 | 5000 |
| Initial Productivity[1], gpd | >1200 | >4000 |
| Initial Rejection[1], % chloride | >99 | >98.5 |
| Operating Pressure, psi | 800 | 800 |
| Temperature Range,°C | 0-30 | 0-30 |
| Shell/End Plate Material | Filament Wound Fiberglass/Fiberglass Epoxy | |
| Permeator Weight Filled With Water | 50 lbs. | 155 lbs. |

[1] 800 psi, 3.5% TDS (seasalt), 25°C, 30% recovery

Adequate rejection has generally been less of a problem than productivity decline. The permeators were typically cleaned once a month with various agents that could include flushes with polyphosphate, urea and enzymatic detergents. Membrane post-treatments to restore rejection can, if needed, be applied in the field. The longest operating experience with an 8 inch permeator is 2000 hours. Daily shock treatments with a sodium bisulfite solution were applied and the productivity has been even more stable than observed with 4 inch permeators (Figure 19).

The trials with single 4 inch permeators helped to consolidate the pretreatment to

chlorine 10-12 ppm,
alum coagulation 30 ppm,
sulfuric acid to pH 6.5,
sand and diatomaceous earth filtration,
carbon treatment,

which maintained water quality in a fouling index range of 3 to 4 with occasional excursions to 5. The scale of field tests was now expanded to multi-permeator operation in a pilot plant (Figure 20). Five 4 inch and two 8 inch B-10 permeators were arranged in the final 1000 hours before shutdown in a staged design in order to achieve high recovery. By that time each permeator had been on test for at least 1000 hours. Arrangement and representative performance data are shown in Figure 21. Total potable water production in these pilot plant trials exceeded 10,000 gallons/day. The brine concentrate contained 5.5 to 6% salt corresponding to at least 30% recovery from seawater at the standard concentration of 3.5% TDS. The permeate from the second and third stage had less than 350 ppm TDS, showing that the B-10 permeators can recover potable water from a feed stream of standard seawater concentration.

## RO in the Seawater Desalination Market

The Membrane Division in the Office of Water Research and Technology has aggressively supported research in seawater RO and as a result several approaches are being developed. The Dow Chemical Company is conducting field tests in one-pass

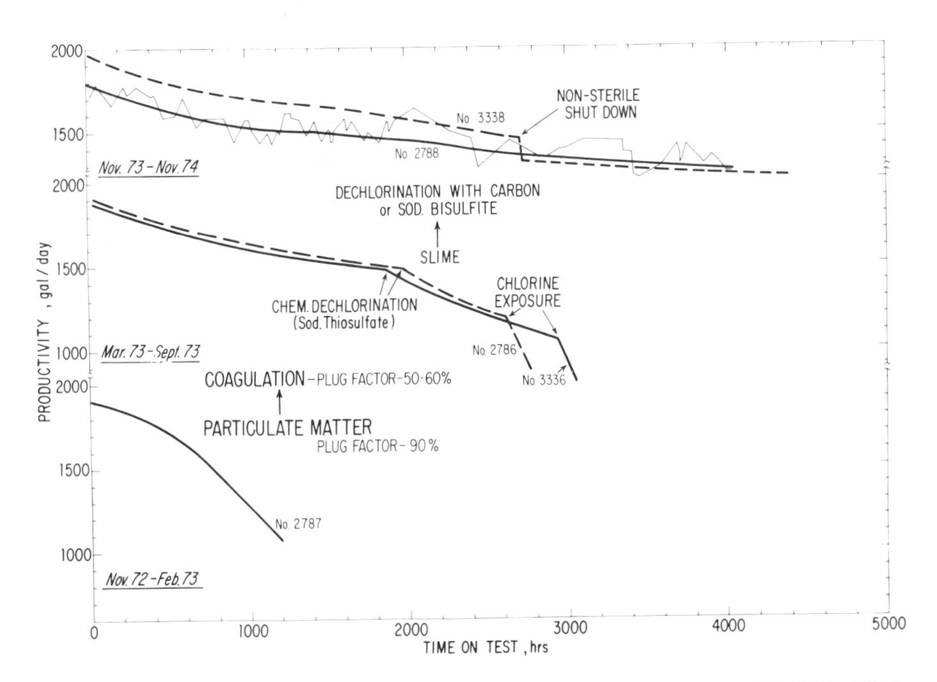


*Figure 18. Field tests—4-in. B-10 permeators (productivity at 800 psi, 2.7% TDS, 25°C, 30% recovery)*

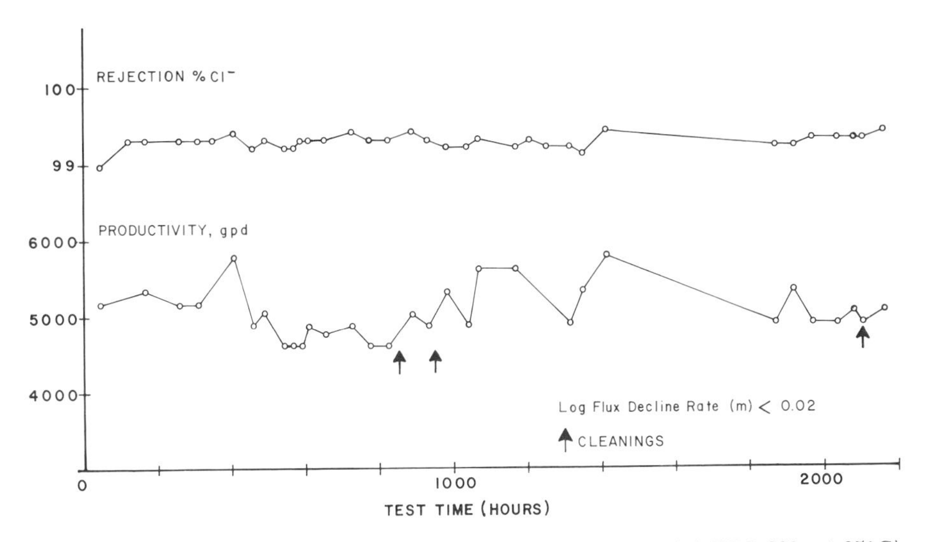


*Figure 19. Performance of 8-in. B-10 permeator (live seawater, 2.7% TDS, 800 psi, 25°C)*

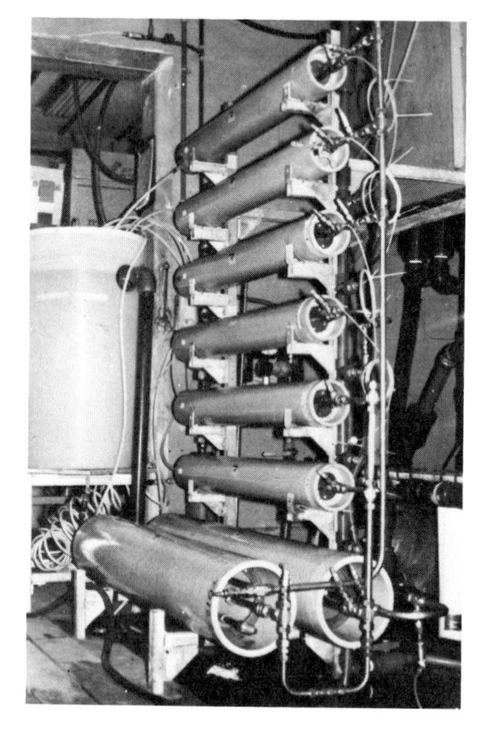


*Figure 20. Pilot plant with "Permasep" permeators*

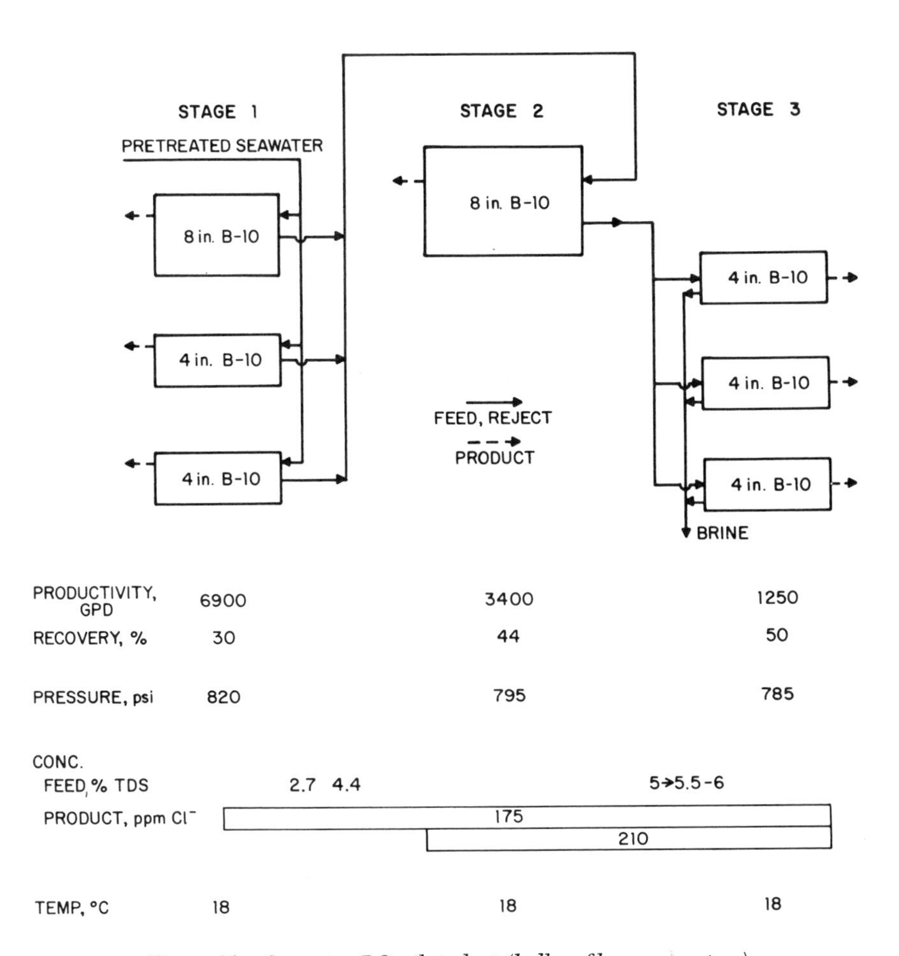


*Figure 21. Seawater RO pilot plant (hollow fiber permeators)*

desalination with triacetate hollow fibers. Universal Oil Products' Roga Division has extended the concept of asymmetry to composite film membranes, where an ultrathin skin is deposited on a porous support structure. Cellulose triacetate thin film composite membranes in small spiral modules (4 sq. ft.) were tested with live seawater and produced 8 gfd/99.7% rejection at 1000 psi (11). Envirogenics Systems Company tested spiral modules with membranes from cellulose acetate blends in one-pass and two-pass operations (9200 East Flair Dr., ElMonte, California 91734).

Commercial desalination is dominated by distillation processes. It is a relatively new technology which grew over the last 20 years from approximately 15 million gallons per day installed capacity in the early fifties to over 50 million gallons per day in the early sixties and to more than 350 million gallons per day in the early seventies. This trend is expected to continue despite increasing cost. Recent cost estimates for new distillation plants with realistic accounting for energy cost predict $2/1000 gallons at the 50,000,000 gpd level, $4-6/1000 gallons at 1,000,000 gpd plant capacity and more than $7/1000 gallons in 100,000 gpd plants. We expect RO to be competitive in the 100,000 to 1,000,000 gpd range, provided membrane life is at least two years and pretreatment cost is not excessive. The Ocean City pretreatment facility was one of the first installations of its kind and had to handle a very difficult water. Several simplifications come readily to mind. Local conditions may also be more favorable, e.g. water from seawells may meet the required quality standards with very limited additional treatments. A major incentive for RO is the lower energy requirements amounting to 25-40% of the thermal processes.

We foresee smaller commercial plants to be in operation in the near future. A small plant will go on stream in the Caribbean area and test programs with individual B-10 permeators are in progress or in preparation in Europe, on the U.S. West Coast and in the Far and Middle East.

## Summary

Du Pont has developed polyamide hollow fiber membranes and polyamidehydrazide film membranes that produce potable water from seawater by the reverse osmosis process. This paper is mainly based on the work done with these non-cellulosic reverse osmosis membranes.

The salt concentration of 3.5% in seawater and the corresponding osmotic pressure of at least 350 psi make the engineering of a pressure driven membrane process to desalt seawater a formidable undertaking. In addition, the interference of membrane foulants that accumulate on the surface and reduce productivity and salt rejection has to be controlled for reliable plant operation. The technical capability to desalt seawater by reverse osmosis has been demonstrated, the leading options being at this time desalination in one-pass with hollow fiber and spiral wound membrane configurations. Practical methods to control membrane fouling by pretreating the seawater exist. "Permasep" hollow fiber permeators have produced potable water from seawater in a 10,000 gallons/day pilot plant. The lower energy requirements of the RO process are an advantage in competition with the conventional thermal (distillation) processes.

## Acknowledgement

The work reported in this paper was carried out by members of a Du Pont task force. Among these, R. A. Halling did the work with DP-1 film membranes. P. P. Goodwyn and S. A. Cope developed the B-10 seawater permeators. P. R. McGinnis was responsible for field tests and J. P. Agrawal worked on fouling control.

The Office of Water Research and Technology (formerly OSW) in the U.S. Department of the Interior supported financially Du Pont's seawater program with film membranes since March 1971 and with hollow fiber permeators since July 1973.

## Literature Cited

1. Panar, M., et al, Macromolecules (1973) 6, 777-780, Sept.-Oct.
2. "Recent Developments in Separation Science", Vol. II, CRD Press, the Chemical Rubber Co., Cleveland, Ohio 44128.
3. S. Sourirajan, "Reverse Osmosis", Academic Press, New York, 1970.
4. Merten, U. (Editor), "Desalination by Reverse Osmosis", The M.I.T. Press, 1966.
5. R. E. Lacey, S. Loeb, (Editors), "Industrial Processing with Membranes", Wiley, New York, 1972.
6. "Why Hollow Fiber Reverse Osmosis Won the Top CE Prize for Du Pont", Chemical Engineering, Nov. 29, 1971, p. 53.
7. Grover, J. R., et al, Desalination (1974) 14, 93-102.
8. Applegate, L. E., Antonson, C. R., "Reverse Osmosis Membrane Research", Plenum Press, New York, London 1972, pp. 243-252.
9. Manjikian, S., Ind. Eng. Chem. Prod. Res. Develop. (1967) 6, 23.
10. Rosenblatt, N. W., 4th Int. Symp. of Fresh Water from the Sea (Heidelberg), (1973) 4, 349-361.
11. Riley, R. L., et al, (Roga Division Universal Oil Products, San Diego), Preprint to be published in "The Permeability of Plastic Films and Coatings To Gases, Vapors and Liquids", Plenum Publishing Corp., New York.

# 33

# Marine Anti-corrosive and Anti-fouling Coatings

MONROE M. WILLEY

E. I. Du Pont Nemours & Co., Marshall R&D Laboratory, Philadelphia, Penn. 19146

One of the oldest and most important resources we obtain from the world's ocean is the low cost, low energy consuming transportation of heavy bulk goods. It's been determined that a ton of coal can often be carried across an ocean at lower cost than across a state. The English term "ton" by which we measure great quantities of nearly everything,originally referred to a maritime container used for the bulk shipment of wine. The ton, or "tunne" was a wooden container of 252 wine gallons. It occupied about 40 cubic feet of hold space, which became the "cargo ton", and weighed about 2240 pounds, or our current "long ton". This, in turn, is the weight of about 35 cubic feet of seawater, or the "displacement ton".

The field of marine coatings is probably only a little younger than the art of navigation, having undoubtedly been born of necessity, when the ancient mariners found that the lives of expensive wooden ships were relatively short in the foreign environment of the sea.

The Old Testament tells us that Noah was instructed to coat his ark within and without with pitch. One of the purposes of a bituminous coating would probably have been to help protect the wood exposed to air from rot, and the underwater hull from the infestation and subsequent destruction by the teredo or ship worm. The effect of the ship worm can often be seen on driftwood in the form of holes or tunnels. While the teredo is a shellfish rather than an insect, it might be viewed as a sea-going termite, that for centuries has been the major "corrosive" agent on wooden ship bottoms in warm waters. A number of preventive measures have been employed over a long period of time. Outside wooden sheathing, metal sheathing, charring the wood to some depth, and a large variety of coating concoctions have all been employed. The ancients in the Mediterranian area who were skilled in working with various metals met success against the ship worm with the use of lead sheathing nailed to that part of the hull below the water line. This left only the problem of the attachment to the surface of the hull of such organisms as barnacles, algae, and the like, which, while relatively non-destructive, impart considerable "drag", or resistance as the vessel moves through the water. This accumulated growth is referred to as "fouling". An example of fouling show on 6" x 12" panel that was immersed in Florida waters for several months. The roughness of a accumulation significantly reduce the ship speed, and had to be removed by scraping every few months.

The use of lead sheathing was successfully carried on for centuries until 1758, when the British Navy apparently ran short of lead and decided to evaluate copper sheathing on the 32 gun frigate, HMS Alarm. After some 20 months at sea, largely in Carribean waters, the Alarm was dry docked, and found not only free of ship worms, but also free of fouling. Why copper sheathing wasn't tried 2000 years earlier is apparently a mystery, but a practical solution to the fouling problem was finally discovered. The use of copper sheathing continued until the advent of steel or iron hull in the mid 1800's. At this time, it was discovered (the hard way) that copper attached to ferrous metals and then immersed in seawater, caused the iron to dissolve in the ensuing electrolysis. H.M.S. Jackal actually foundered as a result of copper sheathing on the iron hull. It was found that copper could be used over iron if the two were insulated from each other with a material such as an intervening wooden sheathing. This rather expensive practice was carried out to some extent (1).

The advent of iron hulls brought with it the problem of severe corrosion of these metals in the marine environment. Standard metal protective paint systems such as red lead pigmented linseed oil primers followed by suitably pigmented oil based topcoats were used to retard corrosion, but were less effective than on land based structural steel. Some improvement was later obtained with the use of synthetic resins such as oil modified phenolic varnishes, and alkyd resins, along with pigmentations that were a little more effective in seawater. Even then, such systems offered limited protection, required constant maintenance by the ships crew, and frequent dry docking for replacement of corroded plates and repainting. Those in the audience who may have served in the World War II Navy or Merchant Marine will no doubt remember "chipping and painting" details with little affection.

The corrosion problem varies in degree and mechanism in three general parts of the ship. These are the "topsides" or the part totally above the water line which is exposed in the atmosphere with occasional wettings in seawater. The "boottop" or splash zone, is that section that remains constantly wet with seawater while simultaneously receiving the high oxygen levels of the atmosphere , and the bottom, or that part which is constantly immersed in the sea but exposed to relatively low oxygen levels.

Steel used in ship construction is received from the mill covered with a blue oxide surface layer called "mill scale" that forms during the hot rolling process. Until recent years, the protective coatings were applied over the mill scale and tightly adhering rust after construction. In the marine environment such systems would rapidly begin to fail by "under-rusting", or the formation of rust under the mill scale and continuing formation in the areas where tight rust existed originally. This situation was held under control by constant chipping away of the under-rusted areas and repainting on voyage. Bottoms, of course, had to be frequently repaired and repainted in dry dock at great expense.

With the development of the modern ships of much greater size, but crews of about the same size as on the smaller vessels, it became impractical or impossible to continue on-voyage maintenance to the extent previously carried out. More efficient coatings systems offering longer term protection and lower maintenance therefore became a necessity.

In the modern day construction mill scale is removed down to bright steel usually prior to fabrication, by the impingment of abrasive grit or steel shot propelled by compressed air, or from centrifugal wheels. This process, known as shot or grit blasting, presents a steel surface free of oxides or other harmful impurities, and imparts a "profile" to the surface that greatly improves the adhesion of the first layer of protective coating. This first coating layer is applied immediately after blasting plates or pre-fabricated sections to prevent the formation of rust, which otherwise would rapidly develop. It is of sufficient thickness and quality to allow the steel to be stored outdoors for several months while awaiting fabrication. It must be compatible with the coating system that later goes on over it, and must not detract from the in-service performance of that system. Pre-construction primers may be of various types, depending on the intended subsequent coating system and area of use; however, the most commom type is probably "inorganic zinc". The latter is a zinc-rich coating consisting of zinc dust particles bound to each other and to the steel substrate by a poly-silica glass. These coatings, being anodic to the steel, provide a high degree of corrosion resistance and are extremely resistant to abrasion from rough handling. They are capable of receiving over them additional coats of inorganic zinc, as well as most other types of coatings that are not easily subject to saponification. The most commonly used inorganic zinc is applied by mixing zinc dust into a vehicle of a partially hydrolized alkyl silicate solution, and spraying the resulting mixture onto the steel. The coating then cures by reaction with atmospheric moisture to split off the alkyl group as alcohol, leaving the silica binder.

Other vehicles such as sodium silicate and lithium silicate have also been used where an aqueous system is desired.

The plates and shapes, or pre-fabricated sections coated with a suitable pre-construction primer are given their complete coating system at a further stage of fabrication, or after completion of construction.

The area above the waterline and the splash zone usually receives a further coating of inorganic zinc over the thin coat of pre-construction primer. Additional protective and decorative organic coatings are then applied over the zinc to provide a long lasting system of good appearance. The organic system over the zinc usually consists of a "tie coat" applied directly to the zinc to provide adhesion over a long period of time, and to build up sufficient thickness to protect against the elements and provide a smooth surface to receive the final topcoats. The tie-coat must be resistant to saponification in the presence of the inorganic zinc, be easily applied in a minimum number of coats, and have the toughness and general resistance properties required by the intended service.

Topcoats are applied over the tie coat to provide the desired color and appearance, as well as resistance to sunlight, rain, and seawater. Special "boottop" coatings are sometimes used at the waterline, or in the case of variable waterline vessels, between the light and deep loadline. These coatings are formulated for maximum resistance to seawater, and may sacrifice some resistance to "chalking", or erosion caused by normal atmosphere weathering.

The types of organic coatings used on the topside are variable in generic type, and are selected with regard to the type of service, production schedules, climatic conditions in the shipyard, and the shipowner's desires and needs. Included are vinyls (polyvinyl chloride), chlorinated rubber, catalyzed epoxies, epoxy esters, alkyds, and their various modifications, and, to an as yet limited extend, urethanes. The reasons for selecting a particular coating are too numerous to go into at this time. It will therefore suffice to say that all have their own particular advantages, and provide good service when properly applied and used.

Moving below the waterline, we encounter another world where corrosion problems are somewhat different, and the function of topcoats is entirely different from those used in the atmosphere. Zinc rich primers other than pre-construction do not seem to be frequently used below the waterline, since it appears to be a common opinion that organic coatings provide at least equally good protection in these surroundings. The anti-corrosive bottom coatings include the same generic types as those used on topsides, but also include bituminous coatings such as coal tar, and tar reinforced with polymeric materials such as epoxy-polyamide or epoxy-amine resins. Metallic aluminum flake finds considerable use as a pigment in bottom coatings to further improve corrosion resistance. Aluminum is anodic to steel, and the flakes may be made to orient themselves in an overlaping fashion to help reduce water permeability. Polyester coatings filled with glass flakes and applied at very high film thicknesses have also made an appearance as bottom coatings. Organic coatings in general that are applied directly to steel are very often pigmented in such a way to further add to the corrosion resistance of the resin itself. Zinc chromate, zinc oxide, and red lead are examples. Various so called inert pigments, or fillers, will also have an effect on corrosion resistance as well as on film properties. Talc, mica, clay and calcium carbonate are typical extender, or filler, pigments.

In addition to providing ordinary corrosion protection, anti-corrosive coatings below the waterline also function as a separator or insulator between the steel hull and copper containing anti-fouling coatings. Being cathodic to steel, the latter coatings have the potential to accelerate corrosion should they come into direct contact with the hull. The use of adequate thicknesses of anti-corrosive coatings on bottoms is therefore doubly necessary.

The protection offered by underwater coatings is often supplemented by a system of "cathodic protection", wherein an electrical potential of less than one volt is imparted to the water immediately surrounding the hull, placing the latter in a cathodic state. This is accomplished either by the attachment of sacrificial anodes of a suitable metal, such as high purity zinc, or by directly impressing a current generated by equipment inside the ship. In the latter case the current is introduced to the water by anodes passing through the hull to the sea. Coatings used in conjunction with cathodic protection systems must be resistant to the alkalinity generaged at the steel surface in order to avoid being lost. Most bottom coatings used today are therefore formulated with cathodic protection being a consideration.

As previously indicated, anti-fouling coatings are applied over the anti-corrosive coatings to prevent the attachment of various marine organisms that reduce speed, increase fuel consumption, and in some cases, contribute to corrosion of the steel plating. The anti-fouling coatings currently in use work by the same principle as copper sheathing; that is, they slowly leach into the water a material toxic to the attaching organisms. Copper sheathing does this simply by slowly corroding and dissolving into the seawater. Anti-fouling paints have the toxin or toxin producing material dispersed through the film from which it is slowly released. Anti-fouling paints may be classified in two general categories. In the first type, the binder, or matrix, is slightly soluble in seawater. It is formulated in such a way that as the toxin at the surface of the film is depleted, the surrounding matrix dissolves at a similar rate, and exposes a fresh toxic surface. In an ideal situation, this process continues until the paint film is completely gone. This type of coating is known as a "soft" or soluble matrix anti-fouling coating, and probably finds it greater use on small pleasure craft. The other general category is the "hard" or insoluble matrix type where the binder is insoluble in seawater, but the toxin is present in sufficient concentration to allow it to be slowly dissolved out of the film leaving a porous, sponge-like "skeleton" (1). There are also intermediate levels where the soluble portion of the film consists of both toxin and a soluble resin.

The most common resin used as the soluble binder is rosin and certain of its derivitives. Other materials such as shellac may be used, but rosin works as well, and is low in cost. Other non-soluble resins may be blended with rosin to help control its rate dissolution, or to form a "hard" coating. Polyvinyl chloride and chlorinated rubber are examples of insoluble resins that may be used.

The most common toxin is probably cuprous oxide. Cuprous oxide, like metallic copper, is slowly dissolved by seawater to provide a surface sufficiently toxic to repel most attaching organisms. Other toxins may also be employed either alone or in conjunction with cuprous oxide. Examples are tributyl tin oxide, tributyl tin fluoride and cupric hydroxide. Organic compounds have also found some use as marine anti-foulants. Combination of biocides are often employed in order to be effective against a wider range of organisms.

Toxic anti-fouling coatings in their present state have certain obvious disadvantages. They have the potential to accumulate in confined waters, such as enclosed harbor areas having little or no current flow, thereby having a possible ill effect on various marine flora and fauna in that location. Secondly, they have by their very nature a limited life span, since their effectiveness depends upon the actual removal of the critical ingredient. Traditionally, we have attempted to obtain maximum effective life of the coating by formulating the matrix in such a way that the toxin is released no faster than is necessary to prevent fouling. In actual practice, a well formulated coating is effective for perhaps 18 months.

In recent years some interesting efforts have been made to extend the life of anti-fouling coatings. One of these is a thin hydrophylic acrylic coating that is applied over the conventional anti-fouling coating that allows the toxin to leach out sufficiently while the ship is at anchor or tied to a dock, but is claimed to prevent the rapid and wasteful extraction that normally occurs while the ship is underway. Since fouling attaches itself only when the vessel is standing still, or nearly so, relative to the surrounding water, such a coating would help save the toxin for the time when it's needed (2).

With many conventional A.F. coatings, the leaching rate of the toxin in the early months of exposure is well above the critical level necessary to prevent fouling. This is done in order to keep the toxin at or above the critical level needed to be effective during the later months. As a result excessive biocide is introduced into the water during the early period, which is both wasteful, and may generate additional pollution to enclosed waters. The U.S. Naval Ship Research and Development Center has announced the development of a series of organometallic polymers consisting of a suitable backbone polymer, such as vinyl, acrylic, etc. chemically combined with an organotin moiety of the following general structure:

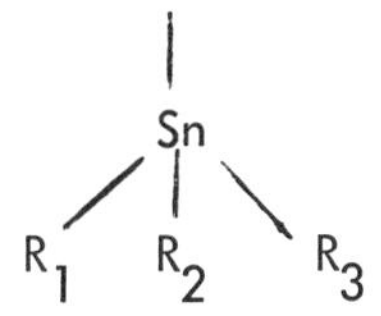


Where the R's may be propyl, butyl, phenyl, or a combination thereof. This in effect makes the polymer toxic in itself, as opposed to the conventional system where the biocide is dispersed through a matrix. It is claimed that these polymers can be formulated to release organotin toxin through hydrolysis at a uniform rate during exposure at or just above the critical level needed to prevent fouling. This, of course, would decrease pollution by avoiding the release of excess toxin, and lengthen the effective life of the coating (3).

A development not in the coatings field, but worthy of mention, is the use of hulls made of copper-nickel alloys that in themselves resist accumulation of fouling. Four shrimp trawlers were recently built with copper-nickel hulls for use in the Indian Ocean. The boats are expected to provide economic advantages in this part of the world, where fouling is heavy, and dry docking facilities are limited (4).

The ideal coating would be one that is not toxic at all, but prevents the attachment of organisms because of its "non-stick", or release properties. Du Pont's Teflon® flourocarbon resins were one of the first candidates to be evaluated in this area. Unfortunately, barnacles and the various other marine organisms found the Teflon® surface to be a firm anchorage. More recently it has been reported by Batelle Memorial Institute that silicone rubber resists the attachment of fouling purely by virtue of its release properties, without the presence of any toxic materials whatever. While anti-fouling release coatings are not yet commercial, such results would indicate that a totally non-toxic, long lasting anti-fouling coating is possible (5).

## Literature Cited

(1) "Marine Fouling and its Prevention", Woods Hole Oceanographic Institution, Woods Hole Mass., Published by U.S. Naval Institute, Annapolis, Md.

(2) Van Londen, A.M., "A Hydrophilic Bottom System to Improve a Ship's Performance", Presented at the 14th Annual Marine Coatings Conference, Williamsburg, Va., (March 1974).

(3) Dykman, E.J. and Monlemarano, J.A., "Performance of Organometallic Polymers as Anti-Fouling Materials", Journal of Paint Technology, (January 1975).

(4) Marine Engineering/Log, "Inland and Offshore", (February 1975).

(5) Mueller, W.J. And Nowacki, L.J., U. S. Patent No. 3,702,778, "Ship's Hull Coated with Anti-Fouling Silicone Rubber".

34

# Applications of Chemistry in Deep Ocean Mining

J. A. OLANDER

Deepsea Ventures, Inc., Gloucester Point, Va. 23062

The Deep Ocean Mining to be discussed here is limited to the mining of manganese nodules. Other deep ocean mining activities such as these relating to the Red Sea muds will not be covered.

While difficult to define, manganese nodules are a form of pelagic agglomeration composed primarily of base metal oxides. Nodules consist of a variety of minerals, hence not of a definite formula, and are therefore most appropriately classified as rocks. The nodules have been described as appearing like potatoes, mammillated cannonballs, marbles, tablets, and other forms, see Figure 1. In size, they range from less than a millimeter to specimens that are many centimeters in diameter. Coloration of nodules is usually earthy black, but this may vary from black to tan. The hardness of nodules is variable, ranging from one to about four on the Moh's scale, with three being an average hardness.

## Occurrence

Manganese nodules occur from the Antarctic to the Arctic, in all major oceans and in deep as well as shallow waters, as shown in Figure 2. They occur in some seas and gulfs and even in some fresh water lakes. They are not, however, distributed as a continuous carpet from shore to shore. Just as land based rocks and ores vary in their occurrence and composition, there is also extensive variation found in nodules. In addition to differences found among samples taken from several locations, there are great variations within a local area in the concentration, size, shape, etc. There are large regions even in the Pacific, which are devoid of nodules, and even in

*Figure 1. Representative nodules showing the variations found in size, shape, and texture*

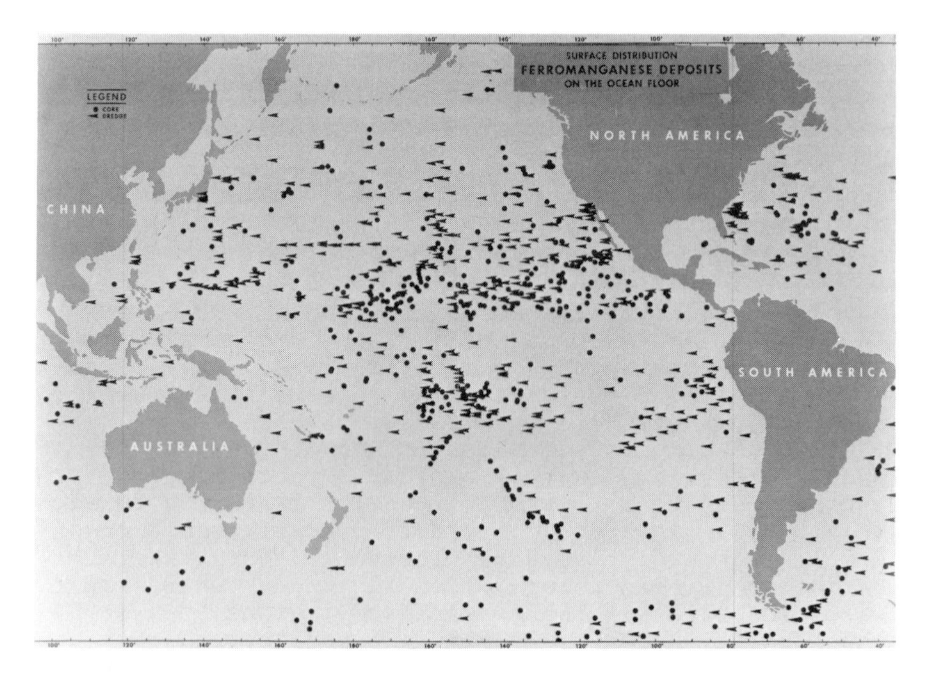


*Figure 2. Location of areas in which nodule deposits have been located thus far (Technical Report No. 1, NSF GX-33616 Office for IDOE; NSF; Washington, D.C., 20550)*

regions where nodules are common, high concentrations are of a local nature.

## Mineralogy

A Chemical analyses of nodules taken from 54 different Pacific Ocean locations are shown in Table I (1,2). In addition to those elements for which analyses are shown, there are also appreciable quantities of cadmium, tin, arsenic, and bismuth present in Pacific nodules.

Examination of nodules reveals that they are comprised mainly of extremely fine grained base metal oxides. The grain size of the oxides varys from less than a micron to only a few microns and rarely exceeds five microns, see Figure 3.

Three manganese minerals have been identified and are present in various ratios in the nodules; todorokite, birnesite, and delta manganese dioxide. Todorokite is quite variable in its chemical composition and can contain significant amounts of other elements substituted for the manganese in the manganese crystal lattice. A typical formula for todorokite is:

$$(Ca,Na,\overset{II}{Mn},K)\ (\overset{IV}{Mn},\overset{II}{Mn},Mg)_6O_{12}\cdot 3H_2O$$

In the oceanographic literature this material has been reported to be 10-A$^o$ manganite.

Birnesite contains less lattice substituted elements than todorokite and has the typical formula:

$$(Na_7Ca_3)Mn_{70}O_{140}\cdot 28H_2O$$

The references in the oceanographic literature to 7-A$^o$ manganite appear to be referring to birnesite. These two "manganites" used in the oceanographic literature should not be confused with the mineral manganite MnOOH which has a very different crystal structure and chemical properties.

In addition to these two manganese minerals, there is also present delta manganese dioxide ($\delta$-$MnO_2$). These three forms of manganese show the degrees of oxidation increasing from todorokite, through birnesite, to delta manganese dioxide.

The only iron mineral which has been recognized in the nodules is goethite, FeOOH. The form in

TABLE I

PACIFIC MANGANESE NODULES (1, 2)

Weight Percentages (Dry Weight Basis)

Statistics on 54 Samples

| Element | Average | Maximum | Minimum |
|---|---|---|---|
| Mn | 24.2 | 50.1 | 8.2 |
| Fe | 14.0 | 26.6 | 2.4 |
| Si | 9.4 | 20.1 | 1.3 |
| Al | 2.9 | 6.9 | 0.8 |
| Na | 2.6 | 4.7 | 1.5 |
| Ca | 1.9 | 4.4 | 0.8 |
| Mg | 1.7 | 2.4 | 1.0 |
| Ni | 0.99 | 2.0 | 0.16 |
| K | 0.8 | 3.1 | 0.3 |
| Ti | 0.67 | 1.7 | 0.11 |
| Cu | 0.53 | 1.6 | 0.028 |
| Co | 0.35 | 2.3 | 0.014 |
| Ba | 0.18 | 0.64 | 0.08 |
| Pb | 0.09 | 0.36 | 0.02 |
| Sr | 0.081 | 0.16 | 0.024 |
| Zr | 0.063 | 0.12 | 0.009 |
| V | 0.054 | 0.11 | 0.021 |
| Mo | 0.052 | 0.15 | 0.01 |
| Zn | 0.047 | 0.08 | 0.04 |
| B | 0.029 | 0.06 | 0.007 |
| Y | 0.016 | 0.045 | 0.033 |
| La | 0.016 | 0.024 | 0.009 |
| Yb | 0.0031 | 0.0066 | 0.0013 |
| Cr | 0.001 | 0.007 | 0.001 |
| Ga | 0.001 | 0.003 | 0.0002 |
| Sc | 0.001 | 0.003 | 0.001 |
| Ag | 0.0003 | 0.0006 | -- |

which the other important metal values are present in the nodules has not been identified.

There are considerable amounts of detrital materials present within the nodules consisting of quartz, calcite, montmorillonite, illite, feldspars, rutile, and barite. These materials are extremely fine grain and are usually distributed throughout the nodule.

Due to the fine grains of the manganese oxide and iron oxide crystals, the porosity and surface area are high. Usually the porosity is better than 50% with the diameters of the majority of pores in the range of 0.1 to 0.01 microns. With porosity of this magnitude, appreciable quantities of seawater and its contained salts are present in the nodules. High porosity with the small pores results in large surface areas of 200 to 300 square meters per gram.

## Formation

Any theory attempting to explain the occurrence of nodules must be compatible with the previous observations. Additionally, other observations about nodules must be considered. For instance, very few buried nodules are found.

A current theory for nodule formation and growth is based upon electrochemical arguments (3). According to this theory, the nodule grows outward from a central nucleus which may be a shark's tooth, or other such material. The precipitation of metal oxides onto the nucleus is a result of the strong oxidizing atmosphere found at the sea floor, and the possibility of forming point charges on the nucleating agent. Hydrous oxides of manganese, iron and other metals will be attracted to the nucleus site, and so be deposited or precipitated. More credence for the electrochemical requirements of the nucleus may be obtained from the fact that nodule encrustation of metalic material is very rapid. Brass shell casings from World War II and pop-top tabs have been found with such accumulations.

The electrochemical explanation for nodule growth and formation may also explain the observation that most nodules are found atop the sea floor sediment. At the seabed, there exists an interface between the highly oxidizing, oxygen enriched sea water, and the less reactive sediment layer with a stagnant and therefore low, oxygen supply. The formation theory presented for nodules require a highly oxidizing environment, and this is found and occurs in the sea water at the ocean floor. The sediment, however, does not present the required oxygen for continued nodule growth. Additionally, it is plausible that decaying marine organisms in the silt layer would produce acids. This combination would not only retard nodule growth, but actually promote dissolution of formed nodule. The net result is that any nodule covered by accumulating silt, or by silt resulting from geological or biological disturbances, would begin a process of dissolution and re-formation at the silt/water interface.

It is also believed that biological activity is important in nodule formation and growth. At least some portion of the iron and manganese in the nodule may be due to the activities of certain marine organisms(4). Perhaps of more importance is the concentration of certain metals through biological processes(3). For instance some biological organisms such as radiolaria require trace quantities of copper and nickel in metabolic processes. Excretions from these organisms and the dead organisms themselves are added to the bottom silt.

Here the contained metal ions are slowly released via a process such as that described for nodule dissolution. The result is a concentration of copper and nickel at the ocean bottom floor interface. Diffusion of the ions through the ooze and into the sea water allows them to be incorporated into the nodule. Although below saturation levels, such phenomenon as adsorption and co-precipitation are believed responsible for the precipitation of these metals.

## Exploration

In any mining program, one of the first requirements is to find and define a suitable mine site. In ocean mining, as with other mining ventures, the location of an ore is only one step in the exploration program. An overall evaluation of the mine site must be made and includes the following criteria (5):

The bathymetry and topography of a potential mine site may be determined via precision depth recordings. Additional information on bottom topography may be obtained from TV viewing of the sea floor, as can estimates of the nodule population. These factors are important in determining the possibility of effectively and economically mining nodules from a given site.

Assay of the nodule material requires actual retrieval and analysis of nodule samples. The gathering of samples is accomplished with towed baskets, "grab-sample" techniques, and any other approach which allows collecting a nodule sample aboard ship.

Assay of samples is carried out on board ship during the exploration cruise. The reasons for this are primarily economic. The cost of operating an oceanographic ship is relatively high, and it is desirable to avoid surveying and exploring areas of low-grade nodule assay. By analyzing the samples on the ship upon retrieval, it is possible to determine if an area warrants further exploration, or if the metal content of the nodule is too low to be attractive.

Analysis of complex ores such as nodules are not particularly difficult in a laboratory ashore, but the project gains new dimensions when it is to be carried out aboard a ship operating on the high seas. Routine laboratory procedures such as the weighing and dissolution of samples become very difficult operations if they are to be used in a shipboard analyses of nodule samples. Therefore, the ideal method of analysis would require little sample preparation and very few manipulation that are best done with firm footing.

The method used on board the R/V PROSPECTOR to analyze nodule samples is x-ray fluorescence analysis (3). A Nuclear-Chicago 9200 Series Analyzer is the instrument used and shown in Figure 4. The sealed radioisotope x-ray source is Plutonium 238 and detection is via a sodium iodide crystal and photomultiplier tube. A sample of Americium 241 is mounted in the instrument to act as a reference energy level.

Calibration curves for the instrument are made using ground nodule samples of known composition. This technique obviates vagaries due to matrix effects as the nodule samples to be analyzed are dried and ground in a fashion similar to the reference samples.

## Mining

After exploration activities have defined a mine site that meets the criteria discussed previously, actual mining of nodules is the next step in a commercial operation. Insofar as is known, no

*Figure 3. Whole nodule and a nodule cross section showing the growth rings and interstitial material*

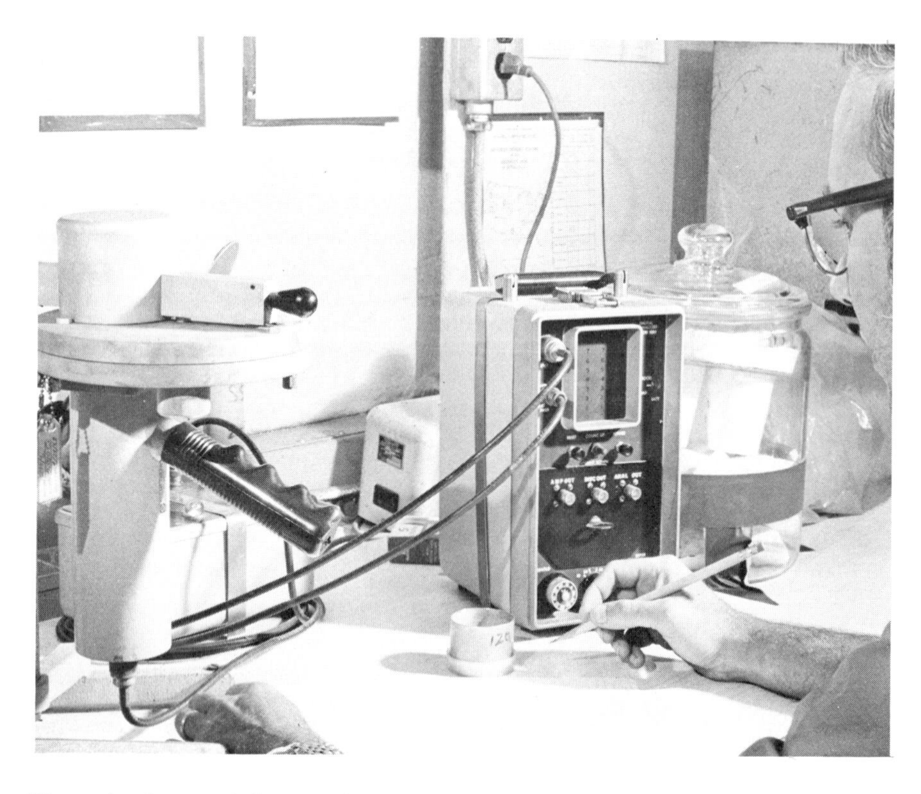

*Figure 4. A view of the x-ray fluorescence apparatus used onboard the R/V Prospector for nodule analysis*

commercial nodule mining operation is underway at this time; therefore, the discussion of mining methods will be restricted to tested prototype systems.

The known mining systems fall into two general catagories (6). The first of these uses a gathering or collecting device connected to the surface by a conduit, and a pumping method to recover the nodules. Mining devices of this type all employ a component which contacts the ocean floor in an effort to make a first separation of the nodule from the surrounding sediment. The second feature common to these systems is control of digging depth into the ocean bottom - usually within the upper few inches of the sediment. The bottom collecting devices employ either a hydraulic or air lift pump to transport the nodules to the surface through a pipe system.

The second type of mining system uses a bucket or basket to drag on the bottom and a mechanical lift procedure. This idea has evolved into the continuous Bucket-Line Dredging System (CLB). This system uses multiple dredge baskets which are connected to a continuous rope rotated by a recovery ship.

Estimates have been made of the environmental impact produced by nodule mining (6). Both the hydraulic and CLB methods will disturb the benthic community in the process of mining, but this is expected to be of little permanent damage as portions of the sea floor throughout a mine site will remain undisturbed. These areas will serve to populate the disturbed communities in the mined portions of the site.

Sediment obtained with nodules in the CLB approach is expected to be washed out of the buckets during their retrieval to the surface. In the hydraulic method of transporting nodules to the surface, both bottom sediment and water will be carried with the nodules. One of the roles of the bottom collector is to minimize the amount of sediment carried to the surface, but any sediment returned to the ocean at or near the surface may cause complications. The cooler, nutrient rich water from the ocean floor is not anticipated to cause any harmful effects and may be beneficial to marine productivity.

Since ocean mining is a new industry and it is being developed in an age of particular awareness of environmental impact, it will be possible to make plans for environmental protection from the beginning. This will begin with establishment of baseline values of existing environment and continue through planning and implementation to maintain these standards.

Processing

Because of the complex mineral and chemical nature of nodules, they do not readily lend themselves to current commercial extractive metallurgical processes. Thus, it becomes necessary to develop new and improved metallurgical techniques for the recovery of the metal metals, and these processes must be tailored for the unique composition of nodules. There has been a recent review of the technical and economic aspects of nodule processing (7).

The nodule processing approaches receiving current serious attention are hydrometallurgical ones. These processes can be divided into those which produce only the associated metals - nickel, copper, cobalt, etc., and those processes which yield these metals in addition to manganese.

One approach for producing the associated metals from nodules is via sulfuric acid leaching. This method takes advantage of the high porosity of the nodules discussed earlier and the fact that manganese dioxide is not soluble in dilute sulfuric acid.

Results for a collection of tests are shown in Table II (1,8,9). The parameters varied were the leaching time, temperature and concentration of sulfuric acid. Of these, it may be seen that acid concentration had the greatest effect on the percentage of metals solubilized. All of the metals were less soluble as the acid strength decreased. Time and temperature of leaching have the greatest effects upon nickel and less so upon cobalt and copper. Results not shown using different particle sizes indicate that the rate of dissolution of copper, nickel, and cobalt is independent of particle size, for particle ranging from 30 to 160 microns. For particle sizes greater than 160 microns, the rate of dissolution markedly decreased - especially for nickel.

The results presented above are for nodules in which the manganese is present as todorokite and the iron content is relatively low. For nodules containing primarily high porosity delta manganese dioxide, there are some differences in the leaching results. The best copper recoveries are from this type of nodule, and the copper recovery is proportional to the amount of manganese dissolved. From

TABLE II

SULFURIC ACID SELECTIVE LEACH(1,8,9)

2P-52 NODULES

Chemical Analysis Wet Basis - 25% Water

| | |
|---|---|
| Mn | 25.6% |
| Fe | 5.0% |
| Co | 0.18% |
| Ni | 1.5% |
| Cu | 1.2% |
| Porosity | 54.3% |
| Manganese Phase | Todorokite |
| Iron Phase | Geothite |

| | Percentage Dissolved | | | | |
|---|---|---|---|---|---|
| Element | Time | Temp. | pH 1 | pH 2 | pH 3 |
| Ni | 6 hrs. | 25°C | 18 | 6 | 1 |
| Ni | 48 hrs. | 25°C | 55 | 24 | 2 |
| Ni | 48 hrs. | 45°C | -- | 60 | -- |
| Ni | 6 hrs. | 94°C | 62 | -- | -- |
| Cu | 6 hrs. | 25°C | 56 | 14 | 3 |
| Cu | 48 hrs. | 25°C | 60 | 17 | 4 |
| Cu | 48 hrs. | 45°C | -- | 23 | -- |
| Cu | 6 hrs. | 94°C | 41 | -- | -- |
| Co | 6 hrs. | 25°C | 58 | 35 | 9 |
| Co | 48 hrs. | 25°C | 70 | 39 | 10 |
| Co | 48 hrs. | 45°C | -- | 44 | -- |
| Mn | 6 hrs. | 94°C | 1.7 | -- | -- |
| Mn | 48 hrs. | 25°C | -- | 0.08 | -- |
| Fe | 6 hrs. | 94°C | 2.9 | -- | -- |
| Fe | 48 hrs. | 25°C | -- | 1.0 | -- |

these and the previous results, it is concluded that copper is present in nodules both as adsorbed and manganese substituted species.

The results for nickel with the delta manganese dioxide nodules also show great sensitivity to pH, and here also the best results are obtained with pH=1 solutions. At pH=3, the best recovery is from the delta manganese dioxide, while at pH=1, the best results were from todorokite nodules. These results indicate that nickel is primarily substituded for manganese in the nodule, and the rate of dissolution is chemically controlled.

Cobalt recovery is best from nodules of low iron content. In nodules of this type, the cobalt is believed to be primarily associated with the manganese via adsorption, and therefore relatively easy to recover. In nodules of higher iron content, the cobalt is evidently substituted in the iron matrix, and therefore somewhat more difficult to leach.

Although these results for sulfuric acid leaching indicate that only partial recoveries of the metals may be obtained, valuable information on the form and composition of nodules may be gleaned therefrom. Also, the low yields found indicate that disrupting of the nodule crystal lattices will be necessary in order to obtain high yields. Such disruption will allow leaching of those metals substituted within the manganese and iron crystals.

One method of effecting this disruption is through the use of a reduction process followed by leaching with ammoniacal solution(7). The reduction process converts some of the oxides to the metal and others to lower oxides. This disrupts the crystal structure of the nodules and allows the leaching to be more effective. Ammoniacal solutions - made from ammonia and an ammonium salt to suppress hydroxide formation - form soluble amine complexes with the so-called associated metals, but not with manganese. Air is added to oxidize metallic elements into soluble compounds. Recoveries of greater than 90% for the associated metals may be obtained with this process.

There are two other hydrometallurgical process that employ disrupting the crystal lattice in order to increase yields of the associated metals, but these processes also solubilize the manganese portion of the nodule. The first of these to be discussed utilizes a sulfur dioxide roast(10). The mechanism in operation here is the oxidation/reduction reaction between $SO_2$ and $MnO_2$. This reaction breaks down the oxide lattices in the nodule, and forms soluble sulfate salts of the metals.

Another hydrometallurgical approach to nodule metal recovery is that of hydrochlorination. Here the classic laboratory method of producing chlorine by allowing hydrogen chloride to react with manganese dioxide is used to solubilize the metals. The oxidation of chloride to chlorine by the $MnO_2$ is the driving force for the reaction. Metal oxides not susceptible to redox reactions with HCl are leached in the acid environment. The yields of metals are very high in this leaching operation - greater than 96% for all valued metals. Water soluble metal chlorides produced in the leaching are then separated to yield the desired products.

In summary, high leaching yields are only obtained from nodules when the crystal lattices of the contained minerals are disrupted. This disruption is most favorably carried out via an oxidation-reduction reaction involving the oxides of manganese. Under these conditions, those metals which are incorporated into the matrix of the manganese dioxide may be exposed to the leaching medium.

The effect upon the environment of any nodule processing plant will be analogous to other mineral processing facilities. Environmental protection measures will be planned into the project at the start rather than added later - as has often been the practice in the past. Solid by-product material from the plant will either be a manganese-rich residue to be stockpiled for future potential use; or the silt-like material from processes leaching manganese. The silt product is generally innocuous, and may be used for landfill or other such purposes.

In addition to the actual emissions produced by the plant, there will be other environmental parameters to consider (11). An operation such as this will require deepwater ports for off-loading of nodules from the transport ships. The increased activity of ships having draughts of 35 to 40 feet in the coastal region will require monitoring to ensure ecological protection.

Acknowledgment

I would like to express my appreciation for the assistance of W. D. Siapno and P. H. Cardwell in the preparation of this material.

Abstract

Manganese nodules are distributed over a large area of the ocean floor and have come to be regarded as an important potential source of manganese, copper, cobalt, nickel, and other metal values. Successful ocean mining will require application of various chemical disciplines. Such applications include analysis of material in the exploration phase, development of recovery methods for nodule processing, understanding of the geochemical processes that led to formation of nodules in order to aid in exploration and processing, and monitoring of the environmental impact of such a venture.

Literature Cited

(1) Hoover, M. P.; "Studies on the Dissolution of Copper, Nickel, and Cobalt from Ocean Manganese Nodules"; Thesis submitted in partial satisfaction of the degree requirements for the degree of Master of Science in Engineering in the Graduate Division of the University of California, Berkeley; (1966).

(2) Mero, J. L.; Economic Geology; (1962); 57; pp. 747-767.

(3) Siapno, W. D.; Private Communication

(4) Greenslate, J.; Nature; (1974); 747.

(5) Kaufman, R. and Siapno, W. D.; "Variability of Pacific Ocean Manganese Nodule Deposits"; Arden House, IDOE; (1972).

(6) Garland, C. and Hagerty, R.; "Environmental Planning Considerations for Deep Ocean Mining"; 8th Annual Conference of Marine Technology Society; (1972).

(7) Agarwal, J. C., Beecher, N., Davies, D. S., Hubred, G. L., Kakaria, V. K. and Kust, R. N.; "Processing of Ocean Nodules - A Technical and Economic Review"; Presented at 104th Annual Meeting of AIME in New York; February 17, 1975.

(8) Fuerstenan, D. W., Herring, A. P. and Hoover, M. P.; "Leaching Manganese Nodules from the Ocean Floor"; Presented at AIME Annual Meeting, Los Angeles; (1967).

(9) Brooks, J. N. and Posser, A. P.; Trans. Inst. Min. Metall.; (1969); 78; c64-73.

(10) Brooks, P. T. and Martin, D. A.; U.S. Department of the Interior, Bureau of Mines; Report of Investigation 7473 (1971).

(11) Kaufman, R.; "Land-Base Requirements for Deep-Ocean Manganese Nodule Mining"; Manganese Nodule Deposits in the Pacific Symposium/Workshop; Honolulu; October, 1972.

# 35

# Nonbiological Degradation and Transformations of Organic Pesticides in Aqueous Systems

SAMUEL D. FAUST

Department of Environmental Sciences, Rutgers, The State University, New Brunswick, N.J. 08903

It may be stated categorically that all organic pesticides are thermodynamically unstable in natural aquatic environments. These so-called recalcitrant compounds have their carbon atoms in a reduced valence state. Consequently, there is an inherent tendency for the carbon to be oxidized to higher valence states in an aerobic environment. If conditions are favorable, $CO_{2(g)}$ (+IVC) is the eventual product.

Many terms are employed to describe the instability of organic pesticides in nonbiological systems: oxidation, transformation, degradation, etc. Definition and interpretation of these terms are, in mose cases, arbitrary for most investigators and authors. In this paper, oxidation and/or degradation refers to a change in the valence state of the organic C. Usually this change occurs whereby the carbon goes to a higher valence state which requires, of course, an electron acceptor. Transformation refers to any alteration in the configuration of an organic molecule. Examples of transformations would be: the hydrolysis of organic phosphorus, carbamates, and urea compounds, replacement of the = S by = O in organic phosphorus compounds, modification of the trichloroethane group in DDT, replacement of the Cl atom by OH in atrazine, etc. These two definitions are somewhat at variance with Kearney and Kaufman (1) who prefer the term "degradation" "to cover all transformations of organic herbicides without particularly trying to ascribe these to enzyme or particulate systems."

## II. Hydrolytic Transformations

Many organic pesticides undergo hydrolytic transformations in aquatic environments. Many of these compounds have been synthesized as an organic ester of some sort. This is especially true for the pesticides that are phosphates, carbamates, ureas, and phenoxyacetates.

A. Organophosphates. The organophosphorus pesticide is a tertiary phosphate or thiophosphate ester:

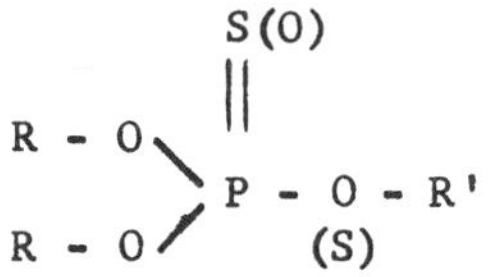

$$\begin{array}{c} \quad\quad\quad S(O) \\ \quad\quad\quad \| \\ R-O \searrow \\ \quad\quad\quad P - O - R' \\ R-O \nearrow \quad\quad (S) \end{array}$$

R is usually a methyl or an ethyl group and R' is an organic moiety. Hydrolytic transformation of these compounds occur from rupture of either the P-O(S) bond or the (S)O-R' bond. Very seldom is the hydrolysis carried to rupture of the R-O bond (tertiary hydrolysis). Once again the nature of the R' group becomes important because the hydrolytic product may be environmentally hazardous.

Whether the P-O(S)-R' bond is ruptured depends upon the structure of the pesticide and the hydrolytic conditions. Various mechanistic studies in $O^{18}$ rich water have shown that, in alkaline solutions, the P-O(S) bond is broken and the (S)O-R' is usually replaced (2). In acidic hydrolysis, however, a rupture of the (S)O-R' bond apparently occurs as an initial step. Secondary esters undergo additional hydrolysis under acid conditions to primary esters which usually does not occur under alkaline conditions. Generalized reactions may be written for these two conditions of hydrolysis:

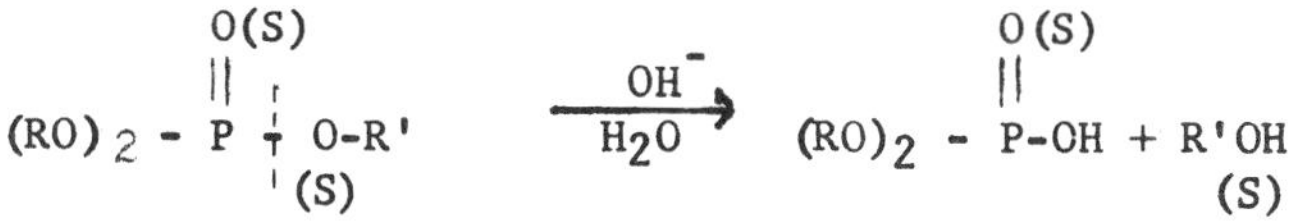

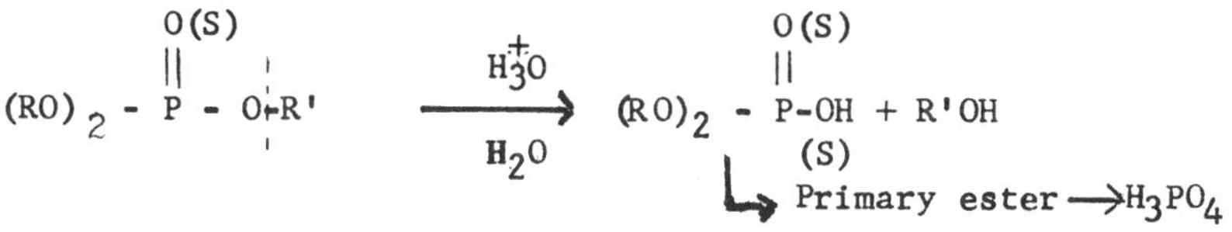

$$(RO)_2-\overset{O(S)}{\overset{\|}{P}} \,\vdots\, \underset{(S)}{O}-R' \xrightarrow[H_2O]{OH^-} (RO)_2-\overset{O(S)}{\overset{\|}{P}}-OH + R'\underset{(S)}{O}H$$

$$(RO)_2-\overset{O(S)}{\overset{\|}{P}}-O \,\vdots\, R' \xrightarrow[H_2O]{H_3O^+} (RO)_2-\overset{O(S)}{\overset{\|}{P}}-OH + R'OH$$

$$\quad\quad (S) \to \text{Primary ester} \longrightarrow H_3PO_4$$

Some organophosphorus compounds, however, do not follow this general rule of acidic and basic hydrolysis. Cleavage of the P-O(S) or the R'O(S) bonds depend, in many cases, upon the nature of the R' group.

Where hydrolysis proceeds under catalysis by $OH^-$, a Sn2 type of reaction occurs (3). That is: (a) the reaction is a nucleophilic substitution in which the $OH^-$ substitutes for the R'O group:

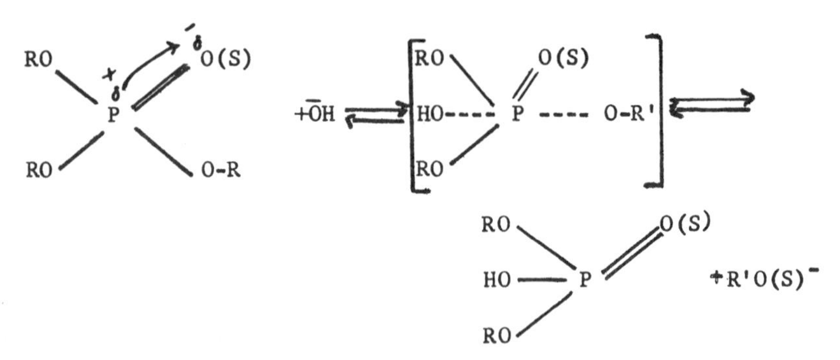


(b) the reaction proceeds by a bimolecular mechanism, and (c) no stable intermediate forms as the $OH^-$ approaches the molecule and attacks the P atom which has been made electrophilic by the inductive effects of the =O or =S atom and the R' group. In all probability, $R'O^-$(S) is released simultaneously as the $OH^-$ attacks the P atom.

The kinetics of hydrolysis was determined for parathion and paraoxon (4) and for diazinon and diazoxon (5). These two studies departed from other investigators wherein the hydrolytic reactions were effected in purely aqueous systems and at temperatures simulating natural water conditions. Also progress of the reactions was determined directly for disappearance of the parent molecule as well as appearance of the products. First-order kinetic behavior was observed for the hydrolytic transformation of these four organophorous compounds. Half-life values are given in Table I at five pH values. Parathion and paraoxon are relatively stable in acidic and neutral conditions. The half-life values observed for parathion and paraoxon at pH values 3.1 to 7.4 are sufficient to permit environmental damage in a natural water body. On the other hand, diazoxon is relatively short-lived especially under highly acidic and alkaline conditions. Diazinon was stable under neutral conditions, but relatively unstable under acidic and alkaline conditions.

Temperature is an environmental factor that affects the rate at which hydrolysis reactions occur. Table II gives the half-life values of parathion, paraoxon, diazinon, and diazoxon at 10° and 40° from which the energy of activation values were calculated (4,5). Reactions which yield the higher Ea values will exhibit the greatest effect of temperature upon their rates of hydrolysis. For example, parathion has an activation energy of 16.4 Kcal/mole at a pH value of 3.1 whereas diazinon has a value of 13.1. Parathion shows the greater difference in the $t_{1/2}$ values at the two temperatures. Two rather important environmental implications evolve from these temperature studies: (a) those compounds with the higher Ea values are more persistent at the lower temperatures of 0-20°C., and (b) hydrolysis is generally slower under acidic conditions.

Table I. HALF-LIFE VALUES IN DAYS OF DIAZINON, DIAZOXON, PARATHION, AND PARAOXON AT 20°C., I=.02M[a]

| pH | Diazinon[b] | Diazoxon[c] | Parathion[d] | Paraoxon[e] |
|---|---|---|---|---|
| 3.1 | .49 | .02 | 174. | 197. |
| 5.0 | 30.8 | 1.28 | 153. | 173. |
| 7.4 | 185. | 29. | 108. | 144. |
| 9.0 | 136. | 18.4 | 22. | 2.9 |
| 10.4 | 6. | .42 | 1.4 | .25 |

a. From references (4,5)
b. Co= $6.61 \times 10^{-5}$M
c. Co= $13.9 \times 10^{-5}$M
d. Co= $3.95 \times 10^{-5}$M
e. Co= $4.81 \times 10^{-5}$M

Table II. EFFECT OF TEMPERATURE ON KINETICS OF HYDROLYSIS[a]

| Compound | Ea Kcal/mole | pH | $t_{1/2}$-days 10°C. | 40°C. |
|---|---|---|---|---|
| Parathion | 16.4 | 3.1 | 555. | 33.7 |
| Paraoxon | 16.4 | 3.1 | 576. | 36. |
| Diazinon | 13.1 | 3.1 | 64. | 7.3 |
| Diazoxon | 12.5 | 3.1 | 2. | .25 |
| Parathion | 14.5 | 9.0 | 53. | 5.3 |
| Paraoxon | 12.0 | 9.0 | 7.5 | .76 |
| Diazinon | 14.3 | 10.4 | 13. | 2. |
| Diazoxon | 13.1 | 10.4 | .9 | .13 |

a. From references (4,5)

Additional information about the hydrolysis of organophosphates may be obtained in the reference of Faust and Gomaa (6).

B. Organocarbamates. Esters of carbamic acid (the carbamate pesticides) also hydrolyze, but mostly under alkaline conditions:

$$R-O-\overset{\overset{O}{\parallel}}{C}-\overset{\overset{R'}{|}}{N}-R'' + H_2O \xrightarrow{(\bar{O}H)} ROH + R'-\overset{H}{N}-R'' + CO_{2(g)}$$

where the products are a hydroxy compound (phenol), an amine, and $CO_{2(g)}$. Aly and El-Dib (7) reported the hydrolytic stability of four carbamates: sevin, baygon, pyrolan, and dimetilan over the pH value range of 2 to 10. First-order kinetics of hydrolysis was observed from which the half-life values were calculated (Table III). Pyrolan and dimetilan did not hydrolyze within this pH range and at 20°C. Baygon resisted hydrolysis at pH values 3 through 7 but did decay under alkaline conditions. Sevin was the least stable of the four carbamates with hydrolysis occurring at pH values 7.0 and above with the rate increasing as the ($OH^-$) was increased. Casida et al. (8) found four p-nitrophenyl N-alkyl carbamates to be quite unstable at a pH value of 7.8. These four compounds are not widely, if at all, used as commercial products.

Table III. HYDROLYTIC HALF-LIFE VALUES FOR SOME CARBAMATE PESTICIDES

| Compound | pH Value | | | | Ea |
|---|---|---|---|---|---|
| | 7 | 8 | 9 | 10 | Kcal/mole |
| Sevin[a] | 10.5 days | 1.3 days | 2.5 hrs. | 15 min. | 16. |
| Baygon[a] | - | 16.0 days | 1.6 days | 4.2 hrs. | 15.8 |
| 1[b] | - | 8.9 min. | - | - | - |
| 2[b] | - | 6.9 min. | - | - | - |
| 3[b] | - | 7.1 min. | - | - | - |
| 4[b] | - | 5.9 min. | - | - | - |

a. After Aly and El-Dib (7) where T=20°C.
b. After Casida et al. (8) where T=22°C. and pH value = 7.8.

Colorimetric and radiometric analyses were used to study the persistence of carbaryl (sevin) in estuarine water and mud in laboratory aquaria held at 8°C. and 20°C. (9). In systems where there was a sharp decrease in the concentration of carbaryl (8°C.), adsorption by the mud was the major reason for the decline. Approximately 90% of the carbaryl added to a control system without mud was present as unchanged insecticide or as 1-naphthol after 38 days. Additional stability experiments performed in the dark in sea water (no mud) showed that, at low temperatures (3.5°C.), no hydrolysis of 10.0 mg/l carbaryl could be detected after 4 days and after 8 days, only 9% of the compound was hydrolyzed (pH=7.8). The amount of carbaryl hydrolyzed in 4 days at 17°C. was 44%, at 20°C. 55%, and at 28°C., 93%. These experiments were presumably performed in autoclaved sea water. In other systems, "complete conversion to non-detectable compounds occurred in 4 days" at 19.5° and 28°C. This is surprising. The authors do not attempt to explain these results and the reader must assume that biological activity is responsible for the rapid disappearance.

The fate of 1-naphthol-1-$^{14}C$ was examined in a simulated estuarine environment by Lamberton and Claeys (10). 1-Naphthol is formed from the hydrolysis of carbaryl (sevin). Figure 1 shows the disappearance of 1-naphthol in light and dark systems

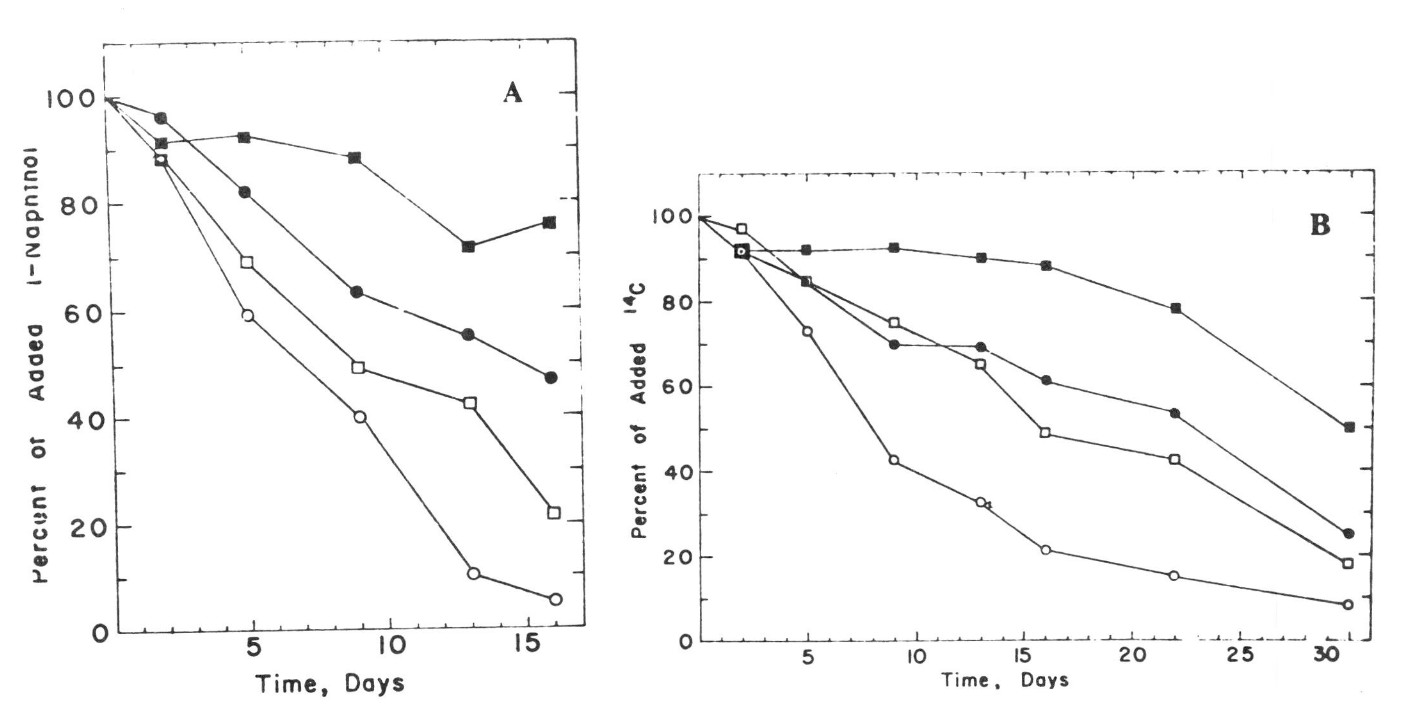


*Figure 1. Effect of light and microorganisms on stability of 1-naphthol in seawater. After Ref. 10. A. Loss of 1-naphthol:* ○ *light, unsterile;* □ *light, sterile. B. Loss of total* $^{14}C$: ● *dark, unsterile;* ■ *dark, sterile.*

and in sterile and unsterile systems. In dark and sterile conditions, this compound is relatively stable with approximately 75% remaining after 16 days. Light and the presence of microorganisms catalyze the disappearance of 1-naphthol in sea water. This observation is frequently the situation where synthetic organic compounds are concerned. Apparently, biological systems can supply the activation energy that is necessary to degrade these compounds.

Wauchope and Hague (11) reported upon the kinetics of hydrolysis of carbaryl (sevin) that was somewhat in disagreement with Aly and El-Dib (7). For example, at pH value of 10.0 and 25°C., Wauchope and Hague reported a $t_{\frac{1}{2}}$ value of 20 minutes whereas Aly and El-Dib reported a value of 15 minutes at the same pH value and at 20°C. The temperature difference does not account for the disagreement which may be due to analytical techniques.

The hydrolysis of zectran (4-dimethylamino-3.5-xylylmethyl carbamate) in alkaline water was reported by Hosler (12). The half-life value at a pH value of 9.5 was approximately two days whereas at pH 7.4, the half-life is approximately 2 weeks at 12-13°C. No order of reaction rates was suggested. Also, Hosler reported a purple colored water solution of zectran under alkaline condtions (pH9-9.5) when exposed to light. This was ascribed to the rapid formation of "xylenol" from hydrolysis which is converted to its "xylenoxide" ionic form that is, in turn, "very sensitive to photooxidation." No proof was offered to substantiate this mechanism. However, Mathews and Faust (13) have made a similar observation.

C. Miscellaneous Compounds. An examination of the hydrolysis of three explosives in sea water was reported by Hoffsommer and Rosen (14). These explosives were: TNT (2,4,6-trinitrotoluene), RDX (1,3,5-triazocyclohexane), and tetryl (N-methyl-N-nitro-2,4,6-triazocyclohexane). Results at 25°C.are summarized below:

| Compound | Time-days | Hydrolysis-% |
|---|---|---|
| TNT | 108 | 0 |
| RDX | 112 | 11.6 |
| Tetryl | 101 | 88. |

It appears that disposal of these compounds (tetryl excepted) at sea is not feasible.

In an attempt to simulate natural conditions, Bailey, et al. (15) studied the hydrolysis of the propylene glycol butyl ether ester of 2-(2,4,5)TP (silvex) in three pond waters. There was an apparent rapid decay of this ester as seen by the $t_{\frac{1}{2}}$ values :

PGBE of 2-(2,4,5) TP: $t_{\frac{1}{2}}$ values in three naturally-occurring pond waters:

| | | |
|---|---|---|
| Pond A | 5 hrs. | (pH=6.25) |
| Pond B | 7 hrs. | (pH=6.09) |
| Pond C | 8 hrs. | (pH=6.07) |

D. Comment. Hydrolysis is, indeed, an important variable affecting the fate of the appropriate organic pesticide in aquatic environments. The rate at which hydrolysis occurs is, of course, unique to the individual compound and is dependent, also, upon such environmental factors as $(H_3O^+)$ and temperature (6). Hydrolytic stability should be viewed with concern about the length of the time period required for complete hydrolysis and about the products that may or may not be more toxic than the parent molecule.

## III. Chemical Systems

A. Thermodynamic Stability. A natural water in equilibrium with the atmosphere should be saturated with dissolved oxygen (neglecting for the moment aerobic biological transformations of organic matter). In this situation, this water has a well-defined redox potential of approximately + 800 mv ($P_{O_2}$ = 0.21 atm, pH = 7.0, 25°C). A simple equilibrium calculation shows that, at this $E_h$ value, all organic carbon should be present as C(+IV) or as $CO_2$, $HCO_3^-$, or $CO_3^{2-}$, S and N should occur in the form of $SO_4^{2-}$ and $NO_3^-$, respectively. That organic compounds are unstable in aquatic environments may be shown from a simple thermodynamic model:

$$CH_2O + H_2O = CO_{2(g)} + 4H^+ + 4e$$
$$O_{2(g)} + 4H^+ + 4e = 2H_2O$$
$$\overline{CH_2O + O_{2(g)} = CO_{2(g)} + H_2O}$$

The free energy change of this reaction is - 117.99 Kcal mole$^{-1}$ which gives a $E_h^o$ value of 1.28 volts. Under the European sign convention, a negative $\Delta G^o$ reaction value and a positive $E_h^o$ values denotes that the left to right reaction is feasible. A similar model may be calculated for the organic herbicide 2,4-D in an oxygenated environment: $2\ C_8H_6O_3\ Cl_2 + 15O_{2(g)} = 16\ CO_{2(g)} + 4H^+ + 4Cl^- + 4H_2O$. The free energy change for this reaction (25°C) is -1600.78 Kcal mole$^{-1}$ which indicates that it is feasible for the carbon in 2,4-D to be oxidized to $CO_{2(g)}$ with $O_{2(g)}$ as an electron acceptor. This model does not, however, indicate the reaction knetics or if, measurably, the reaction will even occur. Only laboratory experimentation will answer these two points. More and detailed information on the thermodynamic stability of organic pesticides in aquatic systems is given by Gomaa and Faust (16).

B. Artificial Systems. There are numerous electron acceptors available for oxidation of organic pesticides in artificially created environments. However, where the natural environment is concerned, the oxidative reactions are extremely slow and are incomplete, in the sense, that toxic reaction products may be formed. It is extremely desireable to effect "complete" degradation to $CO_2$, $H_2O$, etc.

Historically, $KMnO_4$, $ClO_2$, $Cl_2$, and $O_3$ have been employed for the oxidation of organic compounds at water treatment plants. Consequently, these oxidants have been investigated for their capacity to degrade organic pesticides (16, 17, 18, 19, 20). Several basic concepts have evolved from these five studies for the oxidation of organic compounds.

1. Stoichiometry: Oxidative processes at water treatment plants, for example, are often ineffective because inadequate stoichiometries, i.e., chemical dosages are employed. Gomma and Faust (16) proposed the exact stoichiometric relationship for several organic pesticides and inorganic oxidants. For example, parathion and $KMnO_4$ gave:

$$C_{10}H_{14}O_5NSP + 20MnO_4^- + 14H^+ = 20MnO_2 + 10CO_2 + SO_4^{2-} + PO_4^{3-} + 14H_2O + NO_3^-$$

$$3C_{10}H_{14}O_5NSP + 50MnO_4^- = 50MnO_2 + 15C_2O_4^{2-} + 3NO_3^- + 3SO_4^{2-} + 3PO_4^{3-} + 20H_2O + 2\,\bar{O}H$$

These two reaction models suggest rather complex molar ratios of oxidant to compound. The first reaction represents acidic conditions in which the $KMnO_4$/parathion molar ratio is 20:1. In the second reaction, the molar ratio under alkaline conditions is 16.67:1. In order to write these two reactions and others (16), several assumptions were made for the products:

a. $MnO_4^-$ goes to $MnO_{2(s)}$ and not to $Mn^{2+}$
b. organic C goes to $CO_2$ in acid conditions (C+IV)
c. organic C goes to $C_2O_4{}^{2-}$ in alkaline conditions (C+III)
d. nitro group goes to $NO_3^-$, P group goes to $PO_4^{3-}$, and S group goes to $SO_4^{2-}$.

That there is some credibility to the proposed reactions is seen in Table IV where the moles of $KMnO_4$ consumed per mole of compound is in good agreement with the calculated values.

2. Reaction Kinetics. Well-designed kinetic experiments yield data from chemical oxidation systems that will:

a. confirm thermodynamic reactions models with respect to feasibility and stoichiometry

b. determine the order of the disappearance reaction

c. determine the rate and reaction times required to process design

Table IV. POTASSIUM PERMANGANATE CONSUMED BY THE OXIDATION OF PARATHION, PARAOXON, AND p-NITROPHENOL (16)[a]

| Compound | pH | Contact Time (Hours) | Ratio | Mole $MnO_4^-$/Mole Comp. Experimental | Calc. |
|---|---|---|---|---|---|
| Parathion | 3.1 | 120 | 0.386 | 19.48 | 20.00 |
| Paraoxon | 3.1 | 120 | 0.364 | 16.69 | 17.33 |
| p-Nitrophenol | 3.1 | 96 | 0.404 | 9.10 | 9.33 |
| Parathion | 9.0 | 48 | 0.392 | 17.08 | 16.66 |
| Paraoxon | 9.0 | 48 | 0.403 | 14.17 | 14.00 |
| p-Nitrophenol | 9.0 | 96 | 0.367 | 7.62 | 7.33 |

(a) Temperature = $20^o \pm 0.2^oC$.

Determination of the order of the reactions may be accomplished by fitting the data into several rate equations. These reactions are, of course, extremely complex and an exact order is not expected over the entire course of events. Intermediate oxidation products will compete with the parent molecule for the oxidant. Nontheless, Gomma and Faust (21) found that an integrated form of a second-order expression provided reasonably constant rate constants for the $KMnO_4$ oxidation of two dipyridylium quarternary herbicides - Diquat and Paraquat. This expression was:

$$\frac{2.303}{aC^o_{ox} - b\ C^o_{HERB}} \log \frac{C^o_{HERB}\ C_{ox}}{C^o_{ox} C_{HERB}} = K_{ob}t$$

where t is reaction time in minutes, $C^o_{HERB}$ is the initial herbicide molar concentration, $C_{HERB}$ is the herbicide concentration at time t, $C^o_{ox}$ is the initial oxidant concentration, $C_{ox}$ is the oxidant concentration at time t, a is the number of herbicide moles given in the model reaction, and b is the number of oxidant moles given in the model reaction. That a second-order rate expression may be employed to describe the chemical oxidation of Diquat and Paraquat is seen in Table V. The second-order rate constants were confirmed by effecting the reactions at three separate orders of magnitude in initial concentrations of the reactants. This stoichiometric proportions were the same in each system, however.

Table V. VERIFICATION OF ORDER OF THE PERMANGANATE REACTIONS[a]

| System | pH | Mean $K_{ob}$ ($1.\ mole^{-1}\ min^{-1}$) Initial | 10-fold dilution | 100-fold dilution |
|---|---|---|---|---|
| Diquat-$KMnO_4$ | 5.12 | 0.138 | 0.140 | 0.132 |
| | 9.13 | 10.65 | 10.43 | 10.52 |
| Paraquat-$KMnO_4$ | 5.13 | 0.106 | 0.107 | 0.101 |
| | 9.13 | 6.56 | 6.42 | 6.31 |

a. Ionic strength = 0.02M. Temperature = 20°C. Average of three experimental runs at the indicated pH. From reference (21).

Kinetically the oxidative reactions are "slow". Table VI shows the percentage of "complete" oxidation within a specific reaction time. For example, at a pH value of 7.4, only 31% of the initial concentration of parathion was oxidized within a reaction time of 44 hours. Note the very "slow" reaction of 2,4-D toward $KMnO_4$.

Table VI. "SLOW" OXIDATION REACTIONS OF PESTICIDES AND $KMnO_4$[a]

| pH Value | Parathion[b] Time Hours | Oxidation % | pH Value | Paraoxon[b] Hours | Oxidation % |
|---|---|---|---|---|---|
| 3.1 | 44 | 83 | 3.1 | 92 | 48 |
| 5.0 | 44 | 53 | 5.0 | 92 | 25 |
| 7.4 | 44 | 31 | 7.4 | 116 | 14.5 |
| 9.0 | 6 | 91 | 9.0 | 3 | 95 |
| | Diquat[c] | | | Paraquat[c] | |
| 5.1 | 20 | 21 | 5.1 | 24 | 26 |
| 9.1 | 2 | 88 | 9.1 | 6 | 98 |

2,4-D[d] - at pH values of 3.1, 7.4, and 10.1 for 24 hrs. each, % oxidation was zero.

a. Temp. = 20C., I = .02M; b. $[KMnO_4]_o = 8 \times 10^{-4}M$, $[Parathion]_o = 3.95 \times 10^{-5}M$, $[Paraoxon] = 4.81 \times 10^{-5}M$; c. $[KMnO_4]_o = 10^{-3}M$, $[Diquat]_o = 4.16 \times 10^{-5}$, $[Paraquat] = 3.65 \times 10^{-5}M$; d. $[KMnO_4]_o = 158.mg/l$, $[2,4\text{-}D]_o = 25.\ mg/l$

Not all chemical oxidations of organic pesticides are "slow", however. In the decomposition of Diquat and Paraquat by $ClO_2$ at pH values of 8.14, 9.04, and 10.15, the rates were extremely rapid and could not be measured with experimental techniques. That these reactions were complete in less than one minute is seen in Table VII.

Table VII. DIQUAT - PARAQUAT - $ClO_2$ REACTIONS

| pH | Reaction Time min. | Residual[a] Conc. mg/l | Residual[b] $[ClO_2]$ mg/l |
|---|---|---|---|
| | DIQUAT[c] | | |
| 10.15 | 1 | 0.0 | 2.61 |
| 9.04 | 1 | 0.0 | 2.60 |
| 8.14 | 1 | 0.0 | 2.59 |
| 7.12 | 1 | 15.0 | 6.74 |
| 7.12 | 24 hrs. | 15.0 | 6.66 |
| 6.17 | 24 hrs. | 15.0 | 6.65 |
| | PARAQUAT | | |
| 10.15 | 1 | 0.0 | 2.84 |
| 9.04 | 1 | 0.0 | 2.81 |
| 8.14 | 1 | 0.0 | 2.85 |
| 7.12 | 24 hrs. | 15.0 | 6.72 |
| 6.17 | 24 hrs. | 15.0 | 6.73 |

a. Initial [Pesticide]=15.0 mg/l
b. Initial $[ClO_2]$=6.75 mg/l c. Temp. = 20°C., I = .02M

In the design of water treatment processes, sufficient time should be afforded so that the oxidative reaction goes to completion. In this case, completion means the conversion of the organic C to $CO_{2(g)}$. The reaction or contact times required for 50%, 90% and 99% disappearance of an initial concentration of Diquat were calculated for a $KMnO_4$ system. A pseudo first-order rate constant was generated by multiplication of the second-order rate constant by the initial oxidant concentration. An assumption was made whereby this initial oxidant concentration is in excess of the Diquat concentration and little was consumed in the course of the reaction. These contact times are given in Table VIII.

3. <u>Effect of $(H_3{}^+O)$</u>. Apparently, one of the reaction variables affecting the kinetics of chemical oxidation of pesticides is the hydronium ion activity. There is

Table VIII. CONTACT TIMES REQUIRED FOR $KMnO_4$-DIQUAT REACTION (16).

| pH | $K_1$[a] $min.^{-1}$ | $t_{50}$ | $t_{90}$ Hours | $t_{99}$ |
|---|---|---|---|---|
| 5.1 | $1.23 \times 10^{-4}$ | 93.9 | 312.0 | 624.1 |
| 6.0 | $8.69 \times 10^{-5}$ | 132.5 | 441.7 | 883.4 |
| 6.9 | $1.86 \times 10^{-4}$ | 61.9 | 206.4 | 412.7 |
| 8.0 | $1.18 \times 10^{-3}$ | 9.8 | 32.5 | 65.1 |
| 9.1 | $1.48 \times 10^{-2}$ | .8 | 2.6 | 5.2 |
| 10.0 | $5.46 \times 10^{-2}$ | .2 | .7 | 1.4 |

a. Pseduo first-order rate constants obtained by multiplying $K_2$ (second-order rate constant) by $KMnO_4$ conc. ($10^{-3}$M).

sufficient evidence in Tables V, VI, VII, and VIII to demonstrate this point. For one reason or another, these oxidative reactions are faster when alkaline pH values prevail. For example, the rate of reaction of diquat-$KMnO_4$ at pH 9.13 is about 77 times faster than at pH 5.12 wheras the paraquat-$KMnO_4$ reaction is about 62 times faster at pH 9.13 than at 5.12 (21). Similar observations were made with $ClO_2$ and $Cl_2$ as the oxidants (21). In some systems, the effect of increasing ($H_3^+O$) leads to no reaction between the oxidant and pesticide. For example, $ClO_2$ had no oxidative effect on diquat and paraquat at pH values of 5.06, 6.17, and 7.12 (20°C.) (21). It should be obvious that a pH adjustment into the alkaline range is required to effect the fastest reaction and the least contact time.

4. Effect of Temperature. That the rate constants of these chemical oxidative reactions vary greatly with temperature is seen in Table IX (21). It is the general observation that reaction rates increase at higher temperatures. These studies were designed to yield the energies of activation from the Arrhenius equation. These $E_a$ values are given in Table IX also.

5. Formation of Intermediate Products. Ideally, any chemical oxidation of an organic contaminant should carry the degradation of carbon to $CO_{2(g)}$. In the course of the reaction, intermediate products may be formed which, if not completely oxidized, may be more toxic or physiologically more harmful than the parent compound. This is especially important whenever an oxidative process is employed at a potable water treatment plant. It is extremely undesireable to have incomplete oxidation and to have organic toxicants in the potable water.

Throughout the course of a reaction of $KMnO_4$ with parathion and paraoxon (16), it was observed that deviation occurred from the second-order rate equation. This suggested formation of intermediate organic products whose rates of oxidation were, in turn, slower than the parent molecule.

Table IX. POTASSIUM PERMANGANATE AND CHLORINE OXIDATIONS - EFFECT OF TEMPERATURE[a]

Kob (1. $mole^{-1}$ $min^{-1}$)[b]

| | Diquat-$KMnO_4$ | | Paraquat-$KMnO_4$ | |
|---|---|---|---|---|
| Temp. °C. | pH 5.12 | pH 9.13 | pH 5.12 | pH 9.13 |
| 10 | 0.059 | 4.814 | 0.038 | 3.135 |
| 20 | 0.141 | 10.431 | 0.105 | 6.393 |
| 30 | 0.308 | 16.831 | 0.219 | 11.928 |
| 40 | 0.561 | 29.624 | 0.402 | 22.283 |
| | Diquat-$Cl_2$ | | Paraquat-$Cl_2$ | |
| | pH 6.17 | pH 9.04 | pH 8.14 | pH 9.04 |
| 10 | 0.007 | 1.847 | 0.001 | 0.008 |
| 20 | 0.020 | 3.415 | 0.006 | 0.017 |
| 30 | 0.037 | 5.255 | 0.015 | 0.040 |
| 40 | 0.052 | 7.751 | 0.039 | 0.090 |
| | $Ea_{ob}$ Kcalmole$^{-1}$ | | | |
| | 14.2 | 10.8 | 15.1 | 11.5 |
| | 14.4 | 9.0 | 17.5 | 13.9 |

a. From reference (21).

b. Average of three experimental runs. Ionic strength of solution = 0.02M.

Consequently, the $KMnO_4$ - parathion, - paraoxon, - p-nitrophenol, and -2,4 dinitrophenol were studied and were reported in great detail elsewhere (4, 16). Under acidic and neutral conditions in the $KMnO_4$ - parathion system, paraoxon was detected. Under alkaline conditions, p-nitrophenol and traces of paraoxon were detected. These observations led, of course, to kinetic studies of the $KMnO_4$ oxidation of paraoxon and p-nitrophenol. Under acidic and neutral pH values, the paraoxon system did not yield any oxidation products that would have been detected by the analytical techniques of GLC, TLC, and ultraviolet scan. Under alkaline conditions, however, p-nitrophenol was found. In the p-nitrophenol-$KMnO_4$system, 2,4-dinitrophenol appeared as an intermediate product within the pH value range of 3.1 through 9.0. Also two unknown spots appeared on the TLC plate at a pH value of 9.0. These unknowns were identified subsequently by IR and mass spectroscopy as 2-hydroxy-5-nitro benzoic acid and 2,2' dihydroxy-5,5 dinitrophenyl. Figure 2 shows the proposed pathway in the $KMnO_4$ oxidation of parathion (16).

Metabolites and oxidative products of heptachlor were reported by Cochrane and Forbes (22). Using potassium dichromate and chromic oxide in an acidic medium produced heptachlor epoxide that had the same insecticidal activity as the epoxide formed from heptachlor in biological systems. All together, eleven metabolites and derivatives were identified. These systems are, of course atypical in relation to natural environments and to water treatment processes. These observations, however, demonstrate the drastic conditions required to oxidize the chlorinated hydrocarbon pesticide. Heptachlor epoxide has been isolated from biological systems. Whether or not the other 10 metabolites and derivatives are produced, is an unanswered question.

## IV. Photochemical Systems

Among the environmental factors that influence the persistence of organic pesticides in aquatic environments would be decomposition of these compounds under the influence of solar radiation. Many organic pesticides undergo drastic changes upon exposure to ultraviolet light and artificial or natural sunlight (23-29). Sunlight does not necessarily cause degradation of pesticides into their constituent parts or into simpler compounds. In many cases, sunlight transforms pesticides into materials of similar or even greater structural complexity. The toxicology and persistence of these materials are, for the most part, unknown (30). In spite of the difficulties involved in extending laboratory observations to natural environments, much useful information can be obtained from studying the photochemistry of pesticides.

Dieldrin was photoisomerized in sunlight and under laboratory conditions (31, 32) to photodieldrin. Similarly, aldrin was converted to photoaldrin in sunlight. Reaction mixtures obtained from a one-month exposure of an aldrin film contained 2.6% unaltered aldrin and 9.6% photoaldrin. The remainder consisted of 4.5% dieldrin, 24.1% photodieldrin, and 59.7% of the major product, an unidentified material with an average molecular weight of 482. The major product of aldrin photoylsis was approximately five times more toxic to flies than DDT. Yet, this material would not pass through gas chromatographic columns normally employed. If this substance is an actual environmental product, it may go undetected.

Bell (34) detected 2,4-dichlorophenol when a solution of 2,4-D was exposed to UV light. No phenol was produced when a buffered solution at pH 7.0 was irradiated. It was proposed that ultraviolet decomposition of 2,4-D resulted in the formation of phenolic compounds. Also, rupture of the aromatic ring may result in the production of aliphatic products.

Aly and Faust (35) studied the effect of ultraviolet irradiation on three 2,4-D compounds (the Na salt, the isopropyl and butyl esters) in the laboratory. Apparently the initial pH value of the irradiated solution influences the degree of decomposition of each compound. Decomposition was relatively slow at acidic pH values and became faster as the pH value was increased. Irradiation resulted in cleavage of the 2,4-D compounds at the ethereal linkage. 2,4-Dichlorophenol and some unknown free acids were produced as shown by lowered pH values of the irradiated solutions. Under acidic conditions (pH 4.0), 2,4-dichlorophenol accumulated in the systems due to its slow decomposition. On the other hand, no phenol was detected in the irradiated 2,4-D solutions at pH 9.0 because it is decomposed as fast as it is produced at this pH value.

Crosby and Tutass (27) identified the intermediate products resulting from ultraviolet irradiation of 2,4-D. In addition to fission of the etheral linkage this compound underwent step-wise substitution of hydroxyl for chlorine until 1,2,4-trihydroxybenzene was formed. The latter was rapidly air-oxidized to the major product, a polymer whose infrared spectrum was similar to a sample of humic acid prepared from 2-hydroxybenzoquinone.

The herbicide, 3-(p-bromophenyl)-1-methoxy-1-methyl urea (metobromuron), was converted to a 20% yield of a phenolic compound (compound 1) after exposure of a 225 mg/l aqueous solution to sunlight for 17 days (35). Also, oxidation to compounds 2 and 3 occurred:

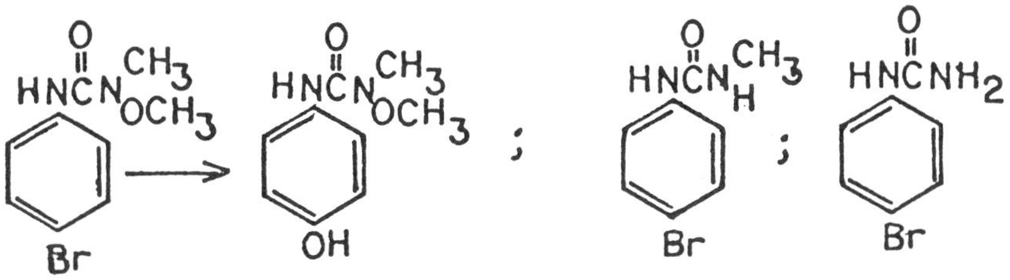

Metobromuron → (1) ; (2) ; (3)

A material whose mass spectrum indicated that it was a condensation product of the parent compound (metobromuron) and product (1) was isolated also. The photolysis of linuron in aqueous solution proceeded in a manner similar to metobromuron as shown in the following reaction (36):

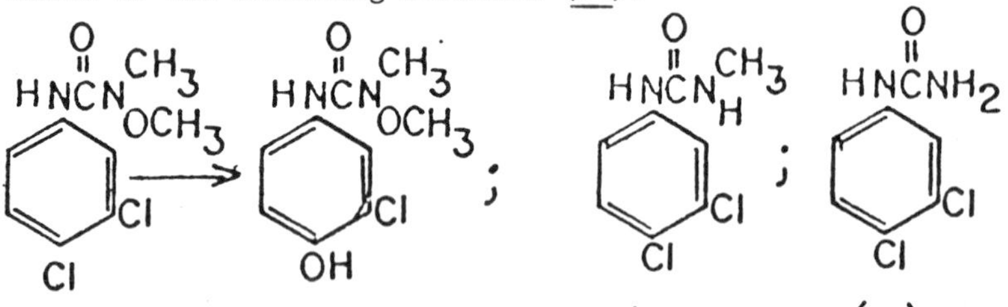

Linuron → (4) ; (5) ; (6)

The effect of ultraviolet irradiation upon the stability of three carbamate insecticides: 1-naphthyl-N-methylcarbamate (sevin), O-isopropoxyphenyl-N-methylcarbamate (baygon), and 1-Phenyl-3-methyl-5-pyrazolyl-dimethylcarbamate (pyrolan) in aqueous solutions was reported by Aly and El-Dib (7). The pH value of the aqueous medium was an important factor in determining rates of photolysis of sevin and baygon. The rates were slow at low pH values and tended to increase with an increase of pH value. The decomposition of pyrolan and dimetilan, however, was not affected by the pH value of the irradiated medium. The primary effect of the ultraviolet light irradiation was cleavage of the ester bond resulting in the phenol or heterocyclic enol of the four carbamate esters. The effect of pH value on the photolysis of sevin and baygon is analogous to the observation from 2,4-D (35) where the photodecomposition was much faster in alkaline than in neutral or acidic media. It was assumed that cleavage of the ester bond is not the only effect of the ultraviolet light but probably other modifications in the molecule tend to occur. Crosby, et al. (26) reported the photodecomposition of some carbamate esters. These investigators reported that photodecomposition of sevin yielded, in addition to 1-naphthol, several cholinesterase inhibitory substances which indicate that these compounds retained the carbamate ester group intact. Also irradiation resulted in changes at other positions in the sevin molecule.

It should be indicated that almost all studies of the photolyses of pesticides have been conducted by exposing these compounds to direct irradiation. Under actual environmental conditions, however, pesticides may be in intimate contact with materials which act as photosensitizers and may absorb light energy which, in turn, is transferred to the pesticide. This alters the products of direct irradiation or causing pesticides which do not absorb light to photolyze. For example, 3,4-dichloroaniline, a metabolite of linuron (37), diuron (38), and propanil (39, 40) is fairly stable to sunlight in aqueous solution. However, in the presence of riboflavin-5-phosphate, a significant amount of photolysis occurs in a few hours. Several photoproducts of the sensitized reactions have been isolated, two of which have been identified as azo compounds. These materials are of interest since they belong to a family of compounds which includes several members with carcinogenic properties (41).

Sunlight energy catalyzed the decomposition of organic compounds in natural waters. An excellent example is provided by Hedlund and Youngson (42) who examined the sunlight-caused photodecomposition of picloram (4-amino-3,5,6-trichloropicolinic acid) under natural conditions. Table X gives a summary of the experimental variables and kinetics of the decomposition reactions. Pseudo-first order kinetics was observed. The experimental conditions were:

Case 1: A hazy sunshine study conducted at Seal Beach, Calif. Thirty days of exposure, 26 of which were actual sunshine.

Case 2: Photodegradation was conducted in 3.65 meter deep containers at Walnut Creek, Calif. Solutions were circulated.

Case 3: Photodecomposition was conducted in distilled and canal water and in 2.54 cm. deep trays at Walnut Creek, Calif.

Case 4: Initial picloram concentrations were varied. Biological and radiochemical assays indicated that the photolysis products were not phytotoxic or in non-detectable concentrations.

Case 5: An effect of depth study. Columns were exposed four hours per day when the sun was most directly overhead.

It appears that picloram does disappear from water via photodecomposition under natural conditions and in a relatively short period of time.

Photodecomposition products may be extremely significant environmental problems. It appears that ultraviolet irradiation of organic pesticides is "incomplete". That is, the parent molecule is not degraded completely to $CO_{2(g)}$. Intermediate or even final decomposition products appear in many systems that, in turn, may have toxic properties or may impart some adverse organoleptic quality to the water, etc. For example, Crosby and Tutass (27) reported several products from the photodecomposition of 2,4-D (XIII in Figure 3). As seen, the major reaction is cleavage of the ether bond to produce 2,4-dichlorophenol which is dehalogenated to 4-chlorocatechol

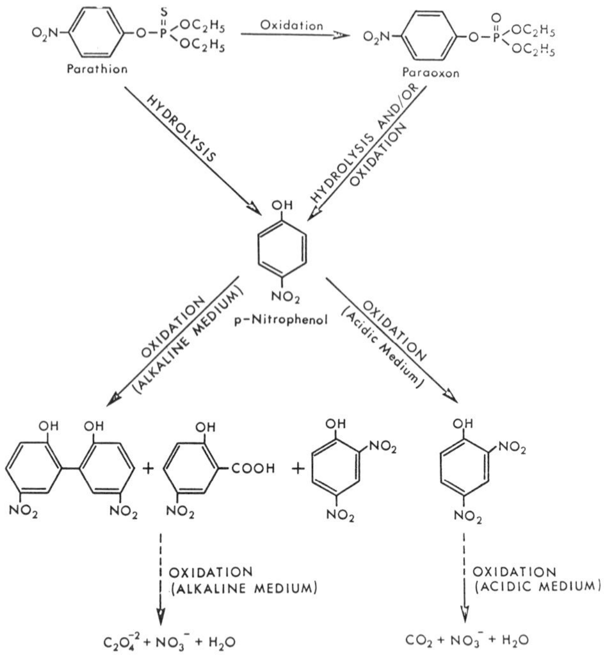

Figure 2. Proposed pathways for the $KMnO_4$ oxidation of parathion. After Ref. 16.

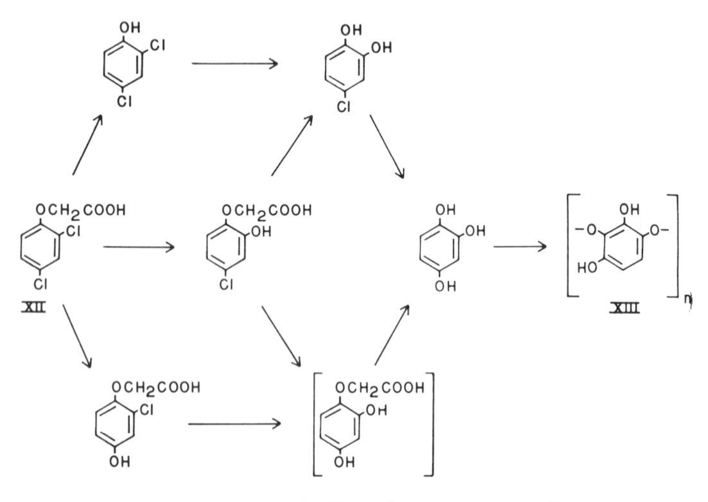


Figure 3. Photolysis of 2,4-dichlorophenoxyacetic acid. After Ref. 27.

Table X. SUMMARY OF EXPERIMENTAL VARIABLES AND KINETICS-PHOTODECOMPOSITION OF PICLORAM[a]

| Case | Initial Conc.[b] ($M \times 10^6$) | Solution Depth | Time of Year | Other Variables | Rate Constant | Half-Life |
|---|---|---|---|---|---|---|
| 1 | 37.3 | 8 cm | March | Actual days | 0.0738 $day^{-1}$ | 9.4 days |
| | | | | Esti. sunshine | 0.0855 $day^{-1}$ | 8.1 days |
| 2 | 4.14 | 3.65 m | Sept.-Oct. | - | 0.0168 $day^{-1}$ | 41.3 days |
| 3 | 20.7 | 2.54 cm | Aug. | Distilled water | 0.306 $day^{-1}$ | 2.3 days |
| | | | | Canal water | 0.280 $day^{-1}$ | 2.5 days |
| 4[c] | 414.1 | 4.6 cm | July | Bioassay | 0.116 $day^{-1}$ | 6.0 days |
| | | | | $BuOH/NH_3$ Chromatography | 0.0956 $day^{-1}$ | 7.3 days |
| | | | | $C_6H_6/C_2H_5COOH/H_2O$ Chromatography | 0.0915 $day^{-1}$ | 7.6 days |
| 5[d] | 0.409 | 0.292 m | Aug.-Sept. | - | $8.83 \times 10^{-3} hr^{-1}$ | 78.5 hr |
| | 0.401 | 1.82 m | | - | $4.49 \times 10^{-3} hr^{-1}$ | 154. hr |
| | 0.421 | 3.65 m | | - | $3.01 \times 10^{-3} hr^{-1}$ | 230. hr |

a. From reference (42).
b. For conversion to other units, $4.14 \times 10^{-6}$M picloram is about equal to 1 ppm or 2.72 pounds per acre-foot of water.
c. Calculations based on construction method.
d. Time values based on noncontinuous exposure, 4 hours per day at midday.

and then to 1,2,4-benzenetriol. In turn, this compound is oxidized rapidly in air to a mixture of polyquinoid humic acids (XIII in Figure 3) by a light independent process. Crosby and Tutass subsequently stated: "the effects observed from irradiation in sunlight were qualitatively very similar to results from laboratory experiments." This research is an excellent example of potentially hazardous products being produced from partial photodecomposition. Similar results were reported by Binkley and Oakes (43). Also, similar photodecomposition products were observed from the irradiation of 2,4,5-T (2,4,5-trichlorophenoxyacetic acid)(45). The major products were: 2,4,5 trichlorophenol and 2, hydroxy-4,5-dichlorophenoxyacetic acid; 4,6-dichlororesorcinol, 4-chlororesorcinol, 2,5-dichlorophenol, and "a dark polymeric product."

Two primary products were observed from the photodecomposition of fenitrothion (O,O-dimethyl-O-(3-methyl-4-nitrophenol) phosphorothioate) in vapor phase and in an ethanolic solution (44). One of these was identified as p-nitrocresol (4-nitro-3-methyl-phenol). The second product was not identified. Photolysis was conducted at 313 nm. These results suggested that fenitrothion is photolyzed by by radiation present in sunlight at the surface of the earth.

It seems reasonable, therefore, to suggest that photodecomposition may account for some loss of the pesticide residues in clear surface waters exposed to long periods of sunlight. However, photolysis may be a minor factor in the decomposition of the pesticide residues in highly turbid waters where the penetration of light will be greatly reduced.

## V. Toxicity of Degradation and Transformation Products

Very little information has been reported about the toxicity values for degradation and transformation products of the organic pesticides cited in this paper. That is, toxicity values were not found in the "Toxic Substances List" compiled by Christensen, et al. (46). In the case of diazinon, the major hydrolysis product, 2-isopropyl-4-methyl-6- hydroxypyrimidine, is considerably less toxic than the parent molecule as seen in Table XI. p-Nitrophenol is the major hydrolysis product from parathion and is considerably less toxic. Paraoxon may be more toxic than parathion whereas 2,4-dinitrophenol is within the same range of toxicity. The hydrolysis products from the carbamates, sevin and dimetilan appear to be less toxic. The information conveyed in Table XI cannot be generalized into the contention that most products are less toxic than the parent molecule. The simple fact is that toxicity values are not known or reported for most degradation

Table XI. TOXICITY VALUES OF SOME DEGRADATION AND TRANSFORMATION PRODUCTS (SOURCE: "TOXIC SUBSTANCES," ANNUAL LIST, H. C. CHRISTENSEN, EDITOR)

| | AO $LD_{50}$ Rats<br>mg/l | $LC_{50}$ Injected Flies |
|---|---|---|
| Parathion | 1.1 - 6.0 | - |
| Paraoxon | 3.0 - 3.5 | - |
| p-Nitrophenol | 350 | - |
| 2,4-Dinitrophenol | 31.2, 13.[a] | - |
| 2,4 Dichlorophenoxyacetic acid | 300 - 1000 | - |
| 2,4-Dichlorophenol | 312.[b] | - |
| Diazinon | 76-435 | $2.04 \times 10^{-4}$M |
| Diazoxon | ? | $6.2 \times 10^{-5}$M |
| 2-isopropyl-4-methyl-6-hydroxypyrimidine | 2700 (mice) | - |
| Sevin (Carbaryl) | 400 - 505 | - |
| 1-Naphthol | 2400 | - |
| Methyl amine | 2500 | - |
| Dimetilan | 25 - 64 | - |
| Dimethylamine | 698 | - |
| 2-Dimethyl carbamyl - | ? | - |
| 3-methyl-5-pyrazolol | | - |
| Baygon | 95 - 175 | - |
| Isopropoxyphenol | ? | - |

a. Oral - bird.
b. Lowest dosage, 39 weeks, intermittent, carcinogenic, skin-muscle.

and transformation products. Diazoxon may be the classic case.

## ACKNOWLEDGMENT

Paper of the Journal Series, New Jersey Agricultural Experiment Station, Rutgers, the State University of New Jersey, Department of Environmental Sciences, New Brunswick, New Jersey.

Literature Cited

1. Kearney, P.C. and D.D. Kaufman, "Degradation of Herbicides", Marcel Dekker, Inc., New York (1969).
2. Blumenthal, E. and J.B. Herbert, Trans. Faraday Soc., 41 611 (1945).
3. Dostrovsky, I. and M. Halmann, J. Chem. Soc., p502 (1953).
4. Gomaa, H.M. and S.D. Faust, Chap. 10 in "Fate of Organic Pesticides in Aquatic Environments", S.D. Faust, Sympos. Chairman, Adv. Chem. Ser.,III, Amer. Chem. Soc., Wash., D.C. (1972).
5. Gomaa, H.M.,et al., Residue Rev., 29, 171 (1969)
6. Faust, S.D. and H.M. Gomaa, Environ. Letters, 3(3), 171 (1972).
7. Aly, O.M. and M.A.El-Dib,Chap 20 in "Organic Compounds in Aquatic Environments", S.D Faust and J.V. Hunter, Eds, Marcel Dekker, Inc., New York (1971).
8. Casida, J.E.,et al., J. Econ. Entomol., 53, 205 (1960).
9. Karinen, J.F., et al., J. Agr. Food Chem., 15(1), 148 (1967).
10. Lamberton, J.G., et al., J. Agr. Food Chem., 18(1), 92 (1970).
11. Wauchope, R.D and R. Hague, Bull. Environ. Contam. & Toxicol., 9 257 (1973).
12. Hosler, C.F., Jr., Bull. Environ. Contam. & Toxicol., 12(5), 599 (1974).
13. Mathews, E.M. and S.D. Faust, Unpublished results (1975).
14. Hoffsommer, J.C. and J.M. Rosen, Bull. Environ. Contam. & Toxicol., 10(2), 78 (1973).
15. Bailey, G.W., et al., Weed Science, 18, 413 (1970).
16. Gomaa, H.M. and S.D. Faust, Chap. 15 in "Organic Compounds in Aquatic Environments", S. D. Faust and J.V. Hunter, Eds, Marcel Dekker, Inc., New York (1971).
17. Cohen, J.M., et al., J. Amer. Water Works Assoc. 52, 1551 (1960).
18. Cohen, J.M., et al., J. Amer. Water Works Assoc., 53 49 (1961).
19. Aly, O.M. and S.D. Faust, J. Amer. Water Works Assoc., 57 221 (1965).
20. Robeck, G., et al., J. Amer. Water Works Assoc., 57 181 (1965).
21. Gomaa, H.M. and S.D. Faust, J. Agr. Food Chem., 19, 302 (1971).
22. Cochrane, W.P. and M. A. Forbes, Chemosphere, 1, 41 (1974).
23. Mitchell, L.C., J. Assoc. Off. Agr. Chemists., 44, 643 (1961).
24. Weldon, L.W. and F.L. Timmons, Weeds, 9, 111 (1961).
25. Jordan, L.S., et al., Weeds, 12, 1(1964).
26. Crosby, D.C., et al., J. Agr. Food Chem., 13, 204 (1965).

27. Crosby, D.C. and H.O. Tutass, J. Agr. Food Chem., 14, 596 (1966).
28. Eberle, D.O. and F. A. Gunther, J. Assoc. Off Agr. Chem., 48, 927 (1965)
29. Abdel-Wahab, A.M. and J.E. Casida, J. Agr. Food Chem., 15 , 479 (1967).
30. Rosen, J.D., Chap. 18 in "Organic Compounds in Aquatic Environments", S. D. Faust and J.V. Hunter, Eds, Marcel Dekker, Inc., New York (1971).
31. Robinson, J., et al., Bull. Environ. Contamin. Toxicol, 1 , 127 (1966).
32. Parsons, A.M. and J. D. Moore, J. Chem. Soc., p.2026 (1966).
33. Rosen, J.D. and D. J. Sutherland, Bull. Environ. Contamin. Toxicol., 2, 1 (1967).
34. Bell, G.R., Bot Gaz., 118,133 (1956).
35. Aly, O. M. and S.D. Faust, J. Agr. Food Chem., 12, 541 (1964).
36. Rosen, J.D. and R.F. Stusz, J. Agr. Food Chem., 16, 568 (1968).
37. Nashed, R.B. and R.D. Ilnicki, Proc. Northeast Weed Control Conf., 21, 564 (1967).
38. Dalton, R.L., et al., Weeds, 14, 31 (1966).
39. Bartha, R. and D. Pramer, Science, 156, 1617, (1967).
40. Still, C.C. and D. Kurzuian, Nature, 216, 799 (1967).
41. Weisburger, J.H. and E. K. Weisburger, Chem. Eng. News, p. 124 (Feb. 7, 1966).
42. Hedlund, R.T. and C. R. Youngson, Chap. 8 in "Fates of Organic Pesticides in Aquatic Environments", S. D. Faust, Symp. Chairman, Adv. Chem. Series, III, Amer. Chem. Soc., Washington, D.C. (1972).
43. Binkley, R.W. and T. R. Oakes, Chemosphere, 1, 3 (1974).
44. Brewer, D.G., et al. Chemosphere, 3, 91 (1974).
45. Grosby, D.G. and A. S. Wong, J. Agr. Food Chem., 21(6) , 1052 (1973).
46. Christensen, H.E., et al., Eds, "The Toxic Substances List", 1973 Edition, U.S. Dept. Health, Educ., and Welfare, Rockville, Md. (June, 1973).

# 36

# Toxins and Bioactive Compounds in the Marine Environment

GEORGE M. PADILLA

Department of Physiology and Pharmacology, Duke University Medical Center, Durham, N.C. 27710

DEAN F. MARTIN

Department of Chemistry, University of South Florida, Tampa, Fla. 33620

Biodynamic compounds are organic products which exert a specific effect on other organisms in the marine environment (1). They may be of natural or artificial origin and may be modified by physical and biological processes. It is fruitful to consider the implications of this concept, particularly as it applies to the coastal areas of the ocean where communities of organisms are highly interdependent. One could say they communicate with one another through the elaboration and release of "active" compounds that form a complex dynamic network. In spite of the fact that these "active" compounds occur at very dilute concentrations, they nevertheless may at times affect the physiological well being of specific organisms, as is the case with sex attractants (2) and ichthyotoxins (3).

To fully understand the role of bioactive compounds in the sea, we must consider how they are distributed throughout the marine environment, the types of organisms which produce them, and finally, the specific modes of action, if any, that these compounds have on target organisms. To achieve all this, the study of biodynamic compounds must be approached from a variety of scientific disciplines. We are dealing not merely with a problem of chemical isolation and identification, but with an analysis of the relationships between organisms, their external environment, and one another. Physiological, ecological, and chemical attributes which define these relationships must be considered and experimentally evaluated.

## Origin and Sources of Biodynamic Compounds

It is clear that there is no single origin of bioactive compounds. Numerous organisms from widely different classes are known to contain or release biologically active chemicals. Table I shows a selection of such compounds. Even if such compounds were produced by a single class of animals (e.g. phytoplankton), it is not clear to what extent these natural products are modified

Table I

Some Unique Organic Compounds
Derived from Marine Organisms

| Compound | Organism | Ref. |
|---|---|---|
| Halogenated Sesquiterpenoid | Red Alga (*Laurecia elata*) | (10) |
| Betaine | Soft Coral (*Pseudoterogorgia americana* | (16) |
| Aromatic terpenoid | Alga (*Taonia atomaria*) | (11) |
| Indoleninones | Mollusc (*Dicathrais orbita*) | (14) |
| Brominated phenols | Red Alga (*Corallina officalis*) | (4) |
| Prostaglandins (15-epi-$PGA_2$) | Coral (*Plexaura homonalla*) | (9) |
| Carageenan (polysaccharide) | Red Seaweed (*Chondrus crispus*) | (15) |
| Saponin | Starfish (*Asterias sp*) | (13) |
| Diethylamine-dithiolanes | Annelid (*Lambriconereis heteropoda*) | (12) |
| Anabaseine | Nemertine Worms (*Paranemertes*) | (17) |

as they pass through the various trophic levels of the whole biotic community. In addition they are mixed with other organic compounds resulting from the continual enrichment and modification of the environment that occurs through the operation of a variety of cycles (e.g. carbon, nitrogen, etc.). These cycles also interact directly with the food chain forming a complex network of energy exchange. In its simplest form, the food chain denotes the passage of nutrients from the "producers" (e.g. phytoplankton) to the "consumers" which include the zooplankton and progressively larger organisms at various trophic levels. The carnivores of the sea, among which we may include ourselves, constitute the last link of the chain.

In a recent article, Benson and Lee (5) stated that the food chain is not only a mechanism for the distribution of organic matter but serves as a dynamic energy depot. They estimated that at least one-half of the organic material produced in the ocean by the primary producers is converted and temporarily stored into waxes by copepods, a type of small, abundant crustaceans. The waxes are not only used as energy sources, they serve as well to regulate these organisms' activity and behavior. For example, a high degree of unsaturation in the wax allows copepods to inhabit polar regions by acting as a sort of biological "anti-freeze". Similarly, changes in an organism's specific gravity, reflecting an altered wax content, permit copepods to migrate vertically for considerable distances. Such vertical migration patterns follow the complex developmental sequences from egg to adult (5).

One usually does not consider food as a specific biodynamic product, but in this instance waxes serve more than a dietary role. Their distribution and modification as they pass through the food chain is an example of the dynamic nature of the roles they play as bioactive compounds.

Recently Banner (6) discussed at some length the food chain theory of Randall as it applies to the origin of ciguatoxin. This toxin will appear sporadically and as yet unpredictably as a contaminant of reef fish in the Pacific. The basic premise of Randall's theory is that fish owe their intoxication to their mode of life. That is to say, the herbivorous fish feed upon certain toxigenic algae on coral reefs. It is thought these algae are the primary producers of the toxin. Even though the toxin does not seem to affect these small herbivorous fish, it is toxic to the larger carnivores which feed upon them. Two possibilities are implicit in this transfer: (1) the principle may become toxin as it is metabolized by the herbivorous fish at the lower trophic levels or (2) the toxic principle is not changed but is merely concentrated in the flesh of the final host fish. It is in this concentrated form lethal to man.

The concentration and storage of ultimately toxic organic compounds by organisms in seemingly "inert" forms is one of the more important aspects of the transfer of bioactive compounds through the food chain. This phenomenon is not limited to biotoxins produced by members of the food chain, it is also applicable to compounds which have been added to the environment by man (e.g. pesticides, organo-metallic compounds, plasticizers, petroleum by-products, etc. [7]). Halstead (8) was one of the first to note the importance of this factor. He was particularly concerned with the impact this would have on the plans to derive a high protein "fish flour" from otherwide inedible "trash" fish. A high level of contamination in the ocean would of course destroy the usefulness of this natural resource.

Marine organisms do not store only harmful products. They have in fact been used as source material for a wide variety of useful industrial and medicinal compounds (1). For example, the prostaglandins are found in soft corals in quantities up to 1.3% of the dry weight (1,9). Included in this class are very potent pharmacological agents with a wide spectrum of physiological activity (1). Figure 1 shows the physiologically active ones.

We have also listed in Table I a selection of some unique compounds obtained from marine organisms. The listing is not comprehensive but is indicative of the diversity of these natural products. Included are halogenated phenols (4), aromatic terpenes (10,11), thiol amines (12), saponins (13), indoleninones (14), polysaccharides (15), betaine (16), and the toxin anabaseine (17). As you can see from Table I, these compounds are derived from diverse organisms. A more extensive list would no doubt increase the diversity of the plants and animals that yield such compounds.

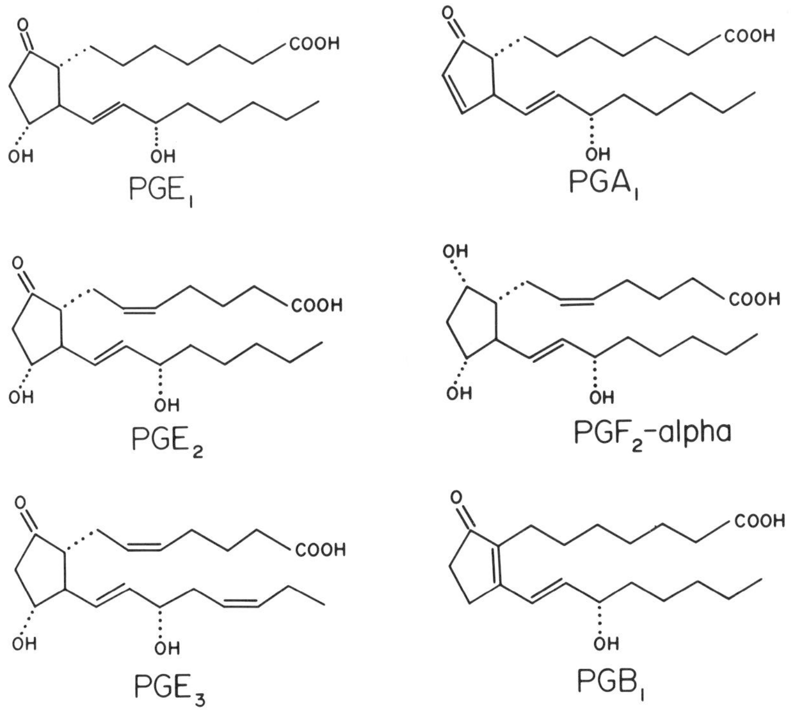

Marine Pharmacognosy

*Figure 1. Chemical structure of prostaglandins with a potential pharmacological activity* (1)

As discussed in this symposium, biodynamic compounds also originate from our highly sophisticated technological activities. Although pollution may not be entirely unavoidable, the impact which our industrial culture has on the environment is now well recognized. The deposition of chemical by-products of technology into the coastal environment is significant not only because of their long term harmful effects on biological communities, as was the case with certain pesticides (7,18), but because of their existence as chemical contaminants. This is of particular concern to those studying natural products from marine sources. There is no guarantee that complex organic molecules from natural sources can be isolated in the absence of contaminating exogenous materials. This is particularly true for widely used industrial compounds such as phthalates which are now reported to be widely distributed in the terrestial and aquatic environment (19,20).

We became aware of this problem during a recent analysis of a red-tide sample off the coast of Florida (21). We were trying to isolate and identify the toxins produced by the red-tide dinoflagellate *Gymnodinium breve*. The isolation procedure involved the extraction of a large volume of sea water with organic solvents. The crude extract, after suitable concentration procedures, was resolved into three toxic fractions by column and thin layer chromatography. The infrared spectra of the toxic fractions ($I_a$, $I_c$, $I_d$) are shown in Fig. 2. The similarity between the spectra of the toxic fractions and that of phthalate-like compounds is readily apparent. To emphasize this point, we have shown the spectrum of a tygon-tubing extract in Fig. 2, as well as the non-toxic fraction ($III_c$) which is usually discarded during the extraction procedure. Note that the tygon extract and fraction $III_c$ have a strong absorption peak at 7.8-7.9 μ (arrows) that is not present in the toxic fractions, $I_a$, $I_c$, $I_d$. The remainder of the spectra are very similar. The presence of the phthalate compounds in all fractions was verified by mass spectroscopy. Fragments with the diagnostic values of 279, 167, and 149 generated by phthalates, particularly di-2-ethyl hexyl phthalate (20), were readily obtained. Further mass spectroscopic analyses, performed by Dr. D. Brent of the Burroughs-Wellcome Laboratories, revealed the presence of additional fragments with m/e values > 700. It is presumed that these were generated by the *G. breve* toxin (unpublished results). We cite this example to illustrate the fact that these man-made additions will add to the technical problems encountered in the study of natural products suspected of having a specific physiological effect.

## Diversity and Distribution of Biodynamic Compounds

All phyla have representative organisms as sources of biodynamic compounds. Because of the pyramidal distribution of organisms in the sea, the phytoplankton, zooplankton, and smaller

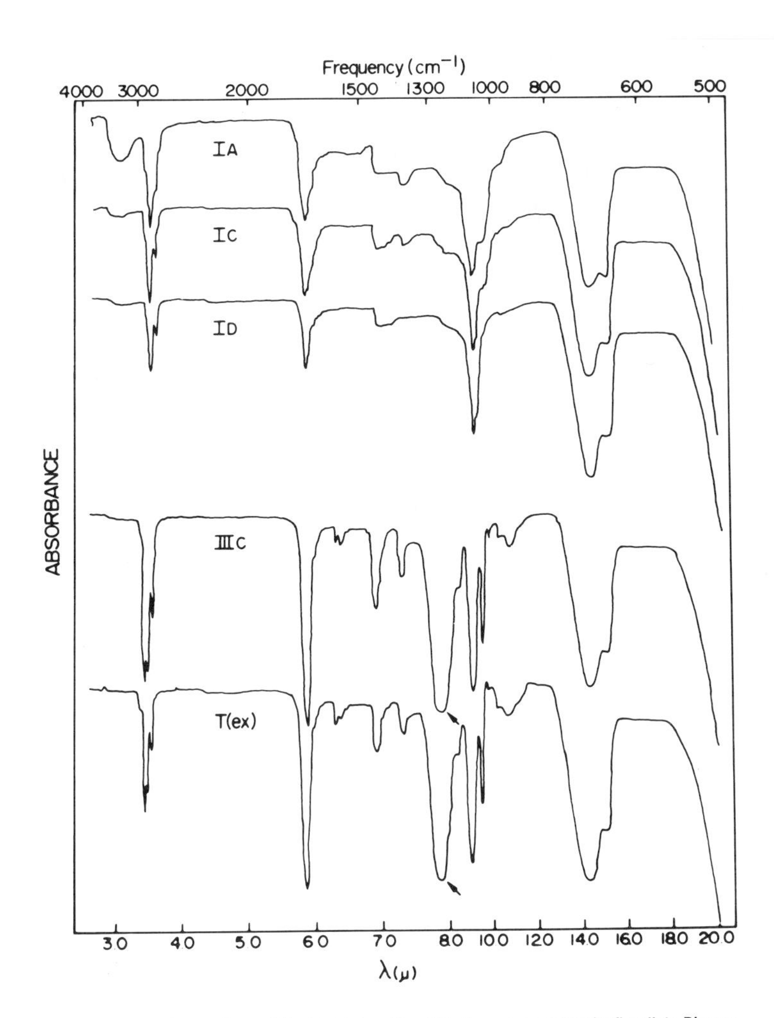


Proceedings of the First International Conference on Toxic Dinoflagellate Blooms

*Figure 2. Infrared spectra of* G. breve *toxins and tygon tubing extract. Upper abscissa, frequency ($cm^{-1}$); lower abscissa, wave length ($\mu$ = microns); ordinate, absorbance. $I_A$ = toxic fraction $I_A$; $I_C$ = toxic fraction $I_C$; $I_D$ = toxic fraction $I_D$; $III_c$ = nontoxic fraction $III_c$; $T_{(ex)}$ = tygon tubing extract* (21).

invertebrates not only occur in larger numbers, but have a higher incidence of speciation and thus may yield a greater variety of bioactive compounds. In other words, they have evolved into specific ecological niches which form a balanced community of organisms.

A second feature which affects the distribution of biodynamic compounds is the seasonal succession of algal blooms, which follow cyclical hydrographic changes in the coastal environment (e.g. changes in salinity during the summer, enrichment of coastal waters from land run-off following seasonal rains, etc.). It has been suggested that seasonal factors play a role in the outbreak of red tides or the incidence of similar phytoplankton-linked toxicities (22). For example, a study of the monthly incidence of ciguatoxin poisonings during 1966-1968 showed that there was a peak of the incidence of toxicity during the summer months of 1968, particularly June-July (6). The level of toxicity was low during the winter months and began to rise during the spring. As discussed earlier, the increased toxicity was due to a parallel increase in the growth of algae. In a study on the incidence of hemolytic agents concentrated by oysters, which were field-collected from 17 stations along the coast of North Carolina, we found a 2 to a 5-fold increase in the hemolytic titer during the months of September and October, coincident with increase in phytoplankton blooms in this region (23). The hemolysin was completely absent during the winter months. The incidence of hemolysin appearance was taken to be an indication that algae containing hemolytic components were being concentrated by the oysters. A similar assay was used to monitor an incipient red-tide outbreak in Florida (24).

The seasonal occurrence of red tides in the coastal environment has been well documented (22,25). For example, Hartwell (25) recently described the hydrographic factors which may set the stage for the red-tide outbreaks of the dinoflagellate *Gonyaulax tamarensis*. Until recently, these organisms were rarely seen in the western gulf of Maine. However, during September 1972, a bloom was seen in the coastal waters off New Hampshire. As many as 4600 μg of toxin/100 g meat of the soft-shelled clam had been accumulated at that time. A second smaller bloom occurred in early June of 1974 and a major one recurred in August 1974. Evidence was presented to show that shifts of a variety of hydrographic factors such as salinity, temperature, water currents, level of nutrients, and dissolved oxygen favored the outbreak of red tides during the summer months of those years.

Steidinger (26) also examined the factors which may influence the incidence of red tides, but from a more ecological point of view. According to her studies, there are at least 3 elements common to red-tide outbreaks: (a) the rate of increase in population size, (b) the presence of favorable hydrographic factors and (c) the maintenance of dinoflagellate populations in discrete "patches" by other hydrologic or meteorologic forces. The interplay between these factors sets the stage for a red-tide outbreak. How such an interplay is brought about is not known at present, but these investigations do show that it is a combination of biological and physical forces that will bring about an expansion or bloom in specific phytoplankton populations. Were it not for the liberation of potent toxins into the marine environment, the presence of these algal blooms would have gone undetected.

## Detection and Identification of Biodynamic Compounds

It is unreasonable to expect that every biodynamic compound will exert dramatic and widespread actions such as the mass fish mortalities which accompany the outbreak of red tides. It would also be an unsatisfactory state of affairs to rely on such ecological disasters before we proceed to identify the toxic principle involved. One's activities would be reduced to a "follow-up" strategy: await a red tide, gather sufficient material, and then discover its mode of action. Whatever the socio-economic implications of this course of action, not only are red tides sporadic and difficult to predict, toxins are produced at extremely low concentrations (i.e. at a few micrograms/liter). It is only because they are specific and potent that they do not escape our attention. Table II shows the relative toxicity of a variety of marine toxins (1). Even when one considers lethality as the index of activity, compounds such as tetrodotoxin, saxitoxin, and *Gymnodinium breve* toxin are

Table II

Relative Toxicities of Selected Marine Biotoxins[a]

| Toxin | Source | Lethal dose[b] (μg/kg) |
|---|---|---|
| Palytoxin | Zoanthid (*Palythoa sp*) | 0.15 |
| Saxitoxin | Dinoflagellate (*G. catenella*) | 3.4 |
| Tetrodotoxin | Puffer fish (*S. rubripes*) | 8.0 |
| Sea snake venom | Sea snake (*L. semifasciata*) | 130 |
| Ciguatoxin | Moray eel (*G. javanicus*) | 500 |
| Prymnesin | Golden alga (*P. parvum*) | 1,400 |
| Holothurin A | Sea cucumber (*A. agassizi*) | 10,000 |
| Ostracitoxin | Boxfish (*O. lengitinosus*) | 200,000 |

[a]See (1) for references; [b]$LD_{50}$ or minimum lethal dose in mice From "Marine Pharmacognosy" with permission of Academic press.

effective at very low concentrations. It has been reported that these toxins inhibit specific physiological activities (i.e. neuronal transmission, active transport, hemolysis, etc.) at very low concentrations ($\sim 10^{-8}$-$10^{-14}$ g/L). Yet lethality is almost the universal bioassay used (1,3,27), and in many cases it is the only one. Most often, it is used with laboratory animals rather than organisms derived from the estuarine environment (7).

Since marine animals communicate with each other through the medium which surrounds them, a biologically active compound must exert its initial stimulatory or noxious effect at the cell membrane. This barrier need not be at the gills or at the outer integument but may reside at a specific locus within the animal, i.e. the nervous system, kidney, hepatopancreas, etc. The biological activities of marine bioactive compounds should thus be examined in terms of effects on membrane-dependent activities. This would include assays based on electrophysiological measurements, transport mechanisms, as well as the binding to specific membrane sites which govern the activity of membrane-bound enzymes. In addition, an affinity for membrane components may result in the disruption of membrane structure. An example of this interaction is seen in the hemolytic activity of toxins from the euryhaline alga Prymnesium parvum (28).

A variety of experimental approaches for determining the activity of compounds on excitable membranes (e.g. the giant axon of the squid, neuromuscular junctions, etc.) are presently available (3). For example, methods using the voltage-clamp technique and recording microelectrodes have defined the specific effects of marine toxins on the ionic conductances of excitable cells (see [3] and [29] for review). The frog skin preparation in which one relates changes in the short circuit current to the active transport of ions has also been used to measure the activity of toxins from Gymnodinium breve (30). Another useful approach employs the frog nerve-muscle preparation to study the effect of biotoxins on the release and activity of neurotransmitters. It has proved valuable in the analysis of toxins derived from dinoflagellates and blue-green algae (3).

It is clear from the above that the study of biodynamic compounds is in essence based on the structure-function relationship which they have with biological membranes. This statement implies that the activity of these compounds is dependent on the existence of membrane sites bearing a unique and specific affinity to effector compounds. This approach is analogous to one based on the receptor concept developed for studies on drug and hormone action. According to this concept, a drug is active because of its interaction with a specific cellular component to which it binds. Subsequent events or a series of events are then triggered by this interaction (32). For example, a specific enzyme may be activated or inhibited, a transmitter substance may be released or a second chemical effector produced. Toxins which are highly reactive must have such a specific interaction by virtue of their chemical structure. Their relative potency is in effect a measure of their affinity to a specific membrane component. This situation is thought to explain the activity of tetrodotoxin. It has been shown to have a unique affinity for the sodium channels in the giant axon of the squid or lobster. The inward movement of sodium is prevented and although the movement of potassium is unaffected, the conduction of electrical pulse along the nerve axon is blocked (see [1,3] for refs.).

A comparison between saxitoxin and tetrodotoxin shows that differences in physico-chemical properties may account for variations in their physiological activity as discussed by Doig et al (1). For example, both toxins possess an N-alkyl-substituted guanidinium ion, $(H_2N)_2C{=}NH^+$, as part of their structure. Such ions are thought be act as current carriers and possible substitutes for sodium ions. However, the bulkiness of the remainder of the molecule offers a steric hindrance which impedes the passage of the guanidinium ion, and thus blocks the movement of sodium ions through "channels" in the cell membrane. Henderson and coworkers (31) recently examined the binding properties of both toxins at the sodium channels of nerve membrane preparations in the presence and absence of a series of monovalent, divalent, and trivalent cations. They provide evidence to show that the sodium channel possesses a negatively charged site with an apparent pKa between 5 and 6. It is at this site that both tetrodotoxin and saxitoxin will bind and block the channel with the guanidinium ion portion of the molecule. Divalent and trivalent cations reversibly compete with the toxins for this site as do some of the monovalent ions. This elegant approach is of necessity possible only with toxins of known chemical structure. The extent to which our understanding of membrane function will be advanced by a parallel understanding of the physico-chemical attributes of marine biotoxins is considerable.

## Concluding Remarks

Marine organisms are linked to each other through a dynamic network of chemical interactions. Bioactive compounds are the elements of this network and as such serve to establish and maintain the physiological and ecological balance of the communities in the coastal environment. It is thought that the specificity and potency of their action is an expression of their affinity to specific membrane sites. It is only through a multidisciplinary approach that we will gain an understanding of the marine environment. We will have at our disposal compounds to be used as probes of membrane function in both research and medicine.

## Acknowledgements

This work was supported in part by a Food and Drug Administration Grant to GMP (FD 00120) and a Research Career Development Award to DFM (1 K04 GM 4269) from NIGMS.

## Abstract

Although our knowledge of the exact chemical structures of bioactive compounds derived from marine sources is limited, it is of value to consider to what extent they originate as by-products of metabolism, constitutents of the organisms that produce them, or as chemically altered compounds introduced by man into the environment. Bioactive compounds were considered in terms of the biological activity they possess, toxicity being one example. The physiological effects of compounds such as prostaglandins and sex attractants were discussed to emphasize the role of bioactive compounds which is not necessarily detrimental to other organisms in the ocean. The structure-function relationship of bioactive compounds was examined at some length and their value as probes of biological activity in research and medicine was evaluated.

## Literature Cited

1. Doig, M.T., Martin, D.F. and Padilla, G.M. In: "Marine Pharmacognosy" (Eds. Martin, D.F. and Padilla, G.M.), pp. 1-35, Academic Press, New York, 1973.
2. Müller, D.G., Jaenicke, L., Donike, M. and Akinobi, T. Science (1971) 171, 815.
3. Sasner, J.J., Jr. In: "Marine Pharmacognosy" (Eds. Martin, D.F. and Padilla, G.M.), pp. 127-177, Academic Press, New York, 1973.
4. Pedersen, M., Senger, P. and Fries, L. Phytochemistry (1974) 13, 2273.
5. Benson, A.A. and Lee, R.F. Scientific American (1975) 232, 77.
6. Banner, A.H. In: "Bioactive Compounds from the Sea" (Eds. Humm, H.J. and Lane, C.E.), pp. 15-36, Marcel Dekker, New York, 1974.
7. Vernberg, F.J. and Vernberg, W.B. (Eds) "Pollution and Physiology of Marine Organisms", 492 pp., Academic Press, New York, 1974.
8. Halstead, B.W. In: "Drugs from the Sea" (Ed. Fruedenthal, H.D.), pp. 229-239, Marine Technology Society, Washington, D.C., 1968.
9. Weinheimer, A.J. and Spraggins, R.L. Tetrahedron Letter (1969) No. 59, 5185.
10. Sims, J.J., Lin, G.H.Y. and Wing, R.M. Tetrahedron Letter (1974) No. 39, 3487.
11. Gonzalez, A.G., Darias, J., Martin, J.D. and Pascual, C. Tetrahedron (1973) 29, 1605.

12. Okaichi, T. and Hashimoto, Y. Bull. Jap. Soc. Sci. Fish. (1962) 28, 930.
13. Yasumoto, T., Watanabe, T. and Hashimoto, Y. Bull. Jap. Soc. Sci. Fish. (1964) 30, 357.
14. Baker, J.T. and Duke, C.C. Aust. J. Chem. (1973) 26, 2153.
15. Mueller, G.P. and Rees, D.A. In: "Drugs from the Sea" (Ed. Fruedenthal, H.D.), pp. 214-255, Marine Technology Society, Washington, D.C., 1968.
16. Weinheimer, A.J., Metzner, E.R. and Mole, M.L., Jr. Tetrahedron (1973) 29, 3135.
17. Kem, W.R. In: "Marine Pharmacognosy" (Eds. Martin, D.F. and Padilla, G.M.), pp. 37-84, Academic Press, New York, 1973.
18. Seba, D.B. and Corcoran, E.F. Science (1973) 191, 925.
19. Ogner, G. and Schnitzer, M. Science (1970) 170, 317.
20. Mayer, F.L., Jr., Salling, D.L. and Johnson, J.L. Nature (1972) 238, 411.
21. Padilla, G.M., Kim, Y.S. and Martin, D.F. Proc. 1st Int. Conf. on Toxic Dinoflagellate Blooms, Massachusetts Science and Technology Foundation, Wakefield, Mass., pp. 299-308, 1975.
22. Prakash, A., Medcof, J.C. and Tennant, A.D. Bulletin 177, Fisheries Research Board of Canada, Ottawa, 1971.
23. Padilla, G.M. and Bragg, R.J. Proc. 4th Ann. Conf. Marine Technology Society, Washington, D.C., pp. 193-204, 1968.
24. Martin, D.F., Martin, B.B. and Padilla, G.M. Environ. Letters (1972) 2, 239.
25. Hartwell, A.D. Proc. 1st Int. Conf. on Toxic Dinoflagellate Blooms, Massachusetts Science and Technology Foundation, Wakefield, Mass., pp. 47-68, 1975.
26. Steidinger, K.A. Proc. 1st Int. Conf. on Toxic Dinoflagellate Blooms, Massachusetts Science and Technology Foundation, Wakefield, Mass., pp. 153-162, 1975.
27. Humm, H.J. and Lane, C.E. (Eds.) "Bioactive Compounds from the Sea", 215 pp., Marcel Dekker, New York, 1974.
28. Martin, D.F. and Padilla, G.M. Biochim. Biophys. Acta (1971) 241, 213.
29. Narahashi, T. Fed. Proc. (1972) 31, 1124.
30. Kim, Y.S., Mandel, L.J., Westerfield, M., Padilla, G.M. and Moore, J.W. Environ. Letters (1975) (In press).
31. Henderson, R., Ritchie, J.M. and Strichartz, G.R. Proc. Nat. Acad. Sci. USA (1974) 71, 3936.
32. Cuatrecasas, P. Ann. Rev. Biochem. (1974) 43, 169.

# 37

# New England Coastal Waters—An Infinite Estuary

CHARLES S. YENTSCH

Bigelow Laboratory for Ocean Sciences, West Boothbay Harbor, Me. 04575

If oceanic regions could be graded using the conventional motives of our society, the coastal waters would get high marks. They host most of the productive fisheries and play a prime role in near-shore recreational activities. Oddly, the study of the processes which cause coastal waters to be productive and aesthetically pleasing, has not received the unified attention of the oceanographic community. The early 1930 studies of H. B. Bigelow represented a giant step towards such a goal; however, what followed was a period where the study of coastal ocean processes was displaced by interests in the deep oceans. Shortly after World War II, interests changed; some oceanographers began to focus interest on the interaction of freshwater and seawater via the estuary. This interest gained momentum largely through the research interests B. H. Ketchum. Armed with techniques largely derived from estuarine study, Ketchum (1) examined the interaction of freshwater and seawater in coastal waters. In essence, he treated these as a gigantic estuary. In the light of our present day concern of the ecological implications of pollution of near-shore waters, this approach must be considered one of the major advancements in the study of the coastal ocean environment.

The approach used by Ketchum considered coastal waters as a mixture of open-ocean water and river inflow. At any distance from either source, the dilution of seawater can be defined by a ratio between the salinity of the open-ocean and the salinity of the coastal water. The approach can be used directly to estimate the dilution of a conservative pollutant. Less direct is the use with a non-conservative pollutant.

It is the intent of this paper to demonstrate a

means of predicting the distribution of non-conservative properties. First, I shall describe the components for the development of the predictive model, and follow by testing the model on data collected in the coastal waters off New England.

My principal aim is to estimate the value of an "estuarine treatment" of coastal waters with reference to one biologically significant compound -- phosphate.

## Models for conservative and non-conservative properties

When freshwater flows into the ocean, a mixture develops that represents a ratio between the two. The amount of freshwater in the mixture can be estimated by measuring the salinity of the mixture ($S_O$) and the salinity of the ocean ($S_R$) neighboring the estuary. The percentage of freshwater (F) is equal to,

$$\frac{S_R - S_o}{S_R} (100) \qquad (1)$$

or the percentage of seawater (S) in the mixture as:

$$\frac{S_O}{S_R} \quad (100) \qquad (2)$$

Units are parts per thousand. By making a series of salinity measurements from the freshwater source to the open ocean, a relationship can be developed (figure 1), which describes the distribution of fresh and oceanwater in the system. If a conservative pollutant (a substance not altered by biochemical activity) enters with freshwater at a uniform rate, then the down stream distribution will follow the curve describing the fresh-oceanwater distribution.

If however, the pollutant is non-conservative (a property whose concentration is altered by biological activity, for example phosphate) then the distribution would not follow the fresh-oceanwater distribution since the concentrations would be altered by biological activity *en route* downstream. The observed concentration of the non-conservative property at any location *en route* would be:

$$N_f = \frac{F\%}{100} \quad (N_{f,r}) - N_u \qquad (3)$$

where $N_f$ would be the concentration at some point down stream from the fresh water source, $N_{f,r}$ would be the concentration of the non-conservative in the fresh-water source and $N_u$ would be the amount removed by biological uptake. $\underline{F}$ is obtained from equation 1. The dimensions of $\underline{N}$ are mass per unit volume.

Thus equation 3 is only valid if there is no contribution by ocean water. The closest example would be a heavily polluted river flowing into phosphate-poor ocean. However the usual case would be for freshwater to carry phosphate to coastal oceanic regions which already have sizable quantities of phosphate present. To counter this situation equation (3) is rewritten for the oceanic contribution ($N_{s,r}$) as;

$$N_s = \frac{S\%}{100}(N_{s,r}) - N_u \qquad (4)$$

where S% is obtained from equation (2).

Thus in a realistic situation any concentration of a non-conservative property ($N_o$) will be determined by fresh and oceanic contributions and biological uptake, hence:

$$N_o = (N_s + N_f) - N_u \qquad (5)$$

and combining with equation (3) and (4) :

$$N_o = \frac{F\%}{100}(N_{f,r}) + \frac{S\%}{100}(N_{s,r}) - N_u \qquad (6)$$

The major difficulty with both, conservative and non-conservative models, is the choice of reference levels, that is, the source concentration. The problem is more severe with the non-conservative model because of the large variability in biological properties. Also, it is difficult to estimate biological uptake $N_u$. In the treatment that follows, I have referenced the phosphate concentrations to the reference salinities of the fresh water and ocean source. $N_u$ values are obtained by a difference between phosphate values computed from the mixing of the two sources and the observed values.

After a short description of the salinity distribution in the coastal waters of New England in April I shall put the model to test.

## Relationship between the distribution of salinity and phosphate

The research of B. H. Ketchum suggested that the coastal mixing processes were similar to those in estuaries and the energy for non-tidal transport was derived from the inflow of freshwater. The question I ask is, can the coastal water distribution of a non-conservative property be explained in the same terms? The equations presented above are the basis for a predictive model of non-conservative distribution which considers the input from two sources. To test the model, I have used data collected in the coastal waters off New England in 1957 (Fig.2) by the Woods Hole Oceanographic Institution (3). To obtain the seasonal extreme, I have used data for the wettest month (April to May, 1957). At this time of year, the water column, thoroughly mixed during the winter, has achieved a degree of thermal stratification and vertical mixing is limited. I have assumed that vertical mixing during this period is unimportant in deciding the level of concentrations of phosphate in the surface waters.

## Salinity

The amount of freshwater in the area seldom exceeded 15% and the source of this water is the major rivers of the region: oceanographic conditions do not permit the entry of freshwater into this area from either the North or the South, and rainfall is equalized by evaporation (1, 3).

Freshwater entry into the coastal region is emphasized by a narrow lens of water of about 31 o/oo (Fig.2), extending from the coast out to 15 miles off shore. As freshwater mixes with oceanic, the shape of the isohaline lines suggests a two layer estuarine flow where the low salinity water at the surface is moving seaward and the high salinity water indrafts near the bottom.

The percentage of freshwater throughout this section (Fig.3) ranges from almost thirteen near-shore to one percent at the oceanic end. The stratification (layering) of the freshwater appears to be pronounced near-shore and off-shore and less pronounced near the middle of the shelf.

## Non-conservative model for phosphate

The reference sources for fresh coastal surface water is 30.8 o/oo and it contains 0.81 ug at./l $PO_4$-P. Curve I (Fig.4) would describe the distribution of phosphate if it were being diluted by a surface

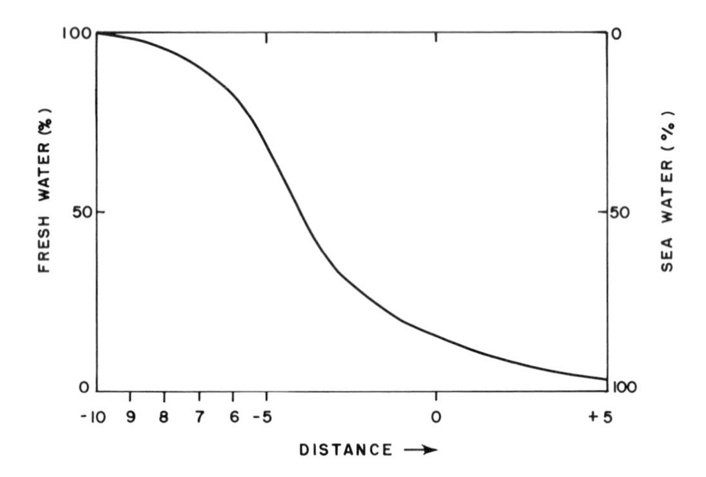


*Figure 1. Schematic of the distribution of fresh and oceanic waters*

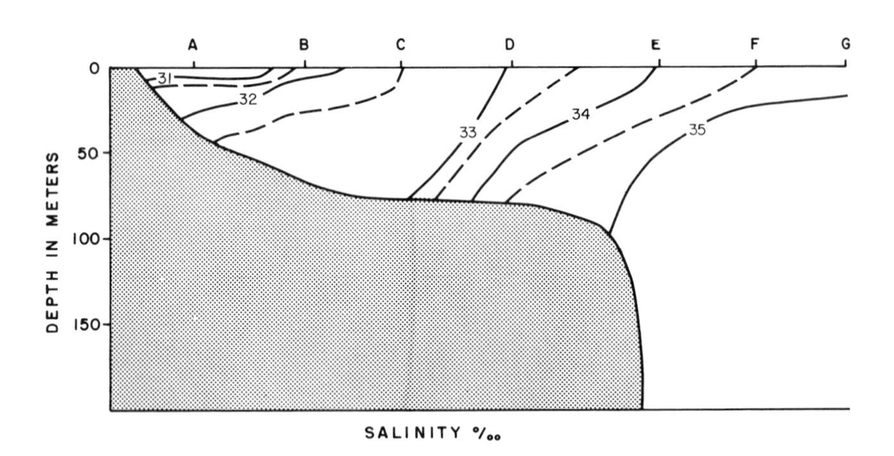


*Figure 2. Distribution of salinity in April–May 1957 across the continental shelf south of Montauk Point Long Island to sea. Reference distance: B → C 15 nautical miles.*

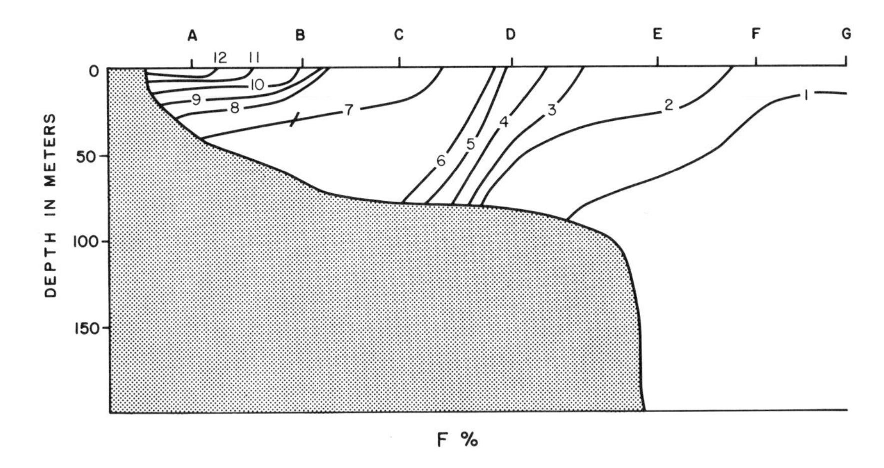


*Figure 3. Distribution of freshwater in April–May 1957 across the continental shelf south of Montauk Point Long Island to sea. Reference distance: B → C 15 nautical miles.*

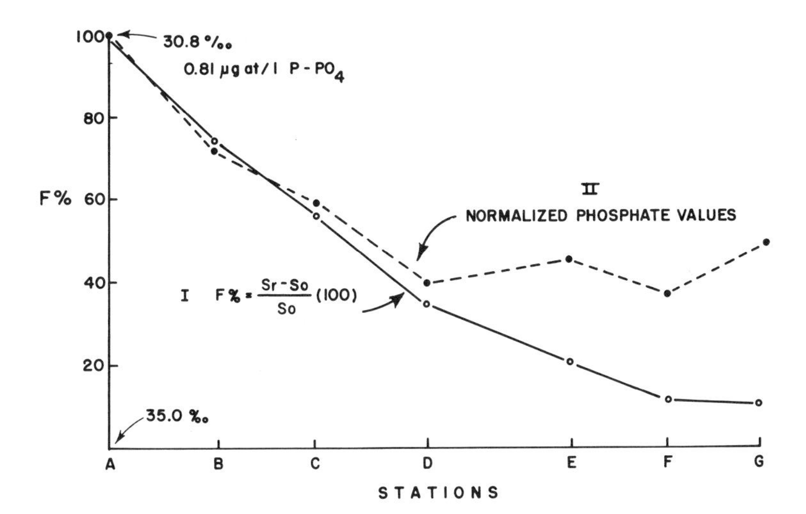


*Figure 4. Distribution of fresh water and phosphate at stations A → G, April–May 1957*

oceanic source having no phosphate and where biological uptake is absent, i. e.

$$N_f = \frac{F\%}{100} (N_{f,r}) \quad (7)$$

where $N_{f,r}$ = 0.81 ug at./l $PO_4$-P

Curve II (Fig. 4) shows the trend of inorganic phosphate measured in surface water at the stations. From the shore to seaward, the initial similarity in the curves suggests that phosphate is behaving as a conservative property: the change in concentration follows the mixing ratio. However, toward the middle of the coastal region, the measured phosphate (curve II) departs from curve I indicating that at the seaward section of the shelf, more phosphate is apparent than can be transported from the coastal freshwater source alone. This can be demonstrated to be an oceanic source as follows. Figure 5, indicates how the two sources interact. From equation (4), the amount of phosphate contributed by the oceanic source declines, moving from station G to A following the percentage of the ocean source in the mixture. Similarly, from equation (3), the phosphate level in the coastal freshwater source decreases moving seaward following the decrease of the fresh coastal source in the mixture. Figure 5 illustrates that the oceanic contribution dominates the concentration at the seaward stations (station E to G) while the coastal fresh source dominates near-shore (stations A to C). At station D, the contributions are approximately equal. By summing the source contributions across the section (Fig.6 curve V) provides an estimate of the distribution of phosphate resulting from the mixture of the two sources but without the effect of biological uptake. From equation 5, the difference between the concentrations predicted by mixing of two sources and that measured by the amount taken up by biological processes (Fig. 6) is estimated.

## Implications

The trend in the concentration of phosphate in the surface waters across the shelf predicted by the model, closely follows the trend in the observed concentration (Fig.6). This argues that spatial changes, i.e. differences between the observed concentrations are largely controlled by the mixing of near-shore and oceanic sources and not to differences in biological

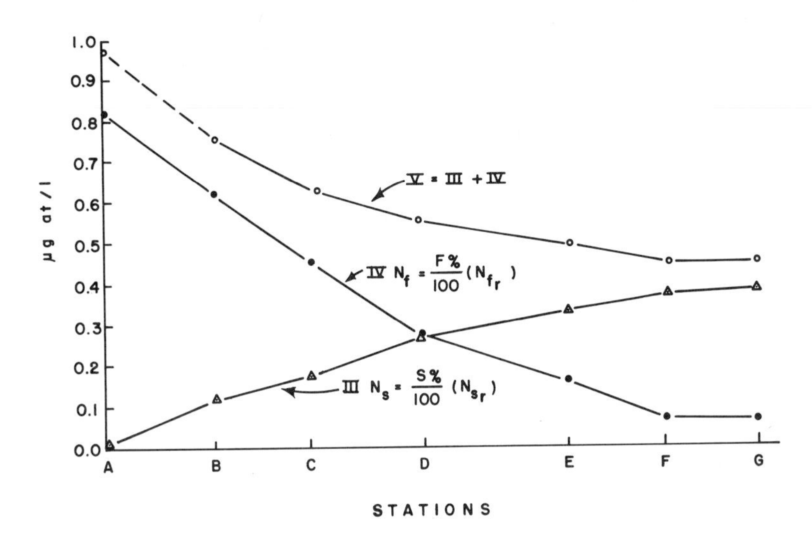


*Figure 5. Distribution of phosphate from near shore and oceanic sources at stations A → G, April–May 1957*

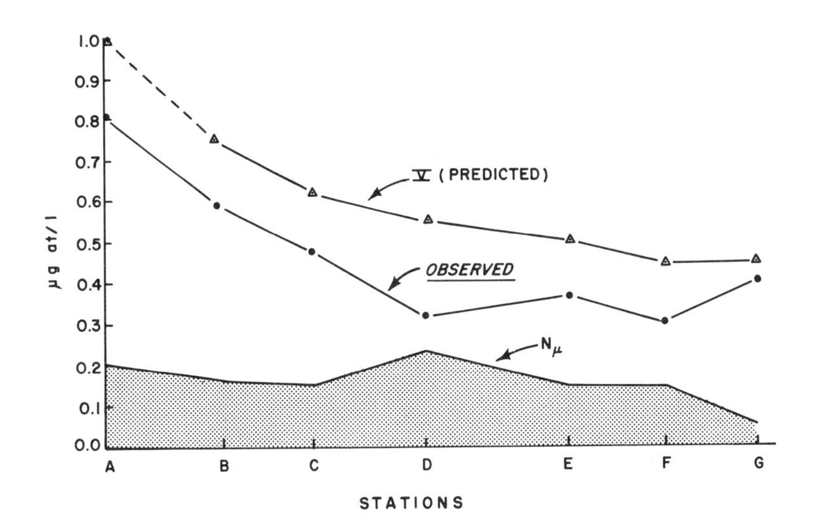


*Figure 6. The concentration of phosphate: V = predicted by sum of phosphate from oceanic and near shore sources; observed = measured phosphate; $N_u$ = concentration taken up by biological processes*

uptake. This is surprising, however it is supported by the rather constant values of estimated phosphate taken up by biology ($N_u$). The consistancy can be explained by the flux of nutrients from the two sources and this is probably one of the reasons for the high productivity of these waters. The validity of the values of $N_u$ are difficult to confirm. They are however, about the same concentration of particulate and dissolved organic phosphate measured in these waters (2).

At this time of year, when freshwater run-off is at its highest, the contribution of phosphate provided by near-shore water is greater than that from the saline ocean source. It seems unlikely that the later can ever become much greater while the water column is thermally stratified (reduced vertical mixing). No similar constraints are imposed on the upper level of phosphate in the near-shore coastal waters. One expects phosphate to increase due to the added sewage load in the rivers. In this case the dominate role of the near-shore waters as major contributor would be firmly decided.

We can ask, what is the tolerable phosphate load for these coastal waters -- What is a reasonable upper limit? Certainly one cannot be satisfied with a limit based on a degree of eutrophication which undersaturates the level of oxygen, things will have gone too far by that time.... What does seem acceptable would be some upper level of water transparency; chlorophyll coupled with species diversity. Deciding limits using these parameters is not easy. To be accurate, the seasonal changes must be understood in detail.

I would argue that if the coastal model which treats mixing from two sources is valid, then the same model can be used to estimate the phosphate load in coastal waters assuming that increases in the freshwater pollutants is forthcoming. Although diversity estimates are difficult, the step from phosphate to a parameter such as transparency or chlorophyll is within the present art of biological oceanography. The errors are large -- but still as a first order estimate, they could provide the basis for the necessary regulation to prevent the coastal oceans from becoming an infinitely polluted estuary.

Abstract

The crux of this paper is to assess the effects of the mixing of near-shore water of low salinity with oceanic waters of high salinity on the distribution of a non-conservative property, namely phosphate. The predictive model uses salinity as a tracer of fresh-water. The proportion of the two sources (inshore and offshore) are expressed as a mixing ratio. It is assumed that phosphate in the two sources mixes following the proportions in each source. The trend of phosphate inshore to offshore predicted, agrees with the concentration measured in the surface, and the model estimates the amount of phosphate taken up by biological processes.

## Literature cited

(1) Ketchum, B.H. and D. J. Keen. 1955. The accumulation of river water over the continental shelf between Cape Cod and Chesapeake Bay. Pap. Mar. Biol. & Oceanogr. Suppl. to Vol. 3, Deep-Sea Res. 346-357.

(2) Bigelow, H. B. 1933. Studies of the waters on the continental shelf, Cape Cod to Chesapeake Bay. I. The cycle of temperature. Papers in Physical Oceanography and Meteorology, Mass. Inst. of Technol. II(4): 1-135.

(3) Woods Hole oceanographic technical report No.61-6 appendix C, 1960.

Bigelow Laboratory Contribution No.75010 4.5.75

This research was supported by the National Science Foundation and the Energy Research Development Administration.

# 38

# Nutrient Budgets in the Hudson River Estuary

H. J. SIMPSON, D. E. HAMMOND, B. L. DECK, and S. C. WILLIAMS

Lamont–Doherty Geological Observatory and Department of Geology, Columbia University, Palisades, N.Y. 10964

The Hudson River forms one of the major estuarine systems of the northeastern United States. The Hudson Estuary, as defined in this paper, includes the entire reach between the mouth of Lower New York Bay and the dam at Green Island a few miles north of Albany (see Figure 1). Sewage from approximately fifteen million people in the New York City area is discharged to the lower Hudson Estuary. Large investments of public funds for sewage treatment in this region have been made over the past half century. By the end of the decade, several billion dollars will have been spent during the 1970s for completion of the few remaining plants in a network of secondary treatment facilities for the City of New York.

The geometry of the lower Hudson Estuary in the zone of major sewage loading is complex, while that upstream of New York City is remarkably simple when compared with a large system such as Chesapeake Bay. The most commonly used location reference system for the Hudson Estuary is statute miles, with the origin located in midchannel opposite the southern tip of Manhattan Island. On this scale, the mouth of the estuary is about mile point (mp) - 15, the seaward entrance of the Narrows leading to Upper New York Bay about mp - 8, and the head of tide about mp + 154 (Figure 1). Semi-diurnal tides of one to two meters occur throughout this reach of approximately 170 miles, but the saline intrusion is always confined below mp 80 even in severe drought years such as those which occurred in the middle 1960s. During high runoff periods typical of early spring or of the first few weeks following major storms, fresh water extends as far south as mp 15 (1).

Nutrient chemistry and the dissolved oxygen distribution throughout much of the saline-intruded reach are dominated by the discharge of enormous volumes of sewage. Levels of sewage treatment range from none for most Manhattan Island outfalls, to crude primary treatment for the major New Jersey outfall, and secondary treatment for most of the rest of New York City. Locations and types of treatment for the major sewage outfalls entering the lower estuary are shown in Figure 2.

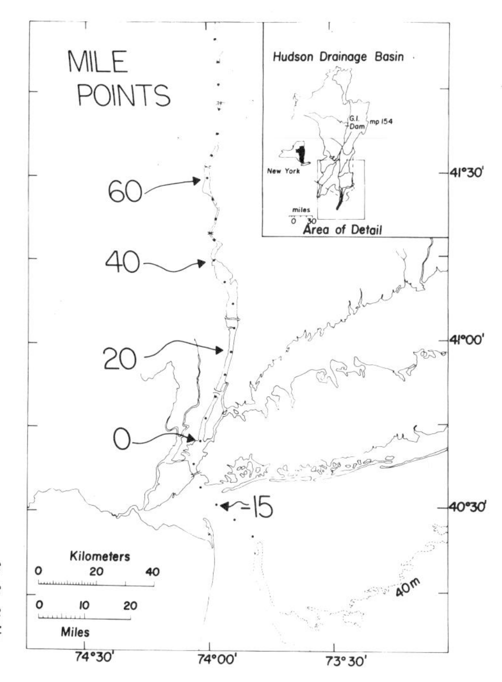


*Figure 1. Reference map for locations within the Hudson Estuary. Distances are given in statute miles with the origin located at the southern tip of Manhattan.*

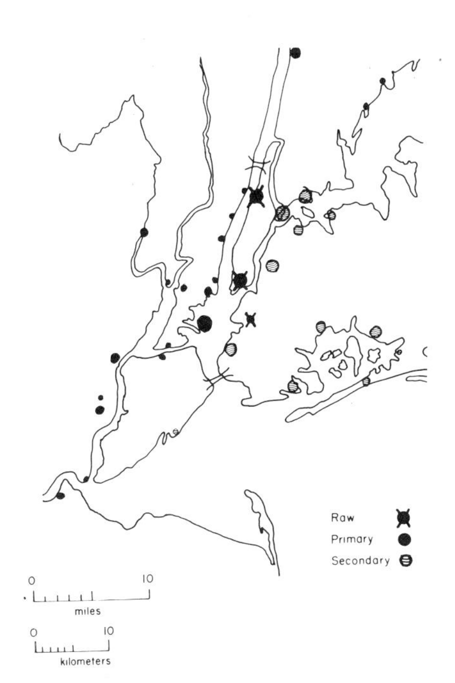


*Figure 2. Major sewage discharges to the Hudson River Estuary near New York City. The area of the circles is proportional to the volume of daily discharge, and the type of sewage treatment is indicated in the legend. The largest discharges are approximately 250 million gal/day, and the smallest discharges shown are approximately 10 million gal/day. The total discharge is somewhat greater than 2 billion gal/day.*

The purpose of this paper is to present a budget description of the loading and flushing characteristics of the Hudson Estuary. This description greatly simplifies the complicated geometry and transport characteristics of the harbor region, and allows first-order behavior of phosphate and silicate to be discussed.

Mixing of sewage with receiving waters is very rapid in the Hudson Estuary and the effects of individual outfalls usually cannot be observed for significant distances from the discharge site. Tidal excursions, particularly through the straits connecting Upper New York Bay with both Long Island Sound (Upper and Lower East River) and Newark Bay (Kill Van Kull) are quite efficient in integrating the effects of the individual outfalls (see Figure 3 for location names). The tidal excursion in the Hudson north of mp 0 is approximately eight miles.

Most of the discharge of sewage-derived pollutants from Upper New York Bay occurs through the Narrows, with transport dominated by the two-layered estuarine circulation primarily focused along the Upper and Lower New York Bay axis. Bottom salinities at least 10-20% greater than surface salinities are maintained throughout the saline-intruded Hudson south of mp 40, even during low fresh water flows. Some pollutants are discharged into western Long Island Sound, largely through the effects of tidal excursions of East River water.

Additional sewage loading occurs seaward of the Narrows, from several large outfalls entering Jamaica Bay and adjacent areas of the Lower Bay, and from Raritan Bay which receives much of the New Jersey industrial wastes discharged to the Arthur Kill and Raritan River.

North of Upper New York Bay, the Hudson has very simple geometry, and can be described reasonably well using one-dimensional presentations. The cross-sectional area of the estuary is nearly constant from mp 0 to mp 80, except for an increase of about 25% between mp 20 and mp 40. Mile point 80 marks the furthest salinity intrusion during drought periods.

Because of this constant cross-sectional area, 4-mile segments of the estuary from mp 80 to mp 0 have comparable volumes. Extending the one-dimensional description of volume as a function of mile point, the harbor between mp 0 and - 8 can be described as two sequential volume elements, with lengths of 4 miles, and volumes of 2 to 2-1/2 times those of the segments north of mp 0. This description represents the Hudson Estuary as a one-dimensional chain of 4-mile boxes, extending from mp 80 to mp - 8, the location where the cross-sectional area increases substantially in Lower New York Bay.

For the purpose of budget calculations, the only sewage outfalls included as inputs to this schematic one-dimensional estuary terminating at the Narrows are those upstream of mp - 8. Approximately two-thirds of the total sewage discharge shown in Figure 2 enters the estuary directly between mp 20 and - 8. Outfalls into Lower New York Bay can influence the estuary upstream of their input, and this effect is included here by specifying the seaward boundary condition to include their contribution. Outfalls to the Upper East River are assumed to leave the system through Long Island Sound. The rate of direct sewage discharge is approximately 60 $m^3$/sec (see Figure 4), which is about 10% of the annual average fresh water discharge of the Hudson River (∿ 550 $m^3$/sec). During low flow, the sewage discharge is more than 30% of the total fresh water entering the lower estuary.

Phosphate Distribution

The spatial distribution of molybdate-reactive phosphate in the Hudson Estuary is simple and indicates the effective mixing of sewage pollutants within the harbor area. Systematic vertical gradients of phosphate within the harbor are usually not observed, except near the seaward end where low-phosphate water enters in the net upstream transport near the bottom. Upstream of the major loading, vertical gradients of phosphate closely follow those in salinity, with lower layer concentrations of both phosphate and salinity 10-40% higher than upper layer concentrations. Figure 5 shows the variation of phosphate along the axis of the Hudson for two types of flow regime, one of high discharge (spring) and one of low discharge (late summer). In each period, the highest values are observed in the harbor (between mp + 4 and - 8), with approximately linear decreases occurring both upstream and downstream of the harbor. Maximum phosphate values observed are 5 - 6 µM during low flow and about 2 µM during high flow.

Plotting phosphate as a function of salinity (Figure 6) removes much of the scatter observed in the plot of phosphate vs. distance. The trend of the phosphate data with salinity in Figure 6 is presented in terms of two line segments intersecting at the maximum phosphate concentration. Using the data points from only a single survey, the trends can sometimes be better represented if a third horizontal line segment (constant $PO_4^{3-}$) in the major loading zone is included, connecting the two primary lines sloping away from the region of maximum concentrations. Because the primary purpose of this paper is to present the first order behavior of phosphate in the Hudson, details of loading geometry and mixing within the harbor region have been purposefully excluded to simplify the discussion.

Considering that a dominant fraction of the total loading above the Narrows is supplied to the Upper Bay between mp 0 and - 4 (Figure 4), and that the observed phosphate-salinity relationship upstream of the major inputs is essentially linear during low-flow summer months (Figure 6), to a first approximation, phosphate is a conservative property in the saline reach of the Hudson upstream of the zone of major loading. Thus, the phosphate distribution is dominated by the balance between sewage inputs and water transport within and out of the system. This situation is substantially

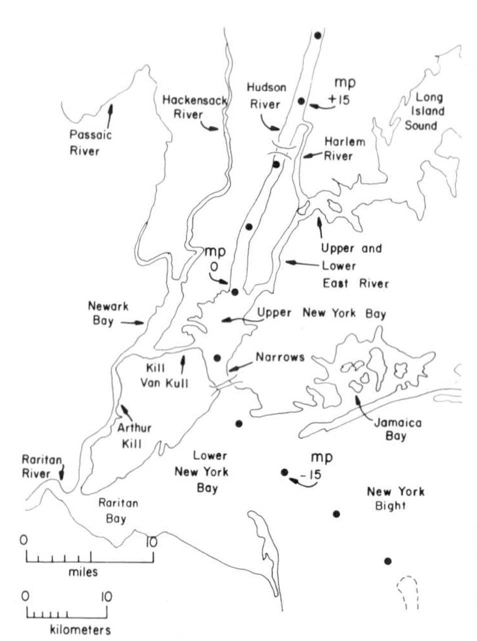


*Figure 3. Location names within the Harbor and Lower Bay region of the Hudson Estuary. More than ¾ of the mean annual fresh water discharge is supplied by the Hudson River. Half of the remainder is supplied by the Passaic and Raritan Rivers and half by sewage discharge.*

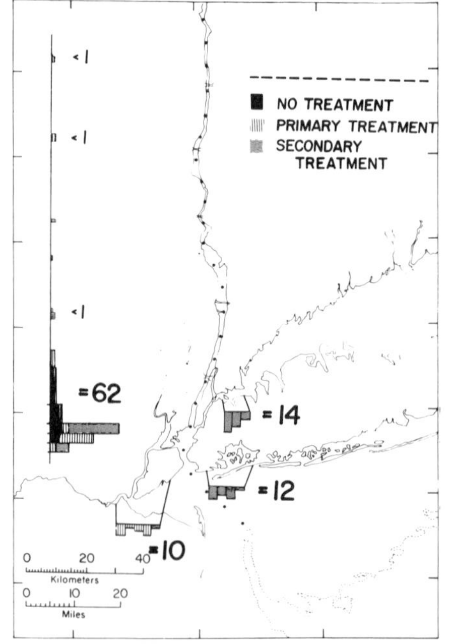


*Figure 4. Volumes of sewage discharged to the Hudson Estuary in cu. m/sec. The total volume included in a one-dimensional model with the seaward end at the Narrows equals 62 $m^3/sec$ while discharges to Jamaica Bay (12 $m^3/sec$), Arthur Kill, and Raritan Bay (10 $m^3/sec$) enter downstream of the Narrows. Discharges to the Upper East River (14 $m^3/sec$) are assumed to leave the system* via *Long Island Sound.*

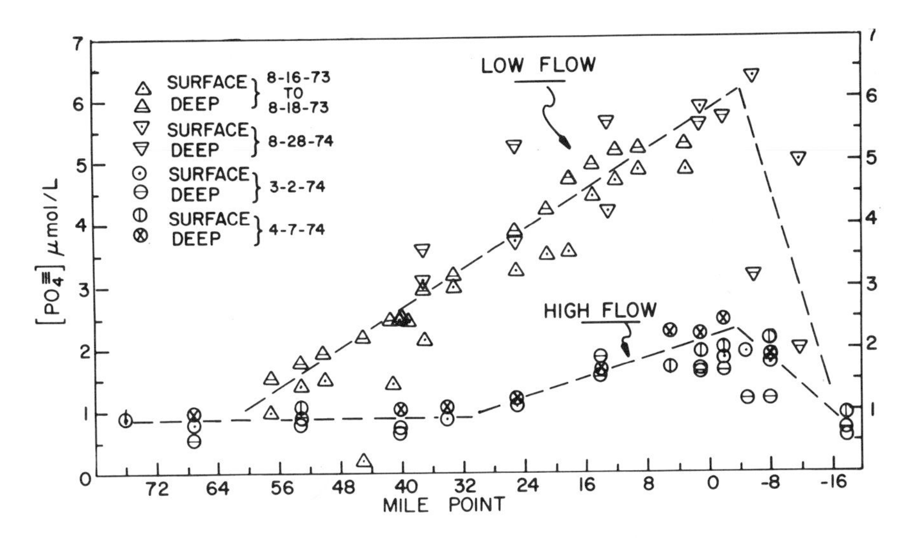


*Figure 5. Molybdate-reactive phosphate as a function of location in the Hudson Estuary. Maximum values during both high flow (March and April) and low flow (August) are within Upper New York Bay (mp 0 to −8) and the first few miles of the Hudson opposite Manhattan (mp 12 to 0).*

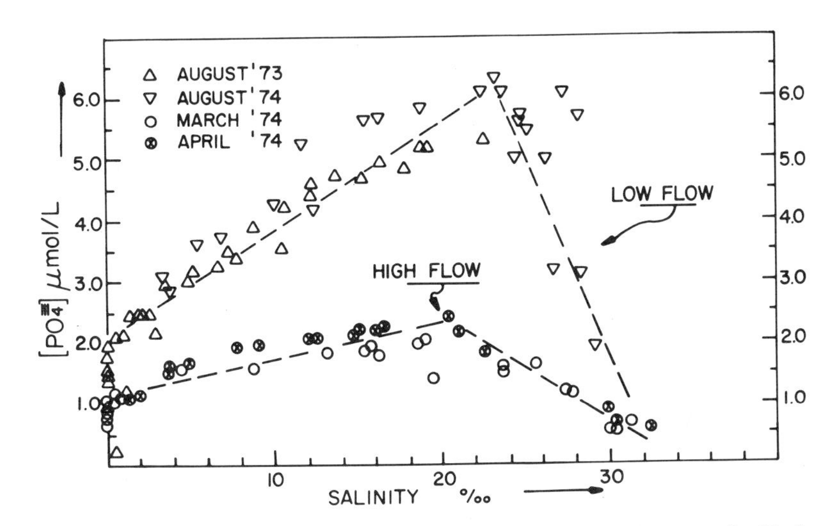


*Figure 6. Molybdate-reactive phosphate as a function of salinity in the Hudson Estuary. These are data from Figure 5 plus samples from mid-depths. Average salinities within Upper New York Bay during high flow (1200 $m^3/sec$) are lower (15–20‰) than during low flow ($\leq$ 300 $m^3/sec$) when they range between 20‰ and 26‰.*

different from that found for most polluted estuaries and lakes where the dominant factors controlling phosphate distribution are sewage inputs and biological activity. Ketchum (2) has previously described total phosphorus as a conservative property in the Hudson on the basis of one set of low flow samples from 1964. Our observation, based on a number of surveys, is that the molybdate-reactive phosphate distribution within the Hudson shows an even simpler and more uniform relationship to salinity than does total phosphorus, due to the variability in particulate phosphate probably resulting from resuspension of bottom sediments.

Typical values of particulate phosphorus in the Upper Bay were 30% of the molybdate-reactive phosphate concentrations, with much greater variability about the mean. In a few near-bottom samples, particulate phosphorus made up about 50% of the total phosphorus content. Dissolved organic phosphorus was always a small fraction of either of the two major phosphorus components discussed above. Considering the evidence that phosphate behaves conservatively during midsummer low-flow conditions (upstream of the major loading) it is reasonable as a first approximation to treat phosphate as conservative within the harbor during all flow conditions.

Our estimate of the total rate of addition of phosphate to the Hudson above the Narrows (excluding the Upper East River) is 6-7 moles/sec. Removal of phosphate from the system can be described in terms of a two-layer advective transport from the seawardmost 4-mile segment of the model Hudson discussed earlier. During both high and low flows, the mean salinity of this segment is approximately 90% of the mean salinity of the bottom water entering the Narrows, which usually varies from 22°/₀₀ to 30°/₀₀ depending on the freshwater discharge. Assuming no net transport of salt through the Narrows, and phosphate concentrations of 1.65 μM in the landward-flowing bottom layer during high flow periods and 3.0 μM during low flow periods, the net seaward fluxes of phosphate are 6.4 moles/sec and 6.7 moles/sec. The volume of this seawardmost 4-mile model segment is 0.22 $km^3$, and the mean residence times of phosphate within the segment are 0.8 and 2.0 days during high and low flows respectively.

A simple one-box description of the harbor can be constructed to illustrate the magnitude of the terms in the phosphate budget (see Figure 7) during both high and low flow periods. The volume of the box is chosen to include the region of maximum phosphate concentration. The flux out of the box is assumed equal to the advective loss rate calculated above. The dominant input (4.7 moles/sec) is sewage discharge, which was computed from a careful inventory of effluent volume and phosphate concentration data for all major sewage inputs (3). Some of the inventory inputs are not well constrained, especially the huge primary-treated outfall from New Jersey which enters Upper New York Bay, but we believe that we have established the sewage loading of phosphate above the Narrows to ± 25%.

Three smaller input fluxes have been included: [1] freshwater transport downstream from the region of the Hudson of sewage loading from the New York City area, [2] an estimate of the phosphate flux from the sediment interstitial waters (0.4 and 0.8 mole/sec), [3] an estimate of the input from particulate organic phosphorus oxidation in the water column (0.4 and 0.8 mole/sec). Neither the sediment flux, nor the particulate oxidation input are well constrained. The former is based on one direct field measurement of sediment phosphate flux and the latter is based on an estimated oxygen budget. Both should be accurate within a factor of two or three which is sufficient to establish them as minor relative to direct sewage input. The difference in sediment and particulate oxidation flux estimates for high and low flow conditions is primarily due to the temperature dependence of molecular diffusion coefficients and biological activity.

Some estimate of the magnitude of the rate of phosphate uptake by phytoplankton can be made from primary productivity data discussed below, using carbon to phosphorus ratios typical of marine plankton. During low flow the estimated uptake rate is comparable to the total sewage phosphate loading rate of $\sim$ 5 moles/sec. In high flow, uptake rates within the harbor region are less than 5% of those estimated for summer low flow conditions. Considering the efficient recycling of phosphate by planktonic systems, and the generation of phosphate from the sewage particulates discharged to the Hudson, it is not surprising that planktonic uptake of phosphate has essentially no measurable impact on plots of phosphate vs. salinity.

On the basis of a one-box harbor model, the steady-state-weighted average residence time of phosphate discharged by sewage to the Hudson Estuary is about 2 days in high flow and 7 days in low flow. Using a similar approach, the times required for the removal through the Narrows of upstream-transported sewage phosphate are given in Figure 8. Thus, phosphate from the upstream saline-intruded region during low flow can take on the order of several months to be removed. In contrast, during high flow the maximum removal time is on the order of a week. In both flow regimes, sewage phosphate near the upstream end of the saline intrusion represents a very small fraction of the total sewage discharge, most of which is removed over much shorter time scales (Figure 7).

## Silicate Distribution

There is extensive literature on the behavior of silicate within estuaries. Two recent treatments, Boyle et al. (4) and Peterson et al. (5), have summarized arguments about the relative importance of conservative and nonconservative mixing between the high-silicate fresh water endmember and low-silicate saline endmember. In general, most of the controversy has focused on [1] if there is, or is not sufficient evidence to indicate a sink for silicate within most estuaries; and [2] if such a sink exists,

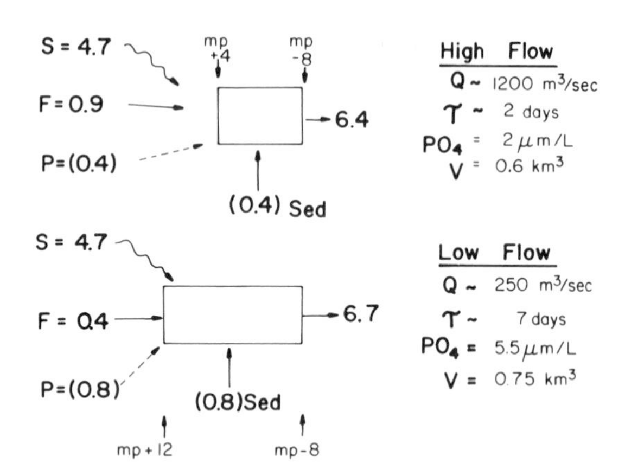


*Figure 7. One-box model for molybdate-reactive phosphate in the lower Hudson Estuary. Input from sewage (S) averages 4.7 mol/sec, and discharge through the Narrows averages 6–7 mol/sec. Lesser contributions are provided by fresh water advective input of phosphate (F), diffusive flux from the harbor sediments (sed), and oxidation of particulate organic phosphorus (P).*

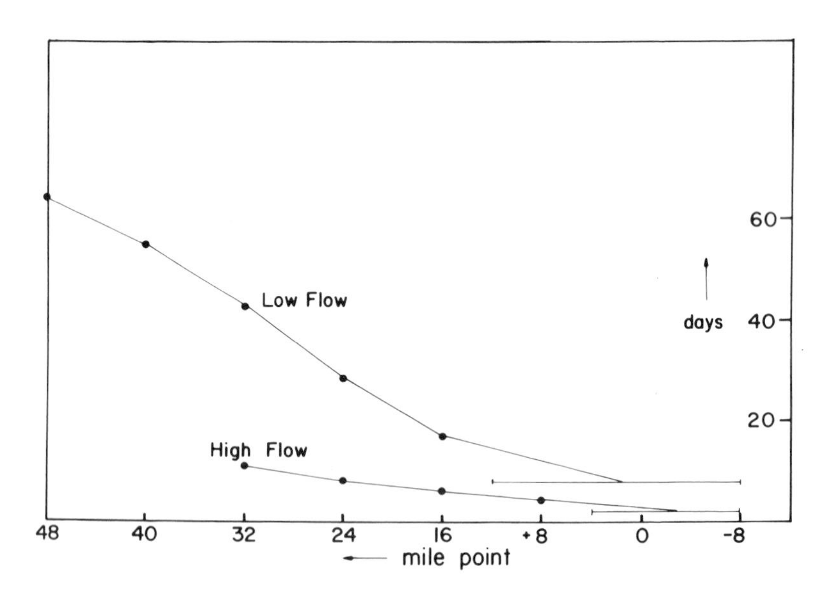


*Figure 8. One-dimensional advective removal times for sewage-derived phosphate within the Hudson Estuary. The points indicate the average time for phosphate within 8-mile segments of the saline-intruded reach to pass beyond the Narrows. The size of the boxes used for the one-box harbor model are indicated at the seaward end of each curve.*

is it dominated by inorganic silicate reactions resulting from the interaction of river-borne clay minerals with high ionic strength saline water, or is the sink primarily due to biological uptake by diatom communities?

The distribution of molybdate-reactive silicate within the Hudson Estuary is very unusual. During a survey in October 1974, silicate was observed to increase with salinity through most of the salt intrusion indicating an internal source rather than a sink (Figure 9). This type of distribution was specifically mentioned by Peterson et al. (5) as the one type of silicate mixing trend for which they had not seen evidence. Figure 10 illustrates the more commonly observed types of silicate profiles in estuaries, indicating either conservative mixing or the existence of an internal sink.

Silicate vs. salinity is plotted again in Figure 11 along with phosphate vs. salinity for the same set of samples. The similarity in distribution indicates spatially similar sources. Two possible sources of this anomaly are the Passaic-Hackensack River flow and sewage. Two samples of the former showed substantially greater values of silicate than the Hudson, as did a sample of one raw sewage outfall on Manhattan's west side. The trend of the phosphate vs. salinity plot for the October survey (Figure 11) is somewhat different than for the low flow August surveys (Figure 6). This may be in part the result of lack of sufficient sampling density in the region of maximum phosphate concentration, but it probably is also the result of distinctly nonequilibrium flow conditions. Fresh water flow gauged at Green Island during the month prior to the survey averaged about twice the August flows and showed four distinct pulses of increased flow, indicating that a fresh water surge was probably passing through the system during our survey.

Aside from the special circumstances of a large internal silicate source, the Hudson Estuary has another unusual feature in its silicate dynamics. The Hudson has an extremely long fresh water tidal reach extending in low flow from about mp + 154 south to mp + 60, and in very high flow to mp + 15. Thus, fresh water has a very long travel time once entering tidal water before it reaches the saline intrusion. As a result, the fresh water tidal Hudson is more like a lake than a river during low-flow conditions (see reference 6 for a similar description), and water may often require substantially more than a month to reach the saline portion of the estuary. During this passage, the diatom community, and whatever other silicate consumers are important in the fresh water reach, have ample time to withdraw significant quantities of dissolved silicate long before salt water is reached.

Figure 12 shows the results of three silicate surveys in the Hudson, each successive survey made during higher fresh water flow from October through January. During each sampling period, there is a major drop in dissolved silicate concentrations along the axis of the tidal Hudson, but the drop is not related to the

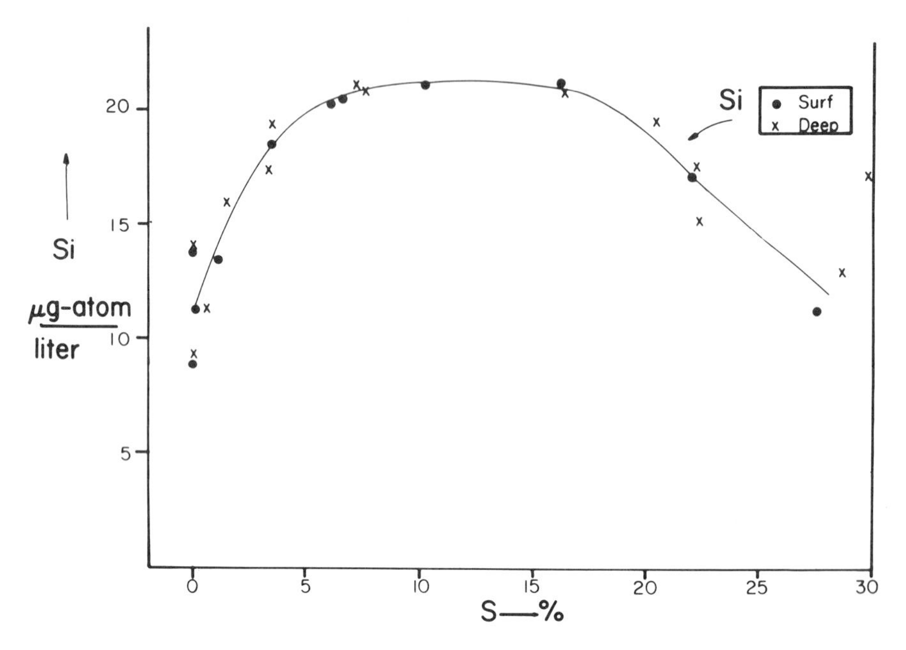


*Figure 9. Silicate as a function of salinity in the Hudson Estuary. The data were collected during early October 1974, a period of moderately low fresh water flow (~ 400 $m^3/sec$). No systematic depletion of surface samples relative to deep samples was observed, and a significant source of dissolved silicate is indicated within the salinity gradient.*

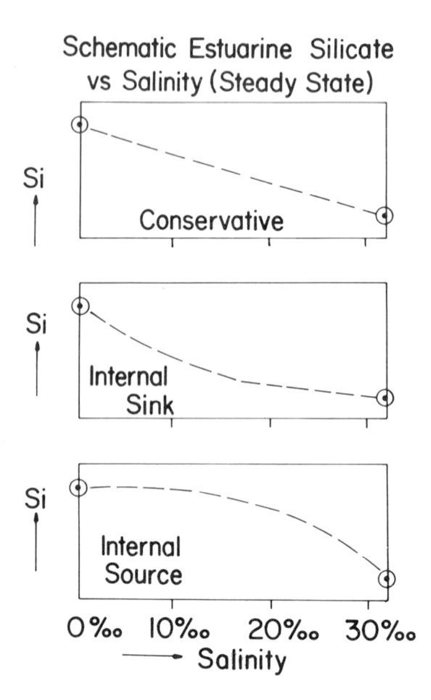


*Figure 10. Possible silicate distributions within estuaries. The most commonly observed patterns are shown in the top two sketches while that shown in the bottom sketch is not typical of most estuarine systems.*

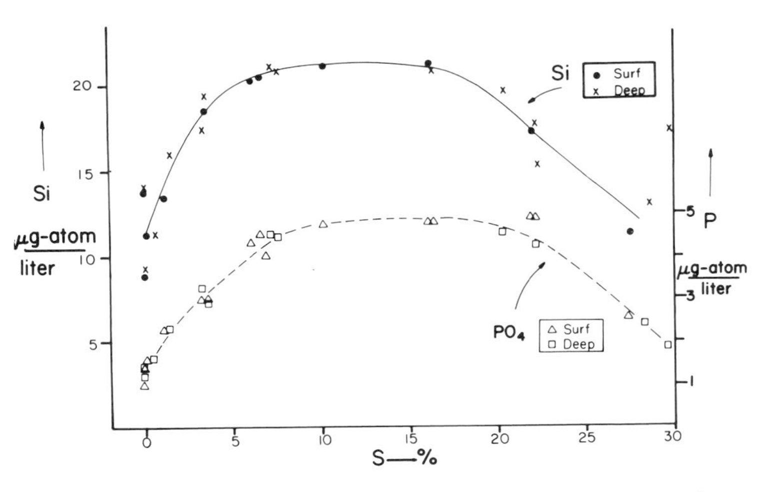


*Figure 11. Phosphate and silicate in the Hudson Estuary. The similarity of the patterns of these two nutrients within the Hudson in October 1974 indicates that sewage discharge can have a substantial impact on the distribution of silicate within polluted estuaries.*

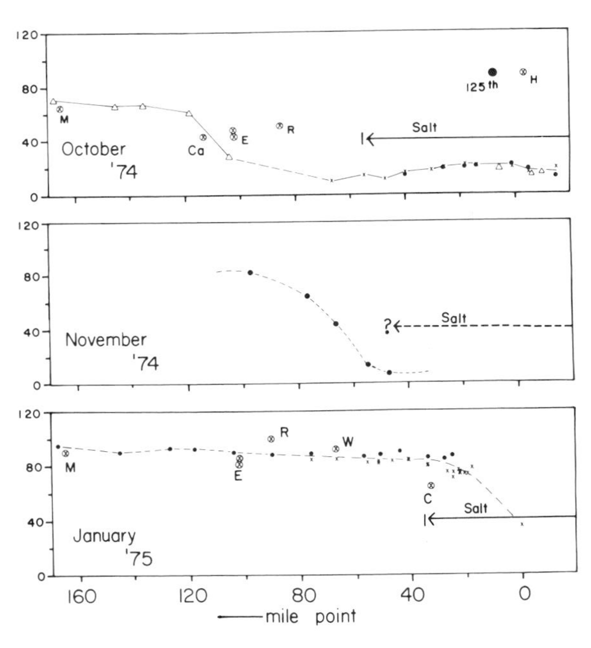


*Figure 12. Silicate distribution as a function of location along the axis of the tidal Hudson River. Points connected by lines represent samples from the Hudson River while those indicated with ◯ are freshwater tributaries. Tributaries samples include the Mohawk (M), Catskill (Ca), Esopus (E), Rondout (R), Walkill (W), Hackensack (H), and a raw sewage outfall from Manhattan (125th Street). These tributaries are all included on the drainage basin insert map in Figure 1.*

presence or absence of salt water. The location of the zone where significant silicate depletion is observed is probably controlled by a combination of longitudinal water transport time and the rate of silicate uptake. During warm low-flow, high-light periods, much of the silicate uptake occurs far upstream of the salinity intrusion. During cold, high-flow, low-light periods, significant depletion may not occur at all resulting in a conservative mixing line between the low-silicate offshore water and high-silicate fresh water.

The effect of sewage-derived silicate is small compared to the magnitude of diatom uptake in the fresh water region during low flow, as indicated by the small "source" anomaly in the October 1974 data when plotted on a more condensed concentration scale (Figure 12) than used in Figures 9 and 11.

## Conclusions

The behavior of molybdate-reactive phosphate and silicate within the Hudson Estuary are quite unusual relative to other large estuarine systems. Phosphate distributions are dominated by the enormous sewage supply rate and by transport within and removal from the system by estuarine circulation. The absolute concentrations of molybdate-reactive phosphate in the Hudson (2 - 6 μM) are intermediate between the low values of Chesapeake Bay (< 0.2 μM (7)) and the high values of the Potomac River downstream of Washington, D.C. (up to > 20 μM (8)). The absolute loading rate of phosphate to the Hudson Estuary is substantially higher than for either the Potomac River or Chesapeake Bay, but the removal rate by physical transport is more rapid than for those two estuarine systems.

The current direction of phosphate management in North America includes reduction or removal of phosphate from detergents on a short-term basis, and the introduction of tertiary sewage treatment for phosphate removal over the longer term. These policies are based on the generally-accepted premise that the state of eutrophy of most fresh water lakes can be clearly related to the loading rate of phosphorus (9). Observations of algal standing crop and primary productivity rates in the Potomac Estuary indicate that similar considerations are probably important for nutrient management for some estuaries (8). However, simple extension of these policies for management of lakes to all estuaries may not be justified.

A comparison between Lake Erie and the Hudson illustrates this point. Both systems receive sewage effluent from about $10^7$ people. The phosphorus loading rates are comparable, although Lake Erie has a somewhat greater rate per capita due to agricultural activities and the existence of a detergent ban in New York State. The budgets for phosphorus are summarized in Table I. One of the major water quality problems in Lake Erie is the existence of nuisance levels of algal standing crops. The high efficiency of

I

Comparison of Phosphate Behavior in the Lower Hudson Estuary and Lake Erie

| | Hudson | Lake Erie |
|---|---|---|
| Volume | 0.6 $km^3$(1) | 500 $km^3$ |
| Area | 60 $km^2$ | 25,000 $km^2$ |
| Mean Depth | 10 m | 20 m |
| Residence Time | 2 days | 1000 days |
| Basin Population | $10^7$(2) | $10^7$ |
| Σ P Loading Rate | - | 26 mole/sec (3) |
| $PO_4^{3-}$ Loading Rate | 6 mole/sec | 13 mole/sec(4) |
| Predicted $PO_4^{3-}$ (5) | 2 μmole/ℓ | 2.5 μmole/ℓ |
| Measured $PO_4^{3-}$ | 2 μmole/ℓ(1) | .03-0.6 μmole/ℓ(6) |
| Measured Σ P | 2.7 μg- atom/ℓ | 0.6-1.2 μg-atom/ℓ (7) |
| $\frac{\text{Measured } PO_4^{3-} \times 100\%}{\text{Predicted } PO_4^{3-} \times 100\%}$ | 100% | 1-25%(6) |

(1) High flow conditions (see Figure 7)
(2) Only populations discharging above the Narrows and not to the upper East River is included (∿ 65% of total in basin).
(3) See reference 10.
(4) Assuming $PO_4^{3-}$ = 50% of Σ P loading rate.
(5) Assuming conservative behavior for $PO_4^{3-}$.
(6) Summer epilimnion near the lower value, winter epilimnion near the upper value. See reference 11.
(7) See reference 11.

phosphorus uptake by algae indicates that reduction of the rate of sewage phosphorus loading is a reasonable way to attack this problem. In the Hudson, algal populations are not usually considered to be a nuisance, and the efficiency of phosphorus removal by algae is quite low (Table I), despite the rather high primary productivity rates (1 - 2 grams $C/M^2$-day) which are observed during low-flow, late-summer periods (T. Malone, personal communication).

The cause of the low removal efficiency of phosphate is poorly understood, but the rate of primary productivity is clearly not phosphate limited in the Hudson Estuary at the present time. High concentrations of sewage-derived nitrogen in the estuary indicate that this nutrient also does not currently limit productivity. Malone (personal communication) has suggested that the standing algal crop may currently be limited by the amount of light penetration through the silt-laden waters or by cropping of algal cells by rapid flushing of the estuary. Control of productivity by these factors does not imply that sewage nutrients have not enhanced productivity nor that the Hudson in its pristine state was not nutrient limited by either N or P. Analysis of the current situation does, however, indicate that introduction of phosphate removal to sewage treatment processes in the lower Hudson should probably not be given first priority at the present time.

There are water quality problems in the Hudson which could be expected to respond immediately to improved sewage treatment. Dissolved oxygen levels are less than 40% of saturation in Upper New York Bay during summer months, and organic detritus contributes to a major siltation problem within the harbor and to the turbid appearance of the water. The most obvious current unresolved sewage treatment problem in the lower Hudson (aside from the raw sewage discharge from Manhattan which will cease during dry weather conditions when present construction projects are completed near the end of the decade) is discharge from the Passaic Valley of New Jersey. This large sewage volume has only very crude primary treatment, and constitutes the largest single-source impact on the oxygen budget of the lower Hudson Estuary.

There is the possibility that completion of secondary treatment plants will reduce suspended particulate levels sufficiently to enhance algal productivity to nuisance levels. If such a situation develops, construction of nutrient removal facilities should then be given serious consideration. Considering the major investments of public funds required, the most reasonable policy would seem to be to complete all major secondary treatment plants and then re-examine the critical factors controlling water quality in the lower Hudson Estuary. In any case, phosphate concentration in the present Hudson Estuary is nearly conservative, in distinct contrast to Lake Erie or the Potomac.

A second major conclusion of this paper concerns the observations of silicate behavior. The removal of silicate within the tidal Hudson is apparently dominated by biological uptake. Because of the long residence time of water upstream of the salt intrusion, organisms can significantly reduce the ambient silicate levels during low-flow periods. Thus, nonconservative behavior of silicate in the Hudson is most clearly related to the length of water transport time and the biological removal rate, and appears to have no significant relationship to ambient salinity. Another unusual feature of silicate distribution in the Hudson is the importance of a source term within the waters of New York Harbor, which may be due to high sewage silicate levels.

## Analytical Methods

Molybdate-reactive phosphate was determined by the method of Murphy and Riley (12), with modifications suggested by Staiton et al. (13). Samples were filtered through washed Whatman GF/C filters and were usually stored at 2°C prior to analysis.

Particulate phosphate was measured by firing glass fiber filters at 700°C for 1/2 hour, followed by leaching with HCl at 104°C, neutralizing with NaOH, and analysis for molybdate-reactive phosphate.

Molybdate-reactive silicate was determined by the method of Strickland and Parsons (14).

## Abstract

The dominant source of dissolved inorganic phosphate to the lower Hudson Estuary is sewage from the New York City area. Essentially linear phosphate-salinity relationships are observed over tens of miles both upstream and downstream of the major sewage loading, indicating conservative phosphate behavior on the time scale of removal from the system, which is a few days during high fresh water flow and a week or more during low flow. Additional phosphate sources and sinks, including release of phosphate from particulates and net phytoplankton uptake, are minor compared with direct sewage discharge. Little immediate improvement in water quality would be expected from the introduction to currently operating secondary plants of tertiary sewage treatment for phosphate removal.

Silicate distributions in the Hudson are also quite unusual, showing substantial silicate removal during low flow in the tidal fresh water Hudson. This suggests that biological uptake rather than a change in ionic strength is primarily responsible for negative deviations from conservative behavior in some estuaries. Silicate-rich sewage may produce a positive deviation from a conservative mixing line on a plot of silicate vs. salinity during some periods in the lower Hudson Estuary.

Literature Cited

1. Simpson, H.J, R. Bopp, and D. Thurber, "Salt Movement Patterns in the Lower Hudson", in Third Symposium on Hudson River Ecology, ed. G.P. Howells and G.J. Lauer, Hudson River Environmental Society, New York, 1974.
2. Ketchum, B.H., "Eutrophication of Estuaries", in Eutrophication Causes, Consequences, Correctives, pp. 197-209, National Academy of Sciences, Washington, D.C., 1969.
3. Hammond, D.E., "Dissolved Gases and Kinetic Processes in the Hudson River Estuary", Ph.D. Thesis, Columbia University, New York, New York, 1975.
4. Boyle, E., R. Collier, A.T. Dengler, J.M. Edmond, A.G. Ng, and R.F. Stallard, Geochim. Cosmochim. Acta, (1974), 38, 1719-1728.
5. Peterson, D.H., T.J. Conomos, W.W. Broenkow, E.P. Scrivan, Recent Advances in Estuarine Research, (to be published in 1975).
6. Howells, G.P., T.J. Kneipe, and M. Eisenbud, Environmental Science and Technology, (1970), 4, 26-35.
7. Carpenter, J.H., D.W. Pritchard, and R.C. Whaley, "Observations of Eutrophication and Nutrient Cycles in Some Coastal Plain Estuaries", in Eutrophication: Causes, Consequences, Correctives, pp. 210-221, National Academy of Sciences, Washington, D.C., 1969.
8. Jaworski, N.A., D.W. Lear, Jr., and O. Villa, Jr., "Nutrient Management in the Potomac Estuary", in Nutrients and Eutrophication, ed. G.E. Likens, Special Symposia, vol. 1, pp. 246-273, American Society of Limnology and Oceanography, 1972.
9. Vollenweider, R.A., Organization for Economic Cooperation and Development (Paris), technical report DAS/CSI 68.27, 1968.
10. Vollenweider, R.A., A. Munawar, and P. Stadelmann, J. Fish. Res. Bd. Can., (1974), 31, 739-762.
11. Dobson, H.F.H., M. Gilbertson, and P. Gisly, J. Fish. Res. Res. Bd. Can., (1974), 31, 731-738.
12. Sirois, D.L., "Community Metabolism and Water Quality in the Lower Hudson River Estuary", Third Symposium on Hudson River Ecology, ed. G.P. Howells and G.J. Lauer, Hudson River Environmental Society, New York, 1974.
13. Murphy, J., and J.P. Riley, Anal. Chim. Acta, (1962), 27, 31-36.
14. Stainton, M.P., M.J. Capel, and F.A.J. Armstrong, "The Chemical Analysis of Fresh Water", Miscellaneous Special Publication, No. 25, Fisheries Research Board of Canada, Ottawa, 1974.
15. Strickland, J.D.H., and T.R. Parsons, "A Practical Handbook of Seawater Analysis", pp. 65-70, Fisheries Research Board of Canada, Ottawa, 1968.

Acknowledgements

A number of ambiguities and inaccuracies were excised from earlier drafts as the result of careful reviewing by R. Bopp, W. Broecker, D. Schindler and T. Takahahsi. Any mistakes and questionable opinions which remain are due to the stubbornness of the authors and approval by the above reviewers is not implied. Ship time for collection of samples during March and April of 1974 was provided by J. Chute of the University Institute of Oceanography of the City University of New York. Sample collection for the January 1975 tributary silicate survey was made by R. Bopp. Financial support for the latter stages of this study was provided by the Environmental Protection Agency under contract number R803113-01.

# 39

# The Seasonal Variation in Sources, Concentrations, and Impacts of Ammonium in the New York Bight Apex

HAROLD B. O'CONNORS and IVER W. DUEDALL

Marine Sciences Research Center, State University of New York, Stony Brook, N.Y. 11794

The flux and concentration of plant nutrients are important factors in controlling the productivity of marine waters (1, 2, 3). Evidence from both laboratory and field measurements have indicated that, while $NH_4^+$, urea, $NO_3^-$, and $NO_2^-$ nitrogen may all be utilized by phytoplankton for uptake and growth, $NH_4^+$ has generally been found to be the preferred form and may inhibit phytoplankton uptake of $NO_3^-$ when present in concentrations above trace amounts (1, 2). Because of the potential impact of added $NH_4^+$ on phytoplankton production, standing crop spatial distribution, and species abundance, it is desirable that information be obtained on its rates of input and concentrations in coastal waters.

In this paper we focus on the input of $NH_4^+$ into the New York Bight apex (Fig. 1). Two sources of $NH_4^+$ input have been identified: (1) sewage effluent which is continuously discharged in large volumes into the receiving waters which surround the New York metropolitan region and (2) sewage sludge which is transported to sea in barges and dumped at a location about 25 km from the entrance to New York Harbor. These wastes introduce $NH_4^+$ to the apex at a combined rate that greatly exceeds the $NH_4^+$ input from the Hudson River.

---

Contribution 123 of the Marine Sciences Research Center (MSRC) of the State University of New York at Stony Brook.

## $NH_4^+$ Input due to Advective Processes

Sewage Effluent. A large number of sewage treatment plants (Fig. 2) are situated throughout the entire New York metropolitan region. Combined, they discharge about 50 $m^3$ $sec^{-1}$ of primary and secondary treated sewage effluent (4). In addition, about 13 $m^3$ $sec^{-1}$ of raw sewage, for which no treatment facilities are available, is directly discharged into receiving waters (T. Glenn, personal communication). Table I summarizes the mean concentrations and rates of inputs of sewage derived nutrients, including organic nitrogen

These concentrations and input rates may show short term variability because the combined sanitary sewer and storm-drain system, which is widely used throughout the metropolitan region, will cause sizable amounts of untreated sewage to bypass treatment plants during periods of heavy rainfall. Under these conditions, the raw sewage is discharged directly into the local receiving waters (5). It is likely, however, that these rain-caused episodes do not introduce much variation in the mean concentration of $NH_4^+$ or the volume of flow from the sewage system for periods longer than several months (see Nutrient Relations, page 4).

A large but unknown fraction of sewage derived nutrients may reach the New York Bight apex (Fig. 1) in a matter of days. The effects of the Hudson River discharge, tidal oscillations, variable coastal circulation patterns, biological processes, and the possibility of variations in rates of input, combine to control the relative composition and rate of flux of nutrients into the Bight apex. To obtain the necessary data needed to estimate the advective flux of $NH_4^+$ and other nutrients into the apex, we (6) investigated the seasonal and tidal variation of nutrients and other oceanographic variables at several stations on a transect (see Fig. 3) between Sandy Hook, New Jersey, and Rockaway Point, New York. The cruises were conducted on November 5, 1973 January 22, March 11, April 20, and June 5, 1974. The fluxes of $NH_4^+$, $NO_2^-$, $NO_3^-$, $PO_4^{3-}$, and $Si(OH)_4$ were computed for the June cruise using previously obtained current data (7). Reported here are only those results that are pertinent to $NH_4^+$, its input and fate in the apex.

Temperature-Salinity Relationships. The strongly linear temperature-salinity (T-S) relationships (see Fig. 4a and b) demonstrate that the water in the transect (Fig. 3) is a mixture of primarily two water

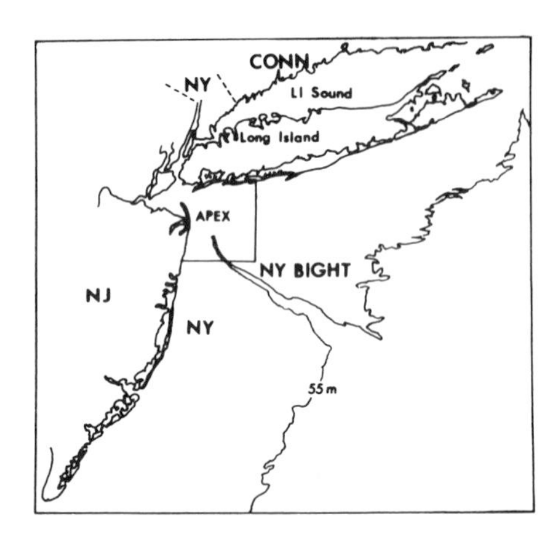


*Figure 1. The New York Bight and its apex*

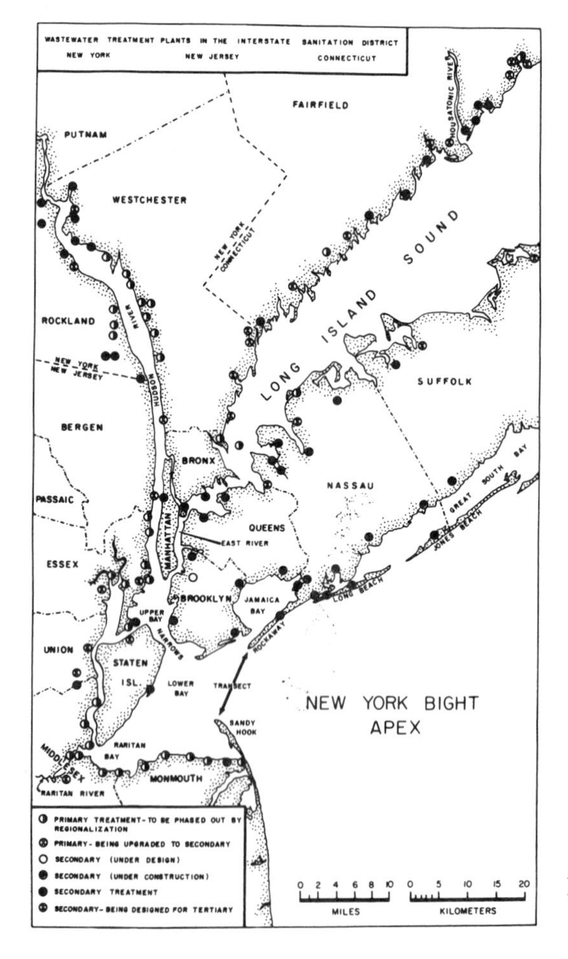


Interstate Sanitation Commission

*Figure 2. Location of sewage treatment plants in the New York metropolitan region (4)*

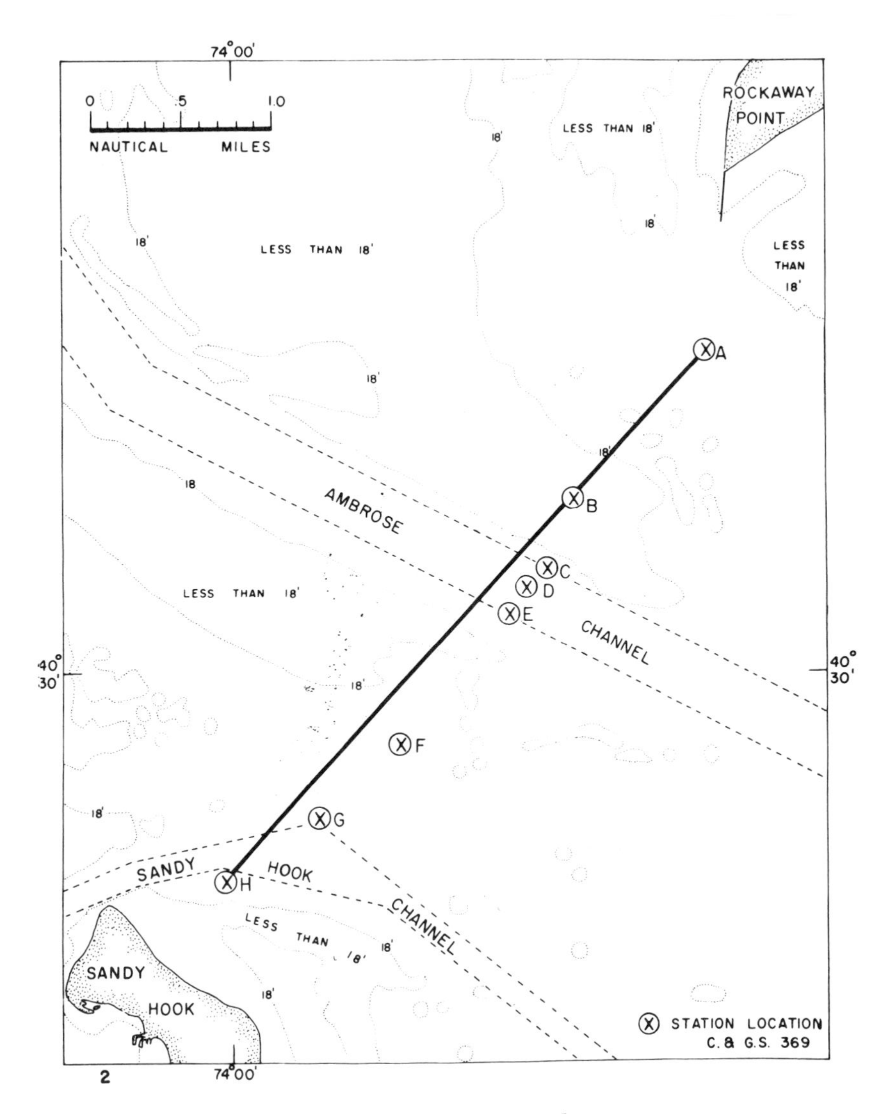


*Figure 3. Station locations in the transect*

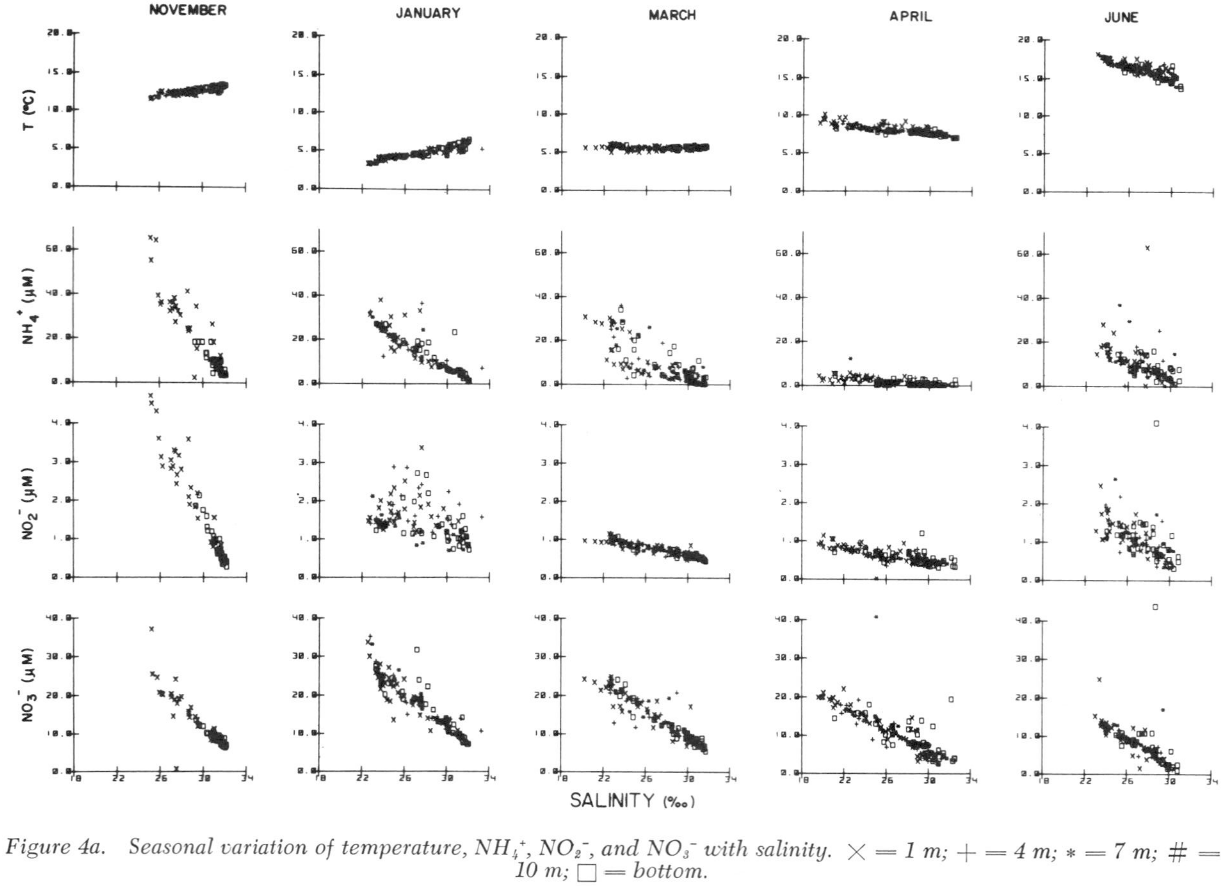


*Figure 4a. Seasonal variation of temperature, $NH_4^+$, $NO_2^-$, and $NO_3^-$ with salinity. $\times$ = 1 m; + = 4 m; * = 7 m; # = 10 m; $\square$ = bottom.*

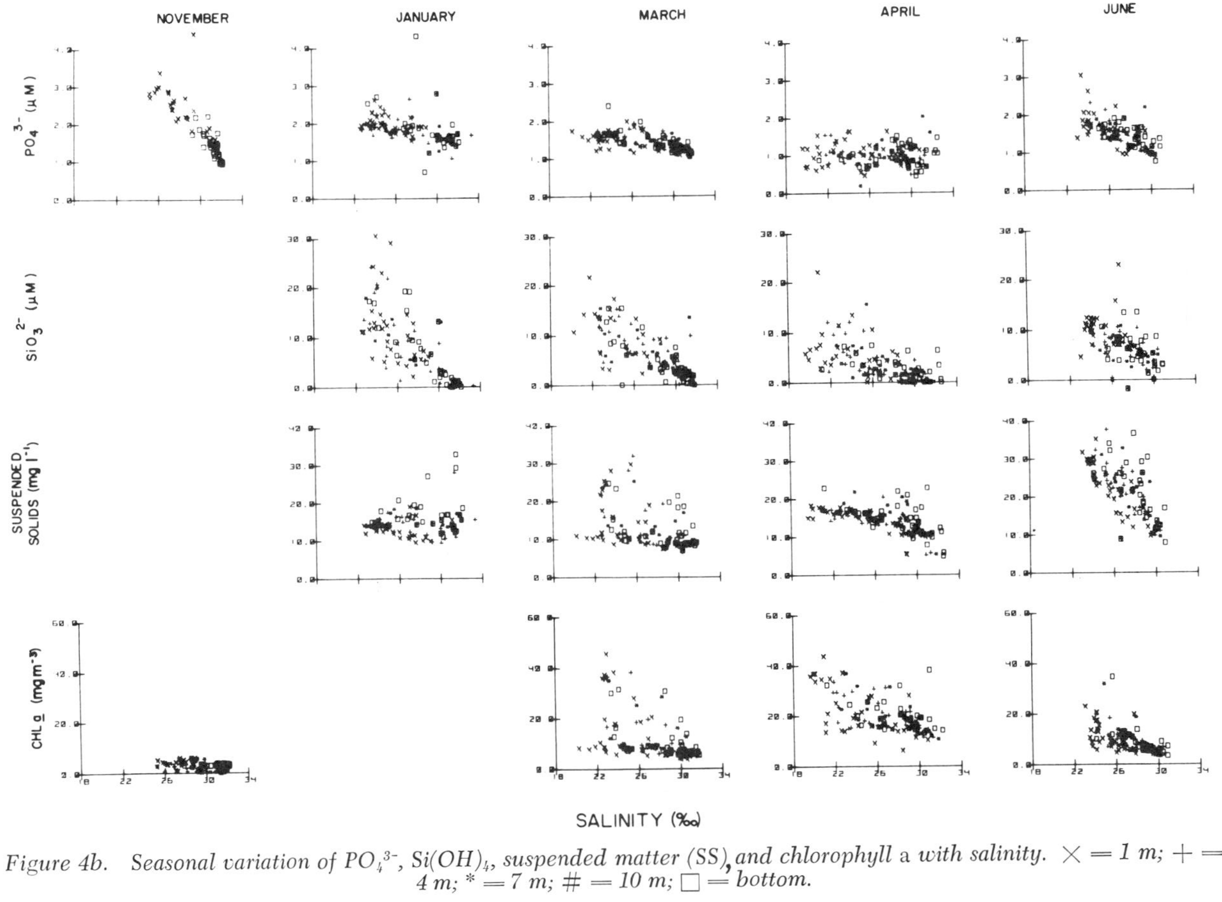


*Figure 4b.* Seasonal variation of $PO_4^{3-}$, $Si(OH)_4$, suspended matter (SS), and chlorophyll a with salinity. × = 1 m; + = 4 m; * = 7 m; # = 10 m; □ = bottom.

Table I. Annual mean concentrations ($\bar{C}$) and annual mean input rates ($\bar{F}$) of sewage effluent contained nutrients discharged through a hypothetical outfall into New York Harbor[a]. Computed for the period July 1973 to July 1974.

| Sewage Component | $\bar{C}$ Concentration[b] in Sewage Effluent (μM) | $\bar{F}$ Input Rate[b] (mole $sec^{-1}$) |
|---|---|---|
| Organic-N | 646 | 63 |
| $NH_4^+$ (+ $NH_3$) | 726 | 71 |
| $NO_2^-$ | 7.7 | 0.8 |
| $NO_3^-$ | 27 | 2.6 |
| $PO_4^{3-}$ | 62 | 6.1 |

[a] Most of the sewage effluent discharged from the New Jersey shore of the harbor has undergone only primary treatment for which only data on mean flows are available (4). Therefore, the values for $c_i$s (see footnote b) for New Jersey effluents were calculated by assuming their concentrations were similar to the Port Richmond primary treatment plant on Staten Island. Calculations exclude effluents from outfalls in Long Island Sound and on the south shore of Long Island.

[b] The concentrations $\bar{C}$ are weighted means calculated from treatment plant annual mean discharges (4) and the mean of monthly concentrations covering the period July 1973 to July 1974. $\bar{C}$ was calculated using the equation:

$$\bar{C} = \frac{\Sigma(c_i x f_i)}{\Sigma f_i}$$

where $c_i$ = annual mean concentration and $f_i$ = annual mean discharge for the $i^{th}$ treatment plant. The input rate $\bar{F}$ was calculated using the equation: $\bar{F} = \Sigma(c_i x f_i)$. The original sewage data used in our calculations were taken from the records of sewage analyses conducted by the Environmental Protection Administration of the City of New York. These records were kindly provided to us by Dr. Seymour Kirschner, Director of the Water Pollution Control Laboratories of the City of New York.

types whose temperatures and salinities show pronounced seasonal changes. During the November cruise, the water mass in the transect was a mixture of two water types: (1) relatively fresher and colder river water and (2) relatively more saline, warmer Bight water. Temperature and salinity ranges were 12-14°C and 24-32°/oo, respectively. During the January cruise, a similar two-component T-S relation was observed except that the water was much colder (3-7°C) and had a wider range of salinities (22-32°/oo). In March the water temperature was nearly uniform at about 5°C and the salinity values (21-32°/oo) had increased slightly from those observed in January. In April the fresher estuarine water was warmer than the more saline Bight water occurring at depth, as shown by the negative slope of the T-S diagram. The salinities (19-32°/oo) exhibited the widest range during the entire study. In June, water temperatures and salinity were both changed substantially from their preceding winter and spring values; temperatures ranged between 13 and 18°C and salinities ranged between 23 and 32°/oo.

The wide range of salinities observed in March and in April demonstrates the strong influence of the spring freshwater discharge. Salinities were more variable near Sandy Hook where the freshwater discharge had the greatest effect on water properties. The most saline water in the transect was always situated near Rockaway Point, due to a strong nontidal transport of Bight water into the harbor.

Nutrient Relations. The approximately linear nutrient-salinity (Nutr-S) relationships (see Figs. 4a and b) also demonstrate that the water in the transect is a two-component system comprised of nutrient impoverished Bight water and nutrient rich estuarine water. The nutrient composition is variable depending on the season and on whether the particular nutrient species originated (1) from sewage effluent or (2) was derived mainly from the freshwater discharge of the Hudson River or a lesser extent, (3) was derived from the discharge of the Raritan and Passaic Rivers.

Most of the $NO_3^-$ measured in the transect comes from Hudson River discharge and not from sewage effluent. Extrapolation of the $NO_3^-$-S results in Fig. 4a to zero salinity gives $NO_3^-$ concentrations that range between 45 and 90 μM. These values are in good agreement with the annual range of 50 to 140 μM reported (8) for $NO_3^-$ in water samples collected from the Hudson and Raritan Rivers upstream from the influence of the New York metropolitan region.

Sewage effluent (Table I) is a principal source of $NH_4^+$, $NO_2^-$, and $PO_4^{3-}$. For November, extrapolation of the $NH_4^+$-S, diagram (see Fig. 4a) to 15°/oo salinity gives a concentration of $NH_4^+$, that compare to some East River values (9) and is about an order of magnitude greater than what is found in Hudson River waters upstream from the metropolitan region.

During the spring freshet, which occurs in April, $NH_4^+$, $NO_2^-$ and $PO_4^{3-}$ concentrations due to sewage discharge would be expected to undergo a greater dilution, with the increased volume of river water in the estuary, than in the earlier winter months. This assumes a relatively constant sewage input of these nutrients. In fact, records of the Environmental Protection Administration of the City of New York (S. Kirschner, personal communication) show that the variability of the flow of sewage effluent and its nutrient concentration is relatively small compared to the amount of dilution. For example, the average coefficients of variation [$V\% = (s\times100)/\bar{X}$] for the mean effluent $NH_4^+$ concentration and flow volume for twelve New York City treatment plants for the period November 1973 to June 1974 were 19.5% and 10.7%, respectively. The April freshwater input from the Hudson River measures about three-fold over that of November (10).

The concentrations of $NO_3^-$ in the transect is primarily a function of the concentration of river water itself and, therefore, may increase or decrease, depending upon whether augmented runoff increases its concentration in the river water or not.

The increased dilution of $NH_4^+$, $NO_2^-$, and $PO_4^{3-}$ takes place at about the same time as the spring phytoplankton bloom, as evidenced by the remarkable increases in the concentration of chlorophyll *a* in the transect, especially near Sandy Hook (Fig. 4b). Therefore, the seasonal variation in the slopes of the $NH_4^+$-S, $NO_2^-$-S and $PO_4^{3-}$-S diagrams (Fig. 4a and b) will not be solely a function of changes in freshwater discharge, but also include the effect of the biological utilization of these chemical species.

For example, assuming that no other processes influence the $NH_4^+$ concentration, the relative effects of (1) phytoplankton uptake during the spring bloom and (2) dilution by river water on the ammonium concentration in the transect can be estimated. This can be accomplished by comparing (1) the measured $NH_4^+$ concentration values with (2) the expected $NH_4^+$ concentration values based solely on dilution. In

November, freshwater discharge in the Hudson estuary is generally about three times less than in April (10). Therefore, it could be expected that dilution would reduce the ammonium concentrations in the transect in April to about one-third of the November values. In Figure 4a, it can be seen that at salinity of 26°/oo, for example, in November the concentration of $NH_4^+$ was about 40 μM. In April, however, the measured $NH_4^+$ concentration at the same salinity ranged from about 5 μM to concentrations below detection, which is about 8 to 13 μM less than the concentration which might be expected from dilution alone. It would appear, therefore, that in April, $NH_4^+$ was being rapidly utilized by the phytoplankton in comparison to November and thus decreased the ammonium concentrations in the transect by about the same amount as that caused by dilution due to the increased freshwater discharge.

Cross-Sections in the Transect. Contoured sections of variables showing the distribution of water properties in the transect were prepared using tidally averaged concentrations (6). Presented here (Figs. 5-7) are sections for salinity, $NH_4^+$, and chlorophyll a.

Two features shown in these sections are the presence of (1) high salinity-low $NH_4^+$ water near Rockaway Point and (2) low salinity-$NH_4^+$ rich water near Sandy Hook. Associated with the low salinity water during April were high concentrations of chlorophyll a (see Fig. 7a through d). Most of the low salinity-$NH_4^+$ rich water probably originates in Raritan Bay where large standing stocks of phytoplankton have been reported (11, 12). The presence of a large but slow moving counterclockwise gyre in Raritan Bay may act to delay some of the nutrient rich Upper Bay water (Fig. 2) that was advected towards the apex but through Raritan Bay by river flow and tidal action. This delay may provide a sufficient amount of time for the development of high chlorophyll a standing crops in Raritan Bay. Tidal current charts (13) show that during ebb tide Raritan Bay waters flow around Sandy Hook into the apex. These low salinity Raritan Bay waters could, during periods of decreased phytoplankton production in the winter, contain elevated concentrations of $NH_4^+$ and other nutrients. However, during the spring when phytoplankton growth is at its maximum in Raritan Bay, the low salinity water flowing from Raritan Bay would

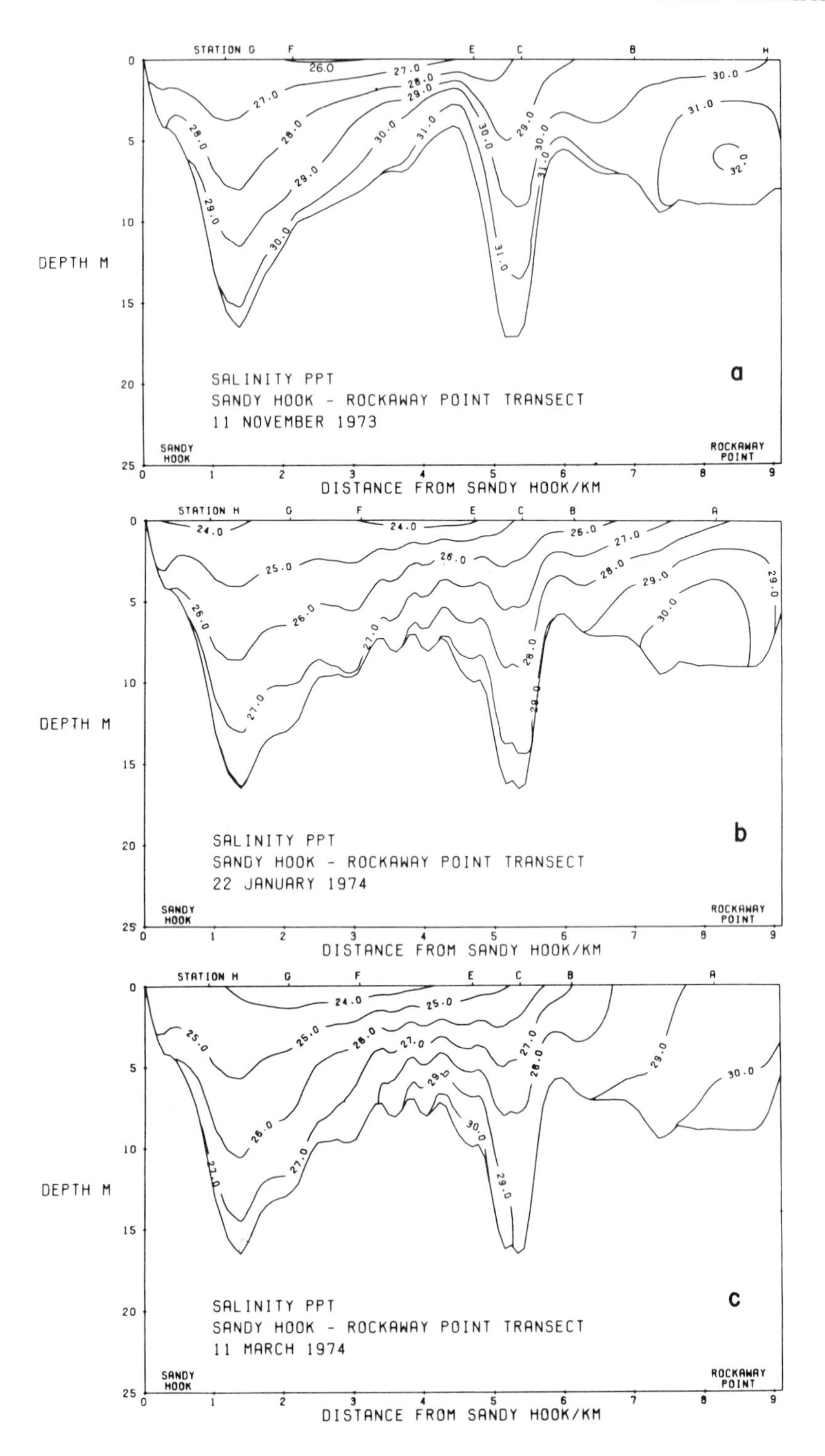

STATION G F E C B H
26.0
27.0
28.0
29.0
30.0
31.0
32.0
DEPTH M
SALINITY PPT
SANDY HOOK - ROCKAWAY POINT TRANSECT
11 NOVEMBER 1973
a
SANDY HOOK
ROCKAWAY POINT
DISTANCE FROM SANDY HOOK/KM
STATION H G F E C B A
24.0
25.0
26.0
27.0
28.0
29.0
30.0
SALINITY PPT
SANDY HOOK - ROCKAWAY POINT TRANSECT
22 JANUARY 1974
b
STATION H G F E C B A
SALINITY PPT
SANDY HOOK - ROCKAWAY POINT TRANSECT
11 MARCH 1974
c

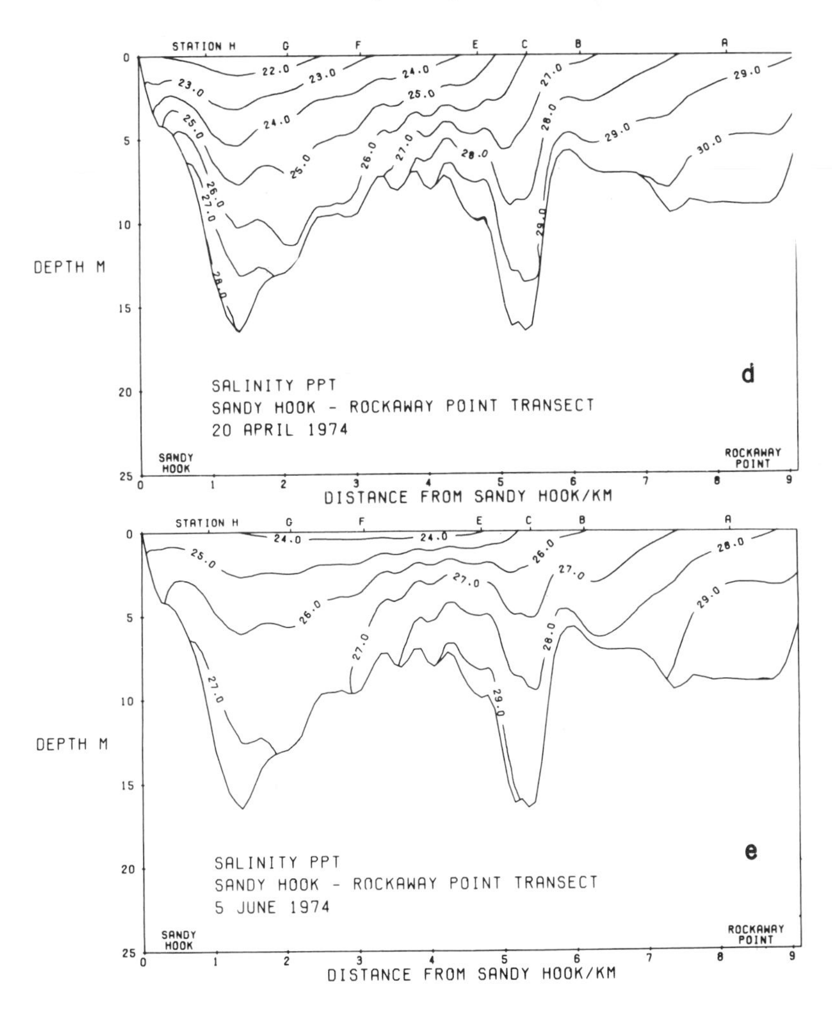


*Figure 5a–e. Tidally-averaged salinity distribution in the transect*

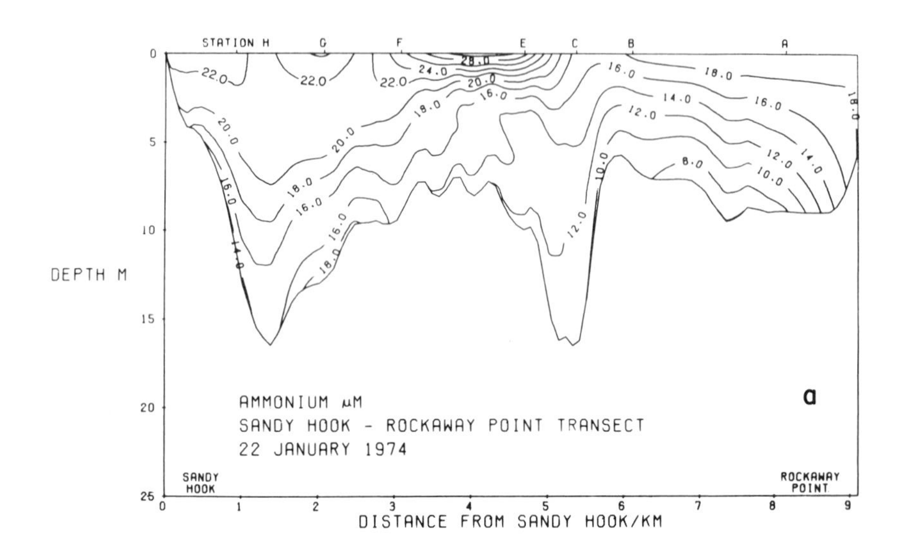


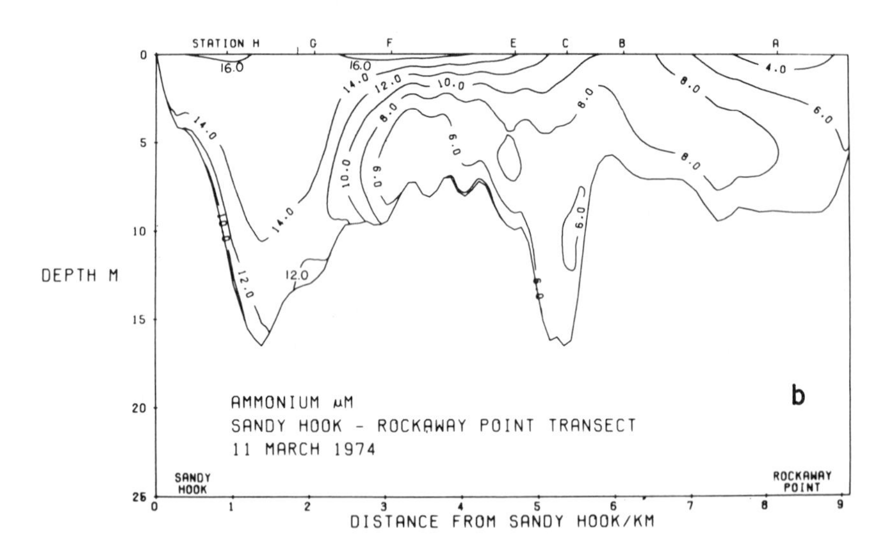


*Figure 6a–d. Tidally-averaged*

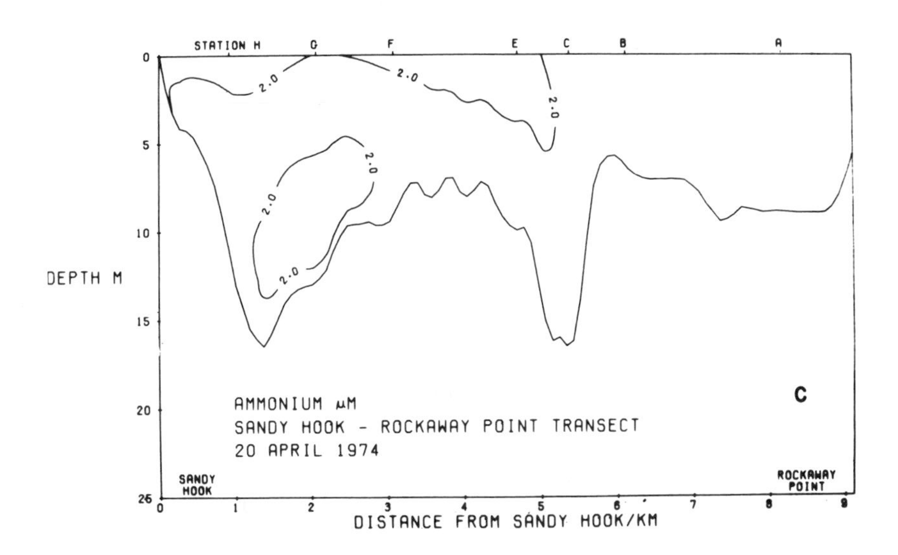


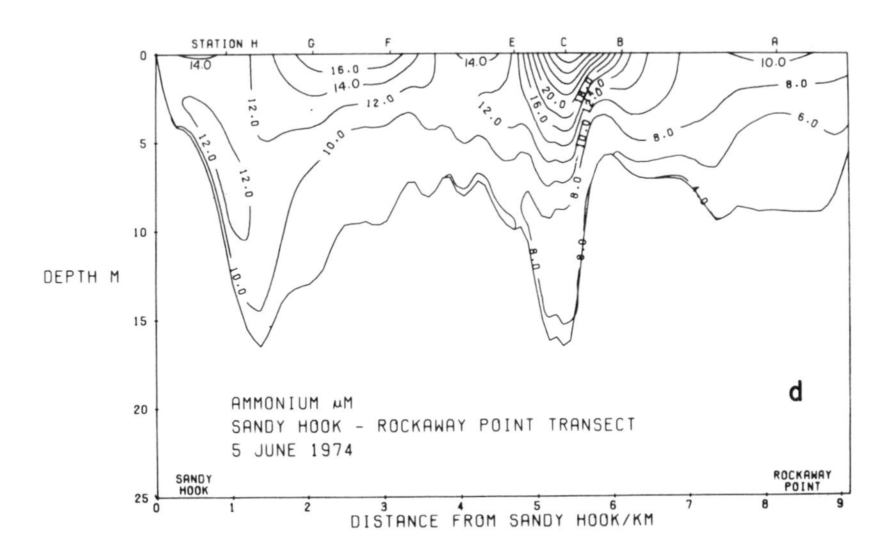


*ammonium distribution in the transect*

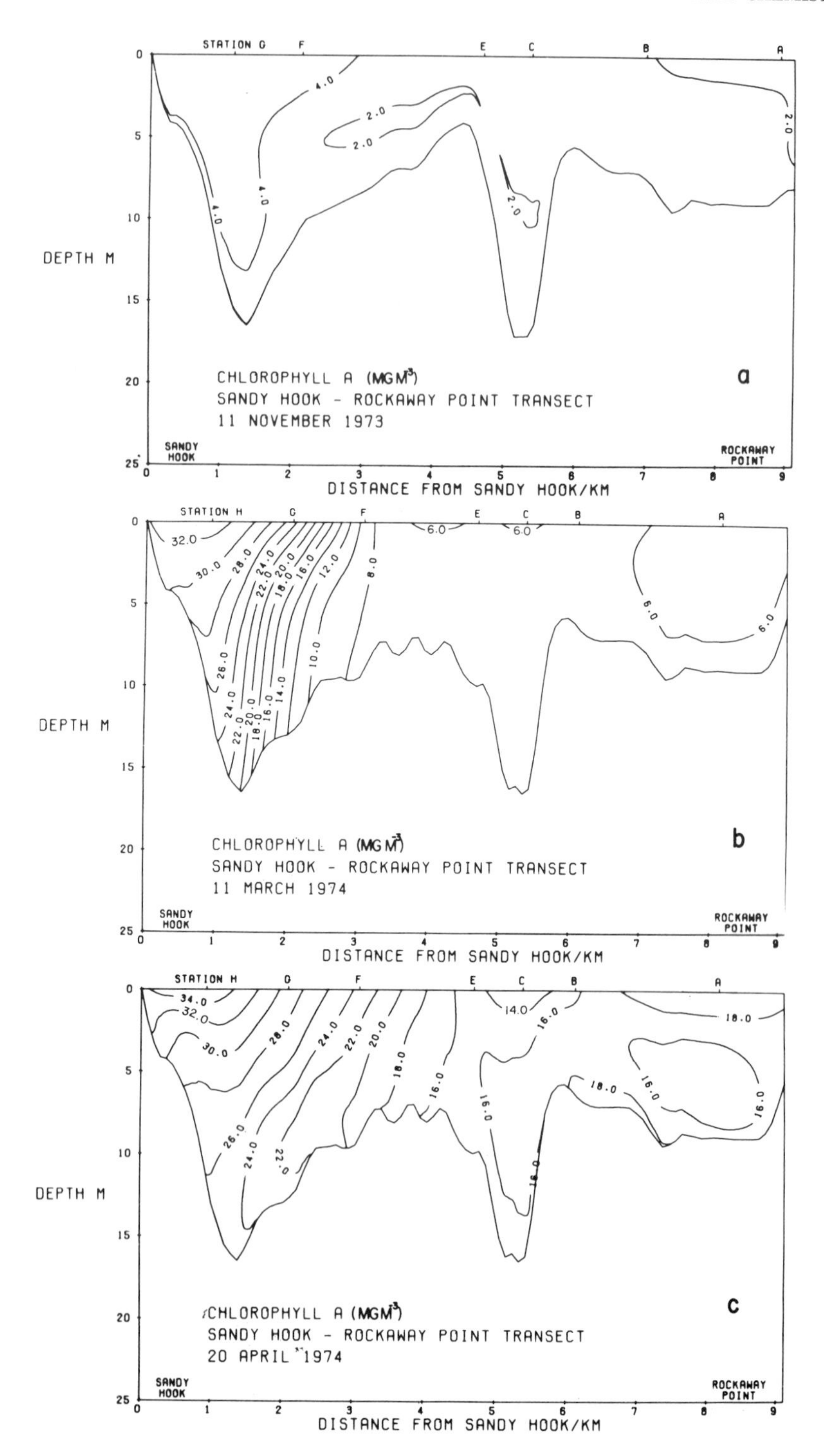

STATION G F E C B A
DEPTH M
CHLOROPHYLL A (MGM$^{-3}$)
SANDY HOOK - ROCKAWAY POINT TRANSECT
11 NOVEMBER 1973
a
SANDY HOOK
ROCKAWAY POINT
DISTANCE FROM SANDY HOOK/KM
STATION H G F E C B A
DEPTH M
CHLOROPHYLL A (MGM$^{-3}$)
SANDY HOOK - ROCKAWAY POINT TRANSECT
11 MARCH 1974
b
SANDY HOOK
ROCKAWAY POINT
DISTANCE FROM SANDY HOOK/KM
STATION H G F E C B A
DEPTH M
CHLOROPHYLL A (MGM$^{-3}$)
SANDY HOOK - ROCKAWAY POINT TRANSECT
20 APRIL 1974
c
SANDY HOOK
ROCKAWAY POINT
DISTANCE FROM SANDY HOOK/KM

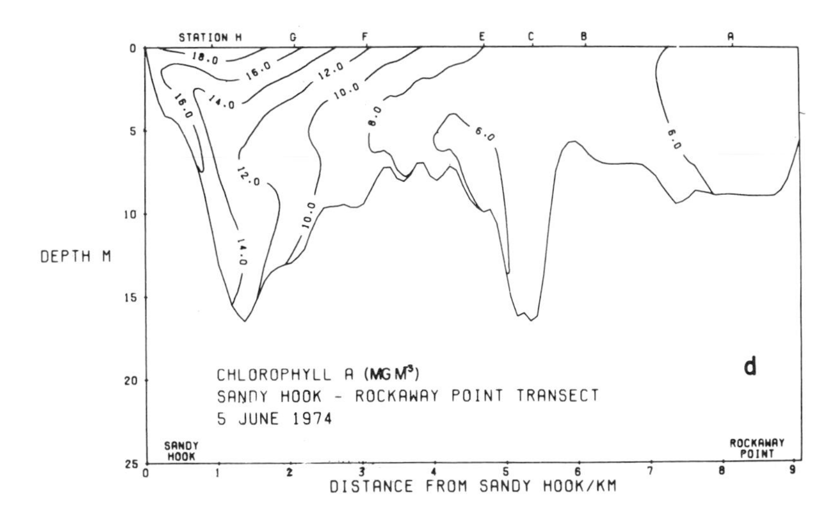


*Figure 7a–d. Tidally-averaged chlorophyll* a *distribution in the transect*

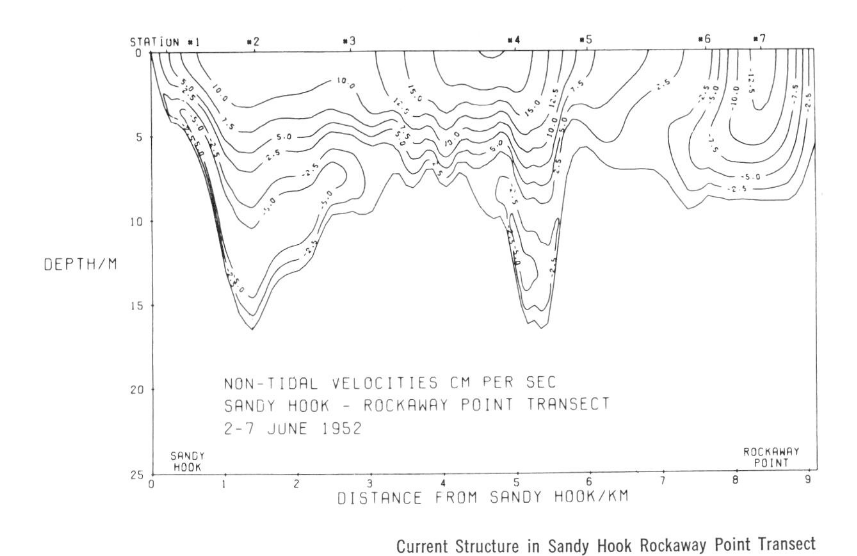


*Figure 8. Nontidal current velocities in the transect* (7). *Negative-sign velocities are into the harbor.*

contain high concentrations of chlorophyll a but low concentrations of nutrients. This is consistent with the features observed in all the nutrient and chlorophyll cross-sections [see Figs. 5, 6, and 7 and also (6)].

The persistence of higher salinity water near Rockaway Point is mainly due to strong non-tidal currents that transport seawater from the apex to the harbor. Figure 8 is a cross-section showing the distribution of non-tidal current velocities normal to the transect during June, 1952 (6). Negative-sign velocities indicate currents moving towards the harbor. Clearly, Rockaway Point is a region where both surface and deep waters are transported into the harbor from the apex. Earlier investigators (14, 15) working in the New York Bight apex postulated, based on salinity, the existence of a non-tidal drift near Rockaway Point which moves towards the harbor. The velocity contours in Figure 8 give the first published quantitative evidence for the presence of this feature. The results of a recent completed seabed drifter study in the apex shows near bottom non-tidal velocities very similar to those in Figure 8 (16).

At other locations, along the transect, non-tidal surface currents move towards the apex, which is consistent with the presence of low salinity-nutrient rich water south of Ambrose channel (see Figs. 5 and 6).

Advective Nutrient Fluxes in the Transect. To obtain estimates of nutrient input to the apex, the advective fluxes of $NH_4^+$, $NO_2^-$, $NO_3^-$, $PO_4^{3-}$, and $Si(OH)_4$ were calculated using the June 5, 1974 nutrient contours [see (6)] and the June 2-8, 1952 velocity profile (Fig. 8). In the calculations, it was assumed that the non-tidal velocity structure in the transect was similar for the two periods, June 2-8, 1952 and June 5, 1974. Storms that might alter non-tidal circulation in the apex are infrequent in June. The non-tidal current structures for May 21-25, 1958, and August 12-16, 1959 are very similar to the non-tidal current structure for June 2-8, 1952 (R. Wilson, personal communication). Advective nutrient fluxes were calculated from the expression $\Sigma(c_i v_i \sigma_i)$.

The $c_i$ and $v_i$ are the tidally averaged nutrient concentrations and non-tidal current velocities interpolated to a grid and contoured in Figures 6 and 8, respectively. $\sigma_i$ are the areas of different grid elements. Table II summarizes our estimates of nutrient fluxes through the transect.

Table II. Calculated Nutrient Fluxes in the Transect.

| Nutrient | FLUX (moles $sec^{-1}$) (a) Into the Harbor | (b) Out of the Harbor | [(b)-(a)] Bight-ward |
|---|---|---|---|
| $NH_4^+$ | 14 | 51 | 37 |
| $NO_2^-$ | 1.6 | 4.6 | 3 |
| $NO_3^-$ | 11 | 37 | 26 |
| $PO_4^{3-}$ | 2.5 | 6.2 | 3.7 |
| $Si(OH)_4$ | 10 | 31 | 21 |

## $NH_4^+$ Inputs due to Sludge Dumping

Sewage sludge, which is dumped in large volumes at sea about 25 km from the entrance to New York Harbor, is also a source of $NH_4^+$ in the apex. For the period 1965 to 1970, it has been estimated that the average annual volume of sludge to the apex is about $3.2x10^6$ $m^3$ (17). The continued construction of large sewage treatment plants in the metropolitan area, including the south shore of Long Island will ensure an increasing input of sludge to the Bight in the future if alternate disposal or recycling schemes are not developed and implemented.

To determine the effect of sludge dumping on the distribution of $NH_4^+$ in apex waters, we conducted a cruise within the sludge dumping grounds during the period July 30-31, 1973 (18). Table III shows that nine dumps of sludge occurred during the period of study.

A rapid sampling grid (Fig. 9) was used in and near the dump site to assess the spatial extent of $NH_4^+$ from sludge dumping. The grid consisted of stations spaced around the perimeter of a 9.4 km radius circle, centered near the designated dump area (see Fig. 9). Surface, ~10m, and bottom water samples were obtained by using a submersible pump and then analysed for $NH_4^+$ onboard the research vessel. In addition, an attempt was made to trace the dispersion of recently dumped sludge by following, with additional water sampling, two parachute drogues set at 4 and 8 m (Fig. 9). Details of the methods used are reported

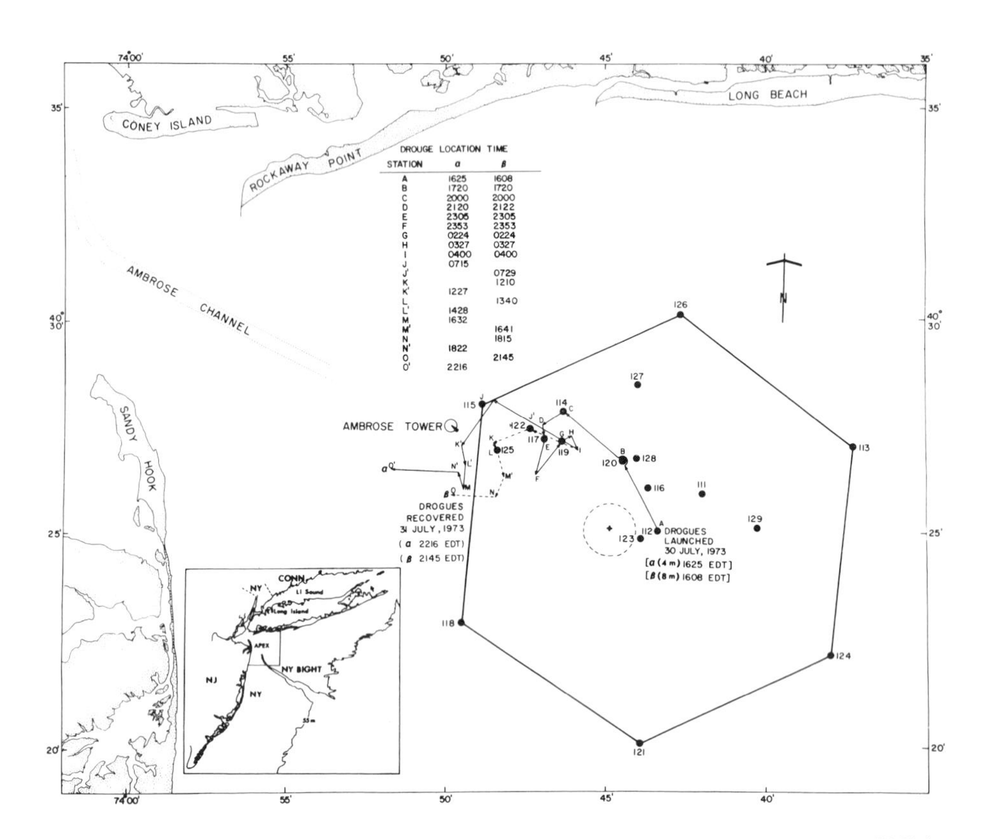


Ocean Dumping in the N.Y. Bight

*Figure 9. Stations occupied and times of drogue release and location (18). Dashed circle near center of perimeter represents the area (3.3 km²) and location (40° 20′ N, 73° 45′ W) of the interim sludge dump area established by the U.S. Environmental Protection Agency.*

Table III. Volumes of sewage sludge (wet) dumped in the New York Bight apex during the period July 30-31, 1973. Data obtained from 1973 Ocean Dumping Logs, Marine Environmental Branch, U. S. Coast Guard, Governors Island, New York. [This tabulation taken from (18).]

| Date | Estimated Time of Arrival (EDT) | Sludge Vessel & Source of Sludge | Volume[a], [b] of Sludge $m^3$ |
|---|---|---|---|
| 30 July | 1055 | "Newtown Creek" (New York City) | 2,860 |
| | 1126 | "Bowery Bay" (New York City) | 1,820 |
| 31 July | 0550 | "Judson K. Stickle" (South Amboy, N.J.) | 152[c] |
| | 1105 | "Newtown Creek" (New York City) | 2,860 |
| | 1111 | "Bowery Bay" (New York City) | 1,870 |
| | 1700 | "Ocean Disposal" (Passaic Valley Sewage Authority, New Jersey) | 155[c] |
| | 1729 | "Newtown Creek" (New York City) | 2,860 |
| | 1740 | "Coney Island" (New York City) | 1,200 |
| | 1900 | "Bowery Bay" (New York City) | 1,890 |

[a] Original logs report volume of wet sludge in units of cubic feet.

[b] The density of wet sludge is about 1 g $cm^{-3}$ (this may be low since some components of sludge sink); at this density 1 $m^3$ equals 1 metric ton.

[c] Original data given in short tons.

elsewhere (18).

The high surface and mid-depth $NH_4^+$ concentrations (Fig. 10) at station 112 confirmed an earlier visual indication of a sludge patch at that station, but the concentrations obtained from the sampling sequence along the trajectories of the drogues are confused. For example, stations 114, 116, and 117, which were sampled subsequent to station 112, showed surface $NH_4^+$ values similar to those measured at stations away from the immediate dump area. It was not until station 119 was sampled that surface and 10 m $NH_4^+$ concentrations similar to those measured at station 112 were found. Extremely high ammonium concentrations (~200 to 500 μM) were observed at station 128 at 1930 eastern daylight time (EDT), July 31. We judged this observation to be attributable to a recent sludge dump, as evidenced by the 1800 EDT sighting of a sludge transport vessel in the vicinity that was returning to New York Harbor (18). It would seem, therefore, that the sludge dump site is characterized by persisting signatures from previous dumping activity which leads to a very non-homogeneous spatial distribution of $NH_4^+$ (Fig. 10). The presence of a pycnocline (Fig. 11) would inhibit vertical mixing of the surface and bottom waters. However, the data obtained at station 128 did show that the effects of dumping can, in some instances, manifest themselves throughout the water column within a very short time.

The concentration of $NH_4^+$ in sewage sludge is variable. Some unpublished analyses performed by the New York City Environmental Protection Administration (S. Kirschner, personal communication) in November and December, 1973, show that the $NH_4^+$ concentration in New York City sludges ranges from 17 to 160 milliM, with a mean value of 70 milliM. Using this mean concentration and the inputs in Table III, we calculate the daily input of $NH_4^+$ to the apex from sludge dumped during our cruise to be $0.3\text{-}0.7 \times 10^6$ moles, which is about one-fifth to one-tenth of the input from the advective processes in the transect. However, unlike the Hudson River's low salinity plume, sludge dumping at sea can produce locally high concentrations of $NH_4^+$ that may be rapidly dispersed by mixing. Further research is needed to determine the spatial distribution, persistence and local biological effects of these elevated $NH_4^+$ concentrations caused by dumped sludge.

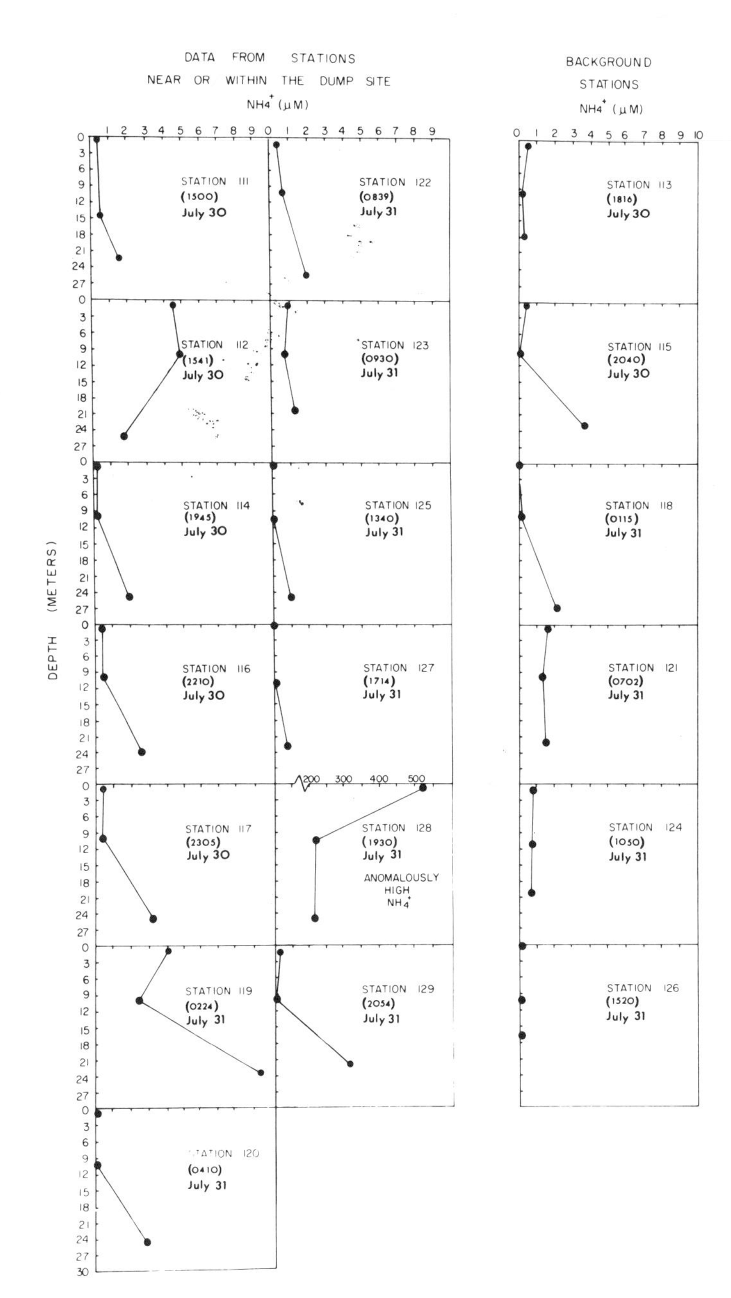


Ocean Dumping in the N.Y. Bight

*Figure 10. Observed ammonium concentrations near or within the dump site and at background stations on the indicated perimeter (18). The number in brackets represents the time (EDT) of arrival on station.*

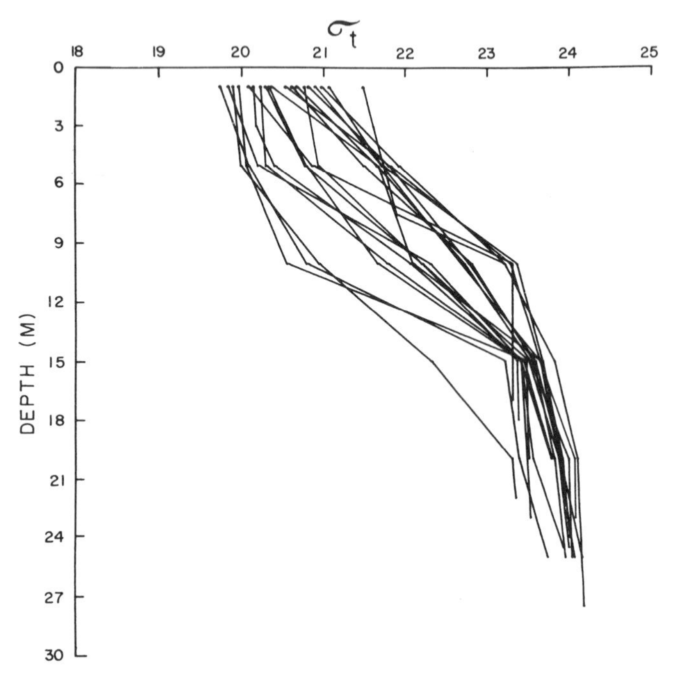


Ocean Dumping in the N.Y. Bight

*Figure 11. Composite plot of $\sigma_t$ vs. depth for all stations occupied in Figure 9 (18)*

## Discussion and Conclusion

The results of our investigations suggest that (1) in June the net advective flux of $NH_4^+$ from New York Harbor to the New York Bight apex probably exceeds the rate of input of $NH_4^+$ from sludge dumping in the Bight apex and (2) the effects of phytoplankton uptake and freshwater dilution are about the same on $NH_4^+$ concentrations in the transect during the spring phytoplankton bloom.

Table IV lists (1) the combined mean daily inputs of nutrients from sewage discharge and river discharge and (2) the flux of nutrients through the transect for June 5, 1974. The relatively large differences between the river plus sewage input and the output through the transect (Table IV, Column C) for $NH_4^+$, $PO_4^{3-}$ and $Si(OH)_4$ suggest that these nutrient species are being utilized by the phytoplankton at rates about one-half the nutrient input rate. The smaller differences for $NO_3^-$ and $NO_2^-$ suggest that these species may behave as qausi-conservative constituents in these eutrophic waters. It has been suggested that $NO_3^-$ is utilized by phytoplankton only in the virtual absence of $NH_4^+$ (1, 2). Most of the $NH_4^+$, $PO_4^{3-}$, and $Si(OH)_4$ utilization probably takes place in Raritan Bay where large standing stocks of phytoplankton have been recorded (11, 12).

We emphasize, however, that the results in Table IV are only approximations and, therefore, have very limited use in any nutrient material-balance calculation because (1) June, 1952 current velocities were used to calculate June 1974 nutrient fluxes in the transect, (2) nutrients may be consumed and recycled before they reach the transect, processes about which we have no direct knowledge. Also, it has been calculated that about 70 per cent of the sewage discharged from the four largest sewage treatment plants on the East River (Fig. 1) is transported to Long Island Sound (20). Furthermore, virtually nothing is known about the fate of the large amount of organic nitrogen (Table I) which in some cases may serve as nitrogen - sources for growth of phytoplankton (21, 2).

The Hudson River Plume and Chlorophyll *a* Input to the Apex. The transect study has demonstrated the Hudson River plume's spatial distribution and temporal permanence across the apex. The plume was most well-developed in the transect between Sandy Hook and Ambrose Channel. Of particular interest are the high concentrations of chlorophyll *a* in the plume and their

Table IV. Calculated daily river and sewage nutrient input and Bight-ward nutrient flux through the transect, June 5, 1974.

| Nutrient | ($x10^6$ moles) (a) Sewage[a] and River[b] Input | (b) Bight-ward Flux | [(b)-(a)] |
|---|---|---|---|
| $NH_4^+$ | 6.5 | 3.2 | -3.3 |
| $NO_2^-$ | .08 | .24 | .16 |
| $NO_3^-$ | 1.9 | 2.2 | .3 |
| $PO_4^{3-}$ | 0.6 | 0.3 | -0.3 |
| $Si(OH)_4$ | 4.9 | 1.8 | -3.1 |

[a] Calculated from Table I. $Si(OH)_4$ inputs in sewage have been neglected for lack of data.

[b] Calculated from May, June and July concentrations from 1972 Water Year data at Chelsea station (8) on the Hudson River. Values adjusted upward by 10% to approximately account for Raritan River. Passaic River input not included.

affect in the apex. Using the June 1952 current velocities (Fig. 8) and June 1974, chlorophyll *a* concentrations, we calculate that about 1.9 metric tons per day of chlorophyll *a* is transported to the Bight apex through the transect. During spring larger daily transports would exist due to development of a more intense phytoplankton bloom (see Fig. 7b, c and d). This seasonal input of large abundances of phytoplankton would be confined to the Hudson plume. The plume, according to previous investigators, remains near to the New Jersey coast (14, 15). These high phytoplankton concentrations undoubtedly represent an important and relatively continuous but seasonally variable supply of particulate food for zooplankton herbivores associated with the river plume. Much remains to be learned about the biological and hydrographic characteristics of the plume. Research should be directed toward developing an adequate understanding of the biological, chemical and physical processes in the plume and to determine what impact the plume has on the region encompassed by the New York Bight.

## Acknowledgement

We gratefully acknowledge the generous and cheerful assistance of Mr. Chris Stuebe, Captain of the *R/V Mic Mac*, Mr. Jeff Parker, party chief, Mr. William Miloski, seawater analyst, Mr. Glen Hulse, instrumentation specialist, and Ms. Sue Oakley and Mr. Alan Robbins for data reduction and computer plotting.

We thank Mrs. Jacqueline Restivo for her generous assistance during the manuscript preparation.

The cruise work would not have been possible without the labors of many graduate students at MSRC, including Mr. Mike White, Ms. Cynthia Marks, and Mr. Robert Olson. Much of the reported work was supported by a grant (No. 04-4-158-19) from the NOAA/MESA New York Bight Project. We thank Mr. H. Stanford for his assistance.

## Abstract

Two sources of $NH_4^+$ input to the New York Bight have been identified: (1) Hudson River-Raritan Bay discharge into the apex of the Bight through the Rockaway Point, New York-Sandy Hook, New Jersey, transect and (2) barge-dumped sewage sludge from the greater New York metropolitan region.

The Bight-ward flux of ammonium through the transect has been calculated for one 24-hour period in

June and was found to be 5 to 10 times greater than the ammonium input from barge-dumped sludge from a typical two-day period in July. During the April bloom, the rate of phytoplankton uptake and the effect of fresh-water dilution were found to decrease the $NH_4^+$ concentration a similar amount in the transect.

The Hudson River plume was observed to be a persistent year-round feature near the New Jersey shore and was responsible for the advection of large amounts of chlorophyll a into the Bight.

## Literature Cited


(1) Pomeroy, L. R., Annual Review of Ecology and Systematics, (1970) 1 171-190.
(2) Stewart, W. D. P., ed., "Algal Physiology and Biochemistry," University of California Press, Berkeley, California, 1974.
(3) Ryther, J. H., Dunstan, W. M., Science, (1971) 171 1008-1013.
(4) Interstate Sanitation Commission, "1974 Annual Report," 10 Columbus Circle, New York, New York, 1975.
(5) Interstate Sanitation Commission, "Combined Sewer Overflow Study for the Hudson River," 10 Columbus Circle, New York, New York, 1972.
(6) Duedall, I. W., O'Connors, H. B., Parker, J. H., Wilson, R., Robbins, A., "The Seasonal and Tidal Distribution of Nutrients and their Fluxes in the New York Bight Apex," Manuscript in Preparation.
(7) Kao, A., "A Study of the Current Structure in the Sandy Hook-Rockaway Point Transect," M. S. Research Paper, Marine Sciences Research Center, State University of New York, Stony Brook, New York, In Preparation.
(8) Water Resources Data for New York, "Part 2. Water Quality Records," U. S. Geological Survey, Albany, New York, 1972.
(9) Duedall, I. W., Unpublished data.
(10) Giese, G. L., Barr, J. W., "The Hudson River Estuary," Bulletin 61, Conservation Department, Water Resources Commission, Albany, New York, 1967.
(11) Jeffries, H. P., Limnology and Oceanography, (1962) 7 21-31.
(12) Patten, B. C., "The Diversity of Species in Net Phytoplankton of the Raritan Estuary," Ph.D. Thesis, Rutgers University, New Brunswick, New Jersey, 1959.

(13) U. S. Coast and Geodetic Survey, "Tidal Current Charts, New York Harbor," 7th Edition, ESSA, Rockville, Maryland, 1956.
(14) Ketchum, B. H., Redfield, A. C., Ayers, J. C., "The Oceanography of the New York Bight," Papers in Physical Oceanography and Meteorology, Massachusetts Institute of Technology and the Woods Hole Oceanographic Institution, Cambridge and Woods Hole, Massachusetts, 1951.
(15) Redfield, A. C., Walford, L. A., "A Study of the Disposal of Chemical Waste at Sea," National Academy of Sciences-National Research Council, Washington, D. C., 1951.
(16) Baylor, E. R., Hardy, C. D., Moskowitz, P., "Bottom Drift over the Continental Shelf of the New York Bight," Manuscript in Preparation.
(17) Pararas-Carayannis, G., "Ocean Dumping in the New York Bight: An Assessment of Environmental Studies," Technical Memorandum No. 39, Coastal Engineering Research Center, U. S. Army Corps of Engineers, Fort Belvoir, Virginia, 1973.
(18) Duedall, I. W., Bowman, M. J., O'Connors, H. B., Estuarine and Coastal Marine Science, In Press.
(19) Ocean Dumping Criteria, Federal Register No. 94, (1973), 38 12872-12877.
(20) Bowman, M. J., "Pollution Prediction Model of Long Island Sound," Ocean Engineering III, American Society of Civil Engineers Specialty Conference, Newark, Delaware, June 9-12, 1975. In press.
(21) Goering, J. J., "The Role of Nitrogen in Eutrophic Processes," In: Water Pollution Microbiology, ed., R. Mitchel, John Wiley and Sons, Inc., New York, New York, 1972.

# 40

# The Dynamics of Nitrogen and Phosphorus Cycling in the Open Waters of the Chesapeake Bay

JAMES J. McCARTHY
Department of Biology, Harvard University, Cambridge, Mass. 02138
W. ROWLAND TAYLOR and JAY L. TAFT
Chesapeake Bay Institute, The Johns Hopkins University, Baltimore, Md. 21218

The use of radioisotopes and rare stable isotopes in field investigations of plankton nutrition both in lakes and the ocean is now widely practiced. The introduction of the $^{14}C$ technique to measure phytoplankton photosynthesis (1) revolutionized the study of aquatic primary productivity. The use of $^{32}P$ in plankton studies has received less attention (2,3,4), and more recently $^{15}N$ has proven to be a useful tool (5,6). The literature of the last two decades which includes various applications of these isotopes in studies of aquatic biology is heavily burdened by discussions of problems in experimental design and data interpretation (7,8,9,10). In general the application of such techniques to aquatic biology has progressed more slowly than in other fields of study, and ecologists have been criticized (11) for their naive and often erroneous use of isotopes in studies of transfer processes within aquatic ecosystems. Recently, innovative applications of isotope tracer studies to plankton nutrition have included tritiated substrates in the study of heterotrophy (12) and $^{29}Si$ (13) and $^{68}Ge$ (14) in the study of the silicious nutrition of diatoms.

Direct and indirect evidence supports the contention that nitrogen is often the element most likely to limit plant productivity in the sea (15,16,17). In addition, studies of the nitrogenous nutrition of plankton are attractive in that nitrogen provides perhaps the simplest perspective from which to view a material balance in a planktonic ecosystem. Whereas cellular nitrogen is primarily bound in structural material, both carbon and phosphorus are largely involved in cellular metabolic activities in addition to their structural roles.

Blue-green algae capable of fixing gaseous nitrogen are considerably more common in fresh-water than in the sea and partially because of this, the availability of nitrogen is considerably less likely to regulate plant productivity in freshwater. Phosphorous is frequently looked to as the most important nutrient in both the regulation of productivity and limitation of biomass in fresh-water and the open waters of estuaries

(18).

In most natural waters dissolved combined nitrogen is present in more than a single form. Concentrations of nitrate, nitrite, and ammonium in the sea vary temporally and spatially (19) and recently, urea has been reported from all areas which have been investigated (20,21,22,23). Amino acids are usually detected (24,25), and although the concentrations are always low, they may at times contribute to phytoplankton nitrogenous nutrition (26,27).

Although we know what forms of combined inorganic nitrogen do and do not commonly exist in the sea, we are largely ignorant of the identity, origin, and fate of a sizeable pool of dissolved organic nitrogen. This material is temporally and spatially ubiquitous (27,28,29) and the implication is that the bulk of it is highly refractory and has little potential in algal nitrogenous nutrition (27,28).

The availability of $^{15}N$ labeled forms of the plant nutrients thought to be of greatest importance, enables one to assess the relative significance of several forms of nitrogen in the nutrition of natural plankton assemblages. A few oceanographic laboratories in this country now have the capacity to rapidly process plankton samples for $^{15}N$ enrichment, and the recent availability of optical emission spectrometers may greatly extend the use of $^{15}N$ in plankton studies. Shipboard studies with large volume cultures established by enriching natural seawater with nutrients have demonstrated that estimates of phytoplankton assimilation derived from short interval measurements of $^{15}N$ uptake agree closely with measurements of both the decrease in dissolved nitrogenous nutrient and increase in particulate nitrogen (30). Field studies of nitrogen fixation have also employed $^{15}N$ (5), but the acetylene reduction assay for nitrogenase activity (31) is far more convenient and it has become widely accepted as the method of choice.

Because of the low natural concentrations of some of the phosphorous and nitrogenous forms of interest, experiments must be initiated quickly. In most cases phytoplankton activity is measured after removal of the zooplankton; such removal certainly reduces the natural supply of the rapidly recycled forms of both phosphorus and nitrogen to the captured phytoplankton. With zooplankton inclusion, phytoplankton are consumed during the experiment, animal metabolites are released, and the interpretation of data is immensely complicated. Since ambient substrate levels should be determined before the initiation of an experiment, there has been much interest in improving the capacity to quantitate phosphorous and nitrogenous nutrients with specific, accurate, precise, and convenient methodology. Most of the techniques of choice are described in both manual and automated forms by Strickland and Parsons (1972) (32). We have streamlined these manual methods to permit us to rapidly process 5ml samples with both precision and sensitivity which exceeds those of the automated methods. A small number of samples can be routinely analyzed by a single analyst for nitrate, nitrite, ammonium, urea, and soluble reactive phosphate within one half hour after sample collection.

In the study of plankton nutrition one has to select an incubation technique which best suits the particular problem under investigation. An *in situ* incubation involves the collection of a water sample, enclosure in a suitable bottle, addition of isotope tracer material, and resuspension of the bottle to the depth from which the sample was collected. The objective is to execute the incubation at near natural light and temperature. Careful and rapid handling are required to minimize physiological shock which may result from exposure of deep water samples to both full incident solar irradiation and great changes in temperature. Any potential problems related to pressure differentials are usually ignored. The *in situ* incubation has been perfected with a single unit which captures, inoculates with material of choice (e.g. isotope stocks), and serves as an incubation chamber (33). After the completion of the incubation the chamber is brought to the surface and the sample is processed. Advantages of such a system are obvious, but the disadvantages are primarily that for nutrient assimilation studies one must make isotope tracer additions without specific knowledge of the ambient nutrient levels. Also, zooplankton which consume phytoplankton and release plant assimilable forms of nitrogen and phosphorus cannot be excluded from the incubation chamber.

Because major sea-going vessels usually cannot remain on stations for extended periods, and there may not be an opportunity to return to a buoy, biological oceanographers frequently use simulated *in situ* incubations. A deckboard incubator is fitted with optical filters to simulate light at depth of sampling, and flowing seawater controls temperature. Some of the problems associated with measuring light and selecting appropriate filters have been investigated and discussed (34).

The Chesapeake Bay is a highly productive coastal plain estuary with a well defined two-layer circulation (35). It is approximately 300 km in length and it is primarily an estuary of the Susquehana River. The nutrient loading resulting from both the Susquehana River drainage from western Pennsylvania and central New York and the metropolitan discharges of the District of Columbia and Baltimore, Maryland, are potentially large. It is indeed interesting to note in the data presented by Carpenter, Pritchard, and Whaley (1969) (36), that the effect of nutrient loading from the adjacent metropolitan areas is virtually undetectable as dissolved inorganic nutrient in the Bay proper below both Baltimore and the junction of the Potomac River with the Bay. It has been demonstrated that the heavy winter-spring runoff from the Susquehana River is the primary source of nitrate in the Bay, but no point sources for either ammonium or phosphate can be identified.

Data from some coastal regions, such as for Buzzards Bay, Massachusetts, suggest that transport of nitrogen remineralized in the sediment interstitial water to the overlying water may contribute significantly to the phytoplankton nitrogen requirement (37). In another study, vertical transport from the sand bottom off La Jolla, California, was shown to provide an insignificant portion of the nitrogen utilized by the phytoplankton in the overlying water (38). Carpenter, Pritchard, and Whaley (1969) (36) concluded from calculated vertical exchange rates for spring and summer that the transport of dissolved nitrogen and phosphorus from the sediments into the euphotic zone was insufficient to account for more than a few per cent of the approximated rates of phytoplankton nutrient utilization. More recently, Bray, Bricker, and Troup (1973) (39) calculated that upward diffusive flux across the sediment-water interface in the Chesapeake Bay would amount to a weekly addition of 5% to the total orthophosphate in the water column. This is considerably less than the rate at which phytoplankton remove phosphorus from the euphotic zone of the Bay (40).

It can be seen, therefore, that the observations relating to nutrient availability and plankton biomass in the Chesapeake Bay are apparently paradoxical: High phytoplankton standing stocks and high phytoplankton productivity persist in the presence of low ambient levels of dissolved inorganic nitrogen and phosphorus throughout much of the year. The short-term temporal stability in these patterns suggests an equilibrium in the biological and chemical processes, and Carpenter, Pritchard, and Whaley (1969) (36) hypothesized that the productivity of the open waters of the Chesapeake Bay is regulated largely by a dynamic nutrient-phytoplankton-zooplankton-nutrient cycle.

Our initial efforts to stimulate primary productivity by enriching natural water samples from various locations within the Chesapeake Bay were unsuccessful. The results clearly demonstrated the unsuitability of the nutrient enrichment-incubation technique in the investigation of a highly dynamic planktonic ecosystem, and supported the hypothesis that only through short-term isotope tracer experiments could one come to an understanding of the plankton nutrition in this estuary.

The phytoplankton may grow at rates sufficient to double their biomass in one or two days and yet such increases in biomass are rarely observed in the Bay. In fact, with the exception of localized blooms, the phytoplankton standing stock varies only by approximately a factor of 5 in the entire main body of the Bay throughout an annual cycle. Consider further a summer condition of nearly constant low levels of plant nutrients in waters containing a population of small herbivorous zooplankters capable of ingesting 2 to 3 times their body mass daily, and one can begin to appreciate both the dynamic nature of estuarine planktonic communities and the great effort required to investigate their nutrition.

What follows is a description of our recent effort to study the plankton nutrition in the Chesapeake Bay with isotope tracer techniques. This program remains in progress, and most of our results to date are either in press or in a state of preparation. Our intention here is to demonstrate through a preview of some of our data the utility of isotope tracers in developing a dynamic image of plankton nutrition which permits insight into the mechanisms by which planktonic productivity is regulated.

We conducted a series of 7 cruises of 2 weeks duration and at 6 week intervals to sample and study 8 geographically fixed stations in the open waters of the Chesapeake Bay and one station on the adjacent continental shelf (Figure 1). Nutrient measurements were made and experiments were initiated to quantitate phytoplankton incorporation rates for carbon, nitrogen, and phosphorus. Numerous other physical, chemical, and biological measurements were made, and the collected biota were also used for additional experiments. Each station required a full day of effort. With the exception of the previously mentioned modifications, all shipboard analytical methods followed the procedures outlined by Strickland and Parsons (1972) (32).

All nutrient, biomass, and nutrient uptake rate measurements which will be reported are averages of two samples from different depths in the euphotic zone, and they are considered to be representative of the upper layer of the estuary. The near constancy of chorophyll *a* concentrations (hereafter referred to as chlorophyll) with depth in the profiles throughout the upper layer supports the notion that this layer, and likewise the euphotic zone, is well mixed. In order to minimize the effect of any small scale inhomogeneities, all individual measurements were made with material collected in a composite sample (the contents of 6 Van Dorn bottles which were cast to the same depth were mixed and aliquots were withdrawn).

$^{14}C$ labeled carbonate, $^{32}P$ labeled phosphoric acid, and $^{15}N$ labeled nitrate, nitrite, ammonium, and urea were added to separate bottles and each was incubated on the deck of the ship under simulated *in situ* conditions. At the termination of the experiments the particulate material was analyzed for isotopic enrichment. $^{14}C$ and $^{32}P$ were determined by liquid scintillation spectrometry and $^{15}N$ by mass spectrometry. The dissolved organic phosphorus and dissolved polyphosphate were also examined for enrichment, and similar measurements for dissolved organic carbon and nitrogen are part of our continuing program. Because of the frequent condition of low nutrients, high biomass, and hence rapid turnover of available nutrient, incubations were of only a few hr duration at midday and extrapolations to daily rates were made using occasional 24 hr sequences of short incubations.

The distribution of plant biomass, as chlorophyll, is rather surprisingly uniform in the main body of the Chesapeake Bay (Figure 2). In April and June higher biomass was observed in the vicinity of the Potomac River discharge, and in June and

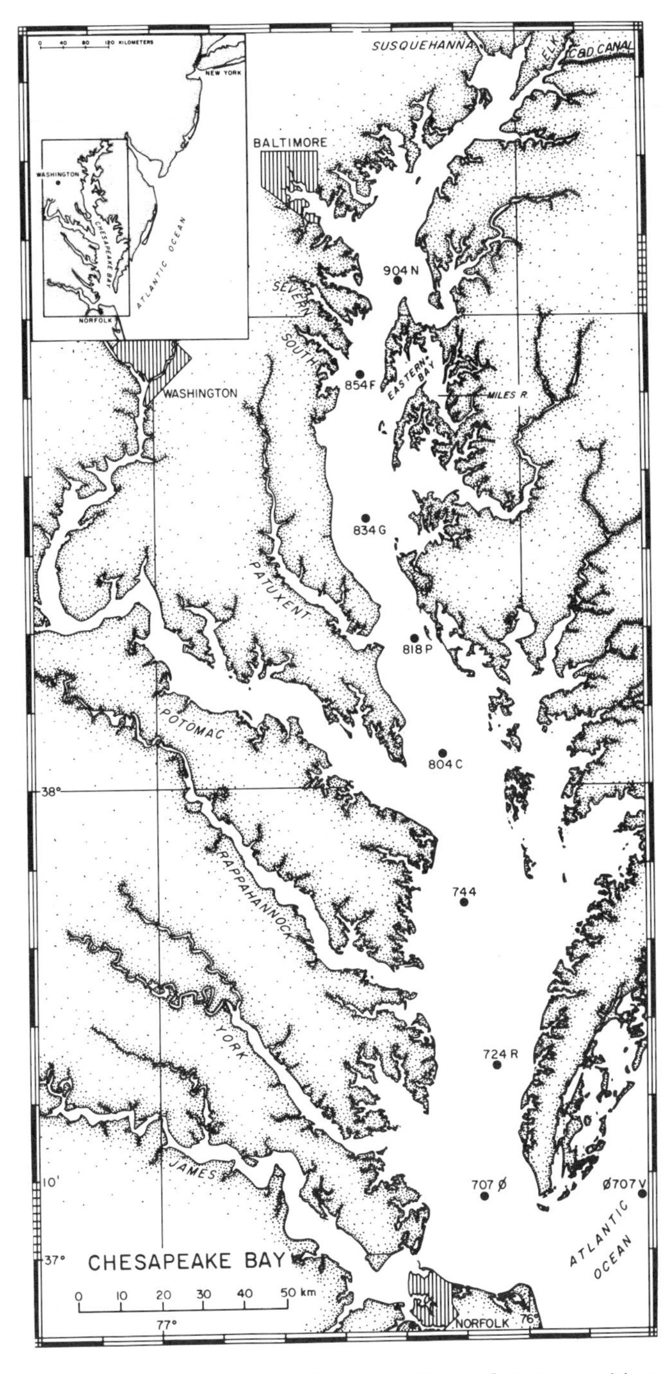


*Figure 1. Location of the Chesapeake Bay and station positions*

August higher biomass was observed below Baltimore. But in general, there is little temporal or spatial heterogenity: $>$50% of the chlorophyll values were between 10 and 20 $\mu g \cdot liter^{-1}$, and 90% were between 5 and 25 $\mu g \cdot liter^{-1}$. The cause of the increased biomass below Baltimore may be the result of greater nutrient availability, but there is not evidence for this in either the nitrogen or phosphorus data throughout most of the year. There is sufficient nitrogen in this region to support a greater plant biomass than is observed and, as demonstrated by Taft, Taylor, and McCarthy (1975) (40), there is no apparent relationship between plant biomass and soluble reactive phosphorus. The same conclusions with respect to nitrogen and phosphorus apply to the area immediately below the Potomac River in August when compared to the stations further north.

There are numerous possible explanations for the occasional association of higher plant biomass with regions downstream from the discharges of Baltimore and District of Columbia, and some of the more obvious include: greater availability of minor nutrients; unidentified stimulating material associated with the discharge of the metropolitan wastes; less herbivorous grazing potential which may or may not be related to the discharges; and reduced thickness of the near surface mixed layer. We have no data to permit evaluation of hypotheses concerning minor nutrients and other stimulating materials. It would be exceedingly difficult to detect the small reduction in herbivorous grazing potential which could within a few days result in plant biomass increases comparable to those observed. The pycnocline was quite shallow below Baltimore in August, and we are further evaluating the relationship of both phytoplankton biomass and phytoplankton productivity on a volume basis to thickness of both the upper mixed layer and the euphotic zone. McCarthy, Taylor and Loftus (1974) (41) demonstrated that throughout the Chesapeake Bay and over an annual cycle, approximately 90% of both the biomass and the productivity of the phytoplankton community is found in the unicellular forms which pass 35 $\mu m$ mesh.

There are well documented occurences of algal blooms of moderate proportions in the Chesapeake Bay. During our study a major visible bloom was never identified on our regular stations. Loftus, Subba Rao, and Seliger (1972) (42) followed the development and dissipation of a bloom which appeared in the open Bay near the Severn River following an intensive rain storm. Their data demonstrate that for the portion of the bloom in the Bay proper there was never adequate nitrogenous nutrient available to permit more than a fractional doubling of plant biomass (1-13%). Therefore, with nutrient limited growth, the bloom was either doomed to dissipate through physical forces or sufficient nutrient had to be delivered through rapid recycling processes within the bloom. *Oxytoxum* sp. was the dominant phytoplankter and the dominant herbivore, a rotifer (*Euchlanis* sp.), reached densities of $10^5$ individuals$\cdot liter^{-1}$. They demonstrated that this density of Euchlanis sp. could almost totally consume maximum observed bloom phytoplankton densities ($4 \times 10^2$ μg chlorophyll·liter$^{-1}$) in 12 hr. A rotifer of similar size (Brachionus sp.) ingests 3 times its mass per day (43), and it can be concluded that the high metabolic rate per unit biomass for such small zooplankters would result in considerable return of ingested nitrogen and phosphorus to the water in dissolved forms.

Time permits the consideration of only a single cruise in any detail, and we have chosen to discuss PROCON 10 of June 1973. It represents an extreme in chlorophyll variations along the major axis of the Bay, and it demonstrates the preferential algal utilization of the more reduced and rapidly recycled forms of nitrogen in the presence of high concentrations of $NO_3^-$.

Time course measurements with multiple samplings during an incubation of a few hours repeatedly demonstrated that $^{32}P$ uptake occurred initially at a rapid rate which could not, from comparisons with photosynthesis as measured by $^{14}C$ fixation, be equated with net uptake associated with growth (40). The subsequent phase of reduced uptake can however, with a few exceptions be shown to approximate phytoplankton synthesis of new cellular material. Table I and Figure 3 demonstrate this phenomenon.

TABLE I. ATOMIC RATIO OF CARBON TO PHOSPHORUS UPTAKE (40)

| Station | Initial Phase | Subsequent Phase |
|---|---|---|
| 834G | 0.16 | 44 |
| 818P | 0.001 | 28 |
| 744 | 0.001 | 60 |
| 724R | 0.002 | 88 |
| 707Ø | 0.015 | 410 |
| Ø707V | 0.45 | 6 |

From the laboratory culture data of Parsons, Stephens, and Strickland (1961) (44), one can calculate atomic ratios for carbon to phosphorus composition of marine phytoplankton which were grown in a phosphorus sufficient medium. For 8 phytoplankters the values ranged from 28 to 102 and had a median of 81. A phosphorus deficient natural phytoplankton population existing in a medium which contains a low constant level of orthophosphate (and no alternate source of phosphorus) would be expected to have less intracellular phosphorus than phytoplankton in a phosphorus sufficient medium (45). The well documented phytoplankton "luxury uptake" of orthophosphate (46) complicates efforts to make meaningful extrapolation from short-term rate measurements. That is to say, that even after taking into consideration the above mentioned dual phase uptake, rate measurements for the second phase may still suggest greater net uptake per unit time than can be reasonably anticipated from independent estimates of

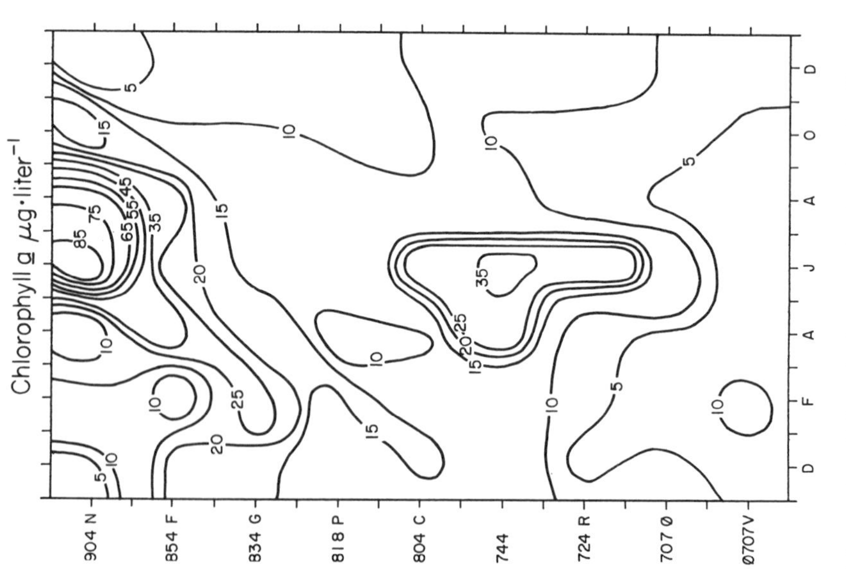


*Figure 2. Chlorophyll concentration in the Chesapeake Bay from December 1972–December 1973*

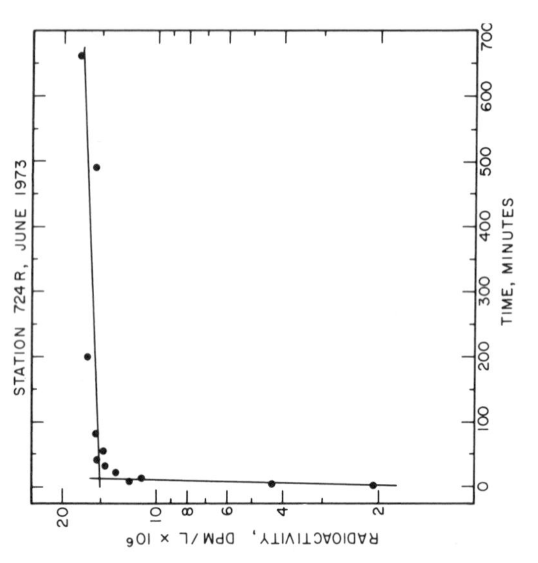


*Figure 3. Uptake of $^{32}P$ orthophosphate into the particulate fraction with time, after Ref. 40.*

phytoplankton growth. If the algae are phosphorus stressed as a result of lengthy exposure to medium with lower than usual levels of orthophosphate, and sufficient phosphorus suddenly becomes available, an extremely rapid uptake would be expected in the second phase. Such a stress would result in a low carbon to phosphorus uptake value such as that observed at station Ø707V (Table 1). Conversely, ample intracellular stores of phosphorus could support high rates of algal carbon fixation concurrent with very slow phosphorus uptake resulting in a high carbon to phosphorus uptake ratio such as that observed at station 707Ø . Stations 707Ø and Ø707V were anomalous in other respects as well.

The high specific activity of $^{32}P$ preparations permits one to make estimates of uptake without measurably increasing the orthophosphate in the medium (radioactive isotope additions contributed $< 2x10^{-4}$ μg atom P • liter $^{-1}$ in these experiments). In that significant changes in substrate were avoided, one is left to conclude that, except for stations 707Ø and Ø707V, our ratios for the subsequent phase of reduced uptake are reasonable approximations for cellular synthesis in Chesapeake Bay phytoplankton.

It should be obvious from the data which we have presented that one cannot make meaningful interpretations from single time determinations of $^{32}P$ uptake by phytoplankton in the Chesapeake Bay. In practice, aquatic biologists are often far too casual with their assumptions that the processes under investigations can be adequately evaluated with a single measurement per experiment.

Taft and Taylor (1975) (47) have described the annual cycle of phosphorus in the Chesapeake Bay, and they note that within the dissolved component, organic phosphorus is often as abundant as orthophosphate. As with dissolved organic nitrogen, much of this phosphorus remains unidentified, but phytoplankton can, with alkaline phosphatase, utilize phosphomonoesters (48,49). Taft, Loftus, and Taylor (50) have determined that throughout the year phosphomonoesters are never more than a small fraction of the total dissolved organic phosphorus in the Bay. The kinetics of algal alkaline phosphatase activity were investigated with $^{32}P$ labeled glucose-6-phosphate, and continuing efforts are concentrating on the nutritional value of other components of the natural dissolved organic phosphorus pool.

Nitrate concentrations (Figure 4) ranged from 0.3 to 33 μg atom N·liter with a rather sharp gradient between 904N and 724R. $NH_4^+$ concentrations (Figure 5) ranged from $<$0.05 to 5.9 μg atom N· -liter$^{-1}$ and showed the sharpest gradient between 834G and 744. $NO_2^-$ concentrations ranged from $<$ 0.05 to 1.0 μg atom N·liter$^{-1}$ and in general varied with $NO_3^-$ concentrations. Urea-N concentrations ranged from $<$ 0.05 to 1.0 μg atom N·liter $^{-1}$ and were highest in mid and upper Bay.

Elsewhere we have given considerable attention to the particulate nitrogen, particulate phosphorus, particulate carbon, chlorophyll and adenosine triphosphate data, and have discussed

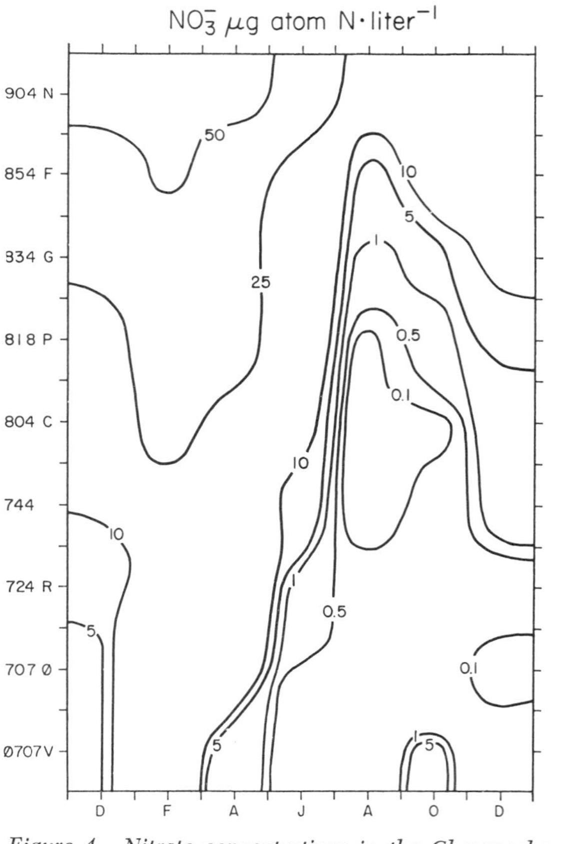


*Figure 4. Nitrate concentrations in the Chesapeake Bay from December 1972–December 1973*

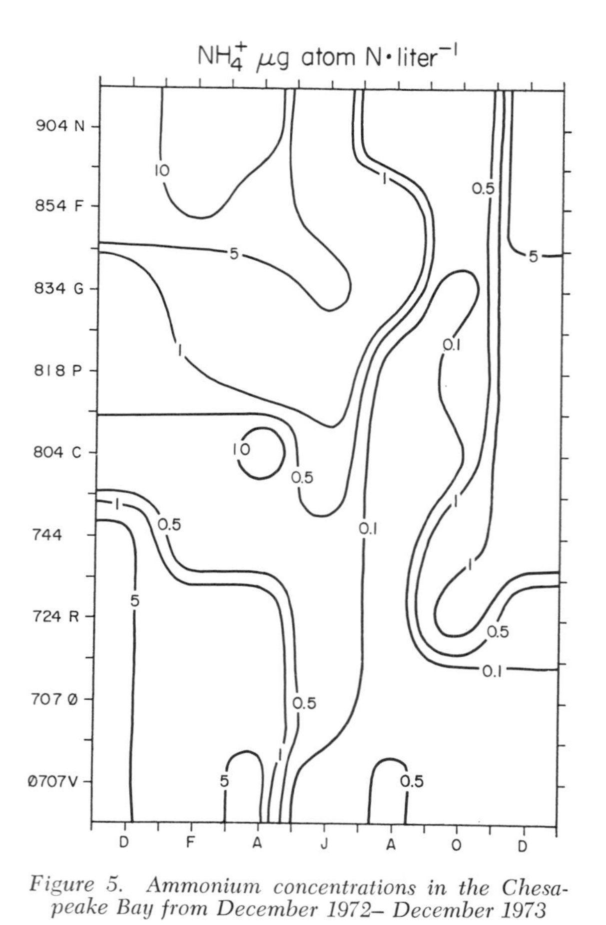


*Figure 5. Ammonium concentrations in the Chesapeake Bay from December 1972– December 1973*

usefulness of ratios from such data in approximating the viable, and specifically the plant,component of the total particulate material (manuscript in preparation). At this point, let it suffice to say that the chlorophyll and particulate nitrogen data (Figure 6) support the notion that the suspended particulate material below Baltimore and at the mouth of the Potomac River is as rich in chlorophyll as that elsewhere in the Bay. This does not rule out the possibility that a major difference in nature of the plant community may to some degree be masking significant detrital input. The salinity (Figure 6) in the upper mixed layer shows the anticipated increase with distance from the head of the Bay.

The stations occupied in this study were selected to represent, at ~ 30 kilometer invervals, various areas of potential interest. High turbidity is frequently encountered in regions of the Bay north of 904N (51), and under such conditions, our experiments could not be executed. The net seaward movement of the upper layer is 0.8 $km \cdot day^{-1}$ in the summer (36) and hence our stations on a particular cruise are dealing approximately with a parcel of water which was located one station north in the Bay on a previous cruise. For the $NO_3^-$ data (Figure 4) one can see that in moving from top to bottom of the plot (down Bay) and from left to right (to a subsequent cruise) the $NO_3^-$ value may be similar but is never much greater. This supports the argument that the Susquehana River (or other supply above station 904N) is the primary source of $NO_3^-$ in the Bay proper.

One can see an interesting pattern to the availability and utilization for each of the 4 nitrogenous nutrients (Figure 7). Although $NO_3^-$ was abundant from 804C to 904N (15-33 μg atom $N \cdot liter^{-1}$) less than 2% of the total nitrogenous ration of the phytoplankton was derived from $NO_3^-$. At the same stations $NH_4^+$ was less available (1-6 μg atom $N \cdot liter^{-1}$), but it invariably accounted for greater than 90% of the nitrogenous ration. From 744 to Ø707V $NH_4^+$ became more scarce (< 0.4 μg atom $N \cdot liter^{-1}$) and although $NO_3^-$ concentrations also became reduced, $NO_3^-$ remained as a major component of the available nitrogenous nutrient pool. It reached maximum importance at Ø707V when a $NO_3^-$ concentration of 0.27 μg atom $N \cdot liter^{-1}$ was the only nitrogenous nutrient detected. The data for both fractional availability and fractional utilization of $NO_3^-$ and $NH_4^+$ are near mirror images, and throughout this series of cruises our data always demonstrated this pattern of preferential uptake of $NH_4^+$. The patterns for $NO_2^-$ and particularly urea-N suggest fractional utilization in proportion to fractional availability.

Calculations for turnover time of $NH_4^+$ in the euphotic zone range from 3 to 20 hours and average 8 hours for PROCON 10. The rapid turnover time and lack of persistent spatial gradients argue further for local origin of $NH_4^+$. If one accepts the Susquehana River as the source of $NO_3^-$, it is clear that at the upper Bay stations there is sufficient $NO_3^-$ for a doubling of

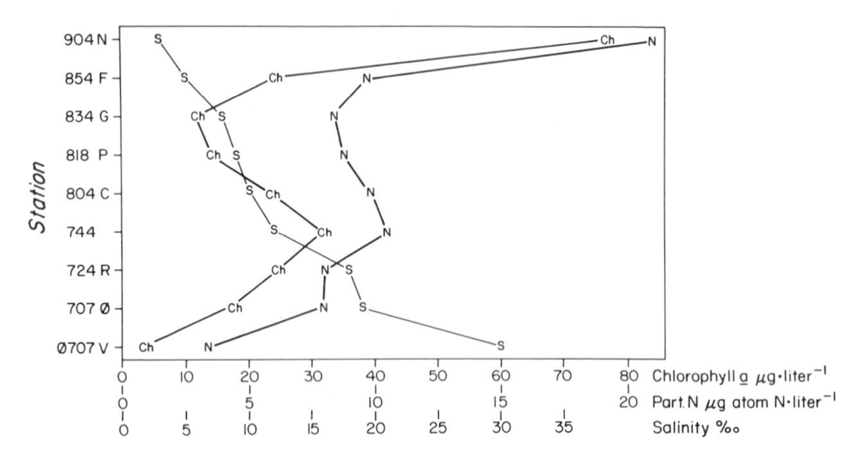


*Figure 6. Salinity, particulate nitrogen, and chlorophyll* a *observed in PROCON 10 in June 1973*

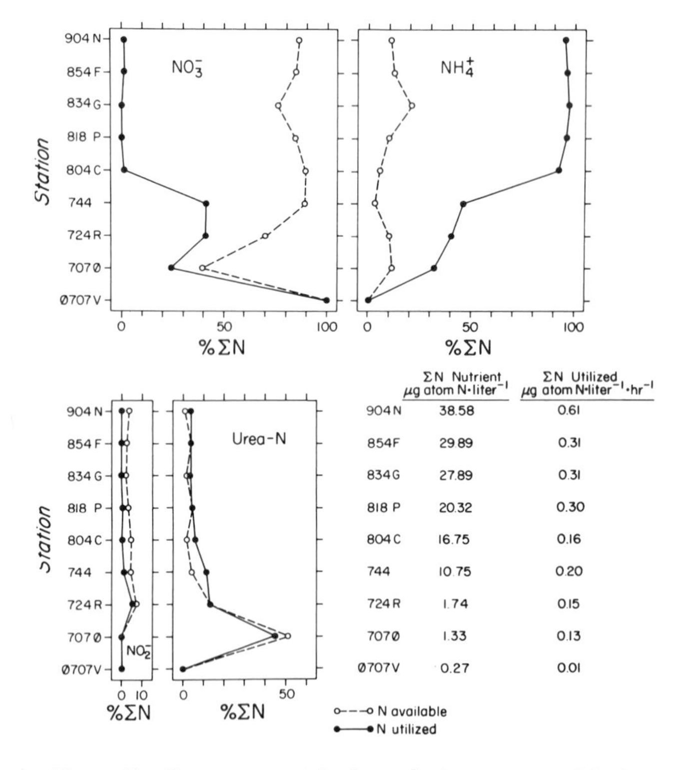


| | ΣN Nutrient μg atom N·liter⁻¹ | ΣN Utilized μg atom N·liter⁻¹·hr⁻¹ |
|---|---|---|
| 904N | 38.58 | 0.61 |
| 854F | 29.89 | 0.31 |
| 834G | 27.89 | 0.31 |
| 818 P | 20.32 | 0.30 |
| 804C | 16.75 | 0.16 |
| 744 | 10.75 | 0.20 |
| 724R | 1.74 | 0.15 |
| 707Ø | 1.33 | 0.13 |
| Ø707V | 0.27 | 0.01 |

*Figure 7. Percentages of both total nitrogenous nutrient available and total nitrogen utilized as nitrate ammonium, nitrite, and urea; total nitrogenous nutrient and rate of phytoplankton nitrogen utilization. Data for PROCON 10 in June 1973.*

of plant biomass (~1-2 days of growth) but the $NO_3^-$ is obviously not utilized at this rate or high levels would not be observed below Baltimore. The standing stock of the phytoplankton changes relatively little with time (although high productivity rates are constantly observed) and one is left to conclude that the herbivorous potential of the zooplankton is sufficient to permit consumption of most of the plant production. Herbivorous zooplankton release nitrogenous metabolites which can be utilized by the plants. These are primarily $NH_4^+$ with some urea (52,53) and under some conditions amino nitrogen (54).

Figure 8 shows the portion of the temporal and spatial coverage represented by our investigation in which there is insufficient nitrogenous nutrient for a single doubling of the particulate nitrogen. And yet one can observe few if any significant daily changes in plant biomass, plant productivity, zooplankton biomass, and nutrient concentrations at the lower and mid Bay stations throughout the summer and autumn.

During PROCON 10 the observed urea concentrations were lower than usual, and the portion of the phytoplankton ration as urea nitrogen was less than our average for the balance of the study; this was probably due to both low urea concentrations and high $NH_4^+$ concentrations. The observation that $NH_4^+$ was selected in preference to $NO_3^-$ was consistently demonstrated, and this has also been noted in laboratory cultures (55) and in a large volume outdoor culture (56). A composite figure of most data from the present study is shown in Figure 9. The universality of the $NH_4^+$ preference phenomenon in the Chesapeake Bay is obvious. The conclusion from this particular analysis is that at $NH_4^+$ concentrations > 1-1.5 μg atom N·liter$^{-1}$, 95% of the nitrogenous ration of the phytoplankton will be met with $NH_4^+$ and urea. This $NH_4^+$ value is similar to that found sufficient to suppress $NO_3^-$ utilization in the culture studies mentioned above.

If one were to determine the effect of ambient $NH_4^+$ concentration on $NO_3^-$ uptake with a unialgal culture, a hyperbolic relationship might be predicted. The scatter in Figure 9 is probably the result of combining data from 120 natural phytoplankton assemblages sampled in waters with salinities ranging from 2 to 30 o/oo and temperatures ranging from 4 to 29°C.

This plot does not necessarily suggest that $NH_4^+$ concentrations < 1-1.5 μg atom N·liter$^{-1}$ will induce $NO_3^-$ uptake, but rather it demonstrates that in excess of this concentration, little if any $NO_3^-$ is utilized when available, and hence the phytoplankton are probably not growth rate limited by nitrogenous nutrient. The kinetics of phytoplankton uptake of $NO_3^-$ and $NH_4^+$ for unialgal cultures are well documented (57), and if one views the uptake of $NO_3^-$ and $NH_4^+$, somewhat paradoxically, as analagous to Holling's "vertebrate" feeding response (58), then the value of 1.5 μg atom N·liter$^{-1}$ may represent a concentration of total nitrogenous nutrient above which there will be no nitrogen limitation of primary productivity.

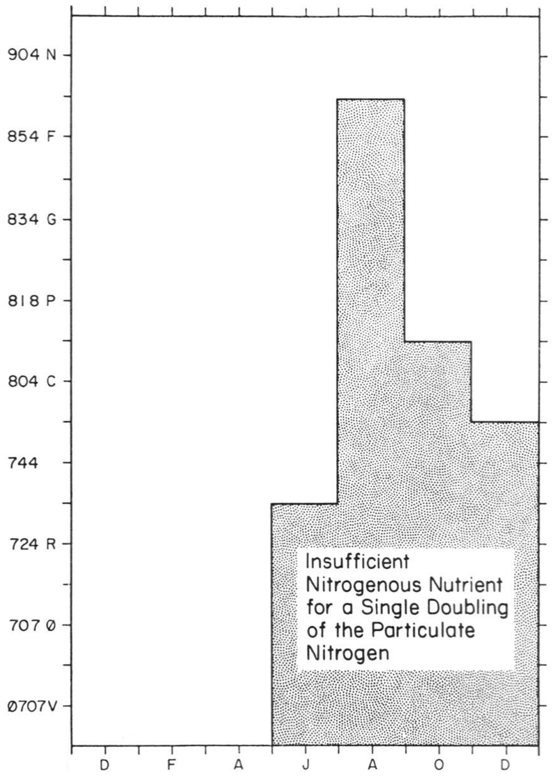

*Figure 8. Portion of the Chesapeake Bay and portion of the 13-month study in which available nitrogenous nutrient was insufficient to permit a single doubling of the particulate nitrogen*

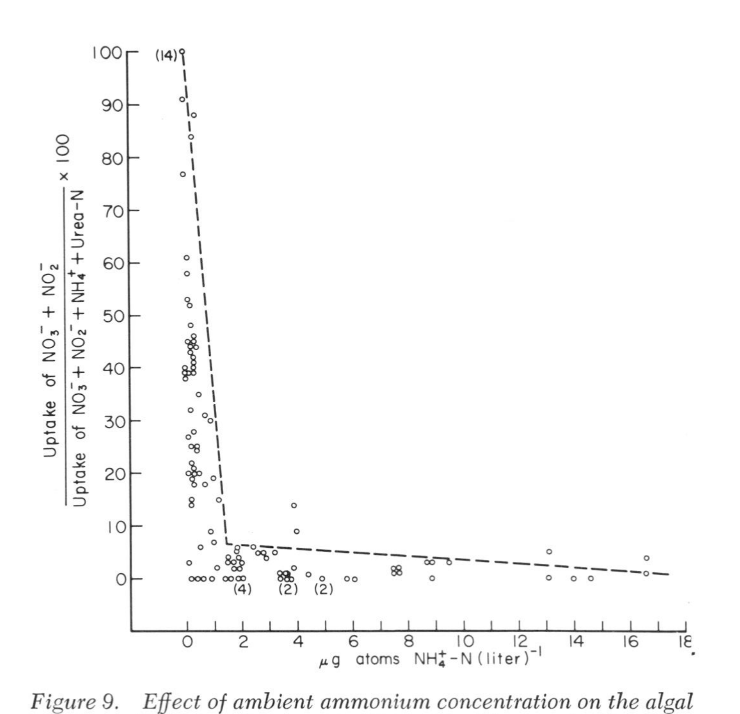


*Figure 9. Effect of ambient ammonium concentration on the algal selection of nitrate. Data are for the entire Chesapeake Bay from December 1972 through December 1973.*

At the present time our greatest uncertainty in the nitrogen and phosphorus budgets of the main body of the Chesapeake Bay rests with the question of local nutrient supply. We have concluded that in general neither vertical transport from the sediments nor rainfall (unpublished data) can be considered as major sources. A large body of data provide both direct and indirect evidence which suggests that herbiverous zooplankters are capable of consuming the phytoplankton productivity. The smaller the zooplankton, the greater the fractional return of ingested nitrogen and phosphorus to the water via excretion. We are in the process of evaluating the importance of this pathway to local nutrient replenishment in the Bay.

Bacteria in the water column, whether free-living or associated with larger particles, may in part be responsible for both supply and loss of the plant nutrients discussed above. Unfortunately, it is extremely difficult to quantitate with even fair accuracy the role of the bacteria, and the significance of bacterial activity in these processes has not been evaluated for any large area of the open Bay. Indirect evidence suggests, however, that the role of bacteria is minor when compared to those of both phytoplankton and zooplankton.

The general impression which we would like you to obtain from this presentation is that plankton nutrition must be viewed as a dynamic process. One can be totally deceived in an effort to understand plankton nutrition solely from measurements of biomass and nutrient concentrations, and, therefore, unless one partitions the nutrient pool and actually measures rates of utilization, little useful information can be obtained from field programs designed to investigate various links in the nutrient-phytoplankton-zooplankton-nutrient cycle.

Acknowledgment

This work was supported by grant GA-33445 and grant DES75-02846 from the National Science Foundation and contract AT (11-1) 3279 with the U.S. Atomic Energy Commission.

LITERATURE CITED

1. Steemann-Nielsen, E., J. Cons. Explor. Mer. (1952) 18, 117-140.
2. Hutchinson, G.E., Bowen, V.T., Proc. Nat. Acad. Sci. (1947) 33, 148-153.
3. Rigler, F.H., Ecology (1956) 37, 550-562.
4. Correll, D.L., Limnol. Oceanogr. (1965) 10, 364-370.
5. Neese, J.C., Dugdale, R.C., Dugdale, V.A., Goering, J., Limnol. Oceanogr. (1962) 7, 163-169.

6. Dugdale, R.C., Goering, J.J., Limnol. Oceanogr. (1967) 12, 196-206.
7. Arthur, C.R., Rigler, F.H., Limnol. Oceanogr. (1967) 12, 121-124.
8. Nalewajko, C., Lean, D.R.S., J. Phycol. (1972) 8, 37-43.
9. Williams, P.J.Le B., Berman, T., Holm-Hansen, O., Nature New Biology (1972) 236, 91-92.
10. Berman, T., J. Phycol. (1973) 9, 327-330.
11. Conover, R.J.,Francis, V., Mar. Biol. (1973) 18, 272-283.
12. Azam, F., Holm-Hansen, O., Mar. Biol. (1973) 23, 191-196.
13. Goering, J.J., Nelson, D.M., Carter, J.A., Deep-Sea Res. (1973) 20, 777-789.
14. Azam, F., Planta. (1974) 121, 205-212.
15. Thomas, W.H., Limnol. Oceanogr. (1966) 11, 393-400.
16. Eppley, R.W., Carlucci, A.F., Holm-Hansen, O., Kiefer, D., McCarthy, J.J., Venrick, E., Williams, P.M., Limnol. Oceanogr. (1971) 16, 741-751.
17. Ryther, J.H., Dunstan, W.M., Science (1971) 171, 1008-1013.
18. Ketchum, B.H., "Eutrophication: Causes, Consequences, Correctives," p. 197-209, Natl. Acad. Sci., Washington, D.C., 1969.
19. Vaccaro, R.F., "Chemical Oceanography," Vol. 1, p. 365-408, J.P. Riley and G. Skirrow, Ed., Academic Press, New York, 1965.
20. Newell, B.S., J. Mar. Biol. Assoc. U.K. (1967) 47, 271-280.
21. McCarthy, J.J., Limnol.Oceanogr. (1970) 15, 309-313.
22. McCarthy, J.J., Kamykowski, D., Fish. Bull. (1972) 70 1261-1274.
23. Remsen, C.C., Limnol. Oceanogr. (1971) 16, 732-740.
24. Chau, Y.K., Riley, J.P., Deep-Sea Res., (1966) 13, 1115-1124.
25. Clark, M.E., Jackson, G.A., North, W.J., Limno. Oceanogr. (1972) 17, 749-758.
26. Wheeler, P.A., North, B.B., Stephens, G.C., Limnol. Oceanogr. (1974) 19, 249-259.
27. Schell, D.M., Limnol. Oceanogr. (1974) 19, 260-270.
28. Thomas, W.H., Renger, E.H., Dodson, A.N., Deep-Sea Res. (1971) 18, 65-71.
29. Gordon, D.C., Sutcliffe, W.H., Marine Chem. (1973) 1, 231-244.
30. McCarthy, J.J., Eppley, R.W., Limnol. Oceanogr. (1972) 17, 371-382.
31. Stewart, W.D.P., Fitzgerald, G.P., Burris, R.H., Proc. Natl. Acad. Sci. U.S. (1967) 58, 2071-2078.
32. Strickland, J.D.H., Parsons, T.R.,"A Practical Handbook of Sea Water Analysis," 310 p., Fish. Res. Bd. Can., Ottawa, 1972.
33. Gundersen, K.,Helgol. Wiss. Meersunters (1973) 24, 465-475.
34. Kiefer, D., Strickland, J.D.H., Limnol. Oceanogr. (1970) 15 408-412.

35. Pritchard, D.W., "Estuaries," p.34-44, G.H. Lauff, Ed., Amer. Assoc. Adv. Sci., Washington, D.C., 1967.
36. Carpenter, J.H., Pritchard, D.W., Whaley, R.C., "Eutrophication; Causes, Consequences, Correctives," p.210-221, Natl. Acad. Sci., Washington, D.C., 1969.
37. Rowe, G.T., personal communication.
38. Hartwig, E.O., "Physical, Chemical, and Biological Aspects of Nutrient Exchange Between the Marine Benthos and the Overlying Water." p.164, Ph.D Dissertation, Univ. Calif. San Diego, 1974.
39. Bray, J.T., Bricker, O.P., Troup, B.N., Science (1973) 180, 1362-1364.
40. Taft, J.L., Taylor, W.R., McCarthy, J.J., Mar. Biol. (1975) in press.
41. McCarthy, J.J., Taylor, W.R., Loftus, M.E., Mar. Biol. (1974) 24, 7-16.
42. Loftus, M.E., Subba Rao, D.V., Seliger, H.H., Ches. Sci. (1972) 13, 282-299.
43. Theilacker, G.H., McMaster, M.F., Mar. Biol. (1971) 10, 183-188.
44. Parsons, T.R., Stephens, K. Strickland, J.D.H., J. Fish. Res. Bd. Can. (1961) 18, 1001-1016.
45. Fuhs, G.W., J. Phycol. (1969) 5, 312-321.
46. Fitzgerald, G.P., Nelson, T.C., J. Phycol. (1966) 2, 32-37.
47. Taft, J.L., Taylor, W.R., Ches. Sci. (1975) in press.
48. Berman, T. Limnol. Oceanogr. (1970) 15, 663-674.
49. Perry, M.J. Mar. Biol. (1972) 15, 113-119.
50. Taft, J.L., Loftus, M.E., Taylor, W.R., unpublished manuscript.
51. Schubel, J.R., Ches. Sci. (1968) 9, 131-135.
52. Corner, E.D.S., Davies, A.G., Adv. Mar. Biol. (1971) 9, 101-204.
53. Eppley, R.W., Renger, E. H., Venrick, E.L., Mullin, M.M. Limnol. Oceanogr. (1973) 18, 534-551.
54. Webb, K.L., Johannes, R.E. Limnol Oceanogr. (1967), 12, 376-382.
55. Eppley, R.W., Coatsworth, J.L., Solórzano, L., Limnol Oceanogr. (1969) 14, 194-205.
56. Strickland, J.D.H., Holm-Hansen, O., Eppley, R.W., Linn, R.J. Limnol. Oceanogr. (1969) 14, 23-34.
57. Eppley, R.W. Rogers, J.N., McCarthy, J.J., Limnol. Oceanogr. (1969) 14, 912-920.
58. Holling, C.S., Mem. ent. Soc. Can. (1965) 45, 5-60.

# 41

# Gross Analyses of Organic Matter in Seawater: Why, How, and from Where

JONATHAN H. SHARP

College of Marine Studies, University of Delaware, Lewes, Del. 19958

The approach in this paper is somewhat simplistic but it seems desirable to first attain a fairly broad knowledge on the scope of marine organic matter before dealing somewhat blindly with the more esoteric or chemically specific aspects. Undoubtedly one needs considerably more information about marine organic matter to better understand a number of areas in chemical oceanography. In this symposium, Millero, Kester, and Wood have stated in their papers that one needs to learn more about organic-inorganic interactions before fully describing ion speciation in coastal waters. Both Jewett and Brinkman have mentioned the problems of misinterpretation of trace metal analyses due to differences between freely ionized and organically-complexed metals. In discussing estuarine nutrient budgets, Yentsch and McCarthy have emphasized the importance of organic matter. Clearly, there is no question of the mandate for considerable research in the field of marine organic chemistry.

In recent years, there has been much activity by workers attempting to measure certain specific pollutants in seawater such as DDT, PCB's, and oil hydrocarbons. However, as Table I shows, very little is known about the total organic pool in seawater. The total organic pool as referenced here does not include volatile organic matter, but as Sackett has demonstrated in this symposium, gaseous hydrocarbons are found in fairly low concentrations and the most abundant of the gases, methane, would contribute only about 0.03 $\mu g\ C \cdot l^{-1}$ (11, 12). Thus, the vast majority of the organic pool is not easily characterized from our present analyses of specific compounds. On the other hand, better than 99% of the total inorganic pool can be characterized with eleven fairly invariant constituents (13). All the organic constituents listed in Table I are quite variable and yet they only make up about 10% of the total. So, in contrast to inorganic chemistry, studies of organic chemistry in seawater must deal with a whole, of which at least 90% is qualitatively unknown and the remainer is at best quite variable.

Table I. Representative values for specific organic compounds measured in seawater. Values are taken from listed references as approximate averages for surface oceanic or not heavily polluted coastal waters and are listed in terms of carbon.

| Compound | Concentration (as $\mu g\ C \cdot l^{-1}$ | Reference |
|---|---|---|
| Total organic pool | 1500 | 1 |
| Glucose | 50 | 2 |
| Free amino acids | 30 | 3 |
| Total hydrocarbons | 30 | 4 |
| Free fatty acids | 15 | 5 |
| Urea | 10 | 6 |
| Glycolic acid | 10 | 7 |
| Creatine | 2 | 8 |
| Vitamins | 0.01 | 9 |
| Chlorinated hydrocarbons | 0.01 | 10 |

The study of organic matter in the sea is very difficult, for the possible array of constituents is enormous. One need only look into the field of petrogenesis (14) to see how complex natural organic deposits can be. Recent studies of organic products in living marine organisms (15) also illustrates the multitudinous compounds and complicated nature of the problem. In addition to the contribution by marine organisms as they go through detrital avenues, excretion of bioactive compounds (16) by healthy organisms presents many interesting constituents to the sea.

The scope of this symposium is the coastal environment while the majority of pertinent research in marine organic chemistry has been pursued in oceanic waters. However, the coastal environment is enough similar to the open ocean that a study of organic chemistry in the former is like an oceanic study with most tools being more easily applied. A basic criterion for this statement and the ensuing discussion is that all organic matter in the sea should originate from plant production. In the oceanic environment, all production takes place in about a 100 meter thick surface veneer (the photic zone) and then by gravity and the food web it falls out in ever diminishing quantities so that the deep sea contains relatively old organic matter in fairly constant and low level proportions. Moving inshore from this environment, one finds that the photic zone production is greater and that the deep water is shallower (containing younger, less constant, and higher levels of organic matter). Additionally, one finds an increasing influence of terriginous input of organic matter as one approaches land. I refer only to plant production (organic matter of animal origin is, of course, indirectly from plant production) and do not include anthropogenic organic matter. I do this for three reasons. First, man's introduction of organic matter into the sea on a global basis is still probably relatively small compared to that from plant origin. Second, this introduction is an inconstant unnatural perturbation to the natural processes of production and subsequent breakdown of marine organic matter. And third, to assess this perturbation one must be able to first describe the natural processes. I do not intend to understate the capability of Homo sapiens to drastically alter his environment; but, in the area of marine organic chemistry, his potential has not yet been fully achieved.

With the above statements in mind, one can see some of the design needed for the analytical methods. If analyses can be used of sufficient sensitivity and thoroughness for deep ocean work, the same methods should be adequate in coastal waters. In the latter case, we expect to find organic matter in larger quantities and of younger age (thus, presumably less difficult to measure in destructive analyses). Changes in other chemical concentrations, such as lower total salt content and higher proportions of some soluble inorganic constituents, should not greatly alter efficacy of the desired methods. In upper estuaries approaching fresh water, some methods may have to be slightly altered. Also, the increased sediment loading in coastal waters as well as higher organic contents often requires the use of smaller sample sizes than when working in oceanic waters. Therefore, in the following discussion I use an oceanic viewpoint both for recommended methodology and for interpretation of the environment.

## Why Gross Organic Analyses

Assuming an oceanic viewpoint, one can see that the vast majority of organic matter in the sea is not within living organisms. The average organic content in surface waters is about 1.5 mg carbon per liter with deep water values being less than 1 mg $C \cdot l^{-1}$ (1). The average concentration of living organic matter is somewhat difficult to assess. In surface waters, organic matter large enough to be classified as particulate makes up about 1 to 7% of the total organic carbon while in deep waters this is usually smaller than 1% (17). Essentially all living organisms fall within this category and estimates of living organic matter show it to be about 10-80% of the particulate matter in surface waters and less than 10% in deep waters (18).

Thus, in the oceanic environment, living organisms probably account for anywhere from less than 0.1% up to 5% of the total organic matter. Moving nearshore, the particulate fraction accounts for a greater portion of the total organic pool (17) and probably the "living fraction" is also larger. Nonetheless, the majority of organic matter in the marine environment is non-living.

A point in discussing gross organic analyses is the overall global balance of elements. Garrels and co-workers (19, 20) calculate that there is a net flux of organic carbon from land to sea for burial in sediments. In a similar fashion, phosphorous and nitrogen pass through the hydrosphere en route to sediments (21). As inorganic ions, they are fixed biologically and are ultimatelly carried to sediments for burial in organic form. Thus, organic carbon, nitrogen, and phosphorous measurements are important in studies of elemental fluxes from inorganic dissolved form into sediments. These constituents can be especially important when viewing coastal waters as a corridor between the land and the deep sea.

Much of this detrital organic matter is important through regenerative cycles as inorganic nitrogen and phosphorous plant nutrients; this regeneration controls subsurface oxygen content (22). Rates of organic degradation are not directly measurable through simple static measurements and considerable effort has been expended to indirectly measure organic utilization in the sea (23, 24, 25). One can see easily that such estimates are made difficult by the fact that both the organic substrates and the organisms effecting the degrading vary from sample to sample. Thus, an attempt is made to measure a rate for which variations occur due to both substrates and to biochemical reaction mechanisms. Observations of differences in oxygen content of water samples can be indicative of organic degradation, but without further inputs, quantification is not feasible.

In waste water analysis, oxygen demand has long been used as an indirect measure of organic loading. The standard biochemical oxygen demand (BOD) is a measure of bacterial use of oxygen in a five-day incubation (26). In the standard procedure (26) it is recommended to seed samples with sewage bacteria when necessary and to detoxify samples when necessary. Comparisons of BOD to total organic carbon analyses have been shown to be poor (27). Clearly, application of this measure to coastal marine chemistry adds very little to our knowledge. It does not tell us anything about the total organic content, but only about that which is degraded under contrived arbitrary conditions. Another standard waste water method is chemical oxygen demand (COD), which is the indirect measure of oxygen utilization by degradation of organic matter in a sulfuric acid-dichromate reflux (26). Comparison of COD to total organic carbon showed poor agreement (28) and the authors of the comparison partially attributed the poor agreement to the differing oxidation states of the organic matter. This is a good point and one that also applies to BOD analyses. An additional fault with the COD method is that the reflux method is probably not an exhaustive oxidation, for evolution of total organic carbon methods shows an interesting history of increasingly stronger oxidizing agents being applied (29). Indeed, recent estimates suggest that wet chemical oxidation methods are not sufficient for total degradation of organic matter in seawater (1,30).

These points summarized are: 1. Even though all the organic matter in the sea has plant production as its origin, the majority of it is in a non-living detrital form; 2. Global elemental balances show a flux of carbon, nitrogen, and phosphorous into the sea from land through organic compounds; 3. Much of the organic matter is important in the inorganic nutrient and oxygen cycles; 4. Currently employed indirect estimates (BOD and COD) of organic loading would be inadequate parameters for marine waters. Thus, the answer to the question in the heading is that gross organic analyses can be valuable tools in understanding biologically-mediated element cycles.

## How To Measure

Measurements of organic matter in seawater are usually preceeded by separation of suspended matter from the sample and then individual measurements of this "particulate organic matter" and the resultant filtrate or "dissolved organic matter". These divisions are quite arbitrary, but rather practical; they are discussed further below. Two types of samples result: dried particulate matter and seawater. The particulate fraction is usually deposited on an "organic-free" microporous glass fiber or silver membrane filter. Usually several hundred to several thousand milliliters must be filtered in the preliminary step for sufficient particulate sample to analyze, while the filtrate samples usually are done on aliquots of 5-10 milliliters or less. The methods given here are this author's estimate of the best routine ones for present-day use, and none are as good as could be hoped for in the future. For the descriptions, the samples on filters will be referred to as "particulate" and those that consist of unfiltered seawater will be referred to as "total". When total organic analysis is made on a filtrate, it fits the much used category of "dissolved" organic matter.

A. Total organic carbon. Two recent papers describe high temperature combustion methods for total organic carbon (1, 30). Neither method is yet at the routine stage, so the best routine method is a wet chemical one (31) in its modified form (1). Comparison of this modified method with the high temperature one for oceanic samples (1) shows the least difference for surface waters. Therefore, this method is probably fairly accurate for nearshore waters. In the method, inorganic carbon is purged from the sample, the sample is sealed in an ampoule with $K_2S_2O_8$, and the $CO_2$ from the oxidation is later measured in a non-dispersive infrared analyzer.

B. Total organic nitrogen. The simplest and best method is that of Armstrong, Williams, and Strickland (32) as described by Strickland and Parsons (33). In it, the organic nitrogen is degraded in solution to nitrate by short-wave ultraviolet light. Then the nitrate is reduced to nitrite by a cadmium-copper amalgam and the resultant nitrite is read colorimetrically. With initial nitrate, nitrite, and ammonium concentrations, organic nitrogen can be calculated.

C. Total organic phosphorous. A method using oxidation by $K_2S_2O_8$ has been proposed (34) and is used in some laboratories. The method suggested here is preferred because it seems a bit easier to perform especially when coupled with the organic nitrogen method. It is the same method as the nitrogen one (32, 33) except that the final readout is phosphate and correction is made for initial phosphate content.

D. Particulate carbon and nitrogen. These two constituents can be analyzed simultaneously making the suggested method superior to most previous independent ones. Several workers have described variations of this method, the one cited (35) is slightly revised with one instrument while two other instruments can also be used (36, 37) and are used in several laboratories. The method uses a commercial carbon-hydrogen-nitrogen analyzer which has a high temperature combustion step and a thermal conductivity gas chromatograph.

E. Particulate phosphorous. This method is again essentially that of Armstrong et. al. (32) as used for particulate matter by Perry and Eppley (38). In it, the filter is placed in distilled water for the ultraviolet exposure and phosphate analysis.

From Where

The title of this section is meant to be somewhat of a double entendre referring both to the sources of the organic matter in the seawater and the type of sample on which to do analyses. It was stated initially that essentially all organic matter in the sea comes from plant production. This being the case, one would expect to find a high concentration of organic matter at the time of high plant production and indeed this does occur (39). In fact, there is the contention that active healthy phytoplankton excrete directly into the sea a significan portion of their photoassimilated carbon (40). Proponents of this theory point out that the relative amount of phytoplankton excretion is less in coastal than in oceanic waters (41). The case for extensive phytoplankton excretion is possibly overstated and circumstantial, being largely based upon experimental artifacts.(42). In lieu of this source, probably most of the organic matter comes from post-flowering plant production and from inefficiency in food-chain transfer. A thought-provoking suggestion, made to me several years ago by Dr. Gordon A.

Riley, is that zooplankton when feeding upon phytoplankton must cause a good deal of the phytoplankton cell fluids to spill out into the seawater. As an analogy, one can visualize a child eating a juicy piece of fruit.

As Duursma suggested (39), probably post-flowering plankton populations contribute considerable amounts of organic matter to the sea. It is convenient, and necessary for understanding dynamic processes, to visualize the oceanic environment in a steady state (43). However, much of the organic matter may come from sporadic plankton blooms. Anyone working in the open sea is impressed by both the normal paucity of life and the extensiveness of blooms. In the central gyre of the North Pacific Ocean, I worked with other researchers on samples taken from the end of a diatom bloom. A description of the phytoplankton population has been published (44) and additional publications are pending. An interesting observation from this occurrence was that a large population of the diatom appeared to be localized immediately above the temporary thermocline at about 45 meters and that the cells did not appear to be very active physiologically. The horizontal extent of this population was probably only several square kilometers. Oxygen anomalies of similar geographical proportions have recently been documented (45) and they may not be uncommon phenomena. So, bloom conditions may be a cause of sporadic injections of organic matter in the oceanic environment as they have been shown to be in nearshore waters (39). Since these sporadic injections probably are a major source of organic matter, sampling on broad time and space scales is necessary to obtain a representative picture of the environment.

Detrital organic matter from higher plants is known to be of utmost importance to salt-marsh food chains (46) and would seem also to be important in general organic phosphorous cycling in estuaries (47). How much organic matter reaches coastal waters from marsh grasses is not known, but it could be appreciable. To get some feeling for this, one can compare data from estuarine and oceanic waters. It is difficult to get truely representative data for the comparison, but an attempt is made here by using data from the mouth of the Patuxent River in the Chesapeake Bay (48) and the Pacific Ocean off California (49). These two sources were chosen because they have similar data listings. Table II shows the comparisons, and although the exercise is somewhat superficial, several interesting points are illustrated.

First, the estuary is richer in all six organic classes than is the ocean. Second, the enrichment of estuary over ocean is far more profound in the particulate than in the dissolved classes. Both these observations meet with logical explanations since the estuary serves as a greater source for the constituents and because as the water passes from the estuary to the ocean, the particulate matter will tend to settle out.

Table II. Comparison of Estuarine and Oceanic Waters. Estuarine values from Flemer et al. (48) and oceanic ones from Holm-Hansen et al. (49).

| | Micrograms per liter | | |
|---|---|---|---|
| Class | Estuarine | Oceanic | Estuarine/ Oceanic |
| Dissolved organic carbon | 2900 | 800 | 3.6 |
| Dissolved organic nitrogen | 280 | 100 | 2.8 |
| Dissolved organic phosphorous | 25 | 15 | 1.7 |
| Particulate organic carbon | 2000 | 150 | 13 |
| Particulate organic nitrogen | 280 | 25 | 11 |
| Particulate organic phosphorous | 40 | 4 | 10 |

One would expect similar trends with inorganic constituents and that this is the case can be illustrated with data from Kester (50). Ratios of estuarine to oceanic concentrations of iron are calculated in the dissolved form at 30 and the particulate form at 1700. The considerably larger ratios for iron than for the gross organic constituents suggests that iron has a large estuarine source and solely a sink in the ocean. In contrast, the organics additionally have a secondary source in the oceanic environment through autochthonous production. This partially vindicates the prior claim that anthropogenic influences are not yet very impressive in the form of gross organic matter.

A third observation from Table II is the trend shown with both the dissolved and particulate classes of decreasing enrichment in going from carbon to nitrogen to phosphorous. This observation is interesting in that it illustrates how gross organic analyses can lead toward important qualitative understandings. Going from estuarine waters seaward, the sources of new organic matter will decrease since both higher productivity and allochthonous inputs pertain to the nearshore waters. Biological breakdown in this same geographic excursion will tend to leave a greater proportion of the organic matter at sea older and thus further from its source on a time axis. As we might expect, the observation of qualitative differences evident in the above comparison therefore suggests differential lability of organic compounds containing carbon, nitrogen, and phosphorous, with compounds containing the latter two elements being less resistant. Another way of viewing this differential degradation is by looking at oceanic depth profiles, since surface waters are the source areas for organic matter which when transferred to deeper waters shows increasing degradation. Using cytochemical staining techniques, it has been shown that particulate matter contains a smaller proportion of protein in deeper waters as compared to shallow ones (29).

In this same fashion, Gordon (51) finds that carbon to nitrogen ratios in particulate matter increase with depth; see

Table III. This trend suggests that nitrogen-containing compounds are generally more labile than non-nitrogenous ones.

Table III. Atomic carbon to nitrogen ratios as a function of depth. Data from Gordon (51).

| Depth Interval | Atomic C/N |
|---|---|
| 0-250 meters | 8.7 |
| 250-500 | 11.3 |
| 500-1000 | 11.1 |
| 1000-2000 | 14.4 |
| 2000-3000 | 15.7 |
| 3000-4000 | 17.7 |

A third way of imposing a time-dependent axis on organic degradation is by looking at size classes. The rationale in this case is that increasingly greater proportions of the organic matter represent more non-living detritus as one looks at smaller and smaller particle sizes until ultimately the size class is below the minimal size of any living cell. This is illustrated in Table IV where the size column represents a downward trend from practically pure living proteinaceous matter to an old detrital organic residue.

Table IV. Atomic carbon to nitrogen ratios as a function of particle size. Data from Sharp and Renger (52).

| Size class | Atomic C/N |
|---|---|
| Greater than 505 microns | 3.9 |
| Greater than 308 microns | 4.7 |
| Greater than 183 microns | 5.8 |
| Greater than 101 microns | 6.8 |
| Greater than 35 microns | 7.4 |
| Greater than 20 microns | 8.0 |
| Greater than 1 micron | 8.8 |
| Greater than 0.003 micron | 18.2 |
| Greater than 0.0012 micron | 18.6 |

The size-class approach brings up an interesting point. As mentioned earlier, the cutoff between particulate and dissolved organic matter is an arbitrary and functional one. It has a long history (53) and the concept of particulate organic matter has been very valuable (54). However, the artificial nature of the particulate-dissolved cutoff must not be forgotten and it should be appreciated that in nature probably a continuum of of particle sizes exists. This is illustrated in Figure 1 which was constructed from carbon analyses done on various size classes. Samples in the particulate classes (greater than 1 μm) were ultimately analyzed on filters after prior screening according to various mesh sizes. Samples in the less than 1 μm classes were analyzed as total organic samples after prior filtration through ultrafiltration membranes. The continuum of sizes of organic matter illustrates the artificiality of the particulate-dissolved cutoff. Not only is the routine particulate class not a discrete category, but also much of the dissolved class is probably not really in solution. The latter observation has lead to the contention that organic matter between 0.001 and 1 μm in size should be called colloidal (17). This colloidal size class is far larger than the particulate one and its distribution in the sea is apparently unlike that of the particulate class as shown in Figure 2. In addition to the importance of the colloidal size class, the accessibility of this class via ultrafiltration membranes makes study of it very desirable.

Using an approach similar to that used for organic carbon, Sharp and Renger (52) found a somewhat similar continuous size spectrum for organic nitrogen, although the quantitative nature of the curve is somewhat different from that of carbon. Protein analyses were run on the retained colloidal organic matter, and the protein expressed in light of the total organic nitrogen spectrum. In this manner, it was found that approximately 30 percent of the total organic nitrogen in the oceanic North Pacific photic zone can be accounted for as protein (or, combined amino acids). This is shown in Table V as are some carbon data. From this table, protein could also account for more than a third of the colloidal organic carbon and for about 7 percent of the total organic carbon.

Table V. Organic carbon and nitrogen from surface oceanic waters. Data from Sharp and Renger (52).

| Size Class | $\mu g\ C \cdot l^{-1}$ | $\mu g\ N \cdot l^{-1}$ | % total C | % total N |
|---|---|---|---|---|
| Total | 1270 | 85 | -- | -- |
| Particulate | 31 | 3.6 | 2.4 | 4.1 |
| Colloidal | 216[1] | 40 | 17 | 47 |
| Colloidal protein | 85[2] | 25 | 6.7 | 29 |

1. Calculated as 17% total carbon from Figure 1.
2. Calculated from protein nitrogen using a weight ratio for C/N of 3.4-value for albumin (55).

This last point takes one back to Table I where all the listed specific compounds summed account for only 10% of the total organic carbon. So, by using gross organic analyses of size classes, one can begin to find qualitative information about significant portions of the total organic pool that were other-

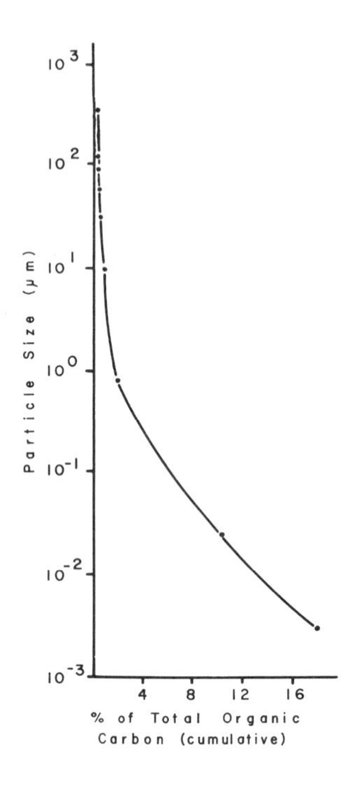


Limnology and Oceanography

*Figure 1. Organic carbon in the North Atlantic Ocean as a function of size class. The apparent particle size classes are determined by various mesh plankton netting an ultrafiltration membranes. The ordinate is a cumulative percentage of the total organic carbon* (17).

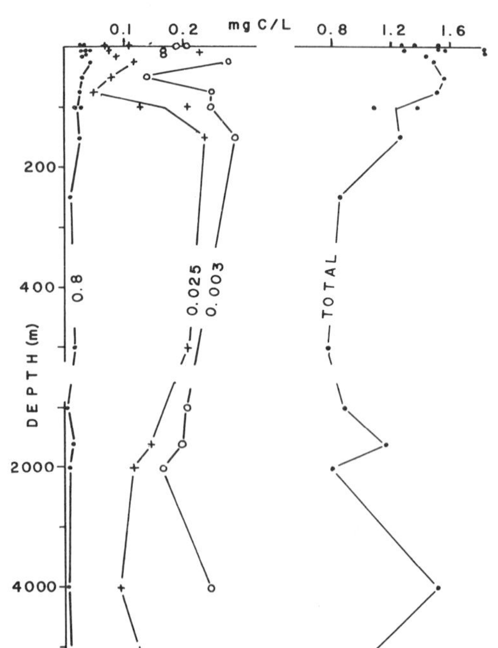


Limnology and Oceanography

*Figure 2. Depth distribution in the North Atlantic Ocean of size classes of organic carbon. Total organic carbon (TOTAL), particulate organic carbon (0.8), and two size groups of colloidal organic carbon (0.008 and 0.025) are plotted* (17).

wise not yet classified on a molecular basis.

To readdress the questions "from where": 1. The organic matter comes from plant production and largely via residual detrital avenues; 2. Samples taken in comparative sequences over geographic or size spectra can begin to provide information on the chemical nature of marine organic matter and its fate in the ocean.

## Abstract

The meanings of indirect measurements of COD and BOD and of direct measurements of particulate organic carbon, nitrogen, and phosphorous, and dissolved organic carbon, nitrogen, and phosphorous are discussed. Methods for these analyses are included in the discussion. The common division between "particulate" and "dissolved" organics is challenged as a somewhat misleading arbitrary concept. A popular consideration among marine scientists is that active healthy phytoplankton excrete appreciable amounts of organics into seawater. This is disputed and the importance of post-flowering plankton populations and macrophytic plant detritus are considered.

## Literature Cited

(1) Sharp, J. H. Total organic carbon in seawater. Comparison of measurements using persulfate oxidation and high temperature combustion. Mar. Chem. (1973) 1:211-229.
(2) Vaccaro, R. F., S. E. Hicks, H. W. Jannasch and F. G. Carey. The occurrence and role of glucose in seawater. Limnol. Oceanogr. (1968) 13:356-360.
(3) Clark, M. E., G. A. Jackson and W. J. North. Dissolved free amino acids in southern California coastal waters. Limnol. Oceanogr. (1972) 17:749-758.
(4) Barbier, M., D. Joly, A. Saliot and D. Tourres. Hydrocarbons from seawater. Deep Sea Res. (1973) 20:305-314.
(5) Treguer, P., P. LeCorre and P. Courlot. A method for determination of the total dissolved free fatty-acid content of seawater. J. Mar. Biol. Assoc. U. K. (1972) 52:1045-1055.
(6) Remsen, C. C. The distribution of urea in coastal and oceanic waters. Limnol. Oceanogr. (1971) 16:732-740.
(7) Shah, N. M. and R. T. Wright. The occurrence of glycolic acid in seawater. Mar. Biol. (1974) 24:121-124.
(8) Whittledge, T. E. and R. C. Dugdale. Creatine in seawater. Limnol. Oceanogr. (1972) 17:304-314.
(9) Provasoli, L. and A. F. Carlucci. Vitamins and growth regulators. In: W. D. P. Stewart-ed. "Algal Physiology and Biochemistry". pp. 741-787. University of California Press (Berkeley). 1974.
(10) Risebrough, R. W. Chlorinated hydrocarbons. In: D. W. Hood-ed. "Impingement of Man on the Oceans" pp. 259-286. Wiley-Interscience. 1971.

(11) Sackett, W. M. and J. M. Brooks. Origin and distribution of low molecular weight hydrocarbons in Gulf of Mexico Coastal Waters. In: T. M. Church-ed. "Marine Chemistry in the Coastal Environment". American Chemical Society. 1975. (Paper 17, this volume).

(12) Seiler, W. and V. Schmidt. Dissolved non-conservative gases in seawater. In: E. D. Goldberg-ed. "The Sea". Vol. 5. Interscience (New York). pp. 219-243. 1974.

(13) Culkin, F. The major constituents of seawater. In: J. P. Riley and G. Skirrow-eds. "Chemical Oceanography" Vol. 1. pp. 121-161. Academic Press. 1965.

(14) Yen, T. F. Genesis and degradation of petroleum hydrocarbons in the marine environment. In: T. M. Church-ed. "Marine Chemistry in the Coastal Environment". American Chemical Society. 1975. (paper 18, this volume).

(15) Faulkner, D. J. and R. J. Anderson. Natural products chemistry of the marine environment. In: E. D. Goldberg-ed. "The Sea". Vol. 5. pp. 679-714. Interscience. 1974.

(16) Padilla, G. M. and D. F. Martin. Toxins and bioactive compounds in the marine environment. In: T. M. Church-ed. "Marine Chemistry in the Coastal Environment". American Chemical Society. 1975. (Paper 40, this volume).

(17) Sharp, J. H. Size classes of organic carbon in seawater. Limnol. Oceanogr. (1973) 18:441-447.

(18) Hobbie, J. E., O. Holm-Hansen, T. T. Packard, L. R. Pomeroy, R. W. Sheldon, J. P. Thomas and W. J. Wiebe. A study of the distribution and activity of microorganisms in ocean water. Limnol. Oceanogr. (1972) 17:544-555.

(19) Garrels, R. M. and F. T. Mackenzie. A quantitative model for the sedimentary rock cycle. Mar. Chem. (1972) 1:27-41.

(20) Garrels, R. M. and E. A. Perry. Cycling of carbon, sulfur, and oxygen through geologic time. In: E. D. Goldberg-ed. "The Sea". Vol. 5 pp.303-336. Interscience. 1974.

(21) Deevey, E. S. Biogeochemistry of lakes: major substances. In: G. E. Likens-ed. "Nutrients and Entrophication: the Limiting Nutrient Controversy". pp. 14-20. American Society of Limnology and Oceanography. 1972.

(22) Redfield, A. C., B. H. Ketchum and F. A. Richards. The influence of organisms on the composition of seawater. In: M. N. Hill-ed. "The Sea". Vol. 2. pp. 26-77. Interscience (New York). 1963.

(23) Andrews, P. and P. J. L.Williams. Heterotrophic utilization of dissolved organic compounds in the sea. III. Measurement of the oxidation rates and concentrations of glucose and amino acids in seawater. JMBAUK (1971) 51:111-125.

(24) Strickland, J. D. H. Microbial activity in aquatic environments. Symp. Soc. Gen. Microbiol. (1971) 21:231-253.

(25) Azam, F. and O. Holm-Hansen. Use of tritated substrates in the study of heterotrophy in seawater. Mar. Biol. (1973)

23:191-196.
(26) APHA, AWWA, AND WPCF. "Standard Methods for the Examination of Water and Wastewater" Thirteenth Edition. pp. 489-499. Amer. Public Health Assoc. (N. Y.) 1971.
(27) Takahashi, Y., R. T. Moore and R. J. Joyce. Direct determination of organic carbon in water by reductive pyrolysis. Amer. Laboratory (1972) 4:31-38.
(28) Stenger, V. A. and C. E. Van Hall. Analyses of municipal and chemical wastewaters by an instrumental method for COD determination. J. Wat. Pol. Contr. Fed. (1968) 40: 1755-1763.
(29) Sharp. J. H. The formation of particulate organic matter in seawater. Ph.D. dissertation (1972) Dalhousie University. Halifax, N. S. pp. 142.
(30) Gordon, D. C. Jr. and W. H. Sutcliffe, Jr. A new dry combustion method for the simultaneous determination of total organic carbon and nitrogen in seawater. Mar. Chem. (1973) 1:231-244.
(31) Menzel, D. W. and R. F. Vaccaro. The measurement of dissolved and particulate carbon in seawater. Limnol. Oceanogr. (1964) 9:138-142.
(32) Armstrong, F. A. J., P. M. Williams and J. D. H. Strickland. Photo-oxidation of organic matter in seawater by ultraviolet radiation, analytical and other applications. Nature (1966) 211:481-483.
(33) Strickland, J. D. H. and T. R. Parsons. "A Practical Handbook of Seawater Analysis". 310 pp. 2nd Edition. Fish. Res. Brd. Can., Bull. 167 (1972).
(34) Menzel, D. W. and N. Corwin. The measurement of total phosphorous in seawater based on the liberation of origanically bound fractions by persulfate oxidation. limnol. Oceanogr. (1965) 10:280-282.
(35) Sharp, J. H. Improved analysis for "particulate" organic carbon and nitrogen from seawater. Limnol. Oceanogr. (1974) 19:984-989.
(36) Gordon, D. C., Jr. and W. H. Sutcliffe, Jr. Filtration of seawater using silver filters for particulate nitrogen and carbon analysis. Limnol. Oceanogr. (1974) 19:989-993.
(37) Pella, E. and B. Colombo. Study of carbon, hydrogen, and nitrogen determination by combustion gas-chromatography. Mikrochim. Acta. (Wien) (1973):697-719.
(38) Perry, M. J. and R. W. Eppley. Dynamics of phosphorous cycling in the euphotic waters of the central North Pacific Ocean. Deep Sea Res. (1975) 22 (in press).
(39) Duursma, E. K. The production of dissolved organic matter in the sea, as related to the primary gross production of organic matter. Neth. J. Sea Res. (1963) 2:85-94.
(40) Fogg, G. E. The extracellular products of algae. Oceanogr. Mar. Biol. Ann. Rev. (1966) 4:195-212.
(41) Thomas, J. P. Release of dissolved organic matter from

natural populations of marine phytoplankton. Mar. Biol. (1971) 11:311-323.
(42) Sharp, J. H. and E. H. Renger. Extracellular production of organic matter by marine phytoplankton - Do Healthy Cells Excrete? (in preparation) 1975.
(43) Eppley, R. W., E. H. Renger, E. L. Venrick and M. M. Mullin. A study of plankton dynamics and nutrient cycling in the central gyre of the North Pacific Ocean. Limnol. Oceanogr. (1973) 18:534-551.
(44) Venrick, E. L. The distribution and significance of Richelia intracellularis Schmidt in the North Pacific Central Gyre. Limnol. Oceanogr. (1974) 19:437-445.
(45) Kester, D. R., K. T. Crocker, and G. R. Miller. Small scale oxygen variations in the thermocline. Deep Sea Res. (1973) 20:409-412.
(46) Odum, E. P. and A. A. de la Cruz. Particulate organic detritus in a Georgia salt marsh-estuarine ecosystem. In: G. H. Lauff-ed. "Estuaries". pp. 383-388. AAAS. 1967.
(47) Pomeroy, L. R., L. R. Shenton, R. D. H. Jones and R. J. Reimold. Nutrient flux in estuaries. In: G. E. Likens-ed. "Nutrients and Entrophication: The Limiting Nutrient Controversy". pp. 274-291. Allen Press. 1972.
(48) Flemer, D. A., D. H. Hamilton, C. W. Keefe and J. A. Mihursky. The effects of thermal loading and water quality on estuarine primary production. 110 p. Natural Resources Institute, Univ. Maryland, 71-6. 1970.
(49) Holm-Hansen, O., J. D. H. Strickland and P. M. Williams. A detailed analysis of biologically important substances in a profile off southern California. Limnol. Oceanogr. (1966) 11:548-561.
(50) Kester, D. R. Redox reactions of iron in marine sediments. In: T. M. Church-ed. "Marine Chemistry in the Coastal Environment". American Chemical Society. 1975. (Paper 3, this volume).
(51) Gordon, D. C. Distribution of particulate organic carbon and nitrogen at an oceanic station in the central Pacific. Deep Sea Res. (1971) 18:1127-1134.
(52) Sharp, J. H. and E. H. Renger. Colloidal organic nitrogen in seawater. (in preparation). 1975.
(53) Goldberg, E. D., M. Baker and D. L. Fox. Microfiltration in oceanographic research: 1. Marine sampling with the molecular filter. J. Mar. Res. (1952) 11:194-204.
(54) Riley, G. A. Particulate organic matter in seawater. Adv. Mar. Biol. (1970) 8:1-118.
(55) Stecher, P. G.-ed. "The Merck Index" Eighth Edition. p. 29 The Merck Company (Rahway, N. J.) 1968.

# INDEX

## C

## D

## E



## F

## P

## R